Electrodynamics of Continua I

A.C. Eringen
G.A. Maugin

Electrodynamics of Continua I

Foundations and Solid Media

With 64 Illustrations

Springer-Verlag
New York Berlin Heidelberg
London Paris Tokyo Hong Kong

A.C. Eringen
Princeton University
Princeton, N.J. 08544
U.S.A.

G.A. Maugin
Laboratoire de Modélisation
en Mécanique
Université Pierre et Marie
Curie et C.N.R.S.
75252 Paris 05
France

Library of Congress Cataloging in Publication Data
Eringen, A. Cemal.
Electrodynamics of continua / A.C. Eringen, G.A. Maugin.
p. cm.
Includes bibliographical references.
Contents: 1. Foundations and solid media — 2. Fluids and complex media.
ISBN-13: 978-1-4612-7923-5 e-ISBN-13: 978-1-4612-3226-1
DOI: 10.1007/978-1-4612-3226-1
1. Fluid mechanics. 2. Electrodynamics. 3. Magnetohydrodynamics. 4. Continuum mechanics. I. Maugin, G. A. (Gerard A.), 1944-
II. Title.
QC151.E&5 1989
537.6—dc20 89-21880
CIP

Printed on acid-free paper

Softcover reprint of the hardcover 1st edition 1990

Typeset by Asco Trade Typesetting Ltd., Hong Kong.

9 8 7 6 5 4 3 2 1

Preface

The electrodynamics of continua is a branch of the physical sciences concerned with the interaction of electromagnetic fields with deformable bodies. Deformable bodies are considered to be continua endowed with continuous distributions of mass and charge. The theory of electromagnetic continua is concerned with the determination of deformations, motions, stress, and electromagnetic fields developed in bodies upon the applications of external loads. External loads may be of mechanical origin (e.g., forces, couples, constraints placed on the surface of the body, and initial and boundary conditions arising from thermal and other changes) and/or electromagnetic origin (e.g., electric, magnetic, and current fields). Because bodies of different constitutions respond to external stimuli in a different way, it is imperative to characterize properly the response functions relevant to a given class of continua. This is done by means of the constitutive theory. For example, an elastic dielectric responds to electromagnetic fields in a totally different way than a magnetic fluid. The present book is intended to present a unified approach to the subject matter, based on the principles of contemporary continuum physics.

In traditional studies, usually the electromagnetic theory is uncoupled from continuum mechanics, so that researchers in the mechanics community seldom deal with electromagnetic phenomena, while physicists and electrical engineers often ignore the mechanical effects. As a result, the extension of classical theories to strong electromagnetic fields and finite deformations has suffered long delays. Yet many interesting physical phenomena which possess important technological applications lie at the intersection of continuum mechanics and electromagnetic theory. Clearly, the electrodynamics of deformable continua is *not* the sum of these two fields. It is a nonlinear composition of them.

The present work (in two volumes) is written in the hope that it will fill a vacuum that exists in the library of other treatises which are devoted principally to special topics in continuum mechanics and electromagnetic theory. While there exist several works covering the subject of classical (rigid body) electromagnetism, piezoelectricity, magnetoelectricity, magnetohydrodynamics, etc., they generally explore a restricted class of phenomena in one or two fields. By-and-large, they employ the linear constitutive theory, ignoring the unified nature of the field. The division of the subject matter into the mechanics and electromagnetism of special bodies is responsible for drawing our attention away from the study and research in the most exciting interdisciplinary areas of continuum physics. As a result, the tradition has been to place electromagnetic theory within the physics or electrical engineering curricula, and continuum mechanics within other engineering science departments. Yet recent rapid advances in engineering technology have placed severe burdens on engineers and physicists in understanding and being able to use the effects of intense fields and large deformations of bodies. Nonlinear optics, equilibrium of magnetic stars, design of space vehicles, fast computers, fusion reactors, plasma devices, and magnetically levitated vehicles are but a few examples of where a deeper understanding and more careful formulations are necessary. It is here that we must return to the foundations and then build various couplings on sound ground.

This book attempts to present a self-contained, finite deformation, and finite electromagnetic field theory from a unified viewpoint—providing ample critical illustrations by way of applications. The constitution of this work may be sketched as follows:

(i) Development of the ten basic balance laws.
(ii) Establishment of the constitutive theory based on nine fundamental axioms.
(iii) Discussion of special theories and applications.

The ultimate goal for (i) is achieved at the end of Chapter 3, on macroscopic electromagnetic theory. To arrive at these results, it was necessary to discuss the kinematics of deformable continua and relevant notions. This is summarized in Chapter 1. In Chapter 2, we present the electromagnetic theory from a microscopic viewpoint. This provides a basis for the macroscopic theory.

The general constitutive theory is contained in Chapter 5. Here, by means of nine axioms, the local nonlinear theory is developed for finite deformations and electromagnetic fields. Magnetic symmetry groups and allied topics essential to the development of constitutive equations for magnetic materials (ferromagnetic, ferrimagnetic, and antiferromagnetic) are given in five appendices. In this way, for the first time, it has been possible to express the nonlinear constitutive equations for the ninety magnetic groups relevant to magnetic crystals. Nonlinear and linear theories are presented for the elastic solids and viscous fluids that are subject to electromagnetic fields.

Out of necessity, and in consideration of the extra degrees of freedom, the constitutive theory appears again in Chapters 9, 11, 12, and 13 (volume II) for more complicated continua: elastic ferromagnets, electrohydrodynamics, ferrofluids, and memory-dependent materials.

Excluding the last two chapters in volume II, the rest of the work explores various special topics and presents solutions to several linear and nonlinear problems in order to demonstrate the uses of the basic theory. Chapter 4 contains an elementary treatment of various classes of electromagnetic phenomena. Here the emphasis is on the physical nature of diverse electromagnetic effects. Chapter 6 discusses rigid dielectrics, including aspects of the potential theory, linear and nonlinear electromagnetic waves, electromagnetic shock waves, and nonlinear optics. Chapter 7 is devoted to the discussion of linear and nonlinear elastic dielectrics, piezoelectricity, surface and body waves, and the general theory of small deformations and electromagnetic fields superposed on large, static deformations and fields. The latter is important to the discussion of the instability of elastic bodies that are subject to electromagnetic fields. Photoelasticity, electro-optical effects, magneto-optical, and other allied phenomena are presented. Chapter 8 develops nonlinear and linear magnetoelasticity, including a plethora of applications to stress concentrations, magnetoelastic waves, and the instability of plates under strong magnetic fields. Fundamental to all treatments are the basic field equations and jump conditions collected at the end of Chapter 3, as well as the constitutive equations given in Chapter 5.

The second volume of this treatise begins with Chapter 9 on elastic ferromagnets. Here it is necessary to incorporate the extra degree of freedom arising from the gyroscopic nature of the spin density. Balance laws are also extended to include spin–spin interactions. Constitutive theory, based on the general axioms, is developed. The theory is then applied to various problems including waves, the Faraday effect, etc.

Chapter 10 is concerned with a discussion of magnetohydrodynamics (MHD). Bernoulli's equation, Alfvén waves, incompressible viscous flows, compressible flow and magnetohydrodynamic shock waves, magnetohydrodynamic stability, and the equilibrium of magnetic stars are some of the problems considered.

Chapter 11 deals with electrohydrodynamics and Chapter 12 with ferrofluids. The latter includes a discussion of extra degrees of freedom for the spin–spin interaction. Several flow problems which are solved include nonviscous and viscous ferrohydrodynamics, shear flow, stagnation-point flow, and interfacial stability.

Chapter 13 presents the theory of memory-dependent electromagnetic continua. It is well known that energy loss, dispersion, and absorption of waves cannot be explained without consideration of memory effects. This requires incorporation of past histories of strain and electromagnetic fields into constitutive equations. Thermodynamics of rate-dependent materials, and materials with continuous memory of past motions, and fields are discussed. Constitutive equations are obtained for memory-dependent electromagnetic elastic solids and electromagnetic viscous fluids. Applications are made to electromagnetic waves, birefringent materials, photoviscoelasticity, and nonlinear optics.

In Chapter 14 we present an account of the recent theory of nonlocal electromagnetic elastic continua. Here the stress, polarization, magnetization, and current at a reference point in the body are considered to be functionals of the strain and electromagnetic fields of *all* points of the body. The nonlocality extends the domain of the application of the theory to microscopic and atomic scales (without the quantum effect). In this way, electromagnetic waves with short wavelengths can be treated. Point charge, anomalous skin effects, eddy currents, infrared dispersion, and lattice vibrations, and even superconductivity, fall into the domain of applicability of the nonlocal theory. This field is rather new, currently in a development stage. Consequently, a short account has been included here, in the hope that it may provide some stimuli for future researchers.

Since electrodynamics properly falls into the domain of relativity, we also include an account of this subject, to close this treatise.

All chapters on special topics and applications are supplied with relevant basic equations and jump conditions in the opening pages. This way, no need will arise for a search of the basic equations in the text.

It is hoped that this work will serve two basic needs:

(a) as a textbook for graduate studies, and
(b) as a guide and reference for researchers.

As a text, this work could be used for several different courses. A first course may be based on Chapters 2, 3, and 5, with problems selected from Chapters 6 through 9. This sequence has been followed successfully by the first author during the last ten years. Another course can be composed by utilizing material selected from various chapters; for example, the topic of elastic dielectrics and magnetoelasticity may be presented, or a discussion of ferrofluids and electrohydrodynamics could take place. Alternatively, the memory-dependent materials discussed in Chapter 13, and the nonlocal electromagnetic continua presented in Chapter 14, could be offered as advanced topics.

As a guide for researchers, there is some material on foundations; e.g., constitutive theory for magnetic crystal classes, memory-dependent continua and nonlocality, formulations and solutions of problems in nonlinear optics of deformable bodies, instability of magnetized bodies, the effects of high electric fields and currents in deformable bodies and conductors, etc. Once the basic principles and the constitutive theory are well understood, extensions are possible in other directions (e.g., topics on mixtures and semiconductors, which fall beyond the scope of the present treatise).

Much of the material presented in this treatise is either new, or newly composed. Presently, there are no other treatises similar in content and composition. For example, Chapters 3 and 5, and parts of Chapters 7, 8, 13, and 14, contain original material and compositions. Some of these are indicated specifically in the appropriate sections.

The beginning of the formal writing of this book goes back to the Spring of 1976, when G.A. Maugin was a visiting Research Associate at Princeton University. Over the years, the project became increasingly ambitious. The mechanical and communication difficulties arising from the three thousand miles of ocean between Princeton and Paris caused many delays. In spite of all our efforts, errors may have crept into the text. Whether they are typographical, or otherwise, we ask the tolerance of our readers and express our deep appreciation, in advance, should they be brought to our attention.

Because of our extensive involvement with research in various sectors of this wide field, many parts of this book have, naturally, been influenced by our research and thought processes. This natural predisposition does not exclude our respect and admiration for other researchers in the field. The large bibliography stands as a testament to this. We hope that the authors of many other contributions that have been unintentionally left out will understand our limitations.

It is a pleasure to acknowledge the help received from various students and colleagues. Without the generous help provided by Dr. Erhan Kiral, Appendix B on the crystallographic magnetic groups could not have been written. A.C. Eringen and E. Kiral completed a monograph on this subject in 1977. This manuscript is now scheduled for publication by Springer-Verlag. Dr. Naşit Ari and Mr. T.M. Cheng checked parts of Chapter 7, and Mrs. (Lois) M. Nase helped in the search for some references. Eleni Maugin prepared the index. A.C.E's daughters, Meva S. Eringen and Lisa M. Eringen edited the manuscript, and we are deeply indebted to them for their invaluable help. A father's love is forever. A.C.E. expresses his thanks to the authorities of the University of Paris VI, "Laboratore de Modélisation en Mécanique", and to his host, G.A.M., for inviting him as a visiting professor in the spring of 1985. This provided an excellent opportunity to discuss several questions and check portions of Chapters 8 and 9. We acknowledge with thanks, the extreme care and attention provided in the preparation of this manuscript by our publisher Springer-Verlag.

Finally, we acknowledge the permission of the following organizations for allowing us to reprint various tables and figures:

Oxford University Press, Oxford, U.K. (Figure 4.5.3)
Elsevier Science Publishers, Amsterdam, The Netherlands (Figure 9.2.1)
Pergamon Press, Oxford, U.K. (Figures 9.6.1, 9.6.2, 9.6.3, 9.6.4, 9.6.5, 9.6.6, 12.10.2) and Tables C1.1–C1.12 and C2.1–C2.12.
American Institute of Physics, New York (Figures 4.5.7, 8.8.1, 8.8.2, 8.8.3, 9.8.1, 9.8.3, 9.8.4, 11.8.4, 12.7.2, 12.8.2, 12.8.3)
Springer-Verlag, Wien, Austria (Figures 10.5.1, 10.5.2) Springer-Verlag, Berlin, Heidelberg (Figures 13.14.1–13.14.5).
John Wiley and Sons, New York (Figures 6.7.1, 7.5.2, 8.4.2, 10.6.2, 13.7.1, 13.8.1)
Annual Reviews, Inc., Palo Alto, CA (Figure 11.7.1)
Gauthier-Villars, Paris, France (Figures 11.8.2, 11.8.3, 12.4.1, 12.4.2, 12.4.3, 12.4.4)
The MIT Press, Cambridge, MA (Figures 11.9.1, 11.9.2, 12.11.2)
Academic Press, New York (Figure 1.2.1)
Kluwer Academic Publishers, Dordrecht, Holland (Figures 4.5.2, 6.17.2, 6.17.3)

Contents (Volume I)

CHAPTER 3

Macroscopic Electromagnetic Theory 47

CHAPTER 4

Elementary Properties of Electromagnetic Continua 91

Contents (Volume II)

CHAPTER 1

Kinematics of Material Continua

1.1. Scope of the Chapter

This chapter is concerned with the kinematics of deformable and fluent bodies. Here we collect the most essential kinematic notions relevant to the later chapters. The notion of material continua with charges is discussed in Section 1.2. The motion and deformation gradients introduced in Section 1.3 are fundamental concepts, upon which the mechanics of deformable continua are built. Section 1.4 discusses strain measures. By means of the polar decomposition theorem, one is led to the concepts of finite rotation and finite strain in Section 1.5. This concept is also fundamental to an understanding of the constitutive theory. Infinitesimal strains and rotations are presented in Section 1.6, and area and volume changes upon deformation are presented in Section 1.7. Section 1.8 discusses the integrability conditions of the strain tensor, i.e., given a set of strain tensors, and the conditions that lead to a single-valued displacement field corresponding to these strains.

In Section 1.9 we discuss time changes. Thus, material time rates of vectors and tensors relevant to motion are introduced. This leads to the concepts of deformation rate tensors, strain rate tensors, spin, vorticity, etc. Rigid body motion is defined by the vanishing of the strain rate tensor (Section 1.10). Scalars, vectors, and tensors, which keep their identities under time-dependent motions of the spatial frame of reference, are fundamental to the constitutive theory. They are called *objective*. The concept of objectivity (or material-frame indifference) is a basic axiom of the constitutive theory discussed in Chapter 5. Objective time rates of vectors and tensors are presented in Sections 1.11 and 1.12. Kinematics of line, surface, and volume integrals, as discussed in Section 1.13, introduce methods for calculating material derivatives of integrals which occur in the global balance laws.

The intention of this chapter is to provide a brief review of the fundamental notions of mechanics of continua. Those readers who are familiar with continuum mechanics may omit this chapter, and those who wish to acquire further knowledge may consult other books on the subject, e.g., Eringen [1980].

1.2. Mass and Charge

Continuum mechanics is a branch of physical science concerned with the deformations and motions of continuous material media under the influence of external agents. Continuum physics is a broader branch which includes the study of such effects as chemical interactions, electromagnetic effects, and other environmental changes. In the present work, we focus our attention on coupled mechanical–electromagnetic effects, of which the mechanical–electric effects and mechanical–magnetic effects are special cases. In this regard, *mass and electric charge* are the basic tenets of the *electrodynamics of continua.*

In reality, physical bodies appear in the form of collections of material points interconnected by some internal forces. Yet, in continuum physics, the subject matter is concerned with macroscopic phenomena in which the smallest characteristic length is much larger than the size of an atom and atomic distance. In support of the corresponding mathematical idealization, the fundamental assumption of continuity of mass, and of some other fields such as electric charge, represents an excellent approximation. The strength of statistical laws in the molecular chaos of particles is basic to this outcome. Furthermore, numerous instruments measure only the statistical averages that characterize global phenomena.

If the continuous mathematical description is accepted, then continuum physics relies on the fact that the balance laws and the axioms of constitution of matter are valid for every part of a material body, no matter how small it may be. Accordingly, a small volume ΔV enclosed within a surface ΔS possesses a mass density defined by the following limit:

$$\rho = \lim_{\Delta V \to 0} \Delta m/\Delta V \geqq 0, \tag{1.2.1}$$

where Δm is the total mass contained in ΔV. Here, ρ is independent of the size of ΔV and depends only on the position $\mathbf{r}$ and the time t.

In order to judge the physical validity of the mathematical definition (1.2.1), consider the following experiment for the measurement of ρ. The mass density of a *homogeneous* material may be calculated approximately by weighing a large number of pieces having different volumes, and calculating the ratio $\Delta m/\Delta V$ for each piece. If the resulting numbers ρ are plotted versus ΔV, one finds that this ratio is nearly constant when ΔV is greater than a certain critical volume ΔV^* which roughly corresponds to a critical length $L = \sim(\Delta V^*)^{1/3}$, which is of course, not precisely defined. The size of ΔV^* depends on the constitution of the material. As ΔV approaches zero, this dependence becomes extreme (see Figure 1.2.1). The classical continuum theory may therefore not be a good mathematical model for the approximation of a physical theory in the range $\Delta V \ll \Delta V^*$. Examples of continuous descriptions, in excellent agreement with experimental data in the range $\Delta V \gg \Delta V^*$, are those corresponding to the classical nonlinear and linear theories of elasticity and Stokesian fluids. More involved physical theories based on the continuum hypothesis have also

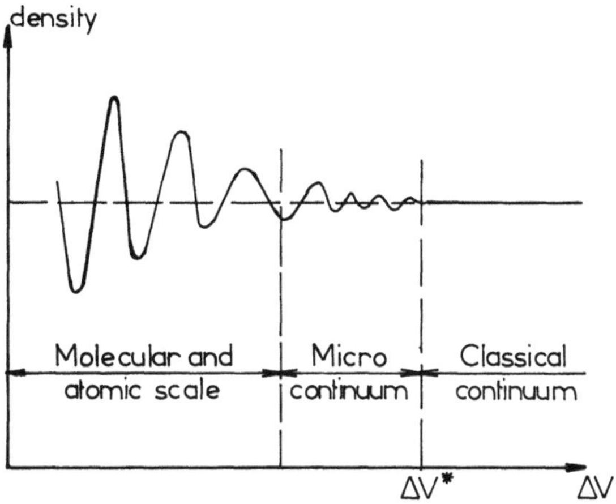

Figure 1.2.1. Mass density versus volume (after Eringen [1968, p. 623]). Reprinted with permission of Academic Press.

been successful. Among many we cite the following: theories of paramagnetic fluids and solids, piezoelectricity in which interactions between motions and deformations and electromagnetic fields play prominent roles; electro-optical, magneto-optical, and photomechanical effects in which the interactions of motion and deformation with light are important.

Equation (1.2.1) defines the basic support for all continuum theories. Accordingly, a *material continuum*, from the mathematical viewpoint, is endowed with a mass *measure*. The same reasoning can be made with regard to the *electric charge continuum.*

The existence of charge is a fundamental postulate of physics, and is based on experimental evidence. The existence of electric currents is attributed to the motion of charges. Accordingly, electric currents are not considered as primitive notions in the subsequent development. In addition to mass,[1] some elementary particles, the aggregates of which compose matter, possess another *measure* called *charge.* The electronic charge, expressed by $e = 1.60 \times 10^{-19}$ coulomb, is the smallest possible division of charge. In contrast to mass which is always positive, electric charge may be positive or negative.

Experimental evidence supports the postulate that the total charge in any isolated system is conserved: that is, if an amount of positive charge appears or disappears within the system, then an equal amount of negative charge does likewise, so that the algebraic sum of the charge remains constant. Charge may also be characterized as being *free* or *bound.* The negative charges carried by free electrons are examples of the former, while the negative charges carried

[1] In addition, particles may possess an intrinsic angular momentum called *spin*, whose importance shows up in the study of magnetic materials (see Section 2.2 and Chapter 9).

by electrons, which make up the inner electron shells of an atom, illustrate the latter. The total charge contained in any spatial volume is some integral multiple of the electronic charge. In the present book, however, we shall assume that, like mass, charge is infinitely divisible in accordance with the continuum hypothesis and, of course, within a volume range defined in a manner similar to that of Figure 1.2.1. Then if ΔQ is the electric charge contained within the volume ΔV, then the charge density per unit volume is defined, analogously to ρ, by

$$q = \lim_{\Delta V \to 0} \Delta Q/\Delta V \gtrless 0. \tag{1.2.2}$$

This quantity may be either positive or negative.

1.3. Motion and Deformation

The material points of a continuous medium, in *the reference state* at time $t=0$, occupy a region B of Euclidean space E^3, which consists of the *material volume* V and its boundary ∂V. The position of a particular point P (P for "particle") in this region may be denoted by a rectangular coordinate system X_1, X_2, X_3, in short X_K, $K = 1, 2, 3$, or simply by $\mathbf{X}$ (Figure 1.3.1). After motion and deformation have taken place, at time t, the material points of V and ∂V occupy a region $\mathscr{B}$ of E^3 consisting of a spatial volume $\mathscr{V}$ and its boundary $\partial\mathscr{V}$. The particular material point or "particle" P is now located at the spatial point p that may be referred to a new rectangular coordinate frame x_k, $k = 1, 2, 3$, which is called the *spatial frame*. The position vector of p referred

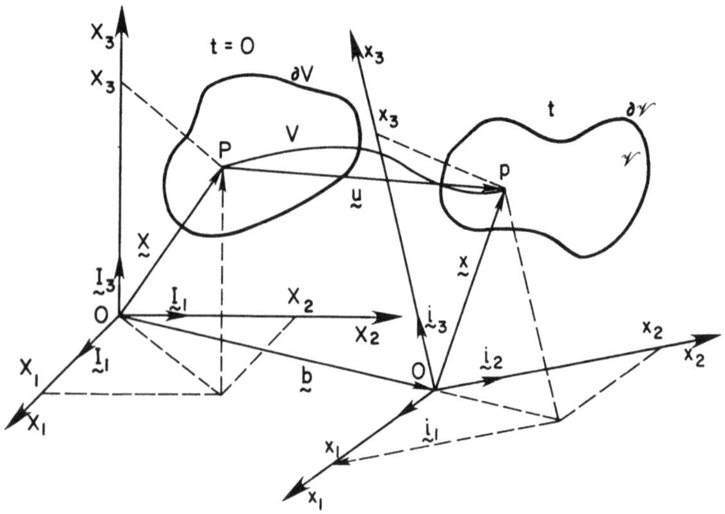

Figure 1.3.1. Motion of a continuum.

to x_k is simply denoted by $\mathbf{x}$. It may be advantageous to select the two reference frames as nonidentical. The choice of two distinct coordinate systems, one for the undeformed body at time $t=0$ and one for the deformed body at time t, is, in fact, particularly suited in the case where curvilinear systems are used.[2]

However, even when both sets of coordinates are chosen to be rectangular, the use of two different sets, in the description of the motion and the deformations of a continuous body, makes many subtle points clearly understood, especially those notions concerned with the *finite* deformations of the body. The coordinates X_K are called *material* or *Lagrangian*, and the coordinates x_k are referred to as spatial or Eulerian. The former describe the *reference* configuration K of the material body at time $t=0$, while the latter describe the present configuraton $\mathscr{K}$ at time t.

Material points of the body in motion will occupy, in general, different spatial points in E^3, resulting in a temporal sequence of configurations. Thus, the motion and deformation of the body carries various material points through spatial positions. Explicitly,

$$\mathbf{x} = \boldsymbol{\chi}(\mathbf{X}, t); \qquad x_k = \chi_k(X_K, t), \tag{1.3.1}$$

where, for each t, $\mathbf{x}$ is a continuously differentiable homeomorphism. Accordingly, the mapping (1.3.1), parametrized by t, is one-to-one and invertible. That is, the inverse motion[3]

$$\mathbf{X} = \mathbf{X}(\mathbf{x}, t); \qquad X_K = X_K(x_k, t), \tag{1.3.2}$$

exists and is unique for all times t. Consequently, the Jacobian

$$J = \det|\mathbf{F}| > 0, \tag{1.3.3}$$

where

$$\mathbf{F} = \{x_{k,K}\} \equiv \{\partial\chi_k/\partial X_K;\ k, K = 1, 2, 3\}, \tag{1.3.4}$$

must not vanish, and is, in fact, strictly positive at all points of the body at all times. It follows that the material "particles" P are *identifiable* and that the *axiom of continuity*, according to which the matter is *indestructible* and impenetrable, is satisfied. In practice, there are cases in which this axiom is violated; for example, materials may fracture, or transmit shock waves or other types of discontinuities. Special attention must be devoted to these cases.

The geometric object $\mathbf{F}$ with components $x_{k,K}$ is called the *deformation gradient*. Obviously, neighboring material points in the configuration K remain so (in both the topological and geometrical sense) in any later configura-

[2] A typical example of the use of such coordinate systems is the finite deformation of a rectangular block into a circular cylinder. The use of rectangular coordinates for the undeformed block and that of cylindrical coordinates for the deformed block proves to be particularly advantageous.

[3] Henceforward, for the sake of economy in notation, we will not distinguish the function and its value, e.g., we shall use x_k for both x_k and χ_k.

tion $\mathscr{X}$. In agreement with (1.3.1) and (1.3.2), for fixed t, we thus have

$$dx_k = x_{k,K}\, dX_K \tag{1.3.5}$$

and

$$dX_K = X_{K,k}\, dx_k, \tag{1.3.6}$$

where

$$\overset{-1}{\mathbf{F}} = \{X_{K,k} = \partial X_K/\partial x_k;\ K, k = 1, 2, 3\} \tag{1.3.7}$$

is the *inverse deformation gradient*. According to the "chain rule" of differentiation

$$x_{k,K} X_{K,l} = \delta_{kl}, \tag{1.3.8}$$

$$X_{K,k} x_{k,L} = \delta_{KL}, \tag{1.3.9}$$

in which δ_{kl} and δ_{KL} are Kronecker symbols in the x_k and X_K systems, respectively. For further use we note the following formulas:

$$\partial/\partial x_i = X_{K,i}\, \partial/\partial X_K, \qquad \partial/\partial X_K = x_{i,K}\, \partial/\partial x_i, \tag{1.3.10}$$

$$J = \frac{1}{3!} \varepsilon_{ijk} \varepsilon_{KLM} x_{i,K} x_{j,L} x_{k,M}, \tag{1.3.11}$$

$$\varepsilon_{ijk} x_{i,K} x_{j,L} x_{k,M} = J \varepsilon_{KLM}, \tag{1.3.12}$$

$$\frac{\partial J}{\partial x_{i,K}} = \text{cofactor}(x_{i,K}) = \tfrac{1}{2} \varepsilon_{ijk} \varepsilon_{KLM} x_{j,L} x_{k,M} = J X_{K,i}, \tag{1.3.13}$$

where ε_{ijk} and ε_{KLM} are the permutation symbols in the x_i and X_K systems, respectively. Thus, for example, we have

$$\varepsilon_{123} = \varepsilon_{231} = \varepsilon_{312} = -\varepsilon_{213} = -\varepsilon_{132} = -\varepsilon_{321} = 1, \qquad \text{all other} \quad \varepsilon_{ijk} = 0.$$

In either one of these systems we have the useful algebraic identities

$$\begin{aligned} \varepsilon_{ijk} \varepsilon_{ipq} &= \delta_{jp} \delta_{kq} - \delta_{jq} \delta_{kp}, \\ \varepsilon_{ijk} \varepsilon_{ijl} &= 2\delta_{kl}, \\ \varepsilon_{ijk} \varepsilon_{ijk} &= 3!. \end{aligned} \tag{1.3.14}$$

We shall often use the following vector notation (the so-called nabla notation)

$$\nabla_0 \equiv \{\partial/\partial X_K;\ K = 1, 2, 3\}, \qquad \nabla \equiv \{\partial/\partial x_k;\ k = 1, 2, 3\}, \tag{1.3.15}$$

to denote gradient operators.

1.4. Strain Measures

The main purpose of the deformation theory in continuum mechanics is to study the variation in magnitude and orientation of two neighboring material points in the course of a motion. In this regard, differential geometry plays a prominent role. Since the physical body is imbedded in Euclidean space in all configurations, the squared infinitesimal distances in the configurations K and $\mathscr{K}$ are given by

$$\begin{aligned} dS^2 &= \delta_{KL}\, dX_K\, dX_L = d\mathbf{X}\cdot d\mathbf{X}, \\ ds^2 &= \delta_{ij}\, dx_i\, dx_j = d\mathbf{x}\cdot d\mathbf{x}. \end{aligned} \tag{1.4.1}$$

Because of (1.3.5) and (1.3.6), we can also write

$$\begin{aligned} dS^2 &= c_{ij}(\mathbf{x}, t)\, dx_i\, dx_j, \\ ds^2 &= C_{KL}(\mathbf{X}, t)\, dX_K\, dX_L, \end{aligned} \tag{1.4.2}$$

where

$$C_{KL} \equiv x_{i,K} x_{i,L} \tag{1.4.3}$$

and

$$c_{ij} \equiv X_{K,i} X_{K,j} \tag{1.4.4}$$

are called the *Green* and *Cauchy* deformation tensors or strain measures, respectively.

The difference $ds^2 - dS^2$ for the same material points in K and $\mathscr{K}$ is a measure of the change in length. This difference vanishes for any two neighboring points if the deformation has not altered the distance between the pair. When this difference is zero for all points in the body, then the material body is said to have undergone only *rigid displacement*.

According to (1.4.1) and (1.4.2), we have

$$ds^2 - dS^2 = 2E_{KL}(\mathbf{X}, t)\, dX_K\, dX_L = 2e_{ij}(\mathbf{x}, t)\, dx_i\, dx_j, \tag{1.4.5}$$

where

$$E_{KL} \equiv \tfrac{1}{2}(C_{KL} - \delta_{KL}) \tag{1.4.6}$$

and

$$e_{ij} \equiv \tfrac{1}{2}(\delta_{ij} - c_{ij}) \tag{1.4.7}$$

are called the *Lagrangian* and *Eulerian* strain tensors, respectively. Clearly,

$$E_{KL} = e_{ij} x_{i,K} x_{j,L}, \qquad e_{ij} = E_{KL} X_{K,i} X_{L,j}. \tag{1.4.8}$$

The *rigid displacement* of a body corresponds to the kinematical constraints

$$C_{KL}(\mathbf{X}, t) = \delta_{KL} \qquad \text{or} \qquad c_{ij}(\mathbf{x}, t) = \delta_{ij} \tag{1.4.9}$$

at all t, and for all points $\mathbf{x}$ and $\mathbf{X}$ related by (1.3.1) or (1.3.2).

It is an easy matter to show that the Jacobian J is also given by

$$J = (\det \mathbf{C})^{1/2} = (\det \mathbf{c})^{-1/2}. \tag{1.4.10}$$

Since $J > 0$, there exist $\overset{-1}{\mathbf{C}}$ and $\overset{-1}{\mathbf{c}}$ such that, in components,

$$\overset{-1}{C}_{KL} \equiv X_{K,i} X_{L,i} \tag{1.4.11}$$

and

$$\overset{-1}{c}_{ij} \equiv x_{i,K} x_{j,K}. \tag{1.4.12}$$

These two tensors are often called the *Piola* and the *Finger strain* tensors, respectively. In certain works[4] $\overset{-1}{\mathbf{c}}$ is denoted by $\mathbf{B}$, so that $\mathbf{c} = \overset{-1}{\mathbf{B}}$.

1.5. Polar Decomposition

Since $\mathbf{F} \equiv \{x_{k,K}\}$ is nonsingular ($J \neq 0$), a theorem due to Cauchy,[5] referred to as the *polar decomposition theorem*, enables us to write $\mathbf{F}$ in the alternative forms

$$\mathbf{F} = \mathbf{R} \cdot \mathbf{U} = \mathbf{V} \cdot \mathbf{R}; \qquad x_{i,K} = R_{iL} U_{LK} = V_{ij} R_{jK}, \tag{1.5.1}$$

with (T $\equiv$ transpose)

$$\mathbf{U} = \mathbf{U}^{\mathrm{T}}, \qquad \mathbf{V} = \mathbf{V}^{\mathrm{T}}, \qquad \mathbf{R} \cdot \mathbf{R}^{\mathrm{T}} = \mathbf{R}^{\mathrm{T}} \cdot \mathbf{R} = \mathbf{1}, \tag{1.5.2}$$

where $\mathbf{R}$ is an orthogonal tensor, while $\mathbf{U}$ and $\mathbf{V}$ are positive-definite symmetric tensors. $\mathbf{R}$, $\mathbf{U}$, and $\mathbf{V}$ can be shown to be unique. Cauchy's polar decomposition theorem tells us that the deformation corresponding locally to $\mathbf{F}$ may be obtained by pure stretches of amount, say λ_α, $\alpha = 1, 2, 3$, along three suitable mutually orthogonal directions $\mathbf{e}_\alpha$, followed by a rigid rotation of these directions. The order of this process is unimportant. Namely, we may perform the rotation first, and then the same stretches along the resulting directions next. The λ_α are called the *principal stretches*. The corresponding unit eigenvectors of $\mathbf{U}$ and $\mathbf{V}$ point along the *principal axes of strain* in the configurations K and $\mathscr{K}$, respectively, since

$$\mathbf{U} \cdot \mathbf{e}_\alpha = \lambda_\alpha \mathbf{e}_\alpha, \tag{1.5.3}$$

so that with (1.5.1), (1.5.2)

$$\mathbf{V} \cdot (\mathbf{R} \cdot \mathbf{e}_\alpha) = (\mathbf{R}\mathbf{U}\mathbf{R}^{\mathrm{T}}) \cdot \mathbf{R}\mathbf{e}_\alpha = \lambda_\alpha \mathbf{R}\mathbf{e}_\alpha. \tag{1.5.4}$$

This shows that $\mathbf{U}$ and $\mathbf{V}$ have common eigenvalues but different principal vectors, and that $\mathbf{R}$ is the *rotation* which carries the principal axes of $\mathbf{U}$ into the principal axes of $\mathbf{V}$, $\mathbf{R}$ is orthogonal but need not be proper orthogonal. That is, we may have $\det \mathbf{R} = \pm 1$ (but $\det \mathbf{R}$ maintains either one of these values by continuity for all $\mathbf{X}$ and t). It follows, by forming the determinant of $\mathbf{F}$, that

$$J = \det \mathbf{U} = \det \mathbf{V}. \tag{1.5.5}$$

The tensor $\mathbf{R}$ is naturally caled the *rotation tensor*, whereas $\mathbf{U}$ and $\mathbf{V}$ are called the right and left *stretch* tensors, respectively. From (1.5.1), (1.5.2), (1.4.3), and

---

[4] See Truesdell and Noll [1965].

[5] See Eringen [1975a, p. 37].

(1.4.12), it follows that

$$\mathbf{C} = \mathbf{U}^2, \qquad \overset{-1}{\mathbf{c}} = \mathbf{V}^2. \tag{1.5.6}$$

These equations show that, while $\mathbf{C}$ and $\overset{-1}{\mathbf{c}}$ in general are computed from their definitions (1.4.3) and (1.4.12), the calculation of $\mathbf{U}$, $\mathbf{V}$, and $\mathbf{R}$, in certain special cases, may be awkward, since irrational operations are usually required. Nevertheless, the fundamental decompositions (1.5.1) play a major role in the proof of general theorems of constitutive theory. In this regard, the fundamental invariants of $\mathbf{C}$ and $\overset{-1}{\mathbf{c}}$ are important:

$$\mathrm{I}_C = \operatorname{tr} \overset{-1}{\mathbf{c}} = \operatorname{tr} \mathbf{C} = \lambda_1^2 + \lambda_2^2 + \lambda_3^2 = \sum_\alpha \lambda_\alpha^2,$$

$$\mathrm{II}_C = \tfrac{1}{2}\left[\left(\operatorname{tr} \overset{-1}{\mathbf{c}}\right)^2 - \operatorname{tr} \overset{-2}{\mathbf{c}}\right] = \tfrac{1}{2}[(\operatorname{tr} \mathbf{C})^2 - \operatorname{tr} \mathbf{C}^2] = \lambda_1^2\lambda_2^2 + \lambda_2^2\lambda_3^2 + \lambda_3^2\lambda_1^2,$$

$$\mathrm{III}_C = \det \overset{-1}{\mathbf{c}} = \det \mathbf{C} = J^2 = \lambda_1^2\lambda_2^2\lambda_3^2 = \prod_\alpha \lambda_\alpha^2, \tag{1.5.7}$$

where tr means trace.

We refer the reader to classical treatises[6] on continuum mechanics for the notion of the *strain ellipsoid of Cauchy* and a detailed discussion of finite rotation.

1.6. Infinitesimal Strains

So far no special hypotheses have been made concerning the magnitude of the deformations. In many applications, however, an approximate deformation theory is sufficient. In order to describe these approximations, the notion of *displacement* must be introduced. For the sake of simplicity, let the x_i and X_K systems have the same origin. Then the *displacement vector* denoted by $\mathbf{u}$, with components u_i in the x_i-system and with components U_K in the X_K-system, is given by (Figure 1.3.1)

$$u_k = x_k - \delta_{kL} X_L, \qquad U_K = \delta_{Ki} x_i - X_K. \tag{1.6.1}$$

The symbol δ_{kL} or δ_{Ki}, often called the *shifter*, plays the same role as the Kronecker symbol when the material and spatial frames are the same.[7]

By differentiation we have

$$x_{i,K} = \delta_{iK} + u_{i,K} \tag{1.6.2}$$

---

[6] For example, Eringen [1967, pp. 28–41] and [1980, Sect. 1.11].

[7] When coordinate frames do not coincide, they represent director cosines of the axes of the two frames.

and

$$X_{K,i} = \delta_{Ki} - U_{K,i}, \tag{1.6.3}$$

where $u_{i,K}$ $(i, K = 1, 2, 3)$ and $U_{K,i}$ $(K, i = 1, 2, 3)$ are the *displacement gradients*. It follows from (1.4.6) and (1.4.7) that

$$E_{KL} = \tfrac{1}{2}(C_{KL} - \delta_{KL}) = \tfrac{1}{2}(U_{K,L} + U_{L,K} + U_{M,K}U_{M,L}) \tag{1.6.4}$$

and

$$e_{ij} = \tfrac{1}{2}(\delta_{ij} - c_{ij}) = \tfrac{1}{2}(u_{i,j} + u_{j,i} - u_{m,i}u_{m,j}). \tag{1.6.5}$$

These are *exact* formulas. However, if $u_{i,K}$ or $U_{K,i}$ is considered as an infinitesimally small quantity, then we can neglect product terms for this quantity. From the above formulas it follows that

$$\begin{aligned} C_{KL} &\approx \delta_{KL} + U_{K,L} + U_{L,K}, \\ R_{iL} &\approx \delta_{iK}[\delta_{KL} + \tfrac{1}{2}(U_{K,L} - U_{L,K})], \\ E_{KL} &\approx \delta_{Ki}e_{ij}\delta_{jL} \approx \tfrac{1}{2}(u_{i,j} + u_{j,i})\delta_{iK}\delta_{jL}, \end{aligned} \tag{1.6.6}$$

and, to the same degree of approximation, (1.3.10) will give

$$\partial/\partial x_i \approx \delta_{Ki}\, \partial/\partial X_K, \qquad \partial/\partial X_K \approx \delta_{iK}\, \partial/\partial x_i. \tag{1.6.7}$$

This means that, in the infinitesimal strain theory, the distinction between the Lagrangian and Eulerian strains disappears. In particular, only one system of coordinates (e.g., x_i) can be used. Then the last two equations of (1.6.6) give

$$R_{ij} \approx \delta_{ij} + \tilde{r}_{ij}, \qquad E_{ij} \approx e_{ij} = \tilde{e}_{ij}, \tag{1.6.8}$$

where

$$\begin{aligned} \tilde{r}_{ij} &= u_{[i,j]} \equiv \tfrac{1}{2}(u_{i,j} - u_{j,i}) = -\tilde{r}_{ji}, \\ \tilde{e}_{ij} &= u_{(i,j)} \equiv \tfrac{1}{2}(u_{i,j} + u_{j,i}) = \tilde{e}_{ji}, \end{aligned} \tag{1.6.9}$$

and $\tilde{r}_{ij}$ and $\tilde{e}_{ij}$ are, respectively, the *infinitesimal rotation* (skew-symmetric) tensor and the *linearized Eulerian strain* tensor. The latter is the symmetric strain tensor used in classical linear elasticity theory.

1.7. Volume and Area Changes

Let dV and dv be the volume material elements in the configurations K and $\mathscr{K}$, respectively. Then, from the transformation of volume integrals we have

$$dv = J\, dV. \tag{1.7.1}$$

In a similar (but a little more involved) way we can establish a relationship between deformed and undeformed area elements: let $d\mathbf{A} = \mathbf{N}\, dA$ and $d\mathbf{a} = \mathbf{n}\, da$ be the oriented surface elements in the configurations K and $\mathscr{K}$, respectively, with corresponding unit normals $\mathbf{N}$ and $\mathbf{n}$. Then, we have the so-called

Nanson formulas:[8]

$$JN_K\,dA = x_{i,K}n_i\,da \tag{1.7.2}$$

and

$$J^{-1}n_i\,da = X_{K,i}N_K\,dA. \tag{1.7.3}$$

The volume analogues of these equations are

$$(J^{-1}x_{i,K})_{;i} = 0, \qquad (JX_{K,i})_{;K} = 0. \tag{1.7.4}$$

1.8. Compatibility Conditions

In E^3 the strain measures E_{KL} and e_{kl}, defined by (1.6.4) and (1.6.5) in general, have six independent components each. However, they are constructed from the displacement vector which has only three independent components. If the six components of a strain tensor are given, then the question arises as to the existence of a single-valued continuous displacement field corresponding to these strains. Clearly, equations such as (1.6.4) and (1.6.5) may not have solutions for the three unknowns u_i or U_K unless certain *integrability* or *compatibility* conditions are satisfied. These conditions, which constitute a set of partial differential equations, must involve only strains. For instance, in the infinitesimal strain theory, the usual compatibility conditions read (Lamé's relations)

$$\varepsilon_{ijk}\varepsilon_{pqr}\tilde{e}_{jq,kr} = 0. \tag{1.8.1}$$

In order to obtain the corresponding compatibility conditions in the finite strain theory, it is sufficient to remark that, in Newtonian continuum mechanics, the three-dimensional *material* manifold in the deformed state is Euclidean (in the absence of dislocations). That is, it is *flat*, hence with vanishing curvature. According to (1.4.2), c_{ij} and C_{KL} are metric tensors for this Euclidean manifold, if and only if, the Riemann–Christoffel curvature tensors formed from them vanish. That is,[9]

$$R_{KLMN}(\mathbf{C}) = 0 \qquad \text{or} \qquad R_{ijkl}(\mathbf{c}) = 0. \tag{1.8.2}$$

In E^3 either one of these equations (with six independent components) generalizes the condition (1.8.1) to finite strain theory. Obviously, in the infinitesimal strain theory, $\tilde{e}_{ij}$ may be considered to be a metric tensor for the deformed material manifold. Hence, the condition (1.8.2) must read $R_{ijkl}(\tilde{\mathbf{e}}) = 0$. However, in (and only in) three-dimensional space, one can associate the tensor R_{ijkl} with a dual symmetric twice-covariant tensor S_{ij} (with six independent components) given by $S_{ij} = \frac{1}{4}\varepsilon_{ikl}\varepsilon_{jmn}R_{klmn}$. Compatibility conditions (1.8.1) follow at once from the usual expression of R_{ijkl} found in all books on differential geometry,[10] and from the definition of S_{ij}.

---

[8] See Eringen [1967, pp. 71–72] and [1975a, p. 40].

[9] See Eringen [1967, p. 58] and [1980, Sect. 1.14 and Appendix C6] or [1975a, p. 42].

[10] See Eringen [1971b, p. 134].

1.9. Deformation Rates

In the kinematics of continua, time rates of various fields play an essential role. In this regard, two types of partial time differentitions may be distinguished, one with **x** fixed, the other with **X** fixed. Thus, for example, the velocity vector $\mathbf{V}(\mathbf{X}, t)$ of a material point is defined by

$$\mathbf{V}(\mathbf{X}, t) = \partial \mathbf{x}(\mathbf{X}, t)/\partial t. \tag{1.9.1}$$

Here, clearly, $\mathbf{X} =$ fixed. Also if we substitute from (1.3.2) we will have

$$\mathbf{V}(\mathbf{X}(\mathbf{x}, t), t) \equiv \mathbf{v}(\mathbf{x}, t). \tag{1.9.2}$$

Here $\mathbf{v}(\mathbf{x}, t)$ is the *velocity field* in $\mathscr{K}$ which by itself does not indicate which particle it belongs to. If we wish to obtain the acceleration vector, we can now use (1.9.1) to obtain

$$\mathbf{A}(\mathbf{X}, t) = \partial V(\mathbf{X}, t)/\partial t = \partial^2 \mathbf{x}(\mathbf{X}, t)/\partial t^2. \tag{1.9.3}$$

But in the use of (1.9.2) we need differentiation with **X** fixed for

$$\begin{aligned} \mathbf{a}(\mathbf{x}, t) \equiv \left.\frac{\partial \mathbf{v}(\mathbf{x}, t)}{\partial t}\right|_{\mathrm{X=fixed}} &= \frac{\partial \mathbf{v}(\mathbf{x}, t)}{\partial t} + \frac{\partial \mathbf{v}}{\partial x_k}\frac{\partial x_k}{\partial t} \\ &= \frac{\partial \mathbf{v}}{\partial t} + (\mathbf{v}\cdot\nabla)\mathbf{v}. \end{aligned} \tag{1.9.4}$$

It follows from this that the material derivative of a tensor field $\mathbf{A}(\mathbf{x}, t)$, denoted by d/dt or by a superposed dot, is given by

$$\frac{d\mathbf{A}}{dt} = \dot{\mathbf{A}} = \frac{\partial \mathbf{A}}{\partial t} + (\mathbf{v}\cdot\nabla)\mathbf{A} \tag{1.9.5}$$

or

$$\frac{dA_{ij\ldots k}}{dt} = \dot{A}_{ij\ldots k} = \frac{\partial A_{ij\ldots k}}{\partial t} + A_{ij\ldots k,l}v_l.$$

Of special interest is the velocity gradient tensor **L** defined in the present configuration $\mathscr{K}$ by

$$\mathbf{L} \equiv \nabla\mathbf{v}; \qquad L_{ij} = v_{i,j}. \tag{1.9.6}$$

This general second-order tensor admits a canonical decomposition into a symmetric part and a skew-symmetric part (the Euler–Cauchy–Stokes decomposition) such that

$$\mathbf{L} = \mathbf{d} + \mathbf{w}; \qquad L_{ij} = d_{ij} + w_{ij}, \tag{1.9.7}$$

where

$$d_{ij} = \tfrac{1}{2}(v_{i,j} + v_{j,i}) \equiv v_{(i,j)}, \tag{1.9.8}$$

$$w_{ij} = \tfrac{1}{2}(v_{i,j} - v_{j,i}) \equiv v_{[i,j]}. \tag{1.9.9}$$

The skew-symmetric tensor w_{ij} can be associated in a unique way with a vector of components w_i, such that

$$w_i = -\tfrac{1}{2}\varepsilon_{ijk}w_{jk} = \tfrac{1}{2}(\nabla \times \mathbf{v})_i; \qquad w_{ij} = -\varepsilon_{ijk}w_k. \tag{1.9.10}$$

For reasons that will become clear soon, $\mathbf{d}$ and $\mathbf{w}$ are called the *rate of deformation* tensor and the *rate of rotation* tensor, respectively. Hence w_i is referred to as the rate-of-rotation vector or *vorticity*. Indeed, let us examine the relations between $\mathbf{L}$, $\mathbf{d}$, and $\mathbf{w}$ and $\dot{\mathbf{F}}$, $\dot{\mathbf{C}}$, and $\dot{\mathbf{R}}$. From (1.3.4), (1.3.5), and (1.3.7), it follows imediatey that

$$\dot{\overline{x_{i,K}}} = v_{i,j}x_{j,K}, \tag{1.9.11}$$

$$\dot{\overline{dx_i}} = v_{i,j}\,dx_j, \tag{1.9.12}$$

$$\dot{\overline{X_{K,i}}} = -v_{j,i}X_{K,j}. \tag{1.9.13}$$

Employing these results and equations $(1.4.1)_2$, (1.4.3), (1.4.4), (1.4.6), and (1.4.7), we compute

$$\dot{\overline{ds^2}} = 2d_{ij}\,dx_i\,dx_j, \tag{1.9.14}$$

$$\dot{C}_{KL} = 2d_{ij}x_{i,K}x_{j,L}, \tag{1.9.15}$$

$$\dot{E}_{KL} = d_{ij}x_{i,K}x_{j,L}, \tag{1.9.16}$$

$$\dot{e}_{ij} = d_{ij} - (e_{ik}v_{k,j} + v_{k,i}e_{kj}). \tag{1.9.17}$$

Furthermore, $\mathbf{L}$ can be expressed in terms of $\dot{\mathbf{F}}$ and $\dot{\overline{\overset{-1}{\mathbf{F}}}}$, with the help of (1.9.11) and (1.9.13), by using (1.3.8) and (1.3.9). We get

$$v_{i,j} = \dot{\overline{x_{i,K}}}X_{K,j} = -x_{i,K}\dot{\overline{X_{K,j}}}. \tag{1.9.18}$$

Clearly, (1.9.11), (1.9.14), and (1.9.15) show that d_{ij} represents the rate of deformation, i.e., the time variation of the distance separating two neighboring material points. In particular, if one defines the reference configuration K instantaneously at time t, then (1.9.16) reads

$$\dot{\mathbf{E}} = \mathbf{d}. \tag{1.9.19}$$

However, it must be remarked that, in the finite strain theory, d_{ij} is not exactly the material time derivative of e_{ij} (see (1.9.17)).

The proof of the following exact relationships is left to the reader as an exercise:

$$\begin{aligned}\mathbf{w} &= \dot{\mathbf{R}}\cdot\mathbf{R}^{\mathrm{T}} + \tfrac{1}{2}\mathbf{R}\left(\dot{\mathbf{U}}\,\overset{-1}{\mathbf{U}} - \overset{-1}{\mathbf{U}}\,\dot{\mathbf{U}}\right)\mathbf{R}^{\mathrm{T}},\\ \mathbf{d} &= \tfrac{1}{2}\mathbf{R}\left(\dot{\mathbf{U}}\,\overset{-1}{\mathbf{U}} + \overset{-1}{\mathbf{U}}\,\dot{\mathbf{U}}\right)\mathbf{R}^{\mathrm{T}}.\end{aligned} \tag{1.9.20}$$

If we define instantaneously the reference configuration K at each time t, we have $\mathbf{R}^{\mathrm{T}} = \mathbf{1}$, $\overset{-1}{\mathbf{U}} = \mathbf{1}$, and eqs. (1.9.20) reduce to

$$\mathbf{w} = \dot{\mathbf{R}}, \qquad \mathbf{d} = \dot{\mathbf{U}}. \tag{1.9.21}$$

The first of these shows that $\mathbf{w}$ is the time rate of rotation.

The following time rates are easily established[11]

$$\dot{J} = \overline{\det F}^{\,\cdot} = \frac{\partial J}{\partial x_{i,K}} \overline{x_{i,K}}^{\,\cdot} = J X_{K,i} \overline{x_{i,K}}^{\,\cdot};$$

thus,

$$\dot{J} = J v_{k,k}. \tag{1.9.22}$$

Also,

$$\overline{dv}^{\,\cdot} = v_{k,k}\, dv \tag{1.9.23}$$

and

$$\overline{n_i\, da}^{\,\cdot} = (v_{k,k} n_i - v_{k,i} n_k)\, da. \tag{1.9.24}$$

In the infinitesimal strain theory, we have the following obvious reduction:

$$v_{i,j} \approx \dot{u}_{i,j}, \tag{1.9.25}$$

$$d_{ij} \approx \dot{\tilde{e}}_{ij}, \tag{1.9.26}$$

$$w_{ij} \approx \dot{\tilde{r}}_{ij}. \tag{1.9.27}$$

1.10. Rigid Body Motion

Motions in which

$$w_{ij}(\mathbf{x}, t) = 0 \tag{1.10.1}$$

for all $\mathbf{x} \in \mathscr{V}$ and all time t are called *irrotational*. They form the main subject of study in classical hydrodynamics.

Motions in which (Killing's theorem)

$$d_{ij}(\mathbf{x}, t) = 0 \tag{1.10.2}$$

for all $\mathbf{x} \in \mathscr{V}$ and all times t are called *rigid body motions*. Clearly, if a material body has undergone only a rigid displacement (see (1.4.9)), then according to (1.9.15), which is valid for all nonsingular $\mathbf{F}$, the condition (1.10.2) is fulfilled, so that the condition (1.10.2) is *necessary* for the velocity field $\mathbf{v}(\mathbf{x}, t)$ of the body to be that of a *rigid solid body*. The sufficient condition is established as follows. From the definitions (1.9.9) and (1.9.10) we have

$$w_{ij,k} = \tfrac{1}{2}(v_{i,jk} - v_{j,ik}) = \tfrac{1}{2}(v_{i,jk} + v_{k,ij} - v_{k,ij} - v_{j,ik}).$$

Hence,

$$w_{ij,k} = d_{ik,j} - d_{jk,i}.$$

[11] See Eringen [1975a, p. 48].

If all d_{ij} are zero, then the w_{ij} are constant in space. That is, $w_{ij,k} = 0$. The $v_{i,j}$ are spatially uniform, since $d_{ij} = 0$. Hence, the v_i are linear affine functions of x_i, which can be written as

$$v_i = w_{ij}x_j + V_i, \tag{1.10.3}$$

where

$$w_{ij} = -w_{ji}, \qquad w_{ij,k} = 0, \qquad V_{i,j} = 0. \tag{1.10.4}$$

w_{ij} and V_i obviously may be dependent upon time. Equation (1.10.3) represents the velocity field of a rigid body. It consists of a spatially uniform rotational velocity and a spatially uniform translational velocity.

1.11. Objectivity

In a subsequent chapter we shall see that the constitutive equations of a continuum must be *form-invariant* under rigid motions of the spatial frame of reference.

In classical mechanics, we think of an observer as being a rigid body (or a system of three rigidly attached orthonormal vectors) carrying a clock. Actually, we do not need an observer as such, but only the concept of *change of observer* or, as we shall say, *change of frame*. An ordered couple $(\mathbf{x}, t)$ where $\mathbf{x} \in E^3$, and $t \in R$ (the real line) is a certain value of time, is called an *event*. Events occupy the *space–time* continuum. By definition, a change of frame is a homeomorphism of space–time onto itself that conserves distances (*isometry*), time intervals, and time orientation. We can show in the space–time of Newtonian physics that the most general change of frame is given by

$$\bar{\mathbf{x}} = \mathbf{Q}(t)(\mathbf{x} - \mathbf{a}) + \mathbf{c}(t), \qquad \bar{t} = t - \alpha, \tag{1.11.1}$$

where $\mathbf{c}(t)$ is a time-dependent vector, $\mathbf{a}$ is a constant vector, $\mathbf{Q}(t)$ is a time-dependent orthogonal tensor, and α is a constant. The class of $\mathbf{Q}$'s is usually limited to that of *proper* orthogonal tensors. That is, $\det \mathbf{Q} = +1$ only. The second equation of (1.11.1) represents only a shift of the origin of time on the clock, since we use the concept of *absolute* Newtonian time. We may describe (1.11.1) as a "change from the unbarred frame to the barred frame," although a frame is not to be confused with a coordinate system (changes of coordinate system do not involve time). With abstraction of the time shift, we recognize in (1.11.1) the description of a rigid body motion. *The invariance of a field under arbitrary rigid motions of the spatial frame of reference is called objectivity or material frame indifference.* The axiom of objectivity is sometimes extended to include the reflections of the spatial frame of reference. In this case

$$\det \mathbf{Q} = \pm 1. \tag{1.11.2}$$

We now formalize these considerations.

Definition. A quantity is said to be frame-indifferent or objective, if and only if, it is form-invariant under all changes of frame (1.11.1).

More precisely, the following requirements are laid down: for objective scalars f, vectors $\mathbf{V}$, and second-order tensors $\mathbf{t}$

$$\begin{aligned}\bar{f} &= f,\\ \bar{\mathbf{V}} &= \mathbf{Q}\mathbf{V},\\ \bar{\mathbf{t}} &= \mathbf{Q}\mathbf{t}\mathbf{Q}^{\mathrm{T}}\\ &\text{etc} \ldots .\end{aligned} \tag{1.11.3}$$

Then the following are easily proved:

$$\begin{aligned}&\bar{\mathbf{F}} = \mathbf{Q}\mathbf{F}, \qquad \bar{\mathbf{R}} = \mathbf{Q}\mathbf{R}, \qquad \bar{\mathbf{U}} = \mathbf{U},\\ &\bar{\mathbf{V}} = \mathbf{Q}\mathbf{V}\mathbf{Q}^{\mathrm{T}}, \qquad \bar{\mathbf{C}} = \mathbf{C}, \qquad \overset{-1}{\bar{\mathbf{c}}} = \mathbf{Q}\overset{-1}{\mathbf{c}}\mathbf{Q}^{\mathrm{T}}.\end{aligned} \tag{1.11.4}$$

We define the *angular velocity* tensor $\boldsymbol{\Omega} = \{\Omega_{ij}\}$ of the barred frame with respect to the unbarred one by

$$\boldsymbol{\Omega} = \dot{\mathbf{Q}}\mathbf{Q}^{\mathrm{T}} = -\boldsymbol{\Omega}^{\mathrm{T}}. \tag{1.11.5}$$

Then the velocity field $\mathbf{v}$, the acceleration $\mathbf{a} \equiv \dot{\mathbf{v}}$, the material derivative $\dot{\mathbf{V}}$ of an *objective* vector field $\mathbf{V}$, the deformation rate $\mathbf{d}$, and the rotation rate tensor $\mathbf{w}$ can be shown to be transformed in the following manner (in a change to (1.11.1)):

$$\begin{aligned}&\bar{\mathbf{v}} - \mathbf{Q}\mathbf{v} = \dot{\mathbf{c}} + \boldsymbol{\Omega}(\bar{\mathbf{x}} - \mathbf{c}),\\ &\dot{\bar{\mathbf{v}}} - \mathbf{Q}\dot{\mathbf{v}} = \ddot{\mathbf{c}} + 2\boldsymbol{\Omega}(\bar{\mathbf{v}} - \dot{\mathbf{c}}) + (\dot{\boldsymbol{\Omega}} - \boldsymbol{\Omega}^2)(\bar{\mathbf{x}} - \mathbf{c}),\\ &\dot{\bar{\mathbf{V}}} - \mathbf{Q}\dot{\mathbf{V}} = \dot{\mathbf{Q}}\mathbf{V},\\ &\bar{\mathbf{d}} = \mathbf{Q}\mathbf{d}\mathbf{Q}^{\mathrm{T}}, \qquad \bar{\mathbf{w}} = \mathbf{Q}\mathbf{w}\mathbf{Q}^{\mathrm{T}} + \boldsymbol{\Omega}.\end{aligned} \tag{1.11.6}$$

The results (1.11.4)–(1.11.6) show that $\mathbf{F}, \mathbf{R}, \mathbf{U}, \mathbf{V}, \mathbf{C}, \overset{-1}{\mathbf{c}}, \mathbf{E}, \mathbf{e}$, and $\mathbf{d}$ are objective whereas the velocity $\mathbf{v}$, the acceleration $\mathbf{a}$, and the rotation-rate tensor $\mathbf{w}$ are not objective. Also, the material derivative of an objective vector (or of tensor fields of higher order) is not objective (see $(1.11.6)_3$). However, it may happen in constructing involved theories of continua that we need *objective time rates* of objective tensors. We examine this question in the following section.

1.12. Objective Time Rates of Tensor Fields

Among the numerous objective time rates of tensor fields that we may construct, two are of the greatest interest because of their applications to electromagnetic continua. The first one is the so-called *co-rotational derivative*, of which the *Jaumann derivative* is a special case; the second one is the *convective-time derivative*.

Co-rotational Derivative

The notion of co-rotational derivative is naively but efficiently introduced as follows. A triad $\mathbf{e}^\alpha$, $\alpha = 1, 2, 3$, of unit orthogonal vectors defined at $(\mathbf{x}, t)$ represents a reference frame. Accordingly, the conditions of unit length and orthogonality of these vectors read as

$$e_i^\alpha e_j^\alpha = \delta_{ij}, \qquad \mathbf{e}^\alpha \cdot \mathbf{e}^\beta = \delta_{\alpha\beta}, \tag{1.12.1}$$

where α is not a tensorial index, but serves only to label the vectors of the triad. Since the $\mathbf{e}^\alpha$ forms a rigid triad at all times t, by taking the time derivative of (1.12.1), we can show that

$$\frac{d\mathbf{e}^\alpha}{dt} = \mathbf{\Omega} \cdot \mathbf{e}^\alpha; \qquad \frac{de_i^\alpha}{dt} = \Omega_{ij} e_j^\alpha, \tag{1.12.2}$$

where $\Omega_{ij} = -\Omega_{ji}$ is the absolute angular velocity of the triad. Let $\mathbf{V}$ and $\mathbf{t}$ be, respectively, a vector field and a second-order tensor field. We can form their components on the triad $\mathbf{e}^\alpha$ in such a way that

$$V^\alpha \equiv V_i e_i^\alpha, \qquad t^{\alpha\beta} \equiv e_i^\alpha t_{ij} e_j^\beta. \tag{1.12.3}$$

Taking the material derivative of both sides of each of these equations, and using (1.12.2), we get

$$\dot{V}^\alpha = (D_\Omega \mathbf{V})_i e_i^\alpha, \qquad \dot{t}^{\alpha\beta} = (D_\Omega \mathbf{t})_{ij} e_i^\alpha e_j^\beta, \tag{1.12.4}$$

where

$$\begin{aligned} (D_\Omega \mathbf{V})_i &= \dot{V}_i - \Omega_{ij} V_j, \\ (D_\Omega \mathbf{t})_{ij} &= \dot{t}_{ij} - \Omega_{ik} t_{kj} - \Omega_{jk} t_{ik}, \end{aligned} \tag{1.12.5}$$

are the *co-rotational* derivatives of $\mathbf{V}$ and $\mathbf{t}$ with respect to the the triad $\mathbf{e}^\alpha$. An important case is that for which Ω_{ij} is none other than the rate of rotation tensor of the deformable matter.[12] Then equations (1.12.5) provide the Jaumann derivatives—noted D_J—of $\mathbf{V}$ and $\mathbf{t}$:

$$\begin{aligned} (D_w \mathbf{V})_i &\equiv (D_J \mathbf{V})_i = \dot{V}_i - w_{ij} V_j, \\ (D_w \mathbf{t})_{ij} &\equiv (D_J \mathbf{t})_{ij} = \dot{t}_{ij} - w_{ik} t_{kj} - w_{jk} t_{ik}. \end{aligned} \tag{1.12.6}$$

From the definitions (1.12.3) it follows that:

Lemma 1. *If* $\mathbf{V}$ *and* $\mathbf{t}$ *are objective, then* $D_J \mathbf{V}$ *and* $D_J \mathbf{t}$ *are objective.*

Clearly, the definition of the co-rotational and Jaumann derivatives places the emphasis on rotational effects, so that these will prove important in the study of the precession of the magnetic spin in ferromagnetism.[13]

---

[12] In that case, it can be shown that the $\mathbf{e}^\alpha$ are none other than, at each time, the unit eigenvectors of the deformation rate tensor. Then $(1.12.2)_2$ expresses Gosiewski's theorem (see Eringen [1975a, Theorem 2, p. 52]).

[13] See Chapter 9 (Vol. II).

Convective-Time Derivative

The notion of the convective-time derivative can be introduced as follows. Let $\mathbf{a}$ be a vector field in the present configuration $\mathscr{K}$. This vector field can be "pulled back" to the reference configuration K in the following fashion:

$$A_K = X_{K,i}a_i. \tag{1.12.7}$$

Computing the time derivative of $\mathbf{A}$ with the help of (1.9.13) we obtain

$$\dot{A}_K = X_{K,i}(\mathbf{D}_C a)_i, \tag{1.12.8}$$

where

$$(D_C\mathbf{a})_i \equiv \dot{a}_i - v_{i,j}a_j = [\dot{\mathbf{a}} - (\mathbf{a}\cdot\nabla)\mathbf{v}]_i \tag{1.12.9}$$

is the contravariant convective-time derivative of $\mathbf{a}$. Since $X_{K,i}$ is objective, it is clear that, according to the process of construction just used, $D_C\mathbf{a}$ is an objective vector field, if $\mathbf{a}$ is an objective vector field. The definition (1.12.9) is readily extended to tensors of any order. For instance, for the second-order tensor of the components t_{ij}, we have

$$(D_C\mathbf{t})_{ij} = \dot{t}_{ij} - v_{i,k}t_{kj} - v_{j,k}t_{ik}. \tag{1.12.10}$$

Finally, another convective-time derivative, which proves essential in the formulation of the electrodynamics of continua, will be denoted by a superposed asterisk. It is defined by

$$\overset{*}{\mathbf{t}} \equiv D_C\mathbf{t} + \mathbf{t}(\nabla\cdot\mathbf{v}). \tag{1.12.11}$$

For instance, for a second-order tensor t_{ij} and a vector field B_i, we have

$$\begin{aligned} \overset{*}{t}_{ij} &= \dot{t}_{ij} - v_{i,k}t_{kj} - v_{j,k}t_{ik} + t_{ij}v_{k,k}, \\ \overset{*}{B}_i &= \dot{B}_i - v_{i,j}B_j + B_i v_{k,k}. \end{aligned} \tag{1.12.12}$$

Furthermore, on account of (1.9.5), the vector form of the second of these, which reads

$$\overset{*}{\mathbf{B}} = \dot{\mathbf{B}} - (\mathbf{B}\cdot\nabla)\mathbf{v} + \mathbf{B}(\nabla\cdot\mathbf{v}), \tag{1.12.13}$$

is easily shown to read

$$\overset{*}{\mathbf{B}} = \frac{\partial\mathbf{B}}{\partial t} + \nabla\times(\mathbf{B}\times\mathbf{v}) + \mathbf{v}(\nabla\cdot\mathbf{B}). \tag{1.12.14}$$

If t_{ji} and B_i are objective fields, then $\overset{*}{t}_{ji}$ and $\overset{*}{B}_i$ are objective, since d_{ij} is objective. Clearly, by accounting for the last term in (1.12.11), the new convective derivative thus defined takes account, of volume changes also. Indeed, if the vector field $\mathbf{B}$ is in $\mathscr{K}$, we define the corresponding vector in K with components

$$B_K \equiv JX_{K,i}B_i, \qquad B_i = J^{-1}x_{i,K}B_K, \tag{1.12.15}$$

Then we can check, with the help of (1.9.22), that

$$\dot{B}_K \equiv \left.\frac{\partial B_K}{\partial t}\right|_{\mathbf{X}} = JX_{K,i}\overset{*}{B}_i. \tag{1.12.16}$$

1.13. Kinematics of Line, Surface, and Volume Integrals

The formulas for line, surface, and volume changes may be used to establish a series of results usually known as *transport theorems*.

Lemma 1. *The material derivative of a line integral of any field ϕ over a material line $\mathscr{C}$ is calculated by*

$$\frac{d}{dt}\int_{\mathscr{C}} \phi\, dx_k = \int_{\mathscr{C}} [\dot{\phi}\, dx_k + \phi v_{k,l}\, dx_l]. \tag{1.13.1}$$

Proof. The integral on the left-hand side of (1.13.1) has fixed limits in the material description because $\mathscr{C}$ is a material line. Thus the operator d/dt commutes with the integral sign. Using (1.9.12), we obtain (1.13.1).

If $\mathbf{t}$ is the unit-oriented tangent to $\mathscr{C}$, we have $dx_i = t_i\, dl$.

Lemma 2. *The material derivative of a surface integral of any field ϕ over a material surface s is given by*

$$\frac{d}{dt}\int_s \phi\, da_k = \int_s [\dot{\phi}\, da_k + \phi(-v_{l,k}\, da_l + v_{p,p}\, da_k)]. \tag{1.13.2}$$

The proof follows by carrying d/dt inside the integral (since s is fixed in the material description) and using (1.9.24).

A particular case of the result (1.13.2) obviously is

$$\frac{d}{dt}\int_s q_k\, da_k = \int_s \overset{*}{\mathbf{q}}\cdot d\mathbf{a}, \tag{1.13.3}$$

where $\overset{*}{\mathbf{q}}$ is the convective-time derivative defined in (1.12.13) or (1.12.14).

Lemma 3. *The material derivative of any field ϕ over a material volume $\mathscr{V}$ is given by*

$$\frac{d}{dt}\int_{\mathscr{V}} \phi\, dv = \int_{\mathscr{V}} (\dot{\phi} + \phi v_{k,k})\, dv = \int_{\mathscr{V}} \left[\frac{\partial \phi}{\partial t} + (\phi v_k)_{,k}\right] dv. \tag{1.13.4}$$

The proof again follows by carrying d/dt inside the integral (since $\mathscr{V}$ is fixed in the material description) and using (1.9.23).

We now extend the results (1.13.3), (1.13.4) to material surfaces and volumes

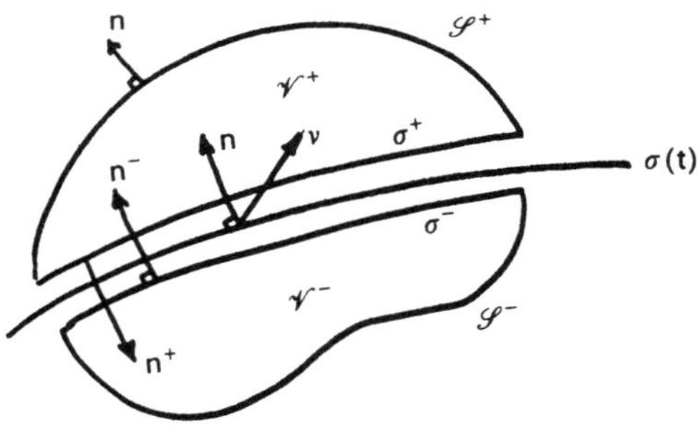

Figure 1.13.1. Discontinuity surface.

containing discontinuity lines and surfaces. For discontinuities sweeping a material manifold are of common occurrence (e.g., shock waves, acceleration waves) in continuum physics. First, we consider the case of a material volume $\mathscr{V}$ intersected by a discontinuity surface $\sigma(t)$ moving with an absolute velocity $\mathbf{v}$ (see Figure 1.13.1). By use of the Green–Gauss theorem, the result (1.13.4) can be rewritten as

$$\frac{d}{dt}\int_v \phi\, dv = \int_v \frac{\partial \phi}{\partial t}\, dv + \int_s \phi v_k\, da_k, \qquad (1.13.5)$$

where $\mathscr{V}$ and its boundary $\partial\mathscr{V}$ have been selected to coincide instantaneously with a spatial volume v and its boundary s. Applying the result (1.13.5) to the two volumes $\mathscr{V}^+$ and $\mathscr{V}^-$ bounded by $s^+ \oplus \sigma^+$ and $s^- \oplus \sigma^-$, respectively, we write

$$\frac{d}{dt}\int_{\mathscr{V}^+} \phi\, dv = \int_{\mathscr{V}^+} \frac{\partial \phi}{\partial t}\, dv + \int_{s^+} \phi \mathbf{v}\cdot d\mathbf{a} - \int_{\sigma^+} \phi\mathbf{v}\cdot d\mathbf{a},$$

$$\frac{d}{dt}\int_{\mathscr{V}^-} \phi\, dv = \int_{\mathscr{V}^-} \frac{\partial \phi}{\partial t}\, dv + \int_{s^-} \phi \mathbf{v}\cdot d\mathbf{a} + \int_{\sigma^-} \phi\mathbf{v}\cdot da.$$

Now add these two equations and let σ^+ and σ^- approach σ. This gives

$$\frac{d}{dt}\int_{\mathscr{V}-\sigma} \phi\, dv = \int_{\mathscr{V}-\sigma} \frac{\partial \phi}{\partial t}\, dv + \int_{\partial\mathscr{V}-\sigma} \phi \mathbf{v}\cdot d\mathbf{a} - \int_{\sigma(t)} [\phi\mathbf{v}]\cdot d\mathbf{a}, \quad (1.13.6)$$

where $\mathscr{V} - \sigma$ and $\partial\mathscr{V} - \sigma$ denote the material volume and surface excluding those points lying on $\sigma(t)$.

A simpler but similar proof[14] yields the following generalized form of the Green–Gauss theorem:

$$\int_{\mathscr{V}-\sigma} \nabla\cdot\mathbf{A}\, dv + \int_{\sigma(t)} [\mathbf{A}]\cdot d\mathbf{a} = \int_{\partial\mathscr{V}-\sigma} \mathbf{A}\cdot d\mathbf{a}. \qquad (1.13.7)$$

By means of this result the second integral on the right-hand side of (1.13.6)

[14] See Eringen [1967, p. 428] and [1980, Appendix A2].

may be converted to a volume integral. Thus,

$$\frac{d}{dt}\int_{\mathscr{V}-\sigma} \phi \, dv = \int_{\mathscr{V}-\sigma} \left\{\frac{\partial \phi}{\partial t} + \nabla \cdot (\mathbf{v}\phi)\right\} dv + \int_{\sigma(t)} [\phi(\mathbf{v} - \boldsymbol{\nu})] \cdot d\mathbf{a}. \quad (1.13.8)$$

In both (1.13.6) and (1.13.8) boldface brackets are used to indicate the jump of its enclosure across $\sigma(t)$, i.e.,

$$[\mathbf{A}] \equiv \mathbf{A}^+ - \mathbf{A}^-, \quad (1.13.9)$$

where $\mathbf{A}^+$ and $\mathbf{A}^-$ are the values of $\mathbf{A}$ on σ as approached from the positive and negative sides of the normal $\mathbf{n}$ of σ (we recall that $d\mathbf{a} = \mathbf{n}\, da$).

A similar argument can be used to extend (1.13.3) to an open material surface s containing a discontinuity line γ moving with an absolute velocity $\boldsymbol{\nu}$ on s. In the absence of a discontinuity line, the result (1.13.3) is also valid for an arbitrary open spatial surface $s(t)$ bounded by a closed curve $c(t)$, the whole moving with a velocity $\mathbf{v}$. Using Stokes' theorem,

$$\int_{s(t)} (\nabla \times \mathbf{A}) \cdot d\mathbf{a} = \int_{c(t)} \mathbf{A} \cdot d\mathbf{x}, \quad (1.13.10)$$

where we assumed that $s(t)$ had no edges (i.e., has a continuous tangent plane), and (1.12.14), we can write, in lieu of (1.13.3),

$$\frac{d}{dt}\int_{s(t)} \mathbf{q} \cdot d\mathbf{a} = \int_{s(t)} \left\{\frac{\partial \mathbf{q}}{\partial t} + \mathbf{v}(\nabla \cdot \mathbf{q})\right\} \cdot d\mathbf{a} + \int_{c(t)} (\mathbf{q} \times \mathbf{v}) \cdot d\mathbf{x}. \quad (1.13.11)$$

Applying (1.13.11) to the surfaces s^+ and s^- (see Figure 1.13.2) and using the theorem (1.13.10) generalized in the form[15]

$$\int_{s-\gamma} (\nabla \times \mathbf{A}) \cdot d\mathbf{a} + \int_{\gamma(t)} [\mathbf{A}] \cdot d\mathbf{x} = \int_{\partial s - \gamma} \mathbf{A} \cdot d\mathbf{x} \quad (1.13.12)$$

to account for the presence of γ (here also it is assumed that $s - \gamma$ has a

[15] See Eringen [1967, p. 428] and [1980, p. 524].

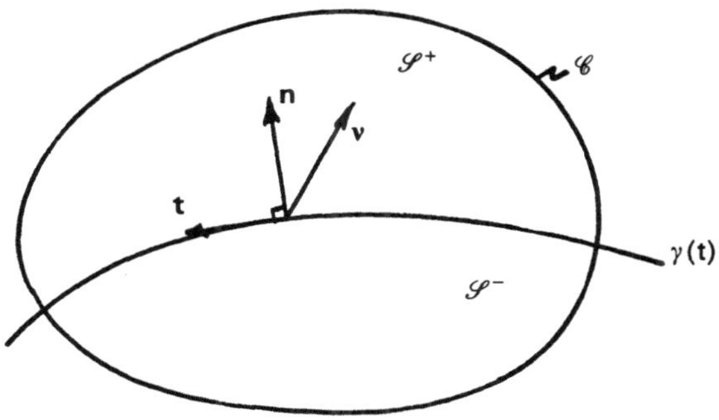

Figure 1.13.2. Discontinuity line.

continuous tangent plane), we obtain

$$\frac{d}{dt}\int_{\mathscr{S}-\gamma} \mathbf{q}\cdot d\mathbf{a} = \int_{\mathscr{S}-\gamma} \overset{*}{\mathbf{q}}\cdot d\mathbf{a} + \int_{\gamma(t)} [\mathbf{q}\times(\mathbf{v}-\mathbf{v})]\cdot d\mathbf{x}. \tag{1.13.13}$$

This last result is particularly important in electromagnetic theory. Three other results or definitions are important in this theory. First, consider Stokes' theorem and consider the case where $\mathbf{A}$ is of the form $\mathbf{q}\times\mathbf{n}$ in which $\mathbf{q}$ is a vector field and $\mathbf{n}$ is the unit outward normal to $s(t)$. Define the operator of projection P_{ij} on the tangent plane to $s(t)$, the surface gradient ∇_s, and the mean curvature Ω of $s(t)$ by

$$\begin{aligned} P_{ij} &\equiv \delta_{ij} - n_i n_j, \\ (\nabla_s)_i &= P_{ij}\nabla_j, \\ 2\Omega &= -\nabla\cdot\mathbf{n} = -P_{ij}n_{i,j}, \end{aligned} \tag{1.13.14}$$

respectively. The unit tangent $\mathbf{t}$ is oriented in the direct sense about $\mathbf{n}$. The unit binormal $\mathbf{b}$ to $c(t)$ is defined by

$$\mathbf{b} = \mathbf{t}\times\mathbf{n}. \tag{1.13.15}$$

Because of (1.13.14) and (1.13.15), equation (1.13.10) with $\mathbf{A}\equiv\mathbf{q}\times\mathbf{n}$ yields

$$\int_{c(t)} \mathbf{q}\cdot\mathbf{b}\,dl = \int_{s(t)} (\nabla_s\mathbf{q} + 2\Omega\mathbf{q}\cdot\mathbf{n})\,da, \tag{1.13.16}$$

where we used $(\nabla\mathbf{n})\cdot\mathbf{n} = 0$, which follows from $\mathbf{n}\cdot\mathbf{n} = 1$. If $\mathbf{q}$ is tangential to $s(t)$, then $\mathbf{q}\cdot\mathbf{n} = 0$, and (1.13.16) reduces to

$$\int_{c(t)} \mathbf{q}\cdot\mathbf{b}\,dl = \int_{s(t)} (\nabla_s\cdot\mathbf{q})\,da. \tag{1.13.17}$$

If on $s(t)$ there is a fixed singular curve γ (an edge where $s(t)$ has a discontinuous tangent plane), then (1.13.17) must be replaced, because of (1.13.10), by

$$\int_{c(t)} \mathbf{q}\cdot\mathbf{b}\,dl = \int_{s(t)-\gamma} (\nabla_s\cdot\mathbf{q} + 2\Omega\mathbf{q}\cdot\mathbf{n})\,da + \int_{\gamma} [\mathbf{q}\cdot\mathbf{b}]\,dl. \tag{1.13.18}$$

If $c(t)$ is *closed*, then the left-hand side of this equation vanishes.

Next, we evaluate integrals of the form

$$\frac{\delta}{\delta t}\int_{s(t)} \phi\,da \tag{1.13.19}$$

on an open surface $s(t)$ moving with velocity $\mathbf{v}$, where ϕ is a scalar and $\delta/\delta t$ is the total time derivative in following the motion of $s(t)$. In the absence of a discontinuity line on $s(t)$, an equation similar to (1.13.11) holds true with $\mathbf{v}$ replaced by $\mathbf{v}$ and $\mathbf{q}\equiv\phi\mathbf{n}$. That is, because of (1.13.15),

$$\frac{\delta}{\delta t}\int_{s(t)} \phi\,da = \int_{s(t)} \left\{\frac{\partial(\phi\mathbf{n})}{\partial t} + \mathbf{v}[\nabla\cdot(\phi\mathbf{n})]\right\}\cdot d\mathbf{a} + \int_{c(t)} \phi\mathbf{v}\cdot\mathbf{b}\,dl. \tag{1.13.20}$$

Upon using (1.13.16), (1.13.14), while noting that $\partial\mathbf{n}/\partial t\cdot\mathbf{n} = 0$, and defining the time derivative (in following the motion of $s(t)$ along its normal) by

$$\frac{\bar{\delta}}{\delta t} \equiv \frac{\partial}{\partial t} + V\frac{\partial}{\partial n},$$
$$V \equiv \mathbf{v}\cdot\mathbf{n}, \qquad \frac{\partial}{\partial n} \equiv \mathbf{n}\cdot\nabla, \tag{1.13.21}$$

we obtain

$$\frac{\delta}{\delta t}\int_{\mathscr{S}} \phi\, da = \int_{\mathscr{S}} \left[\frac{\bar{\delta}\phi}{\delta t} + \nabla_s\cdot(\mathbf{v}\phi)\right] da. \tag{1.13.22}$$

Note the similarity between (1.13.17) and the Green–Gauss theorem on the one hand, and between (1.13.22) and (1.13.4) on the other.

If the surface $s(t)$ possesses a discontinuity line $c(t)$ moving with its own velocity $w(t)$ on $s(t)$ (Figure 1.13.1), then (1.13.22) is generalized to

$$\frac{\delta}{\delta t}\int_{\mathscr{S}-\mathscr{C}} \phi\, da = \int_{\mathscr{S}-\mathscr{C}} \left[\frac{\bar{\delta}\phi}{\delta t} + \nabla_s\cdot(\mathbf{v}\phi)\right] da - \int_{\mathscr{C}} \mathbf{t}\cdot[\mathbf{n}\times(w-\mathbf{v})\phi]\, ds. \tag{1.13.23}$$

This formula remains valid when ϕ is a tensor field.[16]

Finally, we set forth the following useful definition:

Definition. The line integral of the tangential component of a vector field $\mathbf{q}$ along a curve c is called its circulation. Thus,

$$\Gamma_{(q)} = \oint_c \mathbf{q}\cdot d\mathbf{x}. \tag{1.13.24}$$

Using Stokes' theorem (1.13.10), this also reads

$$\Gamma_{(q)} = \int_s (\nabla\times\mathbf{q})\cdot d\mathbf{a}, \tag{1.13.25}$$

where s is any surface bounded by c. This is stated as: *the circulation of* $\mathbf{q}$ *around a closed circuit is equal to the flux of* $\nabla\times\mathbf{q}$ *through any surface bounded by the circuit.* The notion of circulation plays a prominent role in both electromagnetism and fluid dynamics.

Problems

1.1. Prove the following identities:

$$(JX_{K,k})_{,K} = 0, \qquad (J^{-1}x_{k,K})_{,k} = 0.$$

1.2. Show that the Piola and the Finger tensors $\overset{-1}{\mathbf{c}}$ and $\overset{-1}{\mathbf{C}}$ are, respectively, the

---

[16] Private communication from E.S. Şuhubi.

reciprocal tensors to **c** and **C**, e.g.,

$$\overset{-1}{c}_{kl} c_{lm} = \delta_{km}.$$

1.3. The displacement field in a body is given by

$$X_1 = A(r)\cos(\theta - \psi z), \qquad X_2 = A(r)\sin(\theta - \psi z), \qquad X_3 = z/\lambda,$$

where (r, θ, z) are cylindrical polar coordinates in the deformed body. Determine the deformaton tensors c_{kl}, C_{KL} and the infinitesional strain tensor $\tilde{e}_{kl}$.

1.4. In Problem 1.3, for the special case $A(r) = r\lambda^{1/2}$, $\psi = 0$, find the invariants of c_{kl}. Discuss the geometrical picture of deformation.

1.5. Prove the relations (1.5.7).

1.6. Prove the relations (1.7.1), (1.7.2), and $(1.7.4)_2$.

1.7. Prove that a second-order tensor a_{KL} satisfies the equation

$$-\mathbf{a}^3 + \mathrm{I}_a \mathbf{a}^2 - \mathrm{II}_a \mathbf{a} + \mathrm{III}_a \mathbf{1} = \mathbf{0}.$$

This is known as the Cayley–Hamilton theorem.

1.8. The deformaton of a spherical shell is represented by

$$X_1 = A(r)\sin\theta\cos\phi, \qquad X_2 = A(r)\sin\theta\sin\phi, \qquad X_3 = A(r)\cos\theta,$$

where (r, θ, ϕ) are spherical polar coordinates of the deformed body.

(i) For an isochoric deformation, determine the deformation tensors c_{kl} and $\overset{-1}{c}_{kl}$.
(ii) Calculate the invariants of $\overset{-1}{c}_{kl}$.
(iii) Determine the rotation tensor r_{kl}.

1.9. The deformation of a rod is prescribed by

$$X_1 = f(r), \qquad X_2 = g(\theta), \qquad X_3 = z/\lambda,$$

where (r, θ, z) are cylindrical coordinates of the deformed body.
(i) Determine the volume change.
(ii) For an isochoric deformation, calculate the deformation tensor c_{kl} and the rotation tensor r_{kl}.
(iii) Calculate components of the normal vector to the deformed surface $r = r_1$.

1.10. Express compatibility conditions for the plane strain in two-dimensional polar coordinates.

1.11. Calculate the second material time derivative of c_{kl}.

1.12. Calculate the material time derivative of a surface area element da_k.

1.13. The velocity field of a continuum is prescribed by

$$\dot{x}_1 = V(a/r)^2\cos 2\theta, \qquad \dot{x}_2 = V(a/r)^2\sin 2\theta, \qquad \dot{x}_3 = 0,$$

where (r, θ, z) are the cylindrical coordinates. Calculate:
(i) the deformation rate tensor d_{kl};
(ii) the spin tensor w_{kl} and the vorticity vector w_k;
(iii) the invariants of d_{kl}.
(iv) Sketch the path lines (lines $x_k = x_k(\mathbf{X}, t)$ on which $\mathbf{X}$ = fixed).

1.14. (Helmholtz–Zorawski Criterion.) The necessary and sufficient conditions for lines tangential to a vector field $\mathbf{q}$ to be material are

$$\mathbf{q} \times \left[\frac{\partial \mathbf{q}}{\partial t} + \operatorname{curl}(\mathbf{q} \times \mathbf{v}) + \mathbf{v} \operatorname{div} \mathbf{q} \right] = \mathbf{0},$$

where $\mathbf{v}$ is the velocity field. What meaning can be attributed to the special case $\mathbf{q} = \mathbf{v}$?

1.15. Show that $\dot{d}_{kl}$ is not an objective tensor. Find a tensor involving $\dot{d}_{kl}$ that is objective.

1.16. Cauchy's stress tensor t_{kl} obeys the laws of motion

$$t_{kl,k} + \rho(f_1 - \dot{v}_l) = 0, \qquad t_{kl} = t_{lk}, \tag{1}$$

where $\mathbf{f}$ is the body force density and $\mathbf{v}$ is the velocity vector. Piola stress tensors T_{kl} and T_{KL} are defined by

$$T_{KL} = T_{Kl} X_{L,l} = J t_{kl} x_{k,K} x_{l,L}; \tag{2}$$

show that the *pseudo*stress T_{Kl} is the stress at $\mathbf{x}$ measured per unit area of $\mathbf{X} = \mathbf{X}(\mathbf{x}, t)$. Obtain the equation of motion for T_{Kl} and T_{KL}.

1.17. By writing

$$X_{K,k} = (\delta_{lk} - \tilde{e}_{lk} - \tilde{r}_{lk})\delta_{lK}$$

in (2) of Problem 1.16, show that if the strains e_{lk} are small compared to unity, we have

$$T_{Kl} \simeq (t_{ml} - \tilde{r}_{mk} t_{kl})\delta_{mK},$$

$$T_{KL} \simeq (t_{mk} - \tilde{r}_{ml} t_{lk} - \tilde{r}_{kl} t_{lm} + \tilde{r}_{mn} \tilde{r}_{kl} t_{nl})\delta_{mK}\delta_{kL}.$$

Under what conditions will T_{Kl} and T_{KL} be equal to t_{kl}?

1.18. (Short term paper.) Make a literature search and write a paper on the kinematical jump conditions across a shock wave.

CHAPTER 2

Microscopic Electromagnetic Theory

2.1. Scope of the Chapter

In addition to the usual notion of material continuum and the kinematical variables (see Chapter 1), the electrodynamics of continua requires the introduction of new fields. These are the classical electromagnetic fields which exist independently of the fact that space is occupied by a vacuum or is filled by matter. Fields, such as electric charge density, electric polarization, and magnetization reflect the presence of matter. In classical physics, as developed in the nineteenth century, these fields must satisfy a set of partial differential equations known as *Maxwell's equations*. Continuum physics is concerned with the gross, macroscopic behavior of bodies. However, the drastic development of physics during the last eighty years has shown that it is logical to start with the existence of particles at the microscopic level, and to show that Maxwell's equations and electromagnetic theory are but the macroscopic results of the statistical laws that govern them on a finer scale. This line of thought was initiated by H.A. Lorentz in his celebrated theory of electrons. Therefore, in the same way that continuum mechanics is built progressively on the successive notions of mass density, momentum, moment of momentum, and energy, it seems appropriate to introduce those notions (which describe electromagnetic fields in matter) by starting with the initially simple notion of electric point charge (Section 2.2), and then introducing a more realistic but complex picture of assemblies of electric charges and their moments. This is achieved in Sections 2.3 and 2.4, where the notion of electromagnetic multipoles of various orders and macroscopic fields (such as electric charge density, electric polarization, magnetization, and electric current), emerge quite naturally.

In Section 2.5 we discuss briefly the Maxwell–Lorentz theory of microscopic electromagnetic theory, and in Section 2.6 we obtain the expressions of the nonrelativistic force acting on a point particle α with mass m^α and charge e^α.

In Section 2.7 the Maxwell–Lorentz theory is then employed to obtain microscopic field equations for an assembly of charges occupying an atom or a molecule. This process leads naturally to various fundamental concepts, such as atomic and microscopic electric and magnetic moments and currents. In Sections 2.8, 2.9, and 2.10 we derive the expressions of force, couple, and power of electromagnetic fields on a stable assembly of particles (atoms, molecules). These expressions are fundamental to the development of the dynamical electromagnetic theory of deformable bodies.

2.2. Electric Charge

According to modern physics, matter is made up of an aggregate of point-like particles which are endowed with mass and electric charge. All physical phenomena are the result of the interactions of these mass points and charges during their motion. Electronic charge is the smallest possible division of charge having a magnitude of $e = 1.6021917 \times 10^{-19}$ coulomb with $+$ or $-$ signs. An atom is made up of electrons, each carrying a negative charge $-e$, and a nucleus having a positive charge equal to the total charge carried by its electrons. In this book, we are not concerned with the details of the motions of electrons present in a body, but only with those of *stable* groups of charges. In fact, ultimately, we will be concerned only with the behavior of certain crucial statistical moments of stable groups of charges that can describe the macroscopic electromagnetic phenomena, to within the desired accuracy. In this respect, the basic notion is that of a *complex of electric charges*. By this, we understand a *stable* group of electric charges which remain close to one another in the course of their motion in physical space. Thus, for example, an atom or a cluster of atoms, constituting a molecule, may be understood as a complex.

In a microscopic volume element, the number of such stable groups is so large that statistical averages of various physical quantities give extremely high accuracy to the observed phenomena. The continuum theory of electrodynamics, therefore, is justified on the basis of the high accuracy of statistical averaging that smoothes out discontinuities present in the atomic scale.

Many important concepts of the continuum theory take their origins from a discussion of simple models constructed by means of point-like particles. Therefore, we begin our discussion of the point particles as a preparation for the continuum theory.

The simplest point particle is that made up of only one electronic charge e, has mass m, and is situated at a point $\mathbf{x}$, with respect to some Galilean frame of reference R_G—the so-called laboratory frame. Its absolute velocity in R_G is $\mathbf{v} = \dot{\mathbf{x}} \equiv \partial \mathbf{x}/\partial t$. With the help of m and e, the following two basic concepts can be constructed:

(a) *the linear momentum of the point particle*: $m\mathbf{v}$;
(b) *the electric current (convection current) due to the motion of the charge*

$$\mathbf{j} = e\mathbf{v}. \tag{2.2.1}$$

By taking the moment of these two quantities with respect to the origin O of R_G, we define the *angular momentum* $\mathbf{s}_{(O)}$ and the magnetic moment $\mathbf{m}_{(O)}$ of the electric charge about O by

$$\mathbf{s}_{(O)} = \mathbf{x} \times m\mathbf{v} \tag{2.2.2}$$

and

$$\mathbf{m}_{(O)} = \frac{1}{2c}\mathbf{x} \times e\mathbf{v} = \frac{1}{2c}\mathbf{x} \times \mathbf{j}, \tag{2.2.3}$$

respectively, where the constant $c = (2.998925 \pm 0.00006 \times 10^8$ m/sec) is the velocity of light in a vacuum and has been introduced so that the dimension $|\mathbf{m}_{(O)}|$ is that of electric charge times distance. Clearly, from (2.2.2) and (2.2.3), we have

$$\mathbf{m}_{(O)} = \gamma \mathbf{s}_{(O)}, \tag{2.2.4}$$

where

$$\gamma = e/2mc \tag{2.2.5}$$

is called the *gyromagnetic ratio* of the electronic charge e of mass m. The magnetic moment defined by (2.2.3) is due to the *orbital motion* of the charged particle about O.

Consider now the case of the elementary electric quantum called the *electron* with negative electric charge $-e$ and the so-called *rest mass* $m_0 = 9.11 \times 10^{-31}$ kg. Although the electric charge here is considered a point-like particle, it is endowed with an internal angular momentum, called *spin*. The magnetic moment of an electron may be due to two types of motion, its orbital motion about an origin O and its *spin*. If $\mathbf{x}$ and $\mathbf{v}_e = \dot{\mathbf{x}}$ are the position and the velocity vectors of the electron, then in a "stationary" motion, the electron has an angular momentum $\mathbf{s}_{(O)}$ given by

$$\mathbf{s}_{(O)} = \overline{\mathbf{x} \times m_0 \mathbf{v}_e}, \tag{2.2.6}$$

where the bar indicates a time average. Accordingly, its magnetic moment is

$$\mathbf{m}_{(O)} = \frac{1}{2c}\overline{\mathbf{x} \times (-e\mathbf{v}_e)} = -\frac{e}{2m_0 c}\mathbf{s}_{(O)} = \gamma_e \mathbf{s}_{(O)}. \tag{2.2.7}$$

The latter equation, in fact, also holds for a collection of electrons such as those present in an atom. For the pure orbital motion of an electron, we have $\gamma_e = -e/2m_0 c$. More generally, this is modified by a g-factor.

$$\gamma = -\frac{e}{2m_0 c} g, \tag{2.2.8}$$

so that the unit angular momentum is taken as Plank's constant $\hbar\,(= 1.05459 \times 10^{-27}$ erg $\times$ sec) of quantum mechanics. Accordingly, the unit magnetic moment of an electron is the *Bohr magneton* defined by

$$\mu_B = \frac{e\hbar}{2m_0 c} = 0.927 \times 10^{-20} \text{ erg/gauss}. \tag{2.2.9}$$

Clearly $g = 1$ for the orbital motion of an electron. However, the g-factor corresponding to the spin-angular momentum is twice that corresponding to the orbital angular momentum, i.e., $g = 2$. An accurate value is, in fact, 2.0023 instead of 2, as we learn from quantum electrodynamics. If the spin is denoted by $\hbar\mathbf{s}$, then the magnetic moment due to spin is given by

$$\mathbf{m}_s = -\frac{e}{m_0 c}\hbar\mathbf{s} = -2\mu_B\mathbf{s}. \tag{2.2.10}$$

This equation is usually derived from Dirac's relativistic quantum mechanics.

The spins of protons and neutrons also induce magnetic moments. However, the magnetic moment of these nucleons, the nuclear Bohr magneton, is much smaller than the electronic one given by (2.2.9). It is of the order 5.05×10^{-25} erg/gauss. Thus, in general, electrons make the dominant contribution to the magnetic moment of an atom, the nuclei contributing little, so that their effect is practically unnoticeable in most cases. We remark, however, that it is possible to single out and measure nuclear magnetic moments by making use of such dynamic effects as the nuclear spin resonance. The excursion just made into the world of the atom will find its use when we study ferromagnetic materials.

2.3. Electric Moments

The next complex of electric charges is made up of two point electric charges separated by a certain distance d. This represents a physical doublet or dipole, with the help of which, we will introduce the notion of electric moment. Consider two point charges situated at $\mathbf{x}_1 = \mathbf{x} + \boldsymbol{\xi}$ and $\mathbf{x}_2 = \mathbf{x}$, respectively, with opposite electric charges e and $-e$. The total electric charge of this elementary complex is zero. But we can define the *electric moment* of the complex by

$$\mathbf{p} = \mathbf{e}(\mathbf{x} + \boldsymbol{\xi}) - e\mathbf{x} = e\boldsymbol{\xi}, \qquad |\mathbf{p}| = e\xi, \tag{2.3.1}$$

in R_G. $\boldsymbol{\xi}$ is the directed distance from the negative charge to the positive charge. Equation (2.3.1), with $\boldsymbol{\xi} \neq \mathbf{0}$, defines the electric moment of a *physical* doublet. The *ideal* (or mathematical) electric doublet is obtained if we simultaneously let e tend to infinity and let ξ tend to zero, in such a way that $|e\boldsymbol{\xi}|$ retains its constant value $|\mathbf{p}|$. The complex of charges just considered is neutral from the electrical viewpoint.

The next step would be to consider a complex of charges with vanishing electric moment. It can be seen that at least four charges are required to form such a complex. We shall consider this situation as a special case of an arbitrary number of charges. Consider an assembly of point charges e^α, $\alpha = 1, 2, \ldots$, with masses m^α, located at spatial points $\mathbf{x}^\alpha$ in the Galilean frame R_G. The *mass centroid* **C** of this group of charges is given by

$$\mathbf{x} = \sum_\alpha m^\alpha \mathbf{x}^\alpha \Big/ \sum_\alpha m^\alpha. \tag{2.3.2}$$

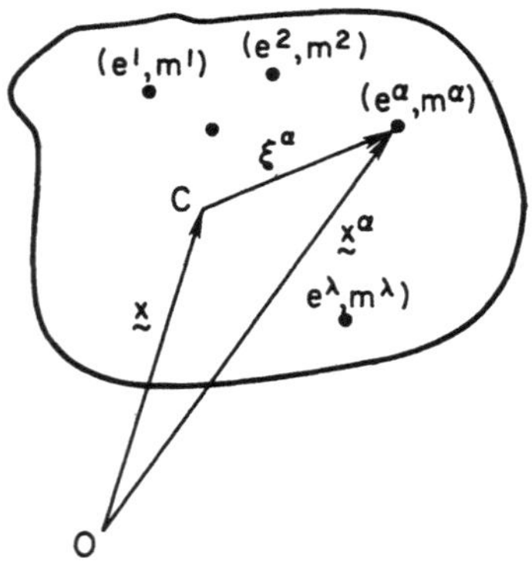

Figure 2.3.1. Assembly of point charges (C is the mass centroid).

We set (see Figure 2.3.1)

$$\mathbf{x}^\alpha = \mathbf{x} + \boldsymbol{\xi}^\alpha, \tag{2.3.3}$$

where $\boldsymbol{\xi}^\alpha$ is the position vector of e^α with respect to the centroid $\mathbf{C}$. From (2.3.2) and (2.3.3) it follows that

$$\sum_\alpha m^\alpha \boldsymbol{\xi}^\alpha = \mathbf{0}. \tag{2.3.4}$$

The total charge of the assembly is given by

$$e = \sum_\alpha e^\alpha. \tag{2.3.5}$$

If the complex of charge considered is electrically neutral, then $e = 0$. The *electric moment* of the complex about the centroid $\mathbf{C}$ is defined by

$$\mathbf{p} = \sum_\alpha e^\alpha \boldsymbol{\xi}^\alpha. \tag{2.3.6}$$

If the complex of charge is electrically neutral, then $e^+ = \sum_{\alpha^+} e^\alpha$ is the total positive electric charge in the complex. We can introduce charge centers $\boldsymbol{\xi}^+$ and $\boldsymbol{\xi}^-$ such that

$$\sum_{\alpha^+} e^\alpha \boldsymbol{\xi}^\alpha = \boldsymbol{\xi}^+ \sum_{\alpha^+} e^\alpha = e^+\boldsymbol{\xi}^+, \qquad \sum_{\alpha^-} e^\alpha \boldsymbol{\xi}^\alpha = \boldsymbol{\xi}^- \sum_{\alpha^-} e^\alpha = -e^+\boldsymbol{\xi}^-,$$

so that (2.3.6) takes the form

$$\mathbf{p} = e^+(\boldsymbol{\xi}^+ - \boldsymbol{\xi}^-) = e^+\boldsymbol{\xi}. \tag{2.3.7}$$

This relation is analogous to that given by (2.3.1) for a physical doublet. A simple example, to which (2.3.7) is applicable, is the case of a single atom with $\boldsymbol{\xi}^+$ as the equivalent center of the nucleus charge and $\boldsymbol{\xi}^-$ as the equivalent center of the electron cloud. If the atom is polarized, these positions do not coincide and $\boldsymbol{\xi}$ has a value different from zero. If the time average of $\boldsymbol{\xi}$ is constant, then the atom is statically polarized. Alternatively, under the influence of a time-harmonic electric field, $\boldsymbol{\xi}$ will be oscillatory.

More generally, the electric moment of the nth order is defined by

$$p_{ij\ldots q} = \frac{1}{n!}\sum_{\alpha} e^{\alpha}\xi^{\alpha}{}_{i}\xi_{j}^{\alpha}, \ldots, \xi_{q}^{\alpha}, \tag{2.3.8}$$

where $p_{ij\ldots q}$ is completely symmetric in all indices. We indicate this by enclosing the indices in parentheses, i.e.,

$$p_{ij\ldots q} = p_{(ij\ldots q)}. \tag{2.3.9}$$

In dyadic notation, (2.3.8) may be abbreviated as

$$\mathbf{p}^{(n)} = \frac{1}{n!}\sum_{\alpha} e^{\alpha}\boldsymbol{\xi}^{\alpha} \otimes \cdots \otimes \boldsymbol{\xi}^{\alpha} = \frac{1}{n!}\sum_{\alpha} e^{\alpha}(\boldsymbol{\xi}^{\alpha})^{n}. \tag{2.3.10}$$

The electric moment of second order is called the *electric quadrupole moment*. This is usually denoted by q_{ij} where

$$q_{ij} = p_{ij}^{(2)} = \frac{1}{2}\sum_{\alpha} e^{\alpha}\xi_{i}^{\alpha}\xi_{j}^{\alpha} = q_{ji}. \tag{2.3.11}$$

Some authors prefer to consider the traceless symmetric tensor called the reduced quadrupole moment. It is defined as the *deviatoric part* of q_{ij}. That is,

$$\begin{aligned}\bar{q}_{ij} &= \frac{1}{2}\sum_{\alpha} e^{\alpha}(\xi_{i}^{\alpha}\xi_{j}^{\alpha} - \tfrac{1}{3}\boldsymbol{\xi}^{\alpha}\cdot\boldsymbol{\xi}^{\alpha}\delta_{ij})\\ &= q_{ij} - \tfrac{1}{3}q_{kk}\delta_{ij},\end{aligned} \tag{2.3.12}$$

so that

$$\bar{q}_{kk} = 0. \tag{2.3.13}$$

The simplest complex of electric charges that gives rise to a quadrupole moment is illustrated by the following example: consider a neutral complex with a vanishing electric (dipole) moment. In this case, the left-hand side of (2.3.6) is zero. At least four charges are required to arrive at such an arrangement. However, the position and value of the charges are not determined uniquely by the given values of a quadrupole moment tensor q_{ij}. A unique problem results when, for instance, we require that all four charges have the same magnitude e, and occupy the four corners of a rhombus (see Figure 2.3.2(a)). The elongated or stretched quadrupole shown in Figure 2.3.2(b) represents an important special case. Here, the rhombus collapses to a straight line which carries charges $+e$ on each of its ends and a charge $-2e$ in the middle. Clearly, the models of Figure 2.3.2 can be made up of two equal dipoles, rotated 180° with respect to one another, and placed at the appropriate dipole separation.

As in the case of the dipole, we can pass to the ideal (or mathematical) electric quadrupole, in which all separations are allowed to approach zero in a constant ratio and, inversely to the square of this ratio, all charges are allowed to go to infinity—all this in such a way that the q_{ij} values remain constant.

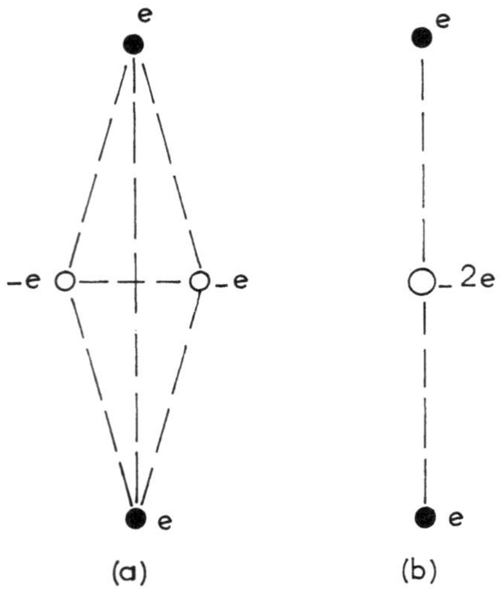

Figure 2.3.2. (a) and (b) Quadrupole moment.

The above discussion shows that many different configurations of the charges e^α (within a complex of charges having the same electric dipole), are entirely equivalent. The same situation does not apply, however, to the electric quadrupole which—like the inertia tensor of a body (see (2.3.11))—contains information about the geometrical configuration of the charges within the complex.

2.4. Current, Magnetic Moments

With each charge e^α, located at $\mathbf{x}^\alpha$ within a stable element, we associate a *current* $\mathbf{j}$ defined by

$$\mathbf{j}^\alpha = e^\alpha \dot{\mathbf{x}}^\alpha, \qquad \dot{\mathbf{x}}^\alpha \equiv \partial \mathbf{x}^\alpha / \partial t. \tag{2.4.1}$$

The center of mass $\mathbf{C}$ of the complex is, as usual, denoted by $\mathbf{x}$ according to (2.3.2) and (2.3.3) so that

$$\mathbf{j}^\alpha = e^\alpha \dot{\mathbf{x}} + \hat{\mathbf{j}}^\alpha, \tag{2.4.2}$$

$$\hat{\mathbf{j}}^\alpha = e^\alpha \dot{\xi}^\alpha. \tag{2.4.3}$$

For the complex, summing over α, we obtain the total current

$$\mathbf{j} = \sum_\alpha \mathbf{j}^\alpha = e\dot{\mathbf{x}} + \hat{\mathbf{j}}, \tag{2.4.4}$$

where $e\dot{\mathbf{x}}$ is the *convection current* carried by the center of mass of the element, and $\hat{\mathbf{j}}$ is the *conduction current* defined by

$$\hat{\mathbf{j}} = \sum_\alpha \hat{\mathbf{j}}^\alpha = \sum_\alpha e^\alpha \dot{\xi}^\alpha. \tag{2.4.5}$$

The linear momentum of e^α corresponding to the internal motion is $m^\alpha \dot{\xi}^\alpha$.

The angular momentum and the magnetic moment of e^α, with respect to **C**, are then given by

$$\mathbf{s}^\alpha{}_{(\mathrm{C})} = \boldsymbol{\xi}^\alpha \times m^\alpha \dot{\boldsymbol{\xi}}^\alpha \tag{2.4.6}$$

and

$$\boldsymbol{\mu}^\alpha{}_{(\mathrm{C})} = \frac{1}{2c} \boldsymbol{\xi}^\alpha \times \hat{\mathbf{j}}^\alpha, \tag{2.4.7}$$

respectively. The total angular momentum and the total magnetic moment (or magnetic dipole) of the complex about **C** are given by

$$\mathbf{s}_{(\mathrm{C})} = \sum_\alpha m^\alpha \boldsymbol{\xi}^\alpha \times \dot{\boldsymbol{\xi}}^\alpha \tag{2.4.8}$$

and

$$\boldsymbol{\mu}_{(\mathrm{C})} = \frac{1}{2c} \sum_\alpha e^\alpha \boldsymbol{\xi}^\alpha \times \dot{\boldsymbol{\xi}}^\alpha, \tag{2.4.9}$$

respectively. Unless all point particles of the complex are of the same nature—i.e., have the same mass-to-charge ratio—the relationship (2.2.4) is no longer valid.

We remark that $\boldsymbol{\mu}_{(\mathrm{C})}$ is not defined with respect to the fixed Galilean frame R_G of origin O, but rather with respect to the centroid **C** of the complex. That is, it is defined with respect to a frame R_C—co-moving with the centroid **C** of the complex.

The magnetic 2^n-*pole* moment is defined by

$$\boldsymbol{\mu}^{(n)}_{(\mathrm{C})} = \frac{n}{(n+1)!\,c} \sum_{\alpha=1}^{n} e^\alpha \underbrace{\boldsymbol{\xi}^\alpha \otimes \boldsymbol{\xi}^\alpha \otimes \cdots \otimes \boldsymbol{\xi}^\alpha}_{(n-1)\text{-fold dyadic product}} (\boldsymbol{\xi}^\alpha \times \dot{\boldsymbol{\xi}}^\alpha)$$

$$= \frac{n}{(n+1)!\,c} \sum_\alpha e^\alpha (\boldsymbol{\xi}^\alpha)^n \times \dot{\boldsymbol{\xi}}^\alpha, \tag{2.4.10}$$

where

$$(\boldsymbol{\xi}^\alpha \otimes \boldsymbol{\xi}^\alpha \otimes \cdots \otimes \boldsymbol{\xi}^\alpha)_{ij\ldots q} = \xi_i^\alpha \xi_j^\alpha \ldots \xi_q^\alpha. \tag{2.4.11}$$

2.5. Microscopic Electromagnetic Fields

The electric and magnetic fields $\mathbf{e}(\mathbf{x}, t)$ and $\mathbf{b}(\mathbf{x}, t)$ at the point $\mathbf{x}$ at time t, generated by a collection of point particles with charges e^α, $\alpha = 1, 2, \ldots$, located at $\mathbf{x}^\alpha$, are obtained by solving the *Maxwell–Lorentz field equations* (see Lorentz [1952])

$$\nabla \cdot \mathbf{e} = \sum_\alpha e^\alpha \delta(\mathbf{x}^\alpha - \mathbf{x}), \tag{2.5.1}$$

$$\nabla \times \mathbf{e} + \frac{1}{c}\frac{\partial \mathbf{b}}{\partial t} = \mathbf{0}, \tag{2.5.2}$$

$$\nabla \cdot \mathbf{b} = 0, \tag{2.5.3}$$

$$\nabla \times \mathbf{b} - \frac{1}{c}\frac{\partial \mathbf{e}}{\partial t} = \frac{1}{c}\sum_{\alpha} e^{\alpha}\dot{\mathbf{x}}^{\alpha}\delta(\mathbf{x}^{\alpha} - \mathbf{x}), \tag{2.5.4}$$

where the operator ∇ is with respect to $\mathbf{x}$ and δ is the Dirac generalized function defined operationally by

$$\int_{R^3} \delta(\mathbf{x} - \mathbf{x}_0) f(\mathbf{x})\, dv = f(\mathbf{x}_0), \qquad \int_{R^3} \delta(\mathbf{x})\, dv = 1,$$

where $f(\mathbf{x})$ is any suitably continuous function.

To obtain a solution of these equations, it is convenient to introduce potentials $\mathbf{a}(\mathbf{x}, t)$ and $\phi(\mathbf{x}, t)$ by

$$\mathbf{e} = -\nabla\phi - \frac{1}{c}\frac{\partial \mathbf{a}}{\partial t}, \qquad \mathbf{b} = \nabla \times \mathbf{a}. \tag{2.5.5}$$

With these, (2.5.2) and (2.5.3) are satisfied identically and (2.5.1) and (2.5.4) give

$$\nabla^2\phi + \frac{1}{c}\frac{\partial}{\partial t}(\nabla \cdot \mathbf{a}) = -\sum_{\alpha} e^{\alpha}\delta(\mathbf{x}^{\alpha} - \mathbf{x}), \tag{2.5.6}$$

$$\nabla^2\mathbf{a} - \nabla\left(\frac{1}{c}\frac{\partial \phi}{\partial t} + \nabla \cdot \mathbf{a}\right) - \frac{1}{c^2}\frac{\partial^2 \mathbf{a}}{\partial t^2} = -\frac{1}{c}\sum_{\alpha} e^{\alpha}\dot{\mathbf{x}}^{\alpha}\delta(\mathbf{x}^{\alpha} - \mathbf{x}). \tag{2.5.7}$$

The potentials $\mathbf{a}$ and ϕ are not unique. If the same electromagnetic fields are described by $\mathbf{a}'$ and ϕ', then they are related to $\mathbf{a}$ and ϕ by a *gauge transformation*

$$\phi' = \phi - \frac{1}{c}\frac{\partial \psi}{\partial t}, \qquad \mathbf{a}' = \mathbf{a} + \nabla\psi, \tag{2.5.8}$$

where ψ is an arbitrary function. By a proper choice of ψ, we arrive at a *Lorentz condition*

$$\frac{1}{c}\frac{\partial \phi}{\partial t} + \nabla \cdot \mathbf{a} = 0. \tag{2.5.9}$$

Employing (2.5.9) in (2.5.7) and dropping terms which contain c^{-2} (*nonrelativistic approximation*) in (2.5.6) and (2.5.7), we arrive at

$$\nabla^2\phi = -\sum_{\alpha} e^{\alpha}\delta(\mathbf{x}^{\alpha} - \mathbf{x}), \tag{2.5.10}$$

$$\nabla^2\mathbf{a} = -c^{-1}\sum_{\alpha} e^{\alpha}\dot{\mathbf{x}}^{\alpha}\delta(\mathbf{x}^{\alpha} - \mathbf{x}). \tag{2.5.11}$$

Remembering a property of the delta function

$$\nabla^2\frac{1}{|\mathbf{x}|} = -4\pi\delta(x), \tag{2.5.12}$$

the nonrelativistic potentials ϕ and $\mathbf{a}$, due to point charges e^{α}, are found to be

$$\phi = \sum_\alpha \frac{e^\alpha}{4\pi|\mathbf{x}^\alpha - \mathbf{x}|}, \tag{2.5.13}$$

$$\mathbf{a} = \frac{1}{c}\sum_\alpha \frac{e^\alpha \dot{\mathbf{x}}^\alpha}{4\pi|\mathbf{x}^\alpha - \mathbf{x}|}. \tag{2.5.14}$$

The nonrelativistic fields (2.5.5) are now given by

$$\mathbf{e} = \sum_\alpha \mathbf{e}^\alpha, \qquad \mathbf{e}^\alpha = -\nabla \frac{e^\alpha}{4\pi|\mathbf{x}^\alpha - \mathbf{x}|}, \tag{2.5.15}$$

$$\mathbf{b} = \sum_\alpha \mathbf{b}^\alpha, \qquad \mathbf{b}^\alpha = \frac{1}{c}\nabla \times \frac{e^\alpha \dot{\mathbf{x}}^\alpha}{4\pi|\mathbf{x}^\alpha - \mathbf{x}|}. \tag{2.5.16}$$

These results indicate that the nonrelativistic **b**-field is of the order c^{-1} while the **e**-field is of the order one. Hence, in the nonrelativistic theory, the second term in (2.5.2) can be neglected.

2.6. Electromagnetic Force on a Point Particle

The electromagnetic force $\mathbf{f}^\alpha$ acting on a point particle with charge e^α and mass m^α located at $\mathbf{x}^\alpha$ is given by Lorentz as

$$\mathbf{f}^\alpha = e^\alpha\left[\mathbf{e}_t(\mathbf{x}^\alpha, t) + \frac{1}{c}\mathbf{x}^\alpha \times \mathbf{b}_t(\mathbf{x}^\alpha, t)\right], \tag{2.6.1}$$

where $\mathbf{e}_t$ and $\mathbf{b}_t$ are the total electric fields and magnetic fields. They are the sums of the external fields $\mathbf{E}_e(\mathbf{x}^\alpha, t)$, $\mathbf{B}_e(\mathbf{x}^\alpha, t)$ and the fields generated by all other point particles on the αth particle, i.e.,

$$\mathbf{e}_t(\mathbf{x}^\alpha, t) = \mathbf{E}_e(\mathbf{x}^\alpha, t) + \sum_{\beta(\neq\alpha)} \mathbf{e}^\beta(\mathbf{x}^\alpha, t), \tag{2.6.2}$$

$$\mathbf{b}_t(\mathbf{x}^\alpha, t) = \mathbf{B}_e(\mathbf{x}^\alpha, t) + \sum_{\beta(\neq\alpha)} \mathbf{b}^\beta(\mathbf{x}^\alpha, t), \tag{2.6.3}$$

where $\mathbf{e}^\beta$ and $\mathbf{b}^\beta$ are given by (2.5.15) and (2.5.16). Since $\mathbf{b}^\beta$ is of the order c^{-1}, and in (2.6.1) $\mathbf{b}_t$ is multiplied by c^{-1}, the second term in (2.6.3) can be dropped so that

$$\mathbf{e}_t(\mathbf{x}^\alpha, t) = \mathbf{E}_e(\mathbf{x}^\alpha, t) - \sum_{\beta(\neq\alpha)} \nabla_\alpha \frac{e^\beta}{4\pi|\mathbf{x}^\alpha - \mathbf{x}^\beta|}, \tag{2.6.4}$$

$$\mathbf{b}_t(\mathbf{x}^\alpha, t) = \mathbf{B}_e(\mathbf{x}^\alpha, t), \tag{2.6.5}$$

where $\nabla_\alpha = \partial/\partial \mathbf{x}_\alpha$.

The equations of motion of the αth particle in the assembly are given by

$$m^\alpha \ddot{\mathbf{x}}^\alpha = \mathbf{f}^\alpha, \tag{2.6.6}$$

where

$$\mathbf{f}^\alpha = e^\alpha\left[\mathbf{E}_e(\mathbf{x}^\alpha, t) + \frac{1}{c}\dot{\mathbf{x}}^\alpha \times \mathbf{B}_e(\mathbf{x}^\alpha, t)\right] - \sum_{\beta(\neq\alpha)} \nabla_\alpha \frac{e^\alpha e^\beta}{4\pi|\mathbf{x}^\alpha - \mathbf{x}^\beta|} \tag{2.6.7}$$

is the nonrelativistic force acting on the αth point particle.

2.7. Microscopic Maxwell's Equations

Here we give microscopic Maxwell's equations for a stable set of particles. We consider an assembly of a large number of stable particles whose elements are electronic charges in motion. Let $\mathbf{x}^\alpha$ be the position vector of the mass center C^α, of the αth atom P^α, and $\mathbf{x}^{\alpha\beta} = \mathbf{x}^\alpha + \xi^{\alpha\beta}$ that of the electronic charge $e^{\alpha\beta}$ of P^α. As usual, ξ^α denotes the relative position of C^α from the center of mass $\mathbf{C}$ of the assembly. Since $\xi^{\alpha\beta}$ is of the order of atomic dimensions, and we are interested only in the fields at points outside the atom, we assume that (Figure 2.7.1)

$$|\xi^{\alpha\beta}| \ll |\xi^\alpha|. \tag{2.7.1}$$

In accordance with the Lorentz theory of electrons, the microscopic electric field $\mathbf{e}$ and the magnetic induction $\mathbf{b}$ must satisfy the Maxwell–Lorentz

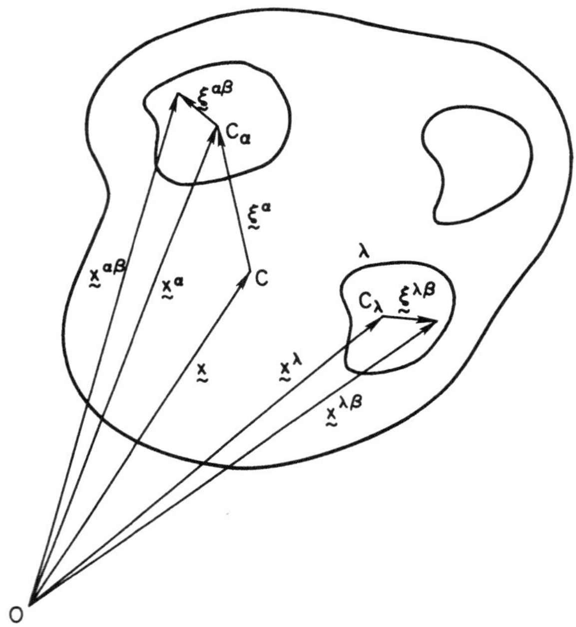

Figure 2.7.1. Stable particles with charges.

equations

$$\nabla \cdot \mathbf{e} = \sum_{\alpha,\beta} e^{\alpha\beta}\delta(\mathbf{x}^{\alpha} + \boldsymbol{\xi}^{\alpha\beta} - \mathbf{x}), \tag{2.7.2}$$

$$\nabla \times \mathbf{e} + \frac{1}{c}\frac{\partial \mathbf{b}}{\partial t} = \mathbf{0}, \tag{2.7.3}$$

$$\nabla \cdot \mathbf{b} = 0, \tag{2.7.4}$$

$$\nabla \times \mathbf{b} - \frac{1}{c}\frac{\partial \mathbf{e}}{\partial t} = \frac{1}{c}\sum_{\alpha,\beta} e^{\alpha\beta}(\dot{\mathbf{x}}^{\alpha} + \dot{\boldsymbol{\xi}}^{\alpha\beta})\delta(\mathbf{x}^{\alpha} + \boldsymbol{\xi}^{\alpha\beta} - \mathbf{x}), \tag{2.7.5}$$

where $\mathbf{e}$ and $\mathbf{b}$ are calculated at a point $\mathbf{x}$ outside the atom P^{α}. Assuming that (2.7.1) holds, the δ-function may be expanded into a power series of $\xi^{\alpha\beta}$ in (2.7.2) and (2.7.5) leading to

$$\nabla \cdot \mathbf{e} = \sum_{\alpha,\beta} e^{\alpha\beta} \sum_{n} \frac{(-)^n}{n!}(\boldsymbol{\xi}^{\alpha\beta} \cdot \nabla)^n \delta(\mathbf{x}^{\alpha} - \mathbf{x}), \tag{2.7.6}$$

$$\nabla \times \mathbf{b} - \frac{1}{c}\frac{\partial \mathbf{e}}{\partial t} = \frac{1}{c}\sum_{\alpha,\beta} e^{\alpha\beta}(\dot{\mathbf{x}}^{\alpha} + \dot{\boldsymbol{\xi}}^{\alpha\beta}) \sum_{n=0}^{\infty} \frac{(-)^n}{n!}(\boldsymbol{\xi}^{\alpha\beta} \cdot \nabla)^n \delta(\mathbf{x}^{\alpha} - \mathbf{x}). \tag{2.7.7}$$

We note that

$$\frac{\partial \delta(\mathbf{x}^{\alpha} - \mathbf{x})}{\partial \mathbf{x}^{\alpha}} = -\frac{\partial \delta(\mathbf{x}^{\alpha} - \mathbf{x})}{\partial \mathbf{x}} = -\nabla\delta(\mathbf{x}^{\alpha} - \mathbf{x}).$$

Using this, (2.7.6) may be written as

$$\nabla \cdot \mathbf{e} = q - \nabla \cdot \mathbf{p}, \tag{2.7.8}$$

where

$$\begin{aligned} q &= \sum_{\alpha} q^{\alpha}, \qquad \mathbf{p} = \sum_{\alpha} \mathbf{p}^{\alpha}, \\ q^{\alpha} &= \sum_{\beta} e^{\alpha\beta}\delta(\mathbf{x}^{\alpha} - \mathbf{x}), \\ \mathbf{p}^{\alpha} &= \sum_{\beta} e^{\alpha\beta} \sum_{n=1}^{\infty} \frac{(-)^{n-1}}{n!} \boldsymbol{\xi}^{\alpha\beta}(\boldsymbol{\xi}^{\alpha\beta} \cdot \nabla)^{n-1}\delta(\mathbf{x}^{\alpha} - \mathbf{x}). \end{aligned} \tag{2.7.9}$$

Equation (2.7.7) can be written as

$$\nabla \times \mathbf{b} - \frac{1}{c}\frac{\partial \mathbf{e}}{\partial t} = \frac{1}{c}\left[\mathbf{j} - \sum_{\alpha} \dot{\mathbf{x}}^{\alpha}\nabla \cdot \mathbf{p}^{\alpha} + \sum_{\alpha,\beta} e^{\alpha\beta} \sum_{n=0}^{\infty} \frac{(-)^n}{n!} \dot{\boldsymbol{\xi}}^{\alpha\beta}(\boldsymbol{\xi}^{\alpha\beta} \cdot \nabla)^n \delta(\mathbf{x}^{\alpha} - \mathbf{x})\right], \tag{2.7.10}$$

where

$$\mathbf{j} = \sum_{\alpha} \mathbf{j}^{\alpha}, \qquad \mathbf{j}^{\alpha} = \sum_{\beta} e^{\alpha\beta}\dot{\mathbf{x}}^{\alpha}\delta(\mathbf{x}^{\alpha} - \mathbf{x}). \tag{2.7.11}$$

Employing the time derivative of $\mathbf{p}^{\alpha}$ in (2.7.9) and replacing n by $n + 1$ in (2.7.10), we obtain

$$\begin{aligned} &\frac{1}{c}\frac{\partial \mathbf{p}^{\alpha}}{\partial t} + \frac{1}{c}(\dot{\mathbf{x}}^{\alpha} \cdot \nabla)\mathbf{p}^{\alpha} - \frac{1}{c}\sum_{\alpha} e^{\alpha\beta} \sum_{n=0}^{\infty} \frac{(-)^n}{(n+1)!} \\ &\qquad \times (\dot{\boldsymbol{\xi}}^{\alpha\beta}\boldsymbol{\xi}^{\alpha\beta} \cdot \nabla + n\boldsymbol{\xi}^{\alpha\beta}\dot{\boldsymbol{\xi}}^{\alpha\beta} \cdot \nabla)(\boldsymbol{\xi}^{\alpha\beta} \cdot \nabla)^{n-1}\delta(\mathbf{x}^{\alpha} - \mathbf{x}) = 0, \end{aligned} \tag{2.7.12}$$

where we used $\nabla\nabla^{-1} = 1$ and

$$\frac{\partial}{\partial t}\delta(\mathbf{x}^\alpha - \mathbf{x}) = -(\dot{\mathbf{x}}^\alpha \cdot \nabla)\delta(\mathbf{x}^\alpha - \mathbf{x}). \tag{2.7.13}$$

Adding (2.7.12) to (2.7.10), we obtain

$$\nabla \times \mathbf{b} - \frac{1}{c}\frac{\partial \mathbf{e}}{\partial t} = \frac{1}{c}\mathbf{j} + \frac{1}{c}\frac{\partial \mathbf{p}}{\partial t} + \nabla \times \mathbf{m}, \tag{2.7.14}$$

where we introduced the abbreviations

$$\mathbf{m} = \sum_\alpha \mathbf{m}^\alpha,$$

$$\mathbf{m}^\alpha = \frac{1}{c}\mathbf{p}^\alpha \times \dot{\mathbf{x}}^\alpha + \frac{1}{c}\sum_\alpha e^{\alpha\beta} \sum_{n=1}^{\infty} \frac{(-)^{n-1}n}{(n+1)!}\xi^{\alpha\beta} \times \dot{\xi}^{\alpha\beta}(\xi^{\alpha\beta}\cdot\nabla)^{n-1}\delta(\mathbf{x}^\alpha - \mathbf{x}). \tag{2.7.15}$$

In deriving (2.7.14), we also employed the identity

$$\nabla \cdot (\mathbf{uv} - \mathbf{vu}) = \nabla \times (\mathbf{v} \times \mathbf{u}). \tag{2.7.16}$$

Summarizing, we have Maxwell's equations for the microscopic fields

$$\begin{aligned}
\nabla \cdot \mathbf{e} &= \mathbf{q} - \nabla \cdot \mathbf{p}, \\
\nabla \times \mathbf{e} + \frac{1}{c}\frac{\partial \mathbf{b}}{\partial t} &= \mathbf{0}, \\
\nabla \cdot \mathbf{b} &= 0, \\
\nabla \times \mathbf{b} - \frac{1}{c}\frac{\partial \mathbf{e}}{\partial t} &= \frac{1}{c}\left(\mathbf{j} + \frac{\partial \mathbf{p}}{\partial t} + c\nabla \times \mathbf{m}\right).
\end{aligned} \tag{2.7.17}$$

The physical meaning of various quantities is clear from their definitions.

Atomic constituents are:

$$e^\alpha = \sum_\beta e^{\alpha\beta} \qquad \text{(atomic charge)},$$

$$\mathbf{p}^{(n)\alpha} = \frac{1}{n!}\sum_\beta e^{\alpha\beta}(\xi^{\alpha\beta})^n \qquad \text{(atomic electric } 2^n\text{-pole moment)},$$

$$\boldsymbol{\mu}^{(n)\alpha} = \frac{n}{(n+1)!\,c}\sum_\beta e^{\alpha\beta}(\xi^{\alpha\beta})^n \times \dot{\xi}^{\alpha\beta} \qquad \text{(atomic magnetic } 2^n\text{-pole moment)}. \tag{2.7.18}$$

Microscopic densities are:

$$q = \sum_\alpha q^\alpha = \sum_\alpha e^\alpha\delta(\mathbf{x}^\alpha - \mathbf{x}) \qquad \text{(microscopic charge density)},$$

$$\mathbf{j} = \sum_\alpha \mathbf{j}^\alpha = \sum_\alpha e^\alpha\dot{\mathbf{x}}^\alpha\delta(\mathbf{x}^\alpha - \mathbf{x}) \qquad \text{(microscopic current density)}$$

$$\mathbf{p} = \sum_{n=1}^{\infty} (-)^{n-1}\boldsymbol{\nabla}^{n-1} \odot \sum_{\alpha} \mathbf{p}^{(n)\alpha}\delta(\mathbf{x}^\alpha - \mathbf{x}) \quad \text{(microscopic electric polarization density)},$$

$$\mathbf{m} = \sum_{n=1}^{\infty} (-)^{n-1}\boldsymbol{\nabla}^{n-1} \odot \sum_{\alpha} \left(\boldsymbol{\mu}^{(n)\alpha} + \frac{1}{c}\mathbf{p}^{(n)\alpha} \times \dot{\mathbf{x}}^\alpha\right)\delta(\mathbf{x}^\alpha - \mathbf{x})$$
$$\text{(microscopic magnetic polarization density)}, \tag{2.7.19}$$

where the symbol $\odot$ represents multifold contraction, e.g.,

$$(\boldsymbol{\nabla}^n \odot A)_p = \frac{\partial}{\partial x_{i_1}} \frac{\partial}{\partial x_{i_2}} \cdots \frac{\partial}{\partial x_{i_n}} A_{i_1 i_2 \ldots i_n p}. \tag{2.7.20}$$

By substituting $\dot{\mathbf{x}}^\alpha = \dot{\mathbf{x}} + \dot{\boldsymbol{\xi}}^\alpha$ into $(2.7.19)_2$, the current vector may be decomposed as

$$\mathbf{j} = q\dot{\mathbf{x}} + \hat{\mathbf{j}}, \tag{2.7.21}$$

where $\hat{\mathbf{j}}$ is the conduction current density, defined by

$$\hat{\mathbf{j}} = \sum_{\alpha} e^\alpha \dot{\boldsymbol{\xi}}^\alpha \delta(\mathbf{x}^\alpha - \mathbf{x}). \tag{2.7.22}$$

In examining $(2.7.19)_4$, we notice that the magnetization of the atom contains contributions from the electric multipoles in motion. A similar term indicating the contribution of moving magnetic moments on the polarization of an atom is not present in this *nonrelativistic* treatment.

If we introduce the atomic displacement vector **d** and the magnetic field **h** by

$$\mathbf{d} = \mathbf{e} + \mathbf{p}, \qquad \mathbf{h} = \mathbf{b} - \mathbf{m}, \tag{2.7.23}$$

then the microscopic field equations (2.7.17) may be written as

$$\begin{aligned} \boldsymbol{\nabla} \cdot \mathbf{d} &= q, \\ \boldsymbol{\nabla} \times \mathbf{e} + \frac{1}{c}\frac{\partial \mathbf{b}}{\partial t} &= 0, \\ \boldsymbol{\nabla} \cdot \mathbf{b} &= 0, \\ \boldsymbol{\nabla} \times \mathbf{h} - \frac{1}{c}\frac{\partial \mathbf{d}}{\partial t} &= \frac{1}{c}\mathbf{j}. \end{aligned} \tag{2.7.24}$$

These are Maxwell's equations for the microscopic fields. Various fields here are expressed in terms of the constituent fields of the atom (stable element) by (2.7.19). Note that no averaging has taken place. Of course, (2.7.24) are valid only outside the atom. The equation of conservation of charge follows from $(2.7.24)_1$ and $(2.7.24)_4$.

$$\frac{\partial q}{\partial t} + \boldsymbol{\nabla} \cdot \mathbf{j} = 0. \tag{2.7.25}$$

2.8. Electromagnetic Force on Composite Particles[1]

We consider a stable composite particle P^α whose constituents are point particles with mass $m^{\alpha\beta}$ and charge $e^{\alpha\beta}$, $\beta = 1, 2, \ldots$. The position vector of $m^{\alpha\beta}$ is denoted, as usual, by $\mathbf{x}^{\alpha\beta}$, in a Galilean frame of reference R_G. The center of mass of P^α is given by

$$\mathbf{x}^\alpha = \sum_\beta m^{\alpha\beta}\mathbf{x}^{\alpha\beta}/m^\alpha, \qquad m^\alpha \equiv \sum_\beta m^{\alpha\beta}, \tag{2.8.1}$$

where m^α is the total mass of P^α.

The force acting on point particles due to external fields $\mathbf{E}_e(\mathbf{x}^{\alpha\beta}, t)$ and $\mathbf{B}_e(\mathbf{x}^{\alpha\beta}, t)$ and those of $e^{\alpha\beta}$, according to (2.6.7), is given by

$$\mathbf{f}^{\kappa\alpha} = e^{\kappa\alpha}\left[\mathbf{e}_t(\mathbf{x}^{\kappa\alpha}, t) + \frac{1}{c}\dot{\mathbf{x}}^{\kappa\alpha} \times \mathbf{b}_t(\mathbf{x}^{\kappa\alpha}, t)\right], \tag{2.8.2}$$

where $\mathbf{e}_t$ and $\mathbf{b}_t$ are the total electric and magnetic fields given by

$$\begin{aligned} \mathbf{e}_t(\mathbf{x}^{\kappa\alpha}, t) &= \mathbf{E}_e(\mathbf{x}^{\kappa\alpha}, t) - \sum_{\beta(\neq\alpha)} \nabla_{\kappa\alpha}\frac{e^{\kappa\beta}}{4\pi|\mathbf{x}^{\kappa\alpha} - \mathbf{x}^{\kappa\beta}|} \\ &\quad - \sum_{\lambda(\neq\kappa),\beta} \nabla_{\kappa\alpha}\frac{e^{\lambda\beta}}{4\pi|\mathbf{x}^{\kappa\alpha} - \mathbf{x}^{\lambda\beta}|}, \\ \mathbf{b}_t(\mathbf{x}^{\kappa\alpha}, t) &= \mathbf{B}_e(\mathbf{x}^{\kappa\alpha}, t). \end{aligned} \tag{2.8.3}$$

The force acting on the composite particle is obtained by summing $\mathbf{f}^{\kappa\alpha}$ over α, i.e.,

$$\mathbf{f}^\kappa(\mathbf{x}^\kappa, t) = \sum_\alpha \mathbf{f}^{\kappa\alpha}. \tag{2.8.4}$$

Substituting from (2.8.2) and (2.8.3), we find that the contribution of the *intra-atomic* fields drops out (since they are central forces) and we obtain

$$\mathbf{f}^\kappa(\mathbf{x}^\kappa, t) = \mathbf{f}_e^\kappa + \mathbf{f}_i^\kappa, \tag{2.8.5}$$

where $\mathbf{f}_e^\kappa$ and $\mathbf{f}_i^\kappa$ are, respectively, the forces due to the external and internal fields given by

$$\mathbf{f}_e^\kappa = \sum_\alpha e^{\kappa\alpha}\left[\mathbf{E}_e(\mathbf{x}^{\kappa\alpha}, t) + \frac{1}{c}\dot{\mathbf{x}}^{\kappa\alpha} \times \mathbf{B}_e(\mathbf{x}^{\kappa\alpha}, t)\right], \tag{2.8.6}$$

$$\mathbf{f}_i^\kappa = -\sum_{\lambda(\neq\kappa),\alpha,\beta} \nabla_{\kappa\alpha}\frac{e^{\kappa\alpha}e^{\lambda\beta}}{4\pi|\mathbf{x}^{\kappa\alpha} - \mathbf{x}^{\lambda\beta}|}. \tag{2.8.7}$$

Introducing the internal coordinates

$$\boldsymbol{\xi}^{\alpha\beta} = \mathbf{x}^{\alpha\beta} - \mathbf{x}^\alpha \tag{2.8.8}$$

---

[1] In this, and in the next two sections, we follow de Groot and Suttorp [1972, Sect. 5.8.6].

we expand $\mathbf{E}_e$ and $\mathbf{B}_e$ into Taylor series, leading to

$$\mathbf{f}_e^\kappa = \sum_\alpha e^{\kappa\alpha} \sum_{n=0}^{\infty} \frac{1}{n!}(\boldsymbol{\xi}^{\kappa\alpha} \cdot \nabla_\kappa)^n \left[\mathbf{E}_e(\mathbf{x}^\kappa, t) + \frac{1}{c}\dot{\mathbf{x}}^\kappa \times \mathbf{B}_e(\mathbf{x}^\kappa, t) + \frac{1}{c}\dot{\boldsymbol{\xi}}^{\kappa\alpha} \times \mathbf{B}_e(\mathbf{x}^\kappa, t) \right], \tag{2.8.9}$$

where the *gradient operator* ∇_κ applies only to the external fields $\mathbf{E}_e$ and $\mathbf{B}_e$. Since the external fields vary slowly over the atom, we may only retain terms involving electric and magnetic dipole moments

$$\mathbf{p}^\kappa = \sum_\alpha e^{\kappa\alpha}\boldsymbol{\xi}^{\kappa\alpha}, \qquad \boldsymbol{\mu}^\kappa = \frac{1}{2c}\sum_\alpha e^{\kappa\alpha}\boldsymbol{\xi}^{\kappa\alpha} \times \dot{\boldsymbol{\xi}}^{\kappa\alpha}. \tag{2.8.10}$$

Using the identity

$$\boldsymbol{\xi}^{\kappa\alpha} \cdot \nabla_\kappa \dot{\boldsymbol{\xi}}^{\kappa\alpha} = \frac{1}{2}\frac{d}{dt}(\boldsymbol{\xi}^{\kappa\alpha}\boldsymbol{\xi}^{\kappa\alpha}) \cdot \nabla_\kappa + \tfrac{1}{2}(\boldsymbol{\xi}^{\kappa\alpha} \times \dot{\boldsymbol{\xi}}^{\kappa\alpha}) \times \nabla_\kappa, \tag{2.8.11}$$

eq. (2.8.9) reduces to

$$\mathbf{f}_e^\kappa = (e^\kappa + \mathbf{p}^\kappa \cdot \nabla_\kappa)\left[\mathbf{E}_e(\mathbf{x}^\kappa, t) + \frac{1}{c}\dot{\mathbf{x}}^\kappa \times \mathbf{B}_e(\mathbf{x}^\kappa, t) \right]$$
$$+ \left(\frac{1}{c}\dot{\mathbf{p}}^\kappa + \boldsymbol{\mu}^\kappa \times \nabla_\kappa \right) \times \mathbf{B}_e(\mathbf{x}^\kappa, t). \tag{2.8.12}$$

The interatomic contribution to the force may also be expanded into Taylor series about $\mathbf{x}^\alpha$, if $|\mathbf{x}^\kappa - \mathbf{x}^\lambda|$ is greater than the sum of the largest $|\boldsymbol{\xi}^{\kappa\alpha}|$ and $|\boldsymbol{\xi}^{\lambda\beta}|$. We denote this series of f_i^κ by ${}_L f_i^\kappa$ where

$${}_L f_i^\kappa = - \sum_{\lambda(\neq\kappa),\alpha,\beta} \sum_{n,m=0}^{\infty} \frac{1}{n!\,m!}(\boldsymbol{\xi}^{\kappa\alpha} \cdot \nabla_\kappa)^n(\boldsymbol{\xi}^{\lambda\beta} \cdot \nabla_\lambda)^m \nabla_\kappa \frac{e^{\kappa\alpha}e^{\lambda\beta}}{4\pi|\mathbf{x}^\kappa - \mathbf{x}^\lambda|}. \tag{2.8.13}$$

Since it is valid under the condition stated above, the difference between ${}_L f_i^\kappa$ and f_i^κ gives the short-range force

$${}_S\mathbf{f}^\kappa = \mathbf{f}_i^\kappa - {}_L\mathbf{f}_i^\kappa. \tag{2.8.14}$$

The force ${}_L\mathbf{f}_i^\kappa$ may be expressed in terms of electric multiple moments $\mathbf{p}^{(n)\alpha}$, so that

$${}_L\mathbf{f}_i^\kappa = - \sum_{\lambda(\neq\kappa)} \sum_{n,m=0}^{\infty} \mathbf{p}^{(n)\kappa} \odot \nabla_\kappa^n \mathbf{p}^{(m)\lambda} \odot \nabla_\lambda^m \nabla_\kappa \frac{1}{4\pi|\mathbf{x}^\kappa - \mathbf{x}^\lambda|}. \tag{2.8.15}$$

The force $\mathbf{f}^\kappa(\mathbf{x}^\alpha)$, acting on the atom P^κ, may thus be decomposed into

$$\mathbf{f}^\kappa = {}_L\mathbf{f}^\kappa + {}_S\mathbf{f}^\kappa, \tag{2.8.16}$$

where ${}_L\mathbf{f}^\kappa$ is the long-range force given by

$${}_L\mathbf{f}^\kappa = \mathbf{f}_e^\kappa + {}_L\mathbf{f}_i^\kappa \tag{2.8.17}$$

and ${}_S\mathbf{f}^\kappa$ is the short-range force, since it vanishes when the atoms are outside each other.

The external fields obey the field equations

$$\nabla \cdot \mathbf{B}_e = 0, \qquad \nabla \times \mathbf{E}_e + \frac{1}{c}\frac{\partial \mathbf{B}_e}{\partial t} = \mathbf{0}. \tag{2.8.18}$$

Employing these, the long-range external force $\mathbf{f}_e^\kappa$ may be expressed as

$$\begin{aligned}\mathbf{f}_e^\kappa = e^\kappa\left[\mathbf{E}_e(\mathbf{x}^\kappa, t) + \frac{1}{c}\dot{\mathbf{x}}^\kappa \times \mathbf{B}_e(\mathbf{x}^\kappa, t)\right] + [\nabla_\kappa \mathbf{E}_e(\mathbf{x}^\kappa, t)]\cdot \mathbf{p}^\kappa \\ + [\nabla_\kappa \mathbf{B}_e(\mathbf{x}^\kappa, t)]\cdot\left(\boldsymbol{\mu}^\kappa + \frac{1}{c}\mathbf{p}^\kappa \times \dot{\mathbf{x}}^\kappa\right) + \frac{1}{c}\frac{d}{dt}[\mathbf{p}^\kappa \times \mathbf{B}_e(\mathbf{x}^\kappa, t)].\end{aligned} \tag{2.8.19}$$

The first term is the *Lorentz force* acting on a charged composite particle. The second term is known as the *Kelvin force* on the electric dipole moment. The third term is the force acting on the magnetic dipole moment and the electric dipole moment in motion. The last term is an electrodynamic effect.

2.9. Couple Acting on a Composite Particle

The couple acting on a point mass $m^{\alpha\beta}$, with charge $e^{\alpha\beta}$, located at $\mathbf{x}^{\alpha\beta}$, is given by

$$\mathbf{c}^{\kappa\alpha} = \boldsymbol{\xi}^{\kappa\alpha} \times \mathbf{f}^{\kappa\alpha}, \tag{2.9.1}$$

where $\mathbf{f}^{\kappa\alpha}$ is the force acting on the point mass, and $\boldsymbol{\xi}^{\kappa\alpha}$ is its relative distance from the center of mass of the composite particle. Substituting (2.8.2), and summing over α, we obtain the couple acting on the composite particle

$$\mathbf{c}^\kappa(\mathbf{x}^\kappa, t) = \mathbf{c}_e^\kappa + \mathbf{c}_i^\kappa, \tag{2.9.2}$$

where $\mathbf{c}_e^\kappa$ and $\mathbf{c}_i^\kappa$ are, respectively, the moments of the force due to external and internal fields about the center of mass $\mathbf{x}^\kappa$, of the composite particle

$$\mathbf{c}_e^\kappa = \sum_\alpha e^{\kappa\alpha}\boldsymbol{\xi}^{\kappa\alpha} \times \left[\mathbf{E}_e(\mathbf{x}^{\kappa\alpha}, t) + \frac{1}{c}\dot{\mathbf{x}}^{\kappa\alpha} \times \mathbf{B}_e(\mathbf{x}^{\kappa\alpha}, t)\right], \tag{2.9.3}$$

$$\mathbf{c}_i^\kappa = -\sum_{\lambda(\neq\kappa),\alpha,\beta} \boldsymbol{\xi}^{\kappa\alpha} \times \nabla_{\kappa\alpha}\frac{e^{\kappa\alpha}e^{\lambda\beta}}{4\pi|\mathbf{x}^{\kappa\alpha} - \mathbf{x}^{\lambda\beta}|}. \tag{2.9.4}$$

For slowly varying external fields $\mathbf{E}_e$ and $\mathbf{B}_e$, (2.9.3) may be expanded into Taylor series in terms of the internal coordinates defined by (2.8.8) so that

$$\mathbf{c}_e^\kappa = \sum_\alpha e^{\kappa\alpha}\boldsymbol{\xi}^{\kappa\alpha} \times \sum_{n=0}^{\infty}\frac{1}{n!}(\boldsymbol{\xi}^{\kappa\alpha}\cdot\nabla_\kappa)^n\left[\mathbf{E}_e(\mathbf{x}^\kappa, t) + \frac{1}{c}(\dot{\mathbf{x}}^\kappa + \dot{\boldsymbol{\xi}}^{\kappa\alpha}) \times \mathbf{B}_e(\mathbf{x}^\kappa, t)\right]. \tag{2.9.5}$$

For sufficiently smooth fields, if we limit ourselves to electric and magnetic dipole moments and use the identity

$$\xi^{\kappa\alpha} \times (\dot{\xi}^{\kappa\alpha} \times \mathbf{B}_e) = \frac{1}{2}(\xi^{\kappa\alpha} \times \dot{\xi}^{\kappa\alpha}) \times \mathbf{B}_e + \frac{1}{2}\frac{d}{dt}(\xi^{\kappa\alpha}\xi^{\kappa\alpha}) \cdot \mathbf{B}_e - \frac{1}{2}\frac{d}{dt}(\xi^{\kappa\alpha})^2 \mathbf{B}_e, \tag{2.9.6}$$

we obtain

$$\mathbf{c}_e^\kappa = \mathbf{p}^\kappa \times \left[\mathbf{E}_e(\mathbf{x}^\kappa, t) + \frac{1}{c}\dot{\mathbf{x}}^\kappa \times \mathbf{B}_e(\mathbf{x}^\kappa, t) \right] + \boldsymbol{\mu}^\kappa \times \mathbf{B}_e(\mathbf{x}^\kappa, t). \tag{2.9.7}$$

In the case of the atoms being outside each other, the interatomic contribution $\mathbf{c}_i^\kappa$ may also be written as a power series of $\xi^{\kappa\alpha}$. We denote this series by ${}_L\mathbf{c}_i^\kappa$, where

$${}_L\mathbf{c}_i^\kappa = -\sum_{\lambda(\neq\kappa),\alpha,\beta} \sum_{n,m=0}^{\infty} \frac{1}{n!\,m!} \xi^{\kappa\alpha} \times \nabla_\kappa (\xi^{\kappa\alpha} \cdot \nabla_\kappa)^n (\xi^{\lambda\beta} \cdot \nabla_\lambda)^m \frac{e^{\kappa\alpha} e^{\lambda\beta}}{4\pi|\mathbf{x}^\kappa - \mathbf{x}^\lambda|}. \tag{2.9.8}$$

In terms of the electric multipole moments, this reads as

$${}_L\mathbf{c}_i = \sum_{\lambda(\neq\kappa)} \sum_{n,m=0}^{\infty} n\nabla_\kappa \times \mathbf{p}^{(n)\kappa} \odot \nabla_\kappa^{n-1} \mathbf{p}^{(m)\lambda} \odot \nabla_\lambda^m \frac{1}{4\pi|\mathbf{x}^\kappa - \mathbf{x}^\lambda|}. \tag{2.9.9}$$

This is the contribution of the interatomic fields to the long-range couple. Consequently, (2.9.2) may be written as

$$\mathbf{c}^\kappa(\mathbf{x}^\kappa, t) = {}_L\mathbf{c}^\kappa + {}_S\mathbf{c}^\kappa, \tag{2.9.10}$$

where ${}_L\mathbf{c}^\kappa$ and ${}_S\mathbf{c}^\kappa$ are, respectively, long- and short-range couples given by

$${}_L\mathbf{c}^\kappa = \mathbf{c}_e^\kappa + {}_L\mathbf{c}_i^\kappa, \qquad {}_S\mathbf{c}^\kappa = \mathbf{c}_i^\kappa - {}_L\mathbf{c}_i^\kappa. \tag{2.9.11}$$

For a single atom in an external field, the couple acting on the atom is given by

$$\mathbf{c}^\kappa(\mathbf{x}^\kappa, t) = \mathbf{c}_e^\kappa. \tag{2.9.12}$$

2.10. Power of Electromagnetic Forces

The power of electromagnetic forces acting on a composite particle is given by

$$w(\mathbf{x}^\kappa, t) = \sum_\alpha \dot{\mathbf{x}}^{\kappa\alpha} \cdot \mathbf{f}^{\kappa\alpha}. \tag{2.10.1}$$

Substituting from (2.8.2) for $\mathbf{f}^{\kappa\alpha}$ we can express (2.10.1) as

$$w(\mathbf{x}^\kappa, t) = w_e + w_i, \tag{2.10.2}$$

where w_e and w_i are, respectively, the powers of the external and internal fields

$$w_e = \sum_\alpha e^{\kappa\alpha} \dot{\mathbf{x}}^{\kappa\alpha} \mathbf{E}_e(\mathbf{x}^{\kappa\alpha}, t),$$
$$w_i = {}_c w_i + \bar{w}_i, \tag{2.10.3}$$

with

$$
\begin{aligned}
{}_{\mathrm{c}}w_{\mathrm{i}} &= -\sum_{\alpha,\beta(\alpha\neq\beta)} (\dot{\mathbf{x}}^{\kappa\alpha}\cdot\nabla_{\kappa\alpha} + \dot{x}^{\kappa\beta}\cdot\nabla_{\kappa\beta})\frac{e^{\kappa\alpha}e^{\kappa\beta}}{8\pi|\mathbf{x}^{\kappa\alpha}-\mathbf{x}^{\kappa\beta}|} \\
&= -\frac{d}{dt}\sum_{\alpha,\beta(\alpha\neq\beta)}\frac{e^{\kappa\alpha}e^{\kappa\beta}}{8\pi|\boldsymbol{\xi}^{\kappa\alpha}-\boldsymbol{\xi}^{\kappa\beta}|} = -\frac{d}{dt}e_{\mathrm{c}}, \qquad (2.10.4)\\
\bar{w}_{\mathrm{i}} &= -\sum_{\lambda(\neq\kappa),\alpha,\beta}\dot{\mathbf{x}}^{\kappa\alpha}\cdot\nabla_{\kappa\alpha}\frac{e^{\kappa\alpha}e^{\lambda\beta}}{4\pi|\mathbf{x}^{\kappa\alpha}-\mathbf{x}^{\lambda\beta}|}.
\end{aligned}
$$

The power ${}_{\mathrm{c}}w_{\mathrm{i}}$ is due to the *intra-atomic* contribution which is expressed as the total time rate of the *Coulomb energy* of the atom.

In a similar way to the force and couple, we express the powers of the external fields w_{e} and the remaining part of the internal fields $\bar{w}_{\mathrm{i}}$ as Taylor series in $\xi^{\alpha\beta}$, so that

$$
w_{\mathrm{e}} = \sum_{\alpha} e^{\kappa\alpha}\sum_{n=0}^{\infty}\frac{1}{n!}(\boldsymbol{\xi}^{\kappa\alpha}\cdot\nabla_{\kappa})^n(\dot{\mathbf{x}}^{\kappa}+\dot{\boldsymbol{\xi}}^{\kappa\alpha})\cdot\mathbf{E}_{\mathrm{e}}(\mathbf{x}^{\kappa},t), \qquad (2.10.5)
$$

$$
{}_{\mathrm{L}}\bar{w}_{\mathrm{i}} = -\sum_{\lambda(\neq\kappa),\alpha,\beta}\sum_{n,m=0}^{\infty}\frac{1}{n!\,m!}(\boldsymbol{\xi}^{\kappa\alpha}\cdot\nabla_{\kappa})^n(\boldsymbol{\xi}^{\lambda\beta}\cdot\nabla_{\lambda})^m(\dot{\mathbf{x}}^{\kappa}+\dot{\boldsymbol{\xi}}^{\kappa\alpha})\cdot\nabla_{\kappa}\frac{e^{\kappa\alpha}e^{\lambda\beta}}{4\pi|\mathbf{x}^{\kappa}-\mathbf{x}^{\lambda}|}. \qquad (2.10.6)
$$

The expansion (2.10.6) is valid where the atoms are outside each other, so we denote this portion of w_{i} by ${}_{\mathrm{L}}\bar{w}_{\mathrm{i}}$. We can express these energy rates in terms of electric and magnetic dipole moments. For sufficiently smooth external fields, we may confine ourselves only to first-order moments so that

$$
w_{\mathrm{e}} = (e^{\kappa}+\mathbf{p}^{\kappa}\cdot\nabla_{\kappa})\dot{\mathbf{x}}^{\kappa}\cdot\mathbf{E}_{\mathrm{e}}(\mathbf{x}^{\kappa},t) + (\dot{\mathbf{p}}^{\kappa}+c\boldsymbol{\mu}^{\kappa}\times\nabla_{\kappa})\cdot\mathbf{E}_{\mathrm{e}}(\mathbf{x}^{\kappa},t). \qquad (2.10.7)
$$

With the use of $(2.8.18)_2$, this may be written as:

$$
\begin{aligned}
w_{\mathrm{e}} &= e^{\kappa}\dot{\mathbf{x}}^{\kappa}\cdot\mathbf{E}_{\mathrm{e}}(\mathbf{x}^{\kappa},t) + \dot{\mathbf{x}}^{\kappa}\cdot[\nabla_{\kappa}\mathbf{E}_{\mathrm{e}}(\mathbf{x}^{\kappa},t)]\cdot\mathbf{p}^{\kappa} + \dot{\mathbf{p}}^{\kappa}\cdot\mathbf{E}_{\mathrm{e}}(\mathbf{x}^{\kappa},t)\\
&\quad -\left(\boldsymbol{\mu}^{\kappa}+\frac{1}{c}\mathbf{p}^{\kappa}\times\dot{\mathbf{x}}^{\kappa}\right)\cdot\frac{\partial\mathbf{B}_{\mathrm{e}}(\mathbf{x}^{\kappa},t)}{\partial t}. \qquad (2.10.8)
\end{aligned}
$$

In terms of moments, ${}_{\mathrm{L}}\bar{w}_{\mathrm{i}}$ may be expressed as

$$
\begin{aligned}
{}_{\mathrm{L}}\bar{w}_{\mathrm{i}} &= -\sum_{\lambda(\neq\kappa)}\sum_{m=0}^{\infty}\left[\sum_{n=0}^{\infty}\mathbf{p}^{(n)\kappa}\odot\nabla_{\kappa}^{n}\dot{\mathbf{x}}^{\kappa}\cdot\nabla_{\kappa} + \sum_{n=1}^{\infty}\dot{\mathbf{p}}^{(n)\kappa}\odot\nabla_{\kappa}^{n}\right]\mathbf{p}^{(m)\lambda}\\
&\quad\odot\nabla_{\lambda}^{m}\frac{1}{4\pi|\mathbf{x}^{\kappa}-\mathbf{x}^{\lambda}|}. \qquad (2.10.9)
\end{aligned}
$$

The time rate of the energy of the composite particle, therefore, is expressed as

$$
w = -\frac{de_{\mathrm{c}}}{dt} + {}_{\mathrm{L}}w + {}_{\mathrm{S}}w, \qquad (2.10.10)
$$

where e_{c} is the Coulomb energy, ${}_{\mathrm{L}}w$ is the energy rate of the long-range forces, and ${}_{\mathrm{S}}w$ is that of the short-range forces. Thus

$$e_{\mathrm{c}} = \sum_{\alpha,\beta(\alpha\neq\beta)} \frac{e^{\kappa\alpha}e^{\kappa\beta}}{8\pi|\xi^{\kappa\alpha} - \xi^{\kappa\beta}|}, \tag{2.10.11}$$

$${}_{\mathrm{L}}w = w_{\mathrm{e}} + {}_{\mathrm{L}}\bar{w}_{\mathrm{i}}. \tag{2.10.12}$$

Energy rates w_{e}, ${}_{\mathrm{L}}\bar{w}_{\mathrm{i}}$, and ${}_{\mathrm{S}}w$ are, respectively, given by (2.10.5), (2.10.9) and

$${}_{\mathrm{S}}w = \bar{w}_{\mathrm{i}} - {}_{\mathrm{L}}\bar{w}_{\mathrm{i}}. \tag{2.10.13}$$

For a single atom in an external field, ${}_{\mathrm{S}}w = 0$ and (2.10.10) reduces to

$$w = -\frac{de_{\mathrm{c}}}{dt} + {}_{\mathrm{L}}w, \tag{2.10.14}$$

with ${}_{\mathrm{L}}w$ given by

$${}_{\mathrm{L}}w = e^{\kappa}\dot{\mathbf{x}}^{\kappa}\cdot\mathbf{E}_{\mathrm{e}} + \dot{\mathbf{x}}^{\kappa}\cdot(\nabla_{\kappa}E_{\mathrm{e}})\cdot\mathbf{p}^{\kappa} + \dot{\mathbf{x}}^{\kappa}\cdot(\nabla_{\kappa}\mathbf{B}_{\mathrm{e}})\cdot\left(\boldsymbol{\mu}^{\kappa} + \frac{1}{c}\mathbf{p}^{\kappa}\times\dot{\mathbf{x}}^{\kappa}\right)$$
$$+ \dot{\mathbf{p}}^{\kappa}\cdot\mathbf{E}_{\mathrm{e}} - \left(\boldsymbol{\mu}^{\kappa} + \frac{1}{c}\mathbf{p}^{\kappa}\times\dot{\mathbf{x}}^{\kappa}\right)\cdot\frac{d\mathbf{B}_{\mathrm{e}}}{dt}, \tag{2.10.15}$$

where $d/dt = (\partial/\partial t) + \dot{\mathbf{x}}^{\kappa}\cdot\nabla_{\kappa}$ is used.

Problems

2.1. Calculate the electric moment of fourth order and the magnetic quadrupole moment. Discuss possibilities for their deviatoric parts.

2.2. Obtain explicit expressions of the microscopic electric and magnetic polarizations of second order.

2.3. Show that the electric and magnetic fields given by

$$\mathbf{e}(\mathbf{x}, t) = -\nabla\int[q(\mathbf{x}', t) - \nabla'\cdot\mathbf{p}(\mathbf{x}', t)]\frac{dv(\mathbf{x}')}{4\pi|\mathbf{x} - \mathbf{x}'|},$$

$$\mathbf{b}(\mathbf{x}, t) = \nabla\times\int\left[c^{-1}\mathbf{j}(\mathbf{x}', t) + \frac{\partial\mathbf{p}(\mathbf{x}', t)}{\partial t} + \nabla'\times\mathbf{m}(\mathbf{x}', t)\right]\frac{dv(\mathbf{x}')}{4\pi|\mathbf{x} - \mathbf{x}'|},$$

are the solutions of the microscopic field equations (2.7.17).

2.4. Show that the general expression for the force on a composite particle is of the form

$$\mathbf{f}^{\kappa} = \left(e^{\kappa} + \sum_{n=1}^{\infty}(\mathbf{p}^{(n)\kappa}\odot\nabla_{\kappa}^{n})(\mathbf{E}_{e} + c^{-1}\mathbf{v}\times\mathbf{B}_{e})\right.$$
$$\left. + \sum_{n=1}^{\infty}(c^{-1}\dot{\mathbf{p}}^{(n)\kappa}\times\nabla_{\kappa}^{n-1} + \nabla_{\kappa}^{n-1}\odot\boldsymbol{\mu}^{(n)\kappa}\times\nabla_{\kappa})\times\mathbf{B}_{e}\right).$$

2.5. The equations of motion of the αth particle given by (2.6.6), where the total force f^{α} is expressed in the terms (2.6.4) and (2.6.5), may be written in Hamiltonian form

$$\frac{\partial H}{\partial \mathbf{P}^\alpha} = \dot{\mathbf{x}}^\alpha, \qquad \frac{\partial H}{\partial \mathbf{x}^\alpha} = -\dot{\mathbf{P}}^\alpha.$$

Show that H is the Hamiltonian given by

$$H = \sum_\alpha \frac{P^{\alpha 2}}{2m^\alpha} + \sum_{\alpha,\beta(\alpha\neq\beta)} \frac{e^\alpha e^\beta}{8\pi|\mathbf{x}^\alpha - \mathbf{x}^\beta|} + \sum_\alpha e^\alpha \left\{\phi_e(\mathbf{x}^\alpha, t) - \frac{1}{c}\frac{\mathbf{P}^\alpha}{m^\alpha}\cdot \mathbf{A}_e(\mathbf{x}^\alpha, t)\right\},$$

where ϕ_e and $\mathbf{A}_e$ are potentials for the external fields.

2.6. Derive equation (2.7.14).

2.7. A bound electron e under an external harmonic electric field $E(t)$ is undergoing damped oscillations. The magnetic field is neglected. Determine the motion $x(t)$ and the dipole moment $\mathbf{p}$.

2.8. An electron gas obeying the Maxwell equation (2.7.24), with a background charge density ρ_0 (i.e., q_e is replaced by $\rho - \rho_0$), also obeys the equations of motion

$$\frac{\partial \mathbf{v}}{\partial t} + (\mathbf{v}\cdot\nabla)\mathbf{v} = -\frac{1}{m\rho}\nabla p - \frac{e}{m}\left(\mathbf{E} + \frac{1}{c}\mathbf{v}\times\mathbf{B}\right),$$

$$\frac{\partial \rho}{\partial t} + \nabla\cdot(\rho\mathbf{v}) = 0.$$

The polarization and the pressure gradient ∇p are assumed to be negligible. Using the perturbation expansions

$$\rho = \rho^{(0)} + \varepsilon\rho^{(1)} + \varepsilon^2\rho^{(2)} + \cdots,$$

$$\mathbf{v} = \varepsilon\mathbf{v}^{(1)} + \varepsilon^2\mathbf{v}^{(2)} + \cdots,$$

determine the current density $\mathbf{j} = \rho\mathbf{v}$ up to and including the second order in the perturbation parameter ε. The electric field is assumed to be of the form $\mathbf{E} = \mathbf{E}_0 \exp(i\mathbf{k}\cdot\mathbf{x} - i\omega t)$.

2.9. For the damped linear oscillations of a bound electron, described in Problem 2.7, obtain the mechanical energy $W_{me} = \frac{1}{2}(m^\alpha\dot{\mathbf{x}}^{\alpha 2} + k^\alpha\mathbf{x}^{\alpha 2})$, where k is the linear spring constant.

2.10. A free electron is subject to a linear spring force and a constant external magnetic field $\mathbf{B}_0$. Determine the motion of the electron under a time-dependent external electric field.

2.11. An electron attached to a core with linear springs and dashpots is subject to harmonic electric and magnetic fields in the (x_1, x_2)-plane. Determine the motion. Show that for zero damping and $B_1 = B_2$, resonance occurs at $\omega = \omega_1 \pm (eB/2m)$; $\omega_1 \equiv (k/m)^{1/2}$, where k is the spring constant, m is the mass of electron, and e is its charge. This splitting of the resonance is called the *Zeeman effect*. Find the expression of polarization $\mathbf{p} = Ne\mathbf{x}_\pm$ away from the resonance peaks.

CHAPTER 3

Macroscopic Electromagnetic Theory

3.1. Scope of the Chapter

The present chapter is of central importance to the whole book. Its principal aim is the formulation of the basic balance laws of mechanical and energetic nature which govern the Galilean-invariant electrodynamics of continua, irrespective of the exact mechanical and electromagnetic responses of bodies. To do this, it is necessary to evaluate the electromagnetic contributions of these balance laws. In order to avoid an arbitrary choice of these contributions, and to keep in touch with the microscopic model developed in Chapter 2, we deduce these contributions from microscopic equations by performing statistical averaging as introduced in Section 3.2. Maxwell's equations for the macroscopic fields are then obtained, in Section 3.3, from the corresponding equations of the microscopic fields derived in Section 2.5. In a natural way, this leads to the notion of macroscopic concepts such as polarization, magnetization, charge, and current, and their expressions in terms of microscopic notions. Galilean invariance of Maxwell's equations is shown in Section 3.4, which leads to the introduction of fields expressed in a moving frame of reference.

In Section 3.5 we obtain expressions for the macroscopic densities of mass, force, couple, and power by means of the statistical averaging procedure. The coupling of the electromagnetic fields to the mechanical field can only be provided when these expressions are found. We obtain the volume contributions of the electromagnetic force, couple, and power which are relevant to the continuum formulation. Expressions of surface and interatomic contributions are given implicitly. Nevertheless, they cannot be evaluated explicitly for real materials. Fortunately, they are not needed in a continuum theory, since the theory of constitutive equations is built precisely to replace these contributions.

In Section 3.6, we establish an identity which relates the electromagnetic force, couple, and power to a surface force density (electromagnetic stress tensor), a time rate of a volume density (electromagnetic momentum), and a surface energy density (Poynting vector). In this way, such secondary quantities as the electromagnetic stress tensor and the electromagnetic momentum are defined (nonuniquely) for practical uses. These notions serve to obtain, in Section 3.7, expressions of the electromagnetic force, couple, and energy at a discontinuity surface.

The global statements of the balance laws of continuum physics are expressed, in general form, in Section 3.8. From these, by means of a localization process, we obtain the local balance laws and jump conditions at a moving discontinuity surface. The global and local forms of the balance laws of electrodynamics (Gauss' law, conservation of charge, Faraday's law, Ampere's law) and the associated jump condition are obtained in Section 3.9, from the general results given in Section 3.8, by means of identification of various fields. The same process leads to the thermomechanical balance laws (conservation of mass, balance of momenta, conservation energy) and the associated jump conditions in Section 3.10.

Various equivalent forms of the Clausius–Duhem (C–D) inequality, fundamental to the thermodynamics of electromagnetic continua, are obtained in Section 3.11. These inequalities are used several times in later chapters of this book for the derivation of constitutive equations.

In Section 3.12 we show the Galilean invariance of the electromagnetic force, couple, and power, and introduce the axiom of *time reversal* which is the result of the microscopic reversibility. These expressions are also fundamental to the formulation of constitutive equations. Finally, in Section 3.13, we formulate the principle of virtual power, a concept which may be used as an alternative way of deriving field equations. It is also used frequently in numerical solutions.

Throughout this book electromagnetic equations are expressed in the Heaviside–Lorentz system of units. For the expressions of the macroscopic Maxwell equations and the Lorentz force in other systems of units, see Appendix F.

3.2. Statistical Averages

In order to express the global balance laws of momentum, moment of momentum, and energy we need to determine the expression of macroscopic force, torque (also referred to as the ponderomotive force, couple), and energy exerted by the electromagnetic fields on a charged, magnetized, and electrically polarized material continuum. Most authors interested in continuum physics postulate the form of the force, torque, and energy under the influence of the classical rigid body electrodynamics. Expressions so derived for these quantities and for those of the *Maxwell stress tensor* and the *electromagnetic momentum density* have created a controversy which has lasted for over fifty years.[1] In our opinion, the proper physical motivation for the choice of a

[1] See Penfield and Haus [1967], Pao [1978], and Hutter and Ven [1978].

particular ponderomotive force, torque, and energy must come from the electron theory of Lorentz, as applied to a collection of atoms.

The expressions of microscopic quantities obtained in Chapter 2 may be used to find the corresponding macroscopic quantities. While atomic quantities change quickly over space and time, macroscopic quantities vary much more slowly. This is because the devices that measure these quantities are of macroscopic sizes, and they record averages over much larger distances and time periods compared to atomic dimensions and time. The macroscopic field can therefore be derived by an averaging process. Dixon and Eringen [1965a], by use of a volume averaging procedure, obtained the expression of macroscopic force, couple, and energy[2] from the basic concept of charges in motion. The notions of magnetic monopoles and infinitesimal Amperian current loops are not used.[3] The method of Dixon and Eringen was used by Maugin and Eringen [1977] to formulate the electrodynamics of deformable continua. For a complete and concise account, see Eringen [1980, Chap. 10].

In this chapter we employ the principles of statistical mechanics to obtain the macroscopic force, torque, and energy densities relevant to continuum theories. In order to calculate averages, and/or various statistical moments of microscopic fields, we need a statistical distribution function f, which depends on the position and velocity (or canonically conjugate momentum) of each point-like particle, and time, i.e.,

$$f = f(\mathbf{x}^{\kappa\alpha}, \dot{\mathbf{x}}^{\kappa\alpha}; t). \tag{3.2.1}$$

The volume element in the space of $(\mathbf{x}^{\kappa\alpha}, \dot{\mathbf{x}}^{\kappa\alpha})$ (the so-called *fluxion space*)[4] is given by

$$d\phi = \prod_{\kappa\alpha} d\mathbf{x}^{\kappa\alpha}\, d\dot{\mathbf{x}}^{\kappa\alpha}. \tag{3.2.2}$$

The probability of finding the system in the volume element $d\phi$ is then given by $f\, d\phi$. Thus, if $a(\mathbf{x}^{\kappa\alpha}, \dot{\mathbf{x}}^{\kappa\alpha}; t)$ is any microscopic quantity, the average of this quantity is given by

$$A(\mathbf{x}, t) = \langle a \rangle \equiv \int af\, d\phi. \tag{3.2.3}$$

The measure of probability $f\, d\phi$ remains constant in time if one follows the points of fluxion space. Therefore, we have two fundamental lemmas:

---

[2] A similar method was used by Masson and Weaver [1929] for the electrostatic case. Independently, Lax and Nelson [1971] repeated the analysis of Dixon and Eringen.

[3] Such approaches are favored by other authors, e.g., Fano et al. [1960], Penfield and Haus [1967], Pao and Hutter [1975], and Grot [1976]. Basic equations of electromagnetic elastic solids may be derived by means of the global energy balance law which is postulated to be invariant under Euclidean transformations (Alblas [1974], Parkus [1972], Van de Ven [1975]). For an excellent comparative discussion of different models, see Hutter and Van de Ven [1978].

[4] In the case of positions and canonical momenta it is called the *phase space*.

Lemma 1. *Time differentiation and averaging commute, i.e.,*

$$\frac{\partial}{\partial t}\langle a \rangle = \frac{\partial}{\partial t}\int af\, d\phi = \int \frac{da}{dt} f\, d\phi = \left\langle \frac{da}{dt} \right\rangle, \tag{3.2.4}$$

where d/dt denotes the total derivative in the fluxion space, i.e.,

$$\frac{da}{dt} = \frac{\partial a}{\partial t} + \sum_{\kappa,\alpha} \left(\dot{\mathbf{x}}^{\kappa\alpha} \cdot \frac{\partial a}{\partial \mathbf{x}^{\kappa\alpha}} + \ddot{\mathbf{x}}^{\kappa\alpha} \cdot \frac{\partial a}{\partial \dot{\mathbf{x}}^{\kappa\alpha}} \right). \tag{3.2.5}$$

As a result of the equations of motion, $\ddot{\mathbf{x}}^{\kappa\alpha}$ can be expressed in terms of $\mathbf{x}^{\kappa\alpha}$ and $\dot{\mathbf{x}}^{\kappa\alpha}$.

Lemma 2. *Space differentiation and averaging commute*

$$\nabla \langle a \rangle = \langle \nabla a \rangle. \tag{3.2.6}$$

This result follows from the fact that f does not depend on the space (centroid) coordinate $\mathbf{x}$.

When a quantity a depends only on the variables pertinent to a few atoms, then the integration of the remaining variables can be carried out, resulting in a distribution function that depends on the coordinates and momenta of these atoms. Thus, for example, if $a = a_1(\mathbf{x}^1, \dot{\mathbf{x}}^1, \xi^{1\alpha}, \dot{\xi}^{1\alpha})$, then

$$A = \int a_1(\mathbf{x}^1, \dot{\mathbf{x}}^1, \xi^{1\alpha}, \dot{\xi}^1) f_1(\mathbf{x}^1, \dot{\mathbf{x}}^1, \xi^{1\alpha}, \dot{\xi}^{1\alpha}, t)\, d\dot{\mathbf{x}}^1 \prod_\alpha d\xi^{1\alpha}\, d\dot{\xi}^{1\alpha}, \tag{3.2.7}$$

or, in shorter notation,

$$A = \int a_1(1) f_1(1; t)\, d\phi_1. \tag{3.2.8}$$

In the following sections, we employ these averaging processes to obtain the expressions of macroscopic fields.

3.3. Maxwell's Equations

Maxwell's equations of macroscopic fields are obtained from the atomic field equations (2.7.24) by the statistical method described in Section 3.2.

To this end, we introduce average fields

$$\begin{aligned} \mathbf{E} &= \langle \mathbf{e} \rangle, & \mathbf{B} &= \langle \mathbf{b} \rangle, \\ \mathbf{H} &= \langle \mathbf{h} \rangle, & \mathbf{D} &= \langle \mathbf{d} \rangle, \\ \mathbf{P} &= \langle \mathbf{p} \rangle, & \mathbf{M} &= \langle \mathbf{m} \rangle, \\ \mathbf{J} &= \langle \mathbf{j} \rangle, & q_e &= \langle q \rangle, \\ \hat{\mathbf{J}} &= \langle \hat{\mathbf{j}} \rangle. \end{aligned} \tag{3.3.1}$$

Employing Lemmas 1 and 2 in (2.7.24) we obtain

$$\begin{aligned} \nabla\cdot\mathbf{D} &= q_e, \\ \nabla\times\mathbf{E} + \frac{1}{c}\frac{\partial\mathbf{B}}{\partial t} &= \mathbf{0}, \\ \nabla\cdot\mathbf{B} &= 0, \\ \nabla\times\mathbf{H} - \frac{1}{c}\frac{\partial\mathbf{D}}{\partial t} &= \frac{1}{c}\mathbf{J}. \end{aligned} \tag{3.3.2}$$

These are macroscopic Maxwell's equations. Note that they are identical in form to the microscopic equations, except that the fields occurring in these equations are smoothed out. From (2.7.23) and (2.7.21), it also follows that

$$\mathbf{D} = \mathbf{E} + \mathbf{P}, \qquad \mathbf{B} = \mathbf{H} + \mathbf{M}, \qquad \mathbf{J} = q_e\mathbf{v} + \hat{\mathbf{J}}, \tag{3.3.3}$$

where $\hat{\mathbf{J}}$ is the conduction current.

The macroscopic equation of conservation of charge is

$$\frac{\partial q_e}{\partial t} + \nabla\cdot\mathbf{J} = 0. \tag{3.3.4}$$

The macroscopic polarization $\mathbf{P}$, magnetization $\mathbf{M}$, charge q_e, and current $\mathbf{J}$ may be expressed in terms of the one-point probability distribution function as

$$\begin{aligned} \mathbf{P}(\mathbf{x}, t) &= \langle\mathbf{p}\rangle = \sum_\alpha \sum_{n=1}^{\infty} (-)^{n-1}\nabla^{n-1} \odot \int \mathbf{p}^{(n)} f_1^\alpha(\mathbf{x}, 1; t)\, d\phi_1, \\ \mathbf{M}(\mathbf{x}, t) &= \langle\mathbf{m}\rangle = \sum_\alpha \sum_{n=1}^{\infty} (-)^{n-1}\nabla^{n-1} \odot \int \left(\boldsymbol{\mu}^{(n)1} + \frac{1}{c}\mathbf{p}^{(n)1} \times \dot{\mathbf{x}}^1\right) f_1^\alpha(\mathbf{x}, 1; t)\, d\phi_1, \\ q_e(x, t) &= \langle q\rangle = \sum_\alpha e^\alpha f_1^\alpha(\mathbf{x}; t), \\ \mathbf{J}(\mathbf{x}, t) &= \langle j\rangle = \sum_\alpha \int e\dot{\mathbf{x}}^1 f_1(\mathbf{x}, \dot{\mathbf{x}}^1; t)\, d\dot{\mathbf{x}}^1. \end{aligned} \tag{3.3.5}$$

In the expression of $\mathbf{P}$ the symbol 1 (in the argument of $f_1(\mathbf{x}, 1; t)$) indicates *all* $\mathbf{p}^{(n)1}$, and in the expression of $\mathbf{M}$ it indicates *all* $\mathbf{p}^{(n)1}$, $\boldsymbol{\mu}^{(n)1}$, and $\dot{\mathbf{x}}^1$. Note that, just like multipole moments, the polarization and magnetization vectors $\mathbf{P}$ and $\mathbf{M}$ may be due to permanent and induced effects.

From (3.3.5) it is clear that $\mathbf{P}$, $\mathbf{M}$, q_e, and $\mathbf{J}$ *can be constructed from the atomic constitutents without any constitutive equations being postulated.* However, such details are generally not available at the outset for real materials—it is beyond our hope to perform these calculations except possibly in some ideal situations. Therefore, it is necessary to appeal to constitutive assumptions in order to close the theory. The continuum theory of electromagnetic materials is justified on this ground as well.

In the same manner as in the atomic theory, we can introduce an electric scalar potential Φ, and a magnetic vector potential $\mathbf{A}$, by

$$\mathbf{B} = \nabla \times \mathbf{A}, \qquad \mathbf{E} = -\nabla\Phi - \frac{1}{c}\frac{\partial \mathbf{A}}{\partial t}, \tag{3.3.6}$$

subject to the Lorentz gauge condition

$$\nabla \cdot \mathbf{A} + \frac{1}{c}\frac{\partial \Phi}{\partial t} = 0, \tag{3.3.7}$$

$(3.3.2)_1$ and $(3.3.2)_4$ yield the wave equations

$$\square\Phi + q^{\text{eff}} = 0, \qquad \square\mathbf{A} + \frac{1}{c}\mathbf{J}^{\text{eff}} = \mathbf{0}, \tag{3.3.8}$$

where $\square$ denotes the *d'Alembertian operator* such that (in rectangular coordinates)

$$\square = \nabla^2 - \frac{1}{c^2}\frac{\partial^2}{\partial t^2} \tag{3.3.9}$$

and we have defined the *effective charge density* and *current* by

$$q^{\text{eff}} = q_e - \nabla \cdot \mathbf{P}, \qquad \mathbf{J}^{\text{eff}} = \mathbf{J} + \frac{\partial \mathbf{P}}{\partial t} + c\nabla \times \mathbf{M}. \tag{3.3.10}$$

Clearly, the vector fields $\partial \mathbf{P}/\partial t$ and $c\nabla \times \mathbf{M}$ may be referred to as the *polarization current* and the *magnetization current* (in the frame R_G), respectively.

If the terms of the order c^{-2} are neglected (the operator $\square$ is replaced by ∇^2 in (3.3.8)), then the solutions of Maxwell's equation (3.3.2), up to order c^{-1}, are obtained as

$$\begin{aligned} \mathbf{E}(\mathbf{x}, t) &= \mathbf{E}_e(\mathbf{x}, t) - \nabla \int [q_e(\mathbf{x}', t) - \nabla' \cdot \mathbf{P}(\mathbf{x}', t)] \frac{d\mathbf{x}'}{4\pi|\mathbf{x} - \mathbf{x}'|}, \\ \mathbf{B}(\mathbf{x}, t) &= \mathbf{B}_e(\mathbf{x}, t) + \frac{1}{c}\nabla \times \int \left[\mathbf{J}(\mathbf{x}', t) + \frac{\partial \mathbf{P}(\mathbf{x}', t)}{\partial t} \right. \\ &\qquad \left. + c\nabla' \times \mathbf{M}(\mathbf{x}', t) \right] \frac{d\mathbf{x}'}{4\pi|\mathbf{x} - \mathbf{x}'|}. \end{aligned} \tag{3.3.11}$$

These solutions may be verified by substituting them into (3.3.2) and using (3.3.4).

3.4. Galilean Invariance of Maxwell's Equations

The Galilean transformations of inertial frames form a group which consists of spatial (time-independent) rotations and pure Galilean transformations. The pure Galilean transformations between two Galilean frames R_G and $R_{G'}$

are expressed as

$$\mathbf{x}' = \mathbf{x} + \mathbf{V}t, \qquad t' = t, \tag{3.4.1}$$

where $\mathbf{V}$ is a uniform velocity in space and time. The form-invariance of Maxwell's equation (3.3.2), under the time-independent rotational and translational members of the Galilean group, is trivial since these equations are written in vector form. The invariance under transformations of the type (3.4.1) is less obvious. Let us superpose a prime on the operators and fields expressed in $R_{G'}$. Thus, for example, we write ∇' and $\partial/\partial t'$, respectively, for the gradient operator and the partial time derivative in $R_{G'}$. Clearly, then we have

$$\nabla' = \nabla, \qquad \frac{\partial}{\partial t'} = \frac{\partial}{\partial t} - \mathbf{V}\cdot\nabla. \tag{3.4.2}$$

Moreover, the distribution functions are invariant

$$f_1^{\alpha'}(1'; t') = f_1^{\alpha}(1, t) \tag{3.4.3}$$

as a result of the probabilistic interpretation. Here $1'$ and 1 refer to the quantities associated with the atom 1 under the transformations (3.4.1). Thus, for example, $\mathbf{x}'^1 = \mathbf{x}^1 + \mathbf{V}t$, $\dot{\mathbf{x}}'^1 = \dot{\mathbf{x}}^1 + \mathbf{V}$, $\mathbf{p}_1' = \mathbf{p}_1$, $\mathbf{m}_1' = \mathbf{m}_1$. With the help of these formulas from (3.3.5) we can show that

$$q_e'(\mathbf{x}', t') = q_e(\mathbf{x}, t), \tag{3.4.4}$$

$$\mathbf{J}'(\mathbf{x}', t') = \mathbf{J}(\mathbf{x}, t) + q_e(\mathbf{x}, t)\mathbf{V}, \tag{3.4.5}$$

$$\mathbf{P}'(\mathbf{x}', t') = \mathbf{P}(\mathbf{x}, t), \tag{3.4.6}$$

$$\mathbf{M}'(\mathbf{x}', t') = \mathbf{M}(\mathbf{x}, t) - \frac{1}{c}\mathbf{V} \times \mathbf{P}(\mathbf{x}, t). \tag{3.4.7}$$

Employing these in Maxwell's equations, (3.3.2) with (3.3.3), it is found that Maxwell's equations are invariant under the transformation (3.4.1) if and only if (3.4.4)–(3.4.7) are valid and

$$\mathbf{E}'(\mathbf{x}', t') = \mathbf{E}(\mathbf{x}, t) - \frac{1}{c}\mathbf{V} \times \mathbf{B}(\mathbf{x}, t), \tag{3.4.8}$$

$$\mathbf{B}'(\mathbf{x}', t') = \mathbf{B}(\mathbf{x}, t) + \frac{1}{c}\mathbf{V} \times \mathbf{E}(\mathbf{x}, t). \tag{3.4.9}$$

Now, at each time, we may consider the motion of the continuum, with a velocity $\dot{\mathbf{x}} = \mathbf{v}$, defining itself *locally* and *instantaneously* as Galilean transformation of the type (3.4.1). Then setting instantaneously $R_{G'} = R_C$, where R_C is the co-moving frame previously introduced, and $\mathbf{V} = -\mathbf{v}$, we will have

$$\mathcal{J} = \mathbf{J} - q_e\mathbf{v} = \hat{\mathbf{J}}, \tag{3.4.10}$$

$$\mathcal{M} = \mathbf{M} + \frac{1}{c}\mathbf{v} \times \mathbf{P}, \tag{3.4.11}$$

$$\mathscr{E} = \mathbf{E} + \frac{1}{c}\mathbf{v} \times \mathbf{B}, \tag{3.4.12}$$

$$\mathscr{B} = \mathbf{B} - \frac{1}{c}\mathbf{v} \times \mathbf{E}, \tag{3.4.13}$$

where, from now on, script symbols indicate fields expressed in the co-moving frame R_C. Defining the field $\mathscr{H}$ by

$$\mathscr{H} = \mathscr{B} - \mathscr{M} \tag{3.4.14}$$

we see that $\mathscr{H}$ and $\mathbf{H}$ are related by

$$\mathscr{H} = \mathbf{H} - \frac{1}{c}\mathbf{v} \times \mathbf{D}. \tag{3.4.15}$$

Equations (3.4.10)–(3.4.15) define the different electromagnetic fields in R_C (the co-moving or, according to the relativistic terminology, the *proper* frame) as functions of the same fields in R_G. We see that whereas $\mathbf{J}$ is the electric current in R_G, the conduction current $\mathscr{J}$ is none other than the electric current in R_C. If the latter is objective, then the former cannot be objective. Similarly, we now see that the objective field $\mathscr{M}$, of course, is the effective magnetization in R_C, whereas $\mathbf{M}$ is the effective magnetization in R_G and, in general, is not objective in the dynamical theory.

Equation $(3.3.2)_2$ and $(3.3.2)_4$ can be transformed to the frame R_C by using the converse of (3.4.10)–(3.4.15)

$$\nabla \times \mathscr{E} + \frac{1}{c}\overset{*}{\mathbf{B}} = \mathbf{0}, \tag{3.4.16}$$

$$\nabla \times \mathscr{H} - \frac{1}{c}\overset{*}{\mathbf{D}} = \frac{1}{c}\mathscr{J}. \tag{3.4.17}$$

The field $\mathscr{E}$ defined in R_C is usually referred to as the *electromotive intensity*. We recall that a superposed asterisk indicates convective time differentiation (see the definition (1.12.12)). Because of (3.4.14) and $(3.3.3)_1$, we can rewrite (3.4.17) in the alternative form

$$\nabla \times \mathscr{B} - \frac{1}{c}\overset{*}{\mathbf{E}} = \frac{1}{c}\mathscr{J}^{\text{eff}}, \tag{3.4.18}$$

where $\mathscr{J}^{\text{eff}}$ is the effective current in R_C, defined by

$$\mathscr{J}^{\text{eff}} = \mathscr{J} + \overset{*}{\mathbf{P}} + c\nabla \times \mathscr{M}. \tag{3.4.19}$$

This field is objective as a consequence of the objective character of $\mathscr{J}$, $\mathbf{P}$, and $\mathscr{M}$.

Finally, it must be emphasized that the transformation formulas obtained above are nonrelativistic. As a rule, they neglect terms of the order of $\beta^2 \equiv v^2/c^2$. The terms in $\beta = |\mathbf{v}|/c$ are those which are neglected in quasi-static theories of magnetized and electrically polarized continua. A typical nonrelativistic feature of the transformation laws obtained is the asymmetry between the formulas (3.4.6) and (3.4.7). That is, according to Galilean relativity, a polarized moving body will appear to be magnetized (this is not surprising since moving charge distributions produce currents), whereas a magnetized moving body will not appear to be polarized. It must be remarked that few observable conclusions can be drawn due to the difficulty of obtaining sufficiently high velocities for material media. However, it is this lack of symmetry which stimulated the study of relativistic electrodynamics at the beginning of this century.

The fully symmetric relativistic laws which replace (3.4.6) and (3.4.7) are[5]

$$\begin{aligned} \mathbf{P}' &= \gamma[\mathbf{P} - \boldsymbol{\beta} \times \mathbf{M} + \boldsymbol{\beta}(\boldsymbol{\beta}\cdot\mathbf{P})\beta^{-2}(\gamma^{-1} - 1)], \\ \mathbf{M}' &= \gamma[\mathbf{M} + \boldsymbol{\beta} \times \mathbf{P} + \boldsymbol{\beta}(\boldsymbol{\beta}\cdot\mathbf{M})\beta^{-2}(\gamma^{-1} - 1)], \end{aligned} \tag{3.4.20}$$

where

$$\gamma = (1 - \beta^2)^{-1/2}, \qquad \boldsymbol{\beta} \equiv \mathbf{v}/c.$$

3.5. Macroscopic Densities of Mass, Force, Couple, and Power

A. Mass Density

The atomic mass density is given by

$$\rho_A = \sum_\alpha m^\alpha \delta(\mathbf{x}^\alpha - \mathbf{x}), \tag{3.5.1}$$

where m^α is the mass of the (identical) atoms in one-component systems. Here, for simplicity, we are interested only in single-component systems and are not concerned with mixtures. The macroscopic mass density is obtained by taking the statistical average of (3.5.1), i.e.,

$$\rho(\mathbf{x}, t) = \left\langle \sum_\alpha m^\alpha \delta(\mathbf{x}^\alpha - \mathbf{x}) \right\rangle = m f_1(\mathbf{x}; t), \tag{3.5.2}$$

where $f_1(\mathbf{x}; t)$ is a one-point distribution function which depends on $\mathbf{x}^1 \equiv \mathbf{x}$ and time. The time derivative of (3.5.2) is given by

$$\frac{\partial \rho}{\partial t} = -\left\langle \sum_\alpha m^\alpha \dot{\mathbf{x}}^\alpha \cdot \nabla\delta(\mathbf{x}^\alpha - \mathbf{x}) \right\rangle, \tag{3.5.3}$$

where we used Lemma 1, Section 3.2, and (2.7.13). Introducing the local barycentric velocity $\dot{\mathbf{x}} = \mathbf{v}(\mathbf{x}, t)$,

$$\rho(\mathbf{x}, t)\mathbf{v}(\mathbf{x}, t) = \left\langle \sum_\alpha m^\alpha \dot{\mathbf{x}}^\alpha \delta(\mathbf{x}^\alpha - \mathbf{x}) \right\rangle = \int m\dot{\mathbf{x}}^1 f_1(\mathbf{x}, \dot{\mathbf{x}}^1; t)\, d\dot{\mathbf{x}}_1 \tag{3.5.4}$$

[5] See Jackson [1975, p. 552].

we express (3.5.3) as

$$\frac{\partial\rho}{\partial t} + \nabla\cdot(\rho\mathbf{v}) = 0. \tag{3.5.5}$$

This is the macroscopic law of conservation of mass.

B. Force Density

To obtain the expression of the macroscopic electromagnetic force, we take the statistical average of the microscopic force given in Section 2.8. The total macroscopic force is the average of (2.8.16), namely,

$$\mathbf{F}(\mathbf{x}, t) = \mathbf{F}_e(\mathbf{x}, t) + {}_L\mathbf{F}_i(\mathbf{x}, t) + {}_S\mathbf{F}(\mathbf{x}, t), \tag{3.5.6}$$

where $\mathbf{F}_e$, ${}_L\mathbf{F}_i$, and ${}_S\mathbf{F}(\mathbf{x}, t)$ are, respectively, the macroscopic force due to long-range external fields, the long-range contribution of internal fields, and the short-range contribution of internal fields

$$\begin{aligned}
\mathbf{F}_e(\mathbf{x}, t) &= \left\langle \sum_\kappa \mathbf{f}_e^\kappa(\mathbf{x}^\alpha - \mathbf{x}) \right\rangle, \\
{}_L\mathbf{F}_i(\mathbf{x}, t) &= \left\langle \sum_\kappa {}_L\mathbf{f}_i^\kappa \delta(\mathbf{x}^\alpha - \mathbf{x}) \right\rangle, \\
{}_S\mathbf{F}(\mathbf{x}, t) &= \left\langle \sum_\kappa {}_S\mathbf{f}^\kappa \delta(\mathbf{x}^\alpha - \mathbf{x}) \right\rangle.
\end{aligned} \tag{3.5.7}$$

In this book we are not interested in mixtures. Considering identical particles ($e^\kappa = e$), and writing

$$\begin{aligned}
\mathbf{J}(\mathbf{x}, t) &= \left\langle \sum_k e^\kappa \dot{\mathbf{x}}^\kappa \delta(\mathbf{x}^\kappa - \mathbf{x}) \right\rangle, \\
\mathbf{P}(\mathbf{x}, t) &= \left\langle \sum_\kappa \mathbf{p}^\kappa \delta(\mathbf{x}^\kappa - \mathbf{x}) \right\rangle, \\
\mathbf{M}(\mathbf{x}, t) &= \left\langle \sum_\kappa \left(\boldsymbol{\mu}^\kappa + \frac{1}{c}\mathbf{p}^\kappa \times \dot{\mathbf{x}} \right) \delta(\mathbf{x}^\kappa - \mathbf{x}) \right\rangle,
\end{aligned} \tag{3.5.8}$$

for contributions of the external fields, we obtain

$$\begin{aligned}
\mathbf{F}_e(\mathbf{x}, t) = {} & \mathbf{q}_e\mathbf{F}_e + \frac{1}{c}\mathbf{J} \times \mathbf{B}_e + (\nabla\mathbf{E}_e)\cdot\mathbf{P} + (\nabla\mathbf{B}_e)\cdot\mathbf{M} \\
& + \frac{1}{c}\left\langle \sum_\kappa \frac{\mathrm{d}}{\mathrm{d}t}[\mathbf{p}^\kappa \times \mathbf{B}_e(\mathbf{x}^\kappa, t)]\delta(\mathbf{x}^\kappa - \mathbf{x}) \right\rangle.
\end{aligned} \tag{3.5.9}$$

The last term, upon using (3.2.4), becomes

$$\frac{1}{c}\frac{\partial}{\partial t}\left\langle \sum_\kappa \mathbf{p}^\kappa \times \mathbf{B}_e\delta(\mathbf{x}^\kappa - \mathbf{x}) \right\rangle + \frac{1}{c}\nabla\cdot\langle \dot{\mathbf{x}}^\kappa\mathbf{p}^\kappa \times \mathbf{B}_e\delta(\mathbf{x}^\kappa - \mathbf{x})\rangle.$$

If (3.5.8) and $\dot{\mathbf{x}}^\kappa = \dot{\mathbf{x}} + \dot{\xi}^\kappa$ are used in these expressions, and the result is substituted into (3.5.9), we obtain

$$\mathbf{F}_e = q_e\mathbf{E}_e + \frac{1}{c}\mathbf{J} \times \mathbf{B}_e + (\nabla\mathbf{E}_e)\cdot\mathbf{P} + (\nabla\mathbf{B}_e)\cdot\mathbf{M} + \frac{1}{c}\frac{\partial}{\partial t}(\mathbf{P} \times \mathbf{B}_e)$$
$$+ \frac{1}{c}\nabla\cdot[\dot{\mathbf{x}}(\mathbf{P} \times \mathbf{B}_e)] - \nabla\cdot\mathbf{P}^F, \tag{3.5.10}$$

where $\mathbf{P}^F$ is a pressure tensor, due to the action of the magnetic field $\mathbf{B}$ on the electric dipoles, given by

$$\mathbf{P}^F = -\frac{1}{c}\int \dot{\xi}^1\mathbf{p}^1 \times \mathbf{B}_e f_1(\mathbf{x}, \dot{\mathbf{x}}^1, \mathbf{p}^1; t)\, d\dot{\mathbf{x}}^1\, d\mathbf{p}^1. \tag{3.5.11}$$

Substituting for $\mathbf{E}_e$ from (3.3.11) and $\mathbf{B}_e = \mathbf{B}$ (up to the order c^{-1}) we obtain for the long-range force

$${}_L\mathbf{F} = \mathbf{F}_e + {}_L\mathbf{F}_i = q_e\mathbf{E} + \frac{1}{c}\mathbf{J} \times \mathbf{B} + (\nabla\mathbf{E})\cdot\mathbf{P} + (\nabla\mathbf{B})\cdot\mathbf{M}$$
$$+ \frac{1}{c}\frac{\partial}{\partial t}(\mathbf{P} \times \mathbf{B}) + \frac{1}{c}\nabla\cdot[\dot{\mathbf{x}}(\mathbf{P} \times \mathbf{B})] - \nabla\cdot\mathbf{P}^F + \mathbf{F}^C, \tag{3.5.12}$$

where

$$\mathbf{F}^C = {}_L\mathbf{F}_i + \int [q_e(\mathbf{x}, t) + \mathbf{P}(\mathbf{x}, t)\cdot\nabla][q_e(\mathbf{x}', t) + \mathbf{P}(\mathbf{x}', t)\cdot\nabla']\nabla\frac{1}{4\pi|\mathbf{x} - \mathbf{x}'|}\, d\mathbf{x}'. \tag{3.5.13}$$

We now introduce a two-point correlation function

$$c_2(1, 2; t) = f_2(1, 2; t) - f_1(1; t)f_1(2; t), \tag{3.5.14}$$

where f_2 is the two-point probability density. Using (2.8.15) we calculate ${}_L\mathbf{F}_i$ by

$${}_L\mathbf{F}_i = -\int \sum_{n,m=0}^{\infty} \mathbf{p}^{(n)1} \odot \nabla_1{}^n\mathbf{p}^{(m)2} \odot \nabla_2{}^m\nabla_1\frac{1}{4\pi|\mathbf{x}_1 - \mathbf{x}_2|}$$
$$\times\, \delta(\mathbf{x} - \mathbf{x}_1)f_2(\mathbf{x}_1, 1, \mathbf{x}_2, 2; t)\, d\mathbf{x}_1\, d\phi_1\, d\mathbf{x}_2\, d\phi_2, \tag{3.5.15}$$

where 1 (and 2) denotes the complete set of electric multipole moments $\mathbf{p}^{(n)1}$ (and $\mathbf{p}^{(m)2}$), $n, m = 1, 2, \ldots$. If we substitute f_2 from (3.5.14), the part of the integrand containing $f_1(1; t)f_1(2; t)$ (which must be multiplied by $\delta(\mathbf{x}' - \mathbf{x})$), can be integrated over $\mathbf{x}'$. If we use (3.5.8) and neglect all multipole densities of an order higher than two, we obtain exactly the negative of the second term in (3.5.13).

Consequently,

$$\mathbf{F}^C = -\int \sum_{n,m=0}^{\infty} \mathbf{p}^{(n)1} \odot \nabla{}^n\mathbf{p}^{(m)2} \odot \nabla'{}^m\nabla\frac{1}{4\pi|\mathbf{x} - \mathbf{x}'|}$$
$$\times\, c_2(\mathbf{x}, 1, \mathbf{x}', 2; t')\, d\mathbf{x}'\, d\phi_1\, d\phi_2, \tag{3.5.16}$$

which may be called "the correlation force." For fluids and polycrystalline solids of neutral atoms, the correlation length is much smaller than the macroscopic characteristic length over which macroscopic fields change appreciably. In this case, expanding c_2 into a Taylor series, and keeping the first two terms (the Irving and Kirkwood [1950] approximation), we have

$$c_2(\mathbf{x}, 1, \mathbf{x} - \mathbf{s}, 2; t) = c_2(\mathbf{x} + \tfrac{1}{2}\mathbf{s}, 1, \mathbf{x} - \tfrac{1}{2}\mathbf{s}, 2; t) - \tfrac{1}{2}\mathbf{s} \cdot \nabla c_2(\mathbf{x} + \tfrac{1}{2}\mathbf{s}, 1, \mathbf{x} - \tfrac{1}{2}\mathbf{s}, 2; t). \quad (3.5.17)$$

The first term in (3.5.16) then gives no contribution, and $\mathbf{F}^{\mathrm{C}}$ becomes a divergence of a pressure tensor $\mathbf{P}^{\mathrm{C}}$, i.e.,

$$\mathbf{F}^{\mathrm{C}} = -\nabla \cdot \mathbf{P}^{\mathrm{C}}, \quad (3.5.18)$$

where

$$\mathbf{P}^{\mathrm{C}} \equiv -\int \sum_{n,m=0}^{\infty} (-)^m \mathbf{s} \nabla_s p^{(n)1} \odot \nabla_s^n \mathbf{p}^{(n)2} \odot \nabla_s^m \frac{1}{8\pi s} \times c_2(\mathbf{x} + \tfrac{1}{2}\mathbf{s}, 1; \mathbf{x} - \tfrac{1}{2}\mathbf{s}, 2; t)\, ds\, d\phi_1\, d\phi_2. \quad (3.5.19)$$

For crystalline solids, the correlation length may be large. But a similar result may be obtained by using a "mean correlation function" (see de Groot and Suttorp [1972, p. 56]).

For the short-range force, with $\mathbf{s} = \mathbf{x} - \mathbf{x}'$ and $s = |\mathbf{s}|$, we have

$${}_{\mathrm{S}}\mathbf{F} = \int\left\{-\sum_{\alpha,\beta} \nabla_s \frac{e^\alpha e^\beta}{4\pi|\mathbf{s} + \mathbf{x}^{1\alpha} - \mathbf{x}^{2\beta}|} + \sum_{n,m=0}^{\infty} (-)^m \mathbf{p}^{(n)1} \odot \nabla_s^n \mathbf{p}^{(m)2} \odot \nabla_s^m \nabla_s \frac{1}{4\pi s}\right\} f_2(\mathbf{x}, 1, \mathbf{x} - \mathbf{s}, 2; t)\, ds\, d\phi_1\, d\phi_2, \quad (3.5.20)$$

where e^α and e^β are charges of the constituent particles of the identical atoms 1 and 2. The integrand vanishes when the atoms are outside each other. Thus, the range of s is small. Expanding f_2 into a Taylor series about $\mathbf{x} = \mathbf{s}/2$, and retaining only the first two terms (Irving–Kirkwood approximation), we have

$$f_2(\mathbf{x}, 1, \mathbf{x} - \mathbf{s}, 2; t) = f_2(\mathbf{x} + \tfrac{1}{2}\mathbf{s}, 1, \mathbf{x} - \tfrac{1}{2}\mathbf{s}, 2; t) - \tfrac{1}{2}\mathbf{s} \cdot \nabla f_2(\mathbf{x} + \tfrac{1}{2}\mathbf{s}, 1, \mathbf{x} - \tfrac{1}{2}\mathbf{s}, 2; t). \quad (3.5.21)$$

The first term is symmetrical with respect to the interchange of the first pair of variables with the second pair and does not contribute to the integral. We now have

$${}_{\mathrm{S}}\mathbf{F} = -\nabla \cdot \mathbf{P}^{\mathrm{S}}, \quad (3.5.22)$$

$$\mathbf{P}^{\mathrm{S}} = -\int\left\{\mathbf{s}\nabla_s\left[\sum_{\alpha,\beta} \frac{e^\alpha e^\beta}{8\pi|\mathbf{s} + \mathbf{x}^{1\alpha} - \mathbf{x}^{2\alpha}|} - \sum_{n,m=0}^{\infty} (-)^m \mathbf{p}^{(n)1} \odot \nabla_s^n \mathbf{p}^{(m)2} \odot \nabla_s^m \frac{1}{8\pi s}\right]\right\} f_2(\mathbf{x} + \tfrac{1}{2}\mathbf{s}, 1, \mathbf{x} - \tfrac{1}{2}\mathbf{s}, 2; t)\, ds\, d\phi_1\, d\phi_2. \quad (3.5.23)$$

For the total force, therefore, we may write

$$\mathbf{F} = q_e\mathbf{E} + \frac{1}{c}\mathbf{J} \times \mathbf{B} + (\nabla\mathbf{E})\cdot\mathbf{P} + (\nabla\mathbf{B})\cdot\mathbf{M} + \frac{1}{c}\nabla\cdot(\mathbf{v}\mathbf{P} \times \mathbf{B})$$

$$+ \frac{1}{c}\frac{\partial}{\partial t}(\mathbf{P} \times \mathbf{B}) - \nabla\cdot\mathbf{P}^{\mathrm{E}}, \tag{3.5.24}$$

where

$$\mathbf{P}^{\mathrm{E}} = \mathbf{P}^{\mathrm{F}} + \mathbf{P}^{\mathrm{C}} + \mathbf{P}^{\mathrm{S}}. \tag{3.5.25}$$

In a continuum theory, $\mathbf{P}^{\mathrm{E}}$ can be incorporated into the stress tensor, for which constitutive equations are constructed. The source term that is relevant to continuum theory therefore is

$$\mathbf{F}^{\mathrm{E}} = q_e\mathbf{E} + \frac{1}{c}\mathbf{J} \times \mathbf{B} + (\nabla\mathbf{E})\cdot\mathbf{P} + (\nabla\mathbf{B})\cdot\mathbf{M} + \frac{1}{c}\nabla\cdot(\mathbf{v}\mathbf{P} \times \mathbf{B}) + \frac{1}{c}\frac{\partial}{\partial t}(\mathbf{P} \times \mathbf{B}). \tag{3.5.26}$$

This is identical in form to $\mathbf{F}_e$ as given by (3.5.10) (excluding the last term in (3.5.10)), and writing $\mathbf{E}$ and $\mathbf{B}$ in place of $\mathbf{E}_e$ and $\mathbf{B}_e$, respectively. This is because the internal fields contribute only to the stress tensor.

C. Couple Density

The macroscopic couple due to electromagnetic fields is obtained by taking the statistical average of (2.9.10)

$$\mathbf{C} = \mathbf{C}_e + {}_L\mathbf{C}_i + {}_S\mathbf{C}, \tag{3.5.27}$$

where $\mathbf{C}_e$, ${}_L\mathbf{C}_i$, and ${}_S\mathbf{C}$ are, respectively, the macroscopic couple due to long-range external fields, long-range contribution of internal fields, and short-range contribution of internal fields, i.e.,

$$\begin{aligned} \mathbf{C}_e(\mathbf{x}, t) &= \left\langle \sum_{\kappa} \mathbf{c}_e^{\kappa}\delta(\mathbf{x}^{\kappa} - \mathbf{x}) \right\rangle, \\ {}_L\mathbf{C}_i(\mathbf{x}, t) &= \left\langle \sum_{\kappa} {}_L\mathbf{C}_i^{\kappa}\delta(\mathbf{x}^{\kappa} - \mathbf{x}) \right\rangle, \\ {}_S\mathbf{C}(\mathbf{x}, t) &= \left\langle \sum_{\kappa s} {}_S\mathbf{C}^{\kappa}\delta(\mathbf{x}^{\kappa} - \mathbf{x}) \right\rangle. \end{aligned} \tag{3.5.28}$$

Using the identity

$$\mathbf{a} \times (\mathbf{b} \times \mathbf{c}) + \mathbf{c} \times (\mathbf{a} \times \mathbf{b}) + \mathbf{b} \times (\mathbf{c} \times \mathbf{a}) = \mathbf{0},$$

we express (2.9.7) as

$$\mathbf{c}_e^{\kappa} = \mathbf{p}^{\kappa} \times \mathbf{E}_e + \left(\boldsymbol{\mu}^{\kappa} + \frac{1}{c}\mathbf{p}^{\kappa} \times \dot{\mathbf{x}}^{\kappa}\right) \times \mathbf{B}_e + \frac{1}{c}\dot{\mathbf{x}}^{\kappa} \times (\mathbf{p}^{\kappa} \times \mathbf{B}_e).$$

Applying the procedure of averaging discussed in Section 3.2 we obtain

$$\mathbf{C}_e = \mathbf{P} \times \mathbf{E}_e + \mathbf{M} \times \mathbf{B}_e + \frac{1}{c}\mathbf{v} \times (\mathbf{P} \times \mathbf{B}_e) + \frac{1}{c}\langle \dot{\boldsymbol{\xi}}^\kappa \times (\mathbf{p}^\kappa \times \mathbf{B}_e)\rangle. \quad (3.5.29)$$

For the external fields, we substitute from (3.3.11) for $\mathbf{E}_e$ and $\mathbf{B}_e = \mathbf{B}$ to obtain, for the total couple,

$$\mathbf{C} = \mathbf{P} \times \mathbf{E} + \mathbf{M} \times \mathbf{B} + \frac{1}{c}\mathbf{v} \times (\mathbf{P} \times \mathbf{B}) + {}_{\mathrm{I}}\mathbf{C}, \quad (3.5.30)$$

where

$${}_{\mathrm{I}}\mathbf{C} = \frac{1}{c}\langle \dot{\boldsymbol{\xi}}^\kappa \times (\mathbf{p}^\kappa \times \mathbf{B}_e)\rangle - \frac{1}{c}\int \mathbf{P}(\mathbf{x}, t) \times \nabla[q_e(\mathbf{x}', t)$$

$$+ \mathbf{P}(\mathbf{x}', t)\cdot\nabla'] \frac{d\mathbf{x}'}{4\pi|\mathbf{x} - \mathbf{x}'|} + {}_{\mathrm{L}}\mathbf{C}_{\mathrm{i}} + {}_{\mathrm{S}}\mathbf{C}. \quad (3.5.31)$$

Just as in the case of force density, though somewhat more elaborately, it can be shown that ${}_{\mathrm{I}}\mathbf{C}$ contributes only to the antisymmetric part of the stress tensor and couple stress (a divergence term) for which constitutive equations will be constructed. Therefore, the electromagnetic couple density given by

$$\mathbf{C}^{\mathrm{E}} = \mathbf{P} \times \mathbf{E} + \mathbf{M} \times \mathbf{B} + \frac{1}{c}\mathbf{v} \times (\mathbf{P} \times \mathbf{B}) \quad (3.5.32)$$

is the source term relevant to the continuum theory. We need not reproduce macroscopic expressions of couples ${}_{\mathrm{I}}\mathbf{C}$, ${}_{\mathrm{L}}\mathbf{C}$, and ${}_{\mathrm{S}}\mathbf{C}$. For these results, the reader is referred to de Groot and Suttorp [1972, Sects. 5 and 6].

D. Electromagnetic Power

The macroscopic expression of power is obtained by taking the statistical average of (2.10.10)

$$W = {}_{\mathrm{C}}W + {}_{\mathrm{L}}W + {}_{\mathrm{S}}W, \quad (3.5.33)$$

where

$$\begin{aligned} {}_{\mathrm{C}}W &= -\left\langle \frac{de_{\mathrm{c}}}{dt}\delta(\mathbf{x}^\kappa - \mathbf{x})\right\rangle, \\ {}_{\mathrm{L}}W &= \langle {}_{\mathrm{L}}w\delta(\mathbf{x}^\kappa - \mathbf{x})\rangle = \langle(\mathbf{w}_e + {}_{\mathrm{L}}\bar{w}_{\mathrm{i}})\delta(\mathbf{x}^\kappa - \mathbf{x})\rangle, \\ {}_{\mathrm{S}}W &= \langle {}_{\mathrm{S}}w\delta(\mathbf{x}^\kappa - \mathbf{x})\rangle. \end{aligned} \quad (3.5.34)$$

Writing

$${}_{\mathrm{C}}W = -\left\langle \frac{d}{dt}[e_{\mathrm{c}}\delta(\mathbf{x}^\kappa - \mathbf{x})]\right\rangle - \langle \nabla\cdot[(\dot{\mathbf{x}} + \dot{\boldsymbol{\xi}}^\kappa)e_{\mathrm{c}}\delta(\mathbf{x}^\kappa - \mathbf{x})]\rangle,$$

with the help of (3.2.4) and (3.2.6) we obtain

$$ {}_{\mathrm{C}}W = -\frac{dE_{\mathrm{c}}}{dt} - \nabla\cdot\mathbf{q}^{\mathrm{c}}, \tag{3.5.35}$$

where

$$\begin{aligned} E_{\mathrm{c}} &= \langle e_{\mathrm{c}}\delta(\mathbf{x}^{\kappa}-\mathbf{x})\rangle, \\ \mathbf{q}^{\mathrm{c}} &= \langle \dot{\xi}^{\kappa} e_{\mathrm{c}}\delta(\mathbf{x}^{\kappa}-\mathbf{x})\rangle, \end{aligned} \tag{3.5.36}$$

where E_{c} is a contribution to the macroscopic internal energy density, and $\mathbf{q}^{\mathrm{c}}$ is contribution to the heat flux.

Next we calculate the power of the external fields. We first note that

$$\begin{aligned} \dot{\mathbf{p}}^{\kappa}\cdot\mathbf{E}_{\mathrm{e}}\delta(\mathbf{x}^{\kappa}-\mathbf{x}) &= \frac{d}{dt}[\mathbf{p}^{\kappa}\delta(\mathbf{x}^{\kappa}-\mathbf{x})]\cdot\mathbf{E}_{\mathrm{e}} - (\mathbf{p}^{\kappa}\cdot\mathbf{E}_{\mathrm{e}})\dot{\mathbf{x}}^{\kappa}\cdot\nabla_{\kappa}\delta(\mathbf{x}^{\kappa}-\mathbf{x}) \\ &= \frac{d}{dt}[\mathbf{p}^{\kappa}\delta(\mathbf{x}^{\kappa}-\mathbf{x})]\cdot\mathbf{E}_{\mathrm{e}} + \nabla\cdot[\dot{\mathbf{x}}^{\kappa}\mathbf{p}^{\kappa}\cdot\mathbf{E}_{\mathrm{e}}\delta(\mathbf{x}^{\kappa}-\mathbf{x})] \\ &\quad - \mathbf{v}^{\kappa}\cdot(\nabla_{\kappa}\mathbf{E}_{\mathrm{e}})\cdot\mathbf{p}^{\kappa}\delta(\mathbf{x}^{\kappa}-\mathbf{x}). \end{aligned} \tag{3.5.37}$$

We multiply (2.10.8) by $\delta(\mathbf{x}^{\kappa}-\mathbf{x})$, employ (3.5.31), and take the average of the resulting expression to obtain

$$\begin{aligned} W_{\mathrm{e}} \equiv \langle w_{\mathrm{e}}\delta(\mathbf{x}^{\kappa}-\mathbf{x})\rangle &= \mathbf{J}\cdot\mathbf{E}_{\mathrm{e}} + \frac{\partial\mathbf{P}}{\partial t}\cdot\mathbf{E}_{\mathrm{e}} - \mathbf{M}\cdot\frac{\partial\mathbf{B}_{\mathrm{e}}}{\partial t} + \nabla\cdot(\mathbf{v}\mathbf{P}\cdot\mathbf{E}_{\mathrm{e}}) \\ &\quad + \nabla\cdot\langle\dot{\xi}^{\kappa}(\mathbf{p}^{\kappa}\cdot\mathbf{E}_{\mathrm{e}})\delta(\mathbf{x}^{\kappa}-\mathbf{x})\rangle. \end{aligned} \tag{3.5.38}$$

Upon substituting $\mathbf{B} = \mathbf{B}_{\mathrm{e}}$ and $\mathbf{E}_{\mathrm{e}}$ from (3.3.11), this gives

$$W_{\mathrm{e}} = \mathbf{J}\cdot\mathbf{E} + \mathbf{E}\cdot\frac{\partial\mathbf{P}}{\partial t} - \mathbf{M}\cdot\frac{\partial\mathbf{B}}{\partial t} + \nabla\cdot(v\mathbf{P}\cdot\mathbf{E}) + {}_{\mathrm{I}}W, \tag{3.5.39}$$

where

$$\begin{aligned} {}_{\mathrm{I}}W = \nabla\cdot[\langle\dot{\xi}^{\kappa}\mathbf{p}^{\kappa}\delta(\mathbf{x}^{\kappa}-\mathbf{x})\rangle\cdot\mathbf{E}_{\mathrm{e}}] + \left(\mathbf{J} + \frac{\partial\mathbf{P}}{\partial t}\right)\cdot\nabla\int[q_{\mathrm{e}}(\mathbf{x}', t) \\ - \nabla'\cdot\mathbf{P}(\mathbf{x}', t)]\frac{d\mathbf{x}'}{4\pi|\mathbf{x}-\mathbf{x}'|}. \end{aligned} \tag{3.5.40}$$

Again, it can be shown that W can be expressed as the sum of $W_{\mathrm{e}} + {}_{\mathrm{I}}W$, a time rate and a divergence, of which last two can be incorporated into the rate of the internal energy density and heat flux. Thus, the macroscopic electromagnetic power source is

$$W^{\mathrm{E}} = \mathbf{J}\cdot\mathbf{E} + \frac{\partial\mathbf{P}}{\partial t}\cdot\mathbf{E} - \mathbf{M}\cdot\frac{\partial\mathbf{B}}{\partial t} + \nabla\cdot[\mathbf{v}(\mathbf{P}\cdot\mathbf{E})]. \tag{3.5.41}$$

For the internal energy and the heat flux, separate constitutive equations are constructed. Therefore, we do not need to reproduce the expressions of these terms. The analysis presented above clearly indicates various contributions from atomic fields to macroscopic fields. In a continuum theory, only the source terms due to the field outside the material point are needed, since the contributions of the internal fields are taken into account through the concepts of the internal energy density and the stress tensor.

The expressions of the macroscopic source terms for the electromagnetic force, couple, and energy were obtained in a simpler way by Dixon and Eringen [1965a]. This method was later used by Maugin and Eringen [1977] and others. For these derivations, see also Eringen [1980, Chap. 10].

The electromagnetic force, couple, and energy sources obtained above may be written in the moving (proper) reference frame by using expressions (3.4.10)–(3.4.13) and

$$\overset{*}{\mathbf{P}} = \dot{\mathbf{P}} - (\mathbf{P}\cdot\nabla)\mathbf{v} + \mathbf{P}(\nabla\cdot\mathbf{v}). \tag{3.5.42}$$

Hence,

$$\mathbf{F}^{\mathrm{E}} = q_{\mathrm{e}}\mathscr{E} + \frac{1}{c}\left(\mathscr{J} + \overset{*}{\mathbf{P}}\right) \times \mathbf{B} + (\mathbf{P}\cdot\nabla)\mathscr{E} + (\nabla\mathbf{B})\cdot\mathscr{M}, \tag{3.5.43}$$

$$\mathbf{C}^{\mathrm{E}} = \mathbf{P} \times \mathscr{E} + \mathscr{M} \times \mathbf{B}, \tag{3.5.44}$$

$$W^{\mathrm{E}} = \mathbf{F}^{\mathrm{E}}\cdot\mathbf{v} + \rho\mathscr{E}\cdot\dot{\boldsymbol{\pi}} - \mathscr{M}\cdot\dot{\mathbf{B}} + \mathscr{J}\cdot\mathscr{E}, \tag{3.5.45}$$

where π is the polarization per unit mass defined by

$$\boldsymbol{\pi} = \mathbf{P}/\rho. \tag{3.5.46}$$

3.6. Electromagnetic Stress Tensor, Momentum, and Poynting Vector

For some purposes it may prove more convenient to express the electromagnetic force, couple, and energy in the form of the sum of a divergence of a tensor field and the time rate of a vector field, that is,

$$\nabla\cdot\mathbf{t}^{\mathrm{E}} - \frac{\partial\mathbf{G}}{\partial t} = \mathbf{F}^{\mathrm{E}}, \tag{3.6.1}$$

$$\varepsilon_{klm}t^{\mathrm{E}}{}_{lm} = C^{\mathrm{E}}_{k}, \tag{3.6.2}$$

$$-\nabla\cdot[\mathbf{S} - \mathbf{v}(\mathbf{E}\cdot\mathbf{P})] - \frac{\partial}{\partial t}[\tfrac{1}{2}(E^2 + B^2)] = W^{\mathrm{E}}, \tag{3.6.3}$$

where

$$\mathbf{t}^{\mathrm{E}} = \mathbf{P}\otimes\mathscr{E} - \mathbf{B}\otimes\mathscr{M} + \mathbf{E}\otimes\mathbf{E} + \mathbf{B}\otimes\mathbf{B} - \tfrac{1}{2}(E^2 + B^2 - 2\mathscr{M}\cdot\mathbf{B})\mathbf{1}, \tag{3.6.4}$$

$$\mathbf{G} = \frac{1}{c}\mathbf{E} \times \mathbf{B}, \tag{3.6.5}$$

$$\mathbf{S} = c\mathbf{E} \times \mathbf{H}, \tag{3.6.6}$$

are, respectively, the *electromagnetic stress tensor*, the *electromagnetic momentum*, and the *Poynting vector*.

The verification of (3.6.1)–(3.6.3) is made by direct substitution of (3.6.4) in to (3.6.6), and by using Maxwell's equation (3.3.2).[6]

It is convenient to introduce the electromagnetic momentum per unit mass $\mathbf{g}$ by

$$\mathbf{g} = \mathbf{G}/\rho. \tag{3.6.7}$$

The identity (3.6.1) can then be written as

$$\nabla \cdot (\mathbf{t}^{\mathrm{E}} + \rho \mathbf{v} \otimes \mathbf{g}) - \rho \dot{\mathbf{g}} = \mathbf{F}^{\mathrm{E}}. \tag{3.6.8}$$

We can also decompose the electromagnetic tensor $\mathbf{t}^{\mathrm{E}}$ into a symmetric tensor and another tensor, by introducing the so-called *Maxwell stress* tensor $\mathbf{t}^{\mathrm{M}}$ so that

$$\mathbf{t}^{\mathrm{E}} = \mathbf{t}^{\mathrm{M}} + \overline{\mathbf{t}}^{\mathrm{E}}, \tag{3.6.9}$$

where

$$\begin{aligned} \mathbf{t}^{\mathrm{M}} &= \mathbf{E} \otimes \mathbf{E} + \mathbf{B} \otimes \mathbf{B} - \tfrac{1}{2}(E^2 + B^2)\mathbf{1}, \\ \overline{\mathbf{t}}^{\mathrm{E}} &= \mathbf{P} \otimes \boldsymbol{\mathscr{E}} - \mathbf{B} \otimes \boldsymbol{\mathscr{M}} + \boldsymbol{\mathscr{M}} \cdot \mathbf{B}\mathbf{1}. \end{aligned} \tag{3.6.10}$$

In terms of these quantities, expressions for the force, couple, and power take the forms

$$\mathbf{F}^{\mathrm{E}} = \mathbf{F}^{\mathrm{L}} + \nabla \cdot \overline{\mathbf{t}}^{\mathrm{E}}, \tag{3.6.11}$$

$$\mathbf{C}^{\mathrm{E}} = \varepsilon_{klm} \overline{t}^{\mathrm{E}}_{lm} \mathbf{i}_k = \mathbf{P} \times \boldsymbol{\mathscr{E}} + \boldsymbol{\mathscr{M}} \times \mathbf{B}, \tag{3.6.12}$$

$$W^{\mathrm{E}} = \mathbf{F}^{\mathrm{E}} \cdot \mathbf{v} + \mathbf{C}^{\mathrm{E}} \cdot \mathbf{w} + \rho h^{\mathrm{E}}, \tag{3.6.13}$$

where $\mathbf{F}^{\mathrm{L}}$ is the *effective Lorentz force*, $\mathbf{w}$ is the vorticity vector, and h^{E} is an energy source term defined by

$$\begin{aligned} \mathbf{F}^{\mathrm{L}} &\equiv q^{\mathrm{eff}} \boldsymbol{\mathscr{E}} + \frac{1}{c} \boldsymbol{\mathscr{J}}^{\mathrm{eff}} \times \mathbf{B}, \\ \mathbf{w} &\equiv \tfrac{1}{2} \nabla \times \mathbf{v}, \\ \rho h^{\mathrm{E}} &\equiv \boldsymbol{\mathscr{J}} \cdot \boldsymbol{\mathscr{E}} + \boldsymbol{\mathscr{E}} \cdot \overset{*}{\mathbf{P}} - \boldsymbol{\mathscr{M}} \cdot \overset{*}{\mathbf{B}} + \overline{t}^{\mathrm{E}}_{kl} d_{lk}, \end{aligned} \tag{3.6.14}$$

with d_{kl} being the deformation rate tensor. From (3.6.1) we also obtain for the effective Lorentz force,

$$\mathbf{F}^{\mathrm{L}} = \nabla \cdot \mathbf{t}^{\mathrm{M}} - \frac{\partial \mathbf{G}}{\partial t}. \tag{3.6.15}$$

In vacuum, $\mathbf{F}^{\mathrm{E}} = \mathbf{F}^{\mathrm{L}} = \mathbf{0}$ and (3.6.15) can be obtained by manipulations of Maxwell's equations for the vacuum.

---

[6] See Maugin and Collet [1974]. Such identities were also obtained by Tiersten and Tsai [1972], for dielectric insulators, by Livens [1962], and by others. See also Eringen [1980, Sects. 10.6 and 10.7].

Finally, we note that the electromagnetic momentum and Poynting vectors defined by (3.6.5) and (3.6.6) are expressed in the fixed Galilean frame. Equally important and useful are the expressions in the co-moving frame R_C

$$\mathscr{G} = \frac{1}{c}\mathscr{E} \times \mathscr{B}, \tag{3.6.16}$$

$$\mathscr{S} = c\mathscr{E} \times \mathscr{H}. \tag{3.6.17}$$

We have the relations

$$c^2(\mathscr{G} - \mathbf{G}) = [\mathbf{t}^{\mathrm{M}} + \mathbf{v} \otimes \mathbf{G} - \tfrac{1}{2}(E^2 + B^2)\mathbf{1}] \cdot \mathbf{v}, \tag{3.6.18}$$

$$\mathscr{S} = \mathbf{S} + [\mathbf{t}^{\mathrm{E}} + \mathbf{v} \otimes \mathbf{G} - \tfrac{1}{2}(E^2 + B^2 + 2\mathbf{E}\cdot\mathbf{P})\mathbf{1}] \cdot \mathbf{v}. \tag{3.6.19}$$

With the help of (3.6.3) we find that

$$W^{\mathrm{E}} = -\rho \frac{d}{dt}\left[\frac{1}{2\rho}(E^2 + B^2)\right] + \nabla \cdot [(\mathbf{t}^{\mathrm{E}} + \mathbf{v} \otimes \mathbf{G}) \cdot \mathbf{v} - \mathscr{S}]. \tag{3.6.20}$$

Some remarks are in order concerning these results: It is clear that the ponderomotive force, couple, and energy supply deduced from the Lorentz theory of electrons are *primary notions*, whereas the electromagnetic stress tensor, electromagnetic momentum, and Poynting vector are *derived notions* via identities (3.6.1)–(3.6.3). The expressions found for $\mathbf{t}^{\mathrm{E}}$ and $\mathbf{G}$ are none of the well-known expressions proposed by various authors in the first decade of this century. For example, some of the celebrated expressions are:

Minkowski [1908]:

$$\mathbf{t}^{\mathrm{Mi}} = \mathbf{D} \otimes \mathbf{E} + \mathbf{B} \otimes \mathbf{H} - \tfrac{1}{2}(\mathbf{E}\cdot\mathbf{D} + \mathbf{H}\cdot\mathbf{B})\mathbf{1}. \tag{3.6.21}$$

Einstein and Laub [1908]:

$$\mathbf{t}^{\mathrm{EL}} = \mathbf{D} \otimes \mathbf{E} + \mathbf{B} \otimes \mathbf{H} - \tfrac{1}{2}(E^2 + B^2)\mathbf{1}. \tag{3.6.22}$$

Abraham [1909], [1910]:

$$\mathbf{t}^{\mathrm{A}} = \tfrac{1}{2}[\mathbf{t}^{\mathrm{EL}} + (\mathbf{t}^{\mathrm{EL}})^{\mathrm{T}}], \qquad \mathbf{G}^{\mathrm{A}} = \frac{1}{c}\mathbf{D} \times \mathbf{B}, \tag{3.6.23}$$

where the superscript T indicates the transpose. We remark that (3.6.21) and (3.6.22) are not expressed in terms of field in the co-moving frame R_C. Therefore, although these stress tensors are nonsymmetric, we would obtain the ponderomotive torque up to the terms of order $\beta = |\mathbf{v}|/c$ by retaining their skew parts, that is, only the torque corresponding to the *statics*. Also, these expressions make use of the fields $\mathbf{H}$ and $\mathbf{D}$ which have only a secondary character. As to the advice of Abraham that the stress tensor must be symmetric (which has been taken too seriously by most workers in magnetohydrodynamics), we have the contradiction observed by the simple example of the alignment of magnetic dipoles with the magnetic field (a basic tenet for the theory of ferromagnetic bodies). It may happen that, in certain circumstances, only the symmetric part of the stress plays a role. Nevertheless, such situations will have to be established by the restrictions emanating from physical laws (e.g., the second law of thermodynamics).[7]

An interpretation of the Poynting vector is revealed if we integrate (3.6.20) over the material volume in the co-moving frame R_C. By use of the Green–Gauss theorem, the divergence terms can be taken to the surface and we obtain

$$\int_{\mathscr{V}} W^{\mathrm{E}}\, dv + \frac{d}{dt}\int_{\mathscr{V}} \tfrac{1}{2}(E^2 + B^2)\, dv = \int_{\partial\mathscr{V}} d\mathbf{a}\cdot(\mathbf{t}^{\mathrm{E}} + \mathbf{v}\otimes\mathbf{G})\cdot\mathbf{v} - \int_{\partial\mathscr{V}} \mathscr{S}\cdot d\mathbf{a}, \tag{3.6.24}$$

where $d\mathbf{a} = \mathbf{n}\, da$ is the area vector over the boundary $\partial\mathscr{V}$ of the material volume $\mathscr{V}$, and $\mathbf{t}^{\mathrm{E}} + \mathbf{n}(\mathbf{v}\otimes\mathbf{G})$ is the total electromagnetic traction acting on $\partial\mathscr{V}$ whose exterior unit normal is $\mathbf{n}$. This expression, which is an identity, is in the form of a balance law which indicates that the Poynting vector $\mathscr{S}$ may be considered as the electromagnetic energy that is lost through the surface $\partial\mathscr{V}$ of the body. The electromagnetic traction is clearly a secondary notion, as in the case of $\mathbf{t}^{\mathrm{E}}$ and $\mathbf{G}$.

3.7. Electromagnetic Force, Couple, and Power at a Discontinuity Surface

Unlike gravitational forces, the electromagnetic force, couple, and power suffer jump discontinuities across a singular surface σ, which may be sweeping the body at a velocity $\mathbf{v}$ in the positive direction of the unit normal to σ. To determine these jumps we evaluate the integrals of electromagnetic force, couple, and power over the volume $\mathscr{V}$, deprived of the singular two-dimensional surface σ, and across σ. Thus, for example, for the force $\mathbf{F}^{\mathrm{E}}$ we write symbolically,

$$\int_{\mathscr{V}} \mathbf{F}^{\mathrm{E}}\, dv = \int_{\mathscr{V}-\sigma} \mathbf{F}^{\mathrm{E}}\, dv + \int_{\sigma} \hat{\mathbf{F}}^{\mathrm{E}}\, da, \tag{3.7.1}$$

where the first integral on the right-hand side is over the open region $\mathscr{V} - \sigma \equiv \mathscr{V} - \mathscr{V} \cap \sigma$ and is valid *almost* everywhere in $\mathscr{V}$; and the second one is over

[7] The controversy about the forms of the electromagnetic stress tensor and the electromagnetic momentum has been discussed, among others, by Møller [1952, Sect. 72], and extensively, by Penfield and Hauss [1967] Hutter and Ven [1978], and Maugin [1980b]. Penfield and Hauss remark (p. 297) that the wrong choice in the expressions of the electromagnetic momentum, once the electromagnetic stress tensor is chosen, "leads to errors in force that are small, that have not time average value, and that are not easily measured." This certainly is true. However, formulations (3.6.1) and (3.6.8) are not subject to this criticism, since they are pure identities, after the ponderomotive force and couple have been evaluated from such a sound physical theory as that of Lorentz.

σ, which contains the jump discontinuity that is being sought. Employing (3.6.8), the first integral to the right-hand side of (3.7.1) reads

$$\int_{\mathscr{V}-\sigma} \mathbf{F}^{\mathrm{E}}\, dv = -\int_{\mathscr{V}-\sigma} \rho\dot{\mathbf{g}}\, dv + \int_{\mathscr{V}-\sigma} \nabla\cdot(\mathbf{t}^{\mathrm{E}} + \rho\mathbf{v}\otimes\mathbf{g})\, dv. \tag{3.7.2}$$

If we now use the transport theorem (1.13.6) and the generalized Green–Gauss theorem (1.13.7), we obtain

$$\int_{\mathscr{V}-\sigma} \mathbf{F}^{\mathrm{E}}\, dv = -\frac{d}{dt}\int_{\mathscr{V}-\sigma} \mathbf{G}\, dv + \int_{\mathscr{V}-\sigma} \mathbf{n}\cdot(\mathbf{t}^{\mathrm{E}} + \mathbf{v}\otimes\mathbf{G})\, da$$
$$-\int_{\sigma} \mathbf{n}\cdot[\mathbf{t}^{\mathrm{E}} + \mathbf{v}\otimes\mathbf{G}]\, da, \tag{3.7.3}$$

where the surface integral is over the bounding surface $\partial\mathscr{V} - \sigma \equiv \partial\mathscr{V} - \partial\mathscr{V}\cap\sigma$ of the body $\mathscr{V}$, excluding the line of intersection of σ with $\partial\mathscr{V}$. From this expression and (3.7.1), it is clear that the electromagnetic force acquires a jump discontinuity across σ in the amount

$$\hat{\mathbf{F}}^{\mathrm{E}}_{(\mathbf{n})} = [\hat{\mathbf{F}}^{\mathrm{E}}_k]n_k, \tag{3.7.4}$$

where

$$\hat{\mathbf{F}}^{\mathrm{E}}_k = (t^{\mathrm{E}}_{kl} + v_k G_l)\mathbf{i}_l. \tag{3.7.5}$$

Similarly, by integrating $\mathbf{x}\times\mathbf{F}^{\mathrm{E}} + \mathbf{C}^{\mathrm{E}}$, with $\mathbf{C}^{\mathrm{E}}$ and W^{E} given by (3.6.12) and (3.6.20) over $\mathscr{V}-\sigma$, we obtain the jump discontinuities in the couple and power

$$\hat{\mathbf{C}}^{\mathrm{E}}_{(\mathbf{n})} = [\hat{\mathbf{C}}^{\mathrm{E}}_k]n_k = \mathbf{x}\times[\hat{\mathbf{F}}^{\mathrm{E}}_k]n_k, \tag{3.7.6}$$

$$\hat{W}^{\mathrm{E}}_{(\mathbf{n})} = [\hat{W}^{\mathrm{E}}_k]n_k, \tag{3.7.7}$$

where

$$\hat{W}^{\mathrm{E}}_k = (t^{\mathrm{E}}_{kl} + v_k G_l)v_l - \mathscr{S}_k - \tfrac{1}{2}(E^2 + B^2)(v_k - \nu_k). \tag{3.7.8}$$

These expressions will be used in formulating the balance laws.

3.8. Balance Laws of Continuum Physics

The balance laws of continuum physics may be divided into the following two basic categories:

(i) *Volume Balance Law*:

The time rate of the total material field ϕ, contained in the body $\mathscr{B}$ with volume $\mathscr{V}$, is balanced by the influx $\tau\{\phi\}$ of ϕ through its surface $\partial\mathscr{V}$ and the total source $g\{\phi\}$ of ϕ in the body, i.e.,

$$\frac{d}{dt}\int_{\mathscr{V}} \phi\, dv = \int_{\partial\mathscr{V}} \tau\{\phi\}\, da + \int_{\mathscr{V}} g\{\phi\}\, dv. \tag{3.8.1}$$

(ii) *Surface Balance Law*:

The time rate of the total field $\boldsymbol{\psi}$, *piercing a material surface* $\mathscr{S}$, *is balanced by the total influx* $\sigma\{\boldsymbol{\psi}\}$ *of* $\boldsymbol{\psi}$ *across the boundary* $\partial\mathscr{S}$ *of* $\mathscr{S}$ *and its source* **h** *on* $\mathscr{S}$. Mathematically,

$$\frac{d}{dt}\int_{\mathscr{S}} \mathbf{n}\cdot\boldsymbol{\psi}\, da = \oint_{\partial\mathscr{S}} \sigma\{\boldsymbol{\psi}\}\, ds + \int_{\mathscr{S}} \mathbf{n}\cdot\mathbf{h}\, da. \tag{3.8.2}$$

In the volume balance law (3.8.1), τ depends on the orientation of the surface $\partial\mathscr{V}$, i.e., its unit exterior normal **n**. This relation is obtained by the localization of (3.8.1) to a volume element Δv in the form of a tetrahedron, whose one face ABC coincides with the surface $\partial\mathscr{V}$ of the body (Figure 3.8.1). Then (3.8.1), by means of the mean-value theorem, gives

$$\frac{d}{dt}\left(\overset{*}{\phi}\,\Delta v\right) = \overset{*}{\tau}_{(\mathbf{n})}\,\Delta a - \overset{*}{\tau}_k\,\Delta a_k + \overset{*}{g}\,\Delta v,$$

where $\overset{*}{\phi}$ and $\overset{*}{g}$ are the values of ϕ and g at some point in Δv; and $\overset{*}{\tau}_{(\mathbf{n})}$ is the value of $\tau_{(\mathbf{n})}$ at some point on Δa. $\overset{*}{\tau}_k$ (which is independent of **n**) is the value of τ on the three coordinate surfaces Δa_k. In the limit, as $\Delta v \to 0$ and $\Delta a \to 0$, $\Delta v/\Delta a \to 0$, and we obtain

$$\tau\{\phi\} = \mathbf{n}\cdot\boldsymbol{\tau}, \tag{3.8.3}$$

since $\mathbf{n} = d\mathbf{a}/da$.

A similar procedure applied to $\sigma\{\boldsymbol{\psi}\}$ on an infinitesimal triangle lying on $\mathscr{S}$, with one of its edges along $\partial\mathscr{S}$, gives

$$\sigma\{\boldsymbol{\psi}\} = t_k\sigma_k, \tag{3.8.4}$$

where σ_k is the value of σ along the coordinate axes, and **t** is the unit tangent vector of $\partial\mathscr{S}$. It makes a right-handed triad with the unit normal of **n** of $\mathscr{S}$

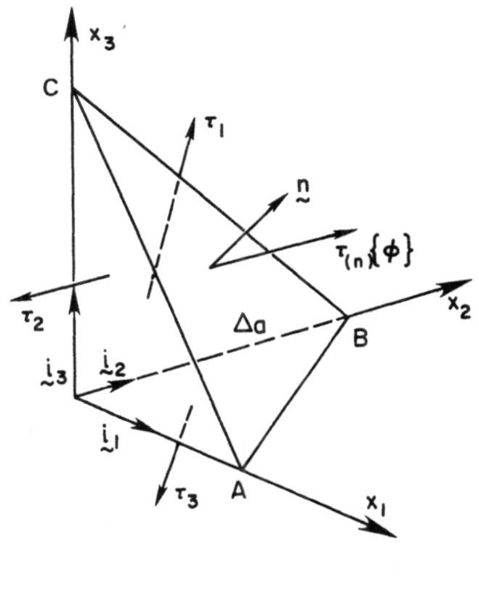

Figure 3.8.1. Infinitesimal tetrahedron.

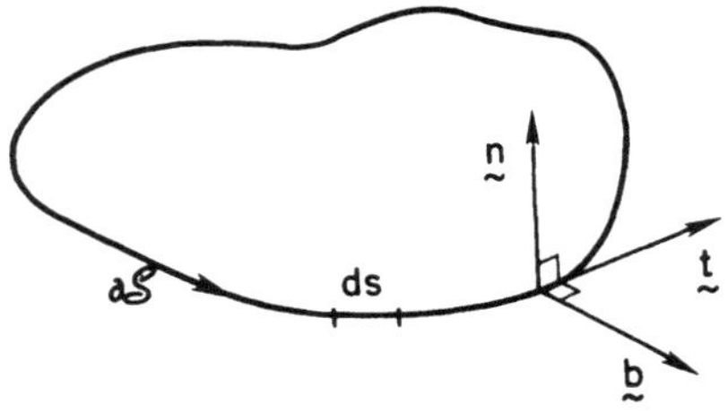

Figure 3.8.2. Unit normal $\mathbf{n}$, binormal $\mathbf{b}$, and tangent vector $\mathbf{t}$.

and the binormal $\mathbf{b}$ defined by

$$\mathbf{b} = \mathbf{t} \times \mathbf{n} \tag{3.8.5}$$

(Figure 3.8.2).

If the body contains a discontinuity surface $\sigma(t)$, which may be sweeping the body with its own velocity $\mathbf{v}$, and $\sigma(t)$ possesses a discontinuity line moving on it by its own velocity $\mathbf{w}$ (Figure 3.8.3), then (3.8.1) is modified to

$$\frac{d}{dt}\int_{\mathscr{V}-\sigma} \phi \, dv + \frac{\delta}{\delta t}\int_{\sigma - C_\sigma} \Phi \, da = \int_{\partial\mathscr{V}-\sigma} \boldsymbol{\tau}\cdot d\mathbf{a} + \int_{\mathscr{V}-\sigma} g \, dv + \int_{\sigma - C_\sigma} G \, da - \oint_{\partial\sigma - C_\sigma} \mathscr{T} \, ds, \tag{3.8.6}$$

where $\delta/\delta t$ denotes the time rate following the surface; Φ and G are, respectively, the surface densities of ϕ and g on σ, excluding the discontinuity line c_σ; and $\mathscr{T} = \mathbf{T}\cdot\mathbf{b}$ is the influx of ϕ through $\partial\sigma$, excluding the points of intersection of c_σ with σ. As usual, we mean

$$\partial\mathscr{V} - \sigma = \partial\mathscr{V} - \sigma \cap \partial\mathscr{V}, \qquad \partial\sigma - c_\sigma = \partial\sigma - c_\sigma \cap \partial\sigma. \tag{3.8.7}$$

In (3.8.6), the expression $\mathscr{T} = \mathbf{T}\cdot\mathbf{b}$ is obtained in a similar fashion to (3.8.4). We also note that, without loss of generality, we can set

$$\mathbf{n}\cdot\mathbf{T} = 0 \quad \text{on } \sigma. \tag{3.8.8}$$

By partitioning the tensor $\mathscr{T}$ on σ as

$$\mathscr{T} = \mathbf{n} \otimes \mathbf{T}_{(\mathbf{n})} + \mathbf{T}_{(\sigma)}, \qquad \mathbf{T}_{(\mathbf{n})} = \mathbf{n}\cdot\mathbf{T}, \qquad \mathbf{n}\cdot\mathbf{T}_{(\sigma)} = 0,$$

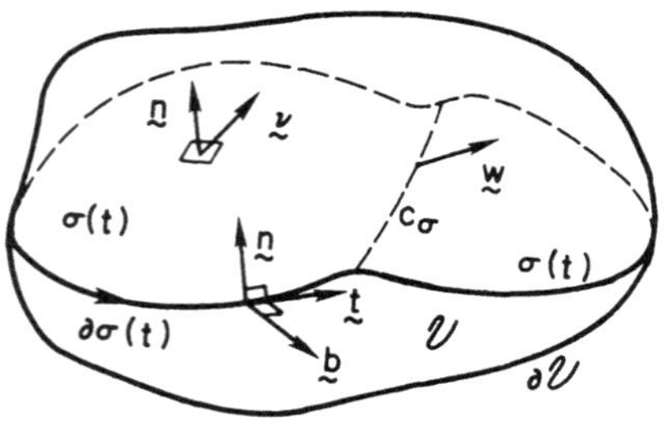

Figure 3.8.3. Moving discontinuity surface $\sigma(t)$.

Figure 3.8.4. Material surface $\mathscr{S}$ with discontinuity line γ.

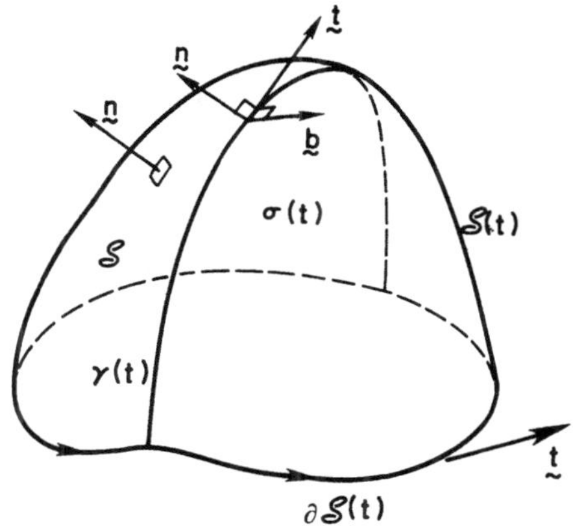

it follows that[8]

$$\mathbf{b}\cdot\mathbf{T} = (\mathbf{b}\cdot\mathbf{n})\mathbf{T}_{(\mathbf{n})} + \mathbf{b}\cdot\mathbf{T}_{(\sigma)} = \mathbf{b}\cdot\mathbf{T}_{(\sigma)}.$$

When the material surface $\mathscr{S}$ with boundary $\partial\mathscr{S}$ contains a discontinuity line $\gamma(t)$ (Figure 3.8.4), then the surface balance law (3.8.2) takes the form

$$\frac{d}{dt}\int_{\mathscr{S}-\gamma} \mathbf{n}\cdot\boldsymbol{\psi}\, da = \oint_{\partial\mathscr{S}-\gamma} \mathbf{t}\cdot\boldsymbol{\sigma}\, ds + \int_{\mathscr{S}-\gamma} \mathbf{n}\cdot\mathbf{h}\, da + \int_{\gamma} \mathbf{b}\cdot\boldsymbol{\Sigma}\, ds, \quad (3.8.9)$$

where $\boldsymbol{\Sigma}$ is the flux density of $\boldsymbol{\psi}$ across $\gamma(t)$ and $\mathbf{b}$ is the binormal of γ, as defined by (3.8.5).

Similar to (3.8.8), we have

$$\mathbf{n}\cdot\boldsymbol{\Sigma} = 0. \quad (3.8.10)$$

The identification of various fields appearing in these expressions is made through the fundamental laws of physics.

The balance laws (3.8.6) and (3.8.9) may be transformed to convenient forms by means of the transport theorems (1.13.8) and (1.13.23) and the Green–Gauss theorems (1.3.7) and (1.13.18), leading to[9]

$$\int_{\mathscr{V}-\sigma}\left[\frac{\partial\phi}{\partial t} + \nabla\cdot(\mathbf{v}\phi) - \nabla\cdot\boldsymbol{\tau} - g\right]dv + \int_{\sigma-c_\sigma}\left\{\frac{\bar{\delta}\boldsymbol{\Phi}}{\delta t} + \nabla_s\cdot(\mathbf{v}\boldsymbol{\Phi})\right.$$

$$\left. - \nabla_s\cdot\mathbf{T} - G - [\mathbf{n}\cdot(\mathbf{v}-\mathbf{v})\phi + \mathbf{n}\cdot\boldsymbol{\tau}]\right\}da + \int_{c_\sigma(t)} [\mathbf{b}\cdot\{\mathbf{T} + (\mathbf{w}-\mathbf{v})\boldsymbol{\Phi}\}]\, ds = 0, \quad (3.8.11)$$

---

[8] E.S. Şuhubi, private communication, 1985. This result was obtained by Kafadar [1971] in connection with the surface polarization.

[9] While equivalent equations, in connection with the electrodynamics of continua and continuum mechanics, were given by Maugin and Eringen [1977] and Eringen [1967], [1980], these elegant formal results are due to E.S. Şuhubi (private communication).

and

$$\int_{\mathscr{S}-\gamma} \mathbf{n}\cdot\left(\overset{*}{\boldsymbol{\psi}} - \nabla \times \boldsymbol{\sigma} - \mathbf{h}\right) da + \int_{\gamma} \mathbf{t}\cdot\{[(\mathbf{v} - \mathbf{v}) \times \boldsymbol{\psi} - \boldsymbol{\sigma}] - \mathbf{n}_\sigma \times \boldsymbol{\Sigma}\}\, ds = \mathbf{0}, \tag{3.8.12}$$

where we used the identity $\mathbf{b}\cdot\boldsymbol{\Sigma} = \mathbf{t}\cdot(\mathbf{n}_\sigma \times \boldsymbol{\Sigma})$. Here $\mathbf{n}_\sigma$ is the unit normal of the surface σ (Figure 3.5.4). Since $\mathbf{t}$ is a unit vector in the surface $\sigma(t)$, it can be expressed in the form $\mathbf{t} = \mathbf{k} \times \mathbf{n}$, where $\mathbf{k}$ is an appropriate vector. Substituting this for $\mathbf{t}$ in the line integral in (3.5.11), and noting that $\mathbf{n}_\sigma\cdot\mathbf{n}_\sigma = 1$, $\mathbf{n}\cdot\boldsymbol{\Sigma} = 0$ (eq. (3.8.10)), we can express (3.8.12) in the form

$$\int_{\mathscr{S}-\gamma} \mathbf{n}\cdot\left(\overset{*}{\boldsymbol{\psi}} - \nabla \times \boldsymbol{\sigma} - \mathbf{h}\right) da + \int_{\gamma} \mathbf{k}\cdot\{\mathbf{n} \times [(\mathbf{v} - v) \times \boldsymbol{\psi} - \sigma] + \boldsymbol{\Sigma}\}\, ds = 0. \tag{3.8.13}$$

These integral balance laws can be written in equivalent "local forms" by introducing appropriate "nonlocal" fields $\{\hat{g}, \hat{G}, \hat{l}\}$ and $\{\hat{h}, \hat{\sigma}\}$. Thus, we may replace (3.8.11) by

$$\begin{aligned} \frac{\partial \phi}{\partial t} + \nabla\cdot(\mathbf{v}\phi) - \nabla\cdot\boldsymbol{\tau} - g &= \hat{g} \qquad \text{in } \sigma, \\ \frac{\bar{\delta}\Phi}{\delta t} + \nabla_s\cdot(\mathbf{v}\Phi) - \nabla_s\cdot\mathbf{T} - G - [\mathbf{n}\cdot(\mathbf{v} - \mathbf{v})\phi + \mathbf{n}\cdot\boldsymbol{\tau}] &= \hat{G} \qquad \text{on } \sigma - c_\sigma, \\ [\mathbf{b}\cdot\{\mathbf{T} + (\mathbf{w} - \mathbf{v})\Phi\}] &= -\hat{l} \qquad \text{on } c_\sigma, \end{aligned} \tag{3.8.14}$$

where $\hat{g}$, $\hat{G}$, and $\hat{l}$ are subject to

$$\int_\sigma \hat{g}\, dv + \int_{\sigma - c_\sigma} \hat{G}\, da + \int_{c_\sigma} \hat{l}\, ds = 0. \tag{3.8.15}$$

Similarly, eq. (3.8.13) is replaced by

$$\begin{aligned} \overset{*}{\boldsymbol{\psi}} - \nabla \times \boldsymbol{\sigma} - \mathbf{h} &= \hat{\mathbf{h}} \qquad \text{in } \mathscr{V} - \hat{\sigma}, \\ \mathbf{n} \times [(\mathbf{v} - \mathbf{v}) \times \boldsymbol{\psi} - \boldsymbol{\sigma}] + \boldsymbol{\Sigma} &= \mathbf{n} \times \hat{\boldsymbol{\sigma}} \qquad \text{on } \sigma, \end{aligned} \tag{3.8.16}$$

where the localization residuals are subject to

$$\int_{\mathscr{S}-\gamma} \mathbf{n}\cdot\hat{\mathbf{h}}\, da + \int_\gamma \mathbf{t}\cdot\hat{\boldsymbol{\sigma}}\, ds = 0. \tag{3.8.17}$$

The basic balance laws (3.8.14) and (3.8.16) are not *closed* by themselves. Various physical fields and nonlocal residuals will have to be determined from constitutional or other physical considerations.

Definition 1. The balance laws are called local if all residuals vanish.

Localization is an important postulate of classical continuum physics. Local forms of the integral balance laws (3.8.11) and (3.8.13) can be obtained by the postulate of localization:

Postulate of Localization. The global laws hold good for any volume element in $\mathscr{V} - \sigma$, any surface element on σ, and any line element on γ.

Physically, it is an assumption which is justified on the basis of the assumption of continuity of matter and the strength of intermolecular forces.

The domain of applicability of the local continuum theory is determined by the ratio of two characteristic lengths. We can associate a characteristic length L with the external excitations or the geometry of the body. Some examples are a wavelength, a length over which external loads are smooth, the thickness of a plate, etc. Equally good is another characteristic length l which is associated with the inner structure of the body, e.g., average atomic or molecular distance, average granular distances (depending on the accuracy desired). The ratio $\varepsilon = l/L$ determines the domain of applicability of physical theory. When $\varepsilon \ll 1$, local continuum physics may provide a reasonably accurate description of the physical phenomena. When $\varepsilon \simeq 1$, the theory breaks down and we may have to resort to atomic or nonlocal theories. Indeed, there exist many important macroscopic phenomena which are cast outside the local continuum theory. The mechanics of granular materials, liquid crystals, stress fields in the vicinity of the tip of a sharp crack and at the core of a dislocation line, dispersion of high frequency electromagnetic waves, turbulence and fatigue, are but a few examples for which classical continuum theories fail to apply.

Recently, a number of continuum theories has been formulated to discuss such physical phenomena. These theories incorporate the notions of microstructure into field equations. They are *nonlocal* in character, i.e., the effect of motions and electromagnetic fields of distant points on the reference point $\mathbf{X}$ in the body is brought in either by means of new kinematical variables (e.g., couple stress, stress moments, intrinsic rotation, rotation gradients, etc.), or in the form of functionals (e.g., the stress at $\mathbf{X}$ as dependent upon strains everywhere in the body). Prominent among the first type are *micropolar* and *micromorphic* theories, which are capable of dealing with granular media, composites, liquid crystals, and suspensions, etc. The micropolar theory incorporates the intrinsic rigid rotation of a microstructure into the constitutive equations. Therefore, it is an ideal vehicle to study liquid crystals and ordered fluids (Eringen [1978b], [1979a, b]), and suspensions (Eringen [1984e], [1985b]). Various expositions of the general theory for elastic solids are to be found in Eringen [1966b, d], [1967a, b], [1968], [1969b], [1970c], [1975b], [1977a] and in Kafadar and Eringen [1971a, b]. For fluids, see Eringen [1966c], [1969a], [1973a]. The general theory and historical reviews of the field are given in Eringen and Kafadar [1976], and Ariman et al. [1973], [1974]. In fact, the first of these (Eringen and Kafadar) discusses the more general theory, namely micromorphic continua. For the original work, see Eringen and Şuhubi [1964], Şuhubi and Eringen [1964], and Eringen [1964], [1968], [1970b], [1972d, e], [1976a, b].

A fully nonlocal theory avoids high order kinematical variables based on spatial moment expansion, but employs constitutive equations involving integrals over the material manifolds. Consequently, it is capable of dealing with physical phenomena involving much smaller internal characteristic lengths. Recent applications show that we can characterize atomic scale phenomena such as the crack tip problem (Eringen and Kim [1974], [1977], Eringen [1977a, b], [1978a, c], Eringen et al. [1977] and dislocations, crack, and fracture (Eringen [1977a, b], [1978c], [1983], [1984a, b], [1985a]). For the development of the theory and various applications, we refer the reader to Eringen [1966a], [1967b], [1972a, b, c], [1974a, b, c], [1984a–d], [1985a], Eringen and Edelen [1972], Kröner [1967], Kunin [1967] and others which are listed in various review articles and books, e.g., Eringen [1975b], [1976b] and Kunin [1967], [1982], [1983]. Nonlocal electromagnetic elastic solids are discussed in Chapter 14.

In this book, we are mostly concerned with macrosopic phenomena which fall within the framework of classical field theories. Excluding excursions made into the polar theories in Chapters 9 and 12 and the nonlocal Electromagnetic theory of Chapter 14, here we posit that the postulate of localization is valid.[10] It then follows that nonlocal residuals vanish *almost everywhere.*

3.9. Balance Laws of Electrodynamics

In Section 3.3 no allusion was made to jump and boundary conditions for Maxwell's equations. These conditions are essential, however, to solving boundary-value problems and studying discontinuities. In order to obtain these conditions in a rational manner, in the case of nonrelativistically moving deformable bodies, we consider the postulational approach to Maxwell's equations, for a macroscopic body of finite extent, without reference to any microscopic description.[11]

A. Global Statement

Let $\mathscr{V}$ be a simply connected, open *material* region of E^3 with regular boundary $\partial\mathscr{V}$. $\sigma(t)$ is a discontinuity surface contained in $\mathscr{V}$ having absolute velocity $\mathbf{v}(t)$, with respect to R_G, and a unit oriented normal $\mathbf{n}$. c_σ is a singular curve on σ (i.e., an edge where σ has a discontinuous tangent plane). c_σ may be moving, with a velocity $\mathbf{w}$, on the surface $\sigma(t)$ (Figure 3.8.3). $\mathscr{S}$ is an open material surface in E^3 enclosed by a closed curve $\partial\mathscr{S}$ and $\gamma(t)$ is a discontinuity curve on $\mathscr{S}$ having absolute velocity $\mathbf{v}$. $d\mathbf{a} = \mathbf{n}\, da$ denotes the oriented surface

---

[10] Even for nonlocal theory, under mild conditions, local balance laws are applicable. The second law of thermodynamics must, however, be postulated for the entire body.

[11] We follow the work of Maugin and Eringen [1977].

element. $d\mathbf{x} = \mathbf{t}\, ds$ denotes the oriented line element. $\mathbf{b}$ is the binormal, as defined by $\mathbf{b} = \mathbf{t} \times \mathbf{n}$ (see Figure 3.8.4).

In describing the electromagnetic properties of moving matter we consider, at most, electric quadrupoles and magnetic dipoles. q_e and w_e are, respectively, volume and surface densities of *free charge*, and $\mathscr{J}$ and $\mathscr{K}$ are volume and surface *conduction* currents, respectively. $\hat{\boldsymbol{\pi}}$, here, is a surface polarization whose existence is justified when the electric quadrupoles are taken into account. $\mathscr{K}$ and $\hat{\boldsymbol{\pi}}$ are supposed to be tangential to the surfaces on which they are defined (see (3.8.8)); i.e.,

$$\mathscr{K} \cdot \mathbf{n} = 0, \qquad \hat{\boldsymbol{\pi}} \cdot \mathbf{n} = 0. \tag{3.9.1}$$

The total volume current $\mathbf{J}$ and surface current $\mathbf{K}$ are defined by

$$\mathbf{J} = \mathscr{J} + q_e \mathbf{v}, \qquad \mathbf{K} = \mathscr{K} + w_e \mathbf{v}. \tag{3.9.2}$$

Maxwell's equations in nonrelativistically moving matter may be stated in global forms, including the effects of moving singular surfaces and curves:

(i) **Volume Balance Laws** (Figure 3.8.3)

Gauss' Law

$$\oint_{\partial\mathscr{V}-\sigma} \mathbf{D} \cdot d\mathbf{a} = \int_{\mathscr{V}-\sigma} q_e\, dv + \int_{\sigma-c_\sigma} w_e\, da + \oint_{\partial\sigma-c_\sigma} \mathbf{b} \cdot \hat{\boldsymbol{\pi}}\, ds. \tag{3.9.3}$$

Conservation of Magnetic Flux:

$$\oint_{\partial\mathscr{V}-\sigma} \mathbf{B} \cdot d\mathbf{a} = 0. \tag{3.9.4}$$

Conservation of Electric Charge:

$$\frac{d}{dt}\int_{\mathscr{V}-\sigma} q_e\, dv + \frac{\delta}{\delta t}\int_{\sigma-c_\sigma} w_e\, da = -\int_{\partial\mathscr{V}-\sigma} \mathscr{J} \cdot d\mathbf{a} - \oint_{\partial\sigma-c_\sigma} \mathbf{b} \cdot \mathscr{K}\, ds. \tag{3.9.5}$$

(ii) **Surface Balance Laws** (Figure 3.8.4)

Faraday's Law:

$$\frac{1}{c}\frac{d}{dt}\int_{\mathscr{S}-\gamma} \mathbf{B} \cdot d\mathbf{a} + \oint_{\partial\mathscr{S}-\gamma} \mathscr{E} \cdot d\mathbf{x} = 0, \tag{3.9.6}$$

Ampère's Law:

$$-\frac{1}{c}\frac{d}{dt}\int_{\mathscr{S}-\gamma} \mathbf{D} \cdot d\mathbf{a} + \oint_{\partial\mathscr{S}-\gamma} \mathscr{H} \cdot d\mathbf{x} = \frac{1}{c}\int_{\mathscr{S}-\gamma} \mathscr{J} \cdot d\mathbf{a} - \frac{1}{c}\int_{\gamma} \mathbf{b} \cdot \mathscr{K}\, ds. \tag{3.9.7}$$

B. Local Balance Laws

If we compare equations of category (i) with (3.8.6) and those of (ii) with (3.8.9), we can easily identify various fields. The localization then follows from (3.8.14) and (3.8.16).

Gauss' Law:

$$\nabla \cdot \mathbf{D} = q_e \qquad \text{in } \mathscr{V} - \sigma, \tag{3.9.8}$$

$$\mathbf{n} \cdot [\mathbf{D}] = w_e - \nabla_\sigma \cdot \hat{\boldsymbol{\pi}} \qquad \text{on } \sigma - c_\sigma, \tag{3.9.9}$$

$$[\mathbf{b} \cdot \hat{\boldsymbol{\pi}}] = 0 \qquad \text{on } c_\sigma. \tag{3.9.10}$$

Conservation of Magnetic Flux:

$$\nabla \cdot \mathbf{B} = 0 \quad \text{in } \mathscr{V} - \sigma, \tag{3.9.11}$$

$$\mathbf{n} \cdot [\mathbf{B}] = 0 \quad \text{on } \sigma - c_\sigma. \tag{3.9.12}$$

Conservation of Charge:

$$\frac{\partial q_e}{\partial t} + \nabla \cdot \mathbf{J} = 0 \quad \text{in } \mathscr{V} - \sigma, \tag{3.9.13}$$

$$\frac{\bar{\delta} w_e}{\delta t} + \nabla_\sigma \cdot \mathbf{K} + \mathbf{n} \cdot [\mathbf{J} - \mathbf{v} q_e] = 0 \quad \text{on } \sigma - c_\sigma, \tag{3.9.14}$$

$$[\mathbf{b} \cdot (w_e \mathbf{v} - \mathbf{K})] = 0 \quad \text{on } c_\sigma, \tag{3.9.15}$$

where we used (3.6.2).

Faraday's Law:

$$\frac{1}{c}\overset{*}{\mathbf{B}} + \nabla \times \mathscr{E} = \mathbf{0} \quad \text{in } \mathscr{V} - \sigma, \tag{3.9.16}$$

$$\mathbf{n} \times \left[\mathscr{E} + \frac{1}{c}\mathbf{B} \times (\mathbf{v} - \mathbf{v}) \right] = \mathbf{0} \quad \text{on } \sigma - c_\sigma. \tag{3.9.17}$$

Ampère's Law:

$$-\frac{1}{c}\overset{*}{\mathbf{D}} + \nabla \times \mathscr{H} = \frac{1}{c}\mathscr{J} \quad \text{in } \mathscr{V} - \sigma, \tag{3.9.18}$$

$$\mathbf{n} \times \left[\mathscr{H} - \frac{1}{c}\mathbf{D} \times (\mathbf{v} - \mathbf{v}) \right] = \frac{1}{c}\mathscr{K} \quad \text{on } \sigma - c_\sigma. \tag{3.9.19}$$

We note that Faraday's and Ampère's local laws are written for $\mathscr{V} - \sigma$ and $\sigma - c_\sigma$, rather than for $\mathscr{S} - \gamma$ and γ. This is because global laws are considered to be valid for all surfaces $\mathscr{S}$ and discontinuity curves γ in the body.

The first equation in each set represents the field equation valid in $\mathscr{V} - \sigma$; the rest are jump conditions across the discontinuity surface σ. Alternative forms of these equations exist and are useful in the treatment of a certain class of problems.

C. Boundary Conditions

If σ is a material surface—i.e., a surface whose geometrical points have an absolute velocity (in R_G) equal to that of the material particles occupying the same place instantaneously—then $\mathbf{v} = \mathbf{v}$. The jump conditions (3.9.9), (3.9.12),

(3.9.14), (3.9.17), and (3.9.19) reduce to

$$
\begin{aligned}
\mathbf{n}\cdot[\mathbf{D}] &= w_e - \nabla_\sigma \cdot \hat{\boldsymbol{\pi}},\\
\mathbf{n}\cdot[\mathbf{B}] &= 0,\\
\frac{\bar{\delta} w_e}{\delta t} + \nabla_\sigma \cdot \mathbf{K} + [\mathscr{J}]\cdot\mathbf{n} &= 0, \qquad (3.9.20)\\
\mathbf{n}\times[\mathscr{E}] &= 0,\\
\mathbf{n}\times[\mathscr{H}] &= \frac{1}{c}\mathscr{K} \quad \text{on } \partial\mathscr{V} - \sigma,
\end{aligned}
$$

where now

$$
\mathbf{K} = \mathscr{K} + w_e \mathbf{v}, \quad \frac{\bar{\delta}}{\delta t} = \frac{\partial}{\partial t} + (\mathbf{v}\cdot\mathbf{n})\frac{\partial}{\partial n}. \qquad (3.9.21)
$$

The boundary $\partial\mathscr{V}$ of a material body is a particular case of material surface. Equations (3.9.20) are the relevant boundary conditions on $\partial\mathscr{V} - \sigma$, if there are a free charge w_e, a surface current $\mathscr{K}$, and a surface polarization π prescribed on it. If these fields are supposedly zero, then (3.9.20) yields the following well-known boundary conditions (with no matter outside $\mathscr{V}$):

$$
\begin{aligned}
\mathbf{n}\times[\mathscr{E}] &= \mathbf{0}, \qquad [\mathbf{B}]\cdot\mathbf{n} = 0,\\
\mathbf{n}\times[\mathscr{H}] &= 0, \qquad [\mathbf{D}]\cdot\mathbf{n} = 0.
\end{aligned} \qquad (3.9.22)
$$

Various situations arising in quasi-electro- or quasi-magnetostatics are easily described as particular cases of (3.9.20) and (3.9.22).

3.10. Thermomechanical Balance Laws

A material body occupies an open region $\mathscr{V}$ of the Euclidean three-dimensional space E^3 at time t. The loads that are exerted on the body by external sources are either through its surface or by action at a distance. The former category of loads is called *contact* (or *surface*) *loads*, and the latter, *body loads*. For body loads we have the examples of gravitational attractions, electromagnetic forces, couples, and power supply; and for surface loads we have the force, couple, and power supply that can be exerted on the body by contact on its surface $\partial\mathscr{V}$. In the previous section we encountered examples of loads on a discontinuity surface by the electromagnetic fields.

Body and surface loads are the cause of motion, deformation, and electromagnetic phenomena that occur in the body. They are subject to certain global balance laws which can be stated in a common form.

The balance laws of the electrodynamics of continua are the union of laws governing electromagnetic fields per se and thermomechanical balance laws. The electrodynamical laws are Maxwell's equations as given in Section 3.9.

The thermomechanical laws consist of:

(i) Conservation of Mass.
(ii) Balance of Momentum.
(iii) Balance of Moment of Momentum.
(iv) Conservation of Energy.
(v) Entropy Inequality.

The last two laws are often referred to as the first and the second laws of thermodynamics.

For the thermomechanical case the global balance law (3.8.6) may be stated in the general form

$$\frac{d}{dt}\int_{\mathscr{V}-\sigma} \phi\, dv + \frac{\delta}{\delta t}\int_{\sigma} \Phi\, da = \int_{\partial\mathscr{V}-\sigma} \tau_k n_k da + \int_{\mathscr{V}-\sigma} (g + g^{\mathrm{E}})\, dv + \int_{\sigma} [\hat{\mathbf{g}}_k] n_k\, da, \qquad (3.10.1)$$

where ϕ is a fundamental volume tensor field that is to be balanced. The tensor field Φ represents an associated surface field per unit area on the discontinuity surface.[12] The surface integral involving Φ represents the time rate following the surface σ. The conservation of electric charge (3.9.5) is an example where such a term occurs. However, this term is seldom encountered in mechanical balance laws. τ_k is the flux of ϕ.

We also note that the balance law (3.10.1) is a special case of the volume balance law (3.8.6). Restricting the discussion to the case where $\Phi = 0$, by means of the transport theorem (1.13.6) and the Green–Gauss theorem (1.13.7), equation (3.10.1) may be transformed to (see also (3.8.11))

$$\int_{\mathscr{V}-\sigma} \left[\frac{\partial\phi}{\partial t} + \nabla\cdot(\mathbf{v}\phi - \boldsymbol{\tau}) - g - g^{\mathrm{E}}\right] dv + \int_{\sigma} [\phi(v_k - \nu_k) - \tau_k - \hat{g}_k^{\mathrm{E}}] n_k\, da = 0. \qquad (3.10.2)$$

The localization process now gives

$$\frac{\partial\phi}{\partial t} + \nabla\cdot(\mathbf{v}\phi - \boldsymbol{\tau}) - g - g^{\mathrm{E}} = 0, \quad \text{in } \mathscr{V} - \sigma,$$
$$[\phi(v_k - \nu_k) - \tau_k - \hat{g}_k^{\mathrm{E}}]\mathbf{n}_k = 0, \quad \text{on } \sigma. \qquad (3.10.3)$$

Here g and g^{E} are, respectively, the sources of ϕ of purely mechanical and electromagnetic origin.[13] The electromagnetic jump across the discontinuity surfaceσ is represented by the term involving $\hat{g}_k^{\mathrm{E}}$, which was discussed in Section 3.7.

---

[12] In the case of viscous subshocks imbedded in a thermal shock, a surface intensity Φ may be encountered.

[13] The jump discontinuity of the electromagnetic body force has been separated out. g^{E}, like g, is continuous throughout $\mathscr{V} - \sigma$.

The global balance law (3.10.1) generalizes the usual balance law of continuum mechanics, by incorporating electromagnetic effects throughout the terms g^{E} and $[\hat{g}_k^{\mathrm{E}}]$ which are of *purely* electromagnetic origin. In this form, ϕ represents a mechanical field (e.g., mechanical momentum). τ_k is the surface flux corresponding to ϕ (e.g., the mechanical part of the surface traction). The total surface load (e.g., the surface traction) includes an electromagnetic part in addition to τ_k. This is clearly indicated by the jump conditions (see also the alternative formulation (3.10.13)–(3.10.15)).

These are the master balance laws validly covering the balance laws (i)–(iv). For the entropy inequality we merely replace the $(=)$ sign with $(\geqslant)$. We now identify the fields ϕ, τ, g, g^{E}, and $\hat{g}^{\mathrm{E}}$ for each balance law:

(i) *Law of Conservation of Mass: The total mass of a material body is unchanged during the motion of the body.*

If ρ denotes the mass density at $\mathbf{x}$, at time t, it is then clear that

$$\phi = \rho, \qquad \tau_k = \hat{g}_k^{\mathrm{E}} = 0, \qquad g = g^{\mathrm{E}} = 0,$$

and (3.10.3) gives

$$\begin{aligned} \frac{\partial \rho}{\partial t} + \nabla \cdot (\rho \mathbf{v}) = 0 \quad \text{in } \mathscr{V} - \sigma, \\ [\rho(\mathbf{v} - \boldsymbol{\nu})] = 0 \quad \text{on } \sigma. \end{aligned} \tag{3.10.4}$$

These are the equations of the local conservation of mass (also called continuity equations). The first one is valid throughout $\mathscr{V} - \sigma$ and the second one represents the jump across σ.

(ii) *Law of Balance of Momentum: The time rate of change of momentum of the material body is equal to the resultant force acting upon the body.*

In this case we identify

$$\phi = \rho \mathbf{v}, \qquad \tau_k = \mathbf{t}_k = t_{kl}\mathbf{i}_l, \qquad g = \rho \mathbf{f},$$
$$g^{\mathrm{E}} = \mathbf{F}^{\mathrm{E}}, \qquad \hat{g}_k^{\mathrm{E}} = \hat{\mathbf{F}}_k^{\mathrm{E}},$$

where $\mathbf{t}_k$ is the surface traction per unit area on the coordinate surface $x_k =$ const.; t_{kl} is Cauchy's stress tensor; $\mathbf{f}$ is the mechanical body force density; and $\mathbf{F}^{\mathrm{E}}$ and $\hat{\mathbf{F}}_k^{\mathrm{E}}$ are, respectively, the electromagnetic force density in the body and the electromagnetic force jump across σ. These latter quantities are given by (3.5.43) and (3.7.5).

The local balance of momentum now follows from (3.10.3), upon use of (3.10.4),

$$\begin{aligned} t_{kl,k} + \rho(f_l - \dot{v}_l) + F_l^{\mathrm{E}} = 0 \quad \text{in } \mathscr{V} - \sigma, \\ [\rho v_l(v_k - \nu_k) - t_{kl} - t_{kl}^{\mathrm{E}} - \nu_k G_l] n_k = 0 \quad \text{on } \sigma. \end{aligned} \tag{3.10.5}$$

(iii) *Law of Balance of Moment of Momentum: The time rate of moment of momentum of the material body is equal to the resultant moment of all forces and the resultant of couples acting upon the body.*

In this case we have

$$\phi = \mathbf{x} \times \rho\mathbf{v}, \qquad \tau_k = \mathbf{x} \times \mathbf{t}_k, \qquad g = \mathbf{x} \times \rho\mathbf{f},$$

$$g^{\mathrm{E}} = \mathbf{x} \times \mathbf{F}^{\mathrm{E}} + \mathbf{C}^{\mathrm{E}}, \qquad \hat{g}_k^{E} = \hat{C}_k^{E},$$

where $\mathbf{C}^{\mathrm{E}}$ and $\hat{\mathbf{C}}_k^{\mathrm{E}}$ are, respectively, electromagnetic couples in the body and on σ. They are given by (3.5.44) and (3.7.6). Through (3.10.3), and upon using (3.10.4) and (3.10.5), we obtain

$$\varepsilon_{klm} t_{lm} + C_k^{\mathrm{E}} = 0 \quad \text{in } \mathscr{V} - \sigma. \tag{3.10.6}$$

The associated jump condition is satisfied identically.[14]

(iv) *Law of Conservation of Energy*: (*The first principle of thermodynamics*): *The time rate of change of the sum of the internal and kinetic energies of a material body, considered as a closed system, is equal to the sum of the rate of work (or power) of all forces and couples and the energies that enter or leave the body per unit time.*

We identify

$$\phi = \rho\varepsilon + \tfrac{1}{2}\rho v^2, \qquad \tau_k = \mathbf{t}_k \cdot \mathbf{v} + q_k, \qquad g = \rho\mathbf{f}\cdot\mathbf{v} + \rho h,$$

$$g^{\mathrm{E}} = W^{\mathrm{E}}, \qquad g_h^{\mathrm{E}} = \hat{W}_k^{\mathrm{E}},$$

where ε is the internal energy density per unit mass, q_k is the heat flux vector per unit area, h is the heat source per unit mass, and W^{E} and $\hat{W}_k^{\mathrm{E}}$ are, respectively, the electromagnetic energy supplies in $\mathscr{V} - \sigma$ and on σ. They are given by (3.5.45) and (3.7.8).

Upon substituting these expressions into (3.10.3), and by using (3.10.4)–(3.10.6), we arrive at the local balance equations of energy

$$\rho\dot{\varepsilon} - t_{kl} v_{l,k} - \nabla\cdot\mathbf{q} - \rho h - \rho\mathscr{E}\cdot\dot{\boldsymbol{\pi}} + \mathscr{M}\cdot\dot{\mathbf{B}} - \mathscr{J}\cdot\mathscr{E} = 0 \quad \text{in } \mathscr{V} - \sigma,$$

$$[\{\rho\varepsilon + \tfrac{1}{2}\rho v^2 + \tfrac{1}{2}(E^2 + B^2)\}(v_k - \nu_k) - (t_{kl} + t_{kl}^{\mathrm{E}} + \nu_k G_l)v_l - q_k + \mathscr{S}_k]n_k = 0$$

$$\text{on } \sigma. \tag{3.10.7}$$

(v) *Law of Entropy*: (*The second principle of thermodynamics*): *The time rate of change of the total entropy is never less than the sum of the entropy supply due to body sources and the entropy influx through the surface of the body.*

We identify

$$\phi = \rho\eta, \qquad \tau_k = q_k/\theta, \qquad g = \rho h/\theta,$$

$$g^{\mathrm{E}} = 0, \qquad \hat{g}_k^{\mathrm{E}} = 0,$$

[14] In polar theories of continua, the mechanical surface body couple densities, and angular momentum must be taken into account. In this case, (3.10.6) is more complicated and the jump condition survives. For the polar theories, see Eringen and Kafadar [1976].

where η is the entropy density per unit mass and $\theta > 0$ (inf. $\theta = 0$) is the absolute temperature.[15]

The second law of thermodynamics is expressed by carrying these expressions into (3.10.3), upon using (3.10.4), and replacing the equality sign $(=)$ by $(\geq)$

$$\begin{aligned} \rho\gamma \equiv \rho\dot{\eta} - \nabla\cdot(\mathbf{q}/\theta) - (\rho h/\theta) \geq 0 \quad &\text{in } \mathscr{V} - \sigma, \\ [\rho\eta(\mathbf{v} - \mathbf{v}) - (\mathbf{q}/\theta)]\cdot\mathbf{n} \geq 0 \quad &\text{on } \sigma. \end{aligned} \tag{3.10.8}$$

Equations (3.10.4)–(3.10.8) are the expressions of the local balance laws. In each set the first equation (valid in $\mathscr{V} - \sigma$) is the balance law at all parts of the body, excluding the points situated on the discontinuity surface; and the second equation (valid on σ) is the jump condition across σ. This latter condition can be used to obtain the boundary conditions on the surface of the body, by merely setting $\mathbf{v} = \mathbf{v}$.

Upon carrying (3.5.44) into (3.10.6), we find that

$$t_{[kl]} = \mathscr{E}_{[k}P_{l]} + B_{[k}\mathscr{M}_{l]}, \tag{3.10.9}$$

where a bracket enclosing indices indicates, as usual, the antisymmetry. We use parentheses to indicate symmetry, e.g.,

$$\begin{aligned} t_{(kl)} &= \tfrac{1}{2}(t_{kl} + t_{lk}), & t_{[kl]} &= \tfrac{1}{2}(t_{kl} - t_{lk}), \\ \mathscr{E}_{(k}P_{l)} &= \tfrac{1}{2}(\mathscr{E}_k P_l + \mathscr{E}_l P_k), & \mathscr{E}_{[k}P_{l]} &= \tfrac{1}{2}(\mathscr{E}_k P_l - \mathscr{E}_l P_k). \end{aligned} \tag{3.10.10}$$

Hence, the stress tensor is *not* symmetric. If it happens that $\mathscr{E}$ is proportional to $\mathbf{P}$ and $\mathbf{B}$ is proportional to $\mathscr{M}$, then the right-hand side of (3.10.9) vanishes and the stress tensor becomes symmetric. This situation prevails in the case of isotropic fluids in both magnetohydrodynamics and electrohydrodynamics. In Voigt's piezoelectricity, the ponderomotive couple is neglected, leading to a symmetric stress tensor by default. The stress tensor is, in general, not symmetric for nonlinear materials and when materials are affected by intrinsic spin and/or electric quadrupoles (Dixon and Eringen [1965a]). Ferroelectric and ferromagnetic materials fall into this class also.

The stress tensor can be expressed in the form

$$t_{kl} = {}_{\mathrm{E}}t_{kl} - P_k\mathscr{E}_l - \mathscr{M}_k B_l, \tag{3.10.11}$$

where ${}_{\mathrm{E}}t_{kl}$ is symmetric, i.e.,

$${}_{\mathrm{E}}t_{kl} = {}_{\mathrm{E}}t_{lk}. \tag{3.10.12}$$

[15] More generally, the entropy flux vector should be taken as $\tau_k = (q_k/\theta) + s_k$, where s_k is the excess over the classical value q_k/θ (Eringen [1966a]). For mixture and nonsimple materials, s_k may not vanish (see Eringen and Ingram [1966]). See Müller [1968] for a discussion of the form to be assumed *a priori* or *a posteriori* by the entropy flux. Application of Müller's idea to the electromagnetic fluids was made by Liu and Müller [1972].

We close this section with some remarks:

(a) It is possible to express the global balance laws of momenta and energy in different forms, leading to the same local equations obtained above. For example, by use of (3.6.8) and (3.6.20), we may incorporate source terms into the volume rate and surface divergence, and express the global laws of balance of momenta and energy in the following forms:

Balance of Momentum:

$$\frac{d}{dt}\int_{\mathscr{V}-\sigma}(\rho\mathbf{v}+\mathbf{G})\,dv=\int_{\partial\mathscr{V}-\sigma}\mathbf{n}\cdot(\mathbf{t}_{(\mathbf{n})}+\mathbf{t}^{E}_{(\mathbf{n})}+\mathbf{v}\otimes\mathbf{G})\,da+\int_{\mathscr{V}-\sigma}\rho\mathbf{f}\,dv. \tag{3.10.13}$$

Balance of Moment of Momentum:

$$\frac{d}{dt}\int_{\mathscr{V}-\sigma}\mathbf{x}\times(\rho\mathbf{v}+\mathbf{G})\,dv=\int_{\partial\mathscr{V}-\sigma}\mathbf{x}\times(\mathbf{t}_{(\mathbf{n})}+\mathbf{t}^{E}_{(n)}+\mathbf{v}\otimes\mathbf{G})\,da+\int_{\mathscr{V}-\sigma}\mathbf{x}\times\rho\mathbf{f}\,dv. \tag{3.10.14}$$

Conservation of Energy:

$$\frac{d}{dt}\int_{\mathscr{V}-\sigma}[\rho\varepsilon+\tfrac{1}{2}\rho v^2+\tfrac{1}{2}(E^2+B^2)]\,dv$$
$$=\int_{\mathscr{V}-\sigma}\{\mathbf{v}\cdot[\mathbf{t}_{(\mathbf{n})}+\mathbf{t}^{E}_{(\mathbf{n})}+(\mathbf{v}\cdot\mathbf{n})\mathbf{G}]+\mathbf{n}\cdot(\mathbf{q}-\mathscr{S})\}\,da+\int_{\mathscr{V}-\sigma}(\rho\mathbf{f}\cdot\mathbf{v}+\rho h)\,dv. \tag{3.10.15}$$

These equations reduce exactly to (3.10.5)–(3.10.7) upon localization. From the physical viewpoint, these forms emphasize the nonelectromagnetic nature of $\mathbf{f}$ and h by attributing an electromagnetic density $\mathbf{G}$ and a free-field electromagnetic energy density $(E^2 + B^2)/2$ to the free space occupied by the matter. As discussed, these notions are rather secondary in nature.[16]

Still another from of the global laws may be written using the notions of effective Lorentz force and effective current, as introduced in (3.6.14) and (3.3.10), respectively. In this case, electromagnetic inertia and free-field energy do not appear, but body and surface forces (including the discontinuity surface contribution) contain equivalent terms involving effective forces, currents, and the Maxwell stress tensor. These equations are less symmetric in appearance and do not involve primitive notions. Therefore we do not produce them here.[17]

All three statements of global balance laws are, in fact, equivalent mathe-

[16] Tiersten ad Tsai [1972] considered global balance laws of the form (3.10.13)–(3.10.15) in the absence of discontinuity surfaces.

[17] See Liu and Müller [1972], Benach [1974], and Benach and Müller [1974].

matically, carrying only different physical interpretations. In the first statement, the nature of electromagnetism (except for the jump discontinuity) is envisioned as an action at a distance, in the same way as gravitation is considered in Newtonian mechanics. This is in the spirit of continuum mechanics. In the second statement, electromagnetic fields are considered to have inertia and contact actions. In the third statement (not written explicitly), electromagnetic fields are considered to produce surface and volume forces, but not inertia. These three possibilities, in fact, embody the well-known statement that the total energy–momentum stress tensor can be split in arbitrarily different fashions.[18]

The second formulation is closer to the relativistic statement in which the notion of stress–energy is primitive.

(b) *Radiation Heat Flux*: The global energy equation (3.10.15) shows that radiation is written in the form of a flux through the surface of the body. It contains the heat vector $\mathbf{q}$ and the Poynting vector $\mathscr{S}$, expressed in the co-moving frame of reference. The Stephan–Boltzmann law may be included in these terms. However, we choose to include it in ρh, thus eliminating the radiation of heat from the electromagnetic terms. Hence, the electromagnetic fields do not contain the high-frequency components accompanying the radiant heat. Some authors, however, include this radiation part in $\mathbf{q}$ (see Hughes and Young [1966, pp. 127–130]).

3.11. Clausius–Duhem Inequality

The second law of thermodynamics places conditions on the state of the body. Constitutive equatons are restricted by the inequality (3.10.8). In order to study these restrictions we employ the Clausius–Duhem (C–D) inequality, generalized for electromechanical bodies, and obtained by eliminating h between (3.10.7) and (3.10.8). It is also convenient to introduce the generalized Helmholtz free energy Ψ by

$$\Psi = \varepsilon - \theta\eta - \rho^{-1}\mathscr{E}_k P_k. \tag{3.11.1}$$

By use of this, we eliminate $\dot{\varepsilon}$ from the energy equation $(3.10.7)_1$, leading to

$$\rho(\dot{\Psi} + \dot{\theta}\eta + \theta\dot{\eta}) - t_{kl}v_{l,k} - \nabla\cdot\mathbf{q} - \rho h + P_k\dot{\mathscr{E}}_k + \mathscr{M}_k\dot{B}_k - \mathscr{J}_k\mathscr{E}_k = 0. \tag{3.11.2}$$

Eliminating ρh, between this equation and $(3.10.8)_1$, gives

$$-\rho(\dot{\Psi} + \dot{\theta}\eta) + t_{kl}v_{l,k} + \theta^{-1}q_k\theta_{,k} - P_k\dot{\mathscr{E}}_k - \mathscr{M}_k\dot{B}_k + \mathscr{J}_k\mathscr{E}_k \geq 0. \tag{3.11.3}$$

A factor θ^{-1} has been discarded. This form of the C–D inequality is useful in dealing with fluid bodies.

---

[18] This fact is especially emphasized by Møller [1952], Maugin [1971a], [1980b], de Groot and Suttorp [1972], and others.

For the elastic solids, a more convenient form is obtained by utilizing fields expressed in the material frame of reference. To this end, we introduce material tensor fields by

$$T_{KL} = JX_{K,k}X_{L,l}t_{kl}, \qquad Q_K = JX_{K,k}q_k,$$
$$\Pi_K = JX_{K,k}P_k, \qquad M_K = JX_{K,k}\mathscr{M}_k, \tag{3.11.4}$$
$$\mathscr{J}_K = JX_{K,k}\mathscr{J}_k,$$

and write

$$C_{KL} = x_{k,K}x_{k,L}, \qquad \theta_{,K} = \theta_{,k}x_{k,K}, \qquad J = \det x_{k,K} = \rho_0/\rho,$$
$$\mathscr{E}_K = \mathscr{E}_k x_{k,K}, \qquad B_K = B_k x_{k,K}. \tag{3.11.5}$$

The generalized Helmholtz free energy (3.11.1) now reads

$$\Psi = \varepsilon - \theta\eta - \rho_0^{-1}\Pi_K\mathscr{E}_K. \tag{3.11.6}$$

The energy equation (3.11.2), in terms of these material fields, is expressed as

$$\rho_0(\dot{\Psi} + \dot{\theta}\eta + \theta\dot{\eta}) - \tfrac{1}{2}\,{}_{\mathrm{E}}T_{KL}\dot{C}_{KL} - Q_{K,K} - \rho h + \Pi_K\dot{\mathscr{E}}_K + M_K\dot{B}_K - \mathscr{J}_K\mathscr{E}_K = 0, \tag{3.11.7}$$

where

$${}_{\mathrm{E}}T_{KL} = JX_{K,k}X_{L,l}\,{}_{\mathrm{E}}t_{kl} \tag{3.11.8}$$

is a symmetric tensor.

The C–D inequality, in terms of the material fields, is obtained by combining (3.11.7) and $(3.10.8)_1$

$$-\rho_0(\dot{\Psi} + \eta\dot{\theta}) + \tfrac{1}{2}\,{}_{\mathrm{E}}T_{KL}\dot{C}_{KL} + \frac{1}{\theta}Q_K\theta_{,K} - \Pi_K\dot{\mathscr{E}}_K - M_K\dot{B}_K + \mathscr{J}_K\mathscr{E}_K \geq 0. \tag{3.11.9}$$

Other forms of the energy equation and the C–D inequality are possible, depending on the forms of various tensors used to represent electromechanical fields.

3.12. Invariance Requirements for Balance Laws

The local balance laws (including Maxwell's equations) are subject to certain invariance requirements. These are:

(i) invariance under Galilean transformations, and
(ii) invariance under time reversal.

(i) Galilean transformations form a group consisting of time-independent spatial rotations and *pure* Galilean transformations, as expressed by

$$\bar{\mathbf{x}} = \mathbf{Q}_0\mathbf{x} + \mathbf{V}t + \mathbf{b}_0, \qquad \bar{t} = t, \tag{3.12.1}$$

where $\mathbf{Q}_0$ is a constant orthogonal matrix (rotation matrix), $\mathbf{V}$ is a constant velocity, and $\mathbf{b}_0$ is a constant vector. $\mathbf{Q}_0$ is subject to

$$\mathbf{Q}_0\mathbf{Q}_0^T = \mathbf{Q}_0^T\mathbf{Q}_0 = \mathbf{1}, \qquad \det \mathbf{Q}_0 = 1. \tag{3.12.2}$$

In Section 3.4 we have already seen that Maxwell's equations are invariant to within an order $(v/c)^2$, under (3.12.1). The rotational invariance ($\bar{\mathbf{x}} = \mathbf{Q}_0\mathbf{x}$) of the mechanical balance laws is verified trivially, since these laws are expressed in vector notation in which a vector $\mathbf{f}$ and a second-order tensor $\mathbf{t}$ transform as

$$\bar{\mathbf{f}} = \mathbf{Q}_0\mathbf{f}, \qquad \bar{\mathbf{t}} = \mathbf{Q}_0\mathbf{t}\mathbf{Q}_0^T. \tag{3.12.3}$$

We also recall expressions of the electromagnetic force, couple, and power, referred to the co-moving reference frame R_C in which $\mathbf{V} = -\mathbf{v}$. Accordingly,

$$\bar{\mathbf{F}}^E = \mathbf{Q}_0\mathbf{F}^E, \qquad \bar{\mathbf{C}}^E = \mathbf{Q}_0\mathbf{C}^E, \qquad \bar{W}^E = W^E, \tag{3.12.4}$$

where quantities carrying a superposed bar are expressed in the co-moving reference frame. Since the expressions of $\mathbf{F}^E$, $\mathbf{C}^E$, and W^E in the co-moving reference frame are given by (3.5.43)–(3.5.45), we see that the balance laws (3.10.4)–(3.10.7) are invariant under pure Galilean transformations also. Hence, they are Galilean invariants.

(ii) On the microscopic scale, the laws of physics obey the invariance requirement under time reversal.[19] "Microscopic reversibility" implies that macroscopic laws (balance laws) must also be invariant under time reversal. In order to see the consequence of this requirement, we must first examine how various fields change under time reversal. Clearly, the density ρ and the charge density q_e are invariant under time reversal. From the discussion given in Chapter 2, it is clear that $\mathbf{P}$ does not change its sign, but $\mathbf{M}$ and $\mathbf{J}$ reverse signs. From Maxwell's equations (3.3.2), and the balance laws (3.10.4)–(3.10.7), it follows that

$$\begin{gathered} \rho \to \rho, \qquad t_{kl} \to t_{kl}, \\ e \to e, \qquad q_k \to q_k, \qquad \nabla\theta \to \nabla\theta, \\ \mathbf{P} \to \mathbf{P}, \qquad \mathscr{E} \to \mathscr{E}, \qquad \mathscr{M} \to -\mathscr{M}, \\ \mathbf{B} \to -\mathbf{B}, \qquad \mathscr{J} \to -\mathscr{J}, \qquad \mathbf{D} \to \mathbf{D}, \\ \mathbf{E} \to \mathbf{E}, \qquad \mathbf{H} \to -\mathbf{H}, \qquad \mathbf{J} \to -\mathbf{J}, \\ \Psi \to \Psi. \end{gathered} \tag{3.12.5}$$

Thus, all balance laws will be form-invariant under time reversal, except the entropy production $\rho\gamma$, given by (3.10.8), which will change its sign. In order for (3.10.8) not to alter its sign, we must therefore impose the condition that

$$\rho\gamma \geq 0 \qquad \text{for} \quad t \to -t. \tag{3.12.6}$$

[19] The recent rejection of parity in nuclear physics cast doubt on the microscopic domain of applicability of the reversibility and the type of physical phenomena covered. But we need not concern ourselves with this question.

In fact, the invariance under time reversal is equivalent to the statement: *The entropy inequality must be form-invariant under the time reversal.*

In Chapter 5, this condition, together with the form-invariance of Ψ, will find important applications.

3.13. Principle of Virtual Power

Consider (3.10.5), with $\mathbf{F}^{\mathrm{E}}$ given by (3.6.1), and take the scalar product of this equation with a vector field $\tilde{\mathbf{v}}$ whose dimension is that of a velocity field. Then

$$\left(\rho\dot{\mathbf{v}} + \frac{\partial \mathbf{G}}{\partial t}\right)\cdot\tilde{\mathbf{v}} = \rho\mathbf{f}\cdot\tilde{\mathbf{v}} + (t_{kl}\tilde{v}_l)_{,k} + (t^{\mathrm{E}}_{kl}\tilde{v}_l)_{,k} - t_{kl}\tilde{v}_{l,k} - t^{\mathrm{E}}_{kl}\tilde{v}_{l,k}. \quad (3.13.1)$$

Let $\dot{\tilde{\pi}}$ and $\dot{\tilde{\boldsymbol{\mu}}}$ be two arbitrary vector fields having the dimension of time rate of polarization and magnetization per unit mass, respectively. Add to the right-hand side of (3.13.1) the vanishing quantity $\rho(\mathbf{B} - \mathbf{B})\cdot\dot{\tilde{\boldsymbol{\mu}}} + \rho(\mathscr{E} - \mathscr{E})\cdot\dot{\tilde{\boldsymbol{\pi}}}$, and integrate the resulting equation over the continuous material volume $\mathscr{V}$ bounded by the surface $\partial\mathscr{V}$. After using the Green–Gauss theorem, we obtain (Collet and Maugin [1974], Maugin [1980a][20])

$$\tilde{\mathscr{P}}_{\mathrm{a}}(\mathscr{V}) = \tilde{\mathscr{P}}_{\mathrm{i}}(\mathscr{V}) + \tilde{\mathscr{P}}_{\mathrm{v}}(\mathscr{V}) + \tilde{\mathscr{P}}_{\mathrm{c}}(\partial\mathscr{V}), \quad (3.13.2)$$

where

$$\begin{aligned}
\tilde{\mathscr{P}}_{\mathrm{a}}(\mathscr{V}) &\equiv \int_{\mathscr{V}} \left(\rho\dot{\mathbf{v}} + \frac{\partial \mathbf{G}}{\partial t}\right)\cdot\tilde{\mathbf{v}}\, dv,\\
\tilde{\mathscr{P}}_{\mathrm{i}}(\mathscr{V}) &\equiv -\int_{\mathscr{V}} (t_{kl}\tilde{v}_{l,k} + \rho\mathscr{E}\cdot\dot{\tilde{\boldsymbol{\pi}}} + \rho\mathbf{B}\cdot\dot{\tilde{\boldsymbol{\mu}}})\, dv,\\
\tilde{\mathscr{P}}_{\mathrm{v}}(\mathscr{V}) &\equiv \int_{\mathscr{V}} (\rho\mathbf{f}\cdot\tilde{\mathbf{v}} - t^{\mathrm{E}}_{kl}\tilde{v}_{l,k} + \rho\mathscr{E}\cdot\dot{\tilde{\boldsymbol{\pi}}} + \rho\mathbf{B}\cdot\dot{\tilde{\boldsymbol{\mu}}})\, dv,\\
\mathscr{P}_{\mathrm{c}}(\partial\mathscr{V}) &\equiv \int_{\partial\mathscr{V}} (\mathbf{t}_{(n)} + \mathbf{t}^{\mathrm{E}}_{(\mathbf{n})})\cdot\mathbf{v}\, da,
\end{aligned} \quad (3.13.3)$$

are, respectively, *the virtual powers of inertial forces, internal forces, volume forces, and contact forces.* Equation (3.13.2) is none other than the mathematical statement of the principle of virtual power, namely:

Principle of Virtual Power. *In a Galilean frame and for an absolute Newtonian chronology, the virtual power of the inertial forces of a mechanical system is balanced by the virtual power of all other forces, internal and external, impressed upon the system for any virtual velocity, polarization, and magnetization fields.*

---

[20] General theory was given by Germain [1973].

If we recall the Jaumann derivatives defined by (1.12.6), and use (3.10.9), then the virtual power of the internal forces may be expressed as

$$\mathscr{P}_{\mathrm{i}}(\mathscr{V}) = -\int_{\mathscr{V}} [t_{(kl)}\, \tilde{d}_{lk} + \rho\mathbf{B}\cdot(D_J\tilde{\boldsymbol{\mu}}) + \rho\mathscr{E}\cdot(D_J\tilde{\boldsymbol{\pi}})]\, dv, \tag{3.13.4}$$

where $\tilde{d}_{kl} \equiv \tilde{v}_{(k,l)}$. In the case where the virtual velocity field $\tilde{\mathbf{v}}$ is that of a rigid-body motion and $\tilde{\boldsymbol{\mu}}$ and $\tilde{\boldsymbol{\pi}}$ rotate at the same rate as the rigid body, we say that the deformable continuum is made rigid with polarization and magnetization fields "frozen" in the body. In this case, we have

$$\tilde{d}_{kl} = 0, \qquad (D_J\tilde{\boldsymbol{\mu}}) = 0, \qquad (D_J\tilde{\boldsymbol{\mu}}) = 0, \tag{3.13.5}$$

and we have: *The virtual power developed by internal forces vanishes in a virtual velocity field, which makes rigid the material continuum and freezes the polarization and magnetization fields.*

3.14. Résumé of Balance Laws

The balance laws of electrodynamics of deformable bodies are listed below:

Maxwells' Equations (in $\mathscr{V} - \sigma$)

$$\nabla\cdot\mathbf{D} - q_{\mathrm{e}} = 0, \tag{3.14.1}$$

$$\nabla\times\mathbf{E} + \frac{1}{c}\frac{\partial\mathbf{B}}{\partial t} = \mathbf{0}, \tag{3.14.2}$$

$$\nabla\cdot\mathbf{B} = 0, \tag{3.14.3}$$

$$\nabla\times\mathbf{H} - \frac{1}{c}\frac{\partial\mathbf{D}}{\partial t} = \frac{1}{c}\mathbf{J}, \tag{3.14.4}$$

$$\frac{\partial q_{\mathrm{e}}}{\partial t} + \nabla\cdot\mathbf{J} = 0. \tag{3.14.5}$$

Thermomechanical Balance Laws (in $\mathscr{V} - \sigma$)

$$\rho_0 = \rho \mathrm{III}_C^{1/2} \qquad \text{or} \qquad \dot{\rho} + \rho\nabla\cdot\mathbf{v} = 0, \tag{3.14.6}$$

$$t_{kl,k} + \rho(f_l - \dot{v}_l) + F_l^{\mathrm{E}} = 0, \tag{3.14.7}$$

$$t_{[kl]} = \mathscr{E}_{[k}P_{l]} + B_{[k}\mathscr{M}_{l]}, \tag{3.14.8}$$

$$\rho(\dot{\Psi} + \dot{\theta}\eta + \theta\dot{\eta}) - t_{kl}v_{l,k} - \nabla\cdot\mathbf{q} - \rho h + P_k\dot{\mathscr{E}}_k + \mathscr{M}_k\dot{B}_k - \mathscr{J}_k\mathscr{E}_k = 0, \tag{3.14.9}$$

$$\rho\gamma \equiv \rho\dot{\eta} - \nabla\cdot(\mathbf{q}/\theta) - (\rho h/\theta) \geq 0. \tag{3.14.10}$$

Jump Conditions (on $\sigma - \gamma$)

$$\mathbf{n}\cdot[\mathbf{D}] = w_{\mathrm{e}}, \tag{3.14.11}$$

$$\mathbf{n}\times\left[\mathbf{E} + \frac{1}{c}\mathbf{v}\times\mathbf{B}\right] = \mathbf{0}, \tag{3.14.12}$$

$$\mathbf{n} \times \left[\mathbf{H} - \frac{1}{c} \mathbf{v} \times \mathbf{D} \right] = \mathbf{0}, \tag{3.14.13}$$

$$\mathbf{n} \cdot [\mathbf{B}] = 0, \tag{3.14.14}$$

$$\mathbf{n} \cdot [\mathbf{J} - q_e \mathbf{v}] = 0, \tag{3.14.15}$$

where surface polarization and surface currents are discarded;

$$[\rho(\mathbf{v} - \boldsymbol{\nu})] \cdot \mathbf{n} = 0, \tag{3.14.16}$$

$$[\rho v_l(v_k - \nu_k) - t_{kl} - t_{kl}^{\mathrm{E}} - \nu_k G_l] n_k = 0, \tag{3.14.17}$$

$$\begin{aligned}[\{\rho(\Psi + \theta\eta) + \mathscr{E} \cdot \mathbf{P} + \tfrac{1}{2}\rho v^2 + \tfrac{1}{2}(E^2 + B^2)\}(v_k - \nu_k) \\ - (t_{kl} + t_{kl}^{\mathrm{E}} + \nu_k G_l) v_l - q_k + \mathscr{S}_k] n_k = 0,\end{aligned} \tag{3.14.18}$$

$$\left[\rho\eta(\mathbf{v} - \boldsymbol{\nu}) - \frac{1}{\theta} \mathbf{q} \right] \cdot \mathbf{n} \geq 0. \tag{3.14.19}$$

Mechanical Surface Traction (on $\partial\mathscr{V}$)

In the absence of the moving discontinuity surface, the mechanical surface traction is given by

$$t_{(\mathbf{n})k} = (t_{kl} - [t_{kl}^{\mathrm{E}}]) n_k. \tag{3.14.20}$$

Definitions of electromagnetic Fields and Loads

$$\mathscr{E} = \mathbf{E} + \frac{1}{c} \mathbf{v} \times \mathbf{B}, \qquad \mathscr{M} = \mathbf{M} + \frac{1}{c} \mathbf{v} \times \mathbf{P},$$

$$\mathscr{J} = \mathbf{J} - q_e \mathbf{v}, \qquad \mathscr{H} = \mathbf{H} - \frac{1}{c} \mathbf{v} \times \mathbf{D},$$

$$\mathbf{D} = \mathbf{E} + \mathbf{P}, \qquad \mathscr{B} = \mathbf{B} - \frac{1}{c} \mathbf{v} \times \mathbf{E},$$

$$\mathbf{B} = \mathbf{H} + \mathbf{M},$$

$$\begin{aligned}
&t_{kl} = {}_{\mathrm{E}}t_{kl} - P_k \mathscr{E}_l - \mathscr{M}_k B_l, \qquad {}_{\mathrm{E}}t_{kl} = {}_{\mathrm{E}}t_{lk}, \\
&\mathbf{F}^{\mathrm{E}} = q_e \mathbf{E} + \frac{1}{c} \mathbf{J} \times \mathbf{B} + (\nabla \mathbf{E}) \cdot \mathbf{P} + (\nabla \mathbf{B}) \cdot \mathbf{M} + \frac{1}{c}[(\mathbf{P} \times \mathbf{B}) v_k]_{,k} + \frac{1}{c} \frac{\partial}{\partial t} (\mathbf{P} \times \mathbf{B}), \\
&\Psi = e - \theta\eta = \varepsilon - \theta\eta - \rho^{-1} \mathscr{E}_k P_k, \\
&t_{kl}^{\mathrm{E}} = P_k \mathscr{E}_l - B_k \mathscr{M}_l + E_k E_l + B_k B_l - \tfrac{1}{2}(E^2 + B^2 - 2\mathscr{M} \cdot \mathbf{B}) \delta_{kl}, \\
&G_k = \frac{1}{c} (\mathbf{E} \times \mathbf{B})_k, \\
&\mathscr{S}_k = c(\mathbf{E} \times \mathbf{H})_k + [t_{kl}^{\mathrm{E}} + v_k G_l - \tfrac{1}{2}(E^2 + B^2 + 2\mathbf{E} \cdot \mathbf{P}) \delta_{kl}] v_l.
\end{aligned} \tag{3.14.21}$$

PROBLEMS

3.1. Show that eqs. (3.3.11) are the solutions of Maxwell's equations up to order c^{-1}.

3.2. Show that two of Maxwell's equations can be expressed as

$$\nabla \times \mathscr{E} + \frac{1}{c}\overset{*}{\mathbf{P}} = \mathbf{0}, \qquad \nabla \times \mathscr{H} - \frac{1}{c}\overset{*}{\mathbf{D}} = \frac{1}{c}\mathscr{J},$$

where a superposed asterisk represents the convective derivative as expressed by (1.12.12).

3.3. The Lorentz force in a microvolume element $\Delta v_\alpha \subset \Delta v$ is given by

$$\delta f^\alpha = \delta q^\alpha \left[\mathbf{e}(\mathbf{x}^\alpha) + \frac{1}{c}\mathbf{v}^\alpha \times \mathbf{b}(\mathbf{x}^\alpha) \right].$$

Writing $\mathbf{x}^\alpha = \mathbf{x} + \xi^\alpha$ and $\mathbf{v}^\alpha = \mathbf{v} + \dot{\xi}^\alpha + \mathring{\xi}^\alpha$, $|\mathring{\xi}^\alpha| \ll |\dot{\xi}^\alpha|$, and expanding $\mathbf{e}(\mathbf{x}^\alpha)$ and $\mathbf{b}(\mathbf{x}^\alpha)$ about $\mathbf{x}^\alpha = \mathbf{x}$, obtain the expressions of macroscopic force $\mathbf{F}^{\mathrm{E}}$ and couple $\mathbf{C}^{\mathrm{E}}$ (Eringen [1980]), where

$$\mathbf{F}^{\mathrm{E}}\Delta v = \sum_\alpha \delta \mathbf{f}^\alpha, \qquad \mathbf{C}^{\mathrm{E}}\Delta v = \sum_0 \mathbf{x}^\alpha \times \delta \mathbf{f}^\alpha.$$

3.4. Verify eqs. (3.5.26)–(3.5.32).

3.5. When the quadrupole moment is included, show that the electromagnetic power takes the form

$$w^{\mathrm{E}} = \mathbf{F}^{\mathrm{E}} \cdot \mathbf{v} + \mathscr{J}^{\mathrm{eff}} \cdot \mathscr{E} + [\mathscr{E}_i(\mathfrak{M}_{mi} + v_{i,j}Q_{jm})]_{,m} + v_{i,m}\, \bar{t}^{\mathrm{E}}_{mi},$$

where Q_{ij} is the quadrupole electric moment and $\mathfrak{M}_{ij} = c\varepsilon_{ijk}\mathscr{M}_k$.

3.6. Show that upon computing the total force and couple exerted by the electromagnetic field on the material volume $\mathscr{V}$ bounded by $\partial\mathscr{V}$, we find

(a) $${}_{\mathrm{M}}\mathbf{F}(\mathscr{V}) = \int_{\mathscr{V}} \mathbf{F}^{\mathrm{E}}\, dv = \int_{\mathscr{V}} {}_{\mathrm{L}}\mathbf{f}\, dv + \int_{\partial\mathscr{V}} \bar{\mathbf{t}}^{\mathrm{E}}_{(\mathbf{n})}\, da,$$

(b) $${}_{\mathrm{M}}C_i(\mathscr{V}) = \int_{\mathscr{V}} C^{\mathrm{E}}_i\, dv = \int_{\mathscr{V}} (\mathbf{r} \times {}_{\mathrm{L}}\mathbf{f})_i\, dv + \int_{\partial\mathscr{V}} \varepsilon_{ijk}(M_{(\mathbf{n})jk} + r_j\, \bar{t}^{\mathrm{E}}_{(\mathbf{n})k})\, da,$$

where

(c) $$\bar{\mathbf{t}}^{\mathrm{E}}_{(\mathbf{n})} \equiv w_{(\mathbf{n})}\mathscr{E} + (\pi_{(\mathbf{n})} \cdot \nabla)\mathscr{E} + \frac{1}{c}\mathfrak{M}_{(\mathbf{n})} \times \mathbf{B},$$

and

(d) $$M_{(\mathbf{n})jk} \equiv \pi_{(n)[j}\mathscr{E}_{k]}.$$

Interpret (a) and (b) and the quantities $w_{(\mathbf{n})}$, $\pi_{(\mathbf{n})}$, and $\mathfrak{M}_{(\mathbf{n})}$.

3.7. Show that three other equivalent expressions of ${}_{\mathrm{M}}\mathbf{f}$ are (compare 3.5.43):

(a) $$\mathbf{F}^{\mathrm{E}} = q_e\mathbf{E} + \frac{1}{c}\mathbf{J} \times \mathbf{B} + (\mathbf{P} \cdot \nabla)\mathbf{E} + (\nabla\mathbf{B}) \cdot \mathscr{M} + \frac{1}{c}[\dot{\mathbf{P}} + \mathbf{P}(\nabla \cdot \mathbf{v})] \times \mathbf{B} + \frac{1}{c}\mathbf{v} \times [(\mathbf{P} \cdot \nabla)\mathbf{B}],$$

(b) $\mathbf{F}^{\mathrm{E}} = q_e \mathscr{E} + \frac{1}{c}(\rho \dot{\pi} + \mathscr{J}) \times \mathbf{B} + (\dot{\mathbf{P}} \cdot \nabla)\mathscr{E} + (\nabla \mathbf{B}) \cdot \mathscr{M} + \frac{1}{c}\mathbf{v} \times [(\mathbf{P} \cdot \nabla)\mathbf{B}],$

(c) $\mathbf{F}^{\mathrm{E}} = q_e \mathbf{E} + \frac{1}{c}\mathbf{J} \times \mathbf{B} + (\nabla \mathbf{E}) \cdot \mathbf{P} + (\nabla \mathbf{B}) \cdot \mathbf{M} + \frac{1}{c}[(\mathbf{P} \times \mathbf{B})v_k]_{,k}$

$$+ \frac{1}{c}\frac{\partial}{\partial t}(\mathbf{P} \times \mathbf{B}).$$

3.8. Show that in the case of electrostatics, including electric quadrupoles where $\mathbf{P} = \bar{\mathbf{P}} - \nabla \cdot \mathbf{Q}$,

$$F_i^{\mathrm{E}} = q_e E_i + \bar{P}_j E_{i,j} + Q_{jp} E_{i,jp} = {}_{\mathrm{M}}t_{ji,j},$$

$$C_i^{\mathrm{E}} = \varepsilon_{ijk}(P_j E_k + Q_{mj} E_{k,m}),$$

$$t_{ji}^{\mathrm{E}} = D_j E_i + Q_{jk} E_{i,k} - \tfrac{1}{2}\mathbf{E}^2 \delta_{ji}.$$

3.9. For a continuous body $\mathscr{V}$ bounded by $\partial\mathscr{V}$, in the general case of quadrupole electric moments, show that the total power of the electromagnetic forces can be written in the form

$$ {}_{\mathrm{M}}W(\mathscr{V}) = \int_{\mathscr{V}} w^{\mathrm{E}}\, dv = \int_{\mathscr{V}} ({}_{\mathrm{L}}\mathbf{f} \cdot \mathbf{v} + \mathscr{J}^{\mathrm{eff}} \cdot \mathscr{E})\, dv + \int_{\partial\mathscr{V}} (\mathbf{t}_{(\mathbf{n})}^{\mathrm{E}} \cdot \mathbf{v} + \mathfrak{M}_{(\mathbf{n})i} \mathscr{E}_i $$

$$ + \eta_{(\mathbf{n})ij}\, d_{ij} + M_{(\mathbf{n})ji} w_{ji})\, da. $$

What is the definition and the interpretation of the different contributions?

3.10. Show that in quasi-electrostatics with quadrupoles the expression obtained in the previous problem reduces to

$$ {}_{\mathrm{M}}W(\mathscr{V}) = \int_{\mathscr{V}} \left[(q_e - \nabla \cdot \mathbf{P})\mathbf{E} \cdot \mathbf{v} + \overset{*}{\mathbf{P}} \cdot \mathbf{E} \right] dv + \int_{\partial\mathscr{V}} (\mathbf{P} \cdot \mathbf{n})(\mathbf{E} \cdot \mathbf{v})\, da, $$

and that in quasi-magnetostatics in insulators

$$ {}_{\mathrm{M}}W(\mathscr{V}) = \int_{\mathscr{V}} [(\nabla \times \mathbf{M}) \times \mathbf{B}] \cdot \mathbf{v}\, dv + \int_{\partial\mathscr{V}} [(\mathbf{M} \times \mathbf{n}) \times \mathbf{B}] \cdot \mathbf{v}\, da. $$

3.11. Show that the global Poynting theorem can be expressed as

$$ \frac{d}{dt}\int_{\mathscr{V}} \rho u\, dv = -\int_{\partial\mathscr{V}} \mathscr{J} \cdot \mathscr{E}\, dv - \int_{\partial\mathscr{V}} \mathscr{S} \cdot \mathbf{n}\, da, $$

where $\mathscr{V}$ is a *material* volume. How is the electromagnetic energy density u defined? Why is this definition awkward?

3.12. Show that in quasi-electrostatics the Poynting theorem takes on the following degenerate form:

$$ \int_{\mathscr{V}} \mathbf{E} \cdot \frac{\partial \mathbf{D}}{\partial t}\, dv + \int_{\partial\mathscr{V}} \phi \mathbf{n} \cdot \frac{\partial \mathbf{D}}{\partial t}\, da = 0, $$

where ϕ is the electrostatic potential. Similarly, show that, in quasi-magnetostatics in insulators, the corresponding degenerate form is

$$\int_{\mathscr{V}} \mathbf{H}\cdot\frac{\partial \mathbf{B}}{\partial t}\,dv + \int_{\partial\mathscr{V}} \Phi\mathbf{n}\cdot\frac{\partial \mathbf{B}}{\partial t}\,da = 0,$$

where Φ is the scalar magnetic potential.

3.13. Show that, equivalent to (3.6.24), we have the following global energetic identities:

(a) $$\frac{d}{dt}\int_{\mathscr{V}} \tfrac{1}{2}(\mathbf{E}^2 + \mathbf{B}^2)\,dv = -\int_{\mathscr{V}} (\mathbf{F}^{\mathrm{E}}\cdot\mathbf{v} + \mathscr{J}\cdot\mathscr{E} + \rho\mathscr{E}\cdot\dot{\boldsymbol{\pi}} - \mathscr{M}\cdot\dot{\mathbf{B}})\,dv + \int_{\partial\mathscr{V}} (\mathbf{t}^{\mathrm{E}}_{(\mathrm{n})}\cdot\mathbf{v} - \mathscr{S}\cdot\mathbf{n})\,da;$$

(b) $$\frac{d}{dt}\int_{\mathscr{V}} \tfrac{1}{2}(\mathbf{E}^2 + \mathbf{B}^2 + 2\mathscr{E}\cdot\mathbf{P})\,dv = -\int_{\mathscr{V}} (\mathbf{F}^{\mathrm{E}}\cdot\mathbf{v} + \mathscr{J}\cdot\mathscr{E} - \mathbf{P}\cdot\dot{\mathscr{E}} - \mathscr{M}\cdot\dot{\mathbf{B}})\,dv + \int_{\partial\mathscr{V}} (t^{\mathrm{E}}_{(\mathrm{n})}\cdot\mathbf{v} - \mathscr{S}\cdot\mathbf{n})\,da.$$

3.14. (a) For the case of quasi-electrostatics with electric quadrupoles, show that:

$$\frac{\partial}{\partial t}\left(\frac{\mathbf{E}^2}{2}\right) + \mathbf{J}^{\mathrm{eff}}\cdot\mathbf{E} = 0 \qquad \text{with} \quad \mathbf{J}^{\mathrm{eff}} = q_{\mathrm{e}}\mathbf{v} + \frac{\partial \mathbf{P}}{\partial t};$$

(b) in insulators, the power developed by the ponderomotive force reads

$$\mathbf{F}^{\mathrm{E}}\cdot\mathbf{v} = \bar{\mathbf{P}}\cdot\dot{\mathbf{E}} + Q_{ij}\dot{\overline{E_{i,j}}} - \frac{\partial}{\partial t}(\tfrac{1}{2}\mathbf{E}^2 + \bar{\mathbf{P}}\cdot\mathbf{E} + Q_{ij}E_{i,j});$$

(c) and we have either one of the following global energetic identities:

(i) $$\frac{d}{dt}\int_{\mathscr{V}} (\tfrac{1}{2}\mathbf{E}^2 + \bar{\mathbf{P}}\cdot\mathbf{E} + Q_{ij}E_{i,j})\,dv = -\int_{\mathscr{V}} (\mathbf{F}^{\mathrm{E}}\cdot\mathbf{v} - \bar{\mathbf{P}}\cdot\dot{\mathbf{E}} - Q_{ij}\dot{\overline{E_{i,j}}})\,dv + \int_{\partial\mathscr{V}} t^{\mathrm{E}}_{(\mathrm{n})}\cdot\mathbf{v}\,da;$$

(ii) $$\frac{d}{dt}\int_{\mathscr{V}} \tfrac{1}{2}\mathbf{E}^2\,dv = -\int_{\mathscr{V}} (\mathbf{F}^{\mathrm{E}}\cdot\mathbf{v} + \rho\mathbf{E}\cdot\dot{\boldsymbol{\pi}} + \rho E_{i,j}\dot{q}_{ij})\,dv + \int_{\partial\mathscr{V}} t^{\mathrm{E}}_{(\mathrm{n})}\cdot\mathbf{v}\,da,$$

where $\mathscr{V}$ is a material volume bounded by $\partial\mathscr{V}$.

3.15. Let $\mathbf{K}$ be a prescribed surface current on a steady discontinuity surface σ in a magnetized body considered in quasi-magnetostatics. Show that the surface force on σ reads

$$\hat{\mathbf{F}}^{\mathrm{E}} = \frac{1}{c}\mathbf{K}\times\langle\mathbf{B}\rangle + (\langle\mathbf{M}\rangle\cdot[\mathbf{B}])\mathbf{n}, \qquad \langle A\rangle \equiv \tfrac{1}{2}(A^+ + A^-).$$

Similarly, if w_{e} is a free charge prescribed on σ, the electrically polarized body being considered in quasi-electrostatics, then show that

$$\hat{\mathbf{F}}^{\mathrm{E}} = w_{\mathrm{e}}\langle\mathbf{E}\rangle + (\langle\mathbf{P}\rangle\cdot[\mathbf{E}])\mathbf{n}.$$

3.16. Show that in a nonmagnetized moving dielectric ($\mathcal{M} = \mathbf{0}$, $q_e = 0$, $\mathcal{J} = \mathbf{0}$), Maxwell's equations $(3.3.2)_1$ and (3.4.18) take on the form

$$\nabla \cdot \mathbf{E} = -\nabla \cdot \mathbf{P}, \qquad \nabla \times \mathbf{B} - \frac{1}{c}\frac{\partial \mathbf{E}}{\partial t} = \frac{1}{c}\left[\overset{*}{\mathbf{P}} - (\nabla \cdot \mathbf{P})\mathbf{v}\right].$$

Interpret the contributions on the right-hand side of these equations.

3.17. Show that Maxwell's equations in material coordinates have the same form as in any other Galilean frame. That is, they can be written as

$$B_{K,K} = 0, \qquad D_{K,K} = Q_e,$$

$$\varepsilon_{KLM}\mathfrak{E}_{M,L} + \frac{1}{c}\frac{\partial B_K}{\partial t} = 0,$$

$$\varepsilon_{KLM}\mathfrak{H}_{M,L} - \frac{1}{c}\frac{\partial D_K}{\partial t} = \frac{1}{c}\mathcal{J}_K,$$

where

$$B_K \equiv JX_{K,i}B_i, \qquad D_K \equiv JX_{K,i}D_i, \qquad Q_e \equiv Jq_e,$$

$$\mathfrak{E}_K \equiv E_i x_{i,K} + \frac{1}{c}\varepsilon_{KLM}X_{L,i}v_i B_M,$$

$$\mathfrak{H}_K \equiv H_i x_{i,K} - \frac{1}{c}\varepsilon_{KLM}X_{L,i}v_i D_M,$$

$$\mathcal{J}_K \equiv JX_{K,i}\mathcal{J}_i.$$

3.18. Using the integral balance law for the energy, obtain a local expression of the energy balance and the jump conditions.

3.19. Give the expressions of momentum and angular momentum balance in the material frame of reference.

CHAPTER 4

Elementary Properties of Electromagnetic Continua

4.1. Scope of the Chapter

This chapter is concerned with the physical nature of electromagnetic bodies. It is intended to provide some fundamental concepts relevant to the electromagnetic properties of matter. We begin with a discussion of dielectric materials in Section 4.2 and, using classical ideas of polarizability in Section 4.3, we obtain expressions of dielectric susceptibility and permittivity. The classical Clausius–Mossoti equation is given. The nature of ferroelectric crystals, their thermal behavior and electrostriction, piezoelectricity, and pyroelectricity are discussed on the basis of simple atomic models. In Section 4.4 we turn our attention to magnetism and briefly discuss the quantum mechanical origin of magnetism, gyromagnetic effects, and various types of magnetism (e.g., paramagnetic, antiferromagnetic, and ferrimagnetic materials). The nature of ferromagnetic materials, the Curie–Weiss law, the Heisenberg model of spin–spin interaction, and the thermal effects on the magnetic susceptibilities of various types of magnetic materials are discussed in Section 4.5. Magnons or spin waves, magnetostrictive effects, piezomagnetism, and magnon–phonon couplings are touched on briefly in the same section.

Section 4.6 is concerned with the phenomena of electric conduction. Various "cross effects," e.g., Hall, Thompson, Peltier, Ettingshausen, and Righi–Leduc effects are second-degree effects of which some are hidden, in the sense that they do not contribute to entropy production. In Section 4.7 we briefly discuss electromagnetic optics, inducement of optical anisotropy, birefringence, photoelastic, electromagneto-optical effects, and streaming birefringence. These topics are taken up in later chapters from the point of view of the mathematical theory of continuous media. The present chapter, therefore, gives a simple overall view of the rich physical phenomena that are nested in deformable electromagnetic bodies.

4.2. Dielectric Materials

Equation $(3.3.3)_1$ signifies that, in matter, the electric field **E** and the electric displacement **D** differ from each other by the vector **P** that represents the *effective electric* polarization. A material in which **P** differs from zero is said to be electrically polarized. Then, when it exists, the ratio

$$\chi^{E} \equiv |\mathbf{P}|/|\mathbf{E}| \tag{4.2.1}$$

defines the *dielectric susceptibility* of this material. The question arises as to the origin of the property of dielectricity. Of course, we already know that polarization arises from the fact that the material body contains bound electric charges—see (2.7.9), (2.7.18), and (2.7.19). However, in a real material, there are, more or less, some ordered arrangements at the atomic, molecular, or crystal level. In fact, with respect to some aspects of their electrical behavior, materials may be classified as conductors, semiconductors, and *dielectrics* or insulators. Here we focus our attention on the last class. An ideal dielectric is a material which possesses no free charges, $q_e = 0$, and thus, completely inhibits the passage of steady electric currents. However, although electrically neutral, all dielectrics are composed of molecules, which in turn are composed of charged particles (nuclei and electrons), and these particles are usually affected by the presence of an electric field. Such a field displaces the positive and negative charges in opposite directions from their equilibrium positions, thus causing the molecule to become electrically *polarized*. The resulting electric moment of the electrically neutral system of charges, thus, may be defined as in (2.3.7). The average relative displacement of charges, denoted by vector $\boldsymbol{\xi}$, is limited by strong restoring forces, so that the charge shift is seldom more than a fraction of a molecular diameter. Thus, the molecule may be viewed as an elementary electric dipole or as several dipoles. When the contributions to all molecular dipoles are summed, the resultant is often found to alter the field distribution significantly from the value it had in the absence of the dipole, both for points inside and outside the dielectric.

The dipole behavior of a molecule and, hence, the dielectric behavior of the material may arise from three distinct causes. First, the electronic cloud of a constituent atom may shift relative to its nucleus due to the presence of an electric field. This induced effect is called *electronic polarization*. Next, the molecular structure may result from an arrangement of oppositely charged ions, which can shift their equilibrium positions under the action of an electric field, thus giving rise to the so-called *ionic polarization*. Finally, the molecule may consist of an arrangement of atoms which, *in the absence* of an electric field, is a randomly oriented, *permanent* electric dipole. The presence of the field then causes a partial orientation of the permanent molecular dipole, causing a net polarization. This phenomenon is called *orientational polarization*. All three effects may be present in a given material. Here we sketch out the basic concepts of the physical theory of dielectrics.[1]

---

[1] For a comprehensive account on the subject, see the books by Böttcher [1952], Fröhlich [1958], Von Hippel [1954], Zheludev [1971], and Chelkowski [1980].

4.3. Physical Theory of Dielectrics

A. Polarizability

A convenient starting point is to assume that the material is *ideal*, i.e., there is no free charge and all regions of the material are electrically neutral. The fundamental "building block" of the material is designated by the generic term *molecule*. This term stands for a neutral atom (e.g., in noble gases such as A, Ne), or several atoms joined in homopolar (e.g., in O_2, N_2), covalent (e.g., in SiC), or ionic (e.g., in HCl) bonds, or a small group of ion pairs (e.g., in solid NaCl), etc.—whichever is characteristic of the material in question. Let $\mathbf{P}_k^\alpha$, $k = \mathrm{e}$ (for electronic), $k = \mathrm{i}$ (for ionic), and $k = \mathrm{o}$ (for orientational), be the dipole moment within a molecule α arising from each of the three sources of dielectric behavior. Then the dipole moment of the molecule α is

$$\mathbf{P}^\alpha = \mathbf{P}_\mathrm{e}^\alpha + \mathbf{P}_\mathrm{i}^\alpha + \mathbf{P}_\mathrm{o}^\alpha. \tag{4.3.1}$$

If there are N molecules in a small volume ΔV, then the total polarization per unit volume will be

$$\mathbf{P} = \frac{1}{\Delta V} \sum_{\alpha=1}^{N} \mathbf{P}^\alpha. \tag{4.3.2}$$

The electric displacement is then defined as in $(3.3.3)_1$.

To develop the connection between $\mathbf{P}$ and its causes it is useful to introduce the concept of the *local field*, $\mathbf{E}_{\mathrm{loc}}$, which will be defined as the average field intensity acting on a given molecule within the dielectric.[2] It is argued that this average field should be proportional to the dipole moment of the molecule, which in turn is proportional to $\mathbf{P}$ by definition. It turns out that $\mathbf{E}_{\mathrm{loc}}$, $\mathbf{E}$, and $\mathbf{P}$ are related by the equation

$$\mathbf{E}_{\mathrm{loc}} = \mathbf{E} + \lambda \mathbf{P}, \tag{4.3.3}$$

where $\lambda \simeq \frac{1}{3}$ is called the internal field constant. Adopting this formulation of the local field enables us to obtain the relationships between $\mathbf{P}$ and its causes for each of the three types of polarization. From these results we can determine a functional (constitutive) relationship between $\mathbf{D}$ and $\mathbf{E}$. Indeed, it is classically shown that, under suitable assumptions, a *linear* relation exists between the strength of the local field and the induced polarization. This linearity is applicable to most dielectrics, and presumes field strengths which are not too excessive. It includes all three polarization mechanisms. In the case of the electronic polarization of an atom we obtain

$$\mathbf{P}_\mathrm{e} = \alpha_\mathrm{e} \mathbf{E}_{\mathrm{loc}}, \tag{4.3.4}$$

---

[2] $\mathbf{E}_{\mathrm{loc}}$ may be determined by removing the molecule in question, maintaining all other molecules in their time-averaged polarized positions, and calculating the space-averaged electrostatic field in the cavity previously occupied by the removed molecule (Lorentz [1952]).

where the *electric polarizability* α_e of the atom is such that $\alpha_e \propto f(0)^{-1}$, where $f(0)$ is the central value $f(\mathbf{r} = \mathbf{0})$ of the normalized absolute charge density in the electronic cloud.[3]

In the case of ionic polarization of a molecule we obtain

$$\mathbf{P}_i = \alpha_i \mathbf{E}_{loc}, \tag{4.3.5}$$

where α_i is the *ionic polarizability*. If the relative sizes of the atomic radii r_a (distance at which the volume charge density in the electronic cloud drops to zero) and of the normal interatomic spacing d_a between centers of differing atoms within the molecule are taken into consideration, then (4.3.4) and (4.3.5) indicate comparable values for electronic and ionic polarizabilities. Thus, ionic polarization manifests itself quantitatively in the same manner as electronic polarization.

The discussion of orientational polarization requires the consideration of *polar* or *nonpolar* molecules. In the presence of an electric field, polar molecules experience a torque, tending to align them with the field. This is orientational polarization. By contrast, nonpolar molecules are not subjected to this aligning torque, but the presence of a field can induce ionic polarization. If a *homogeneous polar* material is considered, all molecules have the *same* permanent dipole moment and, in the absence of an exciting field, these moments are often randomly oriented. In fact, when an electrostatic field is present, each molecule experiences a torque which tends to align its dipole moment with the field. Were it not for thermal agitation, all the molecules would become aligned in such materials. But the collisions between the molecules keep breaking up the pattern, so that, on the time average, there is only partial alignment. A quantitative indication of this effect may be deduced by considering that a permanent dipole $p = qd$ is assigned to the molecule. Then, using a thermodynamical analysis,[4] it is shown that

$$\mathbf{P}_o = \alpha_o \mathbf{E}_{loc}, \qquad \alpha_o = P^2/3k_B\theta, \tag{4.3.6}$$

where θ is the absolute temperature and k_B is Boltzmann's constant. The factor α_o is the *orientational polarizability* of the molecule. The result (4.3.6) assumes a dielectric whose dipole moments are randomly oriented in the absence of an external field.

Thus, it seems that all three types of induced polarization can be linearly proportional to the local field. Note that dielectrics whose molecules possess permanent dipole moments which are randomly oriented in the absence of a field may exhibit all three types of polarizability, so that, for such dielectrics, we may write for the average polarization per molecule, using (4.3.1),

$$\mathbf{p} = \alpha \mathbf{E}_{loc}, \tag{4.3.7}$$

[3] See Elliott [1966, p. 345].

[4] See, for instance, Zheludev [1971, pp. 360–363] or Kittel [1971, pp. 464–465].

where

$$\alpha = \alpha_e + \alpha_i + \alpha_o. \tag{4.3.8}$$

In this equation, α_e and α_i include the effects of all atoms and ions in the molecule, and must be suitably averaged over all orientations of the molecule with respect to the field. If N is the number of molecules per unit volume, then

$$\mathbf{P} = N\mathbf{p} = N\alpha\mathbf{E}_{loc}. \tag{4.3.9}$$

Because of (4.3.3), this may be rewritten as

$$\mathbf{P} = \chi^E\mathbf{E}, \tag{4.3.10}$$

where

$$\chi^E = N\alpha/(1 - \lambda N\alpha) \tag{4.3.11}$$

is the dielectric susceptibility. From (4.3.10) it is clear that, under the conditions of this analysis, the polarization $\mathbf{P}$ is linearly proportional to the macroscopic electric field. It then follows from (4.3.10) and $(3.3.3)_1$ that

$$\mathbf{D} = \varepsilon\mathbf{E}, \tag{4.3.12}$$

in which $\varepsilon \equiv 1 + \chi^E$ is called the permittivity of the medium. Usually, we consider the relative dielectric constant $\varepsilon_r = \varepsilon/\varepsilon_0$, where ε_0 is the dielectric constant of a vacuum. ε_r is an easily measurable quantity. However, in the present Lorentz–Heaviside system of units, ε_0 is taken as equal to unity and we do not distinguish between ε_r and ε.

Prominent among the dielectric materials used in electronic components are the ionic nonpolar solid dielectrics and, more particularly, the *ionic crystals*, such as the alkali halides (e.g., NaCl, NaI, LiCl, LiBr, ...). Ionic nonpolar solid dielectrics generally contain more than one type of atom, but no permanent dipoles, and evidence ionic as well as electronic polarizability. As to the structure of ionic crystals, it is characterized by a regular three-dimensional alternation of positive and negative ions, and, hence, the entire crystal has no permanent dipole moment. However, in the presence of an external field, the positive ion lattice will suffer a displacement relative to the negative ion lattice resulting in ionic polarization. Additionally, both ion types will show electronic polarization, so that the total polarization density may be written as the sum of these contributions in the form $\mathbf{P} = \mathbf{P}_e + \mathbf{P}_i$. The *static* dielectric constant ε_s is related to $\mathbf{P}$ by the expression

$$\mathbf{P}_e + \mathbf{P}_i = (\varepsilon_s - 1)\mathbf{E}. \tag{4.3.13}$$

However, if the dielectric constant is measured at light frequencies (e.g., by a refraction experiment), the ions are too heavy to follow the field oscillations, and we may set $\mathbf{P}_i = \mathbf{0}$. If ε_l is the dielectric constant under these conditions we can write

$$\mathbf{P}_e = (\varepsilon_l - 1)\mathbf{E}. \tag{4.3.14}$$

Combining (4.3.13) and (4.3.14) we have the ratio

$$\frac{|\mathbf{P}_i|}{|\mathbf{P}_e|} = \frac{\varepsilon_s - \varepsilon_l}{\varepsilon_l - 1}. \tag{4.3.15}$$

Therefore, measurements of the dielectric constant under quasi-static and optical conditions provide an indication of the relative strengths of ionic and electronic polarization. Because $\mathbf{P}_i$, like $\mathbf{P}_e$, is dependent on the atomic structure and the particle density (quantities which normally are unaffected by temperature), ionic solids also reveal susceptibilities which are independent of temperature.

B. The Clausius–Mossoti Equation

Consider (4.3.3), as established by Lorentz [1952], with $\lambda = \frac{1}{3}$. Then (4.3.11) may be written in the form

$$\frac{\chi^E}{\chi^E + 3} = \frac{N\alpha}{3} = \frac{\varepsilon - 1}{\varepsilon + 2}. \tag{4.3.16}$$

If ρ is the mass density of the dielectric material, then the molecular density N is given by $N = \rho N_A/M$, in which N_A ($=6.02 \cdot 10^{26}$ in MKS units) is Avogadro's number and M is the molecular weight. Substituting from N in (4.3.16), we obtain

$$\frac{M}{\rho}\frac{\varepsilon - 1}{\varepsilon + 2} = \frac{N_A}{3}[\alpha_e + \alpha_i + (p^2/3k_B\theta)]. \tag{4.3.17}$$

This is Debye's generalization of the celebrated *Clausius–Mossoti equation*. It relates the dielectric constant to the mass density of the material, and is written in such a way that the right-hand side—called the molar polarizability—is a function of temperature only. Therefore, the left-hand side is independent of density. It must be remembered, however, that this equation is valid only for dielectrics of low density, for which inaccuracies in the Lorentz local field expression have little effect. Thus, agreement between (4.3.17) and experiment is found principally for gases at moderate pressures. The expression (4.3.16), which is often called the *Clausius–Mossoti* formula, gives fair agreement with experiments for nonpolar liquids in which short-range forces are negligible, and also for some simple nonpolar solids for which the Lorentz local field is a good approximation.[5]

C. Ferroelectric Crystals

In view of the linear relationship (4.3.10), the above discussion is concerned with linear dielectrics (also referred to as paraelectric bodies). However, in

[5] See Fröhlich [1958] and Böttcher [1952].

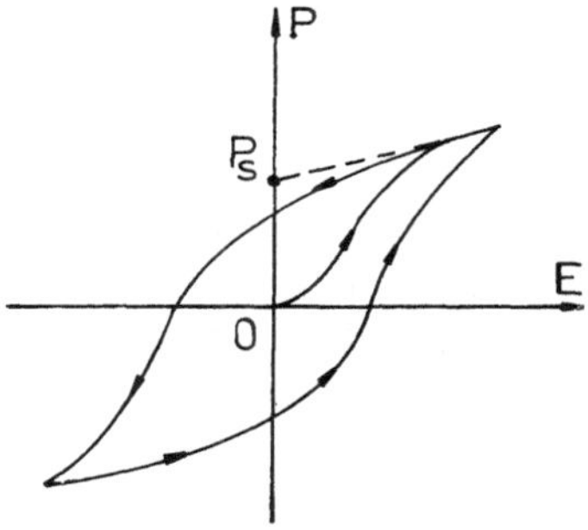

Figure 4.3.1. Hysteresis curve for a ferroelectric crystal.

recent years, several classes of ionic crystals have been discovered to have the following property. When the centers of positive and negative charges have been set apart sufficiently they remain *locked* in their new positions, causing the molecules to become *permanently* polarized. Such crystals then exhibit a polarization, even in the absence of an external field. They are said to be *spontaneously polarized.* It is possible to cause a specimen of such materials to become uniformly polarized. However, a more common situation is the one in which subvolumes of the specimen, called *domains*, are individually polarized, with the direction of polarization varying from one domain to another in a random fashion. These domains can be observed through a microscope using polarized light. When the specimen is in this condition it shows no net bulk polarization. Then, when an external field is applied which tends to align dipole moments along itself, those domains whose polarization is already more or less aligned with the field grow at the expense of other domains. As a result, a hysteresis curve can be drawn for such crystals (see Figure 4.3.1). The similarities between this process and the corresponding one involving ferromagnetic materials[6] explain why dielectrics, which display this hysteresis effect, are called *ferroelectrics.* However, it must be noted that this wording is not entirely appropriate, because the microscopic mechanisms are dissimilar.[7] Of course, the hysteresis curve shows that the polarization is nonlinear in the E-field for ferroelectric materials. Nevertheless, we can define a nonconstant dielectric modulus ε by the differential equation (compare (4.3.10)):

$$\frac{dP}{dE} = \varepsilon - 1. \tag{4.3.18}$$

The value of ε depends on the previous history of the specimen, as well as on the value of E. Usual values of ε may be very high, e.g., values in the range $(500\text{–}5{,}000)\varepsilon_0$, where ε_0 is the dielectric constant of a vacuum (in this work, ε_0 is taken unity).

[6] See Section 5.

[7] In particular, there is no mechanism similar to the exchange forces of ferromagnetism in ferroelectrics.

Spontaneous polarization and orientational polarization have some features in common, but they should not be mistaken for one another. This is particularly true as far as thermal effects are concerned. Orientational polarization—which disappears with the external field—is due to permanent dipole moments possessed by individual molecules, so that molecules orient themselves in a random fashion due to thermal agitation. Hence, orientational polarizability is a temperature-dependent phenomenon. On the other hand, in a ferroelectric crystal, the individual molecules have permanent dipole moments which, within a given domain, are all aligned steadily (thermal agitation not affecting this alignment). Nonetheless, ferroelectric crystals display a different type of temperature effect: the spontaneous polarization usually disappears above a certain characteristic temperature θ_F, called the *Curie temperature*. The reason is that there occurs a change (phase transition) in the crystal structure, such that the individual molecules no longer possess permanent dipole moments (see Figure 4.3.2). Above θ_F, the dielectric constant is found to vary with temperature, in such a way as to obey the *Curie–Weiss law*

$$\varepsilon = \frac{C}{\theta - \theta^*}, \tag{4.3.19}$$

in which C and θ^* are constants characteristic of the crystal. θ^* is usually a few degrees below θ_F.

Several different groups of ferroelectric crystals may be classified on the basis of their chemical composition, structure, and electrical behaviors:

(1) dihydrogen phosphates and arsenates of the alkali metals (e.g., KH_2PO_4 with $\theta_F = 123\,°K$);
(2) tantrates (e.g., Rochelle salt, $NaKC_4H_4O_6 \cdot 4H_2O$, first prepared in 1672 and with two transition temperatures at 255 °K and 296 °K);

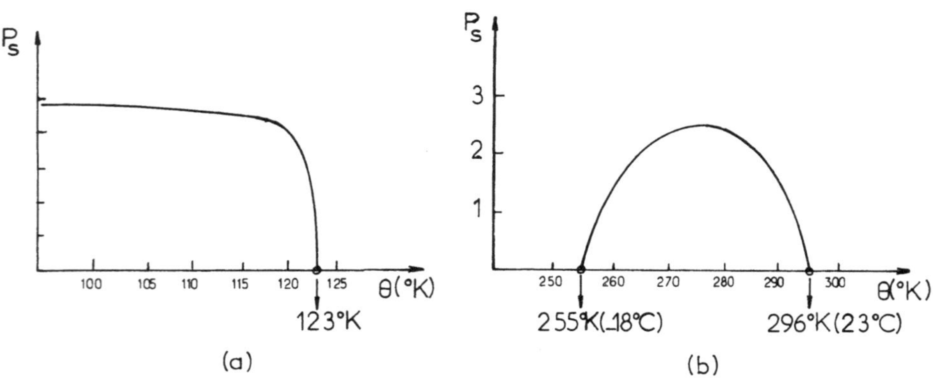

Figure 4.3.2. Spontaneous polarization: (a) potassium dihydrophosphate (KH_2PO_4); (b) Rochelle salt.

(3) the so-called GASH group (so-named after guanidine aluminum sulfate hexahydrate);
(4) the oxygen octahedron group (e.g., barium titanate $BaTiO_3$, which possesses the so-called perovskite structure above $\theta_F = 393\,°K$).

We refer the reader to exhaustive treatises[8] for a comprehensive approach to ferroelectrics. Let us note, however, that the unusual properties of the ferroelectric crystals have led to their use in a variety of practical applications. The nonlinearity between P and E permits the design of devices such as rectifiers, the polarity of spontaneous polarization permits the storage of binary information in a computer memory, and the high dielectric constant permits large capacities in small volumes.

D. Electrostriction and Piezoelectricity

When we consider the polarizability of a crystal such as NaCl, we observe that, at the first-order relative displacements of the negatively and positively charged ionic lattices, a new balance of forces is established, from which an equilibrium displacement is deduced. If second-order effects are also considered, then it is found that the longitudinal shift of the two sets of ions causes transverse force components which result in a slight lateral compression. This, in turn, permits an elongation of the crystal in a direction parallel to the field. If the electric field is reversed, the same deformation occurs. This phenomenon is called *electrostriction* and is common to all dielectric solids. However, if the crystal possesses a center of symmetry for the constituent charges, this is *not* a reciprocal effect, in the sense that a mechanical deformation will not cause a net polarization of the specimen. The clear reason for this is that applied mechanical forces act on elements of mass and do not distinguish between different types of charges (see Figure 4.3.3, in which there is no polarization because, by symmetry, the electrical center of the six Cl-ions still coincides with the position of the Na^+ ion after loading). However, not all crystals contain a center of symmetry, so that the phenomenon, called *piezoelectricity* (after the Greek word "piezein" for press), exists in the classes of crystals which lack a center of symmetry[9] (e.g., quartz and Rochelle salt are materials which exhibit this effect). Such crystals are important because they permit the conversion of mechanical energy into electrical energy and vice-versa. Practical applications include the crystal pickup in a phonograph, stable frequency sources, and sources of ultrasonic vibrations (transducers). For a discussion of the theory of electrostrictive and piezoelectric effects, the reader is referred to Chapter 7.

---

[8] For example, Forsbergh [1956], Cook [1962], Fattuzo and Merz [1967], Burfoot [1967], Jona and Shirane [1962], Megaw [1957], and Zheludev [1971, Volume One].

[9] Piezoelectricity was discovered by Pierre and Jacques Curie in 1881. For a history of piezoelectricity, see Mason [1981].

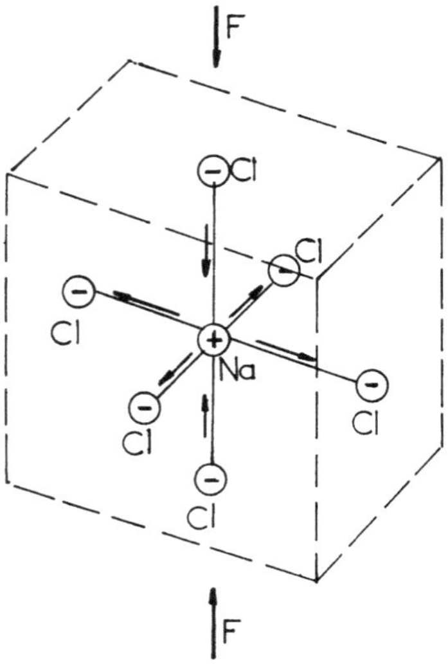

Figure 4.3.3. Mechanical deformation of NaCl crystal.

E. Pyroelectricity

When a piezoelectric crystal is heated or cooled, a charge separation appears across its faces. This effect is known as *pyroelectricity*. The charges are evidence of an internal polarization, caused by the strain associated with thermal expansion or contraction of the crystal.

The optical properties of dielectrics will be reviewed in Section 4.7.

4.4. Magnetic Materials[10]

A. The Origin of Magnetism

In $(3.3.3)_2$, the fields **H**, **B**, and **M** are the magnetic field, the macroscopic magnetic induction, and the *effective magnetization*, respectively (all expressed in the laboratory frame). A material in which the magnetization per unit volume **M** differs from zero is said to be *magnetized*. Then the ratio

$$\chi^{B} \equiv |\mathbf{M}|/|\mathbf{H}|, \tag{4.4.1}$$

when it exists, defines the magnetic susceptibility of this material. More

[10] For an introduction to magnetism, see Kittel [1971]. For a more comprehensive approach, consult the book by Chikazumi [1966]. An exhaustive treatment of the physical properties of magnetic materials is to be found in the encyclopedic work of Vonsovskii [1975].

generally, if there exists a relationship of the type $M = M(a, H)$, then

$$\chi^{B} = \left(\frac{\partial M}{\partial H}\right)_{a}, \tag{4.4.2}$$

where a is any other relevant independent variable, such as the temperature. Clearly, the magnetic susceptibility is a response function analogous to the compressibility, say, of a fluid. Unlike the compressibility, however, there are no stability conditions on its sign. The response of the material may be either to oppose the field or to intensify it. The magnitude and sign of χ^{B} define the different states of magnetism. Moreover, **M** and **H**, in general, are not colinear; it may happen that the *static* susceptibility, as given by (4.4.1) or (4.4.2), is not a well-defined quantity. For instance, **M** may not vanish for vanishing **H**.

The question arises as to the origin of magnetism. The answer to this question varies, depending upon how far we pursue the origin. The standard answer, however, is that magnetism originates from the magnetic moment due to the rotational motion of charged particles. The origin of this concept can be traced back to the molecular current theory of Ampère. We can construct the theory of magnetism from such a basic idea (in this regard, see (2.2.3)). But *quantum mechanics* is an indispensable tool for such a formulation. This can easily be seen from the fact that the existence of a stable microcurrent cannot be demonstrated by classical theory. Magnetism cannot be separated from quantum mechanics because a purely classical system in thermal equilibrium cannot possess a magnetic moment, even in the presence of a magnetic field.[11] In other words, if Planck's constant goes to zero, then a catastrophe occurs: magnetism does not exist in the universe. In this respect the notions of gyromagnetic ratio, the Bohr magneton and magnetic moment due to spin, introduced in Chapter 2, are essential, as well as those of gyromagnetic effects.

B. Gyromagnetic Effects

The Einstein–de Haas [1915] and the Barnett effects (see Barnett [1931]) are gyromagnetic effects. If we hang a cylinder of a magnetic substance with a coil wound around it, and induce a magnetization $\Delta\mathbf{M}$ letting a current flow in the coil, then the angular momentum which induces $\Delta\mathbf{M}$ changes by an amount $\Delta\mathbf{S}_{\text{mag}}$, with $\Delta\mathbf{M} = \gamma\Delta\mathbf{S}_{\text{mag}}$. Since the total angular momentum of the magnetized cylinder is invariant, $\Delta\mathbf{S}_{\text{bulk}}$, given by $\Delta\mathbf{S}_{\text{bulk}} = -\Delta S_{\text{mag}} = -\gamma^{-1}\Delta\mathbf{M}$, is the corresponding change of angular momentum due to rotation of the sample. Hence, a corresponding mechanical torque acts on the entire sample. The gyromagnetic ratio γ can be measured by observing the cancellation of this torque with an externally applied torque or the resonance induced by periodic changes of $\Delta\mathbf{M}$. This is the Einstein–de Haas effect.

---

[11] See the proof of Van Leeuwen's theorem in Van Vleck [1932, pp. 94–104].

Table 4.4.1. Values of g measured by the gyromagnetic effect (after R. Kubo and T. Nagamiya, *Solid State Physics*, McGraw-Hill, New York, 1969, p. 421).

Substance	g	ε	Substance	g	ε
Iron	1.93	0.04	Magnetite	1.93	0.04
Cobalt	1.87	0.07	Heusler's alloy	2.00	0.
Nickel	1.92	0.04	Permalloy	1.91	0.05

The Barnett effect is the magnetization which occurs when a magnetic cylinder is rotated at high speed. If $\mathbf{\Omega}$ is the angular velocity of rotation, each of the angular momenta which induce the magnetic moment is subject to the condition $\dot{\mathbf{s}} + \mathbf{\Omega} \times \mathbf{s} = \mathbf{0}$, or $\dot{\mathbf{s}} = \mathbf{s} \times \mathbf{\Omega}$, in a coordinate system fixed in the rotating sample. The right-hand side of the last equation is equal to the torque $\mathbf{m} \times \mathbf{H}$ which acts on $\mathbf{s}$, so that the field $\mathbf{H} = \gamma^{-1}\mathbf{\Omega} = 2m_0c\mathbf{\Omega}/eg$ is applied to the sample. Therefore, if we measure the magnetization due to the high-speed rotation, and compare it with the magnetization due to the application of the magnetic field, we can find the field $\mathbf{H}$ which is equivalent to $\mathbf{\Omega}$. Then γ or g can be calculated from the ratio of $|\mathbf{H}|$ to $|\mathbf{\Omega}|$. The experiment involves considerable difficulties. Table 4.4.1 shows some experimental results. The quantity $\varepsilon = (2 - g)/2$ is a measure of the difference between the observed value of g and that in the case of spin alone.

4.5. Different Types of Magnetism

Paramagnetic substances are attracted to a magnetic field, whereas diamagnetic substances—with a characteristically small magnetic susceptibility χ^B—are repelled by a magnetic field. Brought below the so-called Curie temperature—which may amount to several hundreds of degrees Centigrade, depending on the material—magnetic materials are subject to an extremely strong type of paramagnetism (large positive χ^B) known as *ferromagnetism*. This can occur in a material containing atoms with net magnetic moments, subject to quantum-mechanical electrostatic "exchange" forces, holding neighboring atomic moments parallel. Ferrimagnetic materials (e.g., ferrites) are magnetic materials capable of combining the resistivity of a good insulator with a high permeability (Snoeck, Néel). *Antiferromagnetic* materials do not present external permanent magnetic moments, because neighboring magnetic spins are antiparallel in such materials.

The set of magnetic moments in a material is often referred to as the *spin system*. Like other statistical systems, spin systems tend to be ordered at low temperatures and disordered at high temperatures. Leaving aside the case of diamagnetism, we focus our attention on the case of materials whose molecules, at the microscopic level, possess permanent magnetic moment.

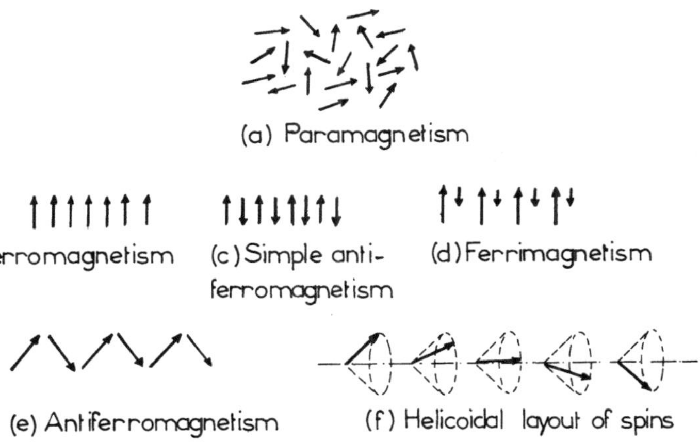

Figure 4.5.1. Possible ordered arrangements of electronic spins.

A. Paramagnetism

At high temperatures, thermal agitation hinders any global ordering of the magnetic moments (see Figure 4.5.1(a)), and the spin system that corresponds to paramagnetism is in complete disorder. The temperature dependence of the magnetic susceptibility is then established as follows: Let the paramagnetic substance have a certain concentration of N ions, of spin S (in $\hbar$ units) per unit volume. Placed in a magnetic field $\mathbf{H}$, the ion of magnetic moment $\mathbf{m}$ develops an energy $\varepsilon = -\mathbf{m}\cdot\mathbf{H} = -mH\cos\phi$, where ϕ is the angle between $\mathbf{m}$ and $\mathbf{H}$. Then the total magnetization per unit volume is $M = Nm\langle\cos\phi\rangle$, where $\langle\ldots\rangle$ indicates the statistical average over the discrete orientations allowed quan-

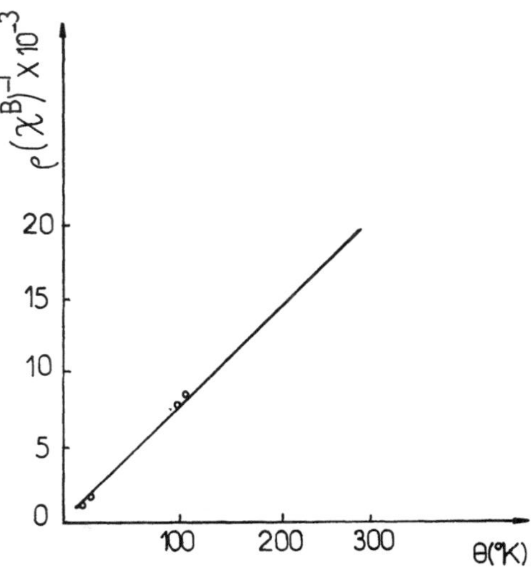

Figure 4.5.2. Reciprocal paramagnetic susceptibility of powdered $CuSO_4 \cdot K_2SO_4 \cdot 6H_2O$ versus temperature (after Hupse, *Physica*, **9**, 633, 1942).

tum mechanically. Then a celebrated calculation performed by Langevin (in the nonquantum case) yields χ^B in the form of *Curie's law*:

$$\chi^B = C/\theta, \tag{4.5.1}$$

where C is Curie's constant such that

$$C = Ng^2S(S+1)\mu_B^2/3k_B, \tag{4.5.2}$$

where μ_B is Bohr's magneton, introduced in Chapter 2. A typical curve for χ^B is given in Figure 4.5.2.

B. Ferromagnetism

When most paramagnetic materials are brought below the critical Curie temperature θ_C, then the thermal agitation is no longer sufficient to destroy the spin ordering that gives rise to ferromagnetism (see Figure 4.5.1(b)). There may exist a *spontaneous magnetization* in the absence of an applied magnetic field. It is an experimental fact, that it is possible to change the overall magnetization of a suitably prepared specimen, from an initial value of zero (in the absence of an applied field) to a saturation value of the order of 1,000 Gauss, by application of a minute field whose strength may be of the order of 0.01 Oe. Magnetic moments and the magnetic field are not necessarily aligned and we cannot have a proportionality relationship of the type $\mathbf{M} = \chi^B\mathbf{H}$. Figures 4.5.3 gives the value of the magnetization as a function of temperature

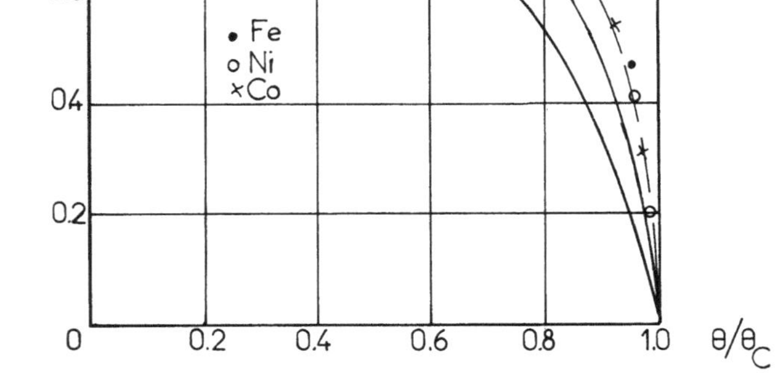

Figure 4.5.3. Spontaneous magnetization versus temperature in ferromagnets (S: angular momentum of atoms in $\hbar$ units). Reprinted with permission from J.H. van Vleck, *The Theory of Electric and Magnetic Susceptibility*, 1931, Oxford University Press.

for ferromagnetic iron, nickel, and cobalt. The phenomenon of spontaneous magnetization can be observed in ferromagnetic samples on a scale corresponding to visual observation with a microscope. Attention, then, is directed at the *local* magnetization over distances of the order of 10^{-5} to 10^{-2} cm, small in comparison with specimen size but still large in comparison with the lattice spacing. On this scale, under a uniform temperature, spontaneous magnetization has a magnitude independent of position, but a direction that can vary from one point to another. We say that the specimen is *magnetically saturated.* At a low temperature, however, some ferromagnetic samples show a tendency to break up into domains, the so-called *magnetic domains* of Weiss [1907]. This fact can explain the possibility of a demagnetization, i.e., the fact that a specimen as a whole presents a negligibly small magnetization. In each domain separated from its neighbors by *walls* (Bloch [1930]) (more realistically discontinuity surfaces which may be considered as transition zones) the spontaneous magnetization is constant in magnitude and direction. However, domain structure is not universal, and we may assume, in a more general way from a phenomenological viewpoint, that the magnetization density (or its director cosines) varies smoothly with position. Domain theory is then replaced by the theory of *micromagnetics* or *micromagnetism,*[12] whose main object is the study of the change in direction of the magnetization under different influences, one of which is that of the interactions responsible for the ferromagnetic behavior. This refers to classical ferromagnetism at low temperatures, much below the Curie transition point.

We now review, briefly, the essentials of the theory of ferromagnetism. The Curie temperature separates the disordered paramagnetic phase ($\theta > \theta_C$) from the ordered ferromagnetic phase ($\theta < \theta_C$) (see Figure 4.5.4). For instance, $\theta_C \simeq 1{,}000$ K for iron, 1,388 K for cobalt, and 627 K for nickel. Pierre Weiss [1907] realized that the difficulty caused by thermal agitation could be circumvented if one postulated that, in ferromagnetic materials, there exists a powerful "*molecular field,*" due to the mutual interaction between electrons, which would tend to line up the magnetic moments parallel to one another. This field also referred to as the exchange field, can be considered to be equivalent to a magnetic field $\mathbf{H}^{ex}$. Note however, that $\mathbf{H}^{ex}$ is *not* a real magnetic field, in that it does not enter Maxwell's equations, and no current density can be associated with it. Its intensity can reach 10^7 gauss and is some 10^4 times greater than the magnetic field of the magnetic dipoles in a ferromagnet. In order to relate $\mathbf{H}^{ex}$ and the magnetization, Pierre Weiss used a very simple idea. The more strongly aligned are the neighbors of a magnetic spin, the greater will be the tendency of this peculiar spin to align parallel with the others. Hence, the molecular field is an increasing function of the magnetization. The simplest relation which can represent this fact is a relation of

[12] See Brown [1963], [1966], Shtrikmann and Treves [1963], Doring [1966], Eringen and Kafadar [1970], Maugin and Eringen [1970a] and Maugin [1976a].

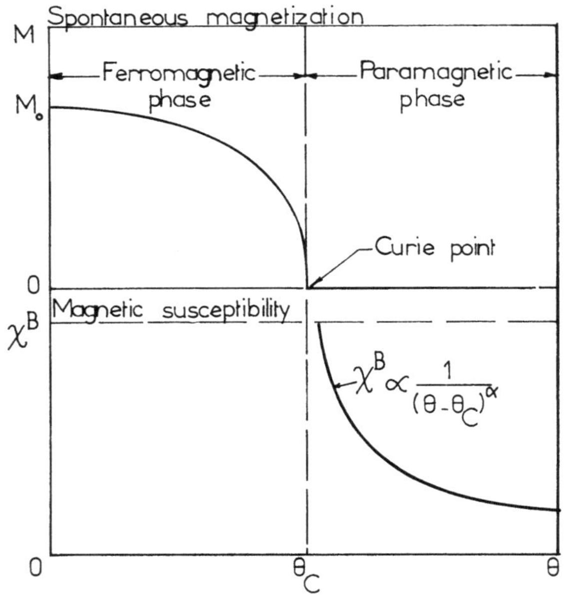

Figure 4.5.4. Schematic curves for the spontaneous magnetization and the magnetic susceptibility versus temperature in ferromagnets.

proportionality

$$\mathbf{H}^{ex} = \lambda \mathbf{M}, \tag{4.5.3}$$

where λ is a constant, independent of temperature. The magnetization field $\mathbf{M}$ used in (4.5.3) is the value at thermal equilibrium in the field $\mathbf{H}^{ex}$. If the body possesses domains, we, of course, refer to the mean value of the magnetization within the domain, so that (4.5.3) is valid within the approximation of the *mean field*. It follows from (4.5.3) that, if there is an applied field $\mathbf{H}$, the effective field acting upon an elementary coil is, according to Weiss,

$$\mathbf{H}^{eff} = \mathbf{H} + \lambda \mathbf{M}. \tag{4.5.4}$$

The main problem, however, is to relate λ to the Curie temperature θ_C. According to the hypothesis of the molecular field, (4.5.1) must be rewritten in the form

$$\chi^B = C/(\theta - C\lambda). \tag{4.5.5}$$

We see that the susceptibility presents a singularity for $\theta = C\lambda$ (see Figure 4.5.4). At this temperature and below, if χ^B is infinite, we may have a finite spontaneous magnetization for vanishing $\mathbf{H}$. Equation (4.5.5) can be rewritten in the form of the so-called *Curie–Weiss law*:

$$\chi^B = \frac{C}{\theta - \theta_C}, \qquad \theta_C = C\lambda. \tag{4.5.6}$$

In fact, more detailed computations using the theory of second-order phase transitions show that we have $\chi^B \propto (\theta - \theta_C)^{-\alpha}$, where the *critical* exponent α is equal, for instance, to 1.33 in iron. Equation $(4.5.6)_1$ provides the susceptibility for values of θ close to, but greater than, θ_C, and is in good agreement with experimental data. The value of λ is found from $(4.5.6)_2$ and (4.5.2). For iron, $\theta_C \simeq 1{,}000$ K, $g \simeq 2$, $S \approx 1$, hence $\lambda \approx 5{,}000$, and $|H^{ex}| \sim k_B\theta/\mu_B \approx 10^{-16} \times 10^3 \times 10^{20} = 10^7$.

If $\theta < \theta_C$, the approximations made to establish (4.5.5) are no longer valid, and the magnetic moment is no longer a linear function of the field $\mathbf{H}$. The molecular field outweights the thermal fluctuations, so that the specimen is ferromagnetic.

The above calculation is purely phenomenological. In fact, Weiss' molecular field gives an approximate representation of the *exchange interaction* of quantum mechanics. Under certain assumptions, it can be shown that the interaction energy between two atoms, located at sites α and β, respectively, and bearing spins $\mathbf{S}^\alpha$ and $\mathbf{S}^\beta$ (in $\hbar$ units), contains a term of the following form:

$$W^{(\alpha\beta)} = -2J\mathbf{S}^\alpha \cdot \mathbf{S}^\beta, \tag{4.5.7}$$

where J is the so-called *exchange integral*, that is related to the overlapping of the charge distributions of atoms α and β (electrostatic energy). Equation (4.5.7) represents *Heisenberg's model*.[13] Of course, J is a function of the radius vector $r^{\alpha\beta}$ joining the sites α and β. However, it must be remarked that the exchange energy has the following important property: in contrast to the Coulomb interaction, it decreases more rapidly (exponentially) with increasing distance between the nuclei, so that J is of appreciable size only for its nearest neighbors. For a system of N spins of value $\mathbf{S}$, set aligned or in a circle, each one linked to its neighbors by the Heisenberg interaction (4.5.7), we have by summation

$$W^{ex} = -2J \sum_{\alpha \neq \beta} \mathbf{S}^\alpha \cdot \mathbf{S}^\beta. \tag{4.5.8}$$

It is possible to establish a simple relationship between λ and J:[14] in iron, with $S \simeq 1$, $\theta_C \simeq 1{,}000$ K, it is found that $J = 1.19 \times 10^{-2}$ eV. A simple "continuous" representation of W^{ex} can be obtained from the above-given expression.[15] It is shown in a *rigid* ferromagnet that the exchange interactions are represented by the contribution of the spatial disuniformities of the magnetization (or of its director cosines at saturation) in the free energy, at any temperature. For instance, it is possible to write this contribution in the form

$$W^{ex} = \int_{\Delta V} \Psi^{ex}\, dv, \qquad \Psi^{ex} = \tfrac{1}{2}\alpha_{ij}\mathbf{M}_{,i} \cdot \mathbf{M}_{,j}, \tag{4.5.9}$$

[13] Heisenberg [1928]; also Dirac [1929].

[14] See Kittel [1971, Chap. 16].

[15] See Akhiezer et al. [1968, p. 15]. For the case of nonlinear elastic solids, see Chapter 9 (vol. II).

where ΔV is a volume small in comparison with usual specimen sizes, but still large in comparison with the atomic scale (microelement). The symmetric second-order tensor α_{ij} defined as

$$\alpha_{ij} = \frac{1}{2(2\mu_B)^2} \int_{\Delta V} \mathscr{J}(\mathbf{r}) x_i x_j \, dv = \alpha_{ji} \tag{4.5.10}$$

may be referred to as the *exchange-modulus* tensor of the ferromagnet. The quantity $\mathscr{J}$ is closely related to the exchange integral J for $\theta \ll \theta_C$ and, like the latter, falls of rapidly with distance. Equation (4.5.9) is the expression introduced by Landau and Lifshitz [1935] in their pioneering phenomenological theory of rigid ferromagnets. We note that the components α_{ij} may depend on the modulus of $\mathbf{M}$ and on the temperature, but the exchange interactions do *not* depend on the direction of $\mathbf{M}$. The order of magnitude of the components α_{ij} can be found from (4.5.10). Let a be a typical crystal lattice constant. Then $|\alpha_{ij}| \simeq Ja^5/\mu_B^2$ if J is the magnitude of the exchange integral between neighboring atoms. Assuming that the magnetization M_S, in the ground state (saturation), is such that $M_S \approx \mu_B/a^3$, and noting that, in energy units ($1 \text{ eV} = 1.1605 \times 10^4$ K), $J \approx \theta_C$, we have

$$|\alpha_{ij}| \approx \theta_C a^2/\mu_B M_S. \tag{4.5.11}$$

Remark finally that J, hence $\mathscr{J}$, must be positive in order that the magnetic spins tend toward parallel alignment (ferromagnetic case). Thus, the quadratic form $(4.5.9)_2$ must be positive definite.

Ferromagnetic bodies also exhibit *directional* properties, expressed by means of the *anisotropy* or *magnetocrystalline* energy, in such a way that the magnetization tends to be directed along certain definite crystallographic axes, the directions of so-called *easy* magnetization. The origin of the anisotropy energy is believed to be the result, at the microscopic level, of the combined effects of *spin–orbit interactions* and the partial quenching of the orbital angular momentum by inhomogeneous crystalline electric fields and orbital exchange interactions with neighboring atoms. In fact, if we examine the magnetic dipole interactions at this level, we find that they consist of: (i) purely magnetic dipole interactions between the magnetic moments of the atoms; and (ii) interactions between the magnetic moments and the electric field of the crystal lattice (spin–orbit interactions). In comparison with the above-described exchange interactions, these interactions have a relatively small magnitude, since they are of the order of $W^{ex} \times (v_e/c)^2$. For this reason, it is often said that they are of "relativistic" origin. However, in spite of their relative weakness in comparison with the exchange interactions, they do play an important role in ferromagnetic materials. The reason for this is two-fold. First, these interactions produce in the crystal a preferred direction of magnetization corresponding to the minimum energy of the ferromagnet. Thus, they give rise to the *anisotropy energy* referred to above, i.e., the dependence of the energy of the ferromagnet upon the direction of the magnetization vector (a fact not accounted for by the exchange energy, as has already been remarked). Second, it is these interactions that provide the coupling between

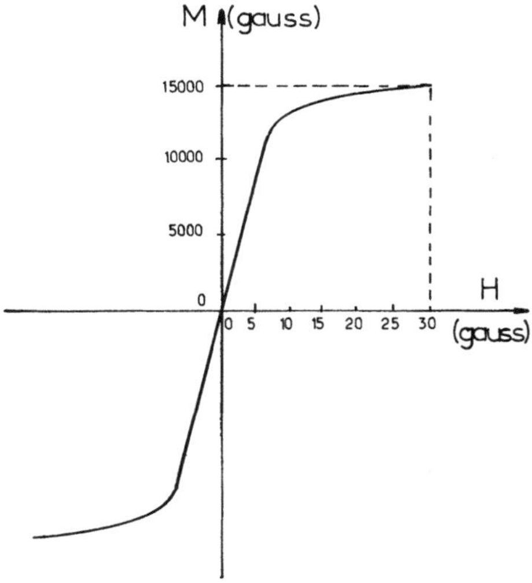

Figure 4.5.5. Magnetization curve for magnetically soft iron (H and M in gauss).

external heat sources and the spin system of the ferromagnet. If such interactions between the spins and the lattice vibrations did not exist, statistical equilibrium could not be established between the heat source and the spin system.

As in the case of ferroelectrics, ferromagnetic materials display a phenomenon of hysteresis, which can be used for the stockage of binary information in computer memories. If the magnetic hysteresis effect is negligible, then we shall say that the material is a *soft ferromagnetic* material (see Figure 4.5.5). Otherwise, the material is said to be *magnetically hard* (see Figure 4.5.6). In

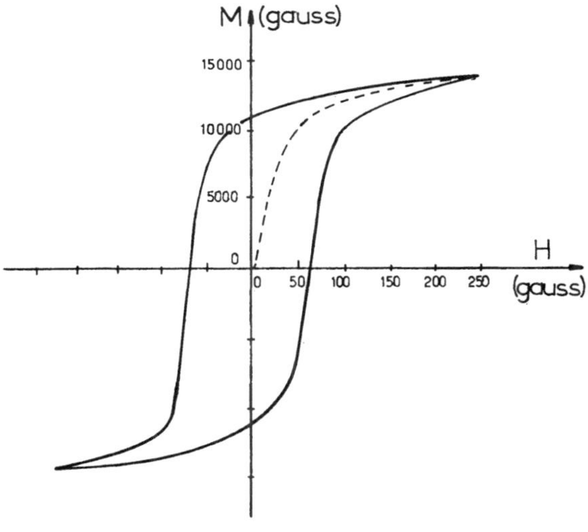

Figure 4.5.6. Magnetization curve for magnetically hard steel.

the latter case, the steep parts of the hysteresis curve are a smooth representation of a semimicroscopic phenomenon called *Barkausen jumps*, which result from successive jerky displacements of domain walls usually anchored on structural defects. The top and bottom parts of the hysteresis curves represent a different process of magnetization (rotation of magnetic dipole). The parameters of the hysteresis curve strongly depend on the state of stresses in the body (see Vonsovskii [1975] and Köszegi and Krönmuller [1984]).

C. Ferrimagnetism and Antiferromagnetism

Ferromagnetism is not the only magnetic order–disorder phenomenon. Indeed, a long-range magnetic ordering is also achieved in ferrimagnets and antiferromagnets (see Figure 4.5.1). Other orders still are possible (for instance, the helicoidal—thus noncoplanar—layout of spins as shown in Figure 4.5.1f).

The word *ferrimagnetism* was coined by Louis Néel [1948] to describe the properties of those substances which, below a certain temperature, exhibit spontaneous magnetization arising from a *nonparallel* alignment of atomic magnetic moments. Among materials possessing this property are the ferrites. The term ferrites denotes a group of iron oxides which have the general formula $MO \cdot Fe_2O_3$, where M is a divalent metal ion. The resultant magnetic moment in a ferrimagnetic sample is partitioned into different magnetic *sublattices*. Simple *antiferromagnetic* samples (see Figure 4.5.1(c)) are not strongly magnetic below the critical temperature because the magnetization densities $\mathbf{M}_A$ and $\mathbf{M}_B$, of the two sublattices A and B, then cancel each other. We have paramagnetism instead of ferromagnetism. However, this paramagnetism is different from the usual one, in that its magnetic susceptibility does not obey Curie's law (see Figure 4.5.7). Unlike those in a ferromagnet, the well-ordered spins in an antiferromagnet cannot be observed as a magnetization; but their existence is demonstrated by neutron diffraction. In fact, antiferromagnetism is the most common form of magnetic order. It is found in almost all inorganic compounds of the transition metals, rare earths, and actinide elements. It is also found in Cr, Mn, Pt, Pd, and rare earth metals and alloys.

The mechanisms responsible for the antiferromagnetic behavior may be sketched out as follows: exchange forces in magnetic materials can operate to hold neighboring moments *antiparallel*, rather than parallel. The sign of the exchange force may depend, among other things, on the atomic spacing. For example, metallic manganese is antiferromagnetic, while many alloys of manganese (in which the average Mn–Mn distance is greater than that of metallic manganese) are ferromagnetic. In some antiferromagnetic compounds the exchange interaction appears to be of the next-nearest neighbor type, taking place through an intervening atom such as oxygen. This type of interaction is termed *superexchange*. This concept of indirect exchange was put forth by Kramers in 1934, and developed in form known as superexchange by several authors (van Vleck, Anderson, ...) in the fifties. Indeed, this indirect exchange mechanism, in which the magnetic ions are coupled through electron transfer with intermediate anions (O, F, S, etc.), arises in both antiferromagnetic and

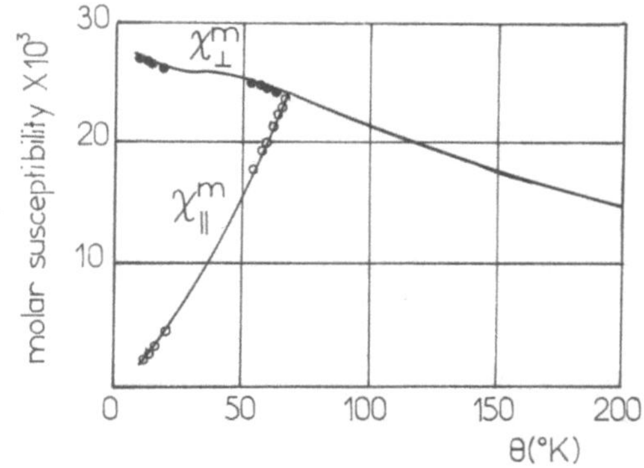

Figure 4.5.7. Magnetic susceptibility of antiferromagnetic MnF_2 (after Griffel and Stout, *J. Chem. Phys.*, **18**, 1455, 1950).

ferrimagnetic insulators. Antiferromagnetic materials, having no net external moment, then show small positive susceptibilities that reach a maximum at the temperature above which the exchange forces and the superexchange mechanisms can no longer hold the moments in ordered arrangements, against thermal agitation. This temperature, Néel's temperature θ_N, corresponds to the Curie temperature of a ferromagnet. Thus, when the temperature decreases from the Néel temperature, the spontaneous antiparallel alignment of neighboring electron spins takes over from the paramagnetic case. The strength of the ordering interaction is characterized by the magnitude of θ_N (e.g., 520 K, 310 K, and 67 K for NiO, Cr_2O_3, and MnF_2, respectively). Since an antiferromagnetic arrangement of spins has no net spontaneous magnetization, the substance, in general, does not exhibit any kind of ferromagnetism. The tendency to maintain the antiferromagnetic order, however, opposes the magnetization, due to the application of an external field and, accordingly, gives rise to a characteristic temperature dependence of the susceptibility. In particular, antiferromagnetic materials show a decrease of susceptibility with a decrease of temperature below θ_N (the curve depends on the orientation of the field relative to the spin direction, see Figure 4.5.7). Above θ_N, the susceptibility always decreases with an increase in temperature as in the case of paramagnetic materials. Thus, a peculiar feature of antiferromagnetism, which distinguishes it from paramagnetism, is that its susceptibility versus temperature curve displays a kink at the transition point $\theta = \theta_N$.

On occasion, the spontaneous magnetization of one spin sublattice has a different intensity from that of another spin sublattice (see Figure 4.5.1). In this case, the spontaneous magnetizations may not cancel each other completely, leaving some net bulk magnetic moment. In effect, this gives a particular type of ferromagnetism, which is called *ferrimagnetism.*

D. Magnons

Because of their magnetic ordering, ferromagnetic (and ferrimagnetic) materials can transmit excitations of a purely magnetic nature, with propagation

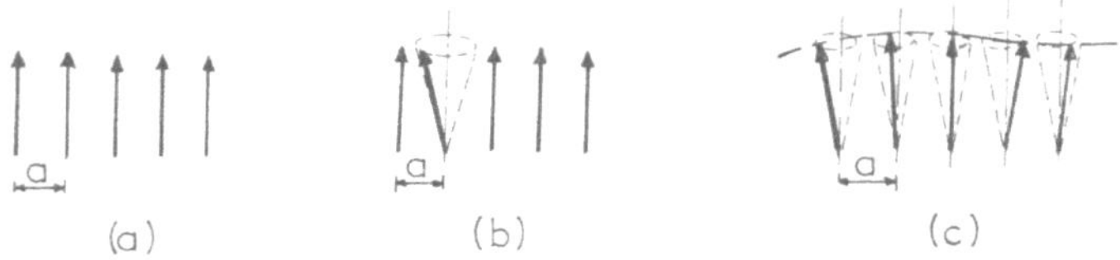

Figure 4.5.8. Elementary mechanism of the excitation of a collective mode of magnons.

effects similar to those of electromagnetic and acoustic excitations. Such excitations are called *spin waves* and were discovered by Bloch [1930]. In the classical representation, magnetic spins precess on a linear chain of atoms, with the same angle, with respect to the internal magnetic field. Spin waves are determined by a spatial shift in the phase of neighboring spins. The existence of this particular type of collective mode in ferromagnetic crystals is best understood, if one recalls the strong correlation between the direction of atomic spins in magnetically ordered crystals. The origin of these modes is easily understood by examining the following simple process. Consider a ferromagnet, in its ground state, at $\theta = 0$ K. All the atomic magnetic moments are then aligned, so that the energy of the ferromagnet is minimized (see Figure 4.5.8(a)). Now deflect the magnetic moment of a particular atom and let it go. It will start to precess about the local effective magnetic field (Figure 4.5.8(b)). However, owing to the presence of the exchange interactions between neighboring spins, the change in direction will not remain localized at the original atom, but will propagate through the crystal, in the form of a mode of wave motion (Figure 4.5.8(c)) called a *spin wave*. We may have transverse or longitudinal spin waves (see Figure 4.5.9). We see that spin waves may be regarded as oscillations in the magnetic moment density, that propagate through a magnetically ordered crystal. Since this ordering is primarily a quantum mechanical effect, the assumptions under which a macroscopic phenomenological study can be carried out ought to be specified. In order that a spin wave propagate through a crystal, it is clear that the spin-wave energy must be equal to the excitation energy of the crystal required to cause the change in the orientation of the spins. Let $\omega_S(\mathbf{k})$ be the frequency of the spin wave, as a function of its wave vector $\mathbf{k}$. The energy of an elementary excitation connected with the spin wave is then given by (quantization of the energy) $\varepsilon_S(\mathbf{k}) = \hbar\omega_S(\mathbf{k})$. The quantity $\mathbf{p} \equiv \hbar\mathbf{k}$, defined in accordance with de Broglie's relation of wave mechanics, can be interpreted as the *momentum* of the spin wave. Hence, a fictitious particle, called *magnon*, can be associated with a spin wave. If the total excitation energy of the ferromagnet is small, then it can be written as the sum of the energies of the individual magnons that propagate through it. That is,

$$E_S(n) = \sum_{\mathbf{k}} \hbar\omega_S(\mathbf{k})n(\mathbf{k}), \tag{4.5.12}$$

where $n(\mathbf{k})$ is the number of magnons with wave vector $\mathbf{k}$, and the summation

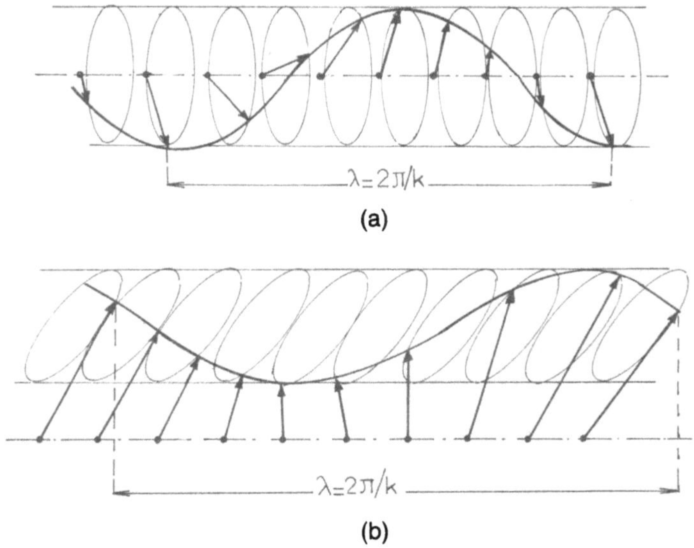

Figure 4.5.9. Schematic view of spin waves: (a) longitudinal spin waves; (b) transverse spin waves.

is carried over all **k**'s. In writing the expression (4.5.12), we have made use of the idea of a *perfect magnon* gas, valid only if the excitation energy of the ferromagnet is small. But low excitation energies correspond to low temperatures, $\theta \ll \theta_C$. That is, (4.5.12) can be used provided $\theta \ll \theta_C$. Using a temperature scale, the mean magnon energy is of the order of the temperature: $\overline{\varepsilon_S(\mathbf{k})} \approx \theta$. Let θ_C^* be the critical temperature for the excitation of the crystal lattice, with corresponding excitation energy of the order $\theta_C^*(ak)^2$, where a is a lattice constant. Since magnons propagate if their excitation energy is equal to that of the crystal lattice, we have $\theta \approx \theta_C^*(ak)^2$. Since $\theta_C \approx \theta_C^*$ and $\theta \ll \theta_C$, it follows that $ak \ll 1$ or $\lambda \gg a$, where $\lambda = k^{-1}$ is the magnon wavelength. Thus, at low temperatures, the magnon wavelength is large compared to the lattice constant. In other words, for $\theta \ll \theta_C$, the magnons excited in a ferromagnet are mainly *long-wave* magnons, and long-wave oscillations of magnetic moments can be described purely phenomenologically, by means of macroscopic electrodynamics.

By the way, we have shown that the dispersion law of magnons in ferromagnets is of the type $\omega_S(\mathbf{k}) \approx \mathbf{k}^2$ (see Figure 4.5.10), so that long-wave magnons are dispersive in ferromagnets. If the same argument is used in antiferromagnets, it is found that $\omega_S(\mathbf{k}) \approx \mathbf{k}$ (Smolenskii [1974], Maugin [1981a]). Finally, it must be remarked that, similar to electromagnetic and acoustic waves, spin waves can suffer a damping. This phenomenon will be examined in Chapter 9 for the case of deformable ferromagnets.

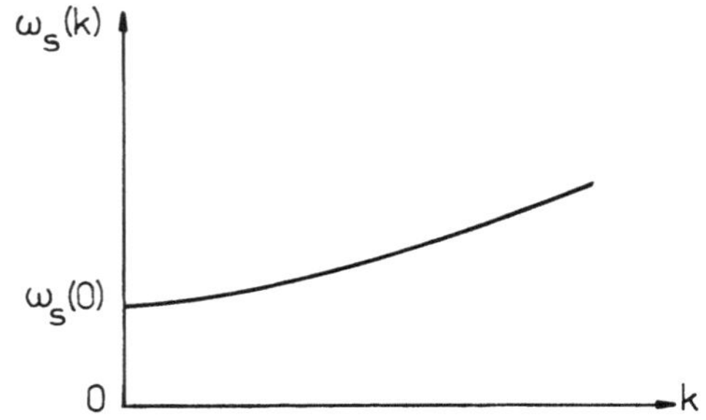

Figure 4.5.10. Dispersion relation for pure spin waves in magnetostatics in ferromagnets.

E. Magnetostriction and Piezomagnetism

Similar to the phenomena of electrostriction and piezoelectricity in deformable dielectrics, magnetic materials can display the properties known as magnetostriction and piezomagnetism. That is, the magnetization of a magnetizable, deformable specimen causes a deformation independent of the sign of the applied magnetic field. This second-order phenomenon (in the field) is called *magnetostriction.* Conversely, in certain non-centrosymmetric crystals, a strain may induce a magnetization. This is piezomagnetism. Moreover, if the specimen is a magnetically ordered crystal (ferromagnet, ferrimagnet), then oscillations in the strain field may be coupled with those in the spin system via the piezomagnetic effect. This coupling phenomenon between elastic and spin waves is known as the *phonon–magnon* coupling.[16] This subject is the main concern of Chapter 9. Finally, we note that the phenomenon of *pyromagnetism,* the analogue of pyroelectricity, is seldom exhibited in magnetic materials.

4.6. Electric Conduction

A. Electron Conduction

Here we discuss briefly electric conduction in metals. Semiconductors are not within the scope of the present work. A short account on superconductivity is to be found in Chapter 14.[17]

The physical model for electrical conduction in metals is the *free electron model*: each electron in motion changes its direction of propagation with an average period of 2τ. This change in direction is not due to the collision of the electron with other electrons, but to the interaction between the electron and a lattice vibration, or a collision with an impurity atom or with a local distortion or defect of the crystal. The time τ is called the mean *life-time* or

[16] *Phonons* are the fictitious particles associated with elastic vibrations via de Broglie's relation. See Kittel [1971, Chap. 5].

[17] For a solid-state approach to these conduction behaviors, see Kittel [1971, Chaps. 11 and 12].

relaxation time of the electronic motion. If an electric field E (in a rigid stationary body) is applied to the electron, then its acceleration is eE/m_0. Thus, the velocity increment of the electron during a time lapse 2τ is $(eE/m_0)2\tau$, and its time average is $(eE/m_0)\tau$. We assume that the electron forgets its history on each collision, so that the velocity increment it acquired before a collision is not maintained after the collision. If n is the density of electrons, then the electron current density is expressed as the product of ne and the average velocity $\bar{v} = (eE/m_0)\tau$. That is,

$$\mathscr{J} = \sigma E, \tag{4.6.1}$$

where $\sigma = ne^2\tau/m_0$ is the *electrical conductivity*. The quantity $1/\sigma$ is called the resistivity. Equation (4.6.1) represents the celebrated *Ohm's law*. If the crystal is free of defects and impurities, nothing obstructs the rectilinear motion of the electron at absolute zero and $\tau \to \infty$, so that $\sigma = \infty$. The medium is then said to be a *perfect* conductor. At high temperatures, if the properties and concentration of the lattice defects are independent of temperature, we have the empirical relation known as *Matthiesen's rule* (a, b: constants),

$$\sigma^{-1} = a + b\theta. \tag{4.6.2}$$

It is well known that the better the electrical conduction of a metal, the better its thermal conduction. This suggests that, in metals, electrons carry heat. In fact, assuming Fourier's law of heat conduction with thermal conductivity κ, we have the *Weidemann–Franz* law:

$$\frac{\kappa}{\sigma} = \frac{\pi^2}{3}\left(\frac{k_B}{e}\right)^2 \theta. \tag{4.6.3}$$

The constant $\kappa/\sigma\theta$ is called the Lorentz number.

The linear relationship (4.6.1) can be written in vectorial form $\mathscr{J} = \sigma\mathbf{E}$. Furthermore, in a moving conductor, $\mathbf{E}$ must be replaced by the electromotive intensity $\mathscr{E}$. That is,

$$\mathscr{J} = \sigma\mathscr{E}. \tag{4.6.4}$$

While this form of Ohm's law is demonstrated to be valid for a large class of conductors (i.e., not only metals, but also fluids), it must be remarked that it represents an *isotropic* property, and assumes no couplings with other effects (i.e., heat conduction, magnetic fields). The latter effects known as thermoelectric, galvanomagnetic, and thermomagnetic require the developement of a complete *phenomenological* constitutive theory.[18] Below we discuss these effects briefly.

B. Cross Effects

In metals, electrical and thermal conductions are intimately related effects. With the absence of an externally applied magnetic field and spontaneous

[18] See Chapters 5 and 7 for such a development.

magnetization, assuming isotropy, we may envisage linear constitutive equations, for the conduction current $\mathscr{J}$ and the heat flux vector $\mathbf{q}$, in the form:[19]

$$\begin{aligned} \mathscr{J} &= \sigma_1 \mathscr{E} + \sigma_2 \nabla\theta, \\ \mathbf{q} &= \kappa_1 \mathscr{E} + \kappa_2 \nabla\theta. \end{aligned} \tag{4.6.5}$$

When $\sigma_2 \neq 0$, a temperature gradient produces a conduction current which is referred to as the *Thompson effect*. Similarly, when $\kappa_1 \neq 0$, an electric field causes a heat flux which is referred to as the *Peltier effect*. The solid-state physics basis for these two thermoelectric effects is given by Wilson [1953]. They can also be related to the phenomenon of the thermocouple (junction of two metals at different temperature) in which a thermoelectric power (or *Seebeck* effect) is developed.[20]

The number of cross effects, resulting from the interference of heat and electric conduction, is increased when an external magnetic field $\mathbf{H}$ is present. The new effects are called *galvanomagnetic effects* when they are caused by an electric field, and *thermomagnetic effects* when they are caused by a temperature gradient. For instance, it is possible that a magnetic field which is not set parallel to the electric field causes both electrical and heat conduction, thus giving rise to the so-called *Hall* and *Ettingshausen effects*, respectively. Similarly, a magnetic field and a temperature gradient which are not parallel to one another may result in the production of electrical and heat conduction, thus giving rise to the so-called *Nernst* and *Righi–Leduc effects*. Finally, we have second-order effects in the magnetic field, which may exist when the electric field and the magnetic field, and the magnetic field and the temperature gradient, respectively, are not orthogonal to one another. If all of the above effects are present, then (4.6.5) are replaced by equations of the form

$$\begin{aligned} \mathscr{J} &= \sigma_1 \mathscr{E} + \sigma_2 \nabla\theta + \sigma_3 \mathscr{E} \times \mathbf{H} + \sigma_4 \nabla\theta \times \mathbf{H} + \sigma_5 (\mathbf{H}\cdot\mathscr{E})\mathbf{H} + \sigma_6 (\mathbf{H}\cdot\nabla\theta)\mathbf{H}, \\ \mathbf{q} &= \kappa_1 \mathscr{E} + \kappa_2 \nabla\theta + \kappa_3 \mathscr{E} \times \mathbf{H} + \kappa_4 \nabla\theta \times \mathbf{H} + \kappa_5 (\mathbf{H}\cdot\mathscr{E})\mathbf{H} + \kappa_6 (\mathbf{H}\cdot\nabla\theta)\mathbf{H}, \end{aligned} \tag{4.6.6}$$

where the σ_α and κ_α, $\alpha = 1, 2, \ldots, 6$, are scalar coefficients which are not necessarily pure constants. Both equations (4.6.6) are jointly linear in the fields $\mathscr{E}$ and $\nabla\theta$, the field $\mathbf{H}$ acting as a parameter. It will be shown in Chapter 5 that (4.6.6) are but special representations, for *isotropy*, of general constitutive equations. Equations (4.6.6) place in evidence the following properties.

For instance, whereas σ_1 corresponds to an isotropic (independent of $\mathbf{H}$) electrical conduction or *Ohm effect*, σ_3 corresponds to an electrical conduction orthogonal to the direction of $\mathbf{H}$, and σ, to an electrical conduction parallel to the direction of $\mathbf{H}$. The same remark holds true with regard to the coeffi-

---

[19] The classical theory of irreversible processes indicates that the coefficients σ_2 and κ_1 are related via the so-called Onsager reciprocity relations (see de Groot and Mazur [1962, p. 350]).

[20] See de Groot and Mazur [1962, p. 351].

cients κ_2, κ_4, and κ_6 for heat conduction, and the coefficients σ_2, σ_4, and σ_6 on the one hand and κ_1, κ_3, and κ_5 on the other. Finally, because the entropy productions associated with electrical and heat conduction are $\theta^{-1}\mathscr{J}\cdot\mathscr{E}$ and $\theta^{-1}\mathbf{q}\cdot\nabla\theta$, respectively, we see that the contributions pertaining to the Hall effect and to the Righi–Leduc effect do not produce entropy. These are *gyroscopic thermodynamically hidden effects.*

Although the Peltier and Thompson effects cannot be deduced from an analysis based only on a simple spherical Fermi distribution of free electrons, such an analysis is possible for the Hall effect. In fact, the following simple model allows us to relate the coefficient σ_3 to microscopic quantities. Let an electric current $\mathscr{J}$ flow in the x-direction, and a magnetic field H be applied in the z-direction. Then, according to $(4.6.6)_1$, an electric field E_y is found to be produced in the y-direction. The quantity R, defined by

$$R = \frac{1}{H\sigma_3} = \frac{E_y}{\mathscr{J}H}, \tag{4.6.7}$$

is called the Hall constant. But the y-component of the Lorentz force acting on an electron in the above conditions is ${}_{\mathrm{L}}f_y = -eE_y + (e/c)v_x H$.

Since there is no current in the y-direction, the average value $\langle {}_{\mathrm{L}}f_y \rangle$ is zero. But, according to a previous calculation, $\mathscr{J} = -en\langle v_x \rangle$, if n is the electron density. Therefore, $\langle {}_{\mathrm{L}}f_y \rangle$ yields $E_y = -\mathscr{J}H/cne$, and (4.6.7) yields

$$R = -\frac{1}{cne}, \qquad \sigma_3 = -\frac{cne}{H}. \tag{4.6.8}$$

Thus, R is inversely proportional to the electron density, and σ_3 is proportional to n and inversely proportional to the strength of the applied magnetic field.

4.7. Interaction Between Light and Electromagnetic Continua

A. Electromagnetic Optics[21]

When light is considered as an electromagnetic oscillation, the subject matter of the interaction between light and electromagnetic continua is called *electromagnetic optics.* By its very nature, optics is a dynamical phenomenon. Electromagnetic bodies which can transmit light without energy loss are said to be *transparent.* Practically all optical properties of an electromagnetic continuum are contained in a relationship between the electric displacement $\mathbf{D}$ and the electric field $\mathbf{E}$ of the form

$$\mathbf{D} = \boldsymbol{\varepsilon}\mathbf{E}, \tag{4.7.1}$$

[21] For a comprehensive approach to electromagnetic optics we refer the reader to classical treatises, such as Landau and Lifshitz [1960], Born [1972], and Klein [1970].

where $\boldsymbol{\varepsilon}$ is a second-order symmetric tensor called the *dielectric tensor*. Thermodynamic considerations show that $\boldsymbol{\varepsilon}$ is associated with a positive definite quadratic form, i.e.,

$$\Phi(\mathbf{V}) = \mathbf{V} \cdot \boldsymbol{\varepsilon}\mathbf{V} > 0$$

for any vector $\mathbf{V}$, so that all components of the tensor $\boldsymbol{\varepsilon}$ are *real* and the eigenvalues of $\boldsymbol{\varepsilon}$ are positive. The fundamental equation (4.7.1) reflects the optical properties of the dielectric medium. For *optically isotropic bodies* $\boldsymbol{\varepsilon}$ is a multiple of the unit dyadic. Crystals, however, are in general optically anisotropic because of their crystalline structure. Then, instead of defining a unique dielectric constant ε by $\varepsilon_{ij} = \varepsilon\delta_{ij}$, as in the case of isotropic media or crystals of the cubic system, it is necessary to take into account the directional dependence of the dielectric tensor. Therefore, for anisotropic crystals, the electric displacement $\mathbf{D}$ will not be aligned with $\mathbf{E}$, so that the general relationship between the two vectors is of the form (4.7.1).[22] A similar relationship holds between $\mathbf{B}$ and $\mathbf{H}$ for magnetized media. In stationary *paramagnetic* solids, for instance, a linear relationship more general than the mere proportionality relationship $\mathbf{B} = \mu\mathbf{H}$ is

$$\mathbf{B} = \boldsymbol{\mu}\mathbf{H}, \tag{4.7.2}$$

where $\boldsymbol{\mu}$ is a second-order symmetric tensor called the *magnetic permeability tensor*. In crystals which possess a symmetry weaker than isotropy or cubic symmetry, it is not possible to define a unique magnetic permeability scalar μ. We must take into account the fact that the magnetic permeability depends on the direction of the magnetic field $\mathbf{H}$ and, therefore, the fields $\mathbf{B}$ and $\mathbf{H}$ are, in general, nonparallel. Furthermore, in ferromagnet or ferrimagnetic materials, we do not have directly a constitutive equation of the type (4.7.2). Nevertheless, an equation similar to (4.7.2) can be obtained in terms of the Fourier components of the perturbing parts of the fields $\mathbf{B}$ and $\mathbf{H}$ (see Chapter 9 in volume II).

If light is composed of a single monochromatic component, the real physical fields $\mathbf{E}$, $\mathbf{B}$, $\mathbf{D}$, and $\mathbf{H}$, which appear in the homogeneous set of Maxwell's equations

$$\begin{gathered} \nabla \cdot \mathbf{B} = 0, \qquad \nabla \cdot \mathbf{D} = 0, \\ \nabla \times \mathbf{E} + \frac{1}{c}\frac{\partial \mathbf{B}}{\partial t} = \mathbf{0}, \qquad \nabla \times \mathbf{H} - \frac{1}{c}\frac{\partial \mathbf{D}}{\partial t} = \mathbf{0}, \end{gathered} \tag{4.7.3}$$

will be the *real* parts of complex fields having an $\exp(-i\omega t)$ dependence, where ω is the circular frequency of the component. Considering a *plane* wave propagating in a transparent medium, all fields are also proportional to

[22] Equation (4.7.1) is linear. A higher-order equation for $\mathbf{D}$, in terms of $\mathbf{E}$, resorts to the theory of nonlinear optics. An account on the subject is to be found in Chapters 7 and 13.

$\exp(i\mathbf{k}\cdot\mathbf{r})$, with a *real wave vector* $\mathbf{k}$. Thus,

$$(\mathbf{E}, \mathbf{D}, \mathbf{H}, \mathbf{B}) \propto \exp[i(\mathbf{kr} - \omega t)]. \tag{4.7.4}$$

For a nonmagnetic body ($\mathbf{B} = \mathbf{H}$), it follows from (4.7.3) that

$$\begin{aligned} &\mathbf{H}\cdot\mathbf{k} = 0, \qquad \mathbf{D}\cdot\mathbf{k} = 0, \\ &\frac{\omega}{c}\mathbf{H} = \mathbf{k}\times\mathbf{E}, \qquad \frac{\omega}{c}\mathbf{D} = -\mathbf{k}\times\mathbf{H}. \end{aligned} \tag{4.7.5}$$

By taking the curl of $(4.7.3)_3$, taking into account $(4.7.3)_4$, and using the dependence (4.7.4), it is also found that $\mathbf{D}$ and $\mathbf{E}$ are related by

$$\mathbf{D} = n^2\mathbf{E} - \mathbf{n}(\mathbf{E}\cdot\mathbf{n}) = n^2[\mathbf{E} - \mathbf{s}(\mathbf{E}\cdot\mathbf{s})], \tag{4.7.6}$$

where $\mathbf{n}$ and $\mathbf{s}$ are defined by

$$\mathbf{k} = \frac{\omega}{c}\mathbf{n}, \qquad \mathbf{n} = n\mathbf{s}, \qquad n = |n|, \qquad \mathbf{s}\cdot\mathbf{s} = 1, \tag{4.7.7}$$

n is the refractive index which, in general, depends on both the direction of $\mathbf{n}$ and the frequency ω. The unit vector $\mathbf{s}$ is normal to the plane surfaces at a constant phase. Upon examining (4.7.5) and (4.7.6), we see that the vectors $\mathbf{D}$, $\mathbf{H}$, and $\mathbf{k}$ are mutually orthogonal. $\mathbf{H}$ is orthogonal to $\mathbf{E}$, so that $\mathbf{D}$, $\mathbf{E}$, and $\mathbf{k}$ (or $\mathbf{n}$) are all orthogonal to $\mathbf{H}$, and thus are coplanar. However, in contrast to what happens in an isotropic medium, $\mathbf{E}$, in general, is not transverse to the direction of propagation. Note, finally, that the derivation of (4.7.5) requires only the use of (4.7.3) and the functional dependence (4.7.4). But a complete study of the optical properties of dielectric crystals requires considering both (4.7.5)–(4.7.6) and (4.7.1). In view of (4.7.4), we may rewrite the latter as

$$\mathbf{D} = \boldsymbol{\varepsilon}(\omega)\mathbf{E}. \tag{4.7.8}$$

Here we have assumed that $\mathbf{B} = \mathbf{H}$, hence $\boldsymbol{\mu} = \mathbf{1}$. The properties derived for $\boldsymbol{\varepsilon}(\omega)$, however, are equally applicable to the magnetic permeability tensor $\boldsymbol{\mu}(\omega)$, in a frequency range where the dispersion of the magnetic permeability is of importance, as in the case of ferromagnetic insulators.

The optical properties of a crystal are, therefore, primarily related to the symmetry properties of the tensor $\boldsymbol{\varepsilon}(\omega)$ and to the geometry of the quadric associated with that tensor. A study along this line of thought leads to the notions of Fresnel's equation, Fresnel's ellipsoid, the optical indicatrix, the wave-vector surface,[23] and the classification of crystals into three classes (cubic or isotropic, uniaxial, and biaxial) according to the properties of the eigenvectors of the tensor $\boldsymbol{\varepsilon}$ or its reciprocal $\overset{-1}{\boldsymbol{\varepsilon}}$. The three eigenvalues of $\boldsymbol{\varepsilon}$, in general, are distinct and greater than one. But the number of distinct eigenvalues may be less than three for certain optical symmetries of the crystal. In

[23] See, e.g., Landau and Lifshitz [1960].

this respect, all crystals fall under three types: cubic, uniaxial, and biaxial. This classification resorts to the optical properties of the crystal and must not be confused with the crystal system to which the crystal belongs in the classical crystallographic classification.[24] *Biaxial* crystals are those for which $\boldsymbol{\varepsilon}$ has three distinct eigenvalues. For instance, crystals of the triclinic, monoclinic, and rhombic systems belong to the class of optically biaxial crystals. *Uniaxial* crystals are those for which $\boldsymbol{\varepsilon}$ has only two distinct eigenvalues. Crystals of the tetragonal, rhombohedral, and hexagonal systems in their *natural* state (i.e., in absence of any *induced* optical anisotropy; see below for this notion) belong to the class of optically uniaxial crystals. Finally, when the three eigenvalues of $\boldsymbol{\varepsilon}$ are equal, it is said that the material belongs to the class of optically *cubic* crystals. This last classification stems from the fact that, for a cubic crystal in its natural state, the tensor $\boldsymbol{\varepsilon}$ can only have the representation $\varepsilon_{ij} = \varepsilon\delta_{ij}$, where ε is the common eigenvalue. In this last case, the principal directions of $\boldsymbol{\varepsilon}$ are left arbitrary, so that, in fact, cubic crystals in their natural state are no different from isotropic bodies, as far as their dielectric and optical properties are concerned.

The case of uniaxial crystals is particularly important, for they occur frequently and have several important uses. Their dielectric tensors have two equal eigenvalues and their dielectric ellipsoids are spheroids. Let $\varepsilon_{\parallel}$ be the eigenvalue in the direction of the greater axis, also called the *optical* axis, of this ellipsoid. The directions of the other two principal axes, in a plane orthogonal to the optical axis, are arbitrary, and the corresponding eigenvalues, denoted $\varepsilon_{\perp}$, are equal. The optical properties of such crystals are invariant with respect to rotations about the direction of the optical axis. If $\mathbf{d}$ is a unit vector in this direction, then $\boldsymbol{\varepsilon}$ has the component expression

$$\varepsilon_{ij} = \varepsilon_{\perp}\delta_{ij} + (\varepsilon_{\parallel} - \varepsilon_{\perp})d_i d_j. \tag{4.7.9}$$

It is then shown that two types of light waves can propagate in uniaxial crystals. With respect to one type, called the *ordinary waves*, the crystal behaves like an isotropic body of refractive index $n_0 = \sqrt{\varepsilon_{\perp}}$. The other type of wave, which does not occur in isotropic bodies, is called *extraordinary waves*. The wave vector $\mathbf{k}$ of extraordinary waves has a magnitude which depends on the angle it makes with the optical axis. The corresponding *ray vector*, usually defined as the vector $\mathbf{R}$ satisfying $\mathbf{n} \cdot \mathbf{R} = 1$, is along the direction of the vector group velocity $\partial\omega/\partial\mathbf{k}$. $\mathbf{R}$ is not in the same direction as the wave vector but is coplanar with that vector and the optical axis. The common plane of the wave vector and the optical axis is called the *principal section* for the prescribed $\mathbf{n}$.

Similar to other anisotropic media, uniaxial crystals may exhibits a phenomenon called *double refraction* or *birefringence*. This occurs in optically anisotropic media (e.g., mica, Iceland spar), that is, media having optical

[24] For this later classification, see Chapter 5 and Appendix A.

properties dependent on direction. The "double" or "bi" in the naming of the effect refers to the two different directions of propagation that a given incident ray can take in such media, depending on the direction of polarization. The birefringence is measured by the difference between the two refractive indices:

$$\Delta n = n_1 - n_2. \tag{4.7.10}$$

When a plane wave is incident on the surface of an optically anisotropic crystal, two different refracted waves are formed. In *uniaxial* crystals ordinary and extraordinary refracted waves are formed. The ordinary (spherical) wave is entirely analogous to the refracted wave in isotropic bodies (its ray vector is in the same direction as its wave vector, and thus lies in the plane of incidence). The extraordinary wave behaves abnormally in that the wave front is not perpendicular to the direction of propagation so that, in general, the ray vector of the extraordinary wave does not lie in the plane of incidence.

B. Inducement of Optical Anisotropy

The symmetry properties of the tensor $\boldsymbol{\varepsilon}$ may be induced by different fields so that, for instance, a transparent medium, which is optically isotropic or cubic in its natural state, may become *optically anisotropic* when these new fields are applied. The effects obtained are called *electromagneto-optical* effects when they result from inducement by an electric or a magnetic field. *Photoelasticity* is an effect resulting from the application of a thermodynamically recoverable strain, whereas *streaming birefringence* is that resulting from viscous flow.

Electromagneto-optical effects in stationary materials are observed when the material is placed in a *strong static* electric or magnetic field and an electromagnetic wave (light) transverses the medium. Clearly, the existence of such effects in material media is direct evidence that the equations of electromagnetic fields in a general material continuum, unlike the equations in a vacuum, are *nonlinear*. An analysis of electromagneto-optical effects, though based on nonlinear equations for which all but the simplest exact solutions are difficult to exhibit, may be simplified and made tractable by assuming that the dynamical part of the solution—the light wave—has such a weak intensity that the vector fields that describe it may be treated as infinitesimal. Thus, in a quite general approach to this class of phenomena, it may generally be assumed that the electric displacement and magnetic induction fields are *functions* of the intense static electric and magnetic fields, and are *linear functionals* in the small dynamic electric and magnetic fields.

For instance, the electro-optical *Kerr effect* results when a strong static electric field $\mathbf{E}^0$ is applied to a solid, such as glass or a liquid such as nitrobenzene, media which can be considered as optically isotropic in their natural state, but behave like uniaxial crystals with optical axis along this electric field. We then have, from (4.7.9),

$$\varepsilon_\perp = \varepsilon, \qquad \varepsilon_\parallel - \varepsilon_\perp \propto (\mathbf{E}^0)^{-2}, \tag{4.7.11}$$

Table 4.7.1. Kerr constant at 20 °C in sodium light ($\lambda = 5890$ Å).

Medium		Refractive index[(a)]	$C_K \times 10^7$ [(b)]
Benzene	C_6H_6	1.482	0.60
Carbon disulfide	CS_2	1.63	3.21
Chloroform	$CHCl_3$	1.446	−3.46
Water	H_2O	1.33	4.7
Nitrobenzene	$C_6H_5NO_2$	1.55	220.0
Nitrotoluene	$C_5H_{71}NO_2$	1.547	123.0
Flint glass	Nr^3 03031	1.60	0.029

(a) From Kaye and Laby [1973, p. 95].
(b) From Born [1972, p. 367].

where ε is the dielectric constant corresponding to the isotropic behavior of the body. The beam of light is supposed to be polarized at about $\pi/2$ to the direction of $\mathbf{E}^0$. It follows from (4.7.11) and (4.7.10), that a birefringence formula follows which is usually written in the form

$$\Delta n = n_\perp - n_\parallel = C_K l(\mathbf{E}^0)^2, \tag{4.7.12}$$

where l is the length of the path of the light beam and C_K is the so-called Kerr constant. Often l is taken to be equal to the wavelength λ of the light in a vacuum. Typical data for C_K are to be found in Kaye and Laby [1973, p. 95] and Born [1972, p. 367] (see also Table 4.7.1). The Kerr effect is a *transverse* effect, which is of second order in $\mathbf{E}^0$. However, the change in optical symmetry may also occur in nonisotropic bodies. In this case, the effect may be of the first order in E^0, but it can occur only in the twenty crystal classes which admit the phenomenon of piezoelectricity (i.e., the classes of crystals which lack a center of symmetry; see Chapter 7).

The magneto-optical effect, known as the *Voigt–Cotton–Mouton* effect after its discovery by Voigt in gases and by Cotton and Mouton in liquids and solids, is the magnetic equivalent of the Kerr effect. Hence (4.7.12) is replaced by

$$\Delta n = n_\parallel - n_\perp = C_M l(\mathbf{H}^0)^2, \tag{4.7.13}$$

where $\mathbf{H}^0$ is the strong static magnetic field (the light propagation is in the direction perpendicular to $\mathbf{H}^0$) and C_M is the so-called Mouton constant[25] (see Table 4.7.2). The sign of this constant cannot be determined from any theory. Materials have been found with positive C_M (aromatic compounds) or negative C_M (saturated alcohols).

While the Voigt–Cotton–Mouton effect is of *second order* in H^0, the magneto-optical phenomenon known as the *Faraday effect*, consisting of a rotation of the plane of polarization of impinging light, is of first order in $\mathbf{H}^0$. This effect disappears if $\mathbf{H}^0$ is orthogonal to $\mathbf{n}$, so that the describing formula

[25] Data concerning C_M can be found in Born [1972, p. 362].

Table 4.7.2. Cotton–Mouton constant in fluids (after Born [1972, p. 362]).

Substance		θ (°C)	λ_0 (Å)	$C_M \times 10^{13}$
Acetone	CH_3COCH_3	20.2	5780	37.6
Aniline	$C_6H_5NH_2$	25.0	5890	5.1
Benzene	C_6H_6	26.5	5800	7.5
Chloroform	$CHCl_3$	17.2	5780	−65.8
Nitrobenzene	$C_6H_5NO_2$	16.3	5780	23.5
Toluene	$C_6H_5CH_3$	19.4	5890	6.71
Carbon disulfide	CS_2	28.0	5800	−4.0

is usually given for a parallel setting in the form[26]

$$\Delta k = C_V H^0, \tag{4.7.14}$$

where Δk is the difference between the wave numbers of the two light oscillations with opposite direction of rotation and C_V is the so-called Verdet constant[27] (see Table 4.7.3).

It is to be noted that, at the beginning of this century, the electromagneto-optical effects played an essential role in the development of the electromagnetic theory of light. At present, the investigation of these effects is undergoing a renewal in relation with fundamental research in solid-state physics, as well as technological development such as electromagneto-optical

[26] A phenomenological treatment of the Faraday effect is given in Landau and Lifshitz [1960, Sect. 82].

[27] Values of C_V for different gases, liquids, and solids are to be found in Kaye and Laby [1973, pp. 98–99].

Table 4.7.3. Verdet constant in rotation angle per ampère (after Kaye and Laby [1973, pp. 98–99]).

Substance		C_V
Gases	Hydrogen	8.867×10^{-6} min $\times A^{-1}$
(0 °C, 1 atom, λ_0 = 5,461 Å)	Oxygen	7.598×10^{-6} min $\times A^{-1}$
	Methane	24.15×10^{-6} min $\times A^{-1}$
	Isobutane	68.12×10^{-6} min $\times A^{-1}$
Liquids	Water (20 °C)	1.645×10^{-2} min $\times A^{-1}$
(λ_0 = 5,893 Å)	Acetone (20 °C)	1.42×10^{-2} min $\times A^{-1}$
	Methyl alcohol (20 °C)	1.17×10^{-2} min $\times A^{-1}$
	Carbon disulfide (18 °C)	5.28×10^{-2} min $\times A^{-1}$
	Chloroform (20 °C)	2.06×10^{-2} min $\times A^{-1}$
	Toluene (28 °C)	3.38×10^{-2} min $\times A^{-1}$
Solids	Crystalline quartz (20 °C)	2.091×10^{-2} min $\times A^{-1}$
(λ_0 = 5,893 Å)	Calcium fluoride	1.127×10^{-2} min $\times A^{-1}$

devices in laser techniques.[28] These effects, and their coupling with the strain field in elastic bodies, are studied on a rational basis in Chapter 7.

Photoelasticity. Elastic deformations can also produce a change in the optical symmetry of a deformable body. In exactly the same way as the Kerr effect, the inducement of strain in a transparent plastic such as perspex (polymethyl metacrylate) causes it to behave as a uniaxial crystal. For example, the distribution of strains in a loaded beam can be investigated in a transparent plastic model by examining it between two polaroids. Indeed, in the case of infinitesimal strains in an elastic body, the components of the dielectric tensor $\boldsymbol{\varepsilon}$ can be written in the form

$$\varepsilon_{ij} = (\varepsilon + \varepsilon_0 \tilde{e}_{kk})\delta_{ij} + 2\varepsilon_1 \tilde{e}_{ij}, \tag{4.7.15}$$

where ε, ε_0, and ε_1 are material constants. Clearly, the dielectric tensor $\boldsymbol{\varepsilon}$ has the same principal directions as the strain tensor $\tilde{\mathbf{e}}$. Unless the latter is proportional to the unit dyadic, the tensor $\boldsymbol{\varepsilon}$, given by (4.7.15), will have unequal eigenvalues. Consequently, the optically isotropic body at the natural state, upon deformation, will behave like a uniaxial or a biaxial crystal, depending on the state of the strains. Thus it will become birefringent. This effect is the basis of the *photoelastic technique* for the analysis of structure on plastic models.[29] Since, in linear elasticity, a linear relationship is assumed between stresses and strains (Hooke's law), it can be shown that the resulting birefringence for a *plane* state of stress is given by

$$\Delta n = n_1 - n_2 = C(t_1 - t_2), \tag{4.7.16}$$

where t_1 and t_2 are the principal stresses and C is the relative *stress-optical* constant of the material. Photoelasticity is a well-developed branch of experimental physics. It is used extensively to measure stress fields in bodies with cracks and holes which are subject to a complicated external load. A rational basis for the theory of photoelasticity in isotropic bodies is given in Chapter 7.

Streaming birefringence. Optical anisotropy can also be induced in other rheological materials, such as moving viscous Newtonian or non-Newtonian fluids, viscoelastic solids, or viscoplastic media. Those media, considered as isotropic in their mechanical properties, may be anisotropic with respect to their optical properties when subjected to a motion. For example, in a moving incompressible Newtonian fluid, neglecting any dragging effect, the components of the dielectric tensor $\boldsymbol{\varepsilon}$ may be written in the form

$$\varepsilon_{ij} = \varepsilon\delta_{ij} + 2\varepsilon_v d_{ij}, \tag{4.7.17}$$

where d_{ij} is the rate-of-strain tensor, ε is the dielectric constant at rest, and ε_v

---

[28] See Lipson and Lipson [1970, p. 371].

[29] For a synthesis and a presentation of modern techniques in photoelasticity, see Jessop [1958] and Kruske and Robertson [1974].

is a new phenomenological constant. The resulting birefringence is due to a shear motion and thus, is referred to as a *streaming birefringence*. The experimental techniques based on streaming birefringence are used to photograph flow patterns in pipes, channels, and around rigid bodies. This phenomenon is studied in Chapter 13 for the case of deformable dielectrics with memory.

PROBLEMS

4.1. *Polarization catastrophe.* Solve (4.3.17) for ε and plot the resulting relationship. What happens for $\rho = 3M/\alpha N_A$? Why does nothing special happen in the case of most real materials?

4.2. (Short term paper.) Study the relevant literature and present a report on the theory of ferroelectric phase transitions.

4.3. (Short term paper.) Study the relevant literature and present a report on the phenomenological theory of magnetic domains.

4.4. (Short term paper.) Study the relevant literature and present a report on the dynamics of Bloch walls in rigid ferromagnets.

4.5. *Equation of telegraphy.* Consider a rigid steady isotropic *electricity conducting* body with constitutive equations (4.7.1), (4.7.2) and $\mathbf{J} = \sigma\mathbf{E}$:

(a) Show that Maxwell's equations yield the so-called *equation of telegraphy*:

$$\nabla^2 \begin{Bmatrix} \mathbf{E} \\ \mathbf{H} \end{Bmatrix} = \frac{\varepsilon\mu}{c^2}\frac{\partial^2}{\partial t^2}\begin{Bmatrix} \mathbf{E} \\ \mathbf{H} \end{Bmatrix} + \frac{\mu\sigma}{c^2}\frac{\partial}{\partial t}\begin{Bmatrix} \mathbf{E} \\ \mathbf{H} \end{Bmatrix};$$

(b) Considering slutions of the plane-harmonic-wave type, show that the corresponding dispersion relation reads

$$k^2 = \frac{\omega^2}{c^2}\eta\mu, \qquad \eta \equiv \varepsilon + \frac{\sigma}{i\omega};$$

(c) Writing $p \equiv \sqrt{\eta\mu} = n - i\kappa$ as the complex index of refraction, where n and κ are real and positive, show that the quantity κ plays the role of an extinction coefficient.

4.6. *Reflection of electromagnetic waves at boundary surfaces.* Consider a plane electromagnetic wave (1) which, having traveled in vacuum, enters perpendicularly into a surface of a rigid steady isotropic paramagnetic electricity conductor. Experimentally, this wave splits at the surface into: (i) a reflected wave (2) of amplitude a' which goes back in vacuum in the direction of the negative x-axis; and into (ii) a penetrating wave (3) of amplitude a'' going in the medium in the positive x-direction. Let $p = \sqrt{\eta\mu} = n - i\kappa$ be the complex index of refraction of the conductor (see Problem 4.5). Show that the *reflectivity* $R \equiv a'/a''$ of the material is given by

$$R = \frac{(n-\mu)^2 + \kappa^2}{(n+\mu)^2 + \kappa^2}.$$

[For a transparent medium $\kappa \equiv 0$ and $\mu = 1$, this reduces to the well-known relationship

$$R \approx (n-1)^2/(n+1)^2.]$$

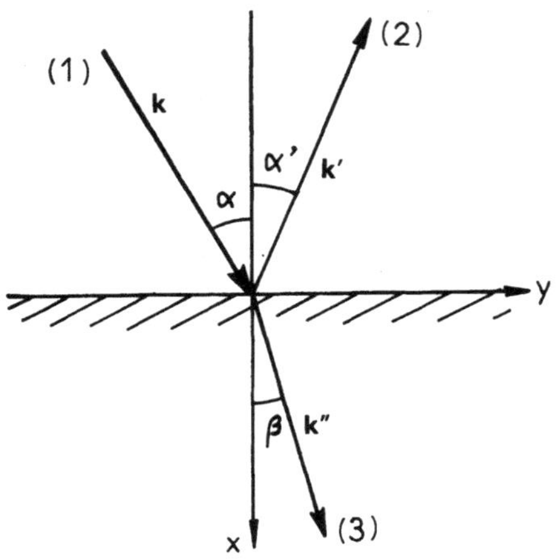

4.7. *Reflection of electromagnetic waves at oblique incidence* (Problem 4.6 continued).

(a) Let $\mathbf{k}$, $\mathbf{k}'$, and $\mathbf{k}''$ be the wave vectors of the incident wave (1), the reflected wave (2), and the penetrating wave (3), respectively. α, α', and β (as indicated in the figure) are, respectively the angle of incidence, the angle of reflection, and the angle of emergence. Then derive the laws of reflection and refraction

$$\alpha = \alpha'; \qquad \sin\alpha = p \sin\beta.$$

(b) If $\mathbf{E}$ is parallel ($\|$) or perpendicular ($\perp$) to the plane of incidence, show that the corresponding reflectivity formulas read

$$R_{\|} = \left|\frac{p\cos\alpha - \mu\cos\beta}{p\cos\alpha + \mu\cos\beta}\right|^2, \qquad R_{\perp} = \left|\frac{p\cos\beta - \mu\cos\alpha}{p\cos\beta + \mu\cos\alpha}\right|^2.$$

(c) For the special case of reflection of light from a *transparent nonmagnetic* medium, show that the above-obtained formulas go over to the *Fresnel formulas*:

$$R_{\|} = \left(\frac{\tan(\alpha-\beta)}{\tan(\alpha+\beta)}\right)^2, \qquad R_{\perp} = \left(\frac{\sin(\alpha-\beta)}{\sin(\alpha+\beta)}\right)^2.$$

(d) Study the variation of $R_{\|}$ and $R_{\perp}$ as functions to α. What happens at the value $\alpha = \tan^{-1}(n)$, known as the *Brewster angle*?

4.8. *Skin effect.* Let a straight metallic wire with circular cross section of radius r_0 carry an alternating current of frequency ω. Only the component E parallel to the wire axis is different from zero. Using the telegraphy equation obtained in Problem 4.5 and considering frequencies so low that the displacement terms can be neglected, find the exact solution in terms of Bessel functions for the field E using the length parameter $d \equiv c/\sqrt{\sigma\omega\mu}$. Then show that for $r_0 \ll d$ (weak skin effect) the approximate solution is

$$E = E_0\left\{1 + \frac{ir^2}{2d^2} + \cdots + \frac{1}{(n!)^2}\left(\frac{ir^2}{2d^2}\right)^n + \cdots\right\},$$

whereas for $r_0 \gg d$ (strong skin effect)

$$E \simeq E_0 \exp[(1 + i)r/d].$$

Interpret both results.

4.9. The Poynting vector, which represents the flux of electromagnetic energy, is defined by $\mathbf{S} = c\mathbf{E} \times \mathbf{H}$, show that for plane waves in an optically anisotropic transparent medium

$$\mathbf{S} = c[E^2\mathbf{n} - (\mathbf{E}\cdot\mathbf{n})\mathbf{E}],$$

where $\mathbf{n}$ is defined as in (4.7.7). Noting that $\mathbf{D} = -\mathbf{n} \times \mathbf{H}$ and $\mathbf{H} = \mathbf{n} \times \mathbf{E}$, show that the energy concentrated in a wave packet moves with the latter or, in other words, that the Poynting vector is in the same direction as the group velocity vector $\mathbf{v}_g = \partial\omega/\partial\mathbf{k}$ (equally, $\mathbf{S}$ is normal to the wave vector surface).

4.10. *Depolarization tensor.*

(a) Show that the electric potential ϕ of a uniformly polarized ellipsoid of principal axes a, b, c is

$$\phi = AxP_x + ByP_y + CzP_z,$$

where the numbers A, B, and C are given

$$A = \frac{abc}{2}\int_0^\infty \frac{ds}{(a^2 + s)R_s}, \qquad B = \frac{abc}{2}\int_0^\infty \frac{ds}{(b^2 + s)R_s},$$

$$C = \frac{abc}{2}\int_0^\infty \frac{ds}{(c^2 + s)R_s}, \qquad R_s \equiv [(a^2 + s)(b^2 + s)(c^2 + s)]^{1/2}.$$

These are the eigenvalues of the depolarization tensor.

(b) Consider a prolate ellipsoid of revolution: $a \geqq b = c$ and set $\delta = a/b$. Show that the values of A for an almost spherical ellipsoid (small $\delta^2 - 1$) and a very long prolate ellipsoid ($\delta \gg 1$) are given by

$$A = \tfrac{1}{3}[1 - \tfrac{2}{5}(\delta^2 - 1) + \cdots]$$

and

$$A \approx \frac{1}{\delta^2}(\ln 2\delta - 1),$$

respectively.

CHAPTER 5

Constitutive Equations

5.1. Scope of the Chapter

This chapter is concerned with the development of the constitutive equations of deformable bodies subject to electromagnetic fields. In Section 5.2 we collect the balance laws of the electrodynamics of continua. An examination of these equations reveals the necessity for additional equations among various fields to characterize the constitution of the body (Section 5.3). Section 5.4 begins a systematic development of the constitutive theory based on nine axioms. After a discussion of the physical foundations of these axioms, mathematical formulations are carried out for the memory-dependent electromagnetic materials. Material symmetry regulations relevant to crystallography are the topic of Section 5.5, where the group-theoretic method is employed to obtain the integrity basis of vectors and tensors.

The method developed, together with the appendices, should enable us to obtain the nonlinear constitutive equations of 32 crystal classes. For magnetic materials, the group symmetry is enlarged from 32 to 90 point groups, due to spin magnetic moments. This is the topic of Section 5.6, which is especially relevant to ferromagnetic, ferrimagnetic, and antiferromagnetic materials. Section 5.7, together with Appendix E, should be sufficient to determine the integrity basis of vectors and tensors for isotropic materials, transverse isotropy included. In Section 5.8 we begin to obtain the constitutive equations for electromagnetic elastic solids, as restricted by the second law of thermodynamics. Special cases of rigid solids, elastic dielectrics, and magnetoelastic solids are given. Nonlinear constitutive equations for isotropic solids are obtained in Section 5.9, where the origins of various physical "cross effects" also are indicated. These include electrostrictive effects, piezoelectricity, piezomagnetism, pyroelectricity, magnetoelectric effects, and nonequilibrium effects such as Seebeck, Righi-Leduc, Ettingshausen, Hall, and Nernst. In Section 5.10 we give some examples from anisotropic solids. Section 5.11 deals with the linear theory. Constitutive equations of electromagnetic thermoelastic solids are given, including all known physical effects for anisotropic and isotropic solids. The effect of the anisotropy clearly comes to surface as a basic property for the existence of many interesting physical phenomenon. Section 5.12 is devoted to the derivation of the constitutive equations of electromagnetic fluids. Finally, Section 5.13 gives a collection of constitutive equations.

The content and composition of this chapter is new and it is an extension of the material presented in Eringen [1967, Chap. 5] and [1980, Chap. 10].

5.2. Résumé of Balance Laws

The local balance laws and entropy inequality were derived in Sections 3.9 to 3.11. Below we reproduce some forms of these expressions which are relevant to a discussion of the constitutive theory.

Conservation of Mass:

$$\rho_0 = \rho \mathrm{III}_C^{1/2} \qquad \text{or} \qquad \dot{\rho} + \rho \nabla \cdot \mathbf{v} = 0 \quad \text{in } \mathscr{V} - \sigma. \tag{5.2.1}$$

Balance of Momentum:

$$t_{kl,k} + \rho(f_l - \dot{v}_l) + F_l^{\mathrm{E}} = 0 \quad \text{in } \mathscr{V} - \sigma. \tag{5.2.2}$$

Balance of Moment of Momentum:

$$t_{[kl]} = \mathscr{E}_{[k} P_{l]} + B_{[k} \mathscr{M}_{l]} \quad \text{in } \mathscr{V} - \sigma. \tag{5.2.3}$$

Conservation of Energy:

$$\rho \dot{e} - t_{kl} v_{l,k} - \nabla \cdot \mathbf{q} - \rho h + \mathbf{P} \cdot \dot{\mathscr{E}} + \mathscr{M} \cdot \dot{\mathbf{B}} - \mathscr{J} \cdot \mathscr{E} = 0 \quad \text{in } \mathscr{V} - \sigma. \tag{5.2.4}$$

Entropy Inequality:

$$\begin{aligned} \rho \gamma \equiv {} & -\rho(\dot{\Psi} + \eta \dot{\theta}) + t_{kl} v_{l,k} + \frac{1}{\theta} \mathbf{q} \cdot \nabla \theta \\ & - \mathbf{P} \cdot \dot{\mathscr{E}} - \mathscr{M} \cdot \dot{\mathbf{B}} + \mathscr{J} \cdot \mathscr{E} \geq 0 \quad \text{in } \mathscr{V} - \sigma. \end{aligned} \tag{5.2.5}$$

Gauss' Law:

$$\nabla \cdot \mathbf{D} - q_e = 0 \quad \text{in } \mathscr{V} - \sigma. \tag{5.2.6}$$

Faraday's Law:

$$\nabla \times \mathbf{E} + \frac{1}{c} \frac{\partial \mathbf{B}}{\partial t} = 0 \quad \text{in } \mathscr{V} - \sigma. \tag{5.2.7}$$

Conservation of Magnetic Flux:

$$\nabla \cdot \mathbf{B} = 0 \quad \text{in } \mathscr{V} - \sigma. \tag{5.2.8}$$

Ampère's Law:

$$\nabla \times \mathbf{H} - \frac{1}{c} \frac{\partial \mathbf{D}}{\partial t} - \frac{1}{c} \mathbf{J} = \mathbf{0} \quad \text{in } \mathscr{V} - \sigma. \tag{5.2.9}$$

Conservation of Charge:

$$\frac{\partial q_e}{\partial t} + \nabla \cdot \mathbf{J} = 0 \quad \text{in } \mathscr{V} - \sigma, \tag{5.2.10}$$

where

ρ = mass density per unit volume,
$\mathbf{v}$ = velocity vector,
t_{kl} = Cauchy stress tensor,
f_l = body force density per unit mass (excluding electromagnetic force),
F_l^{E} = electromagnetic body force density per unit volume,
$\mathscr{E}_k$ = electric field in a co-moving reference frame R_{C},
P_k = polarization vector in the fixed frame R_{G},
B_k = magnetic induction vector in R_G,
$\mathscr{M}_k$ = magnetization in a co-moving reference frame $\mathrm{R_C}$,
e = modified internal energy density per unit mass,
$\mathbf{q}$ = heat vector,
h = heat source per unit mass,
$\mathscr{J}$ = conduction current,
Ψ = modified free energy function,
η = entropy density per unit volume,
θ = absolute temperature,
$\mathbf{D}$ = dielectric displacement vector in R_{G},
q_{e} = free charge density per unit volume,
$\mathbf{E}$ = electric field vector in R_{G},
$\mathbf{H}$ = magnetic field vector in R_{G},
$\mathbf{J}$ = total current in R_{G},
c = speed of light in vacuum.

Definitions of Electromagnetic Fields, Force, and Energy:

$$\mathscr{E} = \mathbf{E} + \frac{1}{c}\mathbf{v} \times \mathbf{B}, \qquad \mathscr{M} = \mathbf{M} + \frac{1}{c}\mathbf{v} \times \mathbf{P}, \tag{5.2.11}$$

$$\mathbf{D} = \mathbf{E} + \mathbf{P}, \qquad \mathbf{B} = \mathbf{H} + \mathbf{M}, \qquad \mathscr{J} = \mathbf{J} - q_{\mathrm{e}}\mathbf{v}, \tag{5.2.12}$$

$$\begin{aligned} t_{kl} &= {}_{\mathrm{E}}t_{kl} - P_k\mathscr{E}_l - \mathscr{M}_k B_l, \\ \mathbf{F}^{\mathrm{E}} &= q_{\mathrm{e}}\mathbf{E} + \frac{1}{c}\mathbf{J} \times \mathbf{B} + (\nabla\mathbf{E})\cdot\mathbf{P} + (\nabla\mathbf{B})\cdot\mathbf{M} + \frac{1}{c}[(\mathbf{P} \times \mathbf{B})v_k]_{,k} + \frac{1}{c}\frac{\partial}{\partial t}(\mathbf{P} \times \mathbf{B}), \end{aligned} \tag{5.2.13}$$

$$e = \varepsilon - \rho^{-1}\mathscr{E}\cdot\mathbf{P}, \qquad \Psi = \varepsilon - \theta\eta - \rho^{-1}\mathscr{E}\cdot\mathbf{P}, \tag{5.2.14}$$

where ε is the internal energy density per unit mass and ${}_{\mathrm{E}}t_{kl}$ is a symmetric tensor, i.e.,

$${}_{\mathrm{E}}t_{[kl]} = 0. \tag{5.2.15}$$

For the discussion of electromagnetic elastic solids, the material form of energy equations and entropy inequality proves to be more convenient. These forms were obtained in Section 3.11.

$$\rho_0\dot{e} - \tfrac{1}{2}{}_{\mathrm{E}}T_{KL}\dot{C}_{KL} - Q_{K,K} - \rho_0 h + \Pi_K\dot{\mathscr{E}}_K + M_K\dot{B}_K - \mathscr{J}_K\mathscr{E}_K = 0, \tag{5.2.16}$$

$$-\rho_0(\dot{\Psi} + \eta\dot{\theta}) + \tfrac{1}{2}{}_{\mathrm{E}}T_{KL}\dot{C}_{KL} + \frac{1}{\theta}Q_K\theta_{,K} - \Pi_K\dot{\mathscr{E}}_K - M_K\dot{B}_K + \mathscr{J}_K\mathscr{E}_K \geq 0, \tag{5.2.17}$$

where

$$\Psi = \varepsilon - \theta\eta - \rho_0^{-1}\Pi_K\mathscr{E}_K. \tag{5.2.18}$$

The Clausius–Duhem (C–D) inequality ((5.2.5) or (5.2.17)) is posited to be valid for all independent variations of the independent variables. Therefore, it is considered a restriction for both the material properties and the thermodynamic evolution of the system. It plays a central role in the development of constitutive equations.

5.3. Raison d'Être

Given the external sources (**f**, h) in the balance laws listed in Section 5.2, the following unknowns must be determined:

$$\begin{gathered} {}_{\mathrm{E}}T_{KL}, \quad Q_K, \quad \Psi, \quad \eta, \quad \mathscr{E}_K, \quad M_K, \quad \mathscr{J}_K, \\ \rho, \quad v_k, \quad \theta, \quad q_e, \quad B_K, \quad \Pi_K. \end{gathered} \tag{5.3.1}$$

These add up to 32 scalar unknowns. To determine these unknowns, we have equatons (5.2.1), (5.2.2), (5.2.4), (5.2.6), (5.2.7), (5.2.8), and (5.2.9)—since (5.2.10) is the result of (5.2.6) and (5.2.9). Thus, we have $1 + 3 + 1 + 1 + 3 + 1 + 3 = 13$ equations. In fact, this number can be reduced by remembering that the divergence of (5.2.7) gives $c^{-1}\partial(\nabla \cdot \mathbf{B})\, \partial t = 0$ or $\nabla \cdot \mathbf{B} = \psi(\mathbf{x})$. Thus, if $\nabla \cdot \mathbf{B} = 0$ initially, it must vanish always. Consequently, (5.2.8) can be disregarded in this counting, except as a restriction to the initial conditions. Therefore, it is clear that for the system to be determinate, we need $32 - 12 = 20$ more equations among the members of the list of unknowns (5.3.1).

In the derivation of balance laws, no consideration was given to the nature of the substance constituting the body. Thus, these laws are valid for all types of substances, e.g., gases, fluids, or solids having a variety of different properties. *The phenomenological properties of materials are brought into play through the constitutive equations.* Naturally, these constitutive equations are limited to be valid in certain domains of their variables. These restrictions arise, first of all, from our inability to characterize mathematically the internal structure of a body. For, to begin with, the inner structures of bodies (excluding some ideal, perfect crystals) are not known in full detail. Moreover, at the microscopic and atomic scale, changes are constantly taking place (e.g., free electrons are moving around, bumping lattice sites, ions are constantly oscillating). Second, the nature of intermolecular forces is not known, except for some ideal configurations and, even for these, it is known only approximately. The existence of submicroscopic faults (e.g., dislocations, impurities, holes, microcracks) makes such a characterization impossible. With the change of the range of applied fields submicroscopic changes occur, often globally affecting the outcome. In fact, these changes take on an important role when the ratio of an external characteristic length, $\mathscr{L}$ (e.g., wavelength, the extent of distribution of loads, the length of a geometrical discontinuity) is compared with the internal characteristic length, l (average molecular or granular distance) of the body. For $\mathscr{L}/l \gg 1$, the *subbodies* that build up the body respond to an external stimulus collectively, so that a properly constructed continuum theory will be successful in the prediction of physical phenomena. However, as $\mathscr{L}$ approaches l the individual behaviors of subbodies become more and more important to arrive at a reasonable prediction. Clearly, then, each constitutive theory will have a limited domain of applicability beyond which more complicated theories need to be constructed. Thus, for example, classical elasticity, as successful as it has been, fails grossly when we are dealing with a sharp crack, or when we transmit waves of small wavelength. For a sharp crack at the crack tip, the curvature of the crack being considered as $\mathscr{L}$, we have $\mathscr{L} \sim l$ where l may be taken as the atomic or the granular distance. This is the reason why the theory predicts infinite stress at the crack tip, a phenomenon which is clearly not possible on physical grounds (Eringen [1978c], [1984a]). In such situations, either we are forced to exclude a small region around the crack tip where the theory of elasticity fails to apply, or we must construct a nonlocal theory that can account for small, external characteristic lengths (see Eringen [1972a], [1974a, b, c], [1977a, b], [1978a], [1985a]; Eringen and Edelen [1972], Eringen et al. [1977]). In this small region the situation is extremely complicated. For some materials, large elastoplastic deformations may take place requiring the use of dislocation and plasticity theories, and for others, we may have to include the effects of interatomic bond failures through an atomic or nonlocal consideration.

Thus, in ascribing a continuum model to a material body, we not only need to select appropriate constitutive dependent and independent variables to characterize the material response, but we also need to investigate or describe the domain of applicability of the constitutive response functions. Of course, this latter restriction is by no means simple to state unequivocally at the outset. To this end, we benefit a great deal from experimental investigations (after the theory is well established and tested, unfortunately!) although we may have some heuristic ideas initially.

The response of electromagnetically active deformable bodies is determined by a collection of 32 variables which we denote by σ_α

$$\sigma_\alpha = \{{}_{\mathrm{E}}T_{KL}, Q_K, \Psi, \eta, \mathscr{E}_K, M_K, \mathscr{J}_K, \rho, x_k, \theta, q_e, B_K, \Pi_K\}, \qquad \alpha = 1, 2, \ldots, 32. \tag{5.3.2}$$

Clearly σ_α depends on the material point $\mathbf{X}$, and time t through its components. The most general constitutive equation that can be constructed from these sets of variables is a set of 20 *functionals* of the form

$$\mathscr{F}_\beta[\sigma_\alpha(\mathbf{X}', t'); \mathbf{X}, t] = 0, \qquad \beta = 1, 2, \ldots, 20. \tag{5.3.3}$$

Here, $\mathscr{F}_\beta$ is a functional of functions $\sigma_\alpha(\mathbf{X}', t')$, and a function of $\mathbf{X}$ and t. The functions σ_α are defined over *all material points* of the body and over all *past times* t', i.e., $X' \in B$, $t' \in [t, -\infty)$. When the 20 functional equations (5.3.2) are known, together with the balance laws, we will have the *necessary* number of *independent* equations to determine (at least in principle) all the unknowns at a material point **X**, at any time t. Thus, (5.3.3) states that *to determine the response of a material body at time t, we need the history of the states σ_α of all points of the body, up to time t.* Clearly, such constitutive equations are too general to be of any utility. Fortunately, for most materials, in a desirable range of variables, much simpler constitutive equations are adequate for a physical description of the outcome. Moreover, gross simplifications are possible by means of certain axioms, which are either verified experimentally or considered obvious. In addition, by noting some obvious characteristics of bodies (e.g., solidity, fluidity, lack of memory dependence), we are able to bring further order to this chaos. When such simple models fail to apply, however, we need to return to the main source (5.3.3) for help.

In the following sections we lay down some axioms for the construction of constitutive equations for thermomechanical and electromagnetically active materials.

5.4. Axioms of Constitutive Theory

For a systematic development of constitutive equations, certain physical and mathematical requirements are necessary. The following axioms, which are the basis for the constitutive theory, have evolved over the past decades (see Eringen [1966a], [1967], [1980]).

(I) Causality.
(II) Determinism.
(III) Equipresence.
(IV) Objectivity.
(V) Time Reversal.
(VI) Material Invariance.
(VII) Neighborhood.
(VIII) Memory.
(IX) Admissibility.

A. Axiom of Causality

The general constitutive equations (5.3.3) do not make any distinction between the sets of dependent and independent variables. Accordingly, any 20 of the sets (5.3.2) may be chosen to be dependent variables. If these variables at (**X**, t) can be solved from (5.3.3) in terms of the rest, we would have explicit functional mappings between the dependent and independent variables. These mappings need not, and in general, will not, be one-to-one. However, for a large class of materials, cause and effect may be recognized within certain domains of definitions at the outset, thus helping to replace (5.3.3) with an explicit set in which the independent and dependent variables are distinguished. In a material body, thermomechanical effects are produced by motion and thermal changes (Eringen [1966a]). Thus, the thermomechanical independent variables are

$$\mathbf{x} = \mathbf{x}(\mathbf{X}, t), \qquad \theta = \theta(\mathbf{X}, t), \tag{5.4.1}$$

the velocity $\mathbf{v} = \dot{\mathbf{x}}$, the acceleration $\mathbf{a} = \dot{\mathbf{v}}$, the velocity gradient $v_{k,l}$, and the mass density $\rho = \rho_0(\det(x_{k,K})$ are all derivable from (5.4.1). In addition, when the body is responsive to external electromagnetic effects, it may be polarized and magnetized, and may conduct electricity. These effects, as we have seen in Chapter 2, are the result of the redistribution and motion of bound and free charges. Thus, it appears reasonable to consider

$$\begin{aligned} \mathbf{\Pi} &= \mathbf{\Pi}(\mathbf{X}, t), \qquad \mathbf{M} = \mathbf{M}(\mathbf{X}, t), \\ \boldsymbol{\mathcal{J}} &= \boldsymbol{\mathcal{J}}(\mathbf{X}, t), \qquad q_e, \end{aligned} \tag{5.4.2}$$

as the electromagnetic independent variables. Given $\boldsymbol{\pi}$, $\mathbf{E}$ is determined by a constitutive equation. Through Gauss' law (5.2.6), q_e is then determined, so that q_e may be exluded from our list of independent variables[1] (see Axiom of Admissibility).

The choice of $\mathbf{\Pi}$ and $\mathbf{M}$ as the electromagnetic independent variables is not unique, nor does it represent the most convenient choice in all cases. Many possibilities exist,[2] e.g.,

$$(\mathbf{P}, \boldsymbol{\mathcal{M}}), \qquad (\mathbf{\Pi}, \mathbf{M}), \qquad (\mathbf{\Pi}, \mathbf{B}),$$
$$(\boldsymbol{\mathcal{E}}, \mathbf{B}), \qquad (\mathbf{E}, \mathbf{B}), \qquad (\mathbf{D}, \mathbf{H}), \ldots .$$

This is adequate to indicate how we may end up with different formulations. Indeed, the literature is full of such choices. While all these formulatons are, in general, equivalent, the passage from one to the other often requires the unique invertibility of nonlinear equations. Approximate theories constructed from each theory differ from one another in a basic way. This situation bears some similarity to the case of nonlinear elasticity theory employing different strain measures (see Eringen [1962], [1975a]). Our approach will be to use the following two sets only:

$$(\mathcal{E}_K, B_K) \qquad \text{or} \qquad (\mathcal{E}_k, B_k), \tag{5.4.3}$$

of which the first set is suitable for solids and the second for fluids. These sets are useful in discussing both conducting and nonconducting materials, while some other sets are not. For example, (Π_K, M_K) or (Π_k, B_k) would be suitable for dielectric materials but they lose their convenience for conducting bodies.

Axiom (I) (Causality). *The motion, temperature, electric field, and magnetic induction of the material points of a body are self-evident and observable in any*

---

[1] If the electronic pressure is not negligible, then q_e must be included in the list of variables.

[2] Jordan and Eringen [1964a, b], employing C_{KL}, E_K, H_K, and $\theta_{,K}$, constructed nonlinear polynomial constitutive equations.

thermoelectromechanical behavior of a body. The remaining quantities (other than those derivable from the motion, temperature, electric field, and magnetic induction) excluding the body force, energy supply, and free charge density that enter the balance laws and the entropy inequality, are the dependent variables.

Thus, the dependent variables are

$$\mathscr{Z} = \{{}_{\mathrm{E}}T_{KL}, Q_K, \Psi, \eta, \Pi_K, M_K, \mathscr{J}_K\}, \tag{5.4.4}$$

and the independent variables are

$$\begin{aligned}\mathscr{Y}' = \mathscr{Y}(\mathbf{X}', t') = \{x'_k = x_k(\mathbf{X}', t'), \theta' = \theta(\mathbf{X}', t'),\\ \mathscr{E}'_K = \mathscr{E}_K(\mathbf{X}', t'), B'_K = B_K(\mathbf{X}', t')\}.\end{aligned} \tag{5.4.5}$$

A prime indicates that the collection includes these functions for *all material points* $\mathbf{X}'$ of the body, and all past times $t' \leq t$. A reference point in the body will be denoted by $\mathbf{X}$, and a specific time by t. The collection (5.4.4) is a composite quantity, in a 20-dimensional Euclidean product space, and (5.4.5) is one, in a 10-dimensional Euclidean product space.

The variables listed in (5.4.4) are considered to be functionals of those listed in (5.4.5) for *all points* $\mathbf{X}'$ of the body, and for all *past time* $t' \leq t$.

B. Axiom of Determinism

Axiom (II) (Determinism). *The value of $\mathscr{Z}$, at a material point $\mathbf{X}$ of the body $\mathscr{B}$ at time t, is determined by the history $\mathscr{Y}'$ of all material points of $\mathscr{B}$.*

This axiom declares merely the basis of science as determinism. In contrast to Newtonian point mechanics and the rigid-body electromagnetism of Maxwell, to determine the future outcome here we need not only information on a particular state of the body (e.g., the initial state), but also on all past histories. This axiom also excludes quantum mechanical concepts of indeterminacy.

According to this axiom, the constitutive equations may be written, symbolically, as

$$\mathscr{Z}(\mathbf{X}, t) = \mathscr{F}[\mathscr{Y}(\mathbf{X}', t'); \mathbf{X}, t], \qquad \mathbf{X}' \in \mathscr{B}, \quad -\infty < t' \leq t. \tag{5.4.6}$$

This means that *each member* of (5.4.4) is a functional of *all members* of (5.4.5), e.g.,

$$\Psi(X, t) = \hat{\Psi}[\mathbf{x}(\mathbf{X}', t'), \theta(\mathbf{X}', t'), \mathscr{E}(\mathbf{X}', t'), \mathbf{B}(\mathbf{X}, t); \mathbf{X}', t],$$

$${}_{\mathrm{E}}T_{KL}(\mathbf{X}, t) = {}_{\mathrm{E}}\hat{T}_{KL}[\mathbf{x}(\mathbf{X}', t'), \theta(\mathbf{X}', t'), \mathscr{E}(\mathbf{X}', t), \mathbf{B}(\mathbf{X}', t'); \mathbf{X}, t].$$

Of course, ${}_{\mathrm{E}}\hat{T}_{KL}$ is a symmetric tensor-valued functional, $\hat{\Psi}$ and $\hat{\eta}$ are scalar-valued functionals, and $\hat{Q}_K$, $\hat{\mathscr{E}}_K$, and $\hat{B}_K$ are vector-valued functionals. They are also functions of $\mathbf{X}$ and t.

We note that (5.4.6) is a special case of (5.3.3), resulting from (5.3.3) when these functions are soluble for the cause set at $(\mathbf{X}, t)$. Naturally, (5.3.3) then includes multiple-valued functional solutions, including rates and gradients of $\mathscr{F}$.

C. Axiom of Equipresence

Axiom (III) (Equipresence). *At the outset, all constitutive response functionals are to be considered to depend on the same list of constitutive variables, until the contrary is deduced.*

This axiom assigns equal rights to all members of the response functionals $\mathscr{F}$, to depend on *all* independent variables. Thus, in the beginning, each member of $\mathscr{Z}$ will be considered to depend on all members of $\mathscr{Y}'$. However, other axioms and balance laws may force the elimination of some of the variables from some of the constitutive equations. Equipresence can hardly be called an axiom. It is, rather, a precautionary measure that is valid in all proper scientific methods.

D. Axiom of Objectivity

Axiom (IV) (Objectivity or Material Frame-Indifference). *The constitutive response functionals are form-invariant under arbitrary rigid motions of the spatial frame of reference and a constant shift of the origin of time.*

From physical considerations, it is evident that the thermomechanical properties of materials cannot depend on the motion of the observer. Excluding large velocities close to the speed of light in vacuum, this observation seems to be valid for all mechanical phenomena. For example, the spring constant of a spring, and the heat conduction coefficient of a solid, do not change when the test is carried out in a moving train. We observe, however, that the balance laws (e.g., Cauchy's equations of motion) are not invariant under *time-dependent* rigid motions of the frame of reference. Maxwell's equations are not invariant either, when referred to a frame of reference that is undergoing *time-dependent* rigid rotatory motions.

If two frames, $\bar{\mathbf{x}}$ and $\mathbf{x}$, can be made to coincide by a rigid motion and a constant time shift, then they must be related to each other by

$$\begin{aligned} &\bar{\mathbf{x}}(\mathbf{X}, t) = \mathbf{Q}(t)\mathbf{x}(\mathbf{X}, t) + \mathbf{b}(t), \\ &\mathbf{Q}\mathbf{Q}^{\mathrm{T}} = \mathbf{Q}^{\mathrm{T}}\mathbf{Q} = 1, \qquad \det \mathbf{Q} = 1; \qquad \bar{t} = t - a, \end{aligned} \tag{5.4.7}$$

where $\mathbf{Q}^{\mathrm{T}}$ denotes the transpose of the matrix $\mathbf{Q}$, $\mathbf{I}$ is the unit matrix, and a is a constant.

A special case of (5.4.7) is the *Galilean transformations*

$$\bar{\mathbf{x}}(\mathbf{X}, t) = \mathbf{Q}_0\mathbf{x}(\mathbf{X}, t) + \mathbf{V}_0 t + \mathbf{b}_0, \tag{5.4.8}$$

where $\mathbf{Q}_0$ is now a *constant* matrix satisfying the orthogonality condition $(5.4.7)_{2,3}$, and $\mathbf{V}_0$ and $\mathbf{b}$ are constant vectors. Referring to (5.4.8), both the balance laws and Maxwell's equations are invariant, the latter to an order of magnitude of $v^2/c^2 \ll 1$ (see Chapter 3).

Galilean transformations are physically equivalent to an arbitrary constant rigid rotation, a constant velocity translation of the frame of reference, and a constant shift of the origin of time. It seems that it would be less restrictive to employ (5.4.8) for electromagnetically active deformable bodies, instead of the more restrictive transformations (5.4.7). However, for fluent and memory-dependent materials, it is necessary to impose the more general invariance requirement (5.4.7) on the response functionals. Presently, no proof of contradiction to the applicability of (5.4.7) exists. We shall, therefore, posit that the response functionals do not violate the objectivity requirement (5.4.7). Often a more restricted form of (5.4.7) is used in which $\det \mathbf{Q} = \pm 1$, thus allowing the invariance under reflection of the spatial frame of reference. With this latter form, (5.4.7) implies the *spatial isotropy* of response functionals.

According to this axiom, the constitutive equations then must obey

$$\mathscr{F}[\mathscr{Y}(\mathbf{X}', t'); \mathbf{X}, t] = \mathscr{F}[\bar{\mathscr{Y}}(\mathbf{X}', t'); \mathbf{X}, \bar{t}], \tag{5.4.9}$$

where

$$\bar{\mathscr{Y}}(\mathbf{X}'t') = \{Q_{kl}(\bar{t}')x_l(\mathbf{X}', \bar{t}) + b_k(\bar{t}'), \theta(\mathbf{X}', \bar{t}'), \mathscr{E}_K(\mathbf{X}', \bar{t}'), B_K(\mathbf{X}', \bar{t}')\}, \tag{5.4.10}$$

$$\mathbf{Q}(\bar{t}')\mathbf{Q}^{\mathrm{T}}(\bar{t}') = \mathbf{Q}^{\mathrm{T}}(\bar{t}')\mathbf{Q}(\bar{t}') = \mathbf{1}, \qquad \det \mathbf{Q}(\bar{t}') = \pm 1. \tag{5.4.11}$$

We now examine (5.4.9) for the following three special cases:

(a) $\mathbf{Q} = \mathbf{1}$, $\quad \mathbf{b} = \mathbf{0}$, $\quad a = t$ (time shift);
(b) $\mathbf{Q} = \mathbf{1}$, $\quad \mathbf{b} = -\mathbf{x}(\mathbf{X}, t')$, $\quad a = 0$ (translation);
(c) $\mathbf{Q} \neq \mathbf{1}$, $\quad \mathbf{b} = \mathbf{0}$, $\quad a = 0$ (rotation);

which are equivalent to the full use of (5.4.7). The use of (a) and (b) shows that

$$\mathscr{Z} = \mathscr{F}[G(\mathbf{X}', t - \tau'); \mathbf{X}], \tag{5.4.12}$$

where

$$\begin{aligned} G(X', t - \tau') = G' = \{&x_k(\mathbf{X}', t - \tau') - x_k(\mathbf{X}, t - \tau'), \theta(\mathbf{X}', t - \tau'), \\ &\mathscr{E}_K(\mathbf{X}', t - \tau'), B_K(\mathbf{X}', t - \tau')\}, \end{aligned} \tag{5.4.13}$$

$$\tau' = t - t' \geq 0, \qquad 0 \leq \tau' < \infty.$$

The condition (c) restricts (5.4.12) further by

$$\mathscr{F}[\bar{G}(\mathbf{X}', t - \tau'); \mathbf{X}] = \mathscr{F}[G(\mathbf{X}', t - \tau'); \mathbf{X}], \tag{5.4.14}$$

where

$$\begin{aligned} \bar{G}(\mathbf{X}', t - \tau') = \{&Q_{kl}(t - \tau')[x_k(\mathbf{X}', t - \tau') - x_k(\mathbf{X}, t - \tau')], \theta(\mathbf{X}', t' - \tau'), \\ &\mathscr{E}_K(\mathbf{X}', t - \tau'), B_K(\mathbf{X}', t - \tau')\} \end{aligned} \tag{5.4.15}$$

for all members of the group $\{\mathbf{Q}\}$.

E. Axiom of Time Reversal

In Section 3.12, we have already seen that invariance of the balance laws under microscopic time reversal places restrictions on the fields, as described by (3.12.5), i.e., as $t \to -t$

$$
\begin{aligned}
(\rho, t_{kl}, q_k, e, \theta_{,k}, \Psi) &\to (\rho, t_{kl}, q_k, e, \theta_{,k}, \Psi), \\
(\mathbf{P}, \mathscr{E}, \mathbf{D}, \mathbf{E}) &\to (\mathbf{P}, \mathscr{E}, \mathbf{D}, \mathbf{E}), \qquad (5.4.16)\\
(\mathscr{M}, \mathbf{B}, \mathbf{H}, \mathscr{J}) &\to -(\mathscr{M}, \mathbf{B}, \mathbf{H}, \mathscr{J}),
\end{aligned}
$$

Axiom (V). *Thermodynamically this is equivalent to: The entropy production must be nonnegative under time reversal.*

$$\rho\gamma \geq 0 \qquad \text{for} \quad t \to -t. \qquad (5.4.17)$$

F. Axiom of Material Invariance

The form of the response functionals, in general, depends on the reference configuration $\boldsymbol{\kappa}$ selected in a neighborhood $N(\mathbf{X})$ of $\mathbf{X}$. If a different reference configuration $\bar{\boldsymbol{\kappa}}$ is selected, which is related to $\boldsymbol{\kappa}$ by

$$\bar{\mathbf{X}} = \boldsymbol{\Lambda}(\mathbf{X}), \qquad \bar{\mathbf{X}}' = \boldsymbol{\Lambda}(\mathbf{X}'), \qquad (5.4.18)$$

the response functionals will, in general, be different. To indicate this, we attach a subscript to the response functionals. Thus, for a purely local thermomechanical medium, we have

$$\{\hat{\mathbf{T}}_{\boldsymbol{\kappa}}, \hat{\mathbf{Q}}_{\boldsymbol{\kappa}}, \hat{\Psi}_{\boldsymbol{\kappa}}, \hat{\eta}_{\boldsymbol{\kappa}}\} = \{\mathbf{S}\hat{\mathbf{T}}_{\bar{\boldsymbol{\kappa}}}\mathbf{S}^{\mathrm{T}}, \mathbf{S}\hat{\mathbf{Q}}_{\bar{\boldsymbol{\kappa}}}, \hat{\Psi}_{\bar{\boldsymbol{\kappa}}}, \hat{\eta}_{\bar{\boldsymbol{\kappa}}}\}, \qquad (5.4.19)$$

where $\mathbf{S}^{\mathrm{T}}$ indicates the transpose of the matrix

$$S_{KL} = \Lambda_{K,L}, \qquad \text{i.e.,} \quad (\mathbf{S}^{\mathrm{T}})_{KL} = S_{LK}. \qquad (5.4.20)$$

If in the neighborhoods of two points $\mathbf{X}$ and $\bar{\mathbf{X}}$ in a body (5.4.19) is valid and the densities are the same ($\rho_{\bar{\boldsymbol{\kappa}}} = \rho_{\bar{\boldsymbol{\kappa}}}$), then it is clear that the thermomechanical material properties of the body at $\mathbf{X}$ and $\bar{\mathbf{X}}$ are indistinguishable. In this case, we say that the material points $\mathbf{X}$ and $\bar{\mathbf{X}}$ are *mechanically isomorphic. A body is called mechanically uniform if all points of the body are mechanically isomorphic to one another.* Such a body may be regarded as consisting of the same kind of material points as far as its mechanical properties are concerned. However, the reference configuration may not be the same for all points of the body. If, however, a *single* reference configuration can be found for all points of the body with (5.4.19) satisfied and $\rho_{\boldsymbol{\kappa}} = \rho_{\bar{\boldsymbol{\kappa}}}$, then the body is said to be *mechanically homogeneous*. Thus, a mechanically homogeneous body is mechanically uniform, but the converse may not be true.

It may happen that, at a material point, there may exist a group of transformations $\{\boldsymbol{\Lambda}(\mathbf{X})\}$ for which the response functionals may have the same functional form. In this case, (5.4.19) gives

$$\{\hat{\mathbf{T}}_{\boldsymbol{\kappa}}, \hat{\mathbf{Q}}_{\boldsymbol{\kappa}}, \Psi_{\boldsymbol{\kappa}}, \eta_{\boldsymbol{\kappa}}\} = \{\mathbf{S}\hat{\mathbf{T}}_{\boldsymbol{\kappa}}\mathbf{S}^{\mathrm{T}}, \mathbf{S}\hat{\mathbf{Q}}_{\boldsymbol{\kappa}}, \Psi_{\boldsymbol{\kappa}}, \eta_{\boldsymbol{\kappa}}\}, \qquad \det \mathbf{S} = \pm 1. \quad (5.4.21)$$

Then, the material possesses symmetry regulations expressed by the group of transformations $\{\mathbf{S}\}$.

One of the simplest of these classes of transformations of $\mathbf{X}$, which leaves density unchanged, is of the form

$$\bar{\mathbf{X}} = \mathbf{S}\mathbf{X} + \mathbf{B}, \tag{5.4.22}$$

where $\mathbf{B}$ is a constant vector and $\mathbf{S}$ is a constant matrix subject to

$$\mathbf{S}\mathbf{S}^{\mathrm{T}} = \mathbf{S}^{\mathrm{T}}\mathbf{S} = \mathbf{I}, \qquad \det \mathbf{S} = \pm 1. \tag{5.4.23}$$

The totality of $\{\mathbf{S}\}$ forms a group, called the *isotropy group*.

Clearly, materials possessing such constant symmetry groups constitute a rather special class in the general materials, with group symmetries that change from point to point. All crystal classes are known to fall into this special case. Homogeneous materials do not depend on $\mathbf{B}$. Since all members of the group $\{\mathbf{S}\}$ can be represented by rotations about an axis, and a reflection in a plane, a point in the body is always fixed. Thus, the group $\{\mathbf{S}\}$ is called a *point group*. The group $\{\mathbf{S}\}$ is, in general, a subgroup of the full orthogonal group $\mathcal{O}$. When $\{\mathbf{S}\}$ is the full group, we say that the material is *isotropic*. Otherwise, the material is said to be anisotropic. If $\{\mathbf{S}\}$ is the **proper orthogonal group** ($\det \mathbf{S} = +1$), we say that the material is *hemitropic*. Including the reflection of the material coordinate axis X_K, there exist 12 members of the full group that characterize the material symmetry of all known 32 classes of crystals.[3] While any subgroup is acceptable to the theory, for real crystals, eleven members of the proper group and one reflection are adequate to describe the mechanical properties of these solids.

Thus far we have avoided mentioning symmetries when electromagnetic effects are present. Magnetic symmetry properties of solids cannot be discussed rationally by means of three-dimensional point groups only. Since magnetism is the result of the (spin) magnetic moment of electrons, we must now include into the symmetry group the spin orientation (a quantum mechanical phenomenon). It is well known that only two possibilities exist for the orientation of the spin. The characteristic values of the spin in any prescribed direction (measured in the units $\hbar/2$ where $\hbar$ is Planck's constant) are $+1$ and -1. A group of rotations $\{\mathbf{S}\}$ may bring a crystal to identical atomic orientations. However, the spins of each atom (for example) may be oriented oppositely.

On the atomic scale component parts of the crystal are in continual motion. Clearly, the structure of the crystal depends on the average positions of its component parts. If, on the atomic scale, there is a relative motion of various parts of the crystal that possess a nonzero time average, then this information is not included in the geometric configuration of the crystal. For example, ferromagnetic, ferrimagnetic, and antiferromagnetic crystals are characterized by an orderly distribution of their spin magnetic moments. On the other hand,

[3] For a discussion of the crystallographic group, see Appendix A.

diamagnetic and paramagnetic crystals do not exhibit such an orderly distribution of their spin magnetic moments. These latter crystals are "time-symmetric" and, therefore, the crystallographic point group is all that is needed for a discussion of the material symmetries of these latter materials. However, for ferromagnetic, ferrimagnetic, and antiferromagnetic materials, we need an additional symmetry operator R which will take care of the time reversal. While R itself may not be a symmetry operator for these materials, when it is combined with the space symmetry, it may become a symmetry operator.

The symmetry group of electromagnetic materials, rationally, should be based on a four-dimensional space–time group. However, since time reversal (or spin angular momentum) possesses only two possibilities, it is possible to study full magnetic group symmetries by means of the point group $\{\mathbf{S}\}$ and the time reversal operator R. Note, for example, that the time reversal operator R leaves the charge unchanged, but changes the sign of magnetization and current vectors. By means of this operator, and using the point group, we can construct the entire 90 magnetic symmetry groups, of which 58 have a magnetic origin.[4]

When the translation symmetry operations, represented by $\{\mathbf{B}\}$, are combined with the point symmetry group, additional symmetry operations are obtained. For example, after a rotation about an axis by translating one part of the crystal lattice with respect to another, we may obtain the original configuration of the crystal. Thus, inclusion of the translatory space operations creates a larger group symmetry. This group is called the *space group* which, for the nonmagnetic crystals, possesses 230 members. Such a symmetry is important, primarily on the atomic scale, and it shows up in X-ray diffractions, as well. The spatial symmetry group generally does not affect the macroscopic properties of materials.[5] For example, the microscopic description that makes use of the space group, distinguishes between symmetry (rotation and translation) and rotational symmetry. On the macroscopic scale, this distinction cannot be detected.

For magnetic crystals, the space group is even larger. The magnetic space group (which is often called Shubnikov's group in honor of its realization by him in 1951) possesses 1,421 members. For a thorough description of the magnetic space group, we refer readers to Shubnikov and Belov [1964].

In its most general form, the axiom of material invariance may be expressed as:

Axiom (VI) (Material Invariance). *The constitutive response functionals must be form-invariant with respect to a group of transformations of the material frame of reference* $\{\mathbf{X} \to \bar{\mathbf{X}}\}$ *and "microscopic time reversal" as* $\{t \to -t\}$ *repre-*

[4] A discussion of the magnetic group is given in Appendix B, For a complete account, see Kiral and Eringen [1976].

[5] An exception to this is, of course, composite materials.

senting the material symmetry conditions. These transformations must leave the density and charge at $(\mathbf{X}, t)$ *unchanged.*

The material symmetry regulations may be curvilinear. For example, automobile tires, with winding steel wires to strengthen them, possess curvilinear anisotropy. In this case, the symmetry will have to be expressed in curvilinear coordinates.

Materials may be subject to other geometrical or internal constraints. For example, rubbery materials are known to be incompressible. This is expressed by the condition that local volume is preserved at all times, i.e., $\mathrm{III}_C = 1$. More generally, internal conditions of materials may not permit deformations, polarization, conduction, etc., in certain directions. For example, materials may contain inextensible strings (fiber composites), conduction may be unidirectional, they may be nonmagnetizable (vanishing magnetization), or nonpolarizable (vanishing polarization), etc. In such cases, additional equations of constraints are written, e.g.,

$$
\begin{aligned}
&f_\alpha(\mathbf{E}) = 0, \quad \alpha = 1, 2, \ldots, n < 6 && \text{(inextensible strings)},\\
&\mathbf{M} = \mathbf{0} && \text{(nonmagnetizable materials)}, \qquad (5.4.24)\\
&\mathbf{\Pi} = \mathbf{0} && \text{(nonpolarizable materials)}.
\end{aligned}
$$

These constraints can also be expressed in the form of the vanishing of the power associated with the constraint, e.g.,

$$T^{Kl} v_{l,K} = 0,$$

where T^{Kl} is the unknown stress of constraints. For example, in the case of the constraint $v_{k,k} = 0$, T^{Kl} reduces to an unknown scalar called hydrostatic pressure. Clearly, the constraints may take the form of functionals involving differential and integral operators in more general situations.

G. Axiom of Neighborhood

Axiom (VII) (Neighborhood). *The values of the response functionals at* $\mathbf{X}$ *are not affected appreciably by the values of the independent constitutive variables at distant points from* $\mathbf{X}$ (Eringen [1966a]).

This axiom states that the independent variable set $\mathscr{X}(\mathbf{X}'t')$, associated with the material point $\mathbf{X}'$, will not contribute to the response functionals $\mathscr{F}(\mathbf{X}, t)$ at $\mathbf{X}$ when $|\mathbf{X}' - \mathbf{X}| > d$ where d is a constant. In other words, the material response is relatively insensitive to the motion, temperature, polarization, and magnetization of distant points. Since physical phenomena are the result of relative changes in the neighborhood of $\mathbf{X}$, the greatest contribution is expected from the points near $\mathbf{X}$. Mathematically, this axiom can be expressed in the form of a continuity requirement for the response functionals. In fact, suppose that $\mathscr{F}$ is a *uniformly* continuous functional of $G(\mathbf{X}')$, i.e., if $G_1(\mathbf{X}')$ and $G_2(\mathbf{X}')$ are any two functions in the class of functions $\{G\}$, then for any $\varepsilon > 0$ there exists a quantity $\delta(\varepsilon)$ such that[6]

$$|\mathscr{F}[G_1(\mathbf{X}')] - \mathscr{F}[G_2(\mathbf{X}')]| < \varepsilon, \tag{5.4.25}$$

whenever

$$\|G_1(\mathbf{X}') - G_2(\mathbf{X}')\| < \delta, \tag{5.4.26}$$

where $\|\cdot\|$ denotes a properly defined norm, e.g., for a vector function $\mathbf{F}(\mathbf{X}')$

$$\|\mathbf{F}\|^2 = \sup[\mathbf{F}(\mathbf{X}')\cdot\mathbf{F}(\mathbf{X}')], \qquad \mathbf{X}' \in V - \Sigma, \tag{5.4.27}$$

or[7]

$$\|\mathbf{F}(\mathbf{X})\|^2 = \int_{V-\Sigma} H(|\mathbf{X}' - \mathbf{X}|)\mathbf{F}(\mathbf{X}')\cdot\mathbf{F}(\mathbf{X}')\,dV(\mathbf{X}'), \tag{5.4.28}$$

where $H(\mathbf{X}')$ (called *alleviator*) is a positive decreasing function of $|\mathbf{X}'|$, $H(0) = 1$. Once the continuity of the response functionals is stated, then we can proceed to obtain various approximations. For our purposes the most important class is that of *simple materials* which is a subclass of *gradient-dependent materials.*

If $G(\mathbf{X}')$ is expressible as a convergent Taylor series about $\mathbf{X}' = \mathbf{X}$, then, according to the Stone–Weierstrass theorem, it may be approximated to any desired degree of accuracy by including a sufficiently large number of terms in the series

$$G^{(M)}(\mathbf{X}') = \sum_{\mu=0}^{M} \frac{1}{\mu!}(\mathbf{X}'_{K_1} - \mathbf{X}_{K_1})\dots(\mathbf{X}'_{K_\mu} - \mathbf{X}_{K_\mu})G_{,K_1\dots K_\mu}(\mathbf{X}). \tag{5.4.29}$$

The functional $\mathscr{F}[G(\mathbf{X}')]$ may now be uniformly approximated by the functions $G_{,K_1\dots K_\mu}$ (still being a functional with respect to τ'), with $\mu = 0, \dots, M$.

Simple Solids. The case in which $M = 1$ is called *simple materials.* These materials are *local* in character, i.e., the effect of long-range interatomic forces is negligible. Thus, for simple memory-dependent materials, the most general form of the constitutive equations is

$$\begin{aligned}\mathscr{Z}(\mathbf{X}, t) = \mathscr{F}[&x_{k,K}(t-\tau'), \theta(t-\tau'), \theta_{,K}(t-\tau'), \mathscr{E}_K(t-\tau'),\\ &\mathscr{E}_{K,L}(t-\tau'), B_K(t-\tau'), B_{K,L}(t-\tau'); \mathbf{X}],\end{aligned} \tag{5.4.30}$$

where, here and henceforth, we avoid indicating the dependence of the argument functions on $\mathbf{X}$, for brevity. We notice that the constitutive equations (5.4.30) include electromagnetic gradient-dependent materials. If $\mathscr{E}_{K,L}$ and $B_{K,L}$ are included in (5.4.30), we need to modify the balance laws also.

---

[6] For brevity, we suppress the dependence of G on $t - \tau'$.

[7] The choice of the norm depends on the internal structure of the body, e.g., the nature of intermolecular forces.

Thus, in general, the dependence on $\mathscr{E}_{K,L}$ and $B_{K,L}$ will also be disregarded presently. In Chapter 9, we shall return to these cases. Thus, for memory-dependent simple deformable electromagnetic solids, we take

$$\mathscr{Z}(\mathbf{X}, t) = \mathscr{F}[x_{k,K}(t-\tau'), \theta(t-\tau'), \theta_{,K}(t-\tau'), \mathscr{E}_K(t-\tau'), B_K(t-\tau'); \mathbf{X}]. \tag{5.4.31}$$

If the axiom of objectivity (5.4.14) is employed we will have

$$\mathscr{F}[\mathbf{Q}(t-\tau')\mathbf{x}_{,K}(t-\tau'), \ldots] = \mathscr{F}[\mathbf{x}_{,K}(t-\tau'), \ldots], \tag{5.4.32}$$

where the only term that is affected by the transformation $\mathbf{Q}$ is $\mathbf{x}_{,K}(t-\tau')$, since all other members of $\mathscr{F}$ are scalars. According to a theorem of Cauchy (see Eringen [1967, App. B-6]), the restriction (5.4.32) for the *full* orthogonal group $\mathbf{Q}$ implies that $\mathscr{F}$ can depend only on the inner products of $\mathbf{x}_{,K}$, i.e.,

$$\mathscr{Z}(\mathbf{X}, t) = \mathscr{F}[C_{KL}(t-\tau'), \theta(t-\tau'), \theta_{,K}(t-\tau'), \mathscr{E}_K(t-\tau'), B_K(t-\tau'); \mathbf{X}]. \tag{5.4.33}$$

This is the final form for the most general memory-dependent simple electromagnetic materials.

H. Axiom of Memory

Axiom (VIII) (Memory). *The values of the constitutive variables, at a distant past from the present, do not affect appreciably the values of the constitutive response functionals at present.*

This axiom is the counterpart of the axiom of neighborhood, in the time domain. Most materials are so constituted that the physical processes that happened to them in a distant past are mostly erased. In other words, a large class of materials possesses a weak memory. In a completely analogous way to the axiom of neighborhood, we can express this requirement as a continuity requirement on the response functionals. These expressions are analogous to (5.4.25)–(5.4.28), with $\mathbf{X}'$ replaced by τ', $\mathbf{X}$ by t and the volume integral in (5.4.28) by an integral over τ from $-\infty$ to t. The rate-dependent materials can then be constructed. To this end, the Taylor expansion (5.4.29) is replaced by an expansion in time, about $\tau' = t$. This results in a *constitutive function* (not functional) in the time rates of C_{KL}, θ, $\theta_{,K}$, $\mathscr{E}_K$, and B_K at time t, i.e.,

$$\mathscr{Z}(\mathbf{X}, t) = \mathscr{F}[\mathbf{C}, \dot{\mathbf{C}}, \ldots, \mathbf{C}^{(m)}, \theta, \dot{\theta}, \ldots, \theta^{(n)}, \theta_{,K}, \dot{\theta}_{,K}, \ldots, \theta^{(p)}_{,K},$$
$$\mathscr{E}_K, \dot{\mathscr{E}}_K, \ldots, \mathscr{E}^{(r)}_K, B_K, \dot{B}_K, \ldots, B^{(s)}_K; \mathbf{X}]. \tag{5.4.34}$$

This is the master equation for all rate-dependent simple materials.

The constitutive functions $\mathscr{F}$ in (5.4.34) express the memory-dependence by simply using a large number of rates at $\tau' = t$. This is tedious and often not acceptable since the constitutive independent variables may possess discontinuities in certain rates (e.g., a suddenly stretched or cooled solid). It is often preferable to employ the general form (5.4.33), with a continuity requirement that only $G(\tau' - t)$ near $\tau' = t$ strongly influences $\mathscr{F}(\mathbf{X}, \mathbf{t})$. The degree of influence is open to our discretion, in view of the absence of any other physical principle or experimental findings. To this end, following Coleman and Noll [1961], we introduce the axiom of *fading memory*. In essence, this is a continuity requirement for the response functionals which may be expressed as follows: For every $\varepsilon > 0$, there exist $\delta(\varepsilon)$ such that

$$|\mathscr{F}[G_1(t')] - \mathscr{F}[G_2(t')]| < \varepsilon, \tag{5.4.35}$$

whenever

$$\|G_1(t') - G_2(t')\| < \delta, \tag{5.4.36}$$

where the norm $\|\cdot\|$ of G is defined by

$$\|G\| = (G, G)_I^{1/2}, \tag{5.4.37}$$

where $(\cdot, \cdot)_I$ indicates the weighted inner product of two such sets of functions, e.g.,

$$(\mathbf{G}_1, \mathbf{G}_2)_I = \int_{-\infty}^{t} I(t - t')\mathbf{G}_1(t') \cdot \mathbf{G}_2(t')\, dt'. \tag{5.4.38}$$

The scalar product $\mathbf{G}_1 \cdot \mathbf{G}_2$ for the set G is defined by a positive definite quadratic form

$$(G, G) = A_{KLMN}C_{KL}C_{MN} + \mathscr{C}\theta^2 + K_{KL}\theta_{,K}\theta_{,L} + \mathscr{E}_{KL}\mathscr{E}_K\mathscr{E}_L + \mu_{KL}B_K B_L. \tag{5.4.39}$$

In (5.4.38), the alleviator $I(t')$ is a positive decreasing function of t' such that $I(0) = 1$. The precise shape of this function is left to our disposal.

The space of functions for G is now a Hilbert space with an influence norm. The continuity requirement (5.4.35) now allows us to state the fading memory axiom as:

Axiom of Fading Memory. *The response functionals are continuous from below in a neighborhood of the history of constitutive independent variables at time t in the Hilbert space defined according to* (5.4.38).

I. Axiom of Admissibility

Axiom (IX) (Admissibility). *Constitutive equations must be consistent with the balance laws and the entropy inequality.*

This seemingly obvious rule, which certainly underlies all scentific disciplines, has far-reaching consequences. For example, since the mass density must obey the equation of conservation of mass

$$\rho_0/\rho(\mathbf{X}, t) = \det[x_{k,K}(\mathbf{X}', t')], \tag{5.4.40}$$

we discarded $\rho(\mathbf{X}', t')$ from the list of constitutive independent variables, unless the dependence on $x_{k,K}$ is through det $x_{k,K}$. A less trivial and deeper consequence of this axiom is through the restrictions placed on the response functionals by the second law of thermodynamics. In fact, since this law must not be violated for all physical processes, we shall find that the thermodynamic admissibility restricts the constitutive equations severely.

5.5. Integrity Basis Under Crystallographic Point Groups

Crystal symmetry places restrictions on the constitutive response functionals. For polynomial constitutive functions, an extensive literature exists for the construction of the polynomial invariants of the independent constitutive variables under a given group of transformations (see Spencer [1971], Smith [1968], [1971], Wang [1969a, b], [1970], Kiral and Smith [1974]). For a comprehensive account on magnetic groups, see Kiral and Eringen [1976]. These invariants form the so-called integrity basis, which constitutes a set of independent variables in the constitutive polynomials restricted by the symmetry of the material. The smallest set of these invariants that are functionally independent is referred to as the *minimal integrity basis*. This set is all that is needed for the construction of the constitutive polynomials, since all other polynomial invariants are expressible in terms of the minimal basis.

The symmetry properties of a given class of materials may be defined by listing a set of matrices $\{\mathbf{S}\} = (\mathbf{S}^1, \ldots, \mathbf{S}^N)$ (see Appendix A, Table A.2), which may be referred to as the crystallographic group $\{\mathbf{S}\}$. As we know, each symmetry transformation, $\alpha = 1, 2, \ldots, N$, is an orthogonal matrix satisfying

$$\mathbf{S}\mathbf{S}^{\mathrm{T}} = \mathbf{S}^{\mathrm{T}}\mathbf{S} = \mathbf{I}, \qquad \det S = \pm 1. \tag{5.5.1}$$

Under each of these transformations, a set of unit Cartesian vectors $\mathbf{I}_K$, $K = 1, 2, 3$, is transformed into another set by

$$\bar{\mathbf{I}}_K = S^{\alpha}_{KL}\mathbf{I}_L, \qquad \alpha = 1, 2, \ldots, N, \quad K, L = 1, 2, 3. \tag{5.5.2}$$

The basic vectors $\bar{\mathbf{I}}^{\alpha}_K$ are the basis vectors of N rectangular frames of reference $\bar{\mathbf{X}}^{\alpha} = \mathbf{S}^{\alpha}\mathbf{X}$, for a given crystal class. Let the independent tensor and vector constitutive variables be denoted by $\mathbf{C}_a$, $\mathbf{p}_b$, $\mathbf{a}_c$ ($a = 1, 2, \ldots, A$; $b = 1, 2, \ldots, B$; $c = 1, 2, \ldots, C$) in which $\mathbf{C}_a$ are symmetric second-order tensors, $\mathbf{p}_b$ are polar vectors, and $\mathbf{a}_c$ are axial vectors. The transformations of these are

$$\bar{\mathbf{C}}_a = \mathbf{S}\mathbf{C}_a\mathbf{S}^{\mathrm{T}}, \qquad \bar{\mathbf{p}}_b = \mathbf{S}\mathbf{p}_b, \qquad a_c = \mathbf{S}\mathbf{a}_c \det \mathbf{S}. \tag{5.5.3}$$

If we wish, we may replace $\mathbf{a}$ by a skew-symmetric tensor $\mathbf{W}$, where

$$W_{KL} = e_{KLM}a_M, \tag{5.5.4}$$

then $\mathbf{W}$ transforms as a second-order absolute tensor

$$\bar{\mathbf{W}}_c = \mathbf{S}\mathbf{W}_c\mathbf{S}^{\mathrm{T}}, \qquad c = 1, 2, \ldots, C. \tag{5.5.5}$$

The symmetry conditions for a scalar function Σ, a vector function $\mathbf{q}$, a tensor function $\mathbf{T}$, and a skew-symmetric tensor function $\mathbf{B}$, of the independent variables $\mathbf{C}_a$, $\mathbf{p}_b$, $\mathbf{W}_c$, require that

$$\begin{aligned}
\Sigma(\mathbf{C}_a, \mathbf{p}_b, \mathbf{W}_c) &= \Sigma(\bar{\mathbf{C}}_a, \bar{\mathbf{p}}_b, \overline{\mathbf{W}}_c),\\
\mathbf{q}(\mathbf{C}_a, \mathbf{p}_b, \mathbf{W}_c) &= \mathbf{S}^{\mathrm{T}}\mathbf{q}(\bar{\mathbf{C}}_a, \bar{\mathbf{p}}_b, \overline{\mathbf{W}}_c),\\
\mathbf{T}(\mathbf{C}_a, \mathbf{p}_b, \mathbf{W}_c) &= \mathbf{S}^{\mathrm{T}}\mathbf{T}(\bar{\mathbf{C}}_a, \bar{\mathbf{p}}_b, \overline{\mathbf{W}}_c)\mathbf{S},\\
\mathbf{B}(\mathbf{C}_a, \mathbf{p}_b, \mathbf{W}_c) &= \mathbf{S}^{\mathrm{T}}\mathbf{B}(\bar{\mathbf{C}}_a, \bar{\mathbf{p}}_b, \overline{\mathbf{W}}_c)\mathbf{S},
\end{aligned} \tag{5.5.6}$$

for all members of $\{\mathbf{S}\} = (\mathbf{S}^1, \mathbf{S}^2, \ldots, \mathbf{S}^N)$.

Introducing

$$Q = q_K \phi_K, \qquad T = T_{KL}\psi_{KL}, \qquad B = B_{KL}\Gamma_{KL}, \tag{5.5.7}$$

where ϕ_K is a polar vector, ψ_{KL} is a symmetric tensor, and Γ_{KL} is a skew-symmetric tensor, we reduce the study of the restriction (5.5.6) to the study of a single scalar form, as described by

$$\begin{aligned}
Q &= q_K\phi_K = Q(\mathbf{C}_a, \mathbf{p}_b, \mathbf{W}_c, \boldsymbol{\phi}),\\
T &= T_{KL}\psi_{KL} = T(\mathbf{C}_a, \mathbf{p}_b, \mathbf{W}_c, \boldsymbol{\psi}),\\
B &= B_{KL}\Gamma_{KL} = B(\mathbf{C}_a, \mathbf{p}_b, \mathbf{W}_c, \Gamma),
\end{aligned}$$

where the functions Q, T, and B are linear in $\boldsymbol{\phi}$, $\boldsymbol{\psi}$, and $\boldsymbol{\Gamma}$, respectively. But these forms are the same as (5.5.6), since each of $\boldsymbol{\phi}$, $\boldsymbol{\psi}$, and $\boldsymbol{\Gamma}$ could be incorporated into each of $\mathbf{p}_b$, $\mathbf{C}_a$, and $\mathbf{W}_c$. Thus, the problem of finding the general forms of (5.5.6) is reduced to that of

$$\Sigma(\mathbf{C}_a, \mathbf{p}_b, \mathbf{W}_c) = \Sigma(\bar{\mathbf{C}}_a, \bar{\mathbf{p}}_b, \overline{\mathbf{W}}_c), \tag{5.5.8}$$

where the range of a, b, c is enlarged by one for $(5.5.6)_2$, $(5.5.6)_3$, and $(5.5.6)_4$. Once this form is determined, then we obtain $\mathbf{q}$, $\mathbf{T}$, and $\mathbf{B}$ by

$$\mathbf{q} = \frac{\partial Q}{\partial \boldsymbol{\phi}}, \qquad \mathbf{T} = \frac{\partial T}{\partial \boldsymbol{\psi}}, \qquad \mathbf{B} = \frac{\partial B}{\partial \boldsymbol{\Gamma}}. \tag{5.5.9}$$

Since Q, T, and B can, at most, be linear in $\boldsymbol{\phi}$, $\boldsymbol{\psi}$, and $\boldsymbol{\Gamma}$, (5.5.9) will not contain these quantities.

The restrictions (5.5.8) constitute N nonlinear equations (one for each member $\mathbf{S}^\alpha$, $\alpha = 1, 2, \ldots, N$, of the group $\{\mathbf{S}\}$) which place severe restrictions on the arguments of Σ. Here we sketch the method of determining these restrictions for the crystallographic point group in the case when Σ is a polynomial in its arguments. For a detailed study of the constitutive equation of electromagnetic materials, the reader is referred to Kiral and Eringen [1976].

The argument set in a constitutive function may possess certain symmetries (e.g., symmetric, skew-symmetric tensors). If this is the case we must first obtain the independent components of these vectors and tensors. We assume that this is done.[8]

Thus, let Z_K, $K = 1, 2, \ldots, P$, denote the independent components of $\mathbf{C}_a$, $\mathbf{p}_b$, $\mathbf{W}_c$. For a given crystal class, all $\mathbf{S}^\alpha\mathbf{X}$, $\alpha = 1, 2, \ldots, N$, constitute N equivalent reference frames. In these coordinate systems, the independent components Z_K of the argument tensors will transform as

$$(\mathbf{S}^\alpha\mathbf{Z})_K = T^\alpha_{KL} Z_L, \tag{5.5.10}$$

where $\mathbf{T}^\alpha = \mathbf{T}(\mathbf{S}^\alpha)$, $\alpha = 1, 2, \ldots, N$, are $NP \times P$ matrices which describe the transformation properties of $\mathbf{Z}$ under the group $\{\mathbf{S}\}$. $\mathbf{T}^\alpha$ forms a *matrix representation* $\boldsymbol{\Gamma}$, *of degree* P, of the group $\{\mathbf{S}\}$, and $Z_1, Z_2, \ldots, Z_P$ are said to form the carrier space for the representation $\boldsymbol{\Gamma}$. By an appropriate change of variables of Z_K, it is possible to decompose $\mathbf{T}^\alpha$ into submatrices of lower degrees. To do this, we change the variable Z_K to Λ_K by

$$\Lambda_K = Q_{KL} Z_L, \tag{5.5.11}$$

where $\mathbf{Q}$ is a nonsingular matrix with an inverse $\mathbf{Q}^{-1}$. Here, Λ_K are the components of $\boldsymbol{\Lambda}$ referred to the reference frame $\mathbf{X}$. The components $(\mathbf{S}^\alpha\boldsymbol{\Lambda})_K$, in the reference frame $\mathbf{S}^\alpha\mathbf{X}$, are therefore calculated by

$$(\mathbf{S}^\alpha\boldsymbol{\Lambda})_K = Q_{KL}(\mathbf{S}^\alpha\mathbf{Z})_L = Q_{KL} T^\alpha_{LM} Q^{-1}_{MN} \Lambda_N. \tag{5.5.12}$$

The set of N matrices $\boldsymbol{\Gamma}^\alpha \equiv \mathbf{Q}\mathbf{T}^\alpha\mathbf{Q}^{-1}$, $\alpha = 1, 2, \ldots, N$, forms a matrix representation of the group $\{\mathbf{S}\}$, which is said to be equivalent to the representation $\boldsymbol{\Gamma} = \{\mathbf{T}(\mathbf{S}^1), \ldots, \mathbf{T}(\mathbf{S}^N)\}$. By a clever choice of $\mathbf{Q}$, however, it is possible to reduce matrices $\mathbf{Q}\mathbf{T}^\alpha\mathbf{Q}^{-1}$ to square matrices, located as blocks on the main diagonal of a matrix with zeros everywhere else, e.g.,

$$\begin{bmatrix} \mathbf{T}_1(\mathbf{S}^\alpha) & 0 & 0 \\ 0 & \mathbf{T}_1(S^\alpha) & 0 \\ 0 & & \mathbf{T}_2(\mathbf{S}^\alpha) \end{bmatrix} \equiv 2\mathbf{T}_1(\mathbf{S}^\alpha) \dot{+} \mathbf{T}_2(\mathbf{S}^\alpha).$$

With the scheme shown, we write this as a direct sum defined as above

$$\mathbf{Q}\mathbf{T}^\alpha\mathbf{Q}^{-1} \equiv \boldsymbol{\Gamma}^\alpha = n_1\mathbf{T}^\alpha_1 \dot{+} n_2\mathbf{T}^\alpha_2 \dot{+} \cdots \dot{+} n_r\mathbf{T}^\alpha_r. \tag{5.5.13}$$

If none of the matrices $\mathbf{T}^\alpha_R$, $R = 1, 2, \ldots, r$, can be written as a direct sum, then it is said that

$$\boldsymbol{\Gamma}_R \equiv \{\mathbf{T}^\alpha_R\} = \{\mathbf{T}_R(\mathbf{S}^\alpha)\}, \qquad \alpha = 1, 2, \ldots, N, \quad R = 1, 2, \ldots, r, \tag{5.5.14}$$

constitute r *irreducible representations* of the group $\{\mathbf{S}\}$. Here, n_R is the number

---

[8] The independent components of a tensor can be obtained systematically by means of Young Tableaux, which is especially useful for the tensors of a higher order than two (see Keown [1975, Chap. 4]). Since we will be dealing with second-order tensors at most, we do not dwell on this subject.

of times the irreducible representation $\{\mathbf{T}_R(\mathbf{S}^\alpha)\}$ is contained in the representation $\boldsymbol{\Gamma} = \{\boldsymbol{\Gamma}^\alpha\}$.

In group theory (see Lomont [1959, p. 55]), it is known that *the number of irreducible representations r of a finite group is equal to its number of classes.* If d_R is the degree (dimension) of the matrices of $\mathbf{T}_R^\alpha$, we also have $\sum_{R=1}^{r} n_R d_R = d$, where d is the degree of the representation $\boldsymbol{\Gamma}$.

The coefficient n_R is given by

$$n_R = \frac{1}{N} \sum_{\alpha=1}^{N} \chi(\mathbf{S}^\alpha) \overset{*}{\chi}_R(\mathbf{S}^\alpha), \qquad R = 1, 2, \ldots, r, \tag{5.5.15}$$

where N is the order of the group $\{\mathbf{S}\}$ and $\chi(\mathbf{S}^\alpha)$ is the *character* (the trace) of the representations $T_{KL}(\mathbf{S}^\alpha)$, $\alpha = 1, 2, \ldots, N$, and $\chi_R(\mathbf{S}^\alpha)$ is that of $\mathbf{T}_R \equiv \mathbf{T}_R(\mathbf{S}^\alpha)$. A superposed asterisk represents the complex conjugate.

With the proper choice of $\mathbf{Q}$, the variable Λ_K will split into the sets of quantities

$$(\Lambda_1, \ldots, \Lambda_K), (\Lambda_{K+1}, \ldots, \Lambda_L), \ldots, (\Lambda_{M+1}, \ldots, \Lambda_P)$$

such that each set forms the carrier space of one of the irreducible representations of the point group $\{\mathbf{S}\}$. These quantities are called *basic quantities.* Components $\Lambda_R(\mathbf{Z})$ of the basic quantities are obtained by using the formula (see Lomont [1959, pp. 75, 76]),

$$\Lambda_K = \sum_{\alpha=1}^{N} \overset{*}{\Gamma}_{1K}(\mathbf{S}^\alpha) T_{PQ}(\mathbf{S}^\alpha) Z_Q. \tag{5.5.16}$$

If, for $P = 1$, this equation gives $\Lambda_1 \neq 0$, then $\Lambda_2, \Lambda_3, \ldots$ are determined from (5.5.16) which, together with Λ_1, form the carrier space for $\boldsymbol{\Gamma}_R$. If $\Lambda_1 = 0$, then we taken $P = 2$, and repeat the process. If $\boldsymbol{\Gamma}_R$ appears n_R times in the decomposition of $\boldsymbol{\Gamma}$, we take $P = 1, 2, \ldots$ in (5.5.16), until we obtain n_R sets of components $(\Lambda_1, \ldots, \Lambda_{d_R}), \ldots, (M_1, \ldots, M_{d_R})$, which form the carrier space for the n_R representations $\boldsymbol{\Gamma}_R$. The degree of $\boldsymbol{\Gamma}_R$ is indicated by d_R. We must require that $n_R d_R$ components of $(\Lambda_1, \ldots, M_{d_R})$ be linearly independent.

Once the basic quantities are found, then the polynomial function will take a form like

$$\Sigma = P(\Lambda_1, \ldots, \Lambda_{d_R}, \ldots, M_1, \ldots, M_{d_R})$$

subject to certain symmetry restrictions, which will be inherent in the nature of the function P due to the symmetry group.

The method outlined above may be used to obtain the integrity basis of arbitrary numbers of vectors and tensors for all crystal classes. The steps for the calculations may be summarized as follows:

1. Given the basic matrices, (A.2), calculate $\{\mathbf{S}^\alpha\}$, $\alpha = 1, 2, \ldots, N$, for the crystal class under consideration. These are listed in Table A.2 for each of the 32 crystal classes.
2. Compute the character table, $\chi(\mathbf{S}^\alpha)$ of the representations $T_{KL}(\mathbf{S}^\alpha)$, $\alpha = 1, 2, \ldots, N$.

3. For the 32 classes of crystallographic point group, the irreducible representations $\boldsymbol{\Gamma}_R(\mathbf{S}^\alpha)$ and their character table $\chi_R(\mathbf{S}^\alpha)$ have been obtained in the literature.
4. Use (5.5.15) to determine n_R.
5. Use (5.5.16) to determine the carrier space $(\Lambda_1, \ldots, \Lambda_{d_R}), \ldots, (M_1, \ldots, M_{d_R})$ for each irreducible representation.
6. Use Theorems 1 to 7, given in Appendix D for easy access, to form the integrity basis for the crystal class under consideration.

Kiral and Smith [1974] and Kiral [1972] have carried out these calculations for a symmetric tensor S_{KL}, a polar vector p_K, and an axial vector a_K. The results of these calculations, from step 1 to step 5, are listed in Tables C1.1 to C1.16. For various crystal classes, entries following $\boldsymbol{\Gamma}_R$ indicate the basic quantities associated with the representation $\boldsymbol{\Gamma}_R$. For example, $\boldsymbol{\Gamma}_3$; ϕ, ψ indicates that ϕ and ψ are basic quantities associated with $\boldsymbol{\Gamma}_3$. The notation $\boldsymbol{\Gamma}_5$: (a_1, a_2), (b_1, b_2) indicates that (a_1, a_2) and (b_1, b_2) are basic quantities associated with the representation $\boldsymbol{\Gamma}_5$ of *degree two*.

The typical elements of the integrity basis for most of the crystal classes are listed following these tables. The typical elements of the integrity basis of degrees 1, 2, 3, ... are denoted by

$$\begin{aligned}
&1.\quad J_1(\phi_1), J_2(\psi_1), \ldots, \\
&2.\quad K_1(\phi_1, \phi_2), K_2(\psi_1, \psi_2), K_3(\phi_1, \psi_1), \qquad\qquad (5.5.17)\\
&3.\quad L_1(\phi_1, \phi_2, \phi_3), L_2(\phi_1, \phi_2, \psi_1), L_3(\phi_1, \psi_1, \psi_2), L_4(\psi_1, \psi_2, \psi_3).
\end{aligned}$$

The integrity basis for the polynomial functions $P(\phi_1, \ldots, \phi_{n_1}; \psi_1, \ldots, \psi_{n_2}; \ldots)$ depending on n_1 basic quantities $\phi_1, \ldots, \phi_{n_1}$; n_2 quantities $\psi_1, \ldots, \psi_{n_2}; \ldots$ can be obtained by replacing ϕ_1 by $\phi_1, \ldots, \phi_n$ and ψ_1 by $\psi_1, \ldots, \psi_{n_2}$ in all possible combinations with repetition included. Thus, the elements of the integrity basis of degrees 1, 2, 3, generated from the typical multilinear elements of the integrity basis (5.5.17), are given by

$$\begin{aligned}
&1.\quad J_1(\phi_i),\quad i = 1, 2, \ldots, n_1; \qquad J_2(\psi_i),\quad i = 1, 2, \ldots, n_2;\\
&2.\quad K_1(\phi_i, \phi_j),\quad i, j = 1, 2, \ldots, n_1; \qquad K_2(\psi_i, \psi_j),\quad i, j = 1, 2, \ldots, n_2;\\
&\qquad K_3(\phi_i, \psi_j),\quad i = 1, 2, \ldots, n_1;\quad j = 1, 2, \ldots, n_2;\\
&3.\quad L_1(\phi_i, \phi_j, \phi_k),\quad i, j, k = 1, 2, \ldots, n_1; \qquad\qquad (5.5.18)\\
&\qquad L_2(\phi_i, \phi_j, \psi_k),\quad i, j = 1, 2, \ldots, n_1;\quad k = 1, 2, \ldots, n_2;\\
&\qquad L_3(\phi_i, \psi_j, \psi_k),\quad i = 1, 2, \ldots, n_1;\quad j, k = 1, 2, \ldots, n_2;\\
&\qquad L_4(\psi_i, \psi_j, \psi_k),\quad i, j, k = 1, 2, \ldots, n_2.
\end{aligned}$$

The following example should illustrate the use of these tables, and the typical multilinear elements of the integrity basis of various crystal classes.

EXAMPLE. Determine the polynomial integrity basis of an absolute scalar that depends on an absolute symmetric tensor C_{KL} and a polar vector $\mathscr{E}_K$ for the ditrigonal-pyramidal class $C_{3v} = 3m$.

From Table C1.9 we see that the space spanned by C_{KL} and $\mathscr{E}_K$ is decomposed into the irreducible representation of C_{3v} as

$$\Gamma = 3\Gamma_1 \dot{+} 3\Gamma_3, \tag{5.5.19}$$

with basic quantities

$$\begin{aligned} &\Gamma_1: \quad \mathscr{E}_3, C_{33}, C_{11} + C_{22}, \\ &\Gamma_3: \quad \mathscr{E}_1 + i\mathscr{E}_2, \qquad S_{13} + iS_{23}, \qquad S_{11} - S_{22} - 2iS_{12}. \end{aligned} \tag{5.5.20}$$

Expressions C2.10, with the use of the basic quantities (5.5.20), give the following 19 members of the minimal integrity basis C_{KL} and $\mathscr{E}_K$ under C_{3v}:

1. $\mathscr{E}_3$; $\mathscr{E}_{33}$; $C_{11} + C_{22}$
2. $\mathscr{E}_1^2 + \mathscr{E}_2^2$; $\quad \mathscr{E}_1 C_{13} + \mathscr{E}_2 C_{23}$; $\quad 2\mathscr{E}_1 C_{12} + \mathscr{E}_2(C_{11} - C_{22})$
   $C_{13}^2 + C_{23}^2$; $\quad 2C_{12}C_{13} + C_{23}(C_{11} - C_{22})$; $\quad 4C_{12}^2 + (C_{11} - C_{22})^2$,
3. $\mathscr{E}_3 C_{23}(C_{11} - C_{22}) - \mathscr{E}_1 C_{13}(C_{11} - C_{22}) - 2C_{13}C_{12}\mathscr{E}_2 - 2C_{12}\mathscr{E}_1 C_{23}$;
   $(\mathscr{E}_2^2 - \mathscr{E}_1^2)(C_{11} - C_{22}) - 4\mathscr{E}_1\mathscr{E}_2 C_{12}$; $\quad (\mathscr{E}_2^2 - \mathscr{E}_1^2)C_{23} - \mathscr{E}_1\mathscr{E}_2 C_{13}$;
   $(C_{23}^2 - C_{13}^2)\mathscr{E}_2 - 2\mathscr{E}_1 C_{13}C_{23}$; $\quad C_{23}^3 - 2C_{12}^2 C_{23}$;
   $(C_{11} - C_{22})^3 - 12C_{12}^2(C_{11} - C_{22})$;
   $(C_{11} - C_{22})^2 C_{23} - 2C_{12}C_{13}(C_{11} - C_{22})$;
   $\mathscr{E}_2^3 - 3\mathscr{E}_1^2\mathscr{E}_2$; $\quad \mathscr{E}_2(C_{11} - C_{22})^2 - 4\mathscr{E}_2 C_{12}^2 - 4\mathscr{E}_1(C_{11} - C_{22})C_{12}$.

(5.5.21)

5.6. Integrity Basis under Magnetic Point Groups

In Section 5.4 we have indicated that for magnetic materials (ferromagnets, antiferromagnets, and ferrimagnets), the crystal symmetry must be supplemented by the spin symmetry. With this, thirty-two (32) crystallographic point groups give rise to ninety (90) magnetic groups (see Appendix B). For example, certain crystals become electrically polarized when placed in a magnetic field and magnetized when placed in an electric field. For small fields, these effects may be characterized by linear equations of the form

$$\begin{aligned} P^K &= \chi_{(\mathrm{E})}^{KL} E_L + \chi_{(\mathrm{EM})}^{KL} B_L, \\ M^K &= \chi_{(\mathrm{ME})}^{KL} E_L + \chi_{(\mathrm{M})}^{KL} B_L, \end{aligned} \tag{5.6.1}$$

where $\mathbf{P}$ is the electric polarization, $\mathbf{E}$ is the electric field, $\mathbf{B}$ is the magnetic induction, and $\mathbf{M}$ is the magnetization. The tensors $\chi_{(\mathrm{E})}$, $\chi_{(\mathrm{M})}$, and $(\chi_{(\mathrm{EM})}, \chi_{(\mathrm{ME})})$ represent electric, magnetic, and magnetoelectric susceptibilities, respectively.

The existence of these tensors depends on the magnetic symmetry of crystals.

The magnetoelectric effect was first proposed by Curie [1908], but its experimental verifications were made much later (Astrov [1960], [1961], Folen et al. [1961], Rado et al. [1962]).[9]

Similarly, on the basis of the erroneous assumption that all crystals are time symmetric, we can conclude that (see Zocher and Török [1953]) pyromagnetism and piezomagnetism are forbidden effects. Since magnetic materials are not time symmetric, this conclusion is wrong. In fact, it was verified (Bhagavantam [1966]) that the piezomagnetic tensor has one independent nonzero component over the crystal class 4/mmm, and two independent components over the magnetic crystal class $\underline{4}/\text{mm}\underline{\text{m}}$. This was observed and measured by Borovik-Romanov [1959] in the antiferromagnetic crystals CoF_2 and MnF_2, which belong to $\underline{4}/\text{mm}\underline{\text{m}}$.

The above considerations indicate the importance of the magnetic symmetry considerations. We now discuss, briefly, a method for the construction of the integrity basis for the magnetic crystals.

A four-dimensional formalism exists to determine the integrity basis of vectors and tensors for the magnetic point group (Mert [1975]). Also, a three-dimensional approach exists similar to the one discussed in Section 5.5, except that the time reversal operator R is adjoined to the 32 nonmagnetic (classical) groups. The establishment of the isomorphism between the classical group $\{\mathbf{S}\}$ and the magnetic group $\{\mathbf{M}\}$ originating from $\{\mathbf{S}\}$ facilitates the process. This was done by Mert [1975], with the conclusion that *classical point groups* $\{\mathbf{S}\}$ *and magnetic groups originated from* $\{\mathbf{S}\}$ *have the same irreducible representations.*

The *four-dimensional* formalism is similar to the case of nonmagnetic groups. Here too, the first task is to obtain the basic quantities associated with the irreducible representation $\mathbf{\Gamma}_R$ of a magnetic group $\{\mathbf{M}\}$. This is again given by (5.5.16), except that $\mathbf{S}^\alpha$ is now replaced by the matrices $\mathbf{M}^\alpha$ of the magnetic group, i.e.,

$$\Lambda_K = \sum_{\alpha=1}^{N} \bar{\Gamma}_{1K}(\mathbf{M}^\alpha) T_{PQ}(\mathbf{M}^\alpha) Z_Q, \tag{5.6.2}$$

where $\bar{\Gamma}_{1K}$ is the complex conjugate of the representations of $\mathbf{M}^\alpha$, and $T_{PQ}(\mathbf{M}^\alpha)$ is defined similarly to (5.5.10). Because of the isomorphism of $\{\mathbf{S}\}$ and $\{\mathbf{M}\}$, the irreducible representations of $\{\mathbf{S}\}$, listed in Tables Appendix C1.1 to C1.16, are also the irreducible representations of $\{\mathbf{M}\}$. Thus, we proceed with $P = 1$, and compute Λ_1 from (5.6.2): If $\Lambda_1 \neq 0$, then eq. (5.6.2) determines the remaining $\Lambda_2, \Lambda_3, \ldots$ which, together with Λ_1, form the carrier space $\mathbf{\Gamma}_R$. If $\Lambda_1 = 0$, then we continue with $P = 2$, and repeat the process. If the irreducible representation $\mathbf{T}_R(\mathbf{M}^\alpha)$ appears n_R times in the representation $\mathbf{\Gamma} = \{\mathbf{\Gamma}^\alpha\}$, as determined by (5.5.15) (of course, with $\mathbf{S}^\alpha$ replaced by $\mathbf{M}^\alpha$ in (5.5.15)), then (5.6.2)

[9] Early positive experimental results of Perrier and Staring on iron and nickel were negated by Perrier and Borel (see O'Dell [1970]). These materials can be shown *not* to allow an appropriate symmetry for the magnetoelectric effect.

will give us n_R sets of components $(\Lambda_1, \ldots, \Lambda_{d_R}), \ldots, (M_1, \ldots, M_{d_R})$ which form the carrier space for the n_R representations $\mathbf{\Gamma}_R$. The degree of $\mathbf{\Gamma}_R$ is given by d_R and, of course, $n_R d_R$ components of $\Lambda_1, \ldots, M_{d_R}$ must be linearly independent. With the basic quantities $\Lambda_1, \ldots, M_{d_R}$ at hand, the integrity basis is then calculated by use of the theorems of Appendix D.[10] Note, however, that here the matrices of the symmetry $\mathbf{M}^\alpha$ are four dimensional.

A three-dimensional treatment is, however, possible, as discussed by Mert [1975]. In this case, we must take notice of two important points.

(a) The nature of argument tensors, e.g., are they absolute or relative tensors?
(b) The form of the symmetry operators, i.e., do they contain the time reversal operator R, or not?

The conventional tensor transformation for the relative tensors of degree 1,

$$P'_{K_1K_2\ldots K_m} = (\det S^\alpha) S^\alpha_{K_1L_1} S^\alpha_{K_2L_2} \cdots S^\alpha_{K_mL_m} P_{L_1L_1\ldots L_m} \tag{5.6.3}$$

may be employed without any modification for 32 of the 90 magnetic point groups, as they are identical to classical point groups and do not contain the time reversal operator R. However, a modification in (5.6.3) is necessary for the remaining 58 additional magnetic point groups. Recall that some of the symmetry operators of the additional magnetic point groups are of the form $R\mathbf{S}^\alpha$. Since the time symmetric tensors (*i-tensors*) are already unaltered under R, the symmetry operator $\mathbf{S}^\alpha$, in $R\mathbf{S}^\alpha$, may be directly substituted into (5.6.3). Hence, i-tensors have the same transformation rule under the corresponding magnetic groups. In the case of time antisymmetric tensors (*c-tensors*), transformations (5.6.3) may be readily used for the elements of magnetic groups which are not in combination with R. However, for symmetry elements of the form $R\mathbf{S}^\alpha$, a modified form of (5.6.3), i.e.,

$$P'_{K_1K_2\ldots K_m} = (-1)(\det S^\alpha) S^\alpha_{K_1L_1} S^\alpha_{K_2L_2} \cdots S^\alpha_{K_mL_m} P_{L_1L_2\ldots L_m} \tag{5.6.4}$$

must be used. The (-1) in the above expression is due to the reversal of c-tensors under R. With these considerations, the remaining procedure is identical to the one discussed for the nonmagnetic case. Following Mert, we illustrate this latter method for the antiferromagnetic crystal Cr_2O_3.

PROBLEM. Determine the integrity basis for an absolute scalar which is a polynomial in an absolute symmetric tensor C_{KL}, an absolute vector $\mathscr{E}_K$, and an axial vector B_K, for the antiferromagnetic crystal Cr_2O_3.

Cr_2O_3 belongs to the magnetic class $\underline{\bar{3}m}$ whose irreducible representations Γ_R are associated with $\bar{3}m = D_{3v}$, and are given in Table Appendix C1.11. The basic quantities for $\mathscr{E}_K$ and C_{KL} are identical to those of the nonmagnetic case,

[10] For a comprehensive account on the method of calculations and tables on the integrity basis, see Kiral and Eringen [1976].

because these are not affected by the time reversal operative R. Thus, C_{KL} and $\mathscr{E}_K$ are decomposed into the irreducible representation of $D_{3v} = \bar{3}m$ as

$$\begin{aligned} \mathscr{E}_K&: \quad \Gamma_4 \dotplus \Gamma_5, \\ C_{KL}&: \quad 2\Gamma_1 \dotplus 2\Gamma_6. \end{aligned} \tag{5.6.5}$$

For B_K, the situation is different, since it is an axial vector and it is affected by the time reversal operator R. Therefore, in calculating $\chi(\mathbf{M}^\alpha)$, we must take notice of R, as present in the magnetic group, e.g., if $\mathbf{M}^\alpha = R\mathbf{S}^\alpha$, for some α (where $\mathbf{S}^\alpha$ is the nonmagnetic group from which $\mathbf{M}^\alpha$ is obtained), then $\chi(R\mathbf{S}^\alpha) = -\chi(\mathbf{S}^\alpha)$. Since, also, $\det(\mathbf{S}^\alpha)$ affects the transformations of B_K altogether, we need to calculate

$$\chi(\mathbf{M}^\alpha) = \chi(R\mathbf{S}^\alpha) = -\chi(\mathbf{S}^\alpha)\det(\mathbf{S}^\alpha) = \chi(\mathbf{S}^\alpha)\det(R\mathbf{S}^\alpha). \tag{5.6.6}$$

Thus, $\det(R\mathbf{S})$ appears as a factor. Since only magnetic elements are in combination with R, (5.6.6) shows that the decomposition of $\mathscr{E}_K$ and B_K are identical, i.e.,

$$B_K: \quad \Gamma_4 \dotplus \Gamma_5. \tag{5.6.7}$$

From Table C1.11 we read the basic quantities

$$\begin{aligned} &\Gamma_1: \quad \varphi \equiv C_{11} + C_{22}; && \varphi' \equiv C_{33}, \\ &\Gamma_4: \quad \eta \equiv \mathscr{E}_3; && \eta' \equiv B_3, \\ &\Gamma_5: \quad \begin{pmatrix} \mathscr{E}_1 \\ \mathscr{E}_2 \end{pmatrix}; && \begin{pmatrix} B_1 \\ B_2 \end{pmatrix}, \\ &\Gamma_6: \quad \begin{pmatrix} 2C_{12} \\ C_{11} - C_{22} \end{pmatrix}; && \begin{pmatrix} C_{13} \\ C_{23} \end{pmatrix}. \end{aligned} \tag{5.6.8}$$

Employing the notation

$$\begin{aligned} &A = \mathscr{E}_1 + i\mathscr{E}_2, && B = B_1 + iB_2, \\ &a = 2C_{12} + i(C_{11} - C_{22}), && b = C_{13} + iC_{23}, \end{aligned} \tag{5.6.9}$$

together with a superposed bar representing the complex conjugate, the typical elements of the integrity basis are

$$\begin{aligned} &1. \quad \varphi, \varphi' \\ &2. \quad a\bar{b} + \bar{a}b, \quad A\bar{B} + \bar{A}B, \quad \eta\eta', \\ &3. \quad aAB - \bar{a}\bar{A}\bar{B}, \quad \eta(a\bar{A} + \bar{a}A), \\ &4. \quad ab\bar{A}\bar{B} + \bar{a}\bar{b}AB, \quad (a\bar{b} - \bar{a}b)(A\bar{B} - \bar{A}B), \\ &\quad\;\; \eta(abA - \bar{a}\bar{b}\bar{A}), \end{aligned} \tag{5.6.10}$$

where in products such as $a\bar{b}$, a and b may be taken equal to a, A and B to include all distinct combinations $a\bar{a}$, $b\bar{b}$, $a\bar{A}$, $b\bar{A}$, $a\bar{B}$, $b\bar{B}$. A similar situation is valid for other products (see Eq. 5.5.21).

The complete set of integrity basis for $\mathbf{C}$, $\boldsymbol{\mathscr{E}}$, and $\mathbf{B}$, over $\bar{3}m$, can then be obtained by using all possible combinations of the basic quantities for C_{KL}, $\mathscr{E}_K$, and B_K, given by (5.6.8) and (5.6.10). The number of elements in this set is 56, and these expressions are too lengthy to list here.[11]

5.7. Integrity Basis of Vectors and Tensors for Isotropic Materials

Let $\mathbf{C}_a$, $\mathbf{p}_b$, and $\mathbf{W}_c$ ($a = 1, 2, \ldots, A$; $b = 1, 2, \ldots, B$; $c = 1, 2, \ldots, C$) denote, respectively, symmetric tensors, vectors, and skew-symmetric tensors. If there exists an orthogonal tensor $\mathbf{S}$ such that

$$\bar{\mathbf{C}}_a = \mathbf{S}\mathbf{C}_a\mathbf{S}^{\mathrm{T}}, \qquad \bar{\mathbf{p}}_b = \mathbf{S}\mathbf{p}_b, \qquad \bar{\mathbf{W}}_c = \mathbf{S}\mathbf{W}\mathbf{S}^{\mathrm{T}}, \tag{5.7.1}$$

we say that $(\mathbf{C}_a, \mathbf{p}_b, \mathbf{W}_c)$ and $(\bar{\mathbf{C}}_a, \bar{\mathbf{p}}_b, \bar{\mathbf{W}}_c)$ are *equivalent*.

A scalar-valued function $\Sigma(\mathbf{C}_a, \mathbf{p}_b, \mathbf{W}_c)$ is called an *isotropic function* or an *invariant*, if it satisfies the condition

$$\Sigma(\bar{\mathbf{C}}_a, \bar{\mathbf{p}}_b, \bar{\mathbf{W}}_c) = \Sigma(\mathbf{C}_a, \mathbf{p}_b, \mathbf{W}_c) \tag{5.7.2}$$

for all orthogonal tensors $\{\mathbf{S}\}$. The symmetry group for isotropic materials is the *full group* $\mathscr{O}$ of orthogonal transformatins of the material frame of reference. Thus, the integrity bases for various tensors and vectors for the full orthogonal group consist of a minimal set out of all possible single and joint invariants of vectors and tensors involved. The complete set must be irreducible, in the sense that no element in the set is functionally dependent on the others. These invariants for polynomials have been tabulated by Spencer [1971]. Here we give the concise results obtained by Wang [1970].

Theorem 1. *For a three-dimensional space, a complete and irreducible representation for any scalar-valued isotropic function* $\Sigma(\mathbf{C}_a, \mathbf{p}_b, \mathbf{W}_c)$, $a = 1, 2, \ldots, A$, $b = 1, 2, \ldots, B$, $c = 1, 2, \ldots, C$, *is given by*

$$\Sigma(\mathbf{C}_a, \mathbf{p}_b, \mathbf{W}_c) = F(I_{\mathbf{cpw}}), \tag{5.7.3}$$

where $I_{\mathbf{cpw}}$ *denotes all invariants given in Table* E.1 (*Appendix* E), *based on any one, two, three, or four variables to within a permutation in the argument*

$$(\mathbf{C}_1, \ldots, \mathbf{C}_A, \mathbf{p}_1, \ldots, \mathbf{p}_B, \mathbf{W}_1, \ldots, \mathbf{W}_C).$$

Thus, the function F does not contain any invariant of more than three symmetric tensors, nor of more than three skew-symmetric tensors.

Example. Determine a complete irreducible representation for a scalar isotropic function Σ, which depends on $\mathbf{C}$, $\mathbf{p}$, and $\mathbf{a}$ where $\mathbf{a}$ is an axial vector.

[11] For this and other tables listing the integrity basis for one symmetric tensor, one polar vector, and one axial vector under a magnetic group, see Kiral and Eringen [1976].

From a_k we form $W_{KL} = e_{KLM}a_M$, so that W_{KL} is an absolute skew-symmetric tensor.

From Table E.1., we read the following 14 quantities:

$$\begin{aligned}
&\text{I.}\quad \operatorname{tr}\mathbf{C},\quad \operatorname{tr}\mathbf{C}^2,\quad \operatorname{tr}\mathbf{C}^3,\quad \mathbf{p}\cdot\mathbf{p},\quad \operatorname{tr}\mathbf{W}^2,\\
&\text{II.}\quad \mathbf{p}\cdot(\mathbf{C}\mathbf{p}),\quad \mathbf{p}\cdot(\mathbf{C}^2\mathbf{p}),\quad \operatorname{tr}(\mathbf{C}\mathbf{W}^2),\quad \operatorname{tr}(\mathbf{C}^2\mathbf{W}^2),\\
&\qquad \operatorname{tr}(\mathbf{C}^2\mathbf{W}^2\mathbf{C}\mathbf{W}),\quad \mathbf{p}\cdot(\mathbf{W}^2\mathbf{p}),\\
&\text{III.}\quad (\mathbf{C}\mathbf{p})(\mathbf{W}\mathbf{p}),\quad (\mathbf{C}^2\mathbf{p})\cdot(\mathbf{W}\mathbf{p}),\quad (\mathbf{C}\mathbf{W}\mathbf{p})\cdot(\mathbf{W}^2\mathbf{p}).
\end{aligned} \tag{5.7.4}$$

To revert to the original variables we substitute $W_{KL} = e_{KLM}a_M$.

The functional basis of vector- and tensor-valued functions can be determined by use of the scalar function basis through equations (5.5.9). However, Wang [1970] gave the following theorem which eliminates some of the tedious algebra.

Theorem 2. *For a three-dimensional space, a complete and irreducible representation for any vector (symmetric tensor, skew-symmetric tensor) valued isotropic function* $f(\mathbf{C}_a, \mathbf{p}_b, \mathbf{W}_c)$ *is given by*

$$\mathbf{f} = \sum_{r=1}^{R} f^r(\mathbf{C}_a, \mathbf{p}_b, \mathbf{W}_c)g_r(\mathbf{C}_a, \mathbf{p}_b, \mathbf{W}_c), \tag{5.7.5}$$

where $f^1, \ldots, f^R$ *are scalar-valued isotropic functions and* $\{g_1, \ldots, g_R\}$ *is a generating set formed by all generators given in Tables* E2, E3, E4, *based on all combinations of none, one, two, or three variables in the argument* $(\mathbf{C}_1, \ldots, \mathbf{C}_A, \mathbf{p}_1, \ldots, \mathbf{p}_B, \mathbf{W}_1, \ldots, \mathbf{W}_c)$ *to within a permutation of entries.*

The representation theory described above applies to all isotropic functions. When the isotropic functions are restricted to polynomial, caution is necessary, in that the functional basis given above is generally a smaller set than the integrity basis for polynomials. For a polynomial basis, a bigger set is generally needed. The functional basis given above, however, is for polynomial functions, so that every integrity basis serves also as a function basis. But generally, an integrity basis cannot be regarded as an *irreducible* functional basis.

Transverse Isotropy. A material that possesses a single preferred direction, which is the same at all points, is said to be *transversely isotropic.* If the preferred direction is the unit vector $\mathbf{I}_3$, directed along the X_3-coordinate axis, then the material properties must be invariant under all rotations about the X_3-axis. Such rotations are represented by matrices of the type:

$$\mathbf{M}_\theta = \begin{bmatrix} \cos\theta & \sin\theta & 0 \\ -\sin\theta & \cos\theta & 0 \\ 0 & 0 & 1 \end{bmatrix}, \qquad \theta \in [0, 2\pi]. \tag{5.7.6}$$

Various other cases of symmetry may also be permitted, as represented by the matrices $\mathbf{R}_1$, $\mathbf{R}_2$, and $\mathbf{D}_2$ (see A3). Five different cases may be distinguished

$$(1)\ \mathbf{M}_\theta, \qquad (2)\ \mathbf{M}_\theta, \mathbf{R}_1, \qquad (3)\ \mathbf{M}_\theta, \mathbf{R}_3,$$
$$(4)\ \mathbf{M}_\theta, \mathbf{D}_2, \qquad (5)\ \mathbf{M}_\theta, \mathbf{R}_1, \mathbf{R}_3, \mathbf{D}_2.$$

The case (2) is the one that is commonly considered in literature.

A complete irreducible representation of transversely isotropic functions may be obtained by the use of Theorem 1 *by including the unit vector* $\mathbf{I}_3$ *among the argument of functions.*

Example. Determine a complete irreducible representation for a scalar transversely isotropic function Σ that depends on a symmetric tensor $\mathbf{C}$ and a vector $\mathbf{p}$.

From the list E.1 with $\mathbf{C}_1 = \mathbf{C}$, $\mathbf{p}_1 = \mathbf{p}$, $\mathbf{p}_2 = \mathbf{I}_3$, we read

$$\begin{aligned}
\text{I.}\quad & \mathbf{C}: && \operatorname{tr}\mathbf{C},\quad \operatorname{tr}\mathbf{C}^2,\quad \operatorname{tr}\mathbf{C}^3,\\
& \mathbf{p}: && \mathbf{p}\cdot\mathbf{p},\\
& \mathbf{I}_3: && \mathbf{I}_3\mathbf{I}_3 = 1.\\
\text{II.}\quad & \mathbf{C}, \mathbf{p}: && \mathbf{p}\cdot(\mathbf{C}\mathbf{p}),\quad \mathbf{p}\cdot(\mathbf{C}^2\mathbf{p}),\\
& \mathbf{C}, \mathbf{I}_3: && C_{33},\quad C_{3K}C_{K3},\\
& \mathbf{p}, \mathbf{I}_3: && p_3.
\end{aligned}$$

Thus, a complete irreducible representation of Σ consists of the following nine (9) members:

$$\begin{gathered}
C_{33},\quad C_{\alpha\alpha},\quad C_{3\alpha}C_{\alpha 3},\quad C_{\alpha\beta}C_{\beta\alpha},\quad C_{\alpha\beta}C_{\beta\gamma}C_{\gamma\alpha} + 3C_{3\alpha}C_{\alpha\beta}C_{\beta 3},\\
p_3,\quad p_\alpha p_\alpha,\quad p_\alpha C_{\alpha\beta}p_\beta,\quad p_\alpha C_{\alpha\beta}C_{\beta\gamma}p_\gamma.
\end{gathered} \tag{5.7.7}$$

where the summation on Greek indices is over (1, 2). Note that this list contains one less invariant than the one for the polynomial integrity basis which may be obtained by using the table of invariants given by Spencer [1971, p. 303] in Eringen [1971b].

5.8. Constitutive Equations of Electromagnetic Elastic Solids

Here we obtain constitutive equations for electromagnetic elastic solids which are heat and electric conducting. The memory-dependence is assumed not present. The constitutive equations of these materials are expressed in the form

$$\rho_0\Psi = \Sigma(C_{MN}, \mathscr{E}_M, B_M, \theta_{,M}, \theta, \mathbf{X}). \tag{5.8.1}$$

Similar expressions containing the same arguments are, of course, valid for η, P_K, M_K, Q_K, J_K, and ${}_{\mathrm{E}}T_{KL}$. We have

$$\rho_0 \dot{\Psi} = \frac{\partial \Sigma}{\partial C_{MN}} \dot{C}_{MN} + \frac{\partial \Sigma}{\partial \mathscr{E}_M} \dot{\mathscr{E}}_M + \frac{\partial \Sigma}{\partial B_M} \dot{B}_M + \frac{\partial \Sigma}{\partial \theta_{,M}} \dot{\theta}_{,M} + \frac{\partial \Sigma}{\partial \theta} \dot{\theta} \tag{5.8.2}$$

Substituting (5.8.2) into the C–D inequality (5.2.17) we have

$$-\left(\frac{\partial \Sigma}{\partial \theta} + \rho_0 \eta\right)\dot{\theta} + \tfrac{1}{2}\left({}_{\mathrm{E}}T_{KL} - 2\frac{\partial \Sigma}{\partial C_{KL}}\right)\dot{C}_{KL} - \rho_0 \frac{\partial \Sigma}{\partial \theta_{,K}} \dot{\theta}_{,K}$$
$$+ \frac{1}{\theta} Q_K \theta_{,K} - \left(\Pi_K + \frac{\partial \Sigma}{\partial \mathscr{E}_K}\right)\dot{\mathscr{E}}_K - \left(M_K + \frac{\partial \Sigma}{\partial B_K}\right)\dot{B}_K + J_K \mathscr{E}_K \geq 0. \tag{5.8.3}$$

This inequality is linear in $\dot{\theta}$, $\dot{C}_{KL}$, $\dot{\theta}_{,K}$, $\dot{\mathscr{E}}_K$, and $\dot{B}_K$ and is posited not to be violated for all independent variations of these quantities. The necessary and sufficient conditions for this are (Eringen [1980, Sect. 10.11])

$$\eta = -\frac{1}{\rho_0}\frac{\partial \Sigma}{\partial \theta}, \qquad \frac{\partial \Sigma}{\partial \theta_{,K}} = 0,$$
$$\Pi_K = -\frac{\partial \Sigma}{\partial \mathscr{E}_K}, \qquad M_K = -\frac{\partial \Sigma}{\partial B_K}, \tag{5.8.4}$$
$${}_{\mathrm{E}}T_{KL} = 2\frac{\partial \Sigma}{\partial C_{KL}},$$

and

$$\rho\gamma \equiv \frac{1}{\theta} Q_K \theta_{,K} + J_K \mathscr{E}_K \geq 0. \tag{5.8.5}$$

From (5.8.4) it is clear that Σ is independent of the temperature gradient. The remaining expressions in (5.8.4) show that η, Π_K, M_K, and ${}_{\mathrm{E}}T_{KL}$ are derivable from the potential $\Sigma = \Sigma(\mathbf{C}, \mathscr{E}, \mathbf{B}, \theta, \mathbf{X})$. If Q_K and J_K are continuous functions of Q_K and $\mathscr{E}_K$, then from (5.8.5) it also follows that

$$Q_K = J_K = 0 \qquad \text{when} \qquad \theta_{,K} = \mathscr{E}_K = 0. \tag{5.8.6}$$

The constitutive equations for $\mathbf{Q}$ and $\mathscr{J}$, which are functions of $\mathbf{C}$, $\mathscr{E}$, $\mathbf{B}$, $\theta_{,M}$, θ, and $\mathbf{X}$, are subject to (5.8.5) and (5.8.6). These equations may be written in the form (5.5.9), i.e.,

$$Q_K = \frac{\partial Q}{\partial \phi_K}, \qquad \mathscr{J}_K = \frac{\partial \mathscr{J}}{\partial \phi_K}, \tag{5.8.7}$$

where the scalar functions Q and $\mathscr{J}$ are linear in ϕ and are defined by

$$Q = Q_K \phi_K = Q(\mathbf{C}, \mathscr{E}, \mathbf{B}, \nabla\theta, \boldsymbol{\phi}; \theta, X),$$
$$\mathscr{J} = \mathscr{J}_K \phi_K = \mathscr{J}(\mathbf{C}, \mathscr{E}, \mathbf{B}, \nabla\theta, \boldsymbol{\phi}; \theta, \mathbf{X}). \tag{5.8.8}$$

The integrity bases of $\mathbf{C}$, $\mathscr{E}$, $\mathbf{B}$, $\nabla\theta$, and $\boldsymbol{\phi}$, under the crystallographic magnetic group, can be obtained by the method described in Sections 5.5 and 5.6. This, then, determines the polynomial functions Q and $\mathscr{J}$, so that, at least, in principle, we can obtain the constitutive equations for Q_K and $\mathcal{J}_K$. Because of the number of vectors and tensors involved, the list of integrity bases for magnetic crystals is too long to be of any utility in practical applications.[12] Even after the expressions of Q_K and $\mathcal{J}_K$ are obtained, we still have two major tasks to complete, namely, to obtain restrictions resulting from the C–D inequality (5.8.5) and those arising from the axiom of time reversal. This latter restriction is, of course, also valid for the free-energy function $\rho_0 \Psi = \Sigma$. Fortunately, the nature of materials and the strength of applied and induced fields allow gross simplifications in the constitutive equations.

(i) *Nature of Electromagnetic Solids.* Electromagnetic materials generally fall into two categories: (a) nonconducting materials, and (b) conducting materials. For nonconducting materials, $\mathcal{J} = \mathbf{0}$. In the case where heat conduction is also negligible, $\nabla\theta$ will not be in the argument, so that (5.8.3) gives $\mathbf{Q} = \mathbf{0}$, and the problem of obtaining constitutive equations for $\mathbf{Q}$ and $\mathcal{J}$ disappears. Moreover, nonconducting solids also fall into two categories, dielectrics and ferrous metals. For dielectrics, the magnetization and electrical conductions are comparatively small. In ferrous metals, electric polarization can be neglected, as compared to other effects.

For conducting metals electric polarization and magnetization can be disregarded. However, heat and electric conductions may be important. Thus, we see that, depending on the nature of materials, the dependence of constitutive functions upon some of the vectors may be ignored.

For *rigid solids* the dependence on the formation tensor $\mathbf{C}$ is dropped.

(ii) The *strength of fields.* If some of the applied and induced fields are not strong, constitutive response functions may be approximated by polynomial expansions containing lower powers and products of these fields. For example, in the linear theory, Q_K and $\mathcal{J}_K$ are linear in the variables $\mathbf{C}$, $\boldsymbol{\mathcal{E}}$, $\mathbf{B}$, and $\nabla\theta$, while Σ is a quadratic function leading to linear equations for $\mathbf{\Pi}$, $\mathbf{M}$, and ${}_E\mathbf{T}$. For the quadratic theory Σ is a third-degree polynomial while $\mathbf{Q}$, $\mathcal{J}$, $\mathbf{\Pi}$, $\mathbf{M}$, and ${}_E\mathbf{T}$ are second-degree in the independent variables. Thus, a great deal of simplification of the constitutive equations is possible since these considerations cover a large variety of materials. Below we discuss some of these special cases.

I. Rigid Materials

For rigid materials, $\mathbf{C} = \mathbf{1}$ so that Σ, Q, and $\mathcal{J}$ are independent of $\mathbf{C}$, i.e.,

$$\begin{aligned} \Sigma &= \Sigma(\mathcal{E}_K, B_K, \theta, X), \\ Q &= Q(\mathcal{E}_K, B_K, \theta_{,K}, \phi_K; \theta, \mathbf{X}), \qquad (5.8.9) \\ \mathcal{J} &= \mathcal{J}(\mathcal{E}_K, B_K, \theta_{,K}, \phi_K; \theta, \mathbf{X}) \qquad \text{(linear in } \phi_K\text{)}. \end{aligned}$$

Excluding ${}_E\mathbf{T}$ and $\mathcal{J}$, equations (5.8.4)–(5.8.7) hold for rigid materials.

[12] Jordan and Eringen [1964a, b] gave nonlinear polynomial constitutive equations employing C_{KL}, E_K, H_K, and $\theta_{,K}$.

II. Elastic Dielectrics

For nonmagnetizable electromagnetic solids the free energy is independent of $\mathbf{B}$. In these materials the electric conduction is small. Hence, we have

$$\Sigma = \Sigma(\mathbf{C}, \mathscr{E}; \theta, \mathbf{X}),$$
$$Q = Q(\mathbf{C}, \mathscr{E}, \theta_{,K}, \phi_K; \theta, \mathbf{X}). \tag{5.8.10}$$

Excluding $\mathbf{M}$ and $\mathscr{J}$, (5.8.4)–(5.8.7) are valid.

III. Magnetoelasticity

For materials that are not polarizable the free energy does not depend on the electric field, so that

$$\Sigma = \Sigma(\mathbf{C}, \mathbf{B}; \theta, \mathbf{X}),$$
$$Q = Q(\mathbf{C}, \mathscr{E}, \mathbf{B}, \theta_{,K}, \phi_K; \theta, \mathbf{X}), \tag{5.8.11}$$
$$\mathscr{J} = \mathscr{J}(\mathbf{C}, \mathscr{E}, \mathbf{B}, \theta_{,K}, \phi_K; \theta, \mathbf{X}).$$

Excluding Π_K, all equations (5.8.4)–(5.8.7) are valid.

Other subspecial cases of these classes of materials are trivially obtained by omitting some of the arguments in the constitutive functions.

The constitutive equations are still subject to the C–D inequality (5.8.5) and the axiom of time reversal. Moreover, the material symmetry regulations place further restrictions on the constitutive response functions. In the following sections we study several typical electromagnetic solids. Following these, we give the linear theory in full detail.

5.9. Isotropic Electromagnetic Elastic Solids

For isotropic solids Σ must be invariant under the full group of orthogonal transformations $\{\mathbf{S}\}$ of the material frame of reference X_K, i.e.,

$$\Sigma(\mathbf{C}, \mathscr{E}, \mathbf{B}) = (\mathbf{SCS}^{\mathrm{T}}, \mathbf{S}\mathscr{E}, \mathbf{SB} \det \mathbf{S}), \tag{5.9.1}$$

where $\mathbf{S}$ is subject to

$$\mathbf{SS}^{\mathrm{T}} = \mathbf{S}^{\mathrm{T}}\mathbf{S} = \mathbf{1}, \qquad \det \mathbf{S} = \pm 1. \tag{5.9.2}$$

Since Σ is a scalar function, (5.9.1) implies that it can depend on $\mathbf{C}$, $\mathscr{E}$, and $\mathbf{B}$, only through the joint invariants of C_{KL}, $\mathscr{E}_K$, and B_K. But these invariants are completely expressible in terms of an equivalent set of spatial tensors and vectors, c_{kl}, $\mathscr{E}_k$, and B_k. Another convenient set is $\overset{-1}{c}_{kl}$, $\mathscr{E}_K$, and B_K. Recalling the relations (3.11.4) and (3.11.5) and the definition

$$\overset{-1}{c}_{kl} = x_{k,K}x_{l,K} \tag{5.9.3}$$

from (5.8.4), we find that

$$\eta = -\frac{1}{\rho_0}\frac{\partial \Sigma}{\partial \theta}, \quad P_k = -\frac{\rho}{\rho_0}\frac{\partial \Sigma}{\partial \mathscr{E}_k}, \quad \mathscr{M}_k = -\frac{\rho}{\rho_0}\frac{\partial \Sigma}{\partial B_k}, \quad {}_{\mathrm{E}}t_{kl} = 2\frac{\rho}{\rho_0}\frac{\partial \Sigma}{\partial \overset{-1}{c}_{kr}}\overset{-1}{c}_{rl}. \tag{5.9.4}$$

The function basis, as discussed in Section 5.7, of $\overset{-1}{c}_{kl}$, $\mathscr{E}_K$, and B_k may read, from Table E.1:[13]

$$
\begin{aligned}
&\mathrm{I}_1 = \operatorname{tr} \overset{-1}{\mathbf{c}}, \quad \mathrm{I}_2 = \operatorname{tr} \overset{-2}{\mathbf{c}}, \quad \mathrm{I}_3 = \operatorname{tr} \overset{-3}{\mathbf{c}}, \quad \mathrm{I}_4 = \mathscr{E} \cdot \mathscr{E}, \quad \mathrm{I}_5 = \mathbf{B} \cdot \mathbf{B}, \\
&\mathrm{I}_6 = \mathscr{E} \overset{-1}{\mathbf{c}} \mathscr{E}, \quad \mathrm{I}_7 = \mathbf{B} \overset{-1}{\mathbf{c}} \mathbf{B}, \quad \mathrm{I}_8 = \mathscr{E} \overset{-2}{c} \mathscr{E}, \quad \mathrm{I}_9 = \mathbf{B} \overset{-2}{\mathbf{c}} \mathbf{B}, \\
&\mathrm{I}_{10} = (\mathscr{E} \cdot \mathbf{B})^2, \quad \mathrm{I}_{11} = \mathscr{E} \cdot [\mathbf{B} \times (\overset{-1}{\mathbf{c}} \mathscr{E})], \quad \mathrm{I}_{12} = \mathscr{E} \cdot [\mathbf{B} \times (\overset{-2}{\mathbf{c}} \mathscr{E})], \\
&\mathrm{I}_{13} = \mathbf{B} \cdot [(\overset{-1}{\mathbf{c}} \mathbf{B}) \times (\overset{-2}{\mathbf{c}} \mathbf{B})], \quad \mathrm{I}_{14} = (\mathscr{E} \cdot \mathbf{B}) \mathscr{E} \cdot [\mathbf{B} \times (\overset{-1}{\mathbf{c}} \mathbf{B})],
\end{aligned}
\tag{5.9.5}
$$

where we employed the notation

$$
\mathbf{a} \cdot \mathbf{a} = a_k a_k, \qquad \mathbf{a}\mathbf{T} = a_k T_{kl} \mathbf{i}_l, \qquad \mathbf{T}\mathbf{a} = T_{kl} a_l \mathbf{i}_k
$$

for a vector $\mathbf{a}$ and a second-order tensor $\mathbf{T}$.

Thus, Σ must be a function of the form

$$
\Sigma = \Sigma(\mathrm{I}_\alpha, \theta, \mathbf{X}), \qquad \alpha = 1, 2, \ldots, 14. \tag{5.9.6}
$$

The use of (5.9.6) in (5.9.4) gives the spatial forms of the constitutive equations in terms of the spatial components of fields

$$
\eta = -\frac{1}{\rho} \frac{\partial \Sigma}{\partial \theta}, \tag{5.9.7}
$$

$$
\begin{aligned}
\mathbf{P} = -2 \frac{\rho}{\rho_0} \bigg\{ & \frac{\partial \Sigma}{\partial \mathrm{I}_4} \mathscr{E} + \frac{\partial \Sigma}{\partial \mathrm{I}_6} \overset{-1}{\mathbf{c}} \mathscr{E} + \frac{\partial \Sigma}{\partial \mathrm{I}_8} \overset{-2}{\mathbf{c}} \mathscr{E} + \frac{\partial \Sigma}{\partial \mathrm{I}_{10}} (\mathscr{E} \cdot \mathbf{B}) \mathbf{B} \\
& + \frac{1}{2} \frac{\partial \Sigma}{\partial \mathrm{I}_{11}} [\mathbf{B} \times (\overset{-1}{\mathbf{c}} \mathscr{E}) + \overset{-1}{\mathbf{c}} (\mathscr{E} \times \mathbf{B})] + \frac{1}{2} \frac{\partial \Sigma}{\partial \mathrm{I}_{12}} [\mathbf{B} \times (\overset{-2}{\mathbf{c}} \mathscr{E}) + \overset{-2}{\mathbf{c}} (\mathscr{E} \times \mathbf{B})] \\
& + \frac{1}{2} \frac{\partial \Sigma}{\partial \mathrm{I}_{14}} [\mathbf{B} \mathscr{E} \cdot (\mathbf{B} \times (\overset{-1}{\mathbf{c}} \mathbf{B})) + (\mathscr{E} \cdot \mathbf{B}) \mathbf{B} \times (\overset{-1}{\mathbf{c}} \mathbf{B})] \bigg\},
\end{aligned}
\tag{5.9.8}
$$

$$
\begin{aligned}
\mathscr{M} = -\frac{2\rho}{\rho_0} \bigg\{ & \frac{\partial \Sigma}{\partial \mathrm{I}_5} \mathbf{B} + \frac{\partial \Sigma}{\partial \mathrm{I}_7} \overset{-1}{\mathbf{c}} \mathbf{B} + \frac{\partial \Sigma}{\partial \mathrm{I}_9} \overset{-2}{\mathbf{c}} \mathbf{B} + \frac{\partial \Sigma}{\partial \mathrm{I}_{10}} (\mathscr{E} \cdot \mathbf{B}) \mathscr{E} \\
& + \frac{1}{2} \frac{\partial \Sigma}{\partial \mathrm{I}_{11}} (\overset{-1}{\mathbf{c}} \mathscr{E}) \times \mathscr{E} + \frac{1}{2} \frac{\partial \Sigma}{\partial \mathrm{I}_{12}} (\overset{-2}{\mathbf{c}} \mathscr{E}) \times \mathscr{E} + \frac{1}{2} \frac{\partial \Sigma}{\partial \mathrm{I}_{13}} \{ (\overset{-1}{\mathbf{c}} \mathbf{B}) \times (\overset{-2}{\mathbf{c}} \mathbf{B}) \\
& + \mathbf{B} \cdot [(\overset{-1}{\mathbf{c}} \mathbf{B}) \times \overset{-2}{\mathbf{c}}] - \mathbf{B} \cdot [\overset{-2}{\mathbf{c}} \mathbf{B} \times \overset{-1}{\mathbf{c}}] \} + \frac{1}{2} \frac{\partial \Sigma}{\partial \mathrm{I}_{14}} [\mathscr{E} \mathscr{E} \cdot (\mathbf{B} \times (\overset{-1}{\mathbf{c}} \mathbf{B}))] \\
& + \mathscr{E} \cdot \mathbf{B} [\overset{-1}{\mathbf{c}} \mathbf{B}) \times \mathscr{E} + \overset{-1}{\mathbf{c}} (\mathscr{E} \times \mathbf{B})] \bigg\},
\end{aligned}
\tag{5.9.9}
$$

---

[13] The polynomial basis (the integrity basis), in this case, contains four more invariants

$$
\begin{aligned}
&\mathrm{I}_{15} = [\mathbf{B} \times (\overset{-1}{\mathbf{c}} \mathbf{B})] \cdot (\overset{-1}{\mathbf{c}} \mathscr{E}), \qquad && \mathrm{I}_{16} = (\mathscr{E} \cdot \mathbf{B})(\mathscr{E} \overset{-1}{\mathbf{c}} \mathbf{B}), \\
&\mathrm{I}_{17} = (\mathscr{E} \cdot \mathbf{B})(\mathscr{E} \overset{-1}{\mathbf{c}} \mathbf{B}), && \mathrm{I}_{18} = (\mathscr{E} \cdot \mathbf{B})(\mathscr{E} \times \mathbf{B}) \cdot (\overset{-1}{\mathbf{c}} \mathbf{B}),
\end{aligned}
$$

(see Pipkin and Rivlin [1966]).

$$
\begin{aligned}
{}_{E}\mathbf{t} = \frac{2\rho}{\rho_0}\Big\{&\frac{\partial\Sigma}{\partial \mathrm{I}}\overset{-1}{\mathbf{c}} + \left(\mathrm{II}\frac{\partial\Sigma}{\partial \mathrm{II}} + \mathrm{III}\frac{\partial\Sigma}{\partial \mathrm{III}}\right)\mathbf{1} - \mathrm{III}\frac{\partial\Sigma}{\partial \mathrm{II}}\mathbf{c} + 2\frac{\partial\Sigma}{\partial \mathrm{I}_6}[\mathscr{E}\otimes(\overset{-1}{\mathbf{c}}\mathscr{E})]_S \\
&+ 2\frac{\partial\Sigma}{\partial \mathrm{I}_7}[\mathbf{B}\otimes(\overset{-1}{\mathbf{c}}\mathbf{B})]_S + 4\frac{\partial\Sigma}{\partial \mathrm{I}_8}[\mathscr{E}\otimes(\overset{-2}{\mathbf{c}}\mathscr{E})]_S + 4\frac{\partial\Sigma}{\partial \mathrm{I}_9}[\mathbf{B}\otimes(\overset{-2}{\mathbf{c}}\mathbf{B})]_S \\
&+ \frac{1}{2}\frac{\partial\Sigma}{\partial \mathrm{I}_{11}}[(\mathscr{E}\times\mathbf{B})\otimes(\overset{-1}{\mathbf{c}}\mathscr{E}) + \mathscr{E}\otimes\overset{-1}{\mathbf{c}}(\mathscr{E}\times\mathbf{B})]_S \\
&+ \frac{1}{2}\frac{\partial\Sigma}{\partial \mathrm{I}_{12}}[(\mathscr{E}\times\mathbf{B})\otimes(\overset{-2}{\mathbf{c}}\mathscr{E}) + \overset{-1}{\mathbf{c}}(\mathscr{E}\times\mathbf{B})\otimes(\overset{-1}{\mathbf{c}}\mathscr{E}) \\
&+ \overset{-1}{\mathbf{c}}\mathscr{E}\otimes\overset{-1}{\mathbf{c}}(\mathscr{E}\times\mathbf{B}) + \mathscr{E}\otimes\overset{-2}{\mathbf{c}}(\mathscr{E}\times\mathbf{B})]_S + \frac{1}{2}\frac{\partial\Sigma}{\partial \mathrm{I}_{13}}\{[(\overset{-1}{\mathbf{c}}\mathbf{B})\times\mathbf{B}\otimes\overset{-1}{\mathbf{c}}\mathbf{B} \\
&+ \mathbf{B}\times(\overset{-1}{\mathbf{c}}\mathbf{B})\otimes\overset{-2}{\mathbf{c}}\mathbf{B} + \overset{-1}{\mathbf{c}}(\mathbf{B}\times\overset{-1}{\mathbf{c}}\mathbf{B})\otimes\overset{-1}{\mathbf{c}}\mathbf{B}]\}_S \\
&+ \frac{1}{2}\frac{\partial\Sigma}{\partial \mathrm{I}_{14}}(\mathscr{E}\cdot\mathbf{B})[(\mathscr{E}\times\mathbf{B})\otimes\overset{-1}{\mathbf{c}}\mathbf{B} + \mathbf{B}\otimes\overset{-1}{\mathbf{c}}(\mathscr{E}\times\mathbf{B})]_S. \qquad (5.9.10)
\end{aligned}
$$

where a subscript S attached to brackets indicates the symmetrization, e.g.,

$$[\mathbf{A}]_S = \tfrac{1}{2}(\mathbf{A} + \mathbf{A}^{\mathrm{T}})$$

In the expression (5.9.10) we also modified the first three terms by using the relations between the principal invariants I, II, and III of $\overset{-1}{\mathbf{c}}$ and of I_1, I_2, and I_3

$$\mathrm{I}_1 = \operatorname{tr}\overset{-1}{\mathbf{c}}, \qquad \mathrm{I}_2 = \operatorname{tr}\overset{-2}{\mathbf{c}} = \mathrm{I}^2 - 2\mathrm{II}, \qquad \mathrm{I}_3 = \operatorname{tr}\overset{-3}{\mathbf{c}} = \mathrm{I}_3 - 2\mathrm{I}.\,\mathrm{II} + 3\mathrm{III}, \qquad (5.9.11)$$

and by using the Cayley–Hamilton theorem

$$\overset{-3}{\mathbf{c}} - \mathrm{I}\,\overset{-2}{\mathbf{c}} + \mathrm{II}\,\overset{-1}{\mathbf{c}} - \mathrm{III}\mathbf{1} = \mathbf{0}. \qquad (5.9.12)$$

Of course, further modification is possible in these expressions, by means of (5.9.12), to eliminate $\overset{-3}{\mathbf{c}}$ from all expressions.

In Appendix E, Table 2, the generators of the vector-valued isotropic functions are listed. The vectors q_k and $\mathscr{J}_k$ depend on one symmetric tensor, $\overset{-1}{\mathbf{c}}$, two absolute vectors, $\theta_{,k}$ and $\mathscr{E}_k$, and one skew-symmetric tensor B^{D}_{kl} defined by

$$B^{\mathrm{D}}_{kl} = \varepsilon_{klm}B_m.$$

There are twelve generators, by use of which we write the constitutive equations for the spatial vectors for q_k and $\mathscr{J}_k$ (Eringen [1980, Sec. 10.14]):

$$
\begin{aligned}
\mathbf{q} = {}&\kappa_1\nabla\theta + \kappa_2\mathscr{E} + \kappa_3\overset{-1}{\mathbf{c}}\nabla\theta + \kappa_4\overset{-1}{\mathbf{c}}\mathscr{E} + \kappa_5\nabla\theta\times\mathbf{B} + \kappa_6\mathscr{E}\times\mathbf{B} \\
&+ \kappa_7\overset{-2}{\mathbf{c}}\nabla\theta + \kappa_8\overset{-2}{\mathbf{c}}\mathscr{E} + \kappa_9(\mathbf{B}\cdot\nabla\theta)\mathbf{B} + \kappa_{10}(\mathbf{B}\cdot\mathscr{E})\mathbf{B} \\
&+ \kappa_{11}[\overset{-1}{\mathbf{c}}(\nabla\theta\times\mathbf{B}) - (\overset{-1}{\mathbf{c}}\nabla\theta)\times\mathbf{B}] + \kappa_{12}[\overset{-1}{\mathbf{c}}(\mathscr{E}\times\mathbf{B}) - (\overset{-1}{\mathbf{c}}\mathscr{E})\times\mathbf{B}], \qquad (5.9.13)
\end{aligned}
$$

$$\begin{aligned}\mathscr{J} = {} & \sigma_1 \mathscr{E} + \sigma_2 \nabla\theta + \sigma_3 \overset{-1}{\mathbf{c}} \mathscr{E} + \sigma_4 \overset{-1}{\mathbf{c}} \nabla\theta + \sigma_5 \mathscr{E} \times \mathbf{B} + \sigma_6 \nabla\theta \times \mathbf{B} \\ & + \sigma_7 \overset{-2}{\mathbf{c}} \mathscr{E} + \sigma_8 \overset{-2}{\mathbf{c}} \nabla\theta + \sigma_9 (\mathbf{B}\cdot\mathscr{E})\mathbf{B} + \sigma_{10}(\mathbf{B}\cdot\nabla\theta)\mathbf{B} \\ & + \sigma_{11}[\overset{-1}{\mathbf{c}}(\mathscr{E} \times \mathbf{B}) - (\overset{-1}{\mathbf{c}}\mathscr{E}) \times \mathbf{B}] + \sigma_{12}[\overset{-1}{\mathbf{c}}(\nabla\boldsymbol{\theta} \times \mathbf{B}) - (\overset{-1}{\mathbf{c}}\nabla\theta) \times \mathbf{B}],\end{aligned} \tag{5.9.14}$$

where κ_1 to κ_{12} and σ_1 to σ_{12} are functions of the joint invariants[14] of $\overset{-1}{\mathbf{c}}$, $\mathscr{E}$, $\mathbf{B}$, and $\nabla\theta$. These invariants can be obtained by use of Table E1.

Constitutive equations (5.9.13) and (5.9.14) must not violate the C–D inequality (5.2.5) which places severe restrictions on the material moduli κ_λ and σ_λ. Moreover, the function Σ and the entropy production $\rho\gamma$, given by (5.8.5), must be invariant under time reversal, i.e., Σ and $\rho\gamma$ must not alter their signs when $t \to -t$. Since $\mathbf{q}$, $\mathscr{J}$, and $\mathbf{B}$ alter their signs, a polynomial function must contain only even degrees of $\mathbf{B}$. For $\rho\gamma$ not to alter its sign with $t \to -t$, the expression of $\rho\gamma$, given by (5.2.5), must be such that time antisymmetric products are cancelled. This process also places restrictions on κ_λ and σ_λ. In the general forms (5.9.8)–(5.9.10), (5.9.13), and (5.9.14) it is not possible to obtain the result of these restrictions. In the special case of the linear theory, these restrictions are discussed in Section 5.11.

A close scrutiny of the nonlinear constitutive equations (5.9.8)–(5.9.10) reveals several interesting physical phenomenon.

1. *The first-order vector and tensor effects*:
   $\partial\Sigma/\partial \mathrm{I}_4 \neq 0$: $\mathscr{E}$-field produces polarization.
   $\partial\Sigma/\partial \mathrm{I}_5 \neq 0$: $\mathbf{B}$-field produces magnetization.
   $\partial\Sigma/\partial \mathrm{I} \neq 0$: Strain produces stress.
2. *The second-order vector and tensor effects*:
   $\partial\Sigma/\partial \mathrm{I}_6 \neq 0$: $\mathscr{E}$-field produces stress, the *electrostrictive effect*.
   $\partial\Sigma/\partial \mathrm{I}_7 \neq 0$: $\mathbf{B}$-field produces stress, the *magnetostrictive effect*.
   $\partial\Sigma/\partial \mathrm{II} \neq 0$: Strain produces the *Poynting effect*.

The third-, fourth-, and fifth-order vector and tensor products complicate the matter further in different combinations. These effects are generally very small, and often not detectable in moderate fields. Important among them is the nonlinear *magnetoelectric* effect ($\partial\Sigma/\partial \mathrm{I}_{10} \neq 0$). If an initial bias field exists, it brings anisotropy to the intermediate state at which incremental fields are applied (see Sections 7.14–7.17). Some of these terms may be important in the case of instability studies (see Section 8.15). Presently, however, the constitutive moduli for these higher-order terms remain to be investigated experimentally.

We now turn our attention to the *nonequilibrium* constitutive equations (5.9.13) and (5.9.14).

---

[14] For an alternative approach to the static, nonlinear theory, see Jordan and Eringen [1964a, b]. For the dynamical theory of polar elastic dielectrics, see Dixon and Eringen [1965a, b] and Demiray and Eringen [1973c].

1. *The first-order vectorial effects*:
   - $\kappa_1 \neq 0$: The temperature gradient produces heat conduction.
   - $\sigma_1 \neq 0$: The electric field gives rise to current.
   - $\kappa_2 \neq 0$: The electric field produces heat conduction. This is the *Peltier effect*.
   - $\sigma_2 \neq 0$: The temperature gradient produces a current. This is the *Seebeck effect*.

2. *The second-order vectorial effects*:
   - $\kappa_4 \neq 0$: Deformation produces an anistropic Peltier effect.
   - $\kappa_5 \neq 0$: Heat flows perpendicular to $\mathbf{B}$ and $\nabla\theta$. This is the *Righi-Leduc* effect.
   - $\kappa_6 \neq 0$: Heat flows perpendicular to $\mathbf{B}$ and $\mathscr{E}$ fields. This is the *Ettingshausen effect*.
   - $\sigma_4 \neq 0$: Deformation produces an anisotropic *Seebeck* effect.
   - $\sigma_5 \neq 0$: Current flows perpendicular to $\mathbf{B}$ and $\mathscr{E}$-fields. This *is the Hall effect*.
   - $\sigma_6 \neq 0$: Current flow perpendicular to $\mathbf{B}$ and $\nabla\theta$. This is the *Nernst effect*.

The third-order effects appearing in (5.9.13) and (5.9.14) further complicate the situation, bringing anisotropies to heat and electric conductions, arising from the cooperative action of electromagnetic fields, strain, and temperature gradients. These third-degree terms are yet to be studied experimentally. We must note that, due to the C–D inequality and the axiom of time reversal, some of these coefficients may be vanishing or expressed in terms of others.

5.10. Anisotropic Electromagnetic Elastic Solids

The material symmetry restrictions for nonmagnetic and magnetic groups were discussed in Sections 5.5 and 5.6. Many interesting physical phenomena that cannot occur in isotropic materials come into play in anisotropic materials. For example, piezoelectricity cannot occur in isotropic solids. For this to take place, the material must be anisotropic and not possess a center of symmetry (or inversion). Similarly, the piezomagnetic effect and other "cross effects" can take place when the material is anisotropic, having the appropriate symmetry. Here we give a few examples of constitutive equations for anisotropic electromagnetic elastic solids for illustrative purposes. For a detailed study of the subject for specific classes, see Kiral and Eringen [1976]. Since the integrity basis for several vectors and tensors is tedious and lengthy to produce, we give here only some important cases involving only one pair out of two vectors and one tensor variable. These include rigid dielectrics ($\mathscr{E}$, $\mathbf{B}$), piezoelectric ($\overset{-1}{\mathbf{c}}$, $\mathscr{E}$), and piezomagnetic ($\overset{-1}{\mathbf{c}}$, $\mathbf{B}$) materials.

I. Rigid Electromagnetic Solids ($m\underline{m}\underline{m}$ magnetic class)

The constitutive equations for nonconducting rigid solids are given by

$$\rho_0\Psi = \Sigma(\mathscr{E}, \mathbf{B}, \theta, \mathbf{X}), \qquad \eta = -\frac{1}{\rho}\frac{\partial\Sigma}{\partial\theta},$$
$$\Pi_K = -\frac{\partial\Sigma}{\partial\mathscr{E}_K}, \qquad M_K = -\frac{\partial\Sigma}{\partial B_K}, \qquad \mathbf{Q} = \mathscr{J} = \mathbf{0}. \tag{5.10.1}$$

Noting that the basic quantities for the ($m\underline{m}\underline{m}$)-class are

$$a = B_3, \qquad b = B_2, \qquad c = B_1, \qquad d = 0,$$
$$B = \mathscr{E}_1, \qquad C = \mathscr{E}_2, \qquad D = \mathscr{E}_3, \qquad \boldsymbol{\Lambda} = \mathbf{0},$$

the integrity basis follows from (C2.3) as:

$$\mathrm{I}_1 = B_3, \qquad \mathrm{I}_2 = B_1^2, \qquad \mathrm{I}_3 = B_2^2,$$
$$\mathrm{I}_4 = \mathscr{E}_1^2, \qquad \mathrm{I}_5 = \mathscr{E}_2^2, \qquad \mathrm{I}_6 = \mathscr{E}_3^2, \tag{5.10.2}$$
$$\mathrm{I}_7 = \mathscr{E}_2\mathscr{E}_3 B_2, \qquad \mathrm{I}_8 = \mathscr{E}_1\mathscr{E}_3 B_1, \qquad \mathrm{I}_9 = \mathscr{E}_1\mathscr{E}_2 B_1 B_2.$$

Hence,

$$\rho_0\Psi = \Sigma(\mathrm{I}_1, \ldots, \mathrm{I}_2, \ldots, \mathrm{I}_9, \theta, \mathbf{X}), \qquad \eta = -\frac{1}{\rho_0}\frac{\partial\Sigma}{\partial\theta}, \qquad \mathbf{Q} = \mathscr{J} = \mathbf{0},$$

$$\begin{aligned}\Pi_K = {} & -2\frac{\partial\Sigma}{\partial\mathrm{I}_4}\mathscr{E}_1\delta_{1K} - 2\frac{\partial\Sigma}{\partial\mathrm{I}_5}\mathscr{E}_2\delta_{2K} - 2\frac{\partial\Sigma}{\partial\mathrm{I}_6}\mathscr{E}_3\delta_{3K} \\ & - \frac{\partial\Sigma}{\partial\mathrm{I}_7}B_2(\mathscr{E}_2\delta_{3K} + \mathscr{E}_3\delta_{2K}) - \frac{\partial\Sigma}{\partial\mathrm{I}_8}B_1(\mathscr{E}_3\delta_{1K} + \mathscr{E}_1\delta_{3K}) \\ & - \frac{\partial\Sigma}{\partial\mathrm{I}_9}B_1B_2(\mathscr{E}_2\delta_{1K} + \mathscr{E}_1\delta_{2K}),\end{aligned}$$

$$\begin{aligned}M_k = {} & -\frac{\partial\Sigma}{\partial\mathrm{I}_1}\delta_{3K} - 2\frac{\partial\Sigma}{\partial\mathrm{I}_2}B_1\delta_{1K} - 2\frac{\partial\Sigma}{\partial\mathrm{I}_3}B_2\delta_{2K} - \frac{\partial\Sigma}{\partial\mathrm{I}_7}\mathscr{E}_2\mathscr{E}_3\delta_{2K} \\ & - \frac{\partial\Sigma}{\partial\mathrm{I}_8}\mathscr{E}_1\mathscr{E}_3\delta_{1K} - \frac{\partial\Sigma}{\partial\mathrm{I}_9}\mathscr{E}_1\mathscr{E}_2(B_1\delta_{2K} + B_2\delta_{1K}).\end{aligned}$$

For example, the antiferromagnetic crystal NiF_2 belongs to this class. It is interesting to note that, for nonvanishing $\partial\Sigma/\partial\mathrm{I}_7$, $\partial\Sigma/\partial\mathrm{I}_8$, and $\partial\Sigma/\partial\mathrm{I}_9$, the magnetic field produces polarization and electric field causes magnetization.[15]

[15] Magnetoelectric coupling plays an important role in the dynamic behavior of certain materials, especially compounds which possess simultaneously ferroelectric and ferromagnetic properties (e.g., the ferroelectric magnet solution of $BaT_2O_3Sr_{0.3}La_{07}MnO_3$). Then the magnetoelectric coupling favors the existence of a resonance-type of coupling between magnons—see Section 4.5.4—and a ferroelectric wave mode (Maugin [1981c]).

II. Elastic Dielectrics ($3m = C_{3v}$)

Some important dielectric materials such as quartz, tellurium, Rochelle salt, MgTe, CdSe, and two recently grown synthetic crystals, lithium tantalate ($LiTaO_3$) and lithium niobate ($LiNbO_3$), belong to this class. The complete, irreducible integrity basis of **C** and Σ, under the group $3m$, is listed in (5.5.21).

The constitutive equations now follow from (5.8.4), with Σ considered as a function of the invariants listed in (5.5.21).

III. Magnetoelastic Solids ($\bar{3}m$)

In Section 5.6 we have already determined the typical elements of the integrity basis of one symmetric tensor C_{KL}, one polar vector $\mathscr{E}_K$, and one axial vector B_K. By setting $\mathscr{E}_K = 0$, by use of this list, we can construct the integrity basis of C_{KL} and B_K for this class. We find that there are 25 invariants.

Chromium oxide (Cr_2O_3) belongs to this class.

5.11. Linear Constitutive Equations

If the electromagnetic fields, strains, and temperature gradients are small, the linear constitutive equations in the independent variables will be adequate for the treatment of physical problems. In this case, Σ is a quadratic polynomial of the form

$$\begin{aligned}\Sigma = \Sigma_0 + \Sigma_{KL}\tilde{E}_{KL} + \tfrac{1}{2}\Sigma_{KLMN}\tilde{E}_{KL}\tilde{E}_{MN} - E_{KLM}\mathscr{E}_K\tilde{E}_{LM} - H_{KLM}B_K\tilde{E}_{LM}\\ - \chi^{\mathrm{E}}_K\mathscr{E}_K - \tfrac{1}{2}\chi^{\mathrm{E}}_{KL}\mathscr{E}_K\mathscr{E}_L - \chi^{\mathrm{B}}_K B_K - \tfrac{1}{2}\chi^{\mathrm{B}}_{KL}B_K B_L - \Lambda_{KL}\mathscr{E}_K B_L, \qquad (5.11.1)\end{aligned}$$

where Σ_0, Σ_{KL}, Σ_{KLMN}, E_{KLM}, H_{KLM}, χ^{E}_{KL}, χ^{B}_K, χ^{B}_{KL}, and Λ_{KL} are functions of θ and **X** only. $\tilde{E}_{KL}$ is the infinitesimal strain tensor. To linearize the thermal effects we further take

$$\begin{gathered}\theta = T_0 + T, \qquad |T| \ll T_0, \qquad T_0 > 0,\\ \Sigma_0 = \sigma_0 - \rho_0\eta_0 T - (\rho_0\gamma/2T_0)T^2, \qquad \Sigma_{KL} = A_{KL} - B_{KL}T, \qquad (5.11.2)\\ \chi^{\mathrm{E}}_K = \Pi^0_K + \tilde{\omega}_K T, \qquad \chi^{\mathrm{B}}_K = M^0_K + \Gamma_K T,\end{gathered}$$

and consider σ_0, η_0, γ, A_{KL}, B_{KL}, Π^0_K, $\tilde{\omega}_K$, M^0_K, and Γ_K as functions of **X**, only for inhomogeneous materials, and constants for homogeneous materials. Substituting (5.11.1) and (5.11.2) into (5.8.4) we obtain

$$\eta = \eta_0 + \frac{\gamma}{T_0}T + \frac{1}{\rho_0}(B_{KL}\tilde{E}_{KL} + \tilde{\omega}_K\mathscr{E}_K + \Gamma_K B_K), \qquad (5.11.3)$$

$$_{\mathrm{E}}T_{KL} = A_{KL} - B_{KL}T + \Sigma_{KLMN}\tilde{E}_{MN} - E_{MKL}\mathscr{E}_M - H_{MKL}B_M, \qquad (5.11.4)$$

$$\Pi_K = \Pi^0_K + \tilde{\omega}_K T + \chi^{\mathrm{E}}_{KL}\mathscr{E}_L + E_{KLM}\tilde{E}_{LM} + \Lambda_{KL}B_L, \qquad (5.11.5)$$

$$M_K = M^0_K + \Gamma_K T + \chi^{\mathrm{B}}_{KL}B_L + H_{KLM}\tilde{E}_{LM} + \Lambda_{LK}\mathscr{E}_L \qquad (5.11.6)$$

To various material moduli, the following terminology may be ascribed:

η_0	entropy at the natural state,
γ	the heat capacity,
A_{KL}	the stress at the natural state,
B_{KL}	thermal stress moduli,
Σ_{KLMN}	elastic moduli,
E_{MKL}	piezoelectric moduli,
H_{MKL}	piezomagnetic moduli,
Π_K^0	polarization at the natural state,
$\tilde{\omega}_K$	pyroelectric polarizability,
χ_{KL}^{E}	dielectric susceptibility,
Λ_{KL}	magnetic polarizability,
M_K^0	magnetization at the natural state,
Γ_K	pyromagnetic moduli,
χ_{KL}^{B}	magnetic susceptibility.

It is important to note the physical significance of various coupled terms: For $E_{MKL} \neq 0$, the electric field causes stress, and strain gives rise to polarization. This is the well-known phenomenon, *piezoelectricity*. Similarly, for $H_{MKL} \neq 0$, the magnetic induction produces stress in the body, and with strain, the body is magnetized. This is the *piezomagnetic effect*. The coupling coefficient Λ_{KL} is responsible for the *magnetoelectric effect*. Since $\mathbf{B}$ changes its sign, but $\mathbf{\Pi}$ does not upon the time reversal, at first sight we may think that piezomagnetic and magnetoelectric effects cannot exist.[16] However, the magnetic symmetry group for certain crystals permits these effects.

For $\tilde{\omega}_K \neq 0$, the temperature produces polarization. This is called *pyroelectricity*. For $\Gamma_K \neq 0$, we will have magnetization with thermal changes. This is the *pyromagnetic effect*.

Under the symmetry group $\{\mathbf{S}\}$, various moduli transform as absolute vectors and tensors, except Γ_K, Λ_{KL}, and H_{KLM} which transform as relative vectors and tensors. Moreover, for the survival of these quantities, a $(-)$ sign is needed incorporating the time reversal so that

$$\begin{aligned} \bar{\Gamma}_K &= -\Gamma_P S_{KP} \det \mathbf{S}, \\ \bar{\Lambda}_{KL} &= -\Lambda_{PQ} S_{KP} S_{LQ} \det \mathbf{S}, \\ \bar{H}_{KLM} &= -H_{PQR} S_{KP} S_{LQ} S_{MR} \det \mathbf{S}. \end{aligned} \tag{5.11.7}$$

We now make several important observations:

(a) *Piezoelectric effect*. This is governed by the third-order coefficient E_{MKL}, which vanishes for crystals having a center of inversion (e.g., $\bar{X}_K = -X_K$). Of the 32 crystals, groups, only 20 allow for a piezoelectric effect.

---

[16] In fact, Zocher and Török [1953] have expressed such a conclusion on the basis of an erroneous assumption that all crystals are time symmetric. Experimental verification of the piezomagnetic effect was made by Borovik-Romanov [1959].

(b) *Pyroelectric effect* ($\tilde{\omega}_K$): The same remark as in (a) above applies to this effect.
(c) *Piezomagnetism* (H_{KLM}). The existence of this effect depends on the combination of the axial character and the time symmetry. Of the 90 magnetic point groups 66 allow for piezomagnetism.
(d) *Magnetoelectric effect* (Λ_{KL}): The same remark as in (c) applies here. There exist 58 members of the magnetic point group that allow for this effect.
(e) *Pyromagnetic effect* (Γ_K): There are 31 members of the magnetic point group which permit the pyromagnetic effect.

The linear constitutive equations for the heat and current vectors are given by

$$Q_K = \kappa_{KL}\theta_{,L} + \kappa_{KLM}\tilde{E}_{LM} + \kappa^{\mathrm{E}}_{KL}\mathscr{E}_L + \kappa^{\mathrm{B}}_{KL}B_L, \tag{5.11.8}$$

$$\mathscr{J}_K = \sigma_{KL}\mathscr{E}_L + \sigma_{KLM}\tilde{E}_{LM} + \sigma^{\theta}_{KL}\theta_{,L} + \sigma^{\mathrm{B}}_{KL}B_L. \tag{5.11.9}$$

Substituting these into the C–D inequality (5.8.5), we find that, because of linearity, κ_{KLM}, $\kappa^{\mathrm{B}}_{KLM}$, σ_{KLM}, and $\sigma^{\mathrm{B}}_{KLM}$ must vanish so that

$$Q_K = \kappa_{KL}\theta_{,L} + \kappa^{\mathrm{E}}_{KL}\mathscr{E}_L, \tag{5.11.10}$$

$$\mathscr{J}_K = \sigma_{KL}\mathscr{E}_L + \sigma^{\theta}_{KL}\theta_{,L}, \tag{5.11.11}$$

$$\theta^{-1}\kappa_{KL}\theta_{,K}\theta_{,L} + \sigma_{KL}\mathscr{E}_K\mathscr{E}_L + (\sigma^{\theta}_{LK} + \theta^{-1}\kappa^{\mathrm{E}}_{KL})\theta_{,K}\mathscr{E}_L \geq 0. \tag{5.11.12}$$

The Onsager principle also tells us that

$$\kappa_{KL} = \kappa_{LK}, \qquad \sigma_{KL} = \sigma_{LK}.$$

The entropy inequality (5.11.12) also shows that κ_{KL} and σ_{KL} are *nonnegative matrices*.

Equations (5.11.10) and (5.11.11) exhibit Peltier and Seebeck effects.

The spatial forms of the constitutive equations are obtained by using

$$\theta = T_0 + T, \qquad |T| \ll T_0, \quad T_0 > 0,$$

$$\rho/\rho_0 \simeq 1 - I_{\tilde{E}}, \qquad x_{k,K} \simeq (\delta_{MK} + \tilde{E}_{MK} + \tilde{R}_{MK})\delta_{Mk},$$

$$\tilde{E}_{KL} \simeq \tilde{e}_{kl}\delta_{kK}\delta_{kL}.$$

When these equations (5.11.1), (5.11.3)–(5.11.6), (5.11.10), and (5.11.11) are combined with (3.11.4) and (3.11.5) we obtain

$$\begin{aligned}\Sigma = \sigma_0 - \rho_0\eta_0 T - \frac{\rho_0\gamma_0}{2T_0}T^2 + (\alpha_{kl} - \beta_{kl}T)\tilde{e}_{kl} + \tfrac{1}{2}\sigma_{klmn}\tilde{e}_{kl}\tilde{e}_{mn} \\ - e_{klm}\mathscr{E}_k\tilde{e}_{lm} - h_{klm}B_k\tilde{e}_{lm} - (\pi^0_k + \tilde{\omega}_k T)\mathscr{E}_k \\ - \tfrac{1}{2}\chi^{\mathrm{E}}_{kl}\mathscr{E}_k\mathscr{E}_l - (m^0_k + \gamma_k T)B_k - \tfrac{1}{2}\chi^{\mathrm{B}}_{kl}B_kB_l - \lambda_{kl}\mathscr{E}_kB_l,\end{aligned} \tag{5.11.13}$$

$$\eta = \eta_0 + \frac{\gamma_0}{T_0}T + \frac{1}{\rho_0}(\beta_{kl}\tilde{e}_{kl} + \tilde{\omega}_k\mathscr{E}_k + \gamma_k B_k), \tag{5.11.14}$$

$$
\begin{aligned}
{}_{\mathrm{E}}t_{kl} &= \alpha_{kl}(1 - \tilde{e}_{rr}) - \beta_{kl} T + \alpha_{km}(\tilde{e}_{lm} + \tilde{r}_{lm}) + \alpha_{ml}(\tilde{e}_{km} + \tilde{r}_{km}) \\
&\quad + \sigma_{klmn}\tilde{e}_{mn} - e_{mkl}\mathscr{E}_m - h_{mkl}B_m, \qquad (5.11.15)
\end{aligned}
$$

$$P_k = [(1 - \tilde{e}_{rr})\delta_{kl} + \tilde{e}_{kl} + \tilde{r}_{kl}]\pi_l^0 + \tilde{\omega}_k T + \chi_{kl}^{\mathrm{E}}\mathscr{E}_l + e_{klm}\tilde{e}_{lm} + \lambda_{kl}B_l, \qquad (5.11.16)$$

$$\mathscr{M}_k = [(1 - \tilde{e}_{rr})\delta_{kl} + \tilde{e}_{kl} + \tilde{r}_{kl}]m_l^0 + \gamma_k T + \chi_{kl}^{\mathrm{B}}B_l + h_{klm}\tilde{e}_{lm} + \lambda_{lk}\mathscr{E}_l, \qquad (5.11.17)$$

$$q_k = \kappa_{kl} T_{,l} + \chi_{kl}^{\mathrm{E}}\mathscr{E}_l. \qquad (5.11.18)$$

$$\mathscr{J}_k = \sigma_{kl}\mathscr{E}_l + \sigma_{kl}^{\theta} T_{,l}, \qquad (5.11.19)$$

where the new material moduli are given by

$$
\begin{aligned}
\{\pi_k^0, m_k^0, \tilde{\omega}_k, \gamma_k\} &= \{\Pi_K^0, M_K^0, \tilde{\omega}_K, \Gamma_K\}\delta_{Kk}, \\
\{\alpha_{kl}, \beta_{kl}, \chi_{kl}^{\mathrm{E}}, \chi_{kl}^{\mathrm{B}}, \lambda_{kl}\} &= \{A_{KL}, B_{KL}, \chi_{KL}^{\mathrm{E}}, \chi_{KL}^{\mathrm{B}}, \Lambda_{KL}\}\delta_{Kk}\delta_{Ll}, \\
\{e_{klm}, h_{klm}\} &= \{E_{KLM}, H_{KLM}\}\delta_{Kk}\delta_{Ll}\delta_{Mm}, \qquad (5.11.20) \\
\sigma_{klmn} &= \Sigma_{KLMN}\delta_{Kk}\delta_{Ll}\delta_{Mm}\delta_{Nn},
\end{aligned}
$$

in which δ_{Kk} is a shifter. It becomes a Kronecker delta when the spatial and material frames coincide.

The symmetry properties of these moduli are apparent from (5.11.13) and the symmetry of $\tilde{e}_{kl}$:

$$
\begin{aligned}
\chi_{kl}^{\mathrm{E}} &= \chi_{lk}^{\mathrm{E}}, \qquad \chi_{kl}^{\mathrm{B}} = \chi_{lk}^{\mathrm{B}}, \\
e_{klm} &= e_{kml}, \qquad h_{klm} = h_{kml}, \qquad (5.11.21) \\
\sigma_{klmn} &= \sigma_{lkmn} = \sigma_{klnm} = \sigma_{mnkl}.
\end{aligned}
$$

Consequently, we have 6 members of each of χ_{kl}^{E} and χ_{kl}^{B}, 18 member of each of e_{klm} and h_{klm}, and 21 members of σ_{klmn}.

Symmetry Group

The symmetry properties of linear anisotropic materials under symmetry groups can be studied in a simple way, eliminating completely the intricate algebraic process required by the nonlinear theory. Moreover, this method is applicable to all magnetic crystal systems as well. This method relies on the basic quantities introduced in Section 5.5 (see (5.5.16) and (5.5.17)). Suppose that the field tensors (e.g., stress, strain, electromagnetic fields, polarization, magnetization, etc.) are decomposed into basic quantities with the appropriate irreducible representations (primitive symmetry species) of a point group. The part of the basic quantities belonging to a given irreducible representation produces an effect belonging to the same representation. Hence, once the basic quantities for the field tensors are obtained for a given class, the linear constitutive equations can be written immediately. The following examples should serve to illustrate the method:

EXAMPLE 1 (Magnetoelectric Effect) (Magnetic class 2). The magnetoelectric effect is described by

$$\mathscr{M}_k = \lambda_{kl}\mathscr{E}_l, \qquad (5.11.22)$$

where $\mathcal{M}$ is the magnetization vector (axial, c-tensor) and $\mathscr{E}$ is the electric field vector (a true i-tensor). For the magnetic crystal class $\underline{2} = \{I, RD_3\}$, we have the irreducible representations

$$\mathcal{M} = 2\Gamma_1 \dotplus \Gamma_2, \qquad \mathscr{E} = \Gamma_1 \dotplus 2\Gamma_2. \tag{5.11.23}$$

The basic quantities are given by (5.6.2) as

Γ_1	$\mathcal{M}_1, \mathcal{M}_2$	$\mathscr{E}_3$
Γ_2	$\mathcal{M}_3$	$\mathscr{E}_1, \mathscr{E}_2$

$$\begin{aligned} \mathcal{M}_1 &= \alpha\mathscr{E}_3 = \lambda_{13}\mathscr{E}_3, \\ \mathcal{M}_2 &= \beta\mathscr{E}_3 = \lambda_{23}\mathscr{E}_3, \\ \mathcal{M}_3 &= \gamma_1\mathscr{E}_1 + \gamma_2\mathscr{E}_2 = \lambda_{31}\mathscr{E}_1 + \lambda_{32}\mathscr{E}_2, \end{aligned} \tag{5.11.24}$$

which shows that the only surviving members of λ_{kl} are $\lambda_{13}, \lambda_{23}, \lambda_{31}$, and λ_{32}.

EXAMPLE 2 (Magnetoelectric Effect) (Magnetic class 3). For the magnetic class $3\underline{m}$, some of the basic quantities form a two-dimensional real carrier space of Γ_5. Noting that

$$\mathcal{M} = \Gamma_4 \dotplus \Gamma_5, \qquad \mathscr{E} = \Gamma_4 \dotplus \Gamma_5, \tag{5.11.25}$$

we have the basic quantities

Γ_4	$\mathcal{M}_3$	$\mathscr{E}_3$
Γ_5	$\begin{pmatrix} \mathcal{M}_1 \\ \mathcal{M}_2 \end{pmatrix}$	$\begin{pmatrix} \mathscr{E}_1 \\ \mathscr{E}_2 \end{pmatrix}$

Hence,

$$\begin{pmatrix} \mathcal{M}_1 \\ \mathcal{M}_2 \end{pmatrix} = \alpha \begin{pmatrix} \mathscr{E}_1 \\ \mathscr{E}_2 \end{pmatrix}, \qquad \mathcal{M}_3 = \beta\mathscr{E}_3,$$

which yields

$$\begin{aligned} \mathcal{M}_1 &= \alpha\mathscr{E}_1 = \lambda_{11}\mathscr{E}_1, \\ \mathcal{M}_2 &= \alpha\mathscr{E}_2 = \lambda_{11}\mathscr{E}_2, \\ \mathcal{M}_3 &= \beta\mathscr{E}_3 = \lambda_{33}\mathscr{E}_3, \end{aligned} \tag{5.11.26}$$

so that $\lambda_{11} = \lambda_{22}$.

EXAMPLE 3 (Piezomagnetic Effect) (Crystal class $C_3 = 3$). Piezomagnetism is described by

$$\mathcal{M}_k = h_{klm}\tilde{e}_{lm}, \tag{5.11.27}$$

where $\mathcal{M}$ is the magnetization and $\tilde{e}_{kl}$ is the strain tensor which is a symmetric, true, i-tensor.

For the crystal class $C_3 = 3$, the relevant basic quantities are given with the aid of Table C1.8:

Γ_1: $\mathscr{M}_3$	$\tilde{e}_{33}, \tilde{e}_{11} + \tilde{e}_{22}$
Γ_2: $\mathscr{M}_1 - i\mathscr{M}_2$	$\tilde{e}_{13} - i\tilde{e}_{23}, \tilde{e}_{11} - \tilde{e}_{22} + 2i\tilde{e}_{12}$
Γ_3: $\mathscr{M}_1 + i\mathscr{M}_2$	$\tilde{e}_{13} + i\tilde{e}_{23}, \tilde{e}_{11} - \tilde{e}_{22} - 2i\tilde{e}_{12}$

from which there follows

$$\begin{aligned}\mathscr{M}_3 &= \alpha_1 \tilde{e}_{33} + \alpha_2(\tilde{e}_{11} + \tilde{e}_{22}),\\ \mathscr{M}_1 - i\mathscr{M}_2 &= \beta(\tilde{e}_{13} - i\tilde{e}_{23}) + \gamma(\tilde{e}_{11} - \tilde{e}_{22} + 2i\tilde{e}_{12})\\ &= (\beta_1 + i\beta_2)(\tilde{e}_{13} - i\tilde{e}_{23}) + (\gamma_1 + i\gamma_2)(\tilde{e}_{11} - \tilde{e}_{22} + 2i\tilde{e}_{12}).\end{aligned} \tag{5.11.28}$$

Consequently, we have the constitutive equations

$$\begin{aligned}\mathscr{M}_1 &= \gamma_1 \tilde{e}_{11} - \gamma_1 \tilde{e}_{22} + \beta_2 \tilde{e}_{23} + \beta_1 \tilde{e}_{13} - 2\gamma_2 \tilde{e}_{12},\\ \mathscr{M}_2 &= -\gamma_2 \tilde{e}_{11} + \gamma_2 \tilde{e}_{22} + \beta_1 \tilde{e}_{23} - \beta_2 \tilde{e}_{13} - 2\gamma_1 \tilde{e}_{12},\\ \mathscr{M}_3 &= \alpha_2 \tilde{e}_{11} + \alpha_2 \tilde{e}_{22} + \alpha_1 \tilde{e}_{33}.\end{aligned} \tag{5.11.29}$$

This is in agreement with the schemes given by Bhagavantam [1966, p. 173].

Isotropic Materials

For the isotropic materials, odd-order tensors vanish and $\lambda_{kl} = 0$ because of the time reversal. For the even-order tensors, we have

$$\begin{aligned}\{\alpha_{kl}, \beta_{kl}, \chi^{\mathrm{E}}_{kl}, \chi^{\mathrm{B}}_{kl}\} &= \{\alpha_{\mathrm{e}}, \beta_{\mathrm{e}}, \chi^{\mathrm{E}}, \chi^{\mathrm{B}}\}\delta_{kl},\\ \sigma_{klmn} &= \lambda_{\mathrm{e}}\delta_{kl}\delta_{mn} + \mu_{\mathrm{e}}(\delta_{km}\delta_{ln} + \delta_{kn}\delta_{lm}),\end{aligned} \tag{5.11.30}$$

and the constitutive equations reduce to

$$\begin{aligned}\Sigma = \sigma_0 - \rho_0\eta_0 T + (\alpha_{\mathrm{e}} - \beta_{\mathrm{e}} T)\operatorname{tr}\tilde{\mathbf{e}} + \tfrac{1}{2}\lambda_{\mathrm{e}}(\operatorname{tr}\tilde{\mathbf{e}})^2 + \mu_{\mathrm{e}}\operatorname{tr}\tilde{\mathbf{e}}^2\\ - \tfrac{1}{2}\chi^{\mathrm{E}}\boldsymbol{\mathscr{E}}\cdot\boldsymbol{\mathscr{E}} - \tfrac{1}{2}\chi^{\mathrm{B}}\mathbf{B}\cdot\mathbf{B},\end{aligned} \tag{5.11.31}$$

$$\eta = \eta_0 + \frac{\gamma_0}{T_0}T + \frac{\beta}{\rho_0}\operatorname{tr}\tilde{\mathbf{e}}, \tag{5.11.32}$$

$${}_{\mathrm{E}}t_{kl} = (\alpha_{\mathrm{e}} - \beta_{\mathrm{e}} T)\delta_{kl} + (\lambda_{\mathrm{e}} - \alpha_{\mathrm{e}})\tilde{e}_{rr}\delta_{kl} + 2(\mu_{\mathrm{e}} + \alpha_{\mathrm{e}})\tilde{e}_{kl}, \tag{5.11.33}$$

$$P_k = \chi^{\mathrm{E}}\mathscr{E}_k, \tag{5.11.34}$$

$$\mathscr{M}_k = \chi^{\mathrm{B}}B_k, \tag{5.11.35}$$

$$q_k = \kappa T_{,k} + \kappa^{\mathrm{E}}\mathscr{E}_k, \tag{5.11.36}$$

$$\mathscr{J}_k = \sigma\mathscr{E}_k + \sigma^\theta T_{,k}. \tag{5.11.37}$$

The entropy inequality (5.11.12) places restrictions on κ, κ^{E}, σ, and σ^θ:

$$\kappa T_0^{-1} T_{,k} T_{,k} + \sigma\mathscr{E}_k\mathscr{E}_k + (\sigma^\theta + T_0^{-1}\kappa^{\mathrm{E}})T_{,k}\mathscr{E}_l \geq 0. \tag{5.11.38}$$

This inequality may be written in the compact form

$$\lambda_{ab}\eta_a\eta_b \geq 0, \qquad a, b = 1, 2, \ldots, 6, \tag{5.11.39}$$

where

$$\begin{aligned}
&\eta_a = \theta_{,a}, \qquad a = 1, 2, 3,\\
&\eta_a = \mathscr{E}_a, \qquad a = 4, 5, 6,\\
&\lambda_{ab} = \kappa T_0^{-1}\delta_{ab}, \qquad a, b = 1, 2, 3,\\
&\lambda_{ab} = \sigma\delta_{ab}, \qquad a, b = 4, 5, 6,\\
&\lambda_{ka} = \lambda_{ak} = \tfrac{1}{2}(\sigma^\theta + T_0^{-1}\kappa^{\mathrm{E}}), \qquad k = 1, 2, 3; \quad a = 4, 5, 6.
\end{aligned} \tag{5.11.40}$$

According to a theorem of algebra, (5.11.39) will not be violated for all η_a, $a = 1, 2, \ldots, 6$, if and only if all sequences of the subdeterminant lying on the main diagonal of a 6×6 matrix λ_{ab}, starting with λ_{11}, are nonnegative. Calculations lead to

$$\kappa \geq 0, \qquad \sigma \geq 0, \qquad 4\kappa\sigma T_0^{-1} - (\kappa^{\mathrm{E}} T_0^{-1} + \sigma^\theta)^2 \geq 0. \tag{5.11.41}$$

5.12. Electromagnetic Fluids

A body is called *fluid* if every instantaneous frame, with fixed density, can be considered a reference frame. Thus, for a fluid, the reference frame $\mathbf{X}$ coincides with the spatial frame $\mathbf{x}$, but the density ρ remains as a constitutive variable.

The constitutive variables of fluids are obtained by considering two motions $\mathbf{x}$ and $\boldsymbol{\xi}$ at times t and τ, i.e.,

$$\mathbf{x} = \boldsymbol{\chi}(\mathbf{X}, t), \qquad \boldsymbol{\xi} = \boldsymbol{\chi}(\mathbf{X}, \tau). \tag{5.12.1}$$

Eliminating $\mathbf{X}$, we obtain

$$\mathbf{x} = \boldsymbol{\chi}(X(\boldsymbol{\xi}, \tau), t) = \boldsymbol{\chi}_{(t)}(\boldsymbol{\xi}, \tau), \tag{5.12.2}$$

where $\boldsymbol{\chi}_{(t)}$ is called the *relative motion*, which is considered to be invertible (axiom of continuity of matter). Thus

$$\boldsymbol{\xi} = \boldsymbol{\xi}_{(\tau)}(\mathbf{x}, t). \tag{5.12.3}$$

The relative strain measures can then be constructed as

$$c_{(t)kl} = \delta_{\alpha\beta}\xi^\alpha_{,k}\xi^\beta_{,l}, \qquad C_{KL}(\tau) = \delta_{\alpha\beta}\xi^\alpha_{,K}\xi^\beta_{,L} = c_{(t)kl}x^k_{,K}x^l_{,L}, \tag{5.12.4}$$

(see Eringen [1976a, Sect. 1.3] and [1980, Sect. 1.4]).

Similar constructions are possible for other constitutive variables. For a fluent body then we let $\tau \to t$, i.e., $\mathbf{x}$ and $\boldsymbol{\xi}$ coincide. This eliminates the strain measures from the list of variables (excluding ρ, which, as posited, stays in), but the rate variables remain. The situation is the same if in (5.4.33) we take $\mathbf{X} \to \mathbf{x}$.

Thus, the general constitutive equations of any rate-dependent fluid may be obtained from (5.4.34) by letting $\mathbf{X} \to \mathbf{x}$ and by including ρ in the set of arguments. For *simple memory-independent fluids*, this leads to the form.[17]

$$\Psi = \Psi(\rho^{-1}, \theta, d_{kl}, \mathscr{E}_k, B_k, \theta_{,k}). \tag{5.12.5}$$

Similar equations (containing the same arguments) are valid for the entropy, stress tensor, heat, polarization, magnetization, and current vectors.

To determine the restrictions arising from the second law of thermodynamics, we now use the C–D inequality (5.2.5), appropriate to the independent variables selected, i.e.,

$$\rho\gamma \equiv -\rho(\dot{\Psi} + \eta\dot{\theta}) + t_{kl}v_{l,k} + \frac{1}{\theta}q_k\theta_{,k} - P_k\dot{\mathscr{E}}_k - \mathscr{M}_k\dot{B}_k + \mathscr{J}_k\mathscr{E}_k \geq 0. \tag{5.12.6}$$

Substituting $\dot{\Psi}$, calculated from (5.12.5), into (5.12.6), we have

$$-\rho\left(\frac{\partial\Psi}{\partial\theta} + \eta\right)\dot{\theta} + \left(t_{kl} - \frac{\partial\Psi}{\partial\rho^{-1}}\right)d_{lk} + t_{kl}w_{lk} - \frac{\partial\Psi}{\partial d_{kl}}\dot{d}_{kl} - \rho\frac{\partial\Psi}{\partial\theta_{,k}}\dot{\overline{\theta_{,k}}}$$
$$+ \frac{1}{\theta}q_k\theta_{,k} - \left(P_k + \rho\frac{\partial\Psi}{\partial\mathscr{E}_k}\right)\dot{\mathscr{E}}_k - \left(\mathscr{M}_k + \rho\frac{\partial\Psi}{\partial B_k}\right)\dot{B}_k + \mathscr{J}_k\mathscr{E}_k \geq 0, \tag{5.12.7}$$

where we used the equation of continuity to replace $\dot{\rho}$ and introduced the spin tensor $w_{kl} = v_{[k,l]}$. This inequality is linear in θ, w_{lk}, $\dot{d}_{kl}$, $\dot{\overline{\theta_{,k}}}$, $\mathscr{E}_k$, and $\dot{B}_k$. The necessary and sufficient conditions for (5.12.7) to remain in one sign are

$$\frac{\partial\Psi}{\partial d_{kl}} = 0, \qquad \frac{\partial\Psi}{\partial\theta_{,k}} = 0, \qquad t_{[kl]} = 0,$$
$$\eta = -\frac{\partial\Psi}{\partial\theta},$$
$$P_k = -\rho\frac{\partial\Psi}{\partial\mathscr{E}_k}, \tag{5.12.8}$$
$$\mathscr{M}_k = -\rho\frac{\partial\Psi}{\partial B_k},$$
$$\rho\gamma \equiv {}_{\mathrm{D}}t_{kl}d_{lk} + \frac{1}{\theta}q_k\theta_{,k} + \mathscr{J}_k\mathscr{E}_k \geq 0, \tag{5.12.9}$$

where

$${}_{\mathrm{D}}t_{kl} = t_{kl} + \pi\delta_{kl}, \qquad \pi = -\frac{\partial\Psi}{\partial\rho^{-1}}. \tag{5.12.10}$$

From the first two of (10.12.8), it is clear that Ψ is independent of $\mathbf{d}$ and $\nabla\theta$ while the third equation indicates that Cauchy's stress tensor must be symmetric. Recalling the expression (5.2.3), we have

$$P_{[k}\mathscr{E}_{l]} + \mathscr{M}_{[k}B_{l]} = 0. \tag{5.12.11}$$

---

[17] Note that discarding the rates of $\mathscr{E}$ and $\mathbf{B}$ eliminates the dissipative electromagnetic effects which are responsible for the absorption of electromagnetic waves.

If ${}_{D}\mathbf{t}$, $\mathbf{q}$, and $\mathscr{J}$ are continuous functions of $\mathbf{d}$, $\nabla\theta$, and $\mathscr{E}$, from (10.12.9) it also follows that

$$ {}_{D}\mathbf{t} = \mathbf{0}, \quad \mathbf{q} = \mathbf{0}, \quad \mathscr{J} = \mathbf{0}, \qquad \text{when} \quad \mathbf{d} = \mathbf{0}, \quad \nabla\theta = \mathbf{0}, \quad \mathscr{E} = \mathbf{0}. \tag{5.12.12} $$

We have, therefore, proved (see Eringen [1980, Sect. 10.24])

Theorem. *The constitutive equations of electromagnetic fluids do not violate the second law of thermodynamics, if they are of the form* (5.12.8) *subject to* (5.12.9) *and* (5.12.12).

If we recall expression (3.6.12) of the electromagnetic couple, (5.12.11) implies that, for *fluids, the electromagnetic couple vanishes.* The free energy function Ψ is an objective function of $\mathscr{E}$ and $\mathbf{B}$. Hence, it can depend on these vectors only through its invariants.

$$ \mathrm{I}_1 = \mathscr{E}\cdot\mathscr{E}, \qquad \mathrm{I}_2 = \mathbf{B}\cdot\mathbf{B}, \qquad \mathrm{I}_3 = (\mathscr{E}\cdot\mathbf{B})^2, \tag{5.12.13} $$

where I_3 is selected so as to satisfy the invariance under time reversal. Thus, from (5.12.8), we obtain

$$ \begin{aligned} \Psi &= \Psi(\mathrm{I}_1, \mathrm{I}_2, \mathrm{I}_3, \theta, \rho^{-1}), \\ \eta &= -\frac{\partial\Psi}{\partial\theta}, \\ \mathbf{P} &= -2\rho\left[\frac{\partial\Psi}{\partial \mathrm{I}_1}\mathscr{E} + \frac{\partial\Psi}{\partial \mathrm{I}_3}(\mathscr{E}\cdot\mathbf{B})\mathbf{B}\right], \\ \mathscr{M} &= -2\rho\left[\frac{\partial\Psi}{\partial \mathrm{I}_2}\mathbf{B} + \frac{\partial\Psi}{\partial \mathrm{I}_3}(\mathscr{E}\cdot\mathbf{B})\right]. \end{aligned} \tag{5.12.14} $$

The constitutive equations for ${}_{D}\mathbf{t}$, $\mathbf{q}$, and $\mathscr{J}$ can be constructed by using Table E2 to obtain the generators of these quantities as functions of the joint invariants of $\mathbf{d}$, $\nabla\sigma$, $\mathscr{E}$, and $\mathbf{B}$. The invariants are too many, and the resulting constitutive equations are too complicated for practical applications. Fortunately, often the linear theory is all that is needed.

Linear Constitutive Equations

The free energy for the linear theory is a quadratic isotropic function, but $\mathbf{P}$, $\mathscr{M}$, ${}_{D}\mathbf{t}$, $\mathbf{q}$, and $\mathscr{J}$ are linear in the independent variables. Hence,

$$ \Psi = \Psi_0 - \frac{1}{2\rho}\chi^{\mathrm{E}}\mathscr{E}\cdot\mathscr{E} - \frac{1}{2\rho}\chi^{\mathrm{B}}\mathbf{B}\cdot\mathbf{B}, \tag{5.12.15} $$

$$ \eta = -\frac{\partial\Psi_0}{\partial\theta} + \frac{1}{2\rho}\frac{\partial\chi^{\mathrm{E}}}{\partial\theta}\mathscr{E}\cdot\mathscr{E} + \frac{1}{2\rho}\frac{\partial\chi^{\mathrm{B}}}{\partial\theta}\mathbf{B}\cdot\mathbf{B}, \tag{5.12.16} $$

$$ \mathbf{P} = \chi^{\mathrm{E}}\mathscr{E}, \tag{5.12.17} $$

$$ \mathscr{M} = \chi^{\mathrm{B}}\mathbf{B}, \tag{5.12.18} $$

$$_{\mathrm{D}}\mathbf{t} = \lambda_v \operatorname{tr} \mathbf{d}\mathbf{1} + 2\mu_v \mathbf{d}, \tag{5.12.19}$$

$$\mathbf{q} = \kappa \nabla\theta + \kappa^{\mathrm{E}} \boldsymbol{\mathscr{E}}, \tag{5.12.20}$$

$$\boldsymbol{\mathscr{J}} = \sigma \boldsymbol{\mathscr{E}} + \sigma^{\theta} \nabla\theta, \tag{5.12.21}$$

where the material moduli $\Psi_0, \chi^{\mathrm{E}}, \chi^{\mathrm{B}}$, the viscosities λ_v, μ_v, the heat conduction coefficients κ, κ^{E}, and the electric conduction coefficients σ and σ^{θ} are functions of ρ^{-1} and θ. Note the appearance of the Peltier and Seebeck effects in eqs. (5.12.20) and (5.12.21). From the C–D inequality (5.12.9), it follows that

$$\begin{gathered} \kappa \geq 0, \qquad \sigma \geq 0, \qquad 3\lambda_v + 2\mu_v \geq 0, \qquad \mu_v \geq 0, \\ 4\kappa\sigma\theta^{-1} - (\kappa^{\mathrm{E}}\theta^{-1} + \sigma^{\theta})^2 \geq 0. \end{gathered} \tag{5.12.22}$$

This completes the theory.

5.13. Résumé of Constitutive Equations

I. Anisotropic Electromagnetic Elastic Solids (Nonlinear)

$$\begin{aligned} \rho_0 \Psi &= \rho_0(\varepsilon - \theta\eta) - \mathscr{E}_K P_K \equiv \Sigma(C_{KL}, \mathscr{E}_K, B_K, \theta, \mathbf{X}), \\ \eta &= -\frac{1}{\rho_0}\frac{\partial \Sigma}{\partial \theta}, \\ P_k &= -\frac{\rho}{\rho_0}\frac{\partial \Sigma}{\partial \mathscr{E}_k} x_{k,K}, \\ \mathscr{M}_k &= -\frac{\rho}{\rho_0}\frac{\partial \Sigma}{\partial B_K} x_{k,K}, \\ {}_{\mathrm{E}}t_{kl} &= 2\frac{\rho}{\rho_0}\frac{\partial \Sigma}{\partial C_{KL}} x_{k,K} x_{l,L}, \\ q_k &= \frac{\rho}{\rho_0}\frac{\partial Q}{\partial \phi_K}\bigg|_{\phi_K = 0} x_{k,K}, \\ \mathscr{J}_k &= \frac{\rho}{\rho_0}\frac{\partial \mathscr{J}}{\partial \phi_K}\bigg|_{\phi_K = 0} x_{k,K}, \\ &\left.\begin{aligned} Q &= Q_K \phi_K = Q(\mathbf{C}, \boldsymbol{\mathscr{E}}, \mathbf{B}, \nabla\theta, \theta, \mathbf{X}; \boldsymbol{\phi}) \\ \mathscr{J} &= \mathscr{J}_K \phi_K = \mathscr{J}(\mathbf{C}, \boldsymbol{\mathscr{E}}, \mathbf{B}, \nabla\theta, \theta, \mathbf{X}; \boldsymbol{\phi}) \end{aligned}\right\} \text{Linear in } \boldsymbol{\phi}, \end{aligned} \tag{5.13.1}$$

where Q_K and $\mathscr{J}_K$ are subject to

$$\rho\gamma \equiv \frac{1}{\theta} Q_K \theta_{,K} + \mathscr{J}_K \mathscr{E}_K \geq 0; \qquad Q_K = \mathscr{J}_K = 0 \qquad \text{when} \quad \theta_{,K} = \mathscr{E}_K = 0.$$

For material symmetry, see Sections 5.5 to 5.7.

II. Isotropic Electromagnetic Elastic Solids (Nonlinear)

$$\Sigma = \Sigma(\mathrm{I}_\alpha, \theta, \mathbf{X}), \qquad \alpha = 1, 2, \ldots, 14,$$

$$\eta = -\frac{1}{\rho}\frac{\partial \Sigma}{\partial \theta},$$

$$\mathbf{P} = -2\frac{\rho}{\rho_0}\left\{\frac{\partial \Sigma}{\partial \mathrm{I}_4}\mathscr{E} + \frac{\partial \Sigma}{\partial \mathrm{I}_6}\overset{-1}{\mathbf{c}}\mathscr{E} + \frac{\partial \Sigma}{\partial \mathrm{I}_8}\overset{-2}{\mathbf{c}}\mathscr{E} + \frac{\partial \Sigma}{\partial \mathrm{I}_{10}}(\mathscr{E}\cdot\mathbf{B})\mathbf{B}\right.$$

$$+ \frac{1}{2}\frac{\partial \Sigma}{\partial \mathrm{I}_{11}}[\mathbf{B}\times(\overset{-1}{\mathbf{c}}\mathscr{E}) + \overset{-1}{\mathbf{c}}(\mathscr{E}\times\mathbf{B})] + \frac{1}{2}\frac{\partial \Sigma}{\partial \mathrm{I}_{12}}[\mathbf{B}\times(\overset{-2}{\mathbf{c}}\mathscr{E}) + \overset{-2}{\mathbf{c}}(\mathscr{E}\times\mathbf{B})]$$

$$\left. + \frac{1}{2}\frac{\partial \Sigma}{\partial \mathrm{I}_{14}}[\mathbf{B}\cdot\mathscr{E}(\mathbf{B}\times(\overset{-1}{\mathbf{c}}\mathbf{B})) + (\mathscr{E}\cdot\mathbf{B})\mathbf{B}\times(\overset{-1}{\mathbf{c}}\mathbf{B})]\right\},$$

$$\mathscr{M} = -\frac{2\rho}{\rho_0}\left\{\frac{\partial \Sigma}{\partial \mathrm{I}_5}\mathbf{B} + \frac{\partial \Sigma}{\partial \mathrm{I}_7}\overset{-1}{\mathbf{c}}\mathbf{B} + \frac{\partial \Sigma}{\partial \mathrm{I}_9}\overset{-2}{\mathbf{c}}\mathbf{B} + \frac{\partial \Sigma}{\partial \mathrm{I}_{10}}(\mathscr{E}\cdot\mathbf{B})\mathscr{E}\right.$$

$$+ \frac{1}{2}\frac{\partial \Sigma}{\partial \mathrm{I}_{11}}(\overset{-1}{\mathbf{c}}\mathscr{E})\times\mathscr{E} + \frac{1}{2}\frac{\partial \Sigma}{\partial \mathrm{I}_{12}}(\overset{-2}{\mathbf{c}}\mathscr{E})\times\mathscr{E} + \frac{1}{2}\frac{\partial \Sigma}{\partial \mathrm{I}_{13}}\{(\overset{-1}{\mathbf{c}}\mathbf{B})\times(\overset{-2}{\mathbf{c}}\mathbf{B})$$

$$+ \mathbf{B}\cdot[(\overset{-1}{\mathbf{c}}\mathbf{B})\times\overset{-2}{\mathbf{c}}] - \mathbf{B}\cdot[(\overset{-2}{\mathbf{c}}\mathbf{B})\times\overset{-1}{\mathbf{c}}]\} + \frac{1}{2}\frac{\partial \Sigma}{\partial \mathrm{I}_{14}}[\mathscr{E}\mathscr{E}\cdot(\mathbf{B}\times(\overset{-1}{\mathbf{c}}\mathbf{B}))]$$

$$\left. + \mathscr{E}\cdot\mathbf{B}[\overset{-1}{\mathbf{c}}\mathbf{B})\times\mathscr{E} + \overset{-1}{\mathbf{c}}(\mathscr{E}\times\mathbf{B})]\right\},$$

$$_{\mathrm{E}}\mathbf{t} = \frac{2\rho}{\rho_0}\left\{\frac{\partial \Sigma}{\partial \mathrm{I}}\overset{-1}{\mathbf{c}} + \left(\mathrm{II}\frac{\partial \Sigma}{\partial \mathrm{II}} + \mathrm{III}\frac{\partial \Sigma}{\partial \mathrm{III}}\right)\mathbf{1} - \mathrm{III}\frac{\partial \Sigma}{\partial \mathrm{II}}\mathbf{c} + 2\frac{\partial \Sigma}{\partial \mathrm{I}_6}[\mathscr{E}\otimes(\overset{-1}{\mathbf{c}}\mathscr{E})]_S\right.$$

$$+ 2\frac{\partial \Sigma}{\partial \mathrm{I}_7}[\mathbf{B}\otimes(\overset{-1}{\mathbf{c}}\mathbf{B})]_S + 4\frac{\partial \Sigma}{\partial \mathrm{I}_8}[\mathscr{E}\otimes(\overset{-2}{\mathbf{c}}\mathscr{E})]_S + 4\frac{\partial \Sigma}{\partial \mathrm{I}_9}[\mathbf{B}\otimes(\overset{-2}{\mathbf{c}}\mathbf{B})]_S \tag{5.13.2}$$

$$+ \frac{1}{2}\frac{\partial \Sigma}{\partial \mathrm{I}_{11}}[(\mathscr{E}\times\mathbf{B})\otimes(\overset{-1}{\mathbf{c}}\mathscr{E}) + \mathscr{E}\otimes\overset{-1}{\mathbf{c}}(\mathscr{E}\times\mathbf{B})]_S$$

$$+ \frac{1}{2}\frac{\partial \Sigma}{\partial \mathrm{I}_{12}}[(\mathscr{E}\times\mathbf{B})\otimes(\overset{-2}{\mathbf{c}}\mathscr{E}) + \overset{-1}{\mathbf{c}}(\mathscr{E}\times\mathbf{B})\otimes(\overset{-1}{\mathbf{c}}\mathscr{E})$$

$$+ \overset{-1}{\mathbf{c}}\mathscr{E}\otimes\overset{-1}{\mathbf{c}}(\mathscr{E}\times\mathbf{B}) + \mathscr{E}\otimes\overset{-2}{\mathbf{c}}(\mathscr{E}\times\mathbf{B})]_S$$

$$+ \frac{1}{2}\frac{\partial \Sigma}{\partial \mathrm{I}_{13}}\{(\overset{-1}{\mathbf{c}}\mathbf{B})\times\mathbf{B}\otimes\overset{-1}{\mathbf{c}}\mathbf{B} + [\mathbf{B}\times(\overset{-1}{\mathbf{c}}\mathbf{B})\otimes\overset{-2}{\mathbf{c}}\mathbf{B}$$

$$+ \overset{-1}{\mathbf{c}}(\mathbf{B}\times\overset{-1}{\mathbf{c}}\mathbf{B})\otimes\overset{-1}{\mathbf{c}}\mathbf{B}]\}_S$$

$$+ \frac{1}{2}\frac{\partial \Sigma}{\partial \mathrm{I}_{14}}(\mathscr{E}\cdot\mathbf{B})[(\mathscr{E}\times\mathbf{B})\otimes\overset{-1}{\mathbf{c}}\mathbf{B} + \mathbf{B}\otimes\overset{-1}{\mathbf{c}}(\mathscr{E}\times\mathbf{B})]_S.$$

$$\mathbf{q} = \kappa_1 \nabla\theta + \kappa_2 \mathscr{E} + \kappa_3 \overset{-1}{\mathbf{c}} \nabla\boldsymbol{\theta} + \kappa_4 \overset{-1}{\mathbf{c}} \mathscr{E} + \kappa_5 \nabla\theta \times \mathbf{B} + \kappa_6 \mathscr{E} \times \mathbf{B}$$
$$+ \kappa_7 \overset{-2}{\mathbf{c}} \nabla\theta + \kappa_8 \overset{-2}{\mathbf{c}} \mathscr{E} + \kappa_9 (\mathbf{B}\cdot\nabla\theta)\mathbf{B} + \kappa_{10}(\mathbf{B}\cdot\mathscr{E})\mathbf{B}$$
$$+ \kappa_{11}[\overset{-1}{\mathbf{c}}(\nabla\boldsymbol{\theta} \times \mathbf{B}) - (\overset{-1}{\mathbf{c}}\nabla\theta) \times \mathbf{B}] + \kappa_{12}[\overset{-1}{\mathbf{c}}(\mathscr{E} \times \mathbf{B}) - (\overset{-1}{\mathbf{c}}\mathscr{E}) \times \mathbf{B}],$$
$$\mathscr{J} = \sigma_1 \mathscr{E} + \sigma_2 \nabla\theta + \sigma_3 \overset{-1}{\mathbf{c}} \mathscr{E} + \sigma_4 \overset{-1}{\mathbf{c}} \nabla\theta + \sigma_5 \mathscr{E} \times \mathbf{B} + \sigma_6 \nabla\theta \times \mathbf{B}$$
$$+ \sigma_7 \overset{-2}{\mathbf{c}} \mathscr{E} + \sigma_8 \overset{-2}{\mathbf{c}} \nabla\theta + \sigma_9 (\mathbf{B}\cdot\mathscr{E})\mathbf{B} + \sigma_{10}(\mathbf{B}\cdot\nabla\theta)\mathbf{B}$$
$$+ \sigma_{11}[\overset{-1}{\mathbf{c}}(\mathscr{E} \times \mathbf{B}) - (\overset{-1}{\mathbf{c}}\mathscr{E}) \times \mathbf{B}] + \sigma_{12}[\overset{-1}{\mathbf{c}}(\nabla\theta \times \mathbf{B}) - (\overset{-1}{\mathbf{c}}\nabla\theta) \times \mathbf{B}],$$

where κ_1 to κ_{12} and σ_1 to σ_{12} are functions of the joint invariants of $\overset{-1}{\mathbf{c}}$, $\mathscr{E}$, $\mathbf{B}$, and $\nabla\theta$. These invariants can be obtained by use of Table E1.

Invariants I_α are given by

$$\mathrm{I}_1 = \operatorname{tr} \overset{-1}{\mathbf{c}}, \quad \mathrm{I}_2 = \operatorname{tr} \overset{-2}{\mathbf{c}}, \quad \mathrm{I}_3 = \operatorname{tr} \overset{-3}{\mathbf{c}}, \quad \mathrm{I}_4 = \mathscr{E}\cdot\mathscr{E}, \quad \mathrm{I}_5 = \mathbf{B}\cdot\mathbf{B},$$
$$\mathrm{I}_6 = \mathscr{E}\overset{-1}{\mathbf{c}}\mathscr{E}, \quad \mathrm{I}_7 = \mathbf{B}\overset{-1}{\mathbf{c}}\mathbf{B}, \quad \mathrm{I}_8 = \mathscr{E}\overset{-2}{\mathbf{c}}\mathscr{E}, \quad \mathrm{I}_9 = \mathbf{B}\overset{-2}{\mathbf{c}}\mathbf{B},$$
$$\mathrm{I}_{10} = (\mathscr{E}\cdot\mathbf{B})^2, \qquad \mathrm{I}_{11} = \mathscr{E}\cdot[\mathbf{B} \times (\overset{-1}{\mathbf{c}}\mathscr{E})],$$
$$\mathrm{I}_{12} = \mathscr{E}\cdot[\mathbf{B} \times (\overset{-2}{\mathbf{c}}\mathscr{E})], \qquad \mathrm{I}_{13} = \mathbf{B}\cdot[(\overset{-1}{\mathbf{c}}\mathbf{B}) \times (\overset{-2}{\mathbf{c}}\mathbf{B})]$$
$$\mathrm{I}_{14} = (\mathscr{E}\cdot\mathbf{B})\mathscr{E}\cdot[\mathbf{B} \times (\overset{-1}{\mathbf{c}}\mathbf{B})], \tag{5.13.3}$$

$\mathbf{q}$ and $\mathscr{J}$ are subject to the dissipation inequality

$$\rho\gamma \equiv \frac{1}{\theta}\mathbf{q}\cdot\nabla\theta + \mathscr{J}\cdot\mathscr{E} \geq 0.$$

III. Anisotropic Electromagnetic Elastic Solids (Linear)

$$\Sigma = \sigma_0 - \rho_0\eta_0 T - \frac{\rho_0\gamma_0}{2T}T^2 + (\alpha_{kl} - \beta_{kl}T)\tilde{e}_{kl} + \tfrac{1}{2}\sigma_{klmn}\tilde{e}_{kl}\tilde{e}_{mn}$$
$$- e_{klm}\mathscr{E}_k\tilde{e}_{lm} - h_{klm}B_k\tilde{e}_{lm} - (\pi_k^0 + \tilde{\omega}_k T)\mathscr{E}_l$$
$$- \tfrac{1}{2}\chi_{kl}^{\mathrm{E}}\mathscr{E}_k\mathscr{E}_l - (m_k^0 + \gamma_k T)B_k - \tfrac{1}{2}\chi_{kl}^{\mathrm{B}}B_kB_l - \lambda_{kl}\mathscr{E}_kB_l,$$
$$\eta = \eta_0 + \frac{\gamma_0}{T_0}T + \frac{1}{\rho_0}(\beta_{kl}\tilde{e}_{kl} + \tilde{\omega}_k\mathscr{E}_k + \gamma_kB_k),$$
$$\begin{aligned} {}_{\mathrm{E}}t_{kl} &= \alpha_{kl}(1 - \tilde{e}_{rr}) - \beta_{kl}T + \alpha_{km}(\tilde{e}_{lm} + \tilde{r}_{lm}) + \alpha_{ml}(\tilde{e}_{km} + \tilde{r}_{km}) \\ &\quad + \sigma_{klmn}\tilde{e}_{mn} - e_{mkl}\mathscr{E}_m - h_{mkl}B_m, \end{aligned} \tag{5.13.4}$$
$$P_k = [(1 - \tilde{e}_{rr})\delta_{kl} + \tilde{e}_{kl} + \tilde{r}_{kl}]\pi_l^0 + \tilde{\omega}_kT + \chi_{kl}^{\mathrm{E}}\mathscr{E}_l + e_{klm}\tilde{e}_{lm} + \lambda_{kl}B_l,$$
$$\mathscr{M}_k = [(1 - \tilde{e}_{rr})\delta_{kl} + \tilde{e}_{kl} + \tilde{r}_{kl}]m_l^0 + \gamma_kT + \chi_{kl}^{\mathrm{B}}B_l + h_{klm}\tilde{e}_{lm} + \lambda_{lk}\mathscr{E}_l,$$
$$q_k = \kappa_{kl}T_{,l} + \chi_{kl}^{\mathrm{E}}\mathscr{E}_l,$$
$$\mathscr{J}_k = \sigma_{kl}\mathscr{E}_l + \sigma_{kl}^{\theta}T_{,l},$$

where q_k and $\mathscr{J}_k$ are subject to the dissipation inequality

$$\rho\gamma \equiv \frac{1}{\theta} q_k \theta_{,k} + \mathscr{J}_k \mathscr{E}_k \geq 0.$$

For the material symmetry regulations, see Section 5.11.

IV. Isotropic Electromagnetic Elastic Solids (Linear)

$$\begin{aligned}
\Sigma &= \sigma_0 - \rho_0 \eta_0 T + (\alpha_e - \beta_e T) \operatorname{tr} \tilde{\mathbf{e}} + \tfrac{1}{2}\lambda_e (\operatorname{tr} \tilde{\mathbf{e}})^2 \\
&\quad + \mu_e \operatorname{tr} \tilde{\mathbf{e}}^2 - \tfrac{1}{2}\chi^E \boldsymbol{\mathscr{E}} \cdot \boldsymbol{\mathscr{E}} - \tfrac{1}{2}\chi^B \mathbf{B} \cdot \mathbf{B}, \\
\eta &= \eta_0 + \frac{\gamma_0}{T_0} T + \frac{\beta}{\rho_0} \operatorname{tr} \tilde{\mathbf{e}}, \\
{}_E t_{kl} &= (\alpha_e - \beta_e T)\delta_{kl} + (\lambda_e - \alpha_e)\tilde{e}_{rr}\delta_{kl} + 2(\mu_e + \alpha_e)\tilde{e}_{kl}, \\
P_k &= \chi^E \mathscr{E}_k, \\
\mathscr{M}_k &= \chi^B B_k, \\
q_k &= \kappa T_{,k} + \kappa^E \mathscr{E}_k, \\
\mathscr{J}_k &= \sigma \mathscr{E}_k + \sigma^\theta T_{,k}, \\
\kappa &\geq 0, \qquad \sigma \geq 0, \qquad 4\kappa\sigma T_0^{-1} - (\kappa^E T_0^{-1} + \sigma^\theta)^2 \geq 0.
\end{aligned} \tag{5.13.5}$$

V. Electromagnetic Fluids (Nonlinear)

$$\begin{aligned}
\Psi &= \Psi(\mathrm{I}_1, \mathrm{I}_2, \mathrm{I}_3, \theta, \rho^{-1}), \\
\eta &= -\frac{\partial \Psi}{\partial \theta}, \qquad \mathrm{I}_1 \equiv \boldsymbol{\mathscr{E}} \cdot \boldsymbol{\mathscr{E}}, \qquad \mathrm{I}_2 \equiv \mathbf{B} \cdot \mathbf{B}, \qquad \mathrm{I}_3 \equiv (\boldsymbol{\mathscr{E}} \cdot \mathbf{B})^2, \\
\mathbf{P} &= -2\rho \left[\frac{\partial \Psi}{\partial \mathrm{I}_1} \boldsymbol{\mathscr{E}} + \frac{\partial \Psi}{\partial \mathrm{I}_3} (\boldsymbol{\mathscr{E}} \cdot \mathbf{B})\mathbf{B} \right], \\
\boldsymbol{\mathscr{M}} &= -2\rho \left[\frac{\partial \Psi}{\partial \mathrm{I}_2} \mathbf{B} + \frac{\partial \Psi}{\partial \mathrm{I}_3} (\boldsymbol{\mathscr{E}} \cdot \mathbf{B})\boldsymbol{\mathscr{E}} \right], \\
\mathbf{t} &= -\pi \mathbf{1} + {}_D t(\rho^{-1}, \mathbf{d}, \boldsymbol{\mathscr{E}}, \mathbf{B}, \theta, \theta_{,k}), \\
\pi &= -\frac{\partial \Psi}{\partial \rho^{-1}},
\end{aligned} \tag{5.13.6}$$

$$\begin{aligned}
{}_D\mathbf{t} &= \alpha_0 \mathbf{1} + \alpha_1 \mathbf{d} + \alpha_2 \mathbf{d}^2 + \alpha_3 \boldsymbol{\mathscr{E}} \otimes \boldsymbol{\mathscr{E}} + \alpha_4 \mathbf{B} \otimes \mathbf{B} + \alpha_5 \nabla\theta \otimes \nabla\theta \\
&\quad + \alpha_6 (\boldsymbol{\mathscr{E}} \otimes \mathbf{d}\boldsymbol{\mathscr{E}})_S + \alpha_7 (\boldsymbol{\mathscr{E}} \otimes \mathbf{d}^2\boldsymbol{\mathscr{E}})_S + \alpha_8 (\nabla\theta \otimes \mathbf{d}\nabla\theta)_S \\
&\quad + \alpha_9 (\nabla\theta \otimes \mathbf{d}^2\nabla\theta)_S + \alpha_{10}(\mathbf{dW} - \mathbf{Wd}) + \alpha_{11}\mathbf{WdW} \\
&\quad + \alpha_{12}(\mathbf{d}^2\mathbf{W} - \mathbf{Wd}^2) + \alpha_{13}(\mathbf{WdW}^2 - \mathbf{W}^2\mathbf{dW}) \\
&\quad + \alpha_{14}(\boldsymbol{\mathscr{E}} \otimes \nabla\theta)_S + \alpha_{15}(\mathbf{W}\boldsymbol{\mathscr{E}} \otimes \boldsymbol{\mathscr{E}}\mathbf{W})_S + \alpha_{16}(\boldsymbol{\mathscr{E}} \otimes \mathbf{W}\boldsymbol{\mathscr{E}})_S
\end{aligned}$$

$$+ \alpha_{17}(\mathbf{W}\mathscr{E} \otimes \mathbf{W}^2\mathscr{E})_S + \alpha_{18}[\mathbf{d}(\mathscr{E} \otimes \nabla\theta - \nabla\theta \otimes \mathscr{E})$$
$$- (\mathscr{E} \otimes \nabla\theta - \nabla\theta \otimes \mathscr{E})\mathbf{d}] + \alpha_{19}[\mathbf{W}(\mathscr{E} \otimes \nabla\theta - \nabla\theta \otimes \mathscr{E}]_S,$$
$$W_{kl} \equiv \varepsilon_{klm} B_m,$$

where α_i are functions of ρ, θ and the invariants of $\mathbf{d}$, $\mathscr{E}$, $\mathbf{B}$, and $\nabla\theta$. These can be read from Table E1. A subscript attached to brackets indicates symmetrization.

Expressions of $\mathbf{q}$ and $\mathscr{J}$ are identical to those of (5.13.2) with $\overset{-1}{\mathbf{c}}$ replaced by $\mathbf{d}$.

${}_D\mathbf{t}$, $\mathbf{q}$, and $\mathscr{J}$ are subject to the C–D inequality

$$\rho\gamma \equiv {}_Dt_{kl}d_{lk} + \frac{1}{\theta}q_k\theta_{,k} + \mathscr{J}_k\mathscr{E}_k \geq 0,$$

$${}_D\mathbf{t} = 0, \qquad \mathbf{q} = \mathbf{0}, \qquad \mathscr{J} = 0 \qquad \text{when} \quad \mathbf{d} = \mathbf{0}, \qquad \nabla\theta = \mathbf{0}, \qquad \mathscr{E} = \mathbf{0}.$$

VI. Electromagnetic Fluids (Linear)

$$\begin{aligned}
\Psi &= \Psi_0 - \frac{1}{2\rho}\chi^E \mathscr{E}\cdot\mathscr{E} - \frac{1}{2\rho}\chi^B \mathbf{B}\cdot\mathbf{B},\\
\eta &= -\frac{\partial\Psi_0}{\partial\theta} + \frac{1}{2\rho}\frac{\partial\chi^E}{\partial\theta}\mathscr{E}\cdot\mathscr{E} + \frac{1}{2\rho}\frac{\partial\chi^B}{\partial\theta}\mathbf{B}\cdot\mathbf{B},\\
\mathbf{P} &= \chi^E\mathscr{E},\\
\mathscr{M} &= \chi^B\mathbf{B},\\
\mathbf{t} &= -\pi\mathbf{1} + {}_D\mathbf{t}; \qquad \pi = -\frac{\partial\Psi}{\partial\theta},\\
{}_D\mathbf{t} &= \lambda_v \operatorname{tr}\mathbf{d}\mathbf{1} + 2\mu_v\mathbf{d},\\
\mathbf{q} &= \kappa\nabla\theta + \kappa^E\mathscr{E},\\
\mathscr{J} &= \sigma\mathscr{E} + \sigma^\theta\nabla\theta,
\end{aligned} \tag{5.13.7}$$

where the material moduli Ψ_0, χ^E, χ^B, the viscosities λ_v, μ_v, the heat conduction coefficients κ, κ^E, and the electric conduction coefficients, σ and σ^θ are functions of ρ^{-1} and θ.

$$\kappa \geq 0, \qquad \sigma \geq 0, \qquad 3\lambda_v + 2\mu_v \geq 0, \qquad \mu_v \geq 0.$$
$$4\kappa\sigma\theta^{-1} - (\kappa^E\theta^{-1} + \sigma^\theta)^2 \geq 0.$$

Problems

5.1. The master constitutive equation of rate-dependent simple materials is given by (5.4.34). Constitutive equations of memory-dependent fluids can be obtained from this equation by letting $\mathbf{X} \to \mathbf{x}$. In this case, which set of variables remains objective? What objective variables can you suggest for those that do not become objective?

5.2. For an electromagnetic fluid, independent constitutive variables are known to be $\{\mathscr{E}_k, \dot{\mathscr{E}}_k, B_k, \dot{B}_k, \rho^{-1}\}$. What objective variable do you suggest to replace $\dot{\mathscr{E}}_k$ and $\dot{B}_k$? For the linear theory, obtain thermodynamically admissible constitutive equations for P_k and $\mathscr{M}_k$.

5.3. In a dielectric fluid, independent constitutive variables are $\{\mathscr{E}_k, \mathscr{E}_{k,l}, \rho^{-1}\}$. What will be the linear constitutive equation for P_k, so that the Clausius–Duhem inequality is not violated? Obtain the nonlinear constitutive equations for P_k.

5.4. An elastic dielectric is governed by $\{x_{k,K}, \mathscr{E}_K, \mathscr{E}_{K,L}, \theta\}$. Obtain constitutive equations for t_{kl} and P_k. Obtain linear constitutive equations for dielectrics with the following symmetry regulations:
(i) isotropic solid;
(ii) cubic symmetry;
(iii) hexagonal symmetry.

5.5. For an electromagnetic elastic dielectric of magnetic class 3, obtain the second-degree constitutive equations for the piezomagnetic effect.

5.6. Cr_2O_3 belongs to the magnetic class $\underline{\bar{3}m}$. Obtain constitutive equations of an electromagnetic elastic solid in this class for t_{kl}, P_k, $\mathscr{M}_k$, and $\mathscr{J}_k$ for the following:
(i) linear theory;
(ii) quadratic theory.

5.7. For the electromagnetic fluids, linear constitutive equations were given in Section 5.12. What forms do these equations take when quadratic terms are included in $\mathscr{E}$ and **B** and their products in **d**?

5.8. If a memory-dependent magnetic fluid is influenced by $\{d_{kl}, B_k, \dot{B}_k, B_{k,l}, \rho^{-1}, \theta\}$, what will be the thermodynamically admissible forms of the constitutive equations for t_{kl}, $\mathscr{M}_k$, and $\mathscr{J}_k$?

CHAPTER 6

Rigid Dielectrics

6.1. Scope of the Chapter

The present chapter is intended as a discussion of various electromagnetic phenomena occurring in rigid continuous bodies. In reality, all bodies are, of course, deformable. However, when the effect of strains are negligible compared with electromagnetic effects, gross mathematical simplifications are achieved. After a summary of basic equations in Section 6.2, we present an account of the potential theory in Section 6.3, including the uniqueness theorem and various methods of solution (e.g., Green's function technique, eigenfunction expansions, mixed boundary-value problems).

Sections 6.4, 6.5, and 6.6 deal, respectively, with the problems of a charged cylinder, potential in a half-plane, and charged disc. Of these, the dual integral equation technique, used in Sections 6.5 and 6.6, is instructive for the mixed boundary-value problems posed.

In Sections 6.7 and 6.8 we present an account of electromagnetic waves. Plane and spherical waves are discussed.

The nonlinear theory of rigid dielectrics is a field of study that deserves more attention than has hitherto been given to it by investigators. We devote Sections 6.9–6.14 to this subject. After the derivation of the field equations in Section 6.9, the uniqueness of the solution is discussed. In Section 6.10 we give solutions for a one-dimensional electrostatic problem, while in Section 6.11 the nature of two-dimensional problems is explored. Section 6.12 presents the solution for a special class of nonlinear dielectric materials. Taking advantage of the special nature of constitutive equations for the polarization vector, we employ a method similar to that pursued for gas dynamics (the Chaplygin transformation).

The subject of Section 6.13 is nonlinear electromagnetic waves, a topic even less amenable to mathematical analysis. Nevertheless, the method of characteristics and special constitutive equations enables us to obtain results that are useful to the theory of nonlinear optics. Section 6.14 is devoted to the subject of electromagnetic shock waves. Section 6.15 gives the solutions of two illustrative problems in moving rigid dielectrics, dragging of light, and rotation in a magnetic field.

Finally, in Sections 6.16 and 6.17, we present accounts of the nonlinear optics and solitary waves which have gained great momentum in recent years. In Section 6.16 we develop the method of *slowly* varying amplitude approximation, and in Section 6.17 the problem of solitary waves. This last problem is but one example of the multitude of technological applications of the theory. Later, in Chapter 13, we shall return to the theory and treatment of other problems.

6.2. Fundamental Equations

Basic equations of the theory of nonconducting rigid materials consist of the following sets:

Maxwell's Equations (Valid in $\mathscr{V} - \sigma$)

$$\nabla \cdot \mathbf{D} = q_e(\mathbf{x}), \tag{6.2.1}$$

$$\nabla \times \mathbf{E} + \frac{1}{c}\frac{\partial \mathbf{B}}{\partial t} = \mathbf{0}, \tag{6.2.2}$$

$$\nabla \cdot \mathbf{B} = 0, \tag{6.2.3}$$

$$\nabla \times \mathbf{H} - \frac{1}{c}\frac{\partial \mathbf{D}}{\partial t} = \mathbf{0}. \tag{6.2.4}$$

Jump Conditions (Valid or σ)

$$[\mathbf{D}]\cdot\mathbf{n} = w_e - \nabla_\sigma \cdot \hat{\boldsymbol{\pi}} \quad \text{on } \sigma - c_\sigma \qquad [\mathbf{b}\cdot\hat{\boldsymbol{\pi}}] = 0 \quad \text{on } c_\sigma, \tag{6.2.5}$$

$$\mathbf{n} \times [\mathbf{E}] - \frac{1}{c}(\mathbf{v}\cdot\mathbf{n})[\mathbf{B}] = \mathbf{0}, \tag{6.2.6}$$

$$[\mathbf{B}]\cdot\mathbf{n} = 0, \tag{6.2.7}$$

$$\mathbf{n} \times [\mathbf{H}] + \frac{1}{c}(\mathbf{v}\cdot\mathbf{n})[\mathbf{D}] = \frac{1}{c}(w_e - \nabla_\sigma\cdot\hat{\boldsymbol{\pi}})\mathbf{v} + \frac{1}{c}\mathscr{K}. \tag{6.2.8}$$

Constitutive Equations

Anisotropic Solids

$$\Sigma = \Sigma(\mathscr{E}_k, B_k, \theta),$$
$$P_k = -\frac{\partial \Sigma}{\partial \mathscr{E}_K}\delta_{kK}, \qquad \mathscr{M}_k = -\frac{\partial \Sigma}{\partial B_K}\delta_{kK}. \tag{6.2.9}$$

The symmetry regulations for magnetic crystals place restrictions on the form of Σ, as discussed in Section 5.6.

Isotropic Solids. Constitutive equations follow from (5.9.8) and (5.9.9) by taking $\overset{-1}{\mathbf{c}} = \mathbf{1}$. Since now $I_4 = I_6 = I_8$ and $I_5 = I_7 = I_9$, Σ can be considered as a function of I_4, I_5, and I_{10}, so that

$$\Sigma = \Sigma(I_4, I_5, I_{10}, \theta),$$
$$\mathbf{P} = \chi^{E}\boldsymbol{\mathscr{E}} + \lambda\mathbf{B}, \tag{6.2.10}$$
$$\boldsymbol{\mathscr{M}} = \chi^{B}\mathbf{B} + \lambda\boldsymbol{\mathscr{E}},$$

where

$$\chi^{E} = -2\frac{\partial\Sigma}{\partial I_4}, \qquad \chi^{B} = -2\frac{\partial\Sigma}{\partial I_5}, \qquad \lambda = -2\frac{\partial\Sigma}{\partial I_{10}}\boldsymbol{\mathscr{E}}\cdot\mathbf{B},$$
$$I_4 = \boldsymbol{\mathscr{E}}\cdot\boldsymbol{\mathscr{E}}, \qquad I_5 = \mathbf{B}\cdot\mathbf{B}, \qquad I_{10} = (\boldsymbol{\mathscr{E}}\cdot\mathbf{B})^2. \tag{6.2.11}$$

Linear Theory. From (5.11.13) to (5.11.17), by setting $\tilde{\mathbf{e}} = \mathbf{0}$ and disregarding the initial fields, heat, and electric conductions, we have

$$\Sigma = -\tfrac{1}{2}\chi^{E}{}_{kl}\mathscr{E}_k\mathscr{E}_l - \tfrac{1}{2}\chi^{B}{}_{kl}B_kB_l - \lambda_{kl}\mathscr{E}_kB_l,$$
$$P_k = \chi^{E}{}_{kl}\mathscr{E}_l + \lambda_{kl}B_l, \tag{6.2.12}$$
$$\mathscr{M}_k = \chi^{B}{}_{kl}B_l + \lambda_{lk}\mathscr{E}_l.$$

For *isotropic solids* these reduce to

$$\Sigma = -\tfrac{1}{2}\chi^{E}\boldsymbol{\mathscr{E}}\cdot\boldsymbol{\mathscr{E}} - \tfrac{1}{2}\chi^{B}\mathbf{B}\cdot\mathbf{B},$$
$$\mathbf{P} = \chi^{E}\boldsymbol{\mathscr{E}}, \qquad \boldsymbol{\mathscr{M}} = \chi^{B}\mathbf{B}. \tag{6.2.13}$$

For static problems the following relations are also useful

$$\mathbf{D} = \mathbf{E} + \mathbf{P} = \varepsilon\mathbf{E}, \qquad \mathbf{B} = \mathbf{H} + \mathbf{M} = \mu\mathbf{H}, \tag{6.2.14}$$

where

$$\varepsilon = 1 + \chi^{E}, \qquad \mu = (1 - \chi^{B})^{-1}. \tag{6.2.15}$$

Boundary and Initial Conditions. The boundary conditions are derivable from jump conditions by considering σ as coinciding with the surface of the body $\partial\mathscr{V}$. In this case, for the jump conditions appearing in (6.2.5)–(6.2.8), we write

$$[A] = A^{+} - A^{-}.$$

Then, either the part containing A^{+} must be specified, or it must be obtained by solving a boundary-value problem for the exterior of the body. Various combinations of (6.2.5)–(6.2.8) are possible but should not violate the existence and uniqueness of the solution.

The initial conditions usually consist of Cauchy data

$$\mathbf{P}(\mathbf{x}, 0) = \mathbf{P}_0(\mathbf{x})$$
$$\boldsymbol{\mathscr{M}}(\mathbf{x}, 0) = \boldsymbol{\mathscr{M}}_0(\mathbf{x}) \quad \text{in } \mathscr{V} - \sigma. \tag{6.2.16}$$

6.3. Potential Theory

In Sections 6.3–6.6 we discuss problems in electrostatics, for which Maxwell's equations (6.2.1) and (6.2.2) reduce to

$$\nabla \cdot \mathbf{D} = q_e(\mathbf{x}), \qquad \nabla \times \mathbf{E} = \mathbf{0}.$$

When the electric field is not very intense, many rigid materials accept a linear constitutive relationship between the polarization and the electric field. In this case, for isotropic solids, we have

$$\mathbf{D} = \varepsilon \mathbf{E}, \qquad \varepsilon = 1 + \chi^E, \tag{6.3.1}$$

where ε is independent of $\mathbf{E}$. The field equations for the electrostatic potential ϕ now read

$$\nabla \cdot (\varepsilon \nabla \phi) = -q_e, \qquad \mathbf{E} = -\nabla \phi \quad \text{in } \mathscr{V} - \sigma. \tag{6.3.2}$$

Jump conditions on the surface of the body may take the form

$$\phi = \phi_0 \qquad \text{on } \partial \mathscr{V}_1, \tag{6.3.3}$$

$$\varepsilon \frac{\partial \phi}{\partial n} = -w_e, \quad \text{on } \partial \mathscr{V}_2 = \partial \mathscr{V} - \partial \mathscr{V}_1, \tag{6.3.4}$$

of which (6.3.3) indicates that the potential is prescribed on a part $\partial \mathscr{V}_1$ of the boundary surface. For example, if $\partial \mathscr{V}_1$ is grounded, $\phi_0 = \text{const.}$, since a conducting surface possesses a constant potential. The boundary condition (6.3.4) expresses the existence of a prescribed surface charge density on $\partial \mathscr{V}_2$. If $\partial \mathscr{V}_1 \equiv \partial \mathscr{V}$, so that (6.3.4) does not apply, the problem is called the *Dirichlet problem*, and if $\partial \mathscr{V}_2 \equiv \partial \mathscr{V}$, then it is called the *Neumann problem*. A more complicated case involving the linear combination

$$\lambda \phi + \varepsilon \frac{\partial \phi}{\partial n} = -w_e, \qquad \lambda > 0 \quad \text{on } \partial \mathscr{V}_3 = \partial \mathscr{V} - \partial \mathscr{V}_1 - \partial \mathscr{V}_2 \tag{6.3.5}$$

is known as the *Robin problem* when $\partial \mathscr{V}_3 \equiv \partial \mathscr{V}$. In the case when the body contains a discontinuity surface σ, then across σ, on sectors of σ, we may prescribe jumps of any one of the quantities ϕ, $\varepsilon(\partial \phi / \partial n)$, and $\lambda \phi + \varepsilon(\partial \phi / \partial n)$. If the body extends to infinity in some or in all directions, then, in the absence of external fields, ϕ vanishes at infinity.

The existence theorem is not given here (see Duvaut and Lions [1972]), but it will be shown that uniqueness is guaranteed under the above conditions. More general boundary and jump conditions are possible and uniqueness can be studied in a similar manner as dealt with here.

The general theory dealing with (6.3.2), subject to (6.3.3)–(6.3.5), is known as the *potential theory*. Accounts of the potential theory are numerous. There exist excellent treatises on the subject (see Kellog [1929], Courant [1965], and Morse and Feschbach [1953]). Here, and in the next three sections, we briefly discuss some aspects of the theory and give a few sample solutions.

A. Green's Theorem

According to the Green–Gauss theorem, for a closed smooth region $\mathscr{V} + \partial\mathscr{V}$ having a discontinuity surface σ, we have

$$\int_{\mathscr{V}-\sigma} \nabla \cdot \mathbf{A}\, dv + \int_{\sigma} [\mathbf{A}] \cdot \mathbf{n}\, da = \int_{\partial\mathscr{V}-\sigma} \mathbf{A} \cdot \mathbf{n}\, da. \qquad (6.3.6)$$

If we take $\mathbf{A} = \psi\varepsilon\nabla\phi$, where ε, ϕ, and ψ are sufficiently smooth scalar fields so as not to violate the application of the Green–Gauss theorem, then (6.3.6) gives

$$\int_{\mathscr{V}-\sigma} [\varepsilon\nabla\phi \cdot \nabla\psi + \psi\nabla \cdot (\varepsilon\nabla\phi)]\, dv + \int_{\sigma} \left[\psi\varepsilon \frac{\partial\phi}{\partial n}\right] da = \int_{\partial\mathscr{V}-\sigma} \varepsilon\psi \frac{\partial\phi}{\partial n}\, da. \qquad (6.3.7)$$

Upon exchanging ϕ and ψ, and substracting, we obtain

$$\int_{\mathscr{V}-\sigma} [\phi\nabla \cdot (\varepsilon\nabla\psi) - \psi\nabla \cdot (\varepsilon\nabla\phi)]\, dv + \int_{\sigma} \left[\varepsilon\left(\phi\frac{\partial\psi}{\partial n} - \psi\frac{\partial\phi}{\partial n}\right)\right] da$$

$$= \int_{\partial\mathscr{V}-\sigma} \varepsilon\left(\phi\frac{\partial\psi}{\partial n} - \psi\frac{\partial\phi}{\partial n}\right) da, \qquad (6.3.8)$$

where $\partial/\partial n$ represents the normal derivative, i.e., $\partial\phi/\partial n = \mathbf{n} \cdot \nabla\phi$. These expressions are fundamental to the potential theory.

By use of (6.3.7) we can prove that if the solution of (6.3.2) exists under the boundary conditions (6.3.3)–(6.3.5) it is *unique*.

B. Uniqueness Theorem

If two solutions $\phi_i(\mathbf{x})$, $i = 1, 2$, *of* (6.3.2) *in* $\mathscr{V}$ *have the same value* $(\partial\phi_i/\partial n) + \lambda\phi_i$ *on* $\partial\mathscr{V}$ *where* $\lambda > 0$, *then both solutions are identical in* $\mathscr{V}$.

To prove this we first take $\psi = \phi$ in (6.3.7) (with no discontinuity surface) leading to

$$\int_{\mathscr{V}} [\varepsilon(\nabla\phi)^2 + \phi\nabla \cdot (\varepsilon\nabla\phi)]\, dv = \int_{\partial\mathscr{V}} \varepsilon\phi\frac{\partial\phi}{\partial n}\, da. \qquad (6.3.9)$$

Since ϕ_i are solutions of (6.3.2), so is $\phi = \phi_1 - \phi_2$. But ϕ satisfies the homogeneous boundary conditions $\varepsilon(\partial\phi/\partial n) + \lambda\phi = 0$ on $\partial\mathscr{V}$. Hence, (6.3.9) reduces to

$$\int_{\mathscr{V}} \varepsilon(\nabla\phi)^2\, dv = -\int_{\partial\mathscr{V}} \varepsilon\lambda\phi^2\, da.$$

Since $\varepsilon \neq 0$ and $\lambda > 0$, the only possibility is $\nabla\phi = \mathbf{0}$ and because of boundary conditions $\phi = 0$ in $\mathscr{V}$. Hence, $\phi_1 = \phi_2$, which means the solution is unique.

For $\lambda = 0$ we have the uniqueness theorem for the Dirichlet problem. In the case of the Neumann problem this procedure leads to $\nabla\phi = 0$ in $\mathscr{V}$. But since $\partial\phi/\partial n = 0$ on $\partial\mathscr{V}$, then $\phi =$ const., so that the solution is again unique to within an arbitrary constant.

C. Representation Theorem

Green's formula can be used to construct the solution of boundary-value problems. For brevity, we consider homogeneous bodies for which $\varepsilon = \text{const.}$ For inhomogeneous bodies, the method of approach requires slight generalization. For an excellent treatment of this case we refer the reader to Bergmann and Schiffer [1953]. Let

$$\psi = \rho^{-1} + F(\mathbf{x}, \mathbf{x}'), \qquad \rho \equiv [(\mathbf{x} - \mathbf{x}')\cdot(\mathbf{x} - \mathbf{x}')]^{1/2}, \tag{6.3.10}$$

where F is a twice-differentiable function

$$\nabla'^2\psi = -4\pi\delta(\mathbf{x} - \mathbf{x}') + \nabla'^2 F, \tag{6.3.11}$$

where $\delta(\mathbf{x} - \mathbf{x}')$ is the Dirac delta function. Upon substituting (6.3.10) into (6.3.8) we obtain

$$\begin{aligned} p\phi(\mathbf{x}) = &\int_{\mathscr{V}-\sigma} \phi\nabla^2 F - F\nabla^2\phi - \rho^{-1}\nabla^2\phi)\, dv(\mathbf{x}') \\ &+ \int_\sigma \left[\phi\frac{\partial\rho^{-1}}{\partial n} - \rho^{-1}\frac{\partial\phi}{\partial n} + \phi\frac{\partial F}{\partial n} - F\frac{\partial\phi}{\partial n}\right] da(\mathbf{x}') \\ &- \int_{\partial\mathscr{V}-\sigma} \left(\phi\frac{\partial\rho^{-1}}{\partial n} - \rho^{-1}\frac{\partial\phi}{\partial n} + \phi\frac{\partial F}{\partial n} - F\frac{\partial\phi}{\partial n}\right) da(\mathbf{x}'), \end{aligned} \tag{6.3.12}$$

where

$$p = \begin{cases} 4\pi & \text{when } \mathbf{x} \text{ is in } \mathscr{V} - \sigma, \\ 2\pi & \text{when } \mathbf{x} \text{ is on } \partial\mathscr{V} - \sigma,^1 \\ 0 & \text{when } \mathbf{x} \text{ is outside } \mathscr{V} - \sigma. \end{cases} \tag{6.3.13}$$

In particular, if we set $F = 0$, we obtain the representation for $\phi(\mathbf{x})$

$$\begin{aligned} \phi(\mathbf{x}) = &-\frac{1}{4\pi}\int_{\mathscr{V}-\sigma} \rho^{-1}\nabla'^2\phi\, dv(x') + \frac{1}{4\pi}\int_\sigma \left[\phi\frac{\partial\rho^{-1}}{\partial n} - \rho^{-1}\frac{\partial\phi}{\partial n}\right] da(\mathbf{x}') \\ &- \frac{1}{4\pi}\int_{\partial\mathscr{V}-\sigma} \left(\phi\frac{\partial\rho^{-1}}{\partial n} - \rho^{-1}\frac{\partial\phi}{\partial n}\right) da(\mathbf{x}') \end{aligned} \tag{6.3.14}$$

valid in the interior of $\mathscr{V} - \sigma$. This is the famous result expressed as:

Every function $\phi(\mathbf{x})$, continuously differentiable in $\mathscr{V} - \sigma$ and possessing continuous second derivatives in $\mathscr{V} + \partial\mathscr{V} - \sigma$, can be considered as the sum of the following potentials:

(a) *a space distribution of density $-\nabla^2\phi/4\pi$ in $\mathscr{V} - \sigma$;*

---

[1] If $\mathbf{x}$ on $\partial\mathscr{V}$ is a conical vertex of the boundary, then at this point, instead of 2π, we set p equal to the vertex angle of the conical vertex.

(b) *single layer surface distributions of densities* $(1/4\pi)\,\partial\phi/\partial n$ *on* $\partial\mathscr{V} - \sigma$ *and* $(1/4\pi)\,\partial\phi/\partial n$ *on* σ;

(c) *double layer distributions of densities* $-\phi/4\pi$ *on* $\partial\mathscr{V} - \sigma$ *and* $\phi/4\pi$ *on* σ.

In two-dimensional space $\mathscr{S}$ the fundamental singularity ρ^{-1} is replaced by $\ln(\rho^{-1})$, so that

$$\psi = \ln(\rho^{-1}) + F(\mathbf{x}, \mathbf{x}'),$$

$$p = \begin{cases} 2\pi & \text{for } \mathbf{x} \text{ on } \mathscr{S}, \\ \pi & \text{for } \mathbf{x} \text{ on } \partial\mathscr{S}, \\ 0 & \text{for } \mathbf{x} \text{ outside } \mathscr{S}. \end{cases} \tag{6.3.15}$$

In this case (6.3.14) is replaced by

$$\phi(\mathbf{x}) = -\frac{1}{2\pi}\int_{\mathscr{S}-\gamma} \ln(1/\rho)\nabla'^2\phi\, da(\mathbf{x}') + \frac{1}{2\pi}\int_{\gamma}\left[\phi\frac{\partial}{\partial n}\ln(1/\rho) - \ln(1/\rho)\frac{\partial\phi}{\partial n}\right] ds(\mathbf{x}') - \frac{1}{2\pi}\int_{\partial\mathscr{S}-\gamma}\left[\phi\frac{\partial}{\partial n}\ln(1/\rho) - \ln(1/\rho)\frac{\partial\phi}{\partial n}\right] ds(\mathbf{x}'). \tag{6.3.16}$$

D. Green's Function

Unfortunately, the representations (6.3.14) and (6.3.16) do not give the solution to our problem, since they require that both ϕ and $\partial\phi/\partial n$ be prescribed on all parts of the boundary of the body and their jumps on the discontinuity surface. In a "well-posed" problem only one of these (or their linear combination in Robin's problem) is prescribed on parts of these surfaces. While these representations can be used to the advantage of various theorems and for obtaining estimates, they do not represent the solution to our problem. The solution can be found only after solving another "simpler" problem for $F(\mathbf{x}, \mathbf{x}')$. For a mixed boundary-value problem involving Dirichlet and Neumann boundary conditions, in a domain without a discontinuity surface, the procedure is as follows: Suppose that we possess a harmonic function $F(\mathbf{x}, \mathbf{x}')$ that satisfies

$$\begin{aligned} \nabla'^2 F &= 0 \quad \text{in } \mathscr{V}, \\ \partial G/\partial n &= 0 \quad \text{on } \partial\mathscr{V}_1, \\ G &= 0 \quad \text{on } \partial\mathscr{V}_2 = \partial\mathscr{V} - \partial\mathscr{V}_1, \end{aligned} \tag{6.3.17}$$

where

$$G(\mathbf{x}, \mathbf{x}') = (4\pi\rho)^{-1} + F(\mathbf{x}, \mathbf{x}') \quad \text{(in three dimensions).} \tag{6.3.18}$$

We substitute $\psi = G(\mathbf{x}, \mathbf{x}')$ in (6.3.8) to obtain the solution. The potential $\phi(\mathbf{x})$, valid for a point $\mathbf{x}$ interior to $\mathscr{V} - \sigma$, is given by

$$\phi(\mathbf{x}) = \frac{1}{\varepsilon}\int_{\mathscr{V}} G(\mathbf{x}, \mathbf{x}')q_e(\mathbf{x}')\, dv(\mathbf{x}') - \int_{\partial\mathscr{V}_2}\phi(\mathbf{x}')\frac{\partial G}{\partial n}\, da(\mathbf{x}') - \frac{1}{\varepsilon}\int_{\partial\mathscr{V}_1} w_e(\mathbf{x}')G\, da(\mathbf{x}'). \tag{6.3.19}$$

Expression (6.3.18) is also the solution to the exterior problem, if $\mathscr{V}$ and $\partial\mathscr{V}_i$ ($i = 1, 2$) are considered to represent the exterior region the body and its surface.

The representation for other mixed boundary-value problems is written similarly, by finding the harmonic function that satisfies appropriate homogeneous boundary conditions. These conditions are those that eliminate the unprescribed parts of the surface integrals in (6.3.8). For the two-dimensional case (6.3.19) is valid, with the understanding that the integrals in (6.3.19) are over $\mathscr{S}$, $\partial\mathscr{S}_1$, and $\partial\mathscr{S}_2$, respectively, and

$$G = \frac{1}{2\pi}\ln(\rho^{-1}) + F(\mathbf{x}, \mathbf{x}'). \tag{6.3.20}$$

To obtain the electric field we must calculate the gradient of (6.3.19). This requires special care, since $G(\mathbf{x}, \mathbf{x}')$ possesses a singularity at $\mathbf{x}' = \mathbf{x}$, and the integrals and the gradient operator may not commute.

The determination of the potential F satisfying (6.3.17) is not always easy. For certain domains it is found by use of the techniques of separation of variables, transforms, integral equations, and in two-dimensional cases, complex functions prove to be powerful.

E. Green's Function for Dirichlet's Problem in Spherical Regions and in Half-Space

Here we give Green's functions, for the spherical and cylindrical regions and for the half-space, in order to formulate solutions to Dirichlet's problem. This requires that we determine Green's function $G(\mathbf{x}, \mathbf{x}')$ satisfying

$$\begin{aligned} \nabla'^2 G &= \delta(\mathbf{x} - \mathbf{x}') \quad \text{in } \mathscr{V}, \\ G &= 0 \qquad\qquad\quad \text{on } \partial\mathscr{V}, \end{aligned} \tag{6.3.21}$$

for a region $\mathscr{V}$. It is not difficult to see that the functions

$$\begin{aligned} g_n(\rho) &= \frac{1}{(n-2)A_n}\rho^{2-n}, \qquad n > 2, \\ g_2(\rho) &= \frac{1}{2\pi}\ln\frac{1}{\rho}, \qquad\qquad n = 2, \end{aligned} \tag{6.3.22}$$

are the fundamental solutions of Laplace's equation in n dimensions, where ρ is the distance of $\mathbf{x}'$ from $\mathbf{x}$ and A_n is the area of the n-dimensional sphere of unit radius, i.e.,

$$\rho^2 = (\mathbf{x} - \mathbf{x}')\cdot(\mathbf{x} - \mathbf{x}'), \qquad A_n = 2\pi^{n/2}/\Gamma\left(\frac{n}{2}\right), \tag{6.3.23}$$

where Γ is the Eulerian function. For example, for $n = 3$, we have

$$g_3(\rho) = 1/4\pi\rho. \tag{6.3.24}$$

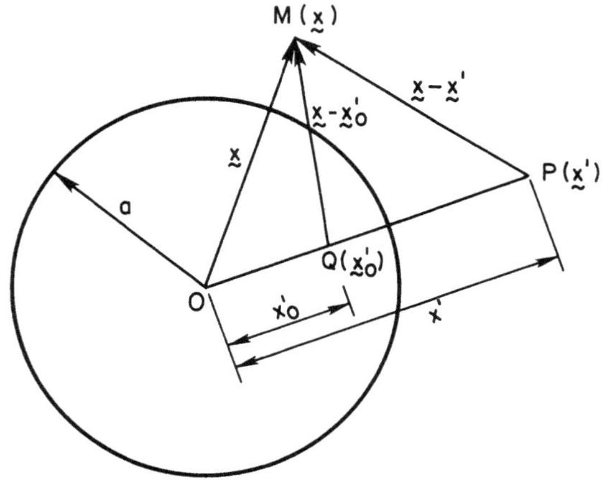

Figure 6.3.1. Image point $Q(\mathbf{x}'_0)$ of point $P(\mathbf{x}')$; $x'_0 = a^2/x'$.

The boundary condition $(6.3.21)_2$ is satisfied if we take

$$G(\mathbf{x}, \mathbf{x}') = g_n(\rho) - g_n(r'\rho_1/a), \tag{6.3.25}$$

where

$$r'^2 = \mathbf{x}' \cdot \mathbf{x}', \qquad \rho_1^2 = [\mathbf{x} - (a/r')^2\mathbf{x}'] \cdot [\mathbf{x} - (a/r')^2\mathbf{x}']. \tag{6.3.26}$$

Note that $g_n(r'\rho_1/a)$ satisfies Laplace's equation and that, at the surface of the n-dimensional surface, G vanishes.

Green's function (6.3.25) has an important physical interpretation. $g_n(\rho)$ represents the field of a positive unit charge (source) located at a point $P(\mathbf{x}')$. The term $g_n(r'\rho_1/a)$ is the field of a negative charge (sink) of magnitude $-a/r'$, located at the reflected image $Q(\mathbf{x}'_0)$ of $P(\mathbf{x}')$ from the sphere (Figure 6.3.1). The potential at $P(\mathbf{x})$ is the sum of these two. If point $P(\mathbf{x})$ is outside the sphere, then P and Q are interchanged. This, however, does not change the form of the Green function given by (6.3.25).

The solution of Dirichlet's problem is now given by

$$\phi(\mathbf{x}) = \frac{1}{\varepsilon} \int_{\mathscr{V}} G(\mathbf{x}, \mathbf{x}') q_e(\mathbf{x}')\, dv(\mathbf{x}') - \int_{\partial\mathscr{V}} \frac{\partial G}{\partial n} \phi_0(\mathbf{x}')\, da(\mathbf{x}'), \tag{6.3.27}$$

where $\mathscr{V}$ is the spherical region and $\partial\mathscr{V}$ is its surface. Substituting from (6.3.22) to (6.3.27) we obtain

$$\phi(r, \theta, \varphi) = \frac{1}{4\pi\varepsilon} \int_0^a \int_{-\pi}^{\pi} \int_0^{\pi} \left(\frac{1}{\rho} - \frac{r'\rho_1}{a}\right) q_e(r', \theta', \varphi') r'^2\, dr' \sin\theta'\, d\theta'\, d\varphi'$$

$$- \frac{a}{4\pi}(a^2 - r^2) \int_{-\pi}^{\pi} \int_0^{\pi} \rho_a^{-3} \phi_0(\theta', \varphi') \sin\theta'\, d\theta'\, d\varphi', \tag{6.3.28}$$

$$\rho^2 = \rho^2(r') = r^2 + r'^2 - 2rr'[\sin\theta \sin\theta' \cos(\varphi - \varphi') + \cos\theta \cos\theta'],$$

$$\rho_a^2 = \rho^2(a), \qquad \rho_1^2 = \rho^2(a^2/r'), \qquad r'^2 = \mathbf{x}' \cdot \mathbf{x}', \qquad r^2 = \mathbf{x} \cdot \mathbf{x},$$

for the spherical region in three dimensions, and

$$\phi(r, \theta) = \frac{1}{2\pi\varepsilon} \int_0^a \int_{-\pi}^{\pi} \ln(r'\rho_1/a\rho) q_e(r', \theta') r' \, dr' \, d\theta' - \frac{1}{2\pi} \int_{-\pi}^{\pi} (a^2 - r^2) \rho_a^{-2} \phi_0(\theta') \, d\theta', \tag{6.3.29}$$

$$\rho^2 = \rho^2(r') = r^2 + r'^2 - 2rr' \cos(\theta - \theta'),$$

$$\rho_a^2 = \rho^2(a), \qquad \rho_1^2 = \rho^2(a^2/r'), \qquad r'^2 = \mathbf{x}' \cdot \mathbf{x}', \qquad r^2 = \mathbf{x} \cdot \mathbf{x},$$

for the circular region in a plane. In calculating $\partial G/\partial n$ on the surface of the sphere we used

$$\frac{\partial G}{\partial n} = \frac{\partial G}{\partial x_k'} n_k' = \frac{\partial G}{\partial x_k'} \frac{x_k'}{a} = g_n(\rho)(a^2 - r^2)/a\rho. \tag{6.3.30}$$

These are Poisson's celebrated formulas for the spherical and circular regions.

Poisson's integral formula for a half-space $z > 0$ is obtained by use of Green's function

$$G(\mathbf{x}, \mathbf{x}') = \frac{1}{2\pi} \left(\frac{1}{\rho_+} - \frac{1}{\rho_-} \right), \tag{6.3.31}$$

where

$$\begin{aligned} \rho_+^2 &= (x - x')^2 + (y - y')^2 + (z - z')^2, \\ \rho_-^2 &= (x - x')^2 + (y - y')^2 + (z + z')^2. \end{aligned} \tag{6.3.32}$$

Expression (6.3.31) is obtained by subtracting the reflection of a unit source $(1/\rho_-)$, in the plane $z = 0$, from the source $(1/\rho_+)$. Hence,

$$\phi(x, y, z) = \frac{1}{4\pi\varepsilon} \int_0^\infty \int_{-\infty}^\infty \int_{-\infty}^\infty \left(\frac{1}{\rho_+} - \frac{1}{\rho_-} \right) q_e \, dx' \, dy' \, dz' - \frac{z}{2\pi} \int_{-\infty}^\infty \int_{-\infty}^\infty \frac{\phi_0(x', y') \, dx' \, dy'}{[(x - x')^2 + (y - y')^2 + z^2]^{3/2}}. \tag{6.3.33}$$

The two-dimensional version of this, for the half-plane $y > 0$, is

$$\phi(x, y) = \frac{1}{2\pi\varepsilon} \int_0^\infty \int_{-\infty}^\infty \ln(\rho_-/\rho_+) q_e(x', y') \, dx' \, dy' - \frac{y}{\pi} \int_{-\infty}^\infty \frac{\phi_0(x') \, dx'}{(x - x')^2 + y^2},$$

$$\rho_+ = [(x - x')^2 + (y - y')^2]^{1/2}, \qquad \rho_- = [(x - x')^2 + (y + y')^2]^{1/2}. \tag{6.3.34}$$

Equation (6.3.34) also may be obtained by integrating (6.3.33) with respect to y' over $(-\infty, \infty)$, when q_e and ϕ_0 are independent of y' and by replacing z by y.

Poisson's formulas can be used to prove various theorems and to obtain estimates of potentials for arbitrary domains. We state some of these without proof.

Theorem 1. *A harmonic function which is regular and positive in every bounded region of the space is a constant.*

Theorem 2. *Every harmonic function that is regular in a domain $\mathscr{V}$ is analytic in $\mathscr{V}$ (it can be expanded into a power series at the interior points of $\mathscr{V}$).*

Theorem 3 (Weierstrass). *A sequence of potential functions ϕ_n, which is regular in $\mathscr{V}$ and continuous in $\mathscr{V} + \partial\mathscr{V}$ and has boundary values ϕ_{0n} converging uniformly on $\partial\mathscr{V}$, converges uniformly in $\mathscr{V}$ to a potential function with the boundary value $\phi_0 = \lim \phi_{0n}$.*

Theorem 4 (Harnack). *If a nondecreasing or nonincreasing sequence of regular harmonic functions in $\mathscr{V}$ converges at a single point of $\mathscr{V}$, then it converges at all points of $\mathscr{V}$, and the convergence is uniform in every closed interior subdomain.*

Theorem 5 (Mean Value). *For every regular potential function ϕ in a region $\mathscr{V}$ the average value, over the surface S_R of a sphere of radius R entirely contained in $\mathscr{V}$, is equal to the value ϕ_0 of the function at the center of the sphere.*

$$\phi_0 = \frac{1}{4\pi R^2} \iint_{S_R} \phi \, da. \tag{6.3.35}$$

A mean value theorem for the interior of the sphere follows from this by integrating (6.3.35) with respect to R after multiplying it by R^2. This is permissible since (6.3.35) holds for all R for which corresponding spheres are inside of $\mathscr{V}$. Hence,

$$\phi_0 = \frac{3}{4\pi a^3} \iiint_{K_a} \phi \, dv,$$

where a is the radius of the solid sphere K_a. All these theorems are for the case $q_e = 0$ in Poisson's formula. For proofs we refer the reader to Courant [1965, Vol. 2, Chap. IV].

F. Eigenfunction Expansions

In the treatment of a large class of problems the eigenfunction expansions of Green's functions prove to be very useful. The product solutions (separation of variables) of Laplace's equation in various coordinate systems can be used to obtain these forms. Here, we briefly present the method based on the expansion of Green's function for the circular region. This is achieved by expanding Green's function into Fourier series of $\cos n(\theta - \theta')$. Alternatively, we introduce complex variables $z = x + iy = re^{i\theta}$, $z' = x' + iy' = r'e^{i\theta'}$, and write

$$2G = \ln(r'\rho_1/a\rho)^2 = \ln\left\{\frac{r'^2}{a^2} \frac{[z - (a^2/r'^2)z'][\bar{z} - (a^2/r'^2)\bar{z}']}{(z - z')(\bar{z} - \bar{z}')}\right\}, \tag{6.3.36}$$

where $\bar{z}$ and $\bar{z}'$ are complex conjugates to z and z', respectively. Now, recalling the series

$$\begin{aligned}\ln|z - z_0| &= \ln z - \sum_{n=1}^{\infty} \frac{1}{n}\left(\frac{z_0}{z}\right)^n, \qquad |z| > |z_0|,\\ &= \ln z_0 - \sum_{n=1}^{\infty} \frac{1}{n}\left(\frac{z}{z_0}\right)^n, \qquad |z| < |z_0|,\end{aligned} \tag{6.3.37}$$

we obtain Green's function for the exterior problem as

$$\begin{aligned}G(r, \theta, r', \theta') &= \ln\frac{r}{a} + \sum_{n=1}^{\infty} \frac{1}{n}\left[\left(\frac{r}{r'}\right)^n - \left(\frac{a^2}{rr'}\right)^n\right] \cos n(\theta' - \theta), \qquad r' > r > a,\\ &= \ln\frac{r'}{a} + \sum_{n=1}^{\infty} \frac{1}{n}\left[\left(\frac{r'}{r}\right)^n - \left(\frac{a^2}{rr'}\right)^n\right] \cos n(\theta - \theta'), \qquad r > r'.\end{aligned} \tag{6.3.38}$$

The expansion (6.3.36), for the interior domain, leads to

$$\begin{aligned}G(r, \theta; r', \theta') &= \ln\frac{r'}{a} + \sum_{n=1}^{\infty} \frac{1}{n}\left[\left(\frac{r}{r'}\right)^n - \left(\frac{rr'}{a^2}\right)^n\right] \cos n(\theta - \theta'), \qquad a > r' > r,\\ &= \ln\frac{a}{r} + \sum_{n=1}^{\infty} \frac{1}{n}\left[\left(\frac{r'}{r}\right)^n - \left(\frac{r'r}{a^2}\right)^n\right] \cos n(\theta' - \theta), \qquad a > r > r'.\end{aligned} \tag{6.3.39}$$

By means of (6.3.7) we can, of course, write the solution of Dirichlet's problem for the circular domain if we note that

$$\frac{\partial G}{\partial n'} = \frac{\partial G}{\partial r'} \quad \text{on } r' = a. \tag{6.3.40}$$

The integrands of both integrals in (6.3.27) are now fully known.

As another example of the eigenfunction expansion of Green's function, we give that of Dirichlet's problem for the half-space $z > 0$. To accomplish this we develop $1/\rho$ into a Fourier series. To this end we note that

$$\begin{aligned}\rho^{-1} &= [r^2 + r'^2 - 2rr' \cos(\theta - \theta') + (z - z')^2]^{-1/2}\\ &= \frac{1}{2\pi} \int_0^{\infty} d\xi \int_0^{\pi} \exp[-\xi|z - z'| + i\xi(r \cos(u - \theta) - r' \cos(u - \theta')] \, du.\end{aligned}$$

This is expandable into a Fourier series where the coefficients are given by

$$\begin{aligned}A_m = \frac{\varepsilon_m}{4\pi^2} \int_0^{\infty} d\xi \int_0^{2\pi} dw \int_0^{2\pi} dt \cos(mt) \exp\{-\xi|z - z'|\\ + i\xi[r \cos w - r' \cos(t + w)]\},\end{aligned}$$

where $\varepsilon_0 = 1$, $\varepsilon_m = 2$, and $w = u - \theta$ and $t = \theta' - \theta$. Now let $t + w = v$ so that

$$A_m = \frac{\varepsilon_m}{4\pi^2} \int_0^\infty d\xi \exp(-\xi(z - z')) \int_0^\infty \exp(i\xi r \cos w) \cos(mw) \, dw$$
$$\cdot \int_0^{2\pi} \exp(-i\xi r \cos v) \cos(mv) \, dv$$
$$= \varepsilon_m i^{2m} \int_0^\infty (-)^m \exp(-\xi|z - z'|) J_m(\xi r) J_m(\xi r') \, d\xi,$$

where $J_m(\xi)$ is the Bessel function of order m. Consequently,

$$1/\rho_\pm = \sum_{m=0}^{\infty} \varepsilon_m \cos m(\theta - \theta') \int_0^\infty J_m(\xi r) J_m(\xi r') \exp(-\xi|z \mp z'|) \, d\xi. \tag{6.3.41}$$

Green's function for the half-space is

$$G(r, \theta, z; r', \theta', z') = \frac{1}{4\pi}\left(\frac{1}{\rho_+} - \frac{1}{\rho_-}\right)$$
$$\frac{\partial G}{\partial n} = \frac{\partial G}{\partial z'}. \tag{6.3.42}$$

The solution of Dirichlet's problem is then given by (6.3.27), i.e.,

$$\phi(r, \theta, z) = \frac{1}{\varepsilon} \int_0^\infty dz' \int_{-\pi}^{\pi} d\theta' \int_0^\infty G q_e r' \, dr'$$
$$- \int_{-\pi}^{\pi} d\theta' \int_0^\infty \frac{\partial G}{\partial z'} \phi_0(r', \theta') r' \, dr'. \tag{6.3.43}$$

In the case of axisymmetric q_e and ϕ_0 distributions the integration over θ' leads to

$$1/\rho_\pm = 2\pi \int_0^\infty J_0(\xi r) J_0(\xi r') \exp(-\xi|z \mp z'|) \, d\xi, \tag{6.3.44}$$

so that the solution to Dirichlet's problem, for the half-space with axisymmetric distribution of potentials, may be calculated by use of (6.3.44).

For spherical and other regions, and for Neumann's problem, eigenfunction expansions are valid. For these we refer the reader to Morse and Feschbach [1953].

6.4. Charged Cylinder

As a simple application of Poisson's formulas, as developed in Section 6.3, we consider the potential due to the field of a charged cylinder. Suppose that the potential ϕ_0 on the surface of an infinite circular cylinder is given by

$$\phi_0(\theta') = \phi^0 \cos \theta', \tag{6.4.1}$$

where ϕ^0 is a constant. With $q_e = 0$, (6.3.29) takes the form

$$\phi(r, \theta) = -\frac{1}{2\pi}\int_{-\pi}^{\pi} \frac{(a^2 - r^2)\phi_0(\theta')}{a^2 - 2\,ar\cos(\theta - \theta') + r^2}\,d\theta'. \tag{6.4.2}$$

Performing the integration we find the following two possible solutions:

$$\phi(r, \theta) = -\phi^0 \frac{r}{a}\cos\theta, \qquad r < a, \tag{6.4.3}$$

$$\phi(r, \theta) = -\phi^0 \frac{r}{a}\cos\theta, \qquad r > a. \tag{6.4.4}$$

Thus the electric field $\mathbf{E} = -\nabla\phi$ is given by

$$E_r = \frac{\phi^0}{a}\cos\theta, \qquad E_\theta = -\frac{\phi^0}{a}\sin\theta, \qquad r < a, \tag{6.4.5}$$

$$E_r = -a\phi^0(a/r^2)\cos\theta, \qquad E_\theta = -\phi^0(a/r^2)\sin\theta, \qquad r > a. \tag{6.4.6}$$

The solution (6.4.3) is for the interior of the cylinder $r < a$, in which case the electric field consists of a constant field parallel to the x-axis inside the cylinder. The solution (6.4.4) is for the exterior problem.

A slightly more involved problem is the potential exterior to a cylinder of radius a when its surface is kept at $\phi_0 =$ const. in the interval $0 < \theta < \pi$ and at $-\phi_0$ in the interval $\pi < \theta < 2\pi$. In this case too, (6.4.2) can be used to evaluate the potential. However, we use the series

$$\left.\frac{\partial G}{\partial r'}\right|_{r'=a} = \frac{1}{a} + \frac{2}{a}\sum_{n=1}^{\infty}\left(\frac{a}{r}\right)^m \cos m(\theta - \theta'), \tag{6.4.7}$$

and

$$\phi(r, \theta) = \frac{\phi_0}{2\pi}\int_0^{\pi} \left.\frac{\partial G}{\partial r'}\right|_{r'=a} d\theta' - \frac{\phi_0}{2\pi}\int_{\pi}^{2\pi} \left.\frac{\partial G}{\partial r'}\right|_{r'=a} d\theta'. \tag{6.4.8}$$

Upon integration, this gives

$$\begin{aligned}\phi(r, \theta) &= -\frac{4\phi_0}{\pi}\sum_{m=0}^{\infty}\frac{1}{2m+1}\left(\frac{a}{r}\right)^{2m+1}\sin(2m+1)\theta \\ &= -\frac{2\phi_0}{\pi}\arctan\left(\frac{2\,\mathrm{arsin}\,\phi}{r^2 - a^2}\right).\end{aligned} \tag{6.4.9}$$

This result would also follow from (6.4.2) by direct integration.

6.5. Potential in Half-Plane

Suppose that a half-plane $y > 0$ is charged to a constant potential ϕ_0 along a segment $|x| < a$, $y = 0$ (Figure 6.5.1), outside of this segment and at infinity

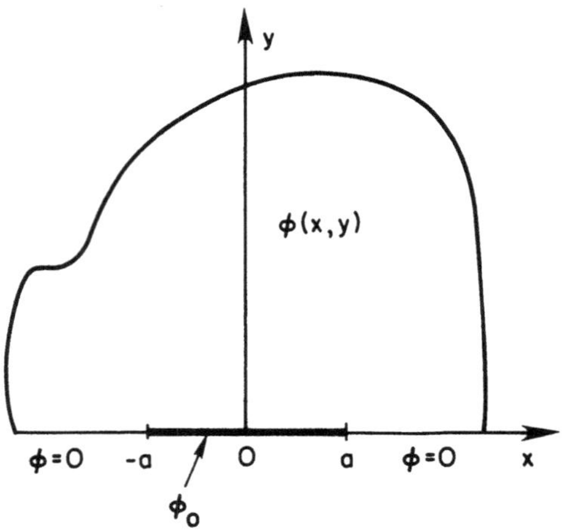

Figure 6.5.1. Potential in half-plane.

the potential vanishes. The boundary conditions are

$$\begin{aligned} \phi(x, 0) &= \phi_0, \qquad |x| < a, \quad y = 0, \\ \phi(x, 0) &= 0, \qquad |x| > a, \quad y = 0, \\ \phi(x, y) &\to 0, \qquad x^2 + y^2 \to \infty. \end{aligned} \tag{6.5.1}$$

The solution of this problem follows from Poisson's formula (6.3.34):

$$\begin{aligned} \phi(x, y) &= -\phi_0 \frac{y}{\pi} \int_{-a}^{a} \frac{dx'}{(x - x')^2 + y^2} \\ &= -\frac{\phi_0}{\pi}\left[\arctan\left(\frac{a - x}{y}\right) + \arctan\left(\frac{a + x}{y}\right)\right]. \end{aligned} \tag{6.5.2}$$

If the potential ϕ_0 on the boundary is a more complicated function, we may employ its Fourier integral form more conveniently. To this end we notice that

$$\frac{1}{\pi}\int_0^\infty e^{-\xi y} \cos[\xi(x - x')]\, d\xi = \frac{y/\pi}{(x - x')^2 + y^2}. \tag{6.5.3}$$

Using this in (6.3.34) Poisson's formula for the half-plane may also be put into the form

$$\phi(x, y) = -\frac{1}{\pi}\int_{-\infty}^{\infty} \phi(x', 0)\, dx' \int_0^\infty e^{-\xi y} \cos[\xi(x - x')]\, d\xi. \tag{6.5.4}$$

Expression (6.5.4) is the Fourier expansion of Green's function for the half-plane. Of course, such expansions are valid for spherical and other regions. Also they can be arrived at by solving Laplace's equation by means of eigenfunction expansions (e.g., separation of variables, transform techniques). In (6.5.4), by using (6.5.1), we again obtain (6.5.2).

Mixed Boundary-Value Problem

A more difficult problem is the following mixed boundary-value problem in which the line segment $|x| < a$, $y = 0$, is charged to a constant potential, but outside this line ($|x| > a$, $y = 0$) the body is a perfect grounded conductor. The boundary conditions are

$$\begin{aligned} \phi(x, 0) &= \phi_0(x), && |x| < a, \quad y = 0, \\ \frac{\partial \phi}{\partial y}(x, 0) &= 0, && |x| > a, \quad y = 0, \\ \phi(x, y) &\to 0 \quad \text{as} && x^2 + y^2 \to \infty. \end{aligned} \tag{6.5.5}$$

In this case Poisson's formula does not provide a direct answer, since $\phi(x, 0)$ at $|x| > a$ is not known. However, we can employ (6.5.4). We treat the problem in which $\phi_0(x)$ is an even function, i.e., $\phi_0(-x) = \phi_0(x)$. A similar treatment can be carried out when $\phi_0(x)$ is an odd function. Further, we put it in non-dimensional form by writing $x_1 = x/a$. Then the interval $|x| < a$ becomes $|x| < 1$. In this case, by expanding $\cos \xi(x - x')$, (6.5.4) may be written as

$$\begin{aligned} \int_0^\infty \bar{\phi}(\xi, 0) \cos(\xi x_1)\, d\xi &= -\phi_0(x_1), \qquad 0 \le x_1 < 1, \\ \int_0^\infty \xi \bar{\phi}(\xi, 0) \cos(\xi x_1)\, d\xi &= 0, \qquad x_1 > 1, \end{aligned} \tag{6.5.6}$$

where we set

$$\begin{aligned} \bar{\phi}(\xi, 0) &= \left(\frac{2}{\pi}\right)^{1/2} \int_0^\infty \phi(x', 0) \cos(x'\xi)\, dx', \\ 0 &= \left(\frac{2}{\pi}\right)^{1/2} \int_0^\infty \phi(x', 0) \sin(x'\xi)\, dx', \end{aligned} \tag{6.5.7}$$

since $\phi(x', 0)$ is an even function. $\bar{\phi}(\xi, 0)$ defined by $(6.5.7)_1$ is, of course, the Fourier transform of $\phi(x, 0)$. The dual integral equations (6.5.6) must be solved to determine the unknown function $\phi(\xi, 0)$. Fortunately, the solution of the set (6.5.6) is known (see Sneddon [1966, p. 98])

$$\bar{\phi}(\xi, 0) = -a_0 \xi J_1(\xi) - 2 \sum_{q=1}^{\infty} q a_q J_{2q}(\xi), \tag{6.5.8}$$

where

$$\begin{aligned} a_0 &= \frac{2}{\pi} \int_0^1 \frac{\phi_0(x_1)}{(1 - x_1^2)^{1/2}}\, dx_1, \\ a_q &= \frac{(-)^q 2^{2-q}}{\Gamma(1/2)\Gamma(q + \frac{1}{2})} \int_0^1 x_1^{2q}(1 - x_1^2)^{q-1/2} \left(\frac{1}{x_1}\frac{\partial}{\partial x_1}\right)^q \phi_0(x_1)\, dx_1, \qquad q > 1. \end{aligned} \tag{6.5.9}$$

Incidentally, the function $\phi_0(x)$ is assumed to accept the expansion

$$\phi_0(x) = a_0 + \sum_{q=1}^{\infty} a_q \mathfrak{F}_q(0, \tfrac{1}{2}, x^2), \tag{6.5.10}$$

where $\mathfrak{F}_q(a, b, x)$ denotes the Jacobi polynomial (Magnus and Oberhettinger [1949, p. 83]).

If we wish that $\partial\phi/\partial y$ remains finite in the interval $|x| < a$, $y = 0$, then we must impose the condition $a_0 = 0$ on our function $\phi_0(x)$. With this we will have

$$\begin{aligned} \phi(x_1, 0) &= -\sum_{q=1}^{\infty} (-)^q [x_1 + (x_1^2 - 1)^{1/2}]^{-2q} a_q, \qquad x_1 > 1, \\ \frac{\partial\phi}{\partial y}(x_1, 0) &= -2(1 - x_1)^{-1/2} \sum_{q=1}^{\infty} q a_q \mathfrak{F}_q(0, \tfrac{1}{2}, x_1^2), \qquad 0 \le x_1 < 1. \end{aligned} \tag{6.5.11}$$

These determine the unknown pieces of the potential and charge density on the line $y = 0$ completing the solution to the problem.

6.6. Charged Disk

Suppose a circular disk of radius a is charged axisymmetrically to a prescribed potential. We wish to determine the potential outside the disk. In cylindrical coordinates (r, θ, z) the boundary conditions read

$$\begin{aligned} [\phi] = 0, \qquad [\partial\phi/\partial z] &= -4\pi w(r) \quad \text{on } z = 0, \\ \phi \to 0 \qquad \text{as} \quad r^2 + z^2 &\to \infty. \end{aligned} \tag{6.6.1}$$

Symmetry conditions reveal that

$$\phi(r, z) = \phi(r, -z),$$

so that the problem is reducible to a problem in the half-space $z > 0$. At the boundary of the half-space for $r \le a$, $z = 0$, the potential is prescribed and outside ($r > a$, $z = 0$), $\partial\phi/\partial z = 0$, i.e.,

$$\begin{aligned} \phi(r, 0) &= \phi_0(r), \qquad r \le a, \quad 0 \le \theta < 2\pi, \\ \frac{\partial\phi}{\partial z}(r, 0) &= 0, \qquad r > a, \quad 0 \le \theta < 2\pi, \\ \phi(r, z) &= 0 \qquad \text{as} \quad r^2 + z^2 \to \infty. \end{aligned} \tag{6.6.2}$$

This is a mixed boundary-value problem. The potential for the half-space is obtained by using (6.3.44). We have

$$\left.\frac{\partial G}{\partial z'}\right|_{z'=0} = -\int_0^{\infty} \xi J_0(\xi r) J_0(\xi r') e^{-\xi z} \, d\xi, \tag{6.6.3}$$

so that

$$\phi(r, z) = \int_0^\infty A(\xi)\xi^{-1}e^{-\xi z}J_0(\xi r)\,d\xi, \qquad z \geq 0, \tag{6.6.4}$$

where we wrote

$$A(\xi) \equiv \int_0^\infty \phi_0(r')\xi^2 J_0(\xi r')\,dr'. \tag{6.6.5}$$

Since $\phi_0(r')$ is not known all along the line $z = 0$, $A(\xi)$ is an unknown function to be determined from the boundary conditions $(6.6.2)_{1,2}$. In nondimensional form these conditions read

$$\begin{aligned} &\int_0^\infty \zeta^{-1}B(\zeta)J_0(\zeta\rho)\,d\zeta = f(\rho), \qquad 0 \leq \rho < 1, \\ &\int_0^\infty B(\zeta)J_0(\zeta\rho)\,d\zeta = 0, \qquad \rho > 1, \end{aligned} \tag{6.6.6}$$

where we set

$$\zeta = a\xi, \qquad \rho = r/a, \qquad A(\xi) = B(\zeta), \qquad \phi_0(r) = f(\rho). \tag{6.6.7}$$

We try a solution of the form

$$\begin{aligned} B(\zeta) &= \int_0^1 \zeta F'(s)\cos(\zeta s)\,ds \\ &= F'(1)\sin\zeta - \int_0^1 F''(s)\sin(\zeta s)\,ds, \end{aligned} \tag{6.6.8}$$

and recall the expressions

$$\begin{aligned} &\int_0^\infty \zeta^{-1}\sin\zeta J_0(\zeta\rho)\,d\zeta = \begin{cases} \pi/2, & 0 \leq \rho < 1, \\ \arcsin(1/\rho), & \rho > 1, \end{cases} \\ &\int_0^\infty \sin\zeta J_0(\rho\zeta)\,d\zeta = \begin{cases} (1-\rho^2)^{-1/2}, & 0 \leq \rho < 1, \\ 0, & \rho > 1, \end{cases} \end{aligned} \tag{6.6.9}$$

Substituting (6.6.8) into $(6.6.6)_2$ we find that the latter is satisfied. On the other hand, $(6.6.6)_1$ gives

$$\int_0^\rho \frac{F'(s)}{(\rho^2 - s^2)^{1/2}}\,ds = f(\rho), \qquad 0 \leq \rho < 1. \tag{6.6.10}$$

This is the celebrated Abel integral equation whose solution is well known (see Tricomi [1957])

$$F'(s) = \frac{2}{\pi}\frac{d}{ds}\int_0^s \frac{tf(t)\,dt}{(s^2 - t^2)^{1/2}}. \tag{6.6.11}$$

With this (6.6.8) gives $B(\zeta)$, thus providing the solution to the problem.

Consider, for instance, $f(t) = 1$. We get $F'(s) = 2/\pi$ and therefore

$$B(\zeta) = \frac{2}{\pi} \sin \zeta, \qquad A(\zeta) = \frac{2}{\pi} \sin a\zeta. \tag{6.6.12}$$

Substituting this into (6.6.4) we obtain the potential

$$\phi(r, z) = \frac{2}{\pi} \int_0^\infty \xi^{-1} e^{-\xi z} J_0(\xi r) \sin(a\xi)\, d\xi, \qquad z \geq 0. \tag{6.6.13}$$

The surface charge density is, therefore,

$$\begin{aligned} w(\rho) &= \frac{1}{\pi^2} \int_0^\infty J_0(\xi r) \sin(a\xi)\, d\xi \\ &= \frac{1}{\pi^2}(a^2 - r^2)^{-1/2} \qquad 0 < r < a. \end{aligned} \tag{6.6.14}$$

For other approaches and for solutions to nonaxisymmetric problems, see Sneddon [1966].

6.7. Plane Electromagnetic Waves in Isotropic Bodies

When the dielectrics contain no charge ($q_e = 0$) or current distributions, Maxwell's equations (6.2.1)–(6.2.4), together with the constitutive equations (6.2.14), are equivalent to two vector wave equations for electric and magnetic fields. To see this we substitute expressions **D** and **H** into (6.2.1)–(6.2.4) to get

$$\nabla \cdot (\varepsilon \mathbf{E}) = 0, \tag{6.7.1}$$

$$\nabla \times \mathbf{E} + \frac{\mu}{c} \frac{\partial \mathbf{H}}{\partial t} = \mathbf{0}, \tag{6.7.2}$$

$$\nabla \cdot (\mu \mathbf{H}) = 0, \tag{6.7.3}$$

$$\nabla \times \mathbf{H} - \frac{\varepsilon}{c} \frac{\partial \mathbf{E}}{\partial t} = \mathbf{0}. \tag{6.7.4}$$

The first of these equations is equivalent to

$$\varepsilon \nabla \cdot \mathbf{E} + \mathbf{E} \cdot \nabla \varepsilon = 0. \tag{6.7.5}$$

From (6.7.2) we write

$$\nabla \times (\mu^{-1} \nabla \times \mathbf{E}) + \frac{1}{c} \frac{\partial}{\partial t} (\nabla \times \mathbf{H}) = \mathbf{0}.$$

Upon using (6.7.4) we eliminate the last term to obtain

$$\nabla \times (\mu^{-1} \nabla \times \mathbf{E}) + \varepsilon c^{-2} \frac{\partial^2 \mathbf{E}}{\partial t^2} = \mathbf{0}.$$

Expanding the first term and using identities

$$\begin{aligned} \nabla \times (\alpha \mathbf{v}) &= \alpha \nabla \times \mathbf{v} + \nabla \alpha \times \mathbf{v}, \\ \nabla \times \nabla \times \mathbf{A} &= \nabla\nabla \cdot \mathbf{A} - \nabla^2 \mathbf{A}, \end{aligned} \tag{6.7.6}$$

we obtain

$$\nabla^2 \mathbf{E} + (\varepsilon\mu/c^2)\frac{\partial^2 \mathbf{E}}{\partial t^2} + \nabla(\ln \mu) \times (\nabla \times \mathbf{E}) - \nabla\nabla \cdot \mathbf{E} = \mathbf{0}. \tag{6.7.7}$$

With the help of (6.7.5) we eliminate the last term so that

$$\nabla^2 \mathbf{E} - (\varepsilon\mu/c^2)\frac{\partial^2 \mathbf{E}}{\partial t^2} + \nabla(\ln \mu) \times (\nabla \times \mathbf{E}) + \nabla(\mathbf{E} \cdot \nabla \ln \varepsilon) = \mathbf{0}. \tag{6.7.8}$$

Similarly, for the **H**-field, we obtain

$$\nabla^2 \mathbf{H} - (\varepsilon\mu/c^2)\frac{\partial^2 \mathbf{H}}{\partial t^2} + \nabla(\ln \varepsilon) \times (\nabla \times \mathbf{H}) + \nabla(\mathbf{H} \cdot \nabla \ln \mu) = \mathbf{0}. \tag{6.7.9}$$

These equations are valid for inhomogeneous solids. For homogeneous solids ε and μ are constants and we have

$$\nabla^2 \mathbf{E} - (\varepsilon\mu/c^2)\frac{\partial^2 \mathbf{E}}{\partial t^2} = \mathbf{0}, \qquad \nabla^2 \mathbf{H} - (\varepsilon\mu/c^2)\frac{\partial^2 \mathbf{H}}{\partial t^2} = \mathbf{0}. \tag{6.7.10}$$

These are the wave equations which suggest the propagation of waves with phase velocity[2]

$$v = (\varepsilon\mu)^{-1/2} c. \tag{6.7.11}$$

The constant c was first determined in 1856, by Kohlrausch and Weber, from the ratio of the values of the capacity of a condenser measured in electrostatic and electromagnetic units. It was found that it was identical to the velocity of light in free-space. Subsequently, Maxwell developed his electromagnetic theory of light which predicted the existence of electromagnetic waves. Hertz elaborated on experiments which confirmed this prediction.

The plane-wave solution of (6.7.10) may be found by considering each rectangular component $V(\mathbf{r}, t)$ of the fields which satisfy the wave equation

$$\nabla^2 V - v^{-2}\frac{\partial^2 V}{\partial t^2} = 0. \tag{6.7.12}$$

Any solution of the form

$$V = V(\mathbf{r} \cdot \mathbf{n}, t) \tag{6.7.13}$$

represents a wave propagating in the direction of unit vectors

$$\mathbf{r} \cdot \mathbf{n} = \xi, \tag{6.7.14}$$

[2] We remind the reader that the Laplacian operator ∇^2 for vectors has the form $(6.7.6)_2$. For rectangular coordinates only $(6.7.6)_2$ reduces to the usual form $A_{i,kk}$.

so that

$$\partial V/\partial x_k = n_k \, \partial V/\partial \xi,$$

and (6.7.12) reduces to a one-dimensional wave equation

$$\frac{\partial^2 V}{\partial \xi^2} - \frac{1}{v^2}\frac{\partial^2 V}{\partial t^2} = 0. \tag{6.7.15}$$

Following d'Alembert, if we change the variables to

$$p = \xi - vt, \qquad q = \xi + vt, \tag{6.7.16}$$

this equation becomes

$$\partial^2 V/\partial p \, \partial q = 0, \tag{6.7.17}$$

whose solution is

$$V = V_1(p) + V_2(q) = V_1(\mathbf{r}\cdot\mathbf{n} - vt) + V_2(\mathbf{r}\cdot\mathbf{n} + vt), \tag{6.7.18}$$

where V_1 and V_2 are arbitrary functions. It is not difficult to see that V_1 represents a disturbance propagating in the positive ξ-direction and V_2 in the negative ξ-direction.

Now, consider propagation in the positive ξ-direction only. The **E** and **H** fields are given by

$$\mathbf{E} = \mathbf{E}(\mathbf{r}\cdot\mathbf{n} - vt), \qquad \mathbf{H} = \mathbf{H}(\mathbf{r}\cdot\mathbf{n} - vt). \tag{6.7.19}$$

Substituting these into (6.7.2) and (6.7.4) we obtain

$$\mathbf{n} \times \mathbf{E}' - (\mu v/c)\mathbf{H}' = \mathbf{0}, \qquad \mathbf{n} \times \mathbf{H}' + (\varepsilon v/c)\mathbf{E}' = \mathbf{0},$$

where primes indicate derivatives with respect to the phase variable $\mathbf{r}\cdot\mathbf{n} - vt$. Setting $v/c = (\varepsilon\mu)^{-1/2}$, and omitting an unimportant constant of integration, these integrate to

$$\mathbf{E} = -(\mu/\varepsilon)^{1/2}\mathbf{n} \times \mathbf{H}, \qquad \mathbf{H} = (\varepsilon/\mu)^{1/2}\mathbf{n} \times \mathbf{E}. \tag{6.7.20}$$

The scalar product of these by **n** shows that

$$\mathbf{E}\cdot\mathbf{n} = 0, \qquad \mathbf{H}\cdot\mathbf{n} = 0. \tag{6.7.21}$$

Thus, electromagnetic waves are "transverse," i.e., the **E** and **H** fields are perpendicular to the direction of propagation (Figure 6.7.1).

From (6.7.20) it is also clear that the **E** and **H** vectors are perpendicular to each other so that **E**, **H**, and **n** form a right-handed orthogonal triad. We also have

$$\mu^{1/2} H = \varepsilon^{1/2} E, \tag{6.7.22}$$

where $E = |\mathbf{E}|$, $H = |\mathbf{H}|$.

For rigid materials the energy equation integrates into

$$\rho_0 e = \frac{\varepsilon}{2}E^2 + \frac{\mu}{2}H^2 \tag{6.7.23}$$

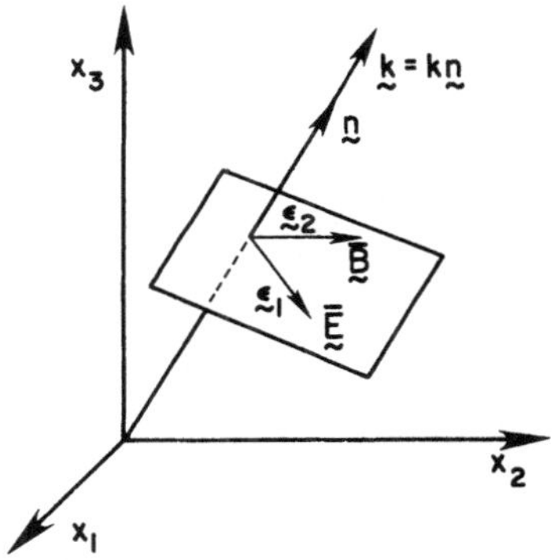

Figure 6.7.1. Plane electromagnetic wave.

and the Poynting vector gives

$$\mathbf{S} = cEH\mathbf{n} = c(\varepsilon/\mu)^{1/2}E^2\mathbf{n} = c(\mu/\varepsilon)^{1/2}H^2\mathbf{n}. \tag{6.7.24}$$

Comparing these two expressions we find

$$\mathbf{S} = \rho_0 e v \mathbf{n}, \tag{6.7.25}$$

where we used (6.7.11). Hence, we see that the Poynting vector in this case represents the flow of energy in the direction of the propagation.

6.8. Spherical Waves

The wave equations $(6.7.10)_1$ in curvilinear coordinates has the form

$$\nabla\nabla\cdot\mathbf{E} - \nabla\times\nabla\times\mathbf{E} - (\varepsilon\mu/c^2)\frac{\partial^2\mathbf{E}}{\partial t^2} = \mathbf{0}.$$

Since ε and μ are constants, in view of (6.7.1) and (6.7.3), $\nabla\cdot\mathbf{E} = \nabla\cdot\mathbf{H} = 0$ and 6.7.10) reads

$$\nabla\times\nabla\times\mathbf{E} + v^{-2}\frac{\partial^2\mathbf{E}}{\partial t^2} = \mathbf{0}, \qquad \nabla\times\nabla\times\mathbf{H} + v^{-2}\frac{\partial^2\mathbf{H}}{\partial t^2} = \mathbf{0}, \tag{6.8.1}$$

where we used (6.7.11). In orthogonal curvilinear coordinates x^k for the curl operator we have Eringen [1967, p. 470])

$$\begin{aligned}\nabla\times\mathbf{A} = {}& (g_{22}g_{33})^{-1/2}\left[\frac{\partial}{\partial x^2}(\sqrt{g_{33}}A^{(3)}) - \frac{\partial}{\partial x^3}(\sqrt{g_{22}}A^{(2)})\right]\mathbf{e}_1\\ &+ (g_{33}g_{11})^{-1/2}\left[\frac{\partial}{\partial x^3}(\sqrt{g_{11}}A^{(1)}) - \frac{\partial}{\partial x^1}(\sqrt{g_{33}}A^{(3)})\right]\mathbf{e}_2\\ &+ (g_{11}g_{22})^{-1/2}\left[\frac{\partial}{\partial x^1}(\sqrt{g_{22}}A^{(2)}) - \frac{\partial}{\partial x^2}(\sqrt{g_{11}}A^{(1)})\right]\mathbf{e}_3,\end{aligned} \tag{6.8.2}$$

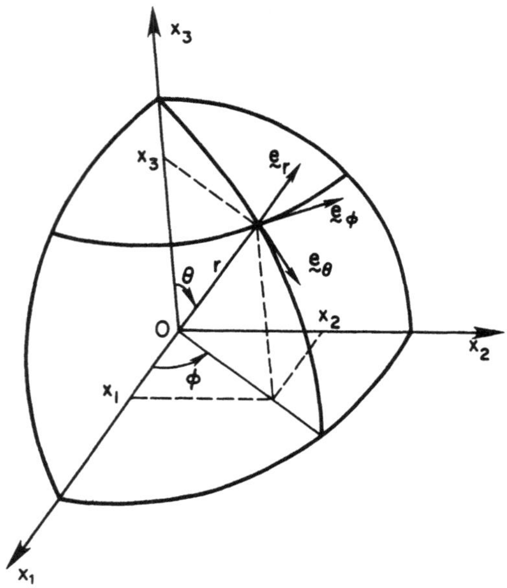

Figure 6.8.1. Spherical polar coordinates.

where g_{ij} is the metric tensor, $A^{(i)}$ are the physical components of the vector $\mathbf{A}$, and $\mathbf{e}_i$ are the unit basis vectors in the curvilinear coordinates x^k. For spherical coordinates we have $x^1 = r$, $x^2 = \theta$, $x^3 = \phi$ (Figure 6.8.1), and

$$g_{11} = 1, \qquad g_{22} = r^2, \qquad g_{33} = r^2 \sin^2 \theta, \tag{6.8.3}$$

$$\begin{aligned}\nabla \times \mathbf{A} &= \frac{1}{r \sin \theta}\left[\frac{\partial}{\partial \theta}(A_\phi \sin \theta) - \frac{\partial A_\theta}{\partial \phi}\right]\mathbf{e}_r \\ &+ \left[\frac{1}{r \sin \theta}\frac{\partial A_r}{\partial \phi} - \frac{1}{r}\frac{\partial}{\partial r}(rA_\phi)\right]\mathbf{e}_\theta + \frac{1}{r}\left[\frac{\partial}{\partial r}(rA_\theta) - \frac{\partial A_r}{\partial \theta}\right]\mathbf{e}_\phi,\end{aligned} \tag{6.8.4}$$

where A_r, A_θ, and A_ϕ are the physical components of $\mathbf{A}$ in spherical coordinates. The spherical electromagnetic wave propagation requires solving (6.8.1) with the curl operators calculated by means of (6.8.4). Here, for simplicity, we consider only scalar waves. To this end, suppose that the only surviving components of $\mathbf{E}$ and $\mathbf{H}$ are $E_r = E(r, t)$ and $H_r = H(r, t)$, and that all other components vanish. In this case, (6.8.1), by use of (6.8.4), gives

$$\frac{\partial^2}{\partial r^2}(rV) - \frac{1}{v^2}\frac{\partial^2 (rV)}{\partial t^2} = 0, \tag{6.8.5}$$

where V represents either E or H. Hence, the solution is (see (6.7.18)).

$$V = \frac{1}{r}V_1(r - vt) + \frac{1}{r}V_2(r + vt), \tag{6.8.6}$$

where V_1 and V_2 are two arbitrary functions, the first of which represents the outgoing (diverging) spherical waves from the origin, and the second one the incoming (converging) spherical waves towards the origin. The velocity of propagation is, of course, v for both types of waves.

6.9. Nonlinear Theory of Rigid Dielectrics

The basic equations listed in Section 6.2 are reduced to the simplest forms for rigid nonmagnetizable bodies ($\mathscr{M} = \mathbf{0}$) in electrostatics. In this case, the only relevant balance laws that survive are (6.2.1), (6.2.2) and the corresponding jump conditions (6.2.5) and (6.2.6), i.e.,

$$\nabla \cdot \mathbf{D} = q_e \quad \text{in } \mathscr{V}, \tag{6.9.1}$$

$$\nabla \times \mathbf{E} = 0 \quad \text{in } \mathscr{V}, \tag{6.9.2}$$

$$[\mathbf{D}] \cdot \mathbf{n} = w_e - \nabla_e \mathfrak{k} \quad \text{on } \partial\mathscr{V} - c_\sigma; \qquad [\mathbf{b} \cdot \mathfrak{k}] = 0, \quad \text{on } c_\sigma, \tag{6.9.3}$$

$$\mathbf{n} \times [\mathbf{E}] = \mathbf{0} \quad \text{on } \partial\mathscr{V}. \tag{6.9.4}$$

For the constitutive equations from (6.2.13) we have

$$\mathbf{P} = \chi \mathbf{E}, \tag{6.9.5}$$

where, for brevity, we drop the suffix E from χ^{E} and write

$$\chi \equiv \chi^{\mathrm{E}} = -2\frac{\partial \Sigma}{\partial J_{\mathrm{E}}}, \qquad J_{\mathrm{E}} \equiv \mathbf{E} \cdot \mathbf{E}. \tag{6.9.6}$$

Introducing the electrostatic potential $\phi(\mathbf{x})$ with

$$\mathbf{E} = -\nabla\phi \tag{6.9.7}$$

we will have (6.9.2) satisfied identically and

$$\mathbf{D} = \mathbf{E} + \mathbf{P} = -(1 + \chi)\nabla\phi. \tag{6.9.8}$$

Substituting this into (6.9.1) we obtain

$$a_{kl}\phi_{,kl} = -q/(1 + \chi), \tag{6.9.9}$$

where

$$a_{kl} \equiv \delta_{kl} + \frac{2\chi'}{1 + \chi}\phi_{,k}\phi_{,l}, \qquad \chi' \equiv \frac{d\chi}{dJ_{\mathrm{E}}}. \tag{6.9.10}$$

The determination of the electrostatic potential, therefore, requires that we solve the nonlinear partial differential equation (6.9.9) subject to the boundary conditions (6.9.3) and (6.9.4):

$$\begin{aligned} &[(1 + \chi)\nabla\phi] \cdot \mathbf{n} = -w_e + \nabla_\sigma \cdot \mathfrak{k} \quad \text{on } \partial\mathscr{V} - c_\sigma, \\ &[\mathbf{n} \times \nabla\phi] = \mathbf{0} \quad \text{on } \partial\mathscr{V}, \\ &[\mathbf{b} \cdot \mathfrak{k}] = 0 \quad \text{on } c_\sigma. \end{aligned} \tag{6.9.11}$$

In general, the solution to this problem has not been found to date, except in special situations (e.g., $\chi = \text{const.}$, the linear theory, see also Section 6.12). The character of the solution, of course, depends on the nature of the differential equation. This, in turn, is related to properties of the matrix a_{kl}. Let Q represent the quadratic form

$$Q = a_{kl}\xi_k\xi_l. \tag{6.9.12}$$

If $Q > 0$ for any vector $\boldsymbol{\xi}$, then (6.9.9) is *elliptic*; if $Q < 0$, it is *hyperbolic*; and if $Q = 0$, it is *parabolic*. We can reduce the quadratic form (6.9.12) into the *normal form* by finding the eigenvalues of a_{kl}. These are the roots of

$$\det(\mathbf{a} - \lambda\mathbf{1}) = 0.$$

The three roots of this equation are

$$\lambda_1 = \lambda_2 = 1, \qquad \lambda_3 = 1 + 2J_E\chi'/(1 + \chi). \tag{6.9.13}$$

Thus, for $\lambda_3 > 0$, (6.9.9) is elliptic; for $\lambda_3 < 0$, it is hyperbolic; and when $\lambda = 0$, it is parabolic. Since χ is a function of $J_E = \phi_{,k}\phi_{,k}$, the nature of (6.9.9) may change depending on the solution $\phi(x)$. λ_3 may be written in the form

$$\lambda_3 = (\ln v)', \qquad v = (1 + \chi)^2 J_E > 0. \tag{6.9.14}$$

Since $J_E > 0$ for $\phi_{,k} \neq 0$, we see that the sign of $v' = dv/dJ_E$ determines the nature of (6.9.9). We first note that $\lambda_3 = 0$ is not a possibility. Suppose that $\lambda_3 = 0$, then

$$(1 + \chi)^2 J_E = \text{const} > 0. \tag{6.9.15}$$

This implies that $\nabla\phi$ is fully determined without reference to the differential equation (6.9.9) and boundary conditions (6.9.11). Equation (6.9.15) cannot be satisfied even for special constitutive functions χ. For a real material, χ is bounded for all J_E. But (6.9.15) implies that for $J_E = 0$, $\chi \to \infty$. Hence, a continuous electric field cannot connect the two regions $\lambda_3 > 0$ and $\lambda_3 < 0$. Assuming that $\chi'/(1 + \chi)$ is bounded for all J_E, we have $\lambda_3 = 1$ for $J_E = 0$. Since, for J_E close to zero, (6.3.10) is elliptic, and by the reasoning that no continuous E-field can span both regions, we see that $\lambda_3 > 0$ for all J_E (in the case of real, rigid dielectrics) and (6.9.9) is elliptic in the entire domain.

For such quasi-linear elliptic equations the solution is known to be unique.

Theorem (Uniqueness). *If* $\phi(\mathbf{x})$ *is a solution of* (6.9.9) *such that the corresponding matrix* a_{kl} *is positive definite everywhere in* V, *then any solution* $\psi(\mathbf{x})$ *of* (6.9.9) *that agrees with* $\phi(x)$ *on the boundary* ∂V *of* V *equals* $\phi(\mathbf{x})$ *throughout* V.

For proof of this theorem see Courant [1965, p. 323].

The determination of the solution of (6.9.9) in this generality is not possible, except for special material functions. Nevertheless, a perturbation scheme can be set up which converges to the solution and (6.9.11). We consider the case

$\hat{\pi} = \mathbf{0}$ and write

$$\begin{aligned} \phi &= \phi_0 + \varepsilon\phi_1 + \varepsilon^2\phi_2 + \cdots, \\ \chi &= \chi_0 + \varepsilon\chi_1 J_E + \varepsilon^2\chi_2 J_E^2 + \cdots, \\ J_E &= \phi_{,k}\phi_{,k}. \end{aligned} \tag{6.9.16}$$

Upon substituting these into (6.9.9) and (6.9.11), and equating the coefficients of equal powers of ε from both sides, we obtain a hierarchy of equations of the forms

$$\begin{aligned} \varepsilon^0: \qquad \nabla^2\phi_0 &= -q_e/(1+\chi_0) \quad \text{in } \mathscr{V}, \\ [\nabla\phi_0]\cdot\mathbf{n} &= -w_e/(1+\chi_0) \quad \text{on } \partial\mathscr{V}, \end{aligned} \tag{6.9.17}$$

$$\begin{aligned} \varepsilon^1: \quad \nabla^2\phi_1 &= (1+\chi_0)^{-2}\chi_1\phi_{0,k}\phi_{0,k}q_e - (1+\chi_0)^{-1}2\chi_1\phi_{0,k}\phi_{0,l}\phi_{0,kl} \\ &\equiv Q_0[\phi_0] \quad \text{in } \mathscr{V}, \\ [\nabla\phi_1]\cdot\mathbf{n} &= -(1+\chi_0)w_e - (1+\chi_0)^{-1}\chi_1[\nabla\phi_0]\cdot\mathbf{n} \\ &\equiv W_0[\phi_0] \quad \text{on } \partial\mathscr{V}, \end{aligned} \tag{6.9.18}$$

....

$$\begin{aligned} \varepsilon^m: \qquad \nabla^2\phi_m &= -Q_m \quad \text{in } \mathscr{V}, \\ [\nabla\phi_m]\cdot\mathbf{n} &= -W_m \quad \text{on } \partial\mathscr{V}, \end{aligned} \tag{6.9.19}$$

where Q_{m-1} and W_{m-1} are all expressible as functions of ϕ_{m-1}.

The solutions to (6.9.17) and (6.9.19) follow from the potential theory for linear elliptic equations, i.e.,

$$\phi_0(\mathbf{x}) = \int_{\mathscr{V}} G(\mathbf{x}, \mathbf{x}')\frac{q_e(\mathbf{x}')}{1+\chi_0}\,dv(\mathbf{x}') - \int_{\partial\mathscr{V}} G(\mathbf{x}, \mathbf{x}')\frac{w_e(\mathbf{x}')}{1+\chi_0}\,da(\mathbf{x}'), \tag{6.9.20}$$

$$\phi_m(\mathbf{x}) = \int_{\mathscr{V}} G(\mathbf{x}, \mathbf{x}')Q_{m-1}\,dv(\mathbf{x}') - \int_{\partial\mathscr{V}} G(\mathbf{x}, \mathbf{x}')w_{m-1}\,da(\mathbf{x}'). \tag{6.9.21}$$

It follows from the Schauder theory (Courant [1965, Chap. IV]) that the solution ϕ_m exists and converges to the solution $\phi(\mathbf{x})$ of (6.9.9). The evaluation of the improper integrals (6.9.20), (6.9.21) and a detailed study of the question of convergence are important problems encountered in potential theory (see Kellog [1929]). The calculation of the electric field from the solution requires calculation of the gradient which needs special care, since at $\mathbf{x} = \mathbf{x}'$, the integrands are singular.

6.10. One-Dimensional Problem

In several technically important cases the electric field varies only in one direction (e.g., a charged slab, parallel plates). Thus, if the electric potential ϕ

depends on only one variable y, then (6.9.9) reduces to

$$\left[1+\chi+2\chi'\left(\frac{d\phi}{dy}\right)^2\right]\frac{d^2\phi}{dy^2}=-q_e(y). \tag{6.10.1}$$

Since χ is a function of $(d\phi/dy)^2$ only we can write this equation as

$$(1+\chi+2\chi'\Phi^2)\frac{d\Phi}{dy}=-q_e(y), \qquad \Phi=\frac{d\phi}{dy}.$$

The integration of this gives

$$\int^{\Phi}(1+\chi+2\chi'\Phi^2)\,d\Phi+C_1=-\int^{y}q_e(y)\,dy. \tag{6.10.2}$$

From this equation we determine $\Phi=\Phi(y)$, and then integrate $d\phi/dy=\Phi(y)$ to obtain the potential $\phi(y)$. Two constants of integration are then determined by use of the boundary condition

$$[(1+\chi)\Phi]=-w_e. \tag{6.10.3}$$

We illustrate the situation for the case in which

$$\chi=\chi_0+\chi_1\left(\frac{d\phi}{dy}\right)^2, \qquad q_e(y)=q_0\,y/h, \tag{6.10.4}$$

where χ_0, χ_1, and q_0 are constants. This corresponds to a rigid slab with quadratic permittivity. On the surfaces of the slab $y=\pm h$ there exist charge densities $\pm w_e$ (Figure 6.10.1). In this case, (6.10.2) gives

$$\Phi^3-3p\Phi-2q=0, \tag{6.10.5}$$

where

$$p=-(1+\chi_0)/3\chi_1, \qquad q=-(2C_1h+q_0y^2)/4\chi_1h. \tag{6.10.6}$$

From energy considerations we have

$$P\cdot E\geq 0\rightarrow(\chi_0+\chi_1\Phi^2)\Phi^2\geq 0. \tag{6.10.7}$$

It follows that $\chi_0\geq 0$. If $\chi_1\geq 0$, then the free energy is nonnegative for all Φ. However, in most materials $\chi_1\leq 0$. In this case the quadratic theory will apply

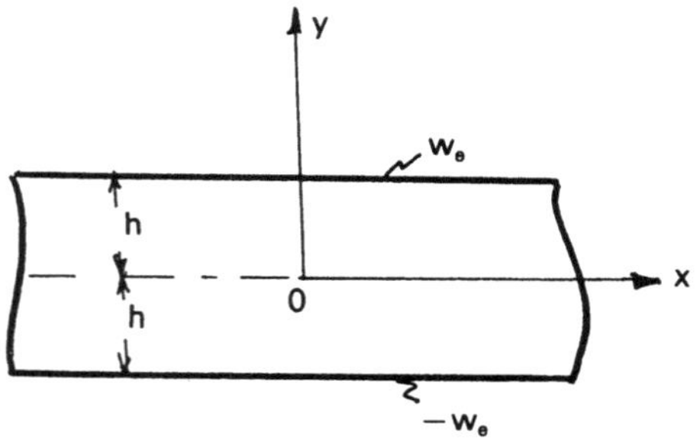

Figure 6.10.1. Slab with surface charges.

only in the region where

$$\Phi^2 \leq -\chi_0/\chi_1 \equiv \Phi_M^2. \tag{6.10.8}$$

Thus, two possibilities exist.

(a) *Hard Polarizers* ($\chi_1 \geq 0$): Hard polarizers possess a concave P-E curve. For these materials $p < 0$, and therefore the cubic equation (6.10.5) possesses a single real root given by

$$\Phi = 2\sqrt{-p}\sinh(\eta/3), \tag{6.10.9}$$

when η is determined from

$$\sinh\eta = q/(-p^3)^{1/2}. \tag{6.10.10}$$

(b) *Soft Polarizers* ($\chi_1 \leq 0$): Soft polarizers possess a convex P-E curve and, of course, the solution is valid in the region $0 < \Phi^2 < \Phi_M^2$. In this case $p > 0$, and an examination of $q^2 - p^3$ shows that $q^2 > p^3$ for all y. Thus, again, the cubic (6.10.5) possesses a single root, given by

$$\Phi = 2\,\mathrm{sgn}(q)\sqrt{p}\cosh(\eta/3), \tag{6.10.11}$$

where η is determined from

$$\cosh\eta = \mathrm{sgn}(q)|q|/p^{3/2}. \tag{6.10.12}$$

It now remains to determine the integration constant C_1 through the boundary condition (6.10.3), i.e.,

$$[1 + \chi_0 + \chi_1\Phi^2(h)]\Phi(h) = w_e. \tag{6.10.13}$$

Using (6.10.9) or (6.10.10) in this equation we obtain a transcendental equation for the determination of C_1. Of course, the electric field is given by

$$\mathbf{E} = (0, -\Phi, 0), \tag{6.10.14}$$

so that the solution is complete, unless we wish to determine the potential $\Phi(y)$. In that case, one more quadrature is necessary. This involves integrating $\Phi = d\phi/dy$.

6.11. Two-Dimensional Nonlinear Problem

In the case where the electric field does not depend on one of the coordinates, the nonlinear problem can be simplified considerably. Suppose that the field and charge are independent of x_3 in a rectangular frame of reference x_k, $k = 1, 2, 3$. In this case, we write $x_1 = x$, $x_2 = y$, and by employing the Legendre transformation

$$\xi = \phi_{,x}, \qquad \eta = \phi_{,y}, \qquad \psi = \xi x + \eta y - \phi, \tag{6.11.1}$$

we can linearize (6.9.9). From (6.11.1) it follows that

$$\psi_{,\xi} = x \qquad \psi_{,\eta} = y. \tag{6.11.2}$$

Differentiating $(6.11.1)_1$ and $(6.11.1)_2$, once with respect to ξ and once with respect to η each, we solve for $\phi_{,kl}$ leading to

$$\phi_{,xx} = \Delta^{-1}\psi_{,\eta\eta}, \qquad \phi_{,yy} = \Delta^{-1}\psi_{,\xi\xi}, \qquad \phi_{,xy} = -\Delta^{-1}\psi_{,\xi\eta}. \tag{6.11.3}$$

Substituting (6.11.1)–(6.11.3) into (6.9.9) we obtain

$$\left(1 + \frac{2\chi'}{1+\chi}\eta^2\right)\psi_{,\xi\xi} - \frac{4\chi'}{1+\chi}\xi\eta\psi_{,\xi\eta} + \left(1 + \frac{2\chi'}{1+\chi}\xi^2\right)\psi_{,\eta\eta}$$
$$= -\frac{q_e}{1+\chi}[\psi_{,\xi\xi}\psi_{,\eta\eta} - (\psi_{,\xi\eta})^2]. \tag{6.11.4}$$

For $q_e = 0$ this equation is linear. The boundary condition for the linearized problem ((6.11.4) with $q_e = 0$) is

$$[(1+\chi)\xi]n_1 + [(1+\chi)\eta]n_2 = -w_e \quad \text{on } \partial\mathscr{V}, \tag{6.11.5}$$

where n_1 and n_2 are functions of x and y depending on the boundary curve. Hence, they depend on $\psi_{,\xi}$ and $\psi_{,\eta}$, i.e.,

$$n_1 = n_1(\psi_{,\xi}, \psi_{,\eta}), \qquad n_2 = n_2(\psi_{,\xi}, \psi_{,\eta}). \tag{6.11.6}$$

Thus, the electrostatic problem is reduced to the solution of (6.11.4) within a region in the *hodograph plane* (ξ, η). The boundary curve of this region has the unit normal given by (6.11.6). We therefore see that, for $q_e = 0$, the Legendre transformation simplifies the differential equations (by reducing it to a linear equation). However, the boundary condition, in general, is complicated and still depends on the solution sought. Hence, the problem is still nonlinear. Nevertheless, the general theory for linear partial differential equations is well developed, so that if a general solution of (6.11.4) is found the arbitrary functions of integration can, in principle, be determined from (6.11.5). In the hodograph plane we can obtain the normal form of the elliptic equation (6.11.4) (with $q_e = 0$). This requires the solution of the system of two first-order equations (called Beltrami equations)

$$\sigma_{,\xi} = \frac{1}{\Delta}\left[-\frac{2\chi'}{1+\chi}\xi\eta\rho_{,\xi} + \left(1 + \frac{2\chi'}{1+\chi}\xi^2\right)\rho_{,\eta}\right],$$
$$\sigma_{,\eta} = -\frac{1}{\Delta}\left[\left(1 + \frac{2\chi'}{1+\chi}\eta^2\right)\rho_{,\xi} - \frac{2\chi'}{1+\chi}\xi\eta\rho_{,\eta}\right], \tag{6.11.7}$$
$$\Delta = 1 + \frac{2\chi'}{1+\chi}J_E.$$

Under certain smoothness conditions [existence of second-order continuous derivatives of the coefficients of (6.11.4) ($q_e = 0$)], the solution of (6.11.7),

$$\sigma = \sigma(\xi, \eta), \qquad \rho = \rho(\xi, \eta), \tag{6.11.8}$$

is known to exist with nonvanishing gradients. Once this solution is found,

(6.11.4) can be written in the normal form

$$\psi_{,\rho\rho} + \psi_{,\sigma\sigma} = 0. \tag{6.11.9}$$

Thus, in the curvilinear coordinates (ρ, σ) in the hodograph plane, the solution of (6.11.9) is an analytic function $F(\zeta)$. Hence,

$$\psi = \operatorname{Re} F(\zeta), \qquad \zeta = \rho + i\sigma. \tag{6.11.10}$$

With the powerful apparatus of the analytic function theory, progress towards the solution of the boundary-value problem is now more likely.

6.12. Solutions of Two-Dimensional Problems for Special Dielectrics

The method of solution presented in Section 6.11 may be employed to obtain the solution of a class of two-dimensional problems in which the constitutive relation $\mathbf{P} = \mathbf{P}(\mathbf{E})$ is of some special forms. These forms are obtained by stipulating that a certain series solution for the potential terminates. Such special solutions contain some parameters which are adjustable for curve fitting purposes, so that the solutions so obtained can have practical applications. Such solutions proved to be useful in other fields of continuum mechanics (see Eringen and Suhubi [1974, Sect. 3.9]) and were pursued by various authors (see Kazakia and Venkataraman [1975], Rogers et al. [1977]). The method pursued here has its parallel in gas dynamics, as discussed by many authors beginning with Chaplygin and others (see Bergmann [1971]).

We express the constitutive equation as

$$\mathbf{D} = \varepsilon_0 k(E^2)\mathbf{E}. \tag{6.12.1}$$

For a charge-free dielectric we have

$$\nabla \cdot \mathbf{D} = 0, \qquad \mathbf{E} = -\nabla\phi. \tag{6.12.2}$$

In two dimensions $(6.12.2)_1$ is satisfied by

$$k(E^2)E_x = +\frac{\partial\psi}{\partial y}, \qquad k(E^2)E_y = -\frac{\partial\psi}{\partial x}, \tag{6.12.3}$$

where ψ is an arbitrary function of x, y.

Using $(6.12.2)_2$ we therefore have

$$\begin{aligned} \psi_{,x} &= k\phi_{,y} = -kE\sin\theta, \\ \psi_{,y} &= -k\phi_{,x} = kE\cos\theta, \end{aligned} \tag{6.12.4}$$

where we wrote $E_x = E\cos\theta$ and $E_y = E\sin\theta$. The hodograph transformation is made by considering

$$x + iy = z(E, \theta). \tag{6.12.5}$$

Using (6.12.4) we obtain

$$\begin{aligned} z_{,E} &= E^{-1}(-\phi_{,E} + ik^{-1}\psi_{,E})e^{i\theta}, \\ z_{,\theta} &= E^{-1}(-\phi_{,\theta} + ik^{-1}\psi_{,\theta})e^{i\theta}. \end{aligned} \tag{6.12.6}$$

The integrability condition $z_{,E\theta} = z_{,\theta E}$ leads to the linear system

$$\psi_{,E} = -kE^{-1}\phi_{,\theta}, \qquad \phi_{,E} = -E\frac{d}{dE}(k^{-1}E^{-1})\psi_{,\theta}. \tag{6.12.7}$$

Introducing

$$E^* = \int_{E_0}^{E} k(\sigma^2)\sigma^{-1}\Lambda(\sigma)\, d\sigma, \tag{6.12.8}$$

where E_0 is an appropriate reference field strength and

$$\Lambda = Ek^{-1}\left[-k\frac{d}{dE}(E^{-1}k^{-1})\right]^{1/2}, \tag{6.12.9}$$

we can write (6.12.7) as

$$\psi_{,E^*} = -\Lambda^{-1}\phi_{,\theta}, \qquad \psi_{,\theta} = \Lambda^{-1}\phi_{,E^*}. \tag{6.12.10}$$

Eliminating ψ between these two equations we obtain

$$\overset{*}{\nabla}{}^2\phi - \Lambda'\Lambda^{-1}\phi_{,E^*} = 0, \tag{6.12.11}$$

where

$$\overset{*}{\nabla}{}^2\phi = \frac{\partial^2\phi}{\partial \overset{*}{E}{}^2} + \frac{\partial^2\phi}{\partial\theta^2}, \qquad \Lambda' = \frac{\partial\Lambda}{\partial E^*}. \tag{6.12.12}$$

By writing

$$\phi^* = \Lambda^{-1/2}\phi, \qquad \Omega = -\Lambda^{1/2}(\Lambda^{-1/2})'', \tag{6.12.13}$$

we may transform (6.12.11) to

$$\nabla^{*2}\phi^* + \Omega\phi^* = 0. \tag{6.12.14}$$

The solution of this equation may be sought in the form of

$$\phi^* = \sum_{n=0}^{\infty} e_n(E^*)F_n(\zeta), \qquad \zeta = E^* + i\theta, \tag{6.12.15}$$

where $F_n(\zeta)$ is a harmonic function. Equation (6.12.14) is satisfied if

$$\begin{aligned} e'_{n+1} &= e''_n + \Omega e_n, \qquad n = 0, 1, 2, \ldots, \\ e_0 &= 1, \\ 2\frac{\partial F_n}{\partial E^*} &+ F_{n-1} = 0, \end{aligned} \tag{6.12.16}$$

where F_0 is an arbitrary harmonic function. Thus, we may represent the solution for F_n as

$$F_n = \mathrm{Re}\{\Phi_n\}, \qquad \Phi_n(\zeta) = -\frac{1}{2}\int_0^{\zeta}\Phi_{n-1}(t)\, dt. \tag{6.12.17}$$

By repeated integration this gives

$$F_n(E^*, \theta) = \frac{(-)^n}{(n-1)!\,2^n} \operatorname{Re} \int_0^\zeta \Phi_0(t)(\zeta - t)^{n-1}\,dt. \tag{6.12.18}$$

Consequently,

$$\phi^*(E^*, \theta) = F_0(E^*, \theta) + \operatorname{Re} \int_0^\zeta \phi_0(t) K(E^*, \zeta - t)\,dt, \tag{6.12.19}$$

where

$$K(E^*, \zeta - t) = \sum_{n=1}^{\infty} \frac{(-)^n e_n(E^*)(\zeta - t)^{n-1}}{2^n (n-1)!}. \tag{6.12.20}$$

The solution is formally complete, provided (6.12.20) converges uniformly in a complex t-domain. The arbitrary harmonic function $F_0(E^*, \theta)$ can be determined to satisfy the boundary conditions. As pointed out in Section 6.11 this may still be a difficult task, since the boundary conditions in the physical domain generally become complicated in the hodograph plane. Alternatively, we may express ϕ as a function of x and y by transforming variables. This process may not be simple to achieve explicitly.

The function $e_n(E^*)$ is not determined as yet, since Ω is a complicated function of E with $k(E^2)$ not specified. By imposing a condition for the termination of the series in (6.12.15) we can determine special forms for $k(E^2)$, which can be used in dealings with certain classes of nonlinear dielectric solids. To this end, we set

$$e_{N+r} = 0, \qquad r = 1, 2, \ldots.$$

From (6.12.16) it then follows that

$$e_N = \Lambda^{-1/2}\left(-2\bar{\alpha}_1 + 2\bar{\beta}_1 \int \Lambda\, dE^*\right), \tag{6.12.21}$$

where $\bar{\alpha}_1$ and $\bar{\beta}_1$ are constants of integration. The case where $\bar{\beta}_1 = 0$ and $N = 1$ was studied by Rogers et al [1977]. In this simple case the only remaining equation is for $n = 0$

$$(\Lambda^{1/2})' + \alpha_2 \Lambda + \alpha_1 = 0, \tag{6.12.22}$$

where α_2 is an integration constant. Thus, Λ is determined to be

$$\begin{aligned}
\Lambda^{1/2} &= (\alpha_2 E^* + \beta_2)^{-1}, && \text{if } \alpha_1 = 0,\\
\Lambda^{1/2} &= -\alpha_1 E^* + \beta_3, && \text{if } \alpha_2 = 0,\\
\Lambda^{1/2} &= (\alpha_1/\alpha_2)^{1/2} \cot[(\alpha_1/\alpha_2)^{1/2}(\alpha_1 E^* + \beta_3)], && \text{if } \alpha_1/\alpha_2 > 0,\\
\Lambda^{1/2} &= (-\alpha_1/\alpha_2)^{1/2} \tanh[(-\alpha_1/\alpha_2)^{1/2}(\alpha_2 E^* + \beta_4)], && \text{if } \alpha_1/\alpha_2 < 0,
\end{aligned} \tag{6.12.23}$$

where β_2 to β_4 are integration constants. By use of the expression (6.12.9) of Λ and (6.12.8) of E^*, it is now possible (in principle) to obtain expressions of $k(E^2)$ corresponding to each case listed in (6.12.23), and therefore a $\mathbf{P} - \mathbf{E}$ relationship. To this end, by differentiating (6.12.8) with respect to E and using

(6.12.9), we obtain the system of differential equations

$$\frac{dE}{dE^*} = \frac{E}{\Lambda k}, \qquad \frac{dk}{dE^*} = \frac{\Lambda^2 k^2 - 1}{\Lambda}. \tag{6.12.24}$$

The integrations of these equations determine E and k as functions of E^*. Upon elimination of E^*, the function $k(E^2)$ is determined. The explicit solutions of these equations are found for the first two cases listed in (6.12.23). The remaining two cases may be done numerically. The result for the first case is

$$\begin{aligned} k &= -E^{-1}(\alpha_2 E^* + \beta_2)(ae^{E^*} + be^{-E^*})^{-1}, \\ E &= [\alpha_2(ae^{E^*} + be^{E^*}) - (\alpha_2 E^* + \beta_2)(ae^{E^*} - be^{-E^*})]^{-1}, \end{aligned} \tag{6.12.25}$$

where a and b are constants of integration. For the second case ($\alpha_2 = 0$) we have the parametric form

$$\begin{aligned} k &= -E^{-1}[-\alpha_1(ae^{E^*} + be^{-E^*}) - (-\alpha_1 E^* + \beta_3)(ae^{E^*} - be^{-E^*})]^{-1}, \\ E &= (-\alpha_1 E^* + \beta_3)(ae^{E^*} + be^{-E^*})^{-1}, \end{aligned} \tag{6.12.26}$$

where a and b are constants of integration.

A uniformly convergent series representation of the potential ϕ is possible for a special Ω, which, in the special case, reduces to the first case. Thus, consider the special form

$$\Omega = C(\beta_2 - E^*)^{-2}, \qquad C > 0, \quad \beta_2 < 0. \tag{6.12.27}$$

In this case, the solution of $(6.12.16)_1$ is given by

$$e_n = n!\, \mu_n(\beta_2 - E^*)^{-n}, \tag{6.12.28}$$

where μ_n is determined by the following recurrence relation:

$$\mu_0 = 1, \qquad \mu_{n+1} = \mu_n(n + \alpha)(n + \beta)(n + 1)^{-1}(n + \gamma)^{-1}, \tag{6.12.29}$$

in which

$$\begin{gathered} \alpha = \tfrac{1}{2} - (\tfrac{1}{4} - C)^{1/2}, \qquad \beta = \tfrac{1}{2} + (\tfrac{1}{4} - C)^{1/2}, \\ \gamma = 1, \qquad 0 < C \leq \tfrac{1}{4}. \end{gathered} \tag{6.12.30}$$

The above recurrence relation is the same as the one that prevails for the hypergeometric series

$$H(\alpha, \beta, \gamma; x) = \sum_{n=0}^{\infty} \mu_n x^n. \tag{6.12.31}$$

Under the conditions of (6.12.30), this series is uniformly convergent for $|x| < 1$. But

$$K(E^*, \zeta - t) = \tfrac{1}{2}(\beta_2 - E^*)^{-1} H'\left(\alpha, \beta, 1; \frac{1}{2}\left|\frac{\zeta - t}{\beta_2 - E^*}\right|\right). \tag{6.12.32}$$

Thus, a uniformly convergent potential is assured in the region

$$\frac{1}{2}\left|\frac{E^* + i\theta - t}{\beta_2 - E^*}\right| < 1. \tag{6.12.33}$$

Using (6.12.27) and (6.12.13) we obtain

$$\Lambda^{-1/2} = \gamma_1|\beta_2 - E^*|^\alpha + \gamma_2|\beta_2 - E^*|^\beta, \tag{6.12.34}$$

where γ_1 and γ_2 are constants of integration.

In the special case $C = 0$, $\alpha = 0$, $\beta = 1$, we have the first case listed in (6.12.23). The electric potential ϕ now is given by

$$\phi = (\gamma_1|\beta_2 - E^*|^\alpha + \gamma_2|\beta_2 - E^*|^\beta)^{-1}$$
$$\times\left\{F_0(E^*, \theta) + \operatorname{Re}\int_0^\zeta\left[1 + \tfrac{1}{2}(\beta_2 - E^*)^{-1}H'\left(\alpha, \beta, 1; \frac{1}{2}\left|\frac{\zeta - t}{\beta_2 - E^*}\right|\right)\right]\phi_0(t)\,dt\right\}. \tag{6.12.35}$$

Since Re $\phi_0(\zeta)$ corresponds to the potential for the linear theory, (6.12.35) links the linear theory to the nonlinear case corresponding to a special constitutive $\mathbf{P} - \mathbf{E}$ relation inherent in the phenomenological relation corresponding to (6.12.34).

6.13. Nonlinear Electromagnetic Waves

The basic tool of nonlinear optics is the Fourier analysis (see Bloembergen [1965] and Zernike and Midwinter [1973]). The method of Fourier analysis, while powerful in the linear theory, precludes a discussion of harmonics higher than the first few, and relies heavily on the assumption of slowly varying amplitudes (see Section 6.16). A few problems have been studied by means of the theory of characteristics and hodograph transformations. These are concerned with nonconducting media and problems that are of a one-dimensional nature (see Kazakia and Vankataraman [1975] and Rogers et al. [1977]). The method of analysis, based on the theory of characteristics, is similar to the one used in the discussion of one-dimensional wave propagation problems (see Eringen and Suhubi [1975]). The hodograph transformation, on the other hand, has already been used in Section 6.12. For wave propagation problems, the same method, together with the linear-operator method of Bergman, can give some exact solutions. For special classes of nonlinear dielectrics, results may be used to test the nature of approximate methods. Moreover, certain nonlinear dielectric constitutive equations are constructed which, by curve-fitting, can represent the real materials within a reasonably wide range of fields.

Suppose that electromagnetic waves are polarized in the $x_3 = $ const. plane and propagate in the x_3-direction with amplitudes which depend on x_3 and t

only, i.e.,

$$\mathbf{E} = E_1(x_3, t)\mathbf{e}_1, \qquad \mathbf{H} = H_2(x_3, t)\mathbf{e}_2,$$
$$\mathbf{D} = D_1(x_3, t)\mathbf{e}_1, \qquad \mathbf{B} = B_2(x_3, t)\mathbf{e}_2, \tag{6.13.1}$$

where $\mathbf{e}_1$ and $\mathbf{e}_2$ are unit vectors in the x_1- and x_2-directions, respectively.

The constitutive equations are of the form

$$D_1 = D(E_1), \qquad B_2 = B(H_2). \tag{6.13.2}$$

Substituting these into Maxwell's equations (6.2.2) and (6.2.4), we obtain

$$\frac{\partial E}{\partial x} + B'(H)\frac{\partial H}{\partial t} = 0, \qquad \frac{\partial H}{\partial x} + D'(E)\frac{\partial E}{\partial t} = 0, \tag{6.13.3}$$

where $B'(H) \equiv dB/dH$, etc., and where we drop the subscripts from E_1, H_2 and x_3. If E and H are considered as independent variables and x and t as dependent variables, then

$$E_{,x} = Jt_{,H}, \qquad E_{,t} = -Jx_{,H},$$
$$H_{,x} = -Jt_{,E}, \qquad H_{,t} = Jx_{,E}, \qquad J \equiv E_{,x}H_{,t} - E_{,t}H_{,x}, \tag{6.13.4}$$

where $0 < |J| < \infty$. Equations $(6.13.3)_1$, $(6.13.3)_2$ now read

$$x_{,H} = -t_{,E}/D'(E), \qquad t_{,H} = -x_{,E}B'(H). \tag{6.13.5}$$

Introducing

$$h = \int_{H_0}^{H} [B'(\zeta)]^{1/2}\, d\zeta, \qquad e = \int_{E_0}^{E} [D'(\zeta)]^{1/2}\, d\zeta,$$
$$A(e, h) = [B'(H)D'(E)]^{-1/2}, \qquad t^* = A^{1/2}t, \qquad x^* = A^{-1/2}x, \tag{6.13.6}$$

we obtain the linear set

$$\Box t^* - \{\Box A^{1/2}/A^{1/2}\}t^* = 0,$$
$$\Box x^* - \{A^{1/2}\Box A^{-1/2}\}x^* = 0, \tag{6.13.7}$$

where $\Box = \partial^2/\partial e^2 - \partial^2/\partial h^2$ and we assumed that $A(e, h) \neq 0$.

Bergmann [1971] discusses the solution of such equations by means of a linear integral-operator technique. To this end, we introduce the Riemann invariants

$$r = \tfrac{1}{2}(e + h), \qquad s = \tfrac{1}{2}(e - h). \tag{6.13.8}$$

By virtue of (6.13.6), (6.13.3) becomes

$$\frac{\partial h}{\partial t} = -A(e, h)\frac{\partial e}{\partial x}, \qquad \frac{\partial e}{\partial t} = -A(e, h)\frac{\partial h}{\partial x}. \tag{6.13.9}$$

Consequently, r is constant along the right propagating characteristics $\alpha(x, t) = \text{const.}$ defined by

$$\frac{dx}{dt} = A \quad \text{along } \alpha = \text{const.}, \tag{6.13.10}$$

and s is constant along the left propagating characteristics $\beta(x, t) = \text{const.}$ defined by

$$\frac{dx}{dt} = -A \quad \text{along } \beta = \text{const.} \tag{6.13.11}$$

Thus, we have

$$\begin{aligned} r &= F(\alpha), & s &= G(\beta), \\ e &= F(\alpha) + G(\beta), & h &= F(\alpha) - G(\beta). \end{aligned} \tag{6.13.12}$$

Sustituting (6.13.8) into (6.13.7) we will have

$$\begin{aligned} &\frac{\partial^2 t^*}{\partial r\,\partial s} - \left(A^{-1/2}\frac{\partial^2 A^{1/2}}{\partial r\,\partial s}\right)t^* = 0, \\ &\frac{\partial^2 x^*}{\partial r\,\partial s} - \left(A^{1/2}\frac{\partial^2 A^{-1/2}}{\partial r\,\partial s}\right)x^* = 0. \end{aligned} \tag{6.13.13}$$

These equations have the common form

$$\frac{\partial^2 f}{\partial r\,\partial s} + \lambda f = 0. \tag{6.13.14}$$

Once $\lambda(e, h) = \lambda(r + s, r - s)$ is specified, this may be integrated. To this end, the solution may be sought in the form

$$f = \sum_{n=0}^{\infty} [\alpha_n(r, s)\Phi_n(r) + \beta_n(r, s)\Psi_n(s)], \tag{6.13.15}$$

where α_n and β_n are to be determined and Φ_n and Ψ_n satisfy the recurrence relations

$$\frac{1}{2}\frac{d\Phi_n}{dr} = \Phi_{n-1}, \qquad \frac{1}{2}\frac{d\Psi_n}{ds} = \Psi_{n-1}. \tag{6.13.16}$$

Substitution of (6.13.15) into (6.13.14) now gives

$$\begin{aligned} &\frac{\partial\beta_0}{\partial r} = 0, \qquad \frac{\partial\alpha_0}{\partial s} = 0, \qquad \alpha_0\Phi_0 + \beta_0\Psi_0 = 0, \\ &2\frac{\partial\beta_n}{\partial r} + \frac{\partial^2\beta_{n-1}}{\partial r\,\partial s} + \lambda\beta_{n-1} = 0, \qquad 2\frac{\partial\alpha_n}{\partial s} + \frac{\partial^2\alpha_{n-1}}{\partial r\,\partial s} + \lambda\alpha_{n-1} = 0. \end{aligned} \tag{6.13.17}$$

When λ is specified, these equations can be integrated recursively. The recurrence relations (6.13.16) are satisfied by

$$\begin{aligned} \Phi_n &= \frac{1}{n!}\int_0^{2r} (2r - \xi)^n \Phi_0'(\xi)\,d\xi, \\ \Psi_n &= \frac{1}{n!}\int_0^{2s} (2s - \zeta)^n \Psi_0'(\zeta)\,d\zeta, \end{aligned} \tag{6.13.18}$$

where $\Phi_0(\xi)$ and $\Psi_0(\zeta)$ are two arbitrary functions to be determined from the initial conditions. Thus, the general solution for f is given by

$$f = \sum_{n=0}^{\infty} \frac{1}{n!}\left[\alpha_n \int_0^{2r} (2r-\xi)^n \Phi_0'(\xi)\, d\xi + \beta_n \int_0^{2s} (2s-\zeta)^n \Psi_0'(\zeta)\, d\zeta\right]. \tag{6.13.19}$$

The formal solution is now complete. The investigation of the convergence of this solution, the determination of α_n and β_n by solving (6.13.17), and finally the determination of the arbitrary functions Φ_0 and Ψ_0 are formidable for arbitrary nonlinear dielectrics. Rogers et al. [1977] investigated in some detail the case $\lambda = 0$. This led to constitutive equations for a class of nonlinear dielectrics for which an exact solution may be found. Among these are the dielectric models previously discussed by Kazakia and Venkataraman [1975].

For $\lambda = 0$ we have

$$\square A^{\pm 1/2} = 0. \tag{6.13.20}$$

Using $(6.13.6)_3$ we obtain

$$A^{\pm 1/2} = f(e)g(h), \tag{6.13.21}$$

where

$$\begin{aligned}
&\text{(i)} \quad f = f_1 e + f_0, && g = g_1 h + g_0,\\
&\text{(ii)} \quad f = f_1 \sinh(ke + f_0), && g = g_1 \sinh(kh + g_0),\\
&\text{(iii)} \quad f = f_1 \sin(ke + f_0), && g = g_1 \sin(kh + f_0),
\end{aligned} \tag{6.13.22}$$

where f_i, g_i, and k are constants. By means of (6.13.6) we determine

$$\begin{aligned}
E - E^0 &= c_1^{\pm 1} \int f^{\pm 2}(e)\, de,\\
H - H_0 &= c_1^{\mp 1} \int g^{\pm 2}(h)\, dh,\\
\bar{D} - D_0 &= c_1{}^{\mp 1} \int f^{\mp 2}(e)\, de,\\
\bar{B} - B_0 &= c_1^{\pm 1} \int g^{\mp 2}(h)\, dh,
\end{aligned} \tag{6.13.23}$$

where $\bar{D}(e) = D(E)$, $\bar{\mathrm{B}}(h) = B(H)$, and c_1, E_0, H_0, D_0, and B_0 are constants.

Carrying (6.13.22) into (6.13.23) we obtain, for the upper signs,

$$\begin{aligned}
\bar{D} - D_0 &= -(c_1 f_1)^{-1}[3f_1 c_1{}^{-1}(E - E_0)]^{-1/3},\\
\bar{B} - B_0 &= -(c_1/g_1)[3g_1 c_1(H - H_0)]^{-1/3},
\end{aligned} \tag{6.13.24}$$

for the case (i),

$$\begin{aligned}
\bar{D} - D_0 &= -c_1^{-1} f_1^{-2} k^{-1} \coth(ke + f_0),\\
\bar{E} - E_0 &= -\tfrac{1}{4} c_1 f_1^2 k^{-1}[2ke - \sinh 2(ke + f_0)],\\
\bar{B} - B_0 &= -c_1 g_1^{-2} k^{-1} \coth(kh + g_0),\\
\bar{H} - H_0 &= -\tfrac{1}{4} c_1^{-1} g_1^2 k^{-1}[2kh - \sinh 2(kh + g_0)],
\end{aligned}$$

in parametric form for the case (ii), and

$$\begin{aligned}
\bar{D} - D_0 &= -c_1^{-1} f_1^{-2} k^{-1} \cot(ke + f_0), \\
\bar{E} - E_0 &= -\tfrac{1}{4} c_1 f_1^2 k^{-1} [2ke - \sin 2(ke + f_0)], \\
\bar{B} - B_0 &= -c_1 g_1^{-2} k^{-1} \cot(kh + g_0), \\
\bar{H} - H_0 &= \tfrac{1}{4} c_1^{-1} g_1^2 k^{-1} [2kh - \sin 2(kh + g_0)],
\end{aligned}$$

in parametric form for the case (iii)

6.14. Electromagnetic Shock Waves in Rigid Stationary Dielectrics

Nonlinear wave propagation can be envisaged as a nondestructive means to test the nonlinearity of a dielectric material. The occurrence of particular shock solutions (i.e., discontinuities in the electromagnetic fields) in such a material strongly depends on the convexity of the $\mathbf{P} - \mathbf{E}$ curve, in the same way that shocks in nonlinear elastic bodies depend on the convexity of the stress–strain relation (see Eringen and Şuhubi [1974]). To illustrate this class of problems, we study the propagation of electromagnetic shock waves in a *rigid nonlinear isotropic* dielectric *at rest*. We intend to determine the speed of propagation and the electromagnetic vectors behind the shock front, in terms of those ahead of the shock front.

Jump Relations. Let $S(\mathbf{x}, t) = 0$ be the equation of the discontinuity front $\sigma(t)$ sweeping the body with an absolute velocity ν. Then, without charges and defining a nondimensional speed v, by

$$v = \frac{V}{c} = \frac{1}{c}\mathbf{v}\cdot\mathbf{n}, \tag{6.14.1}$$

where $\mathbf{n}$ is the unit normal to $\sigma(t)$ in its direction of propagation, the jump conditions (6.2.5)–(6.2.8), with $\hat{\boldsymbol{\pi}} = \boldsymbol{\mathscr{K}} = \mathbf{0}$ and $w_e = 0$, read

$$\begin{aligned}
[\mathbf{D}]\cdot\mathbf{n} &= 0, \\
\mathbf{n} \times [\mathbf{E}] - v[\mathbf{B}] &= \mathbf{0}, \\
[\mathbf{B}]\cdot\mathbf{n} &= 0, \\
\mathbf{n} \times [\mathbf{H}] + v[\mathbf{D}] &= \mathbf{0},
\end{aligned} \tag{6.14.2}$$

across $\sigma(t)$.

These equations are supplemented by the jump of the energy equation and the entropy jump across σ. For *rigid stationary* dielectrics these are given by (3.14.18) and (3.14.19) as

$$[\bar{\Sigma}\mathbf{v} + \tfrac{1}{2}(\mathbf{E}^2 + \mathbf{B}^2)\mathbf{v} - \mathbf{S}]\cdot\mathbf{n} = 0 \tag{6.14.3}$$

and

$$[\rho\eta\mathbf{v}]\cdot\mathbf{n} \leqq 0, \tag{6.14.4}$$

where $\bar{\Sigma} = \rho\varepsilon = \rho(\Psi + \theta\eta) + \mathbf{E}\cdot\mathbf{P}$ and the Poynting vector $\mathbf{S}$ is given by

$$\mathbf{S} = c\mathbf{E} \times \mathbf{H}. \tag{6.14.5}$$

Hence, (6.14.3) takes on the form

$$[\bar{\Sigma} + \tfrac{1}{2}(\mathbf{E}^2 + \mathbf{B}^2)]v = [(\mathbf{E} \times \mathbf{H})\cdot\mathbf{n}]. \tag{6.14.6}$$

For nonmagnetizable dielectrics $\mathbf{H} = \mathbf{B}$, and (6.14.6) can be transformed as follows. Denoting the fields ahead and behind the wave front σ by subscripts 1 and 2, respectively, we have the following definitions and identities:

$$\begin{aligned} &[A] \equiv A_2 - A_1, \qquad \langle A\rangle = \tfrac{1}{2}(A_2 + A_1),\\ &[AB] = \langle A\rangle[B] + [A]\langle B\rangle, \qquad [A^2] = 2\langle A\rangle[A]. \end{aligned} \tag{6.14.7}$$

Because of $(6.14.7)_3$ we can write the right-hand side of (6.14.6) as

$$[(\mathbf{E} \times \mathbf{H})\cdot\mathbf{n}] = \langle\mathbf{B}\rangle\cdot[\mathbf{n} \times \mathbf{E}] + [\mathbf{B}]\cdot\langle\mathbf{n} \times \mathbf{E}\rangle. \tag{6.14.8}$$

But, from $(6.14.2)_2$ and $(6.14.2)_4$,

$$[\mathbf{n} \times \mathbf{E}] = v[\mathbf{B}] \tag{6.14.9}$$

and

$$[\mathbf{B}]\cdot\langle\mathbf{n} \times \mathbf{E}\rangle = \langle\mathbf{E}\rangle\cdot[\mathbf{B} \times \mathbf{n}] = v[\mathbf{D}]\cdot\langle\mathbf{E}\rangle. \tag{6.14.10}$$

Substituting then from (6.14.9) and (6.14.10) in (6.14.8), recognizing $(6.14.7)_4$, with $v \neq 0$, we transform (6.14.6) to

$$[\bar{\Sigma} + \tfrac{1}{2}\mathbf{E}^2] = [\mathbf{D}]\cdot\langle\mathbf{E}\rangle. \tag{6.14.11}$$

This is the so-called *Hugoniot condition*[3] *for rigid, nonmagnetizable, nonlinear dielectrics.*

Constitutive Equations. For rigid, isotropic, nonmagnetizable dielectrics we will have, from (6.2.10), the constitutive equations

$$\mathbf{D} = \varepsilon(J_{\mathrm{E}}, \theta)\mathbf{E}, \qquad \mathbf{H} = \mathbf{B}, \qquad J_{\mathrm{E}} \equiv \mathbf{E}^2, \tag{6.14.12}$$

where

$$\varepsilon(J_{\mathrm{E}}, \theta) = 1 - 2\frac{\partial\Sigma}{\partial J_{\mathrm{E}}}, \qquad \Sigma = \Sigma(J_{\mathrm{E}}, \theta) = \bar{\Sigma} - \rho_0\theta\eta - \mathbf{E}\cdot\mathbf{P}. \tag{6.14.13}$$

Equivalently to $(6.14.12)_1$ and $(6.14.13)_1$, we can write

$$\mathbf{P} = \chi(J_{\mathrm{E}}, \theta)\mathbf{E}, \qquad \chi(J_{\mathrm{E}}, \theta) = -2\frac{\partial\Sigma}{\partial J_{\mathrm{E}}}. \tag{6.14.14}$$

A special form of $(6.14.14)_2$, convenient for curve-fitting with experimental data, is [compare with eq. (6.10.4)]

$$\chi(J_{\mathrm{E}}, \theta) = \chi_0(\theta) + \chi_1(\theta)J_{\mathrm{E}}. \tag{6.14.15}$$

---

[3] Compare with the magnetohydrodynamic case in Chapter 10.

Then $(6.14.13)_2$ yields

$$\bar{\Sigma} = \Sigma\rho_0\theta\eta + \mathbf{E}\cdot\mathbf{P} = \tfrac{1}{2}\chi_0 J_{\mathrm{E}} + \tfrac{3}{4}\chi_1 J_{\mathrm{E}}^2. \tag{6.14.16}$$

Usually

$$\chi_0 \geqq 0 \tag{6.14.17}$$

and

$$\begin{cases} \chi_1 = 0 & \text{for linear dielectrics,} \\ \chi_1 > 0 & \text{for hard polarizers (concave } \mathbf{P} - \mathbf{E} \text{ curve),} \\ \chi_1 < 0 & \text{for soft polarizers (convex } \mathbf{P} - \mathbf{E} \text{ curve).} \end{cases} \tag{6.14.18}$$

Because of (6.14.12), the Hugoniot condition (6.14.11) is shown to take either one of the following forms:

$$\left[\bar{\Sigma} + (1 - \langle\varepsilon\rangle)\frac{\mathbf{E}^2}{2}\right] = [\varepsilon]\langle\mathbf{E}\rangle^2 \tag{6.14.19}$$

or

$$[2\bar{\Sigma}] = [(\varepsilon - 1)\mathbf{E}^2] + [\varepsilon](\mathbf{E}_2\cdot\mathbf{E}_1). \tag{6.14.20}$$

For the special case (6.14.16), this gives

$$[\chi_0 J_{\mathrm{E}} + \tfrac{3}{2}\chi_1 J_{\mathrm{E}}^2] = (\varepsilon_2 - 1)E_2^2 - (\varepsilon_1 - 1)E_1^2 + (\varepsilon_2 - \varepsilon_1)(\mathbf{E}_2\cdot\mathbf{E}_1), \tag{6.14.21}$$

where

$$\varepsilon_\alpha = \varepsilon(E_\alpha^2) \qquad \alpha = 1, 2. \tag{6.14.22}$$

From here on, we ignore thermal influences. For further use, we set

$$n^2 = 1 + \chi_0, \qquad e = \chi_1 E_0^2, \tag{6.14.23}$$

where n is the *index of refraction* of the linear case, and E_0 is a characteristic electric field so that the parameter e, which measures deviations from the linear case, is nondimensional.

General Shock Solutions. We solve the system (6.14.2) for E_2, B_2, and v; these fields being constrained to satisfy (6.14.4) and (6.14.20). Taking the vector product of $(6.14.2)_2$ with $\mathbf{n}$, and substituting from the result in $(6.14.2)_4$ with $v \neq 0$, yields

$$\mathbf{n} \times [\mathbf{n} \times \mathbf{E}] + v^2[\mathbf{D}] = \mathbf{0}. \tag{6.14.24}$$

By expanding the double vector product, we obtain

$$[\mathbf{n}\cdot\mathbf{E}]\mathbf{n} + [(v^2\varepsilon - 1)\mathbf{E}] = \mathbf{0}. \tag{6.14.25}$$

From $(6.14.2)_1$, $(6.14.12)_1$, and $(6.14.7)_1$, we have

$$\mathbf{n}\cdot\mathbf{E}_2 = (\varepsilon_1/\varepsilon_2)\mathbf{n}\cdot\mathbf{E}_1. \tag{6.14.26}$$

Introducing this in (6.14.25), we solve for $\mathbf{E}_2$:

$$\mathbf{E}_2 = Q\mathbf{E}_1 + R\mathbf{n}, \tag{6.14.27}$$

where

$$Q = (v^2\varepsilon_1 - 1)/(v^2\varepsilon_2 - 1),$$
$$R = (\mathbf{n}\cdot\mathbf{E}_1)\left(1 - \frac{\varepsilon_1}{\varepsilon_2}\right)/(v^2\varepsilon_2 - 1). \tag{6.14.28}$$

By means of (6.14.27) and $(6.14.2)_2$, we can determine the magnetic induction $\mathbf{B}_2$ behind the shock front as

$$\mathbf{B}_2 = \mathbf{B}_1 - \left(\frac{v[\varepsilon]}{v^2\varepsilon_2 - 1}\right)\mathbf{n}\times\mathbf{E}_1. \tag{6.14.29}$$

It is immediately noticable from this equation that $(6.14.2)_3$ is identically satisfied. Equations (6.14.27) and (6.14.29) determine the electric field and magnetic induction behind the shock, if $\mathbf{E}_1$, $\mathbf{B}_1$, $\mathbf{E}_2^2$, and v are given, since $\varepsilon_2 = \varepsilon(E_2^2)$. But $\mathbf{E}_2^2$ and v are also unknowns in the problem, so that we need two more equations to solve the problem in its entirety. One of these equations is none other than the Hugoniot condition (6.14.20). The other is obtained by evaluating $\mathbf{E}_2^2$ from (6.14.27). This provides a biquadratic equation for v

$$a_1 v^4 + a_2 v^2 + a_3 = 0, \tag{6.14.30}$$

where

$$a_1 = [\varepsilon^2\mathbf{E}^2],$$
$$a_2 = -2\left\{[\varepsilon\mathbf{E}^2] + \frac{\varepsilon_1}{\varepsilon_2}[\varepsilon](\mathbf{n}\cdot\mathbf{E}_1)^2\right\}, \tag{6.14.31}$$
$$a_3 = [\mathbf{E}^2] + [1 - (\varepsilon_1/\varepsilon_2)^2](\mathbf{n}\cdot\mathbf{E}_1)^2.$$

For physically realistic situations, it is expected that two of the roots of (6.14.30) will be real positive (corresponding to propagation in the positive direction of $\mathbf{n}$) and two will be negative (corresponding to propagation in the negative direction of $\mathbf{n}$). The sign of the realistic v must be such that entropy grows across the shock. Equation (6.14.4) yields

$$[\rho\eta]v \leqq 0. \tag{6.14.32}$$

Since $\rho = \text{const.} > 0$, and necessarily $[\eta] \geqq 0$ across physically admissible shocks, we must have $v < 0$.

The roots of (6.14.30) are given by

$$v^2 = \frac{1}{2a_1}[-a_2 \pm (a_2^2 - 4a_1a_3)^{1/2}]. \tag{6.14.33}$$

This equation gives four real roots if and only if

$$a_2^2 - 4a_1a_3 > 0, \qquad \frac{-a_2}{2a_1} > \frac{1}{2a_1}(a_2^2 - 4a_1a_3)^{1/2} > \frac{a_2}{2a_1}. \tag{6.14.34}$$

This completes the formal solution of the problem, since (6.14.27), (6.14.29),

(6.14.20), and (6.14.30) form a system of eight scalar equations for the eight scalar unknowns of the problem (v, $\mathbf{E}_2^2$, and the components of $\mathbf{E}_2$ and $\mathbf{B}_2$), if $\mathbf{E}_1$ and $\mathbf{B}_1$ are prescribed and the constitutive relation $\varepsilon(\mathbf{E}^2)$ is specified.

Clearly, $\mathbf{B}_1$ plays a passive role in the above formulation. In general, $\mathbf{E}_1$ can be decomposed in components parallel and orthogonal to $\mathbf{n}$ by

$$\begin{aligned} \mathbf{E}_1 &= \mathbf{E}_1^{\parallel} + \mathbf{E}_1^{\perp}, \\ \mathbf{E}_1^{\parallel} &= (\mathbf{n}\cdot\mathbf{E}_1)\mathbf{n}, \qquad \mathbf{E}_1^{\perp} \equiv \mathbf{n}\times(\mathbf{E}_1\times\mathbf{n}), \end{aligned} \tag{6.14.35}$$

so that two cases of interest can be considered:

(i) $\mathbf{E}_1^{\parallel} = \mathbf{0}$, i.e., $\mathbf{n}\cdot\mathbf{E}_1 = 0$ and $\mathbf{E}_1^{\perp} \neq 0$ (transverse setting).
(ii) $\mathbf{E}_1^{\parallel} \neq 0$, $\mathbf{E}_1^{\perp} = \mathbf{0}$, i.e., $\mathbf{n}\times\mathbf{E}_1 = \mathbf{0}$ (parallel setting).

We need consider only these two particular cases. In this state of generality, we can examine the possibility of *switch-on* and *switch-off shocks*.

Equation (6.14.29) shows that in a *nonlinear* dielectric, with $[\varepsilon] \neq 0$, we may have $\mathbf{B}_2 \neq \mathbf{0}$ for $\mathbf{B}_1 = \mathbf{0}$, unless we have a transverse setting. Then $\mathbf{B}_2$, necessarily, will lie in the plane of the shock, so that we may, *a priori*, have *magnetic switch-on* shocks. However, we cannot have $\mathbf{E}_2 \neq \mathbf{0}$ for $\mathbf{E}_1 = \mathbf{0}$, as is shown by (6.14.27) and (6.14.28), even though $\mathbf{B}_1 \neq \mathbf{0}$. Also, we cannot have *electric switch-off* shocks, for $\mathbf{E}_2 = 0$ requires from (6.14.26)–(6.14.28) that $\mathbf{E}_1 = \mathbf{0}$. Finally, we may have a magnetic switch-off shock, if $\mathbf{B}_2 = \mathbf{0}$ for $\mathbf{B}_1 \neq \mathbf{0}$. In this case, (6.14.29) requires that the initial setting of the electromagnetic field and the solution of the shock equations be such that

$$\mathbf{B}_1 = \frac{v[\varepsilon]}{v^2\varepsilon_2 - 1}\mathbf{n}\times\mathbf{E}_1. \tag{6.14.36}$$

Since $\mathbf{n}\cdot\mathbf{B}_1 = 0$ from $(6.14.2)_3$ in this case, $\mathbf{B}_1$ must then be in the plane of the shock, and (6.14.36) will have a solution only in *nonlinear* dielectrics (for $[\varepsilon] \neq 0$) and for a transverse setting.

Transverse Electric-Field Setting. We take $\mathbf{B}_1 = B\mathbf{n}$, so that the magnetic induction ahead of the shock is in the direction of $\mathbf{n}$. We also assume that $\mathbf{n}\cdot\mathbf{E}_1 = 0$ (case (i)), so that we can theoretically have only magnetic switch-on shocks according to the discussion above. Then, (6.14.27) and (6.14.29) reduce to

$$\begin{aligned} \mathbf{E}_2 &= \frac{v^2\varepsilon_1 - 1}{v^2\varepsilon_2 - 1}\mathbf{E}_1, \\ \mathbf{B}_2 &= \mathbf{B}n - \frac{v[\varepsilon]}{v^2\varepsilon_2 - 1}(\mathbf{n}\times\mathbf{E}_1), \end{aligned} \tag{6.14.37}$$

so that

$$\mathbf{E}_2\cdot\mathbf{E}_1 = \frac{v^2\varepsilon_1 - 1}{v^2\varepsilon_2 - 1}E_1^2, \tag{6.14.38}$$

while (6.14.30) provides the roots

$$v^2 = \frac{[\varepsilon \mathbf{E}^2]}{[\varepsilon^2 \mathbf{E}^2]} \pm \left\{ \left(\frac{[\varepsilon \mathbf{E}^2]}{[\varepsilon^2 \mathbf{E}^2]} \right)^2 - \frac{[\mathbf{E}^2]}{[\varepsilon^2 \mathbf{E}^2]} \right\}^{1/2}. \tag{6.14.39}$$

For a *linear* dielectric, (6.14.39) yields the classical solution

$$v^2 = 1/n^2, \qquad \mathbf{E}_2 = \mathbf{E}_1, \qquad \mathbf{B}_2 = B\mathbf{n}. \tag{6.14.40}$$

Parallel Electric Field Setting. We take $\mathbf{B}_1 \neq \mathbf{0}$ and $\mathbf{E}_1 = \mathbf{E}_1^{\|}$, so that $\mathbf{n} \times \mathbf{E}_1 = \mathbf{0}$. Then, (6.14.27) and (6.14.29) reduce to

$$\begin{aligned} \mathbf{E}_2 &= \frac{v^2 \varepsilon_1 - (\varepsilon_1/\varepsilon_2)}{v^2 \varepsilon_2 - 1} (\mathbf{E}_1 \cdot \mathbf{n})\mathbf{n} = \mathbf{E}_2^{\|}, \\ \mathbf{B}_2 &= \mathbf{B}_1, \end{aligned} \tag{6.14.41}$$

so that

$$\mathbf{E}_2 \cdot \mathbf{E}_1 = \frac{v^2 \varepsilon_1 - (\varepsilon_1/\varepsilon_2)}{v^2 \varepsilon_2 - 1} (\mathbf{E}_1 \cdot \mathbf{n})^2. \tag{6.14.42}$$

For a *linear* dielectric, this again yields the classical solution (6.14.40). To proceed further, we need consider a special $\mathbf{P} - \mathbf{E}$ constitutive equation, namely, that given by (6.14.15)

Shocks in Soft Ferroelectrics [constitutive equation (6.14.15)].
Case (i): With $\mathbf{n} \cdot \mathbf{E}_1 = 0$, we define the strength of the shock by

$$\alpha = E_2/E_1, \qquad E_\alpha \equiv |E_\alpha|, \tag{6.14.43}$$

and use the definitions (6.14.23), where $E_0 = E_1$, in the second of these. Then $(6.14.37)_1$, (6.14.38), and (6.14.39) read

$$\begin{aligned} \mathbf{E}_2 &= \pm \alpha \mathbf{E}_1, \\ \mathbf{E}_2 \cdot \mathbf{E}_1 &= \pm \alpha \mathbf{E}_1^2, \end{aligned} \tag{6.14.44}$$

$$v^2 = \frac{1}{n^2} \cdot \left[\frac{\alpha^2\left(1+\frac{e\alpha^2}{n^2}\right) - \left(1+\frac{e}{n^2}\right) \pm \left\{ \left[\alpha^2\left(1+\frac{e\alpha^2}{n^2}\right) - \left(1+\frac{e}{n^2}\right)\right]^2 - (\alpha^2-1)\left[\left(1+\frac{e\alpha^2}{n^2}\right)^2\alpha^2 - \left(1+\frac{e}{n^2}\right)\right] \right\}^{1/2}}{\alpha^2\left(1+\frac{e\alpha^2}{n^2}\right)^2 - \left(1+\frac{e}{n^2}\right)^2} \right]. \tag{6.14.45}$$

Noting

$$\varepsilon_2 = n^2 + e\alpha^2, \qquad \varepsilon_1 = n^2 + e, \tag{6.14.46}$$

we calculate both sides of (6.14.21). This yields

$$(\alpha^2 - 1)(\alpha \mp 1)^2 = 0. \tag{6.14.47}$$

The only two real roots of this equation are the classical solutions $\alpha = \pm 1$, which yield (6.14.40). Hence, we cannot have any nontrivial shock solutions for the transverse electric-field setting. This means the nonlinearity in the dielectric cannot be detected for this initial setting.

Case (ii): With $\mathbf{n} \times \mathbf{E}_1 = \mathbf{0}$, it proves more convenient to define the strength of the shock, β, by

$$\beta = \varepsilon_2/\varepsilon_1. \tag{6.14.48}$$

and to define n^2 by

$$\varepsilon_1 = n^2 + e, \qquad \chi_0 = n^2 - 1. \tag{6.14.49}$$

Then (6.14.44) and (6.14.47) are replaced by

$$\begin{aligned} \mathbf{E}_2 &= \frac{1}{\beta}(\mathbf{E}_1 \cdot \mathbf{n})\mathbf{n}, \\ \mathbf{E}_2 \cdot \mathbf{E}_1 &= \frac{1}{\beta} E_1^2, \end{aligned} \tag{6.14.50}$$

and

$$(1 - \beta^2)\left[\beta^2\left(n^2 - \frac{3e}{2}\right) - \frac{3e}{2}\right] = 0. \tag{6.14.51}$$

The first two roots, $\beta = \pm 1$, of this equation again yield the trivial solutions for which $v^2 = 1/n^2$. The remaining factor in (6.14.51) yields real nontrivial shock solutions if and only if

$$e < 0, \qquad |e| < \tfrac{2}{3}n^2. \tag{6.14.52}$$

That is, this occurs only if the nonlinear dielectric is a soft polarizer (since $e < 0$ implies $\chi_1 < 0$), and, given $\chi_1 < 0$, only for electric fields $\mathbf{E}_1$ ahead of the shock, whose magnitude satisfies the inequality

$$|\mathbf{E}_1| < \frac{2(1 + \chi_0)}{3|\chi_1|}. \tag{6.14.53}$$

This, in turn, means that the $\mathbf{P} - \mathbf{E}$ curve must be markedly convex for small $|\mathbf{E}|$. If this is realized, the nontrivial solutions of (6.14.51) are given by

$$\beta^2 = \frac{1}{(2n^2/3|e|) - 1}. \tag{6.14.54}$$

As $|\mathbf{E}_1|$ is very large, then this tends towards the trivial solutions $\beta = \pm 1$. It is clear that the value of $\mathbf{E}_2$ behind the shock depends only on β, and not on v, and is simply given by (6.14.44). We have $\mathbf{B}_2 = \mathbf{B}_1$. As to v, it is immediately conclusive that all a_α, $\alpha = 1, 2, 3$, of eq. (6.14.31) are zero for this case, so that the speed of propagation is left undetermined (insofar as it does not exceed the light velocity in the body). In other words, it is independent of the shock solution and keeps its initial value. More on electromagnetic shocks may be found in Katayev [1966] and Sedova [1978–1982].

6.15. Moving Rigid Dielectrics

A. Dragging of Light by a Moving Dielectric

Here we study the propagation of an electromagnetic shock wave in a *moving linear* isotropic dielectric. The medium is not necessarily rigid, but it behaves as such, insofar as the present problem is concerned. For instance, it may be water. It is a *transparent* (i.e., with no electromagnetic absorption) *nonmagnetizable*, *incompressible*, isotropic dielectric. Its electromagnetic constitutive equation, therefore, is in a co-moving frame R_C

$$\mathbf{P} = \chi \mathscr{E}, \qquad \mathscr{M} = \mathbf{0}, \tag{6.15.1}$$

where χ is the constant dielectric susceptibility of the matter. From $(6.15.1)_2$, (3.4.13), and (3.4.14), it follows that

$$\mathscr{H} = \mathbf{B} - \frac{1}{c}\mathbf{v} \times \mathbf{E}. \tag{6.15.2}$$

This shows that we are dealing with a true problem of *electrodynamics*. For an electromagnetic shock wave moving at an absolute velocity $\mathbf{v}$ through a dielectric, we set

$$V = \mathbf{v} \cdot n, \tag{6.15.3}$$

where $\mathbf{n}$ is the unit normal of the shock surface. The jump conditions to be satisfied by the electromagnetic fields across σ are those given by (6.2.6), (6.2.7), and (6.2.8), in the absence of source terms, i.e., without right-hand sides. Because of (6.15.3) and (6.15.2), these jump conditions reduce to

$$\mathbf{n} \times [\mathbf{E}] = \frac{V}{c}[\mathbf{B}] \tag{6.15.4}$$

and

$$\mathbf{n} \times [\mathbf{B}] = \mathbf{n} \times \left[\mathbf{P} \times \frac{\mathbf{v}}{c}\right] - \frac{V}{c}[\mathbf{D}]. \tag{6.15.5}$$

Setting $\varepsilon = 1 + \chi$ and neglecting contributions of the order of $\beta^2 (= v^2/c^2) \ll 1$, since we assume a nonrelativistic velocity for the medium, (6.15.5) can be rewritten as

$$\mathbf{n} \times \left[\mathbf{B} + \chi \frac{\mathbf{v}}{c} \times \mathbf{E}\right] + \frac{V}{c}\left[\varepsilon \mathbf{E} + \chi \frac{\mathbf{v}}{c} \times \mathbf{B}\right] = \mathbf{0}. \tag{6.15.6}$$

Supposing that $[\mathbf{v}] = \mathbf{0}$ across σ (i.e., σ is *not* a hydrodynamic shock), (6.15.4) and (6.15.6) form a system of two vectorial equations for the unknowns $[\mathbf{B}]$ and $[\mathbf{E}]$. A single vectorial equation is obtained for $[\mathbf{E}]$ by substituting from (6.15.4) into (6.15.6). We obtain

$$\mathbf{n} \times (\mathbf{n} \times [\mathbf{E}]) + \chi \frac{V}{c} \mathbf{n} \times \left(\frac{\mathbf{v}}{c} \times [\mathbf{E}]\right) + \chi \frac{V}{c}\left[\frac{\mathbf{v}}{c} \times (\mathbf{n} \times [\mathbf{E}])\right] + \varepsilon \frac{V^2}{c^2}[\mathbf{E}] = \mathbf{0}. \tag{6.15.7}$$

Developing various terms and considering the special case that corresponds to motions of the wave front and of the dielectric along the same axis, with $[\mathbf{E}]\cdot\mathbf{n} = 0$, (6.15.7) yields

$$\left(\varepsilon\frac{V^2}{c^2} - 2\chi\frac{V}{c}\frac{U}{c} - 1\right)[\mathbf{E}] + \chi\frac{V}{c}\left(\frac{\mathbf{v}}{c}\cdot[\mathbf{E}]\right)\mathbf{n} = \mathbf{0}; \qquad U \equiv \mathbf{v}\cdot\mathbf{n}.$$

The scalar product of this equation with $\mathbf{n}$ gives

$$\chi\frac{V}{c}\frac{\mathbf{v}}{c}\cdot[\mathbf{E}] = 0.$$

Consequently, the jump in the electric field is normal to the velocity, and we have

$$\left(\varepsilon\frac{V^2}{c^2} - 2\chi\frac{V}{c}\frac{U}{c} - 1\right)[\mathbf{E}] = \mathbf{0}. \tag{6.15.8}$$

For a nonzero jump in the electric field across σ, the factor of $[\mathbf{E}]$ in (6.15.8) must necessarily vanish. This *compatibility condition* corresponds to the dispersion relation in the study of harmonic-wave propagation. Defining the *refractive index* by $n = \sqrt{\varepsilon}$, so that $\chi = n^2 - 1$, we find that, up to relativistic terms, the solutions of this compatibility condition are

$$V = \pm\frac{c}{n} + \alpha U, \tag{6.15.9}$$

where

$$\alpha \equiv 1 - \frac{1}{n^2} \tag{6.15.10}$$

is Fresnel's dragging coefficient.

The historical importance of the result (6.15.9) and (6.15.10) must be emphasized. Fresnel originally derived this result in 18[illegible]8, from his aether theory,[4] by postulating a partial dragging of the aether by refringent media. These results were confirmed by the early experimental investigations of Fizeau [1859], which were devoted to the propagation of light in flowing water. Fizeau's experimental apparatus is sketched in Figure 6.15.1: the light is emitted by the source S, and the displacement of the interference fringes (obtained with the help of a semimirror) is observed at A. This constituted a *crucial experiment* for the electrodynamics of moving media.[5] The experience was repeated with Young's fringes, using a different device by Michelson and Morley [1886], and later by Zeeman [1914] with light of different colors, and again by Zeeman with moving dielectric solids (quartz rod set in a rapid motion). It must be noted that (6.15.9) is an approximation of the first order

---

[4] For the obsolete notion of luminiferous aether, the *ad hoc* substratum of electromagnetic vibrations, see Whittaker [1951].

[5] See Whittaker [1951] and Tonnélat [1971, pp. 82–84].

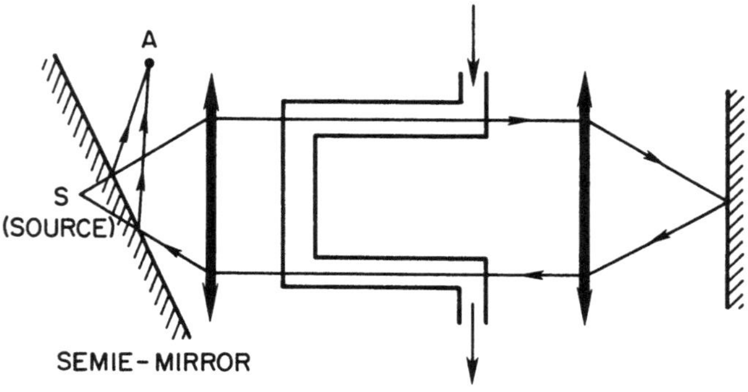

Figure 6.15.1. Fizeau's experiment.

of the rigorous law of addition of velocities. Indeed, V is the "absolute" speed of the electromagnetic wave front, c is its velocity in a vacuum, c/n is its velocity with respect to the moving dielectric in a co-moving frame, and U is the speed of the material dielectric.

For the case of a light-absorbing dielectric medium, we refer the reader to Voigt [1899], Born [1972, Sect. 69], and Boulanger et al. [1971, pp. 23–25 and App. II]. The method above, using shock fronts, was employed by Toupin [1963] and, subsequently, by other authors to study the photoelastic effect (see, e.g., Morris and Nariboli [1972]).

B. The Rotation of a Rigid Dielectric in a Magnetic Field

In order to exhibit the effect of rotation in a magnetic field on a rigid nonmagnetizable dielectric insulator, we consider the experiment of Wilson [1905], later repeated by M. Wilson and H.A. Wilson [1914]. In this experiment a hollow dielectric cylinder is set rotating in a magnetic field directed along its axis (see Figure 6.15.2). Historically, this constitutes another *crucial experiment* for the electrodynamics of moving bodies. The cylinder has radii r_1 and r_2, $r_1 < r_2$, and axis O_z. Its length is L. The inner and outer faces of the cylinder are coated with a conducting material (e.g., copper) and connected through brushes with an electrometer of capacitance C. When the magnetic field is turned on and the cylinder rotated about its axis, a voltage V is developed between the brushes, and the electrometer becomes charged. The experiment is made at constant angular velocity Ω, and is repeated for different values of Ω. Thus, the problem is stationary (no explicit dependence of the variables on the time t). The constitutive equations are those given by (6.15.1). The velocity field is that of a rigid-body motion. The symmetry of the problem dictates the use of cylindrical coordinates (r, θ, z), as shown in Figure 6.15.2. Thus, $\mathbf{v} = \Omega r \mathbf{e}_\theta$. Let E, P, and B denote the so-called physical radial com-

Figure 6.15.2. Wilson's experiments.

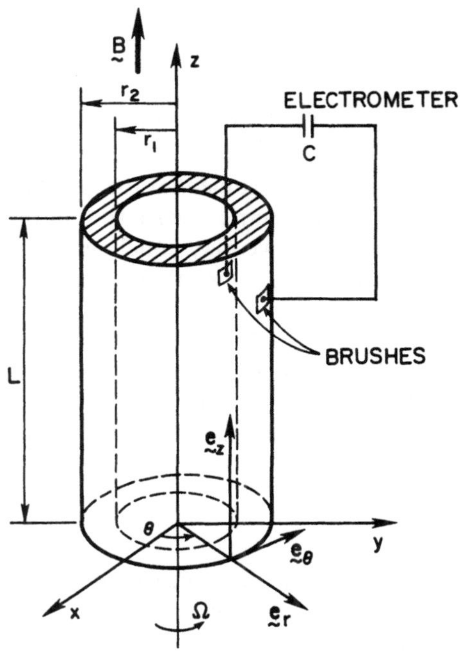

ponents[6] of the fields **E**, **P**, and **B**, respectively. We have

$$\mathbf{E} = E\mathbf{e}_r, \qquad \mathbf{P} = P\mathbf{e}_r, \qquad \mathbf{B} = B\mathbf{e}_z, \tag{6.15.11}$$

with all fields depending only on the radial coordinate r because of the cylindrical symmetry. Equation (6.15.1) reads

$$P = \chi\left(E + \frac{\Omega r}{c} B\right). \tag{6.15.12}$$

The nonvanishing components of Maxwell's equations (3.9.8) and (3.9.18) for a nonmagnetic dielectric insulator read

$$\begin{gathered} \frac{d}{dr}[r(E + P)] = 0, \\ \frac{d}{dr}\left(B + \frac{\Omega r}{c} E\right) - \frac{\Omega}{c}(E + P) = 0. \end{gathered} \tag{6.15.13}$$

These integrate immediately to

$$r(E + P) = k_1, \qquad B + \frac{\Omega r}{c} E - \frac{\Omega k_1}{c} \ln r + k_2 = 0, \tag{6.15.14}$$

[6] See Eringen [1967, p. 463].

where k_1 and k_2 are constants to be evaluated from the boundary data of the experiment. Eliminating P between (6.15.12) and (6.15.14) we obtain

$$E = \left(\frac{k_1}{\chi r} + \frac{\Omega r}{c} k_2 - \frac{\Omega^2 r k_1}{c^2} \ln r\right)\left[\frac{1+\chi}{\chi} - \left(\frac{\Omega r}{c}\right)^2\right]^{-1}. \tag{6.15.15}$$

To evaluate k_1, assume that the net charge on the inner and outer surfaces of the rotating dielectric is zero, so that the electric field exterior to the cylinder vanishes. Then, we may conclude, with the boundary condition (6.2.5) with $\hat{\pi} = 0$, applied at $r = r_2$, that

$$2\pi r_2[E(r_2) + P(r_2)] = 2\pi w_e = 2\pi k_1 = W, \tag{6.15.16}$$

where $W = 2\pi w_e$ is the free charge per unit length of the cylinder, deposited by the electrometer. Then, the total charge C of one sign which has flowed through the electrometer is given by $Q = WL$, from which we conclude with $Q = CV$ that

$$k_1 = \frac{W}{2\pi} = \frac{CV}{2\pi L}, \tag{6.15.17}$$

where V is the voltage registered by the electrometer. The value of k_2 can be determined from the magnetic field at $r = r_2$.

Ignoring the relativistic term $(\Omega r/c)^2$, eq. (6.15.15) may be integrated to give

$$-\varepsilon V = \varepsilon \int_{r_1}^{r_2} E\, dr = k_1 \ln\frac{r_2}{r_1} + \frac{\Omega k_2}{c}\chi(r_2^2 - r_1^2). \tag{6.15.18}$$

Substituting from (6.15.17), we obtain

$$-\varepsilon V = \alpha f; \qquad \alpha \equiv \frac{2k_2\chi}{c\varepsilon}\left(1 + \frac{C}{2\pi\varepsilon L}\ln\frac{r_2}{r_1}\right)^{-1} A, \tag{6.15.19}$$

where

$$A \equiv \pi(r_2^2 - r_1^2), \qquad f \equiv \frac{\Omega}{2\pi}, \tag{6.15.20}$$

are, respectively, the area of a cross section of the cylinder and the rate of rotation, measured per unit time. This result shows that the voltage V is a linear function of A and Ω.

The conditions of the Wilson–Wilson [1914] experiments are the following ones:

$$\chi = 0.17, \qquad B_0 \simeq 4{,}200\text{–}5{,}000 \text{ A}^{-1}, \qquad r_2 = 3.73 \text{ cm}, \qquad r_1 = 2 \text{ cm},$$

$$L = 9.5 \text{ cm},$$

the dielectric sample being made of sealing wax and embedded steel balls. The range of value of f and V is restricted as a consequence of the simple characteristic of the experimental device: 99 rps $\leq f \leq$ 200 rps and 0.0306 volts $\leq$ V $\leq$ 0.1160 volts. In this range, the observed dependence of V on f is approximately linear. Toupin [1963] has conducted the same analysis as that given above, taking account of the polarization inertia besides. Then, a sensible nonlinearity occurs for sufficiently high rotation rates

C. Unipolar Induction

Let us finally examine a third phenomenon, known as *unipolar induction*, which occurs when a magnetized conductor rotates. To that purpose, consider the magnetizing body $\mathscr{B}$ rotating about the axis O_z, with a *time-constant* angular velocity vector $\mathbf{\Omega}$ (see Figure 6.15.3). In a stationary wire ACB, connected to the rotating body by means of two sliding contacts (brushes) A and B, a current flows. Indeed, the electromotive force (e.m.f.) Φ which produces this current can be calculated from Faraday's law (3.9.6) applied to the open circuit ACB. That is, noting that the electric field vanishes in the fixed Galilean frame R_G

$$\Phi = \oint_{ACB} \mathscr{E} \cdot d\mathbf{x} = \frac{1}{c} \int_{ACB} (\mathbf{v} \times \mathbf{B}) \cdot d\mathbf{x}. \tag{6.15.21}$$

It is convenient to use a coordinate system rotating with the body $\mathscr{B}$. Then the wire ACB rotates with angular velocity $-\mathbf{\Omega}$ with respect to this coordinate system. In other words, we have a conductor moving in a given magnetic induction $\mathbf{B}$ due to a fixed magnet, if we assume that the presence of the wire practically does not distort this field. Then, eq. (6.15.21) gives the required solution

$$\Phi = \frac{1}{c} \int_{ACB} [\mathbf{B} \times (\mathbf{r} \times \mathbf{\Omega})] \cdot d\mathbf{x} \tag{6.15.22}$$

since $\mathbf{v} = -\mathbf{r} \times \mathbf{\Omega}$.

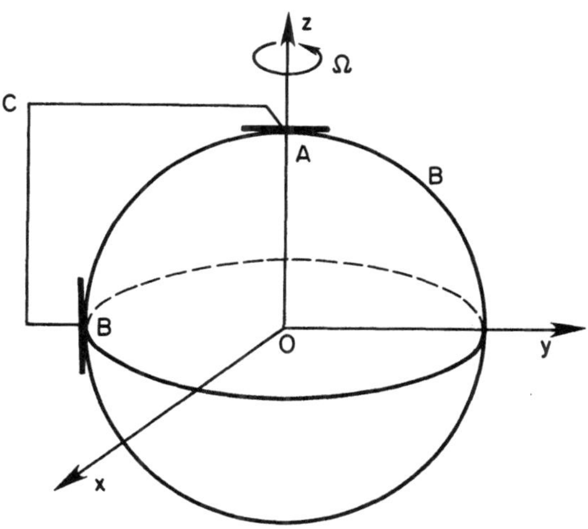

Figure 6.15.3. Unipolar induction.

To materialize the above considerations, consider $\mathscr{B}$ to be a uniformly magnetized sphere of radius r_0 rotating uniformly about the direction of magnetization O_z. The brush A is located at a pole, and the brush B at the equator (Figure 6.15.3). For a steady state rotation with constant angular velocity $\mathbf{\Omega}$, we have $\mathbf{B}^* = \mathbf{0}$ and (3.4.16) reduces to

$$\nabla \times \mathscr{E} = \mathbf{0}.$$

Since $\mathbf{E} = \mathbf{0}$, this gives

$$\nabla \times (\mathbf{v} \times \mathbf{B}) = \mathbf{0}. \tag{6.15.23}$$

This means that the circulation of $\mathbf{v} \times \mathbf{B}$ along any closed circuit vanishes, so that the integral in (6.15.22) along the circuit ACB can be replaced by that along AOB, which circuit lies within the sphere. The line integral along OA does not contribute, since $\mathbf{\Omega}$ and $\mathbf{r}$ are parallel. From (6.15.22) we have

$$\Phi = -\frac{1}{c}\int_0^{r_0} B_0 \Omega r\, dr = -\frac{B_0 \Omega r_0^2}{2c}, \tag{6.15.24}$$

where B_0 is the magnetic induction within the sphere. The latter having demagnetizing factors $\frac{1}{3}$, we have $B_0 = H_0 + M_0 = -\frac{1}{3}M_0 + M_0 = \frac{2}{3}M_0$, where M_0 is the uniform magnetization per unit volume. Let $\mathfrak{M}_0$ be the total magnetic moment of the sphere, then we obtain finally

$$\Phi = -\frac{\mathfrak{M}_0}{4\pi r_0 c}\Omega. \tag{6.15.25}$$

This result shows that the e.m.f. is linear in both $\mathfrak{M}_0$ and Ω.

6.16. Nonlinear Optics

The calculation of the propagation and interaction of electromagnetic waves in nonlinear media is the main task of the field of nonlinear optics. In this sense, nonlinear optics is none other than the dynamical theory of electromagnetism, with special features relevant to the physics of optics. For a large class of optical phenomena, magnetization of the medium and the dependence of polarization on the magnetic field are negligible. Moreover, free charge and current densities are assumed to vanish. Then, Maxwell's equations take the form

$$\nabla \times \mathbf{E} + \frac{1}{c}\frac{\partial \mathbf{H}}{\partial t} = \mathbf{0}, \qquad \nabla \cdot \mathbf{H} = 0, \tag{6.16.1}$$

$$\nabla \times \mathbf{H} - \frac{1}{c}\frac{\partial \mathbf{D}}{\partial t} = \mathbf{0}, \qquad \nabla \cdot \mathbf{D} = 0, \tag{6.16.2}$$

where, for **D**, we have a constitutive equations of the form (see (6.2.10)):[7]

$$\mathbf{D} = \varepsilon(|\mathbf{E}|^2)\mathbf{E}, \qquad \varepsilon \equiv 1 + \chi(|\mathbf{E}|^2). \tag{6.16.3}$$

Eliminating **H**, between (6.16.1) and (6.16.2), we obtain

$$\nabla \times \nabla \times \mathbf{E} + \frac{1}{c^2}\frac{\partial^2 \mathbf{D}}{\partial t^2} = \mathbf{0}. \tag{6.16.4}$$

The major difficulty in this equation stems from the term $\partial^2 \mathbf{D}/\partial t^2$, which is nonlinear in **E**. When the variation of the amplitude $|\mathbf{E}|$ is small over several wavelengths, in some sense, it is possible to replace this equation by a more tractable one. This is the *slowly varying amplitude* approximation. The essence of this idea is as follows: A plane wave propagating in the x-direction may be represented by

$$\mathbf{E}(\mathbf{x}, t) = \mathscr{E}(\mathbf{r}, t)\exp[i(k_0 x - \omega_0 t)], \tag{6.16.5}$$

where k_0 is the wave number and ω_0 is the circular frequency. The amplitude function $\mathscr{E}$ is supposed to be varying slowly over a distance of several wavelengths. The real part of **E** describes the waves. In the linear theory

$$\mathbf{D} = \varepsilon_0 \mathbf{E}, \tag{6.16.6}$$

where ε_0 is constant for nondispersive media. For dispersive media (e.g., memory-dependent materials), ε_0 depends on the frequency. For nonlinear materials, (6.16.6) is replaced by

$$\mathbf{D} = \varepsilon_0 \mathbf{E} + \mathbf{D}^{\mathrm{NL}} = (\varepsilon_0 + \chi)\mathbf{E}, \tag{6.16.7}$$

where χ is a function of $|\mathbf{E}|^2$.

In the linear theory, dispersion relations have the form

$$k^2 = \omega^2 \varepsilon / c^2 = F_0(\omega) \tag{6.16.8}$$

which would be valid for memory-dependent materials for which $\varepsilon = \varepsilon(\omega)$. Note that if we identify

$$\hat{\mathbf{k}} = -i\nabla, \qquad \hat{\omega} = i\frac{\partial}{\partial t}, \tag{6.16.9}$$

then replacing k and ω by $\hat{\mathbf{k}}$ and $\hat{\omega}$ in (6.16.8) and operating on **E**, we return to the wave equation (6.16.4) of the linear theory. We take advantage of this

[7] For an alternative approach, see Bloembergen [1965]. A full discussion of nonlinear optics involves both memory effects and spatial nonlocality. In this generality, **D** is considered to be a nonlinear functional of $\mathbf{E}(\mathbf{x}', t - \tau')$, where $\mathbf{x}'$ covers the entire volume of the body, and $0 \leq \tau' < \infty$ (see Chapter 14). Usually, for **D**, a Volterra series involving multiple space–time integrals, is written in terms of $\mathbf{E}(\mathbf{x}', t - \tau')$ (see Shen [1984] and Schubert and Wilhelmi [1986]). We return to this question later in Chapters 13 and 14.

property to replace (6.16.4) by a more tractable equation. To this end, we expand F_0 into a power series about $\omega = \omega_0$.

$$F_0(\hat{\omega})\mathbf{E} = [F_0(\omega_0) + iF_0'(\omega_0)\frac{\partial \mathscr{E}}{\partial t} - \frac{1}{2}F_0''(\omega_0)\frac{\partial^2 \mathscr{E}}{\partial t^2}$$

$$- \frac{i}{6}F_0'''(\omega_0)\frac{\partial^2 \mathscr{E}}{\partial t^3} + \cdots]e^{i(k_0 x - \omega_0 t)}, \tag{6.16.10}$$

where a prime represents the partial derivative with respect to ω_0. From (6.16.8), we have

$$\begin{aligned} k_0^2 &= \omega_0^2 \varepsilon_0/c^2 = F_0(\omega_0), \qquad F_0'(\omega_0) = 2k_0 k_0' = 2k_0/u_0, \\ F_0''(\omega_0) &= 2(k_0'^2 + k_0 k_0'') = 2(u_0 - k_0 u_0')u_0^{-3}, \\ F_0'''(\omega_0) &= 6k_0' k_0'' + 2k_0 k_0''' = 2(-3u_0 u_0' + 3k_0 u_0'^2 - k_0 u_0 u_0'')u_0^{-3}, \end{aligned} \tag{6.16.11}$$

where $u_0 = 1/k_0' = \partial\omega_0/\partial k_0$ denotes the group velocity and primes denote differentiations, e.g.,

$$u_0' = \frac{\partial u_0}{\partial k_0}, \qquad k_0' = \frac{\partial k_0}{\partial \omega_0}. \tag{6.16.12}$$

Substitution of (6.16.9) into

$$\hat{\mathbf{k}}^2\mathbf{E} = F_0(\hat{\omega})\mathbf{E} \tag{6.16.13}$$

yields an equation for the linear part of (6.16.4). Together with the nonlinear part, we have

$$L\mathscr{E} \equiv i(\mathscr{E}_{,x} + k_0'\mathscr{E}_{,t}) + \frac{1}{2k_0}\nabla_\perp^2\mathscr{E} - \tfrac{1}{2}k_0''\mathscr{E}_{,tt}$$

$$-i\left(\frac{k_0' k_0'''}{2k_0} + \tfrac{1}{6}k_0'''\right)\mathscr{E}_{,ttt} = \frac{1}{2k_0 c^2}[(\chi\mathscr{E})_{,tt} - 2i\omega_0(\chi\mathscr{E})_{,t} - \omega_0^2\chi\mathscr{E}]. \tag{6.16.14}$$

This nonlinear differential equation replaces (6.16.4) in the slowly varying amplitude approximation. In this approximation, a term $\mathscr{E}_{,xx}$ has been cancelled out, with the replacement of $\mathscr{E}_{,tt}$ with $u_0^2\mathscr{E}_{,xx}$. In fact, a further approximation is possible by dropping $(\chi\mathscr{E})_{,tt}$ from the right-hand side of this equation. In nonlinear optics, the slowly varying amplitude approximation is referred to an even higher-order approximation, reducing (6.16.14) to

$$i(\mathscr{E}_{,x} + k_0'\mathscr{E}_{,t}) = \frac{-\omega_0^2}{2k_0 c^2}\chi\mathscr{E} \tag{6.16.15}$$

(see Schubert and Wilhelmi [1986, §1.4.4] and Shen [1984, §3.5]). In fact, often the term $\mathscr{E}_{,t}$ is also dropped (Shen [1984, §3.3]). However, large numbers of scientists have employed equations of the form (6.16.14) involving up to third-order time derivatives (Hasegawa and Kodama [1981], [1982], Boardman and Cooper [1984], and Crosignani et al. [1981]).

Equation (6.16.15) should be used for short-pulse propagation. The time derivative term can be neglected, only if the amplitude variations are not important during the time $T = l\sqrt{\varepsilon_0}/e$, for the light to traverse l. In fact, these approximations require that

$$|\partial^2\mathscr{E}/\partial x^2| \ll k_0|\partial\mathscr{E}/\partial x|, \tag{6.16.16}$$

implying that the energy transfer is not significant, until the waves travel over a distance much longer than the wavelengths. We employ several simplified forms of (6.16.14) in the discussion of nonlinear optical effects with short pulses. Equation (6.16.14) does not contain a loss term. This could be brought in by replacing $\mathscr{E}$ with $\mathscr{E}\exp(-\gamma_0 t)$ in (6.16.14) where $\gamma_0 \geq 0$.

6.17. Solitary Waves

Experiments on nonlinear pulse propagation in optical fibers have demonstrated a variety of nonlinear optical effects which have a plethora of interesting technological applications. In particular, the appearance of optical solitons (Mollenauer et al. [1980]), has promise for distortionless signal transmission with a bit rate improved from 10 to 100 times. As is well known, solitons can not only travel without distortions in "perfect" fibers, but can also cross each other without distortion. Naturally, damping and higher-order dispersions can cause some alterations in their shapes, but it is possible to restore their original forms by amplification and reshaping them at periodic intervals (see Hasekawa and Kodama [1981], [1982]). Presently, a fairly large literature exists on this subject. We mention here only a few: Karpman and Krushkal [1969], Karpman [1975], Hasegawa and Brinkman [1980], Crosignani et al. [1981], Hasegawa et al. [1981], [1982], Boardman and Cooper [1984], and Kikuchi and Hirota [1985].

Consider the propagation of a pulse in a circular fiber. In this case, only a component of the electric field, transverse to the axis of the fiber, is nonzero, i.e.,

$$\mathscr{E}(\mathbf{r}, t) = R(r)\phi(x, t), \tag{6.17.1}$$

where $R(r)$ is an eigenfunction of the linear equation ($\chi \equiv 0$) which depends on the radial coordinate r. Below the threshold of self-focusing of the electric field (which is often 10^2 times higher than the electric field intensity), the use of a linear eigenfunction $R(r)$ is justified. In the lowest polynomial approximation, we may express[8]

$$\chi = \chi_0|\mathscr{E}|^2. \tag{6.17.2}$$

[8] Often the constitutive equation is expressed by means of the index of refraction $ck/\omega = n_0 + n_2|\mathbf{E}|^2$, where $n_0 = 1.5$, and the Kerr coefficient $n_2 = 3 \times 10^{-22}\ (\mathrm{V/m})^{-2}$ for SiO_2.

With these, (6.16.14) reduces to

$$i(\phi_{,x} + k_0'\phi_{,t}) - \frac{k_0''}{2}\phi_{,tt} - i\left(\frac{k_0'k_0''}{2k_0} + \frac{k_0'''}{6}\right)\phi_{,ttt}$$

$$= -\frac{\omega_0^2}{2k_0c^2}\left[\chi_0|\phi|^2\phi + \frac{2i}{\omega_0}(\chi_0|\phi|\phi)_{,t}\right]. \quad (6.17.3)$$

We transform this equation by setting

$$\xi = k_0x/a^2, \qquad \tau = \frac{1}{a}(-k_0/k_0'')^{1/2}\left(t - \frac{x}{u}\right),$$

$$q = \frac{a\omega_0}{k_0c}(\chi_0/2)^{1/2}\phi(x, t), \quad (6.17.4)$$

to obtain

$$i\frac{\partial q}{\partial \xi} + \frac{1}{2}\frac{\partial^2 q}{\partial \tau^2} + |q|^2 q - i\frac{b}{a}\frac{\partial^3 q}{\partial \tau^3} + i\frac{f}{a}\frac{\partial}{\partial \tau}(|q|^2 q) = 0, \quad (6.17.5)$$

where a is a scaling constant and

$$b = \frac{1}{k_0}\left(\frac{k_0'k_0''}{2k_0} + \frac{k_0'''}{6}\right)\left(\frac{k_0}{-k_0''}\right)^{3/2}, \qquad f = \frac{2}{\omega_0}\left(\frac{\beta_0}{-\beta_0''}\right)^{1/2}. \quad (6.17.6)$$

Equation (6.17.5), aside from the loss term $(ia^2r/k_0)q$, is identical to that given by Boardman and Cooper [1984], in a different way.

The equation with $b = f = 0$ has been studied extensively. This equation

$$i\frac{\partial q}{\partial \xi} + \frac{1}{2}\frac{\partial^2 q}{\partial \tau^2} + |q|^2 q = 0 \quad (6.17.7)$$

is known as the nonlinear Schrödinger equation. Zakharov and Shabat [1972] obtained the solution of this equation for any Cauchy data $q(0, \tau)$, by means of the *inverse scattering* technique. For the anomalous dispersion $k_0'' < 0$ (SiO_2 has anomalous group dispersion), the general response $q(\xi, \tau)$ can be described by a set of N solitons and a continuous mode, which die off asymptotically as $\xi \to \infty$. Hasegawa and Tappert [1973] gave a stationary solution of (6.17.7), which corresponds to one soliton solution

$$q(\xi, \tau) = q_0 \operatorname{sech}(q_0\tau)\exp(iq_0\xi/2). \quad (6.17.8)$$

This solution shows that a stationary pulse of the form $|q| = q_0 \operatorname{sech}(q_0\tau)$ is possible, even in the presence of group dispersion. Such a result is due to the balance of group dispersion with the nonlinear response of the fiber. The solution possesses a common feature of nonlinear waves, namely that the amplitude and velocity of pulse are related. The larger the amplitude, the faster it moves. It follows from the integration

$$A = \int_{-\infty}^{\infty} |q(\xi, \tau)|\, d\tau = \pi \quad (6.17.9)$$

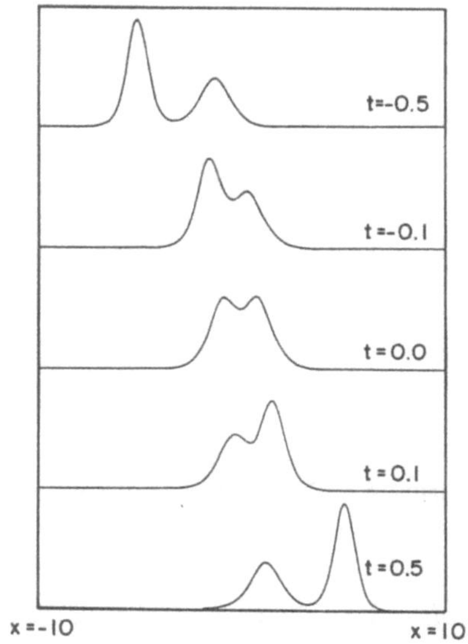

Figure 6.17.1. Interaction of two solitons in which single peak is not formed (from Lamb [1980]). Reprinted with permission from G.L. Lamb, *Elements of Soliton Theory*, © 1980, John Wiley and Sons.

that the width and amplitude of the pulse are related. In fact, the number of solitons depends on A. For the above single soliton, $A = \pi$.

The fundamental nature of the soliton is realized when two solitons interact. They will cross each other, reappearing in their original forms, without change of shape, but with only a shift of phase (see Figure 6.17.1). This subject matter has been studied extensively in connection with Korteweg–de Vries and other equations. Elaborate methods, such as inverse scattering techniques, have been developed for the solution of such evolution equations (see Lamb [1980], Ablowitz and Segur [1981], and Calogero and Degasperis [1982]).

The foregoing soliton solutions are sometimes referred to as the *Bright*

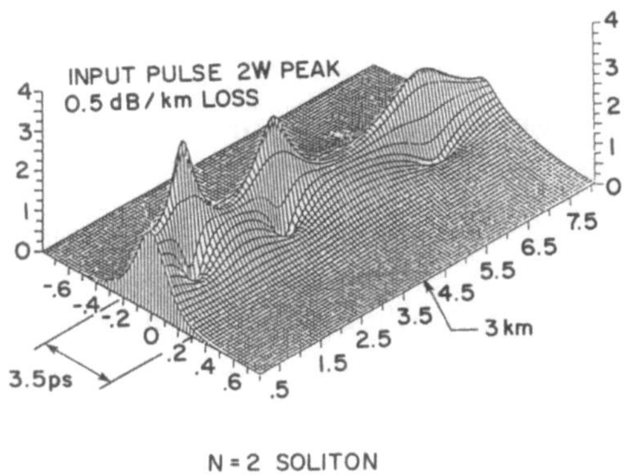

Figure 6.17.2. Solitons in nonlinear optics (after Boardman and Cooper [1984]). (From Boardman and Cooper [1984]. Permission granted by Kluwer Academic Publishers.)

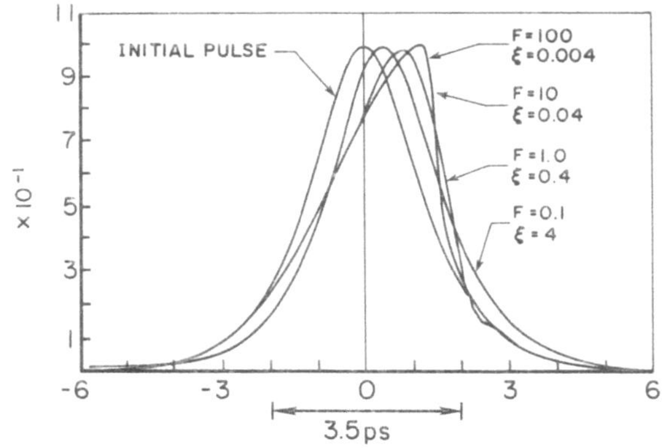

Figure 6.17.3. Amplitudes of solitons (after Boardman and Cooper [1984]). (From Boardman and Cooper [1984]. Permission granted by Kluwer Academic Publishers.)

solitons. In the case of $k_0'' > 0$, the solution is of the form

$$q(\xi, \tau) = q_0 \tanh(q_0 \tau) \exp(i q_0 \xi/2) \tag{6.17.10}$$

and it is called a *Dark soliton*.

The full evolution equation has other terms which produce third-order dispersion and optical shocks. Internal damping can cause diminution of the pulse height. The discussion of these phenomena requires computer solutions of (6.17.5), with an additional loss term $ia^2\, dq$ (Boardman and Cooper [1984] and Figures 6.17.2 and 6.17.3).

Problems

6.1. Obtain constitutive equations for $\mathbf{P}$ and $\mathcal{M}$ which are second degree in $\mathcal{E}$ and $\mathbf{B}$.

6.2. A conducting plane with a circular hole is subject to constant vertical electric fields E_0 and E_1 on the upper and lower sides of the plane far from the hole. The electric potential ϕ is specified by

$$\phi = \begin{cases} E_0 x_3 + \Phi, & x_3 > 0, \\ E_1 x_3 + \Phi, & x_3 < 0, \end{cases}$$

where Φ is the potential due to the presence of the hole. Show that $\Phi(r, x_3)$, where $r = (x_1^2 + x_2^2)^{1/2}$ can be expressed as

$$\Phi(r, x_3) = \int_0^\infty dk\, A(k) e^{-k|x_3|} J_0(kr),$$

where J_0 is the Bessel's function. Determine $A(k)$.

6.3. (a) Show that the electric potential ϕ of a uniformly polarized ellipsoid of principal axes a, b, c is

$$\phi = AxP_x + ByP_y + CzP_z,$$

where the numbers A, B, C are given

$$A = \frac{abc}{2}\int_0^\infty \frac{ds}{(a^2+s)R_s}, \qquad B = \frac{abc}{2}\int_0^\infty \frac{ds}{(b^2+s)R_s},$$

$$C = \frac{abc}{2}\int_0^\infty \frac{ds}{(c^2+s)R_s}, \qquad R_s = [(a^2+s)(b^2+s)(c^2+s)]^{1/2}$$

These are the eigenvalues of the depolarization tensor.

(b) Consider a prolate ellipsoid of revolution: $a \geqq b \geqq c$ and set $\delta = a/b$. Show that the values of A for an almost spherical ellipsoid (small $\delta^2 - 1$) and a very long prolate ellipsoid ($\delta \gg 1$) are given by

$$A = \tfrac{1}{3}[1 - \tfrac{2}{5}(\delta^2 - 1) + \cdots],$$

and

$$A \approx \frac{1}{\delta^2}(\ln 2\delta - 1),$$

respectively.

6.4. Discuss the reflection and refraction of a plane wave incident to a flat boundary between two homogeneous isotropic media with different optical properties. Determine the angle of transmission in terms of the angle of incidence. Find the intensity of light as a function of the angle of incidence.

6.5. For plane harmonic electromagnetic waves in conducting isotropic solids, show that the dispersion relation is given by

$$k^2 = (\mu\omega/c^2)\left(\varepsilon + \frac{i\sigma}{\omega}\right),$$

where k is the wave vector, μ, ε, and σ are material constants, and $\omega/2\pi$ is the frequency.

6.6. Given 4×4 matrices

$$F_{kl} = \begin{bmatrix} 0 & B_3 & -B_2 & -iE_1 \\ -B_2 & 0 & B_1 & -iE_2 \\ B_2 & -B_1 & 0 & -iE_3 \\ iE_1 & iE_2 & iE_3 & 0 \end{bmatrix}, \qquad H_{kl} = \begin{bmatrix} 0 & H_3 & -H_2 & -iD_1 \\ -H_3 & 0 & H_1 & -iD_2 \\ H_2 & -H_1 & 0 & -iD_3 \\ iD_1 & iD_2 & iD_3 & 0 \end{bmatrix}.$$

show that Maxwell's equations reduce to

$$\frac{\partial F_{kl}}{\partial x_m} + \frac{\partial F_{lm}}{\partial x_k} + \frac{\partial F_{mb}}{\partial x_l} = 0, \qquad \frac{\partial H_{kl}}{\partial x_l} = 0.$$

What form do these equations take for conducting solids?

6.7. Consider time-harmonic fields

$$\{\mathbf{E}(\mathbf{x}, t), \mathbf{H}(\mathbf{x}, t)\} = \{\mathbf{E}_0(\mathbf{x}), \mathbf{H}_0(\mathbf{x})\}e^{-i\omega t},$$

where $\mathbf{E}_0$ and $\mathbf{H}_0$ are given by asymptotic series of the form

$$\{\mathbf{E}_0, \mathbf{H}_0\} = e^{ik_0 S(\mathbf{x})}\left\{\sum_{m\geq 0} \mathbf{e}^{(m)}(\mathbf{x})(ik_0)^{-m}, \sum_{m\geq 0} h^{(m)}(\mathbf{x})(ik_0)^{-m}\right\}.$$

Show that for large k_0, $S(x)$ satisfies *eiconal equation*

$$\nabla S \cdot \nabla S = n^2,$$

where n is the refractive index.

6.8. The constitutive equation of a nonlinear dielectric is given by

$$\mathbf{D} = \varepsilon_0(1 + \alpha E^2)\mathbf{E}; \qquad E^2 = \mathbf{E}\cdot\mathbf{E},$$

where ε_0 and α are constants. By means of perturbation, obtain the solution up to second degree for two interacting harmonic waves of the form

$$\mathbf{E} = \mathbf{E}_1 e^{i(\mathbf{k}_1\cdot\mathbf{x} - \omega_1 t)} + \mathbf{E}_2 e^{i(\mathbf{k}_2\cdot\mathbf{x} - \omega_2 t)},$$

where E_1, E_2, k_1, k_2, ω_1, and ω_2 are constants.

6.9. (Short term paper.) Search the literature and write a paper on a two-soliton solution of the nonlinear Schrödinger equation. Discuss the interactions of solitons.

6.10. (Short term paper.) Explain the mathematical methods underlying the *inverse scattering* technique.

CHAPTER 7

Elastic Dielectrics

7.1. Scope of the Chapter

Chapter 7 is concerned with the continuum theory of elastic dielectrics. In Section 7.2 we list basic equations. The uniqueness theorem for the linear theory is presented in Section 7.3. Piezoelectric material moduli and their symmetry regulations for all thirty-two crystal classes are discussed in Section 7.4. In Section 7.5 we start to give solutions for various problems such as the thickness vibrations of piezoelectric plates (Section 7.5), the extensional vibrations of piezoelectric rods (Section 7.6), surface waves (Section 7.7), the radially symmetric vibrations of thin ceramic rings (Section 7.8), the spherically symmetric vibrations of ceramic shells (Section 7.9), and the piezoelectrically generated electric field (Section 7.10). These problems are useful for both experimental research and technological applicaton.[1] In contrast to these solutions in the linear theory, nonlinear solutions are rare and deserve special attention, especially since the recent development of piezoelectric polymers (Sessler [1981]). These problems are concerned with the behavior of incompressible dielectric solids under "controllable surface" loads. Thus, for example, *homogeneous deformations* of a slab under a uniform electric field, such as *simple shear*, are discussed in Section 7.11. Cylindrically symmetric deformations of a thick tube under a radial field are covered in Section 7.12. Various nonlinear effects (e.g., Kelvin, Poynting, the electrostrictive stress field due to the interaction of large strains and fields) are some of the physical phenomena that are not present in a linear theory. Section 7.13 deals with the nonlinear axisymmetric oscillations of an isotropic elastic dielectric.

Many interesting "cross effects" which result from the coupling of electro-

[1] The problems presented here do not represent an exhaustive study. Linear piezoelectricity (Voigt [1928]) is a flourishing field. For further study reference may be made to Cady [1946], Dieulesaint and Royer [1980], Mason [1950], Nelson [1979], Ristic [1983], Tiersten [1969], and Narasimhamurty [1981]. Nonlinear piezoelectric and other electromagnetic effects in crystals are dealt with by Maugin [1985] and by Eringen [1980, Chap. 10] see also Toupin [1956].

magnetic fields with the deformation of a solid can be observed by applying small deformations and fields on a dielectric solid, subject to large static deformation and fields, initially. For example, photoelastic, electroelastic, and magneto-optical effects are among such phenomena which possess important technological applications. The general theory is presented in Section 7.14. Applications of the theory to photoelasticity are discussed in Section 7.15, and to the electro-optical effects in Section 7.16. Section 7.17 deals with the propagation of waves in an initially polarized body under a uniform electric field. This enables us to discuss such cross phenomena as the transverse Kerr effect and the longitudinal Pockels effect. In Section 7.17 we discuss magneto-optical effects. By means of an initially applied magnetic field on an isotropic dielectric, anisotropy is induced. The propagation of waves in a driection normal to the bias magnetic field is known as the *Voigt–Cotton–Mouton effect*, and the propagation of waves in the direction of the bias field is known as the *Faraday effect*. The nature of such waves and the transverse birefringence are the subjects of Section 7.17. In Section 7.18 some ideas concerning elastic ionic crystals (e.g., NaCl) and elastic ferroelectrics (e.g., $BaTiO_3$ and $NaNO_2$) are given, and which require a continuum model more elaborate than the present one.

7.2. Résumé of Basic Equations

Below we collect the basic equations of nonmagnetizable elastic dielectrics that are not heat and electric conducting. In ultrasonics and many other fields (except when light is involved), where the wavelengths of interest are shorter than electromagnetic wavelengths of the same frequency, a quasi-electrostatic approximation is sufficient.

Equations of Balance (Valid in $\mathscr{V} - \sigma$)

$$\rho/\rho_0 = \mathrm{III}^{-1/2}, \tag{7.2.1}$$

$$_{\mathrm{E}}t_{kl,l} + (q_e - \nabla\cdot\mathbf{P})E_l + \rho(f_l - \dot{v}_l) = 0, \tag{7.2.2}$$

$$\nabla\cdot\mathbf{D} = q_e \tag{7.2.3}$$

$$\nabla\times\mathbf{E} = \mathbf{0}, \tag{7.2.4}$$

Equations of Jump (Valid on σ)

$$[{}_{\mathrm{E}}t_{kl} + E_k E_l - \tfrac{1}{2}\mathbf{E}^2\delta_{kl}]n_k = 0, \tag{7.2.5}$$

$$[\mathbf{D}]\cdot\mathbf{n} = w_e, \tag{7.2.6}$$

$$\mathbf{n}\times[\mathbf{E}] = \mathbf{0}, \tag{7.2.7}$$

where q_e and w_e are the prescribed body and surface charges.

Constitutive Equations

Anisotropic Solids

$$\rho_0 \Psi = \Sigma(C_{KL}, E_K, \theta, \mathbf{X}), \qquad \eta = -\frac{1}{\rho}\frac{\partial \Sigma}{\partial \theta}, \tag{7.2.8}$$

$$_{\mathrm{E}}t_{kl} = \frac{2\rho}{\rho_0}\frac{\partial \Sigma}{\partial C_{KL}} x_{k,K} x_{l,L}, \tag{7.2.9}$$

$$P_k = -\frac{\rho}{\rho_0}\frac{\partial \Sigma}{\partial E_K} x_{k,K}. \tag{7.2.10}$$

For material symmetry regulation, see Section 5.10.

Isotropic Solids. The constitutive equations of isotropic solids follow from (5.9.8) and (5.9.10) by dropping all terms that contain $\mathbf{B}$:

$$\Sigma = \Sigma(\mathrm{I}, \mathrm{II}, \mathrm{III}, \mathrm{I}_4, \mathrm{I}_6, \mathrm{I}_8, \theta, \mathbf{X}), \qquad \eta = -\frac{1}{\rho}\frac{\partial \Sigma}{\partial \theta}, \tag{7.2.11}$$

$$\begin{aligned} {}_{\mathrm{E}}\mathbf{t} = b_{-1} \overset{-1}{\mathbf{c}} + b_0 \mathbf{1} + b_1 \mathbf{c} - \chi_1 \left[\mathbf{E} \otimes \left(\overset{-1}{\mathbf{c}}\, \mathbf{E} \right) + \left(\overset{-1}{\mathbf{c}}\, \mathbf{E} \right) \otimes \mathbf{E} \right] \\ - 2\chi_2 \left[\mathbf{E} \otimes \left(\overset{-2}{\mathbf{c}}\, \mathbf{E} \right) + \left(\overset{-2}{\mathbf{c}}\, \mathbf{E} \right) \otimes \mathbf{E} \right], \end{aligned} \tag{7.2.12}$$

$$\mathbf{P} = \chi \mathbf{E} + \chi_1 \overset{-1}{\mathbf{c}}\, \mathbf{E} + \chi_2 \overset{-2}{\mathbf{c}}\, \mathbf{E}, \tag{7.2.13}$$

where

$$b_{-1} \equiv -2\frac{\rho}{\rho_0}\frac{\partial \Sigma}{\partial \mathrm{I}}, \qquad b_0 = 2\frac{\rho}{\rho_0}\left(\mathrm{II}\frac{\partial \Sigma}{\partial \mathrm{II}} + \mathrm{III}\frac{\partial \Sigma}{\partial \mathrm{III}} \right), \qquad b_1 = -2\frac{\rho}{\rho_0}\mathrm{III}\frac{\partial \Sigma}{\partial \mathrm{II}},$$

$$\chi = 2\frac{\rho}{\rho_0}\frac{\partial \Sigma}{\partial \mathrm{I}_4}, \qquad \chi_1 = -2\frac{\rho}{\rho_0}\frac{\partial \Sigma}{\partial \mathrm{I}_6}, \qquad \chi_2 = -2\frac{\rho}{\rho_0}\frac{\partial \Sigma}{\partial \mathrm{I}_8},$$

$$\mathrm{I} = \operatorname{tr} \overset{-1}{c}, \qquad \mathrm{II} = \frac{1}{2}\left[\left(\operatorname{tr} \overset{-1}{c} \right)^2 - \operatorname{tr} \overset{-2}{\mathbf{c}} \right], \qquad \mathrm{III} = \det \overset{-1}{\mathbf{c}},$$

$$\mathrm{I}_4 = \mathbf{E} \cdot \mathbf{E}, \qquad \mathrm{I}_6 = \mathbf{E}\, \overset{-1}{\mathbf{c}}\, \mathbf{E}, \qquad \mathrm{I}_8 = \mathbf{E}\, \overset{-2}{\mathbf{c}}\, \mathbf{E},$$

$$\overset{-1}{c}_{kl} = x_{k,K} x_{l,K}, \qquad c_{kl} = X_{K,k} X_{K,l}, \qquad C_{KL} = x_{k,K} x_{k,L},$$

$$t_{kl} = {}_{\mathrm{E}}t_{kl} - P_k E_l. \tag{7.2.14}$$

Isotropic, Incompressible Solids. For the incompressible solids, $\rho_0 = \rho$, since Σ is subject to the constraint of $\mathrm{III} = 1$, the constitutive equations are modified to the following:

$$\rho_0\psi = \Sigma(\mathrm{I}, \mathrm{II}, \mathrm{I}_4, \mathrm{I}_6, \mathrm{I}_8, \theta, \mathbf{X}), \qquad \eta = -\frac{1}{\rho}\frac{\partial\Sigma}{\partial\theta}, \tag{7.2.15}$$

$$\begin{aligned} {}_{\mathrm{E}}\mathbf{t} = -p\mathbf{1} + b_{-1}\overset{-1}{\mathbf{c}} + b_1\mathbf{c} - \chi_1\left[\mathbf{E}\otimes\left(\overset{-1}{\mathbf{c}}\mathbf{E}\right) + \left(\overset{-1}{\mathbf{c}}\mathbf{E}\right)\otimes\mathbf{E}\right] \\ - 2\chi_2\left[\mathbf{E}\otimes\left(\overset{-2}{\mathbf{c}}\mathbf{E}\right) + \left(\overset{-2}{\mathbf{c}}\mathbf{E}\right)\otimes\mathbf{E}\right], \end{aligned} \tag{7.2.16}$$

$$\mathbf{P} = \chi\mathbf{E} + \chi_1\overset{-1}{\mathbf{c}}\mathbf{E} + \chi_2\overset{-2}{\mathbf{c}}\mathbf{E}, \tag{7.2.17}$$

where $p(\mathbf{x}, t)$ is an unknown pressure and

$$\begin{aligned} b_{-1} = 2\frac{\partial\Sigma}{\partial\mathrm{I}}, \qquad b_1 = -2\frac{\partial\Sigma}{\partial\mathrm{II}}, \\ \chi = -2\frac{\partial\Sigma}{\partial\mathrm{I}}, \qquad \chi_1 = -2\frac{\partial\Sigma}{\partial\mathrm{I}_6}, \qquad \chi_2 = -2\frac{\partial\Sigma}{\partial\mathrm{I}_8}. \end{aligned} \tag{7.2.18}$$

Linear Theory of Piezoelectricity. Constitutive equations of the linear theory are extracted from (5.11.13)–(5.11.16). Without the initial fields, and in the absence of heat and electric conduction, we have

$$\begin{aligned} \Sigma &= \sigma_0 + \tfrac{1}{2}\sigma_{klmn}\tilde{e}_{kl}\tilde{e}_{mn} - \tilde{e}_{klm}E_k\tilde{e}_{lm} - \tfrac{1}{2}\chi^{\mathrm{E}}{}_{kl}E_kE_l, \\ \eta &= \eta_0 + \frac{1}{\rho_0}\beta_{kl}\tilde{e}_{kl}, \end{aligned} \tag{7.2.19}$$

$${}_{\mathrm{E}}t_{kl} = \sigma_{klmn}\tilde{e}_{mn} - e_{mkl}E_m, \tag{7.2.20}$$

$$D_k = \varepsilon_{kl}E_l + e_{klm}\tilde{e}_{lm}, \tag{7.2.21}$$

where we introduced dielectric constants by[2]

$$\varepsilon_{kl} = \varepsilon_0\delta_{kl} + \chi^{\mathrm{E}}{}_{kl}. \tag{7.2.22}$$

Linear Field Equations

If we introduce the electrostatic potential $\phi(\mathbf{x}, t)$ by

$$\mathbf{E} = -\nabla\phi, \tag{7.2.23}$$

then (7.2.4) is satisfied identically. Using this and the strain tensor

$$\tilde{e}_{kl} = \tfrac{1}{2}(u_{k,l} + u_{l,k}) \tag{7.2.24}$$

---

[2] Here we reinstated the vacuum dielectric constant, ε_0, which was taken as unity in the development. In Institute of Radio Engineers Standards [1949], $\varepsilon_0 = 8{,}854 \times 10^{-12}$ farad/m.

in (7.2.20) and (7.2.21), we have

$$ {}_{\mathrm{E}}t_{kl} = \sigma_{klmn} u_{m,n} + e_{mkl}\phi_{,m}, \tag{7.2.25} $$

$$ D_k = -\varepsilon_{kl}\phi_{,l} + e_{klm} u_{l,m}. \tag{7.2.26} $$

Substituting these into (7.2.2) and (7.2.3), and dropping the nonlinear terms in (7.2.2), we obtain the field equations of the linear theory:

$$ \sigma_{klmn} u_{m,nk} + e_{mlk}\phi_{,mk} + \rho(f_l - \ddot{u}_l) = 0, \tag{7.2.27} $$

$$ e_{klm} u_{l,mk} - \varepsilon_{kl}\phi_{,lk} = q_e. \tag{7.2.28} $$

These partial differential equations must be solved under boundary and initial conditions to determine $u_k(\mathbf{x}, t)$ and $\phi(\mathbf{x}, t)$.

Boundary Conditions[3]

$$ \begin{aligned} {}_{\mathrm{E}}t_{kl} n_k &= t_{(\mathbf{n})l} \quad \text{on } \partial\mathscr{V}_{\mathbf{t}}, \\ u_k &= \bar{u}_k \quad \text{on } \partial\mathscr{V}_{\mathbf{u}} = \partial\mathscr{V} - \partial\mathscr{V}_{\mathbf{t}}, \end{aligned} \tag{7.2.29} $$

$$ \begin{aligned} [\mathbf{D}]\cdot\mathbf{n} &= w_e \quad \text{on } \partial\mathscr{V}_{\mathbf{D}}, \\ [\phi] &= 0 \quad \text{on } \partial\mathscr{V}_{\phi} = \partial\mathscr{V} - \partial\mathscr{V}_{\mathbf{D}}. \end{aligned} \tag{7.2.30} $$

where $\partial\mathscr{V}_{\mathbf{t}}$ and $\partial\mathscr{V}_{\mathbf{D}}$ are parts of the boundary $\partial\mathscr{V}$, and $\partial\mathscr{V}_{\mathbf{u}}$ and $\partial\mathscr{V}_{\phi}$ are the remaining parts of $\partial\mathscr{V}$ after either $\partial\mathscr{V}_{\mathbf{t}}$ or $\partial\mathscr{V}_{\mathbf{D}}$ is excluded. Some of these parts may be absent or cover the entire surface. Of course, other possibilities exist. However, in all cases, boundary conditions must not violate the uniqueness theorem.

Initial Conditions. Initial conditions usually consist of the Cauchy data

$$ \begin{aligned} u_k(\mathbf{x}, 0) &= u^0{}_k(\mathbf{x}) \quad \text{in } \mathscr{V}, \\ \dot{u}_k(\mathbf{x}, 0) &= v^0{}_k(\mathbf{x}) \quad \text{in } \mathscr{V}. \end{aligned} \tag{7.2.31} $$

In the electrostatic equation (7.2.28) we notice the absence of time rates. This is due to the absence of magnetization. Clearly, in a moving body a magnetic field is induced. However, for the nonmagnetizable dielectrics, this effect is small. Therefore, the time in ϕ appears as a parameter, and dependence on time is induced by the displacement field.

7.3. Uniqueness Theorem

Definition 1. Let $\mathscr{V}$ be a spatial region with boundary $\partial\mathscr{V}$ and let T be a time interval. If $\mathbf{u}$ and $\mathbf{t}$ are, respectively, a vector-valued and a symmetric tensor-valued function, and ϕ and $\mathbf{D}$ are, respectively, a scalar- and a vector-valued function, all defined on $\mathscr{V} \times T$, then we call the two ordered pair $\mathscr{P}[(\mathbf{u}, \mathbf{t}),$

[3] Note that for the linear theory $t_{kl} = {}_{\mathrm{E}}t_{kl}$.

$(\phi, \mathbf{D})]$ a piezoelectric state on $\mathscr{V} \times T$ provided[4]

$$\begin{aligned} \text{(a)} \quad & \mathbf{u} \in C^{2,2}(\mathscr{V} \times T), \qquad \mathbf{u} \in C^{1,1}(\partial\mathscr{V} \times T), \\ & \mathbf{t} \in C^{1,0}(\mathscr{V} \times T), \\ & \mathbf{D} \in C^1, \qquad \phi \in C^2(\mathscr{V} \times T), \\ & \mathbf{D} \in C^0, \qquad \phi \in C^1(\partial\mathscr{V} \times T) \end{aligned} \tag{7.3.1}$$

(b) $(\mathbf{u}, \mathbf{x})$ and $(\phi, \mathbf{D})$ satisfy the equations

$$\begin{aligned} & t_{kl,k} + \rho(f_l - \ddot{u}_l) = 0, \qquad D_{k,k} = q_e, \\ & t_{kl} = \sigma_{klmn} u_{m,n} + e_{mlk}\phi_{,m}, \\ & D_k = e_{klm} u_{l,m} - \varepsilon_{kl}\phi_{,l} \end{aligned} \tag{7.3.2}$$

or the set

$$\begin{aligned} \sigma_{klmn} u_{m,nk} + e_{mlk}\phi_{,mk} + \rho(f_l - \ddot{u}_l) = 0, \\ e_{klm} u_{l,mk} - \varepsilon_{kl}\phi_{,lk} = q_e, \end{aligned} \tag{7.3.3}$$

where

$$\mathbf{f} \in C^{0,0}(\mathscr{V} \times T), \qquad \rho > 0$$

and

$$\sigma_{klmn} a_k a_m b_l b_m \geq 0, \qquad \varepsilon_{kl} a_k a_l \geq 0, \tag{7.3.4}$$

for arbitrary vectors **a** and **b**.

We denote the class of all piezoelectric states satisfying the above conditions by $\mathscr{E}$ and write

$$\mathscr{P} \in \mathscr{E}(\mathbf{f}, q_e, \rho, \sigma_{klmn}, e_{klm}, \varepsilon_{kl}; \mathscr{V} \times T), \tag{7.3.5}$$

we say that $\mathscr{P}$ is a *piezoelectric state of quiescent past*, and write

$$\mathscr{P} = \mathscr{E}_0(\mathbf{f}, q_e, \rho, \boldsymbol{\sigma}, \mathbf{e}, \varepsilon; \mathscr{V}). \tag{7.3.6}$$

The conditions (7.3.4) express that the constitutive moduli (tensors) are semi-strongly elliptic. Actually, these conditions and the smoothness conditions (7.3.1), are not all necessary and can be weakened.

Uniqueness Theorem. *Let $\mathscr{V}$ be a regular region*[5] *with a boundary $\partial\mathscr{V}$, and let $\mathscr{P}'$ and $\mathscr{P}''$ be two piezoelectric states with the following properties:*

(a)

$$\begin{aligned} \mathscr{P}' &= [(\mathbf{u}', \mathbf{t}'), (\phi', \mathbf{D}')] = \mathscr{E}'(f, q_e, \rho, \boldsymbol{\sigma}, \mathbf{e}, \boldsymbol{\varepsilon}; \mathscr{V} \times T^+), \\ \mathscr{P}'' &= [\mathbf{u}'', \mathbf{t}''), (\phi'', \mathbf{D}'')] = \mathscr{E}''(f, q_e, \rho, \boldsymbol{\sigma}, \mathbf{e}, \boldsymbol{\varepsilon}; \mathscr{V} \times T^+). \end{aligned} \tag{7.3.7}$$

---

[4] $F \in C^{i,j}$ denotes the continuity class of the function $F(\mathbf{x}, t)$, which posseses continuous partial derivatives of a total order i with respect to x_k, and of order j with respect to t. If j is missing, then the statement applies only to x_k.

[5] In the sense of Kellog [1929, p. 112].

(b) *The initial conditions for both states are identical*:

$$\mathbf{u}'(\mathbf{x}, 0) = \mathbf{u}''(\mathbf{x}, 0) \qquad \dot{\mathbf{u}}'(\mathbf{x}, 0) = \dot{\mathbf{u}}''(\mathbf{x}, 0) \quad \text{in } \mathscr{V}. \tag{7.3.8}$$

(c) *The boundary conditions for both states are identical*:

$$\begin{aligned} \mathbf{u}' &= \mathbf{u}'' && \text{on } \partial\mathscr{V}_{\mathbf{u}} \times T^+, \\ \mathbf{t}'_{(\mathbf{n})} &= \mathbf{t}''_{(\mathbf{n})} && \text{on } \partial\mathscr{V}_{\mathbf{t}} \times T^+; \qquad \partial\mathscr{V}_{\mathbf{u}} \cup \partial\mathscr{V}_{\mathbf{t}} = \partial\mathscr{V}, \\ \phi' &= \phi'' && \text{on } \partial\mathscr{V}_{\phi} \times T^+, \\ \mathbf{D}' &= \mathbf{D}'' && \text{on } \partial\mathscr{V}_{\mathbf{D}} \times T^+; \qquad \partial\mathscr{V}_{\phi} \cup \partial\mathscr{V}_{\mathbf{D}} = \partial\mathscr{V}, \end{aligned} \tag{7.3.9}$$

then the states $\mathscr{P}'$ and $\mathscr{P}''$ are identical in $\mathscr{V} \times T$ to within a static rigid body displacement and a constant potential.

Proof. Suppose that the contrary is valid, and two piezoelectric states $\mathscr{P}'$ and $\mathscr{P}''$ exist satisfying (7.3.2) or (7.3.3) and the identical initial and boundary conditions (7.3.8) and (7.3.9). Then, clearly,

$$\mathscr{P}[(\mathbf{u}, \mathbf{t}), (\phi, \mathbf{D})] = \mathscr{P}' - \mathscr{P}'', \tag{7.3.10}$$

also satisfies (7.3.2) or (7.3.3) with $\mathbf{f} = \mathbf{0}$, $q_e = 0$, and homogeneous initial and boundary conditions

$$\begin{aligned} \mathbf{u} = \dot{\mathbf{u}} &= \mathbf{0}, && \mathscr{V}, t = 0, \\ \mathbf{u} &= \mathbf{0}, && \partial\mathscr{V}_u \times T^+, \\ \mathbf{t}_{(\mathbf{n})} &= \mathbf{0}, && \partial\mathscr{V}_{\mathbf{t}} \times T^+, \end{aligned} \tag{7.3.11}$$

$$\begin{aligned} \phi &= 0, && \partial\mathscr{V}_{\phi} \times T^+, \\ \mathbf{D} &= \mathbf{0}, && \partial\mathscr{V}_{\mathbf{D}} \times T^+. \end{aligned} \tag{7.3.12}$$

Multiplying $(7.3.2)_1$ by $\dot{\mathbf{u}}_l$, and integrating the result over $\mathscr{V}$, we will have

$$\int_{\mathscr{V}} (t_{kl,k}\dot{u}_l - \rho\ddot{u}_l\dot{u}_l)\, dv = 0,$$

alternatively,

$$\int_{\mathscr{V}} [(t_{kl,k}v_l)_{,k} - t_{kl}v_{l,k}]\, dv - \frac{d}{dt}\int_{\mathscr{V}} \tfrac{1}{2}\rho v^2\, dv = 0.$$

By means of the Green–Gauss theorem, the first term is converted to a surface integral, so that

$$\int_{\partial\mathscr{V}} t_{kl}v_l\, da_k - \int_{\mathscr{V}} t_{kl}v_{l,k}\, dv - \frac{d\mathscr{K}}{dt} = 0, \tag{7.3.13}$$

where we also wrote

$$\mathscr{K} = \frac{1}{2}\int_{\mathscr{V}} \rho v^2\, dv \tag{7.3.14}$$

for the total kinetic energy. Employing (7.3.2) we have

$$\int_{\partial\mathscr{V}} t_{kl}v_l\,da_k - \int_{\mathscr{V}} (\sigma_{klmn}\dot{\tilde{e}}_{kl}\tilde{e}_{mn} + e_{mlk}\phi_{,m}\dot{\tilde{e}}_{kl} + \dot{\mathscr{K}})\,dv = 0.$$

But from $(7.3.2)_4$ we have

$$\begin{aligned} e_{mlk}\phi_{,m}\dot{\tilde{e}}_{kl} &= \dot{D}_k\phi_{,k} + \varepsilon_{kl}\dot{\phi}_{,k}\phi_{,l}, \\ &= (\dot{D}_k\phi)_{,k} + \varepsilon_{kl}\dot{\phi}_k\phi_{,l}, \end{aligned}$$

where we used $D_{k,k} = 0$, consequently,

$$\int_{\partial\mathscr{V}} (t_{kl}v_l - \phi\dot{D}_k)\,da - \frac{d}{dt}\int_{\mathscr{V}} (\Sigma + \mathscr{K})\,dv = 0, \tag{7.3.15}$$

with

$$\Sigma = \tfrac{1}{2}\sigma_{klmn}\tilde{e}_{kl}\tilde{e}_{mn} + \tfrac{1}{2}\varepsilon_{kl}E_kE_l. \tag{7.3.16}$$

But on the boundary, according to (7.3.12), the first term vanishes. Hence

$$\frac{d}{dt}\int_{\mathscr{V}} (\Sigma + \mathscr{K})\,dv = 0, \tag{7.3.17}$$

$$\tilde{e}_{kl} = 0, \qquad E_k = 0, \qquad \mathscr{V} \times T^+. \tag{7.3.18}$$

Thus, the two solutions are identical to within a static rigid body displacement and a constant potential.

Analogous to the uniqueness theorem of elastodynamics (see Eringen and Şuhubi [1975, p. 365]), the uniqueness theorem, so proved, can be extended to the infinite region.

Essential to the proof is the vanishing of the first term in (7.3.15), i.e.,

$$\int_{\partial\mathscr{V}} (t_{kl}v_l - \phi\dot{D}_k)\,da_k = 0. \tag{7.3.19}$$

Therefore it is clear that, in place of the mixed boundary conditions (7.3.12), other possibilities exist. For example, the vanishing of some components of t_{k1} on a part of the boundary and other components of v_k, e.g.,

$$t_{k1} = 0, \qquad v_2 = v_3 = 0.$$

Following Eringen [1955] such boundary conditions may be called *mixed–mixed*. Among other possible boundary conditions are the linear combinations of t_{kl}, v_l, ϕ, and $\mathbf{D}$ which make the integrand vanish on the boundary.

7.4. Piezoelectric Moduli

In the study and determination of piezoelectric constitutive moduli, it is often useful to employ a compressed notation which reduces the third- and fourth-order tensors in three dimensions to second-order matrices in higher dimensions. To this end, we replace a pair of indices with one index. Thus, we write a, b for each pair (k, l) and (m, n). For example, we write

$$\begin{aligned} &t_{11} = t_1, && t_{22} = t_2, && t_{33} = t_3, \\ &t_{23} = t_{32} = t_4, && t_{31} = t_{13} = t_5, && t_{12} = t_{21} = t_6, \end{aligned} \tag{7.4.1}$$

and similarly, for others. In this notation, due to Voigt, the constitutive equations (7.2.20) and (7.2.21) read

$$\begin{aligned} t_a &= \sigma_{ab}\tilde{e}_b - e_{la}E_l, \qquad && a, b = 1, 2, \ldots, 6, \\ D_k &= e_{ka}\tilde{e}_a + \varepsilon_{kl}E_l, && k, l = 1, 2, 3. \end{aligned} \tag{7.4.2}$$

The material moduli can now be written as matrices

$$\sigma^{ab} = \begin{pmatrix} \sigma_{11} & \sigma_{12} & \sigma_{13} & \sigma_{14} & \sigma_{15} & \sigma_{16} \\ & \sigma_{22} & \sigma_{23} & \sigma_{24} & \sigma_{25} & \sigma_{26} \\ & & \sigma_{33} & \sigma_{34} & \sigma_{35} & \sigma_{36} \\ & & & \sigma_{44} & \sigma_{45} & \sigma_{46} \\ & & & & \sigma_{55} & \sigma_{56} \\ & & & & & \sigma_{66} \end{pmatrix}, \tag{7.4.3}$$

$$e_{ka} = \begin{pmatrix} e_{11} & e_{12} & e_{13} & e_{14} & e_{15} & e_{16} \\ e_{21} & e_{22} & e_{23} & e_{24} & e_{25} & e_{26} \\ e_{31} & e_{32} & e_{33} & e_{34} & e_{35} & e_{36} \end{pmatrix}, \tag{7.4.4}$$

$$\varepsilon_{kl} = \begin{pmatrix} \varepsilon_{11} & \varepsilon_{12} & \varepsilon_{13} \\ & \varepsilon_{22} & \varepsilon_{23} \\ & & \varepsilon_{33} \end{pmatrix}. \tag{7.4.5}$$

The matrices σ_{ab} and ε_{kl} are symmetric with respect to the main diagonal, so we have $21 + 18 + 6 = 45$ independent moduli for the most general piezoelectric crystals. With material symmetry the number of independent components decreases further.

Experimentalists, apparently preferring stress and electric fields as independent variables, have often employed the expressions

$$\begin{aligned} \tilde{e}_a &= s_{ab}t_b + d_{ka}E_k, \\ D_k &= d_{ka}t_a + \bar{\varepsilon}_{kl}E_l. \end{aligned} \tag{7.4.6}$$

By carrying $(7.4.6)_1$ into (7.4.2) we find that

$$\sigma_{ab} = \overset{-1}{\mathbf{s}}_{ab}, \qquad e_{ka} = d_{kb}\,\overset{-1}{\mathbf{s}}_{ab}, \qquad \varepsilon_{kl} = \bar{\varepsilon}_{kl} - d_{ka}\,d_{lb}\,\overset{-1}{\mathbf{s}}_{ab}, \tag{7.4.7}$$

where $\overset{-1}{\mathbf{s}}$ is the inverse matrix to $\mathbf{s}$.

The constitutive equations (7.4.6) are often found to be more convenient for direct measurements. while (7.4.2) are better suited for theoretical work.

Table 7.4.1. Elastoelectric matrices for the 32 crystal classes.[a]

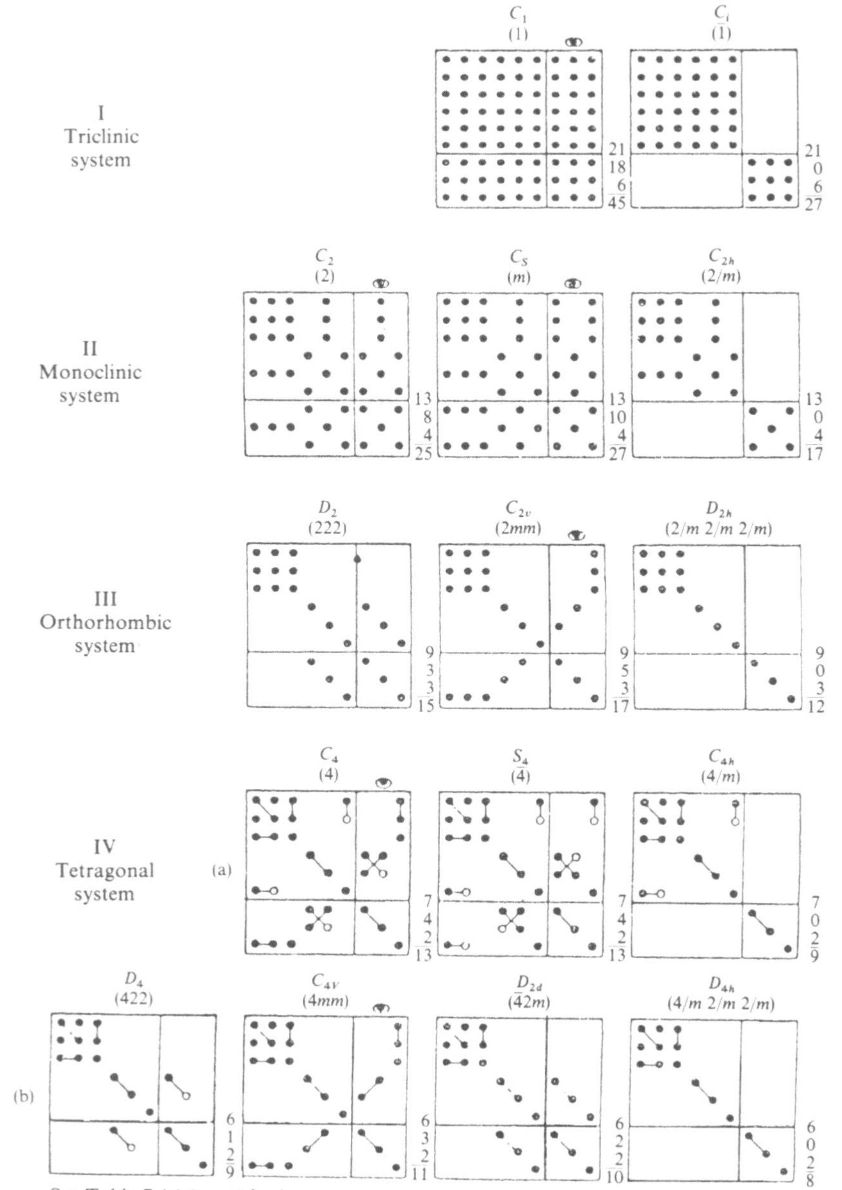

See Table 7.4.1 (*cont.*) for key

[a] The numbers on the right-hand side of each scheme indicate, from top to bottom, the number of the independent elastic, piezoelectric, and dielectric constants.

Table 7.4.1 (*continued*)

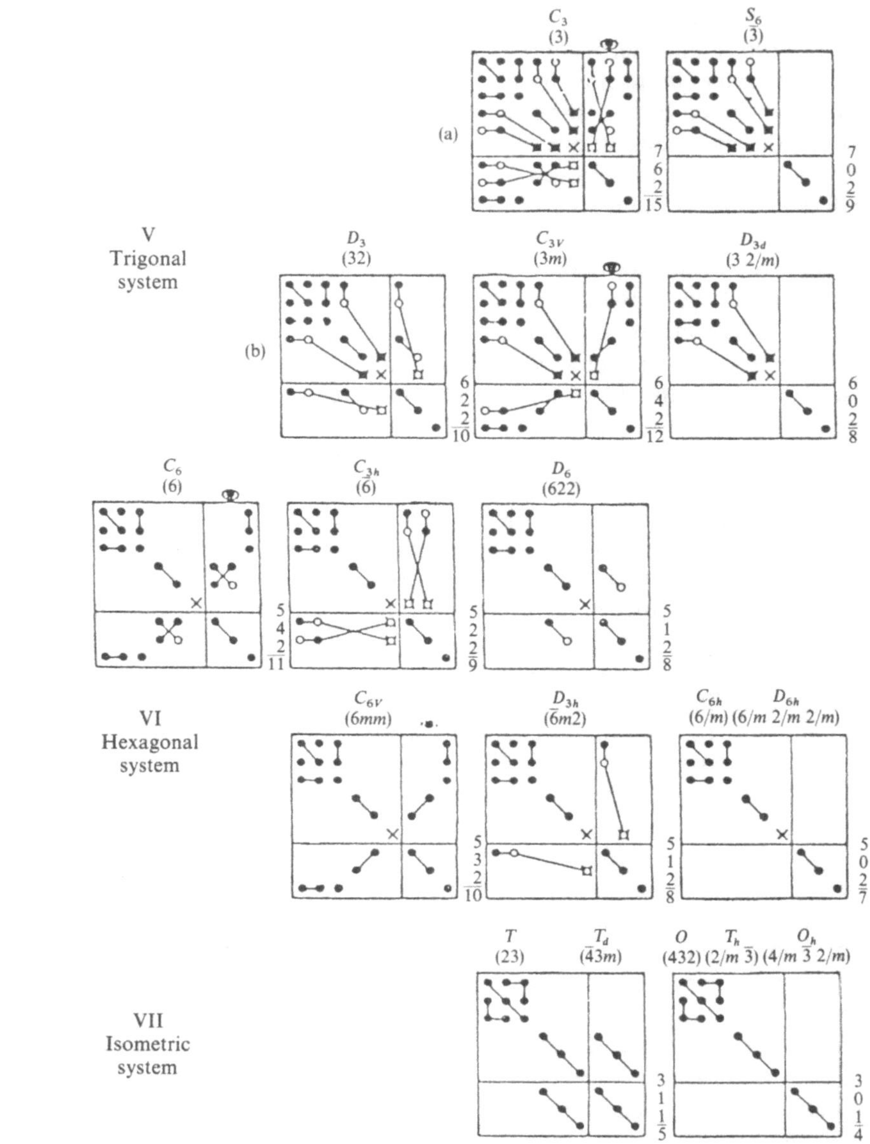

Key: Lines join numerical equalities except for complete reciprocity across principal diagonal which holds for all classes:
(○) indicates negative of ●;
(■) These classes are piezoelectric on hydrostatic compression and have pyroelectric properties.
In the trigonal and hexagonal system
■ □ indicates for s, d, d_1, g or g_1 twice the numerical equalities;
× indicates $2(s_{11} - s_{12})$ or $1/2(c_{11} - c_{12})$.

A thorough collection of numerical data on piezoelectric crystals up to 1956 is given by R. Bechmann in Landolt–Börnstein Tables [1959]. Shorter tables are to be found in the *American Institute of Physics Handbook* [1957], Brown et al. [1962], and Berlincourt et al. [1964]. Early work, particularly useful with experimental techniques and on the properties of quartz and Rochelle salt, is included in the authoritative work of Cady [1946]. Mason's [1950] book contains extensive discussions on these, and on man-made piezoelectric crystals such as a dihydrogen phosphate (ADP), lithium sulphate, etc. These references also contain other electromagnetic elastic moduli which are related to those displayed by (5.11.13)–(5.11.19).

To provide an idea, we reproduce here several tables on the material constants of the most commonly used piezoelectric crystals for transducer design. These tables are taken from Berlincourt et al. [1964]. Table 7.4.1 summarizes, in a graphical form, the nonvanishing components of the constitutive constants. The blank spaces indicate vanishing coefficients. The tie-lines indicate the equality required by the symmetry; and open circles equal the absolute value but with an opposite sign. This representation is attributed to Professor K.S. Van Dyke. The table is reprinted from the Institute of Radio Engineers Standards on piezoelectric crystals [1958].

Table 7.4.2 presents the material moduli of Rochelle salt and other monoclinic crystals. Rochelle salt has two Curie points (see Figure 4.3.2)—one at −18 °C and the other at 23 °C. Between these two points it possesses ferroelectricity, where the crystal symmetry is lowered to monoclinic class 2, requiring additional elastoelectric constants. Nevertheless, the value of piezoelectric coupling around both Curie points, and especially the inclusion of the room temperture region, has kept Rochelle salt in a prominent position for transducer design. Lithium sulfate, $Li_2SO_4 \cdot H_2O$ (class 2), has a very high piezoelectric response to hydrostatic pressure ($d_{21} + d_{23} + d_{22}$) and a strong response to strain, parallel to field d_{22}. With a moderate dielectric constant, this coupling makes lithium sulfate valuable as a transducer for investigating compressional waves in the 0.5–5 Mc range. In Table 7.4.3 we present material constants of some cubic crystals, in Table 7.4.4 those of hexagonal and tetragonal crystals, and in Table 7.4.5 properties of quartz. The data in all of these tables refer to room temperature, unless otherwise specified. Quartz is one of the most widely used materials for transducer design. It has a low piezoelectric coupling, which becomes advantageous at frequencies of about 15 Mc. Because of its excellent mechanical properties, very thin plates can be made. At high frequencies the hexagonal crystals, cadmium sulfide or selenide and zinc oxide, may become competitive because of their semiconductor properties for sound wave generation and amplification. Structurally related cubic semiconductor crystals (Table 7.4.3) are weaker in piezoelectric effects, but offer simplicity in theoretical work. The tables are based on rationalized mks units for the piezoelectric costants, 1 C/N (Coulomb/Newton) = $3 \cdot 10^4$ esu/dyne. Dielectric constants K_{kl} are relative to the vacuum $K_{kl} = \varepsilon_{kl}/\varepsilon_0$ where $\varepsilon_0 = 8.854 \cdot 10^{-12}$ F/M.

Table 7.4.2. Elastoelectric constants of rochelle salt and monoclinic crystals. (Tables from Berlincourt et al. [1964].)

Crystal	Density (kg/m^3)	Elastic compliance s_{ij}^E (10^{-12} m^2/N)												
		s_{11}	s_{22}	s_{33}	s_{12}	s_{13}	s_{23}	s_{44}	s_{55}	s_{66}	s_{15}	s_{25}	s_{35}	s_{46}
Rochelle salt (34 °C)	1767	52.0	36.8	35.9	−16.3	−11.6	−12.2	150.2	350.3	104.2	0	0	0	0
Potassium tartrate (DKT)	1987	23.6	35.3	47.7	−6.1	−4.2	−17.5	122.7	96.1	113.3	−13.4	7.7	−7.4	−6.7
Ethylene diamine tartrate (EDT)	1538	33.4	36.5	100.2	−3	−32.8	−18	191.8	122	191.4	−17	15	−26.5	3.8
Lithium sulfate	2052	26.2	22.5	23.9	−7.0	−9.8	−3.0	38.4	50.6	69.0	4.0	−7.3	−0.5	−8.8

	Piezoelectric constants (10^{-12} C/N)								Dielectric constants K_{ij}^T (relative)			
	d_{14}	d_{25}	d_{36}	d_{16}	d_{34}	d_{21}	d_{22}	d_{23}	K_{11}	K_{22}	K_{33}	K_{13}
Rochelle salt (34 °C)	345	54	12	0	0	0	0	0	205	9.6	9.5	0
Potassium tartrate (DKT)	7.9	−6.5	−23.2	3.5	−12.3	−0.8	4.5	−5.3	6.44	5.80	6.49	0.005
Ethylene diamine tartrate (EDT)	−10.0	−17.9	−18.4	−12.2	−17.0	10.1	2.2	−11.3	4.8	7.74	5.25	0.25
Lithium sulfate	0.8	−5.0	−4.2	−2.0	−2.1	−3.6	18.3	1.7	5.6	10.3	6.5	0.07

Table 7.4.3. Elastoelectric constants of cubic crystals (Tables from Berlincourt et al. [1964].)

Crystal	Density (kg/m^3)	Elastic compliance (10^{-12} m^2/N)			Piezoelectric constants (10^{-12} C/N)	Dielectric constants (relative)
		s_{11}	s_{12}	s_{44}^E	d_{14}	K^T
ZnS	4088	18.39	−7.07	21.68	3.18	8.27
ZnSe	5262	22.6	−8.5	22.7	1.10	9.12
ZnTe	5636	24.0	−8.73	32.1	0.91	10.1
CdTe (−196 °C)	5840	38.3	−15.8	51.1	1.68	9.65

Table 7.4.4. Elastoelectric constants of hexagonal and tetragonal crystals. (Table from Berlincourt et al. [1964].)

Crystal	Density kg/m^3	Elastic compliance s_{ij}^E (10^{-12} m^2/N)					
		s_{11}	s_{33}	s_{12}	s_{13}	s_{44}	s_{66}
BeO	3009	2.30	—	—	—	—	a
ZnO	5675	—	8.3	—	—	—	a
CdS	4819	20.69	16.97	−9.99	−5.81	66.49	a
CdSe	5684	23.38	17.35	−11.22	−5.72	75.95	a
$BaTiO_3$	6020	8.05	15.7	−2.35	−5.24	18.4	8.84

Crystal	Density kg/m^3	Piezoelectric constants (10^{-12} C/N)			Dielectric constants (relative)	
		d_{15}	d_{33}	d_{31}	K_{11}^T	K_{33}^T
BeO	3009	—	+0.24	−0.12	—	7.66
ZnO	5675	−12	+12	−4.7	—	8.2
CdS	4819	−13.98	+10.32	−5.18	9.35	10.33
CdSe	5684	−10.51	+7.84	−3.92	9.70	10.65
$BaTiO_3$	6020	392	85.6	−34.5	2920	168

[a] $s_{66} = 2(s_{11} - s_{12})$.

Table 7.4.5. Elastoelectric constants of quartz. (Table from Berlincourt et al. [1964].)

Density (kg/m^3)	Piezoelectric Elastic compliance s_{ij}^E (10^{-12} m^2/N)							Dielectric constants (10^{-12} C/N)		constants (relative)	
	s_{11}	s_{33}	s_{12}	s_{13}	s_{44}	s_{66}	s_{14}	d_{11}	d_{14}	K_{11}^T	K_{33}^T
2649	12.77	9.60	−1.79	−1.22	20.04	29.12	4.50	2.21	0.727	4.52	4.68

For some technological reasons, different cuts of crystals, not coinciding with the coordinate planes, are preferred. In such cases, material constants in the new coordinate system x'_k are given, in terms of those in x_k, by the tensor transformations

$$s'_{klmn} = s_{pqrs} \frac{\partial x'_k}{\partial x_p} \frac{\partial x_q}{\partial x'_l} \frac{\partial x'_m}{\partial x_r} \frac{\partial x_s}{\partial x'_n},$$

$$d'_{klm} = d_{pqr} \frac{\partial x'_k}{\partial x_p} \frac{\partial x'_l}{\partial x_q} \frac{\partial x_r}{\partial x'_m}, \tag{7.4.8}$$

$$\varepsilon'_{kl} = \varepsilon_{pq} \frac{\partial x'_k}{\partial x_p} \frac{\partial x_q}{\partial x'_l}, \qquad p'_k = p_q \frac{\partial x'_k}{\partial x_q},$$

where $\partial x'_k/\partial x_p$ and $\partial x_r/\partial x'_m$ are calculated from the coordinate transformations

$$x'_k = x'_k(x_p). \tag{7.4.9}$$

For a rigid rotation of the coordinate system, $\partial x'_k/\partial x_p$ represents the direction cosines between the two frames of reference.

7.5. Piezoelectrically Excited Thickness Vibrations of Plates

Here we present the solution of the thickness vibrations[6] of an infinite plate subjected to an oscillating electric potential applied by two large electrodes to its surfaces (Figure 7.5.1). The thickness vibrations consist of waves traveling in the direction of the z-axis which is perpendicular to the surfaces of the plate. The displacement vector can, of course, have an arbitrary direction. The thickness of the plate, h, is considered small, compared to the free-space electromagnetic wavelength, so that the quasi-static approximation is valid. Thus, the differential field equations (7.2.27) and (7.2.28), with $\mathbf{f} = 0$ and $q_e = 0$,

[6] Tiersten [1969].

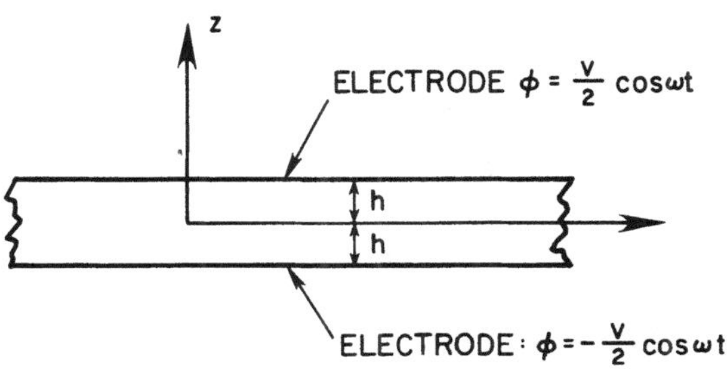

Figure 7.5.1. Piezoelectric thickness vibrations of the electroded plate.

are valid. Let $\mathbf{n}$ be the unit normal to the plate surface directed in the positive direction of the z-axis. The surface of the plate is traction-free. Hence, the boundary conditions are

$$\begin{aligned} (\sigma_{klmn}u_{m,n} + \varepsilon_{mkl}\phi_{,m})n_k &= 0 \qquad \text{at } z = \pm h/2, \\ \phi &= \pm\tfrac{1}{2}V\cos\omega t \quad \text{at } z = \pm h/2, \end{aligned} \tag{7.5.1}$$

of which the second prescribes the applied potential. Because of the one-dimensional nature of the problem, we set $\partial/\partial x_k = n_k\,\partial/\partial z$, so that (7.2.27) and (7.2.28) reduce to

$$\begin{aligned} \sigma_{klmn}n_n n_k u''_m + e_{mkl}n_m\phi'' &= \rho\ddot{u}_l, \\ e_{klm}n_m n_k u''_l - \varepsilon_{kl}n_k n_l\phi'' &= 0, \end{aligned} \tag{7.5.2}$$

where a prime denotes differentiation with respect to z. Eliminating ϕ'' between these two equations we obtain

$$\rho\ddot{u}_l = \bar{\sigma}_{klmn}n_k n_n u''_m, \tag{7.5.3}$$

where the *piezoelectrically stiffened* stiffness tensor $\bar{\sigma}$ is defined by

$$\bar{\sigma}_{klmn} = \sigma_{klmn} + (\mathbf{n}\cdot\boldsymbol{\varepsilon}\cdot\mathbf{n})^{-1}e_{ikl}n_i e_{jmn}n_j, \tag{7.5.4}$$

The standing-wave solution of (7.5.3) is of the form

$$\mathbf{u} = \sum_{\alpha=1}^{3} b^\alpha(A^\alpha\cos k^\alpha z + B^\alpha\sin k^\alpha z)\cos\omega t, \tag{7.5.5}$$

which upon substitution into (7.5.3) gives

$$(\bar{\sigma}_{klmn}n_k n_n - \rho v_\alpha{}^2\delta_{lm})b^\alpha{}_m = 0, \tag{7.5.6}$$

where v_α is the phase velocity defined by

$$v_\alpha = \omega/k^\alpha. \tag{7.5.7}$$

Hence, it is one of the three roots of the characteristic equation

$$\det(\bar{\sigma}_{klmn}n_k n_m - \rho v^2\delta_{lm}) = 0. \tag{7.5.8}$$

Substituting (7.5.5) into $(7.5.2)_2$, we determine the potential

$$\phi = (\mathbf{n}\cdot\boldsymbol{\varepsilon}\cdot\mathbf{n})^{-1}e_{jkl}n_j n_l u_k + (C_1 z + C_2)\cos\omega t. \tag{7.5.9}$$

The integration constants C_1 and C_2 are determined from the boundary conditions $(7.5.1)_2$

$$\begin{aligned} C_1 &= \frac{V}{h} - \frac{2}{h}(\mathbf{n}\cdot\boldsymbol{\varepsilon}\cdot\mathbf{n})^{-1}e_{jkl}n_j n_l\sum_\alpha b^\alpha{}_k B^\alpha\sin(k^\alpha h/2), \\ C_2 &= -(\mathbf{n}\cdot\boldsymbol{\varepsilon}\cdot\mathbf{n})^{-1}e_{jkl}n_j n_l\sum_\alpha b^\alpha{}_k A^\alpha\cos(k^\alpha h/2). \end{aligned} \tag{7.5.10}$$

The remaining boundary conditions (7.5.1) now give

$$\sum_{\alpha=1}^{3} A^\alpha k^\alpha b^\alpha{}_m\bar{\sigma}_{klmn}n_k n_n\sin(k^\alpha\,h/2) = 0,$$

$$\sum_{\alpha=1}^{3} B^\alpha k^\alpha b^\alpha{}_m n_k n_n[\bar{\sigma}_{klmn} \cos(k^\alpha\, h/2) + (\sigma_{klmn} - \bar{\sigma}_{klmn})(2/k^\alpha h) \sin(k^\alpha h/2)]$$

$$= -e_{ikl} n_i n_k\, V/h. \quad (7.5.11)$$

The first of these equations shows that even vibrations of $\mathbf{u}$ in (7.5.5) are not driven and, therefore, can be dropped. The three components of $(7.5.11)_2$ determine B^α, $\alpha = 1, 2, 3$. To obtain B^α, we substitute the eigenvectors $\mathbf{b}^\alpha$ given by (7.5.6). The scalar product of the resulting equation with $\mathbf{b}^\alpha$, because of the orthogonality of the $\mathbf{b}^\alpha$s, gives

$$\rho B^\beta k^\beta v_\beta{}^2 \cos(k^\beta\, h/2) - \sum_\alpha \frac{2}{h}(\mathbf{n}\cdot\boldsymbol{\varepsilon}\cdot\mathbf{n})^{-1} B^\alpha\, d^\alpha\, d^\beta \sin(k^\alpha\, h/2) = -d^\beta\, V/h, \quad (7.5.12)$$

where

$$d^\alpha \equiv e_{klm} n_k b^\alpha{}_l n_m. \quad (7.5.13)$$

The solution for B^α now follows from (7.5.12)

$$B^\alpha = -[\rho D h k^\alpha v_\alpha{}^2 \cos(k^\alpha h/2)]^{-1}\, d^\alpha V, \quad (7.5.14)$$

where D is the determinant given by

$$D = \det\left[\delta_{\beta\gamma} - \frac{d^\beta\, d^\gamma}{\rho(\mathbf{n}\cdot\boldsymbol{\varepsilon}\cdot\mathbf{n})v_\gamma{}^2}\, \frac{2\sin(k^\beta\, h/2)}{k^\gamma h \cos(k^\gamma\, h/2)}\right]$$

$$= 1 - \sum_{\lambda=1}^{3} (K^\lambda)^2 \frac{\tan(k^\lambda\, h/2)}{(k^\lambda\, h/2)}, \quad (7.5.15)$$

where the constant K^λ, defined by

$$K^\lambda = d^\lambda[\rho(\mathbf{n}\cdot\boldsymbol{\varepsilon}\cdot\mathbf{n})v_\lambda{}^2]^{-1/2}, \quad (7.5.16)$$

may be referred to as the *electromechanical coupling* factor for the λth eigenvalue.[7] Because of (7.5.14), the final solution for $\mathbf{u}$ and ϕ can be written as

$$\mathbf{u} = -\sum_{\alpha=1}^{3} [\rho h k^\alpha v_\alpha{}^2 \cos(k^\alpha\, h/2) D]^{-1} \mathbf{b}^\alpha\, d^\alpha V \sin(k^\alpha z) \cos \omega t.$$

$$\phi = \frac{V}{h}\left\{z - \sum_{\alpha=1}^{3} [k^\alpha \cos(k^\alpha\, h/2) D]^{-1} (K^\alpha)^2 [\sin(k^\alpha z) - (2z/h)\sin(k^\alpha\, h/2)]\right\} \cos \omega t. \quad (7.5.17)$$

Frequency Dependence of the Dielectric Tensor

The solution (7.5.17) can be exploited to find the frequency dependence of the dielectric tensor in the piezoelectric resonance region. To this end, following Nelson [1979], we use the electric displacement boundary condition

$$[\mathbf{D}]\cdot\mathbf{n} = w, \qquad z = \pm h/2. \quad (7.5.18)$$

[7] See Nelson [1979, Sect. 11.7].

Since **D** is given by (7.2.26), and the electric field vanishes in the electrode, the total surface charge on the electrode area is given by

$$Q = A^e w = A^e n_i(\varepsilon_{ij} n_j \phi' - e_{ijk} n_k u'_j), \qquad z = \pm h/2. \tag{7.5.19}$$

Substituting from (7.5.17), we can write this as

$$Q = CV, \tag{7.5.20}$$

where C is the capacitance given by

$$C = A^e \varepsilon / h, \tag{7.5.21}$$

and ε is the *effective dielectric* constant given by

$$\varepsilon = (\mathbf{n} \cdot \boldsymbol{\varepsilon} \cdot \mathbf{n}) \left[1 - \sum_\alpha (K^\alpha)^2 (k^\alpha h/2)^{-1} \tan(k^\alpha h/2) \right]^{-1}. \tag{7.5.22}$$

The resonance occurs when the denominator of this equation vanishes. Since $(k^\alpha)^{-1} \sim \omega^{-1}$, the high-frequency limit of (7.5.22) is

$$\varepsilon_\infty = \mathbf{n} \cdot \boldsymbol{\varepsilon} \cdot \mathbf{n} \qquad \text{as} \quad \omega \to \infty. \tag{7.5.23}$$

The low-frequency limit is given by

$$\varepsilon_0 = (\mathbf{n} \cdot \boldsymbol{\varepsilon} \cdot \mathbf{n}) \left[1 - \sum_\alpha (K^\alpha)^2 \right]^{-1} \qquad \text{as} \quad \omega \to 0. \tag{7.5.24}$$

To study the behavior of $\varepsilon(\omega)$ between these limits, we consider a single acoustic mode, e.g., β, to be driven piezoelectrically, because of certain combinations of crystal symmetry and orientation. In such a case, the denominator of (7.5.24) contains only one term so that the resonance occurs when

$$(K^\beta)^2 \frac{\sin(k^\beta h/2)}{(k^\beta h/2)} - \cos(k^\beta h/2) = 0. \tag{7.5.25}$$

For frequencies higher than the lowest resonance frequency, this is approximated by

$$\cos(k^\beta h/2) = 0, \tag{7.5.26}$$

which is valid when the coupling factor K^β is small. The solution (7.5.26) gives evenly distributed frequencies

$$\frac{K^\beta h}{2} = \frac{\omega h}{2 v_\beta} = (2m - 1)\pi/2. \tag{7.5.27}$$

Introducing the acoustic wavelength $\lambda_\beta = 2\pi v_\beta/\omega$, (7.5.27) yields

$$\tfrac{1}{2}(2m - 1)\lambda_\beta = h, \qquad m = 1, 2, 3, \ldots. \tag{7.5.28}$$

From this it is clear that resonance occurs whenever an odd integral number of acoustic wavelengths equals twice the thickness of the plate. The resonance frequencies can be measured experimentally. Consequently, the acoustic velocity v_β, and, therefore, the stiffened elastic moduli $\bar{\sigma}_{klmn}$ given by (7.5.4), can

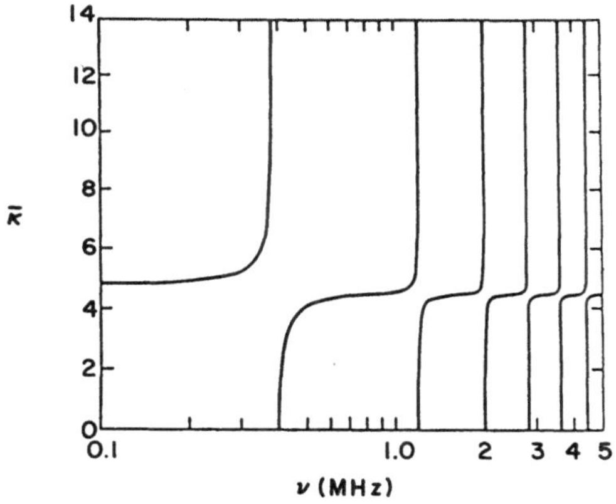

Figure 7.5.2. Piezoelectric resonance effect on the effective dielectric constant ε (7.5.22) of a Y-cut quartz plate 0.5 mm thick versus frequency. The lowest six resonances governed by (7.5.25) are shown. No damping of them is included. For this orientation of quartz (point group 32) only the pure shear wave with the displacement in the X-direction can be piezoelectrically driven. Its electromechanical coupling coefficient is -0.267 and its velocity is 4.20×10^3 m/sec. The high-frequency value for ε_{22} of 4.435 is used in the plot (after Nelson [1979]). Reprinted with permission from D.F. Nelson, *Electric, Optic and Acoustic Interactions in Dielectrics*, © 1979, John Wiley and Sons.

be determined. For the special case of a Y-cut quartz plate, the electromechanical coupling constant K^λ vanishes for two of the three eigenmodes that propagate in the Y-direction, so that (7.5.5) contains only one term. This term represents a pure shear wave, with the displacement vector in the X-crystallographic direction. A plot of ε versus the frequency $\nu = \omega/2\pi$ is given in Figure 7.5.2.

Measurement of the piezoelectric constant requires several experiments in which we must excite unstiffened modes. Some of these modes can be excited by placing electrodes on two opposite edges of the plate. In this case, edge-excited thickness vibrations need to be studied. For these, we refer the reader to Tiersten [1969] and Nelson [1979].

7.6. Extensional Vibrations of Piezoelectric Rods

Vibrations of a piezoelectric rod can be used to determine some of the piezoelectric moduli. They also have industrial applications requiring knowledge of vibration characteristics.

A rod is a body whose two transverse dimensions are much smaller than its length. We consider a rod of rectangular cross section of width W, thickness h, and length L, such that

$$L \gg W > h. \tag{7.6.1}$$

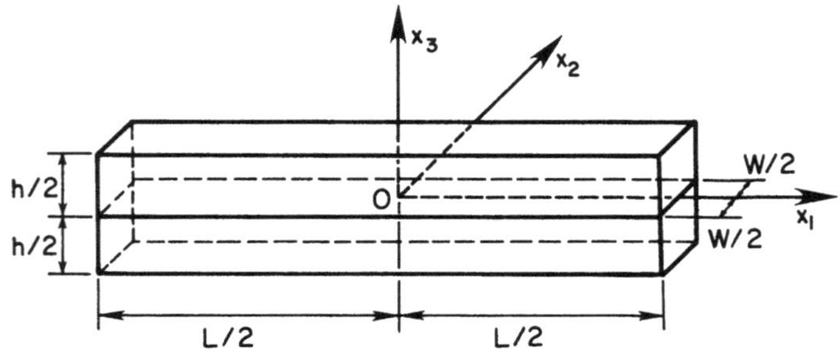

Figure 7.6.1. Geometry of a rod.

Let the x_1-axis of a rectangular frame be directed along the length of the rod, x_2 is taken in the direction of the width, and x_3 in the thickness direction (Figure 7.6.1). Suppose that the two surfaces, $x_3 = \pm h/2$, are electroded. We allow both arbitrary crystal symmetry and crystallographic orientation. The two faces $x_3 = \pm h/2$ and $x_2 = \pm W/2$ are free of tractions, so that we have the boundary conditions

$$\begin{aligned} t_{2k} &= 0, \qquad x_2 = \pm W/2, \\ t_{3k} &= 0, \qquad x_3 = \pm h/2. \end{aligned} \tag{7.6.2}$$

Since the rod is very thin, in both the x_2- and x_3-directions, we assume that (7.6.2) is valid throughout the cross section of the rod. The only remaining nonvanishing component of the stress tensor is t_{11}, which can be solved from $(7.4.6)_1$ as

$$t_{11} = \frac{1}{s_{11}}(u_{1,1} - d_{k1}E_k). \tag{7.6.3}$$

Electrical boundary conditions require that, on the electroded faces $\mathbf{n} \times [\mathbf{E}] = \mathbf{0}$, i.e.,

$$E_1 = E_2 = 0, \qquad x_3 = \pm h/2. \tag{7.6.4}$$

Again, these conditions may be assumed to be valid throughout the rod, so that only the E_3 component of the electric field survives and is given by

$$E_3 = \frac{V}{h}\exp(-i\omega t) \tag{7.6.5}$$

for an oscillatory applied field. Inherent in (7.6.5) is also the assumption that the E_3-field is uniform throughout the rod. The foregoing approximation implies that any acoustic wave traveling along the rod will not induce an appreciable longitudinal field.

Substituting (7.6.3) into Cauchy's equation of motion, we obtain

$$\rho\frac{\partial^2 u_1}{\partial t^2} = \frac{1}{s_{11}}\frac{\partial^2 u_1}{\partial {x_1}^2}. \tag{7.6.6}$$

The general solution of this equation is

$$u_1(x_1, t) = (A \sin kx_1 + B \cos kx_1) \exp(-i\omega t), \tag{7.6.7}$$

where

$$k = (\rho s_{11})^{1/2}\omega. \tag{7.6.8}$$

The vanishing tractions at the free ends, $x_1 = \pm L/2$, determine the constants A and B

$$A = d_{311}\, V/kh \cos(kL/2), \qquad B = 0,$$

so that (7.6.7) reads

$$u_1(x_1, t) = \frac{d_{311} V}{kh} \frac{\sin kx_1}{\cos(kL/2)} \exp(-i\omega t). \tag{7.6.9}$$

Resonance occurs whenever

$$kL/2 = (2m - 1)\pi/2 \qquad m = 1, 2, 3, \ldots.$$

In terms of frequency, this reads

$$\nu = \omega/2\pi = (2m - 1)/2L(\rho s_{1111})^{1/2}, \qquad m = 1, 2, 3, \ldots. \tag{7.6.10}$$

By measuring the resonance frequencies we can determine the compliance $s_{11} = s_{1111}$ of the rod. Note, however, that s_{1111} may be any combination of the crystallographic compliances in the coordinate system which is not a crystallographic system in general.

7.7. Surface Waves

Piezoelectric surface waves can be studied in an approach parallel to that for the Rayleigh surface waves of elastodynamics (Eringen and Suhubi [1975, p. 518]). Such studies, in fact, exist (see Tiersten [1963], Coquin and Tiersten [1967], Berlincourt [1981], Farnell [1978], Foster [1981], Tiersten [1981], and Viktorov [1979]), and are modifications of elastic surface waves for anisotropic bodies. In piezoelectric crystals another type of surface wave exists, one which has no counterpart in a purely elastic homogeneous medium. This type of wave may prove to be useful in high-frequency surface-wave devices. Following Bleustein [1968], we now examine such a solution.

Consider a piezoelectric half-space, $x_2 \geq 0$, whose boundary surface, $x_2 = 0$, is coated with an infinitesimally thin, perfectly conducting electrode (Figure 7.7.1). Suppose that the half-space consists of a material in the crystallographic schoenflies class C_6 ($\equiv$ 6 mm in the international system), for example, zinc oxide or cadmium sulfide. Such a material possesses five elastic constants, three piezoelectric constants, and two dielectric constants. Let the x_3-axis be oriented in the direction of the six-fold axis of the crystal. For motions that are independent of x_3, the four coupled, partial differential equations (7.2.27) and (7.2.28) reduce to two sets and two coupled equations. Here we study an

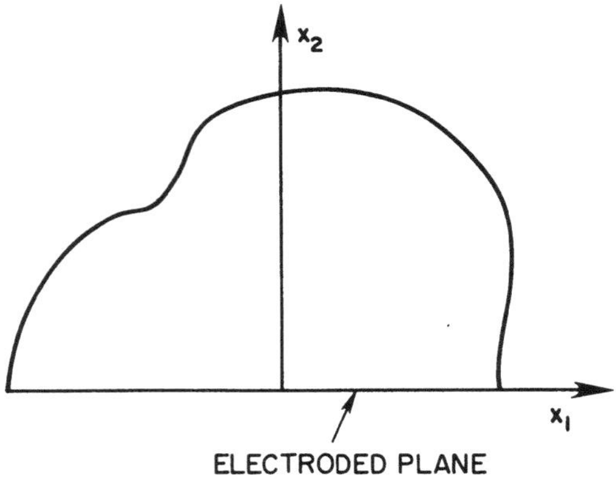

Figure 7.7.1. Antiplane surface waves.

antiplane motion, which is governed by the u_3-components of the displacement field and the electric potential ϕ:

$$\sigma_{44}\nabla^2 u_3 + e_{15}\nabla^2\phi = \rho\frac{\partial^2 u_3}{\partial t^2},$$
$$e_{15}\nabla^2 u_3 - \varepsilon_{11}\nabla^2\phi = 0, \tag{7.7.1}$$

where σ_{ab}, e_{ka}, and e_{kl} denote the matrix elements introduced by (7.4.3)–(7.4.5) and ∇^2 is the two-dimensional Laplacian in x_1- and x_2-coordinates. The constitutive equations (7.4.2) for this case reduce to

$$t_{11} = t_{22} = t_{33} = t_{12} = 0, \qquad D_3 = 0,$$
$$t_{23} = \sigma_{44}u_{3,2} + e_{15}\phi_{,2}, \qquad t_{31} = \sigma_{44}u_{3,1} + e_{15}\phi_{,1}, \tag{7.7.2}$$
$$D_1 = e_{15}u_{3,1} - \varepsilon_{11}\phi_{,1}, \qquad D_2 = e_{15}u_{3,2} - \varepsilon_{11}\phi_{,2}.$$

The boundary conditions are

$$t_{22} = t_{23} = t_{21} = 0 \quad \text{(stress-free)}, \qquad x_2 = 0,$$
$$\phi = 0 \quad \text{(grounded electrode)}, \qquad x_2 = 0, \tag{7.7.3}$$
$$u_3, \phi \to 0 \qquad \text{as} \quad x_2 \to \infty.$$

If we write

$$\psi = \phi - (e_{15}/\varepsilon_{11})u_3, \tag{7.7.4}$$

the system (7.7.1) reduces to

$$\bar{\sigma}_{44}\nabla^2 u_3 = \rho\frac{\partial^2 u_3}{\partial t^2}, \qquad \nabla^2\psi = 0, \tag{7.7.5}$$

where

$$\bar{\sigma}_{44} = \sigma_{44} + e_{15}{}^2/\varepsilon_{11} \tag{7.7.6}$$

is the piezoelectrically stiffened elastic constant. The relevant boundary conditions now read

$$\begin{aligned} t_{23} &= \bar{\sigma}_{44}u_{3,2} + e_{15}\psi_{,2} = 0, & x_2 &= 0, \\ \phi &= (e_{15}/\varepsilon_{11})u_3 + \psi = 0, & x_2 &= 0, \\ u_3, \phi &\to 0 & \text{as}\quad x_2 &\to \infty. \end{aligned} \tag{7.7.7}$$

A solution of (7.7.5), which vanishes at $x_2 = \infty$, has the form

$$\begin{aligned} u_3 &= A\exp(-\xi_2 x_2)\cos(\xi_1 x_1 - \omega t), \\ \psi &= B\exp(-\xi_1 x_2)\cos(\xi_1 x_1 - \omega t), \end{aligned} \tag{7.7.8}$$

provided ξ_1 and ξ_2 are real and positive and

$$\bar{\sigma}_{44}(\xi_1{}^2 - \xi_2{}^2) = \rho\omega^2. \tag{7.7.9}$$

The two boundary conditions on $x_2 = 0$ require that

$$\bar{\sigma}_{44}\xi_2 A + e_{15}\xi_1 B = 0, \qquad (e_{15}/\varepsilon_{11})A + B = 0. \tag{7.7.10}$$

Nontrivial solutions for A and B exist if the determinant of the system vanishes. Hence,

$$\xi_2 = K^2\xi_1, \qquad K^2 = \varepsilon_{15}{}^2/\varepsilon_{11}\bar{\sigma}_{44}. \tag{7.7.11}$$

Here K is the piezoelectric coupling factor. The phase velocity of the surface wave $v_s = \omega/\xi_1$ is given by

$$v_s{}^2 = (\bar{\sigma}_{44}/\rho)(1 - K^4). \tag{7.7.12}$$

This type of surface wave does not exist in a purely elastic solid. For example, in a pure, elastic, homogeneous material with the same hexagonal symmetry, the antiplane motion u_3 vanishes identically throughout the half-space $x \geq 0$.[8]

7.8. Radially Symmetric Vibrations

Radially symmetric vibrations can be set up in a thin ceramic ring, electroded on its inner and outer surfaces and poled using these electrodes. The appropriate coordinate system is the polar one (r, θ, z). Let a be the mean radius, w the width, and h the thickness of the ring. Since h and w are very small compared to a, and the surfaces of the ring are free of tractions, we may assume that $t_{rr} = t_{r\theta} = t_{rz} = t_{\theta z} = 0$ throughout the crosss section of the ring. The boundary conditions $\mathbf{n} \times [\mathbf{E}] = \mathbf{0}$ on the inner and outer surfaces give $E_\theta = E_z = 0$, so that again we take these to be true across the cross section. With these, the

[8] However, such a surface-wave motion may exist in a purely elastic but *nonhomogeneous* material (Bakirtas and Maugin [1982]), and in elastic ferroelectrics for certain conditions of Bias fields (Maugin [1983]).

constitutive equations (7.4.6) for $\tilde{e}_{\theta\theta}$ and D_r reduce to

$$\begin{aligned} \tilde{e}_{\theta\theta} &= s_{11} t_{\theta\theta} + d_{31} E_r, \\ D_r &= d_{31} t_{\theta\theta} + \bar{\varepsilon}_{33} E_r, \end{aligned} \tag{7.8.1}$$

where the crystal x_1-axis is oriented in the tangential θ-direction and the x_3-axis is in the radial direction. For several piezoelectric ceramics, many components of s_{ab} and $\bar{\varepsilon}_{kl}$ vanish. For example, for PZT-4, PZT-5, and barium titanate, the only surviving components are (s_{11}, s_{33}, s_{13}, s_{44}, s_{66}), (d_{33}, d_{31}, d_{15}), and $\bar{\varepsilon}_{11}$ and $\bar{\varepsilon}_{33}$. For these materials then, the only remaining non-vanishing components of the constitutive equations are

$$\tilde{e}_{zz} = s_{31} t_{\theta\theta} + d_{33} E_r. \tag{7.8.2}$$

From (7.8.1), we solve for $t_{\theta\theta}$, so that

$$\begin{aligned} t_{\theta\theta} &= \frac{1}{s_{11}}(\tilde{e}_{\theta\theta} - d_{31} E_r), \\ D_r &= \frac{1}{s_{11}}[d_{31}\tilde{e}_{\theta\theta} + (\bar{\varepsilon}_{33} s_{11} - d_{31}{}^2) E_r]. \end{aligned} \tag{7.8.3}$$

The strain component $\tilde{e}_{\theta\theta}$ is given by

$$\bar{e}_{\theta\theta} = u(t)/a, \tag{7.8.4}$$

where u is the radial displacement. Now, $\nabla \cdot \mathbf{D} = 0$ gives $E_r = E(t)$, and Cauchy's equations of motion in cylindrical coordinates reduce to

$$\frac{1}{a} t_{\theta\theta} + \rho \ddot{u} = 0. \tag{7.8.5}$$

Using $(7.8.3)_1$, this becomes

$$\ddot{u} + \omega_0{}^2 u = a\, d_{31} \omega_0{}^2 E(t), \tag{7.8.6}$$

where

$$\omega_0{}^2 = (a^2 \rho s_{11})^{-1}. \tag{7.8.7}$$

The general solution of (7.8.6) is

$$u(t) = A \cos \omega_0 t + B \sin \omega_0 t + a d_{31} \omega_0 \int^t E(\tau) \sin \omega_0 (t - \tau)\, d\tau, \tag{7.8.8}$$

where A and B are constants of integration to be determined from the initial conditions. For harmonic excitation, $E(t) = E_0 \exp(i\omega t)$, and (7.8.8) gives

$$u(t) = A \cos \omega_0 t + B \sin \omega_0 t + \frac{\mathrm{ad}_{31}\, E_0}{1 - (\omega/\omega_0)^2} \exp(i\omega t). \tag{7.8.9}$$

Once u is given by (7.8.8), $\tilde{e}_{\theta\theta}$ is determined by (7.8.4), and the electric displacement by $(7.8.3)_2$. At $\omega = \omega_0$, the resonance occurs. By exciting the ceramic ring to the resonance frequency, we can determine the compliance s_{11}.

7.9. Spherically Symmetric Vibrations of Thin Ceramic Shells

A thin spherical shell transducer, poled in the thickness direction, with fully electroded inner and outer surfaces, can be treated in a similar fashion to a ceramic ring. In this case, the boundary conditions give $E_1 = E_2 = 0$, $D_1 = D_2 = 0$, and $t_3 = t_4 = t_5 = t_6 = 0$, where the radial direction is called the 3-direction. Symmetry regulation gives $t_1 = t_2 = t$, $\tilde{e}_1 = \tilde{e}_2 = \tilde{e}$, and $s = \frac{1}{2}(s_{11} + s_{22})$. Under these conditions the relevant constitutive equations become

$$\tilde{e} = 2st + d_{31}E_3, \qquad D_3 = 2d_{31}t + \bar{\varepsilon}_{33}E_3. \tag{7.9.1}$$

Since the strain $\tilde{e} = u/a$, where $u(t)$ is the radial displacement and a is the mean radius, Cauchy's equation gives the same equation as (7.8.6), i.e.,

$$\ddot{u} + \omega_0{}^2 u = d_{31}\omega_0{}^2 E_3, \qquad D_3 = 2d_{31}t + \bar{\varepsilon}_{33}E_3. \tag{7.9.2}$$

The equation $\nabla \cdot \mathbf{D} = 0$ gives $E_3 = E_3(t)$. The solution of $u(t)$ is again given by (7.8.8). The electric displacement is obtained from $(7.9.1)_2$, i.e.,

$$D_3 = \frac{1}{s}\left[d_{31}\frac{u}{a} + (\bar{\varepsilon}_{33}s - d_{31}{}^2)E_3\right]. \tag{7.9.3}$$

The resonance frequency is again given by

$$\omega = \omega_0 = (a^2\rho s)^{-1/2}. \tag{7.9.4}$$

Spherically symmetric vibrations of transducers may be used in sonars for underwater detection.

7.10. Piezoelectrically Generated Electric Field

In Sections 7.5–7.9 we studied mechanical vibrations generated by electric fields. Conversely, it is known that mechanical vibrations produce electric fields in dielectric crystals. However, the electrostatic equations used do not contain this coupling. In order to account for the full coupling of the strain and electric fields, we need the fully dynamical set of Maxwell's equations. For nonmagnetizable dielectrics ($\mu = 1$), these are given by

$$\begin{aligned} \nabla \times \mathbf{E} + \frac{1}{c}\frac{\partial \mathbf{B}}{\partial t} &= \mathbf{0}, \qquad & \nabla \times \mathbf{B} - \frac{1}{c}\frac{\partial \mathbf{D}}{\partial t} &= \mathbf{0}, \\ \nabla \cdot \mathbf{B} &= 0, & \nabla \cdot \mathbf{D} &= 0. \end{aligned} \tag{7.10.1}$$

By taking the curl of the first of these, we eliminate $\mathbf{B}$. Upon using the constitutive equation (7.2.21) for $\mathbf{D}$ this results in

$$E_{l,kl} - E_{k,ll} + c^{-2}\varepsilon_{kl}\frac{\partial^2 E_l}{\partial t^2} = -c^{-2}e_{klm}\frac{\partial^2 u_{l,m}}{\partial t^2}. \tag{7.10.2}$$

This set of equations, together with Cauchy's equation $(7.3.3)_1$, in the absence of the body force **f**,

$$\sigma_{klmn} u_{m,nk} - e_{mnl} E_{m,n} = \rho \frac{\partial^2 u_l}{\partial t^2}, \tag{7.10.3}$$

must be solved under appropriate boundary and initial conditions in order to determine the displacement field **u** and the electric field **E**. It is now clear from these equations that the **E**-field causes displacement and, in turn, a displacement causes an **E**-field. The formal solution may be obtained by considering harmonic solutions of the form

$$\begin{aligned} \mathbf{E} &= \mathbf{E}^0 \exp\left[i\omega\left(\frac{n}{c}\mathbf{s}\cdot\mathbf{r} - t\right)\right], \\ \mathbf{u} &= \mathbf{u}^0 \exp\left[i\omega\left(\frac{n}{c}\mathbf{s}\cdot\mathbf{r} - t\right)\right], \end{aligned} \tag{7.10.4}$$

where $\mathbf{E}^0$ and $\mathbf{u}^0$ are constant vectors, n is the refractive index, and **s** is a unit vector defined by

$$\mathbf{k} = \frac{\omega}{c}\mathbf{n}, \qquad \mathbf{n} = n\mathbf{s}, \qquad n = |\mathbf{n}|, \qquad \mathbf{s}\cdot\mathbf{s} = 1, \tag{7.10.5}$$

where **k** is the wave vector. Substituting (7.10.4) into (7.10.2) and (7.10.3) we obtain

$$\begin{aligned} M_{kl} E_l{}^0 - (i\omega n/c)\bar{e}_{kl} u_l{}^0 &= 0, \\ (inc/\omega)\bar{e}_{kl} E_l{}^0 + \bar{\sigma}_{kl} u_l{}^0 &= 0, \end{aligned} \tag{7.10.6}$$

where

$$\begin{aligned} M_{kl} &= n^2(\delta_{kl} - s_k s_l) - \varepsilon_{kl}, \qquad \bar{e}_{kl} = e_{klm} s_m, \\ \bar{\sigma}_{kl} &= c_{jklm} n^2 s_j s_m - \rho c^2 \delta_{kl}. \end{aligned} \tag{7.10.7}$$

Assuming that the inverse of $\bar{\boldsymbol{\sigma}}$, $\overset{-1}{\bar{\boldsymbol{\sigma}}}$, exists, we can solve for $\mathbf{u}^0$ from $(7.10.6)_2$

$$u^0{}_m = -(inc/\omega)\bar{e}_{kl} \overset{-1}{\bar{\sigma}}_{km} E^0{}_l. \tag{7.10.8}$$

Substituting this into $(7.10.6)_1$ we obtain

$$(\varepsilon_{kl} - n^2 \lambda_{kl}) E^0{}_l = 0, \tag{7.10.9}$$

where

$$\lambda_{kl} = \delta_{kl} - s_k s_l - \bar{e}_{km} \bar{e}_{pl} \overset{-1}{\bar{\sigma}}_{pm}. \tag{7.10.10}$$

For a nontrivial solution of the system (7.7.9) to exist, we must have

$$\det(\varepsilon_{kl} - n^2 \lambda_{kl}) = 0. \tag{7.10.11}$$

Once the roots of this equation $n_\alpha{}^2$, $\alpha = 1, 2, 3$, are found, then (7.10.9) can be solved for $\mathbf{E}^0{}_\alpha$ corresponding to each root. These eigenvectors are determined to within an arbitrary factor for each α. With $E^0{}_\alpha$ so determined, (7.10.8) gives the displacement vector $\mathbf{u}^0{}_\alpha$. The boundary conditions are then used to determine the arbitrary amplitude factors.

7.11. Elastic Dielectrics Subject to Finite Deformations and Fields

Very few genuinely nonlinear problems have been treated. Most of these problems are for incompressible solids and they fall into the category of *controllable states* of elastic dielectrics. A controllable state is defined as that which can be maintained by the application of surface loads only (no body load distribution). This is achieved by assuming specific displacement and electric (or polarization) fields which satisfy the field equations without source terms ($\mathbf{f} = \mathbf{0}$, $q_e = 0$). Then the surface tractions and fields are determined to support the assumed fields. Other simple solutions exist in which the electric field or the dielectric displacement is specified. In this and the next few sections, we present solutions to several nonlinear problems in these categories.

(i) *Homogeneous Deformation of a Slab in a Uniform Field* (Figure 7.11.1) The simplest controllable states are obtained if the components of the strain tensor and the polarization vector are constants. We consider an infinite slab, bounded by the surfaces $x_3 = \pm H$, in the undeformed state. Suppose that the slab is subjected to extensions in the coordinate directions, with extension ratios λ_1, λ_2, and λ_3, respectively, and then sheared by the amounts τ_1 and τ_2 in the x_1- and x_2-directions. The total deformation is given by

$$x_1 = \lambda_1 X_1 + \tau_1 \lambda_3 X_3, \qquad x_2 = \lambda_2 X_2 + \tau_2 \lambda_3 X_3, \qquad x_3 = \lambda_3 X_3. \tag{7.11.1}$$

If the material is incompressible, then $\lambda_1 \lambda_2 \lambda_3 = 1$. Suppose that the electric field is a constant vector field

$$\mathbf{E} = (E_1, E_2, E_3). \tag{7.11.2}$$

From (7.2.12) and (7.2.13) it follows that ${}_{\mathrm{E}}\mathbf{t}$ and $\mathbf{P}$ are constant. With $\mathbf{f} = 0$ and $q_e = 0$ the field equations (7.2.2)–(7.2.4) are satisfied. Using (7.2.12) and (7.2.13) we calculate

$$\begin{aligned} {}_{\mathrm{E}}\mathbf{t} = b_{-1} \overset{-1}{\mathbf{c}} + b_0 \mathbf{1} + b_1 \mathbf{c} - \chi_1 \left[\mathbf{E} \otimes \left(\overset{-1}{\mathbf{c}} \mathbf{E} \right) + \left(\overset{-1}{\mathbf{c}} \mathbf{E} \right) \otimes \mathbf{E} \right] \\ - 2\chi_2 \left[\mathbf{E} \otimes \left(\overset{-2}{\mathbf{c}} \mathbf{E} \right) + \left(\overset{-2}{\mathbf{c}} \mathbf{E} \right) \otimes \mathbf{E} \right], \end{aligned} \tag{7.11.3}$$

$$\mathbf{P} = \chi \mathbf{E} + \chi_1 \overset{-1}{\mathbf{c}} \mathbf{E} + \chi_2 \overset{-2}{\mathbf{c}} \mathbf{E}. \tag{7.11.4}$$

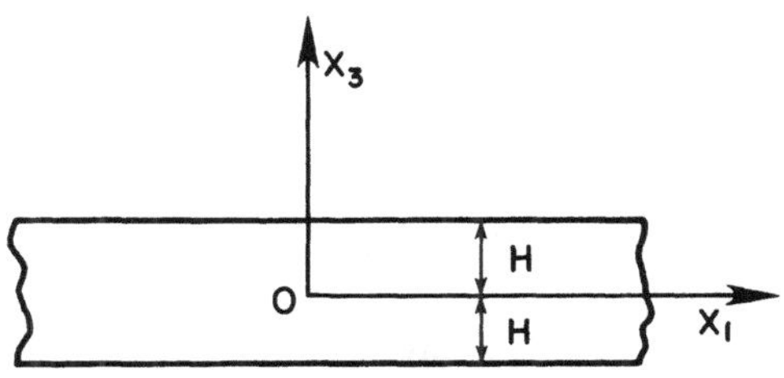

Figure 7.11.1. Slab in a uniform field.

The jump conditions (7.2.5)–(7.2.7) are satisfied on the surfaces where $x_3 = \pm H$, if

$$t_{(3)k} = {}_{E}t_{3k} + E_3 E_k - \tfrac{1}{2}E^2\delta_{3k} - \overset{+}{t}{}^{E}_{3k},$$
$$E_3 + P_3 = \overset{+}{E}_3 + w_e, \qquad E_1 = \overset{+}{E}_1, \qquad E_2 = \overset{+}{E}_2, \tag{7.11.5}$$

where $\overset{+}{t}{}^{E}_{3k}$ is the electrical traction given by

$$\overset{+}{t}{}^{E}_{3k} = \overset{+}{E}_3\,\overset{+}{E}_k - \tfrac{1}{2}\left(\overset{+}{E}\right)^2\delta_{3k}. \tag{7.11.6}$$

Here $t_{(3)k}$ are mechanical tractions applied to the surfaces of the slab, and $\overset{+}{\mathbf{E}}$ is the externally applied electric field. Thus, when the surface charge w and the external electric field $\overset{+}{\mathbf{E}}$ are known, the electric field, polarization, and stress field are fully determined. From $(7.11.5)_1$ it is clear that the slab is subject to surface tractions, which are prescribed on the surfaces $x_3 = \pm H$ of the slab, in order to maintain the assumed displacement field (7.11.1) and the external electric field $\overset{+}{E}$. Among the several special cases included in this solution, the following two may be of some interest.

(ii) *Extensions Without Shearing*

If the slab is assumed to undergo only extensions without shearing, $\tau_1 = \tau_2 = 0$, from the formulas given in (7.2.14), we calculate

$$\overset{-1}{\mathbf{c}} = \begin{pmatrix} \lambda_1{}^2 & 0 & 0 \\ \cdot & \lambda_2{}^2 & 0 \\ \cdot & \cdot & \lambda_3{}^2 \end{pmatrix}, \qquad \mathbf{c} = \begin{pmatrix} \lambda_1{}^{-2} & 0 & 0 \\ \cdot & \lambda_2{}^{-2} & 0 \\ \cdot & \cdot & \lambda_3{}^{-2} \end{pmatrix}, \tag{7.11.7}$$

$$\begin{gathered} \mathrm{I} = \lambda_1{}^2 + \lambda_2{}^2 + \lambda_3{}^2, \qquad \mathrm{II} = \lambda_1{}^2\lambda_2{}^2 + \lambda_2{}^2\lambda_3{}^2 + \lambda_3{}^2\lambda_1{}^2, \\ \mathrm{III} = \lambda_1{}^2\lambda_2{}^2\lambda_3{}^2, \qquad \mathrm{I}_4 = \mathbf{E}\cdot\mathbf{E}, \\ \mathrm{I}_6 = \lambda_1{}^2E_1{}^2 + \lambda_2{}^2E_2{}^2 + \lambda_3{}^2E_3{}^2, \\ \mathrm{I}_8 = \lambda_1{}^4E_1{}^2 + \lambda_2{}^4E_2{}^2 + \lambda_3{}^4E_3{}^2. \end{gathered} \tag{7.11.8}$$

Constitutive equations (7.11.3) and (7.11.4) become

$$\begin{aligned} {}_{E}t_{\underline{kk}} &= b_{-1}\lambda_k{}^2 + b_0 + b_1\lambda_k{}^{-2} - (2\chi_1\lambda^2{}_k + 4\chi_2\lambda_k{}^4)E_k{}^2, \\ {}_{E}t_{kl} &= -[\chi_1(\lambda_k{}^2 + \lambda_l{}^2) + 2\chi_2(\lambda_k{}^4 + \lambda_l{}^4)]E_kE_l, \\ P_k &= (\chi + \chi_1\lambda_{\underline{k}}{}^2 + \chi_2\lambda_{\underline{k}}{}^4)E_k, \end{aligned} \tag{7.11.9}$$

where underlined indices are not summed. From the definitions (7.2.14) of material moduli, it is clear that they are constants. The electric field $\mathbf{E}$ is determined by the last three equations of (7.11.5), of which the equation determining E_3 is highly complicated, since it involves P_3. The surface tractions given by $(7.11.5)_1$ show that the applied external electric field is oblique to the surfaces of the slab so that shearing stresses must be applied to the surfaces, in order to keep them from shearing. These are given by

$$t_{(3)l} = -[\chi_1(\lambda_l^2 + \lambda_3^2) + 2\chi_2(\lambda_l^4 + \lambda_3^4)]E_3 E_l + \left(E_3 - \overset{+}{E}_3\right)E_l, \qquad l = 1, 2. \tag{7.11.10}$$

If, however, $E_1 = E_2 = 0$ but $E_3 \neq 0$, then ${}_{\mathrm{E}}t_{kl} = 0$ for $k \neq l$, and no surface shear is necessary. In this case, the electric field is determined by

$$E_1 = E_2 = 0, \qquad E_3 + P_3 = E_3^0 + w_e. \tag{7.11.11}$$

(iii) *Simple Shear*[9]

The slab undergoes a simple shear deformation in the $x_2 = 0$-plane, if $\lambda_1 = \lambda_2 = \lambda_3 = 1$ and $\tau_2 = 0$. We also assume that the applied electric field is orthogonal to the $x_2 = 0$-plane ($E_2 = 0$). In this case, we have

$$x_1 = X_1 + \tau X_3, \qquad x_2 = X_2, \qquad x_3 = X_3, \tag{7.11.12}$$

$$\overset{-1}{\mathbf{c}} = \begin{pmatrix} 1+\tau^2 & 0 & \tau \\ \cdot & 1 & 0 \\ \cdot & \cdot & 1 \end{pmatrix}, \qquad \mathbf{c} = \begin{pmatrix} 1 & 0 & -\tau \\ \cdot & 1 & 0 \\ \cdot & \cdot & 1+\tau^2 \end{pmatrix}, \tag{7.11.13}$$

$$\begin{aligned} &\mathrm{I} = \mathrm{II} = 3 + \tau^2, \qquad \mathrm{III} = 1, \qquad \mathrm{I}_4 = \mathrm{E}_1{}^2 + \mathrm{E}_3{}^2, \\ &\mathrm{I}_6 = (1+\tau^2)E_1{}^2 + E_3{}^2 + 2\chi E_1 E_3, \\ &\mathrm{I}_8 = [(1+\tau^2) + \tau^2]E_1{}^2 + (1+\tau^2)E_3{}^2 + 2(2\tau + \tau^3)E_1 E_3. \end{aligned} \tag{7.11.14}$$

Since III = 1, the deformation is isochoric. By use of (7.2.12) and (7.2.13), we obtain the stress and polarization fields

$$\begin{aligned} {}_{\mathrm{E}}t_{11} &= b_{-1}(1+\tau^2) + b_0 + b_1 - 2\chi_1[(1+\tau^2)E_1{}^2 + \tau_1 E_1 E_3] \\ &\quad - 4\chi_2[(1 + 3\tau^2 + \tau^4)E_1 + (2\tau + \tau^3)E_1 E_3], \\ {}_{\mathrm{E}}t_{22} &= b_{-1} + b_0 + b_1, \\ {}_{\mathrm{E}}t_{33} &= b_{-1} + b_0 + b_1(1+\tau^2) - 2\chi_1(\tau E_1 + E_3)E_3 \\ &\quad - 4\chi_2[(2\tau + \tau^3)E_1 + (1+\tau^2)E_3]E_3, \\ {}_{\mathrm{E}}t_{13} &= b_{-1}\tau - b_1\tau - \chi_1[\tau(E_1{}^2 + E_3{}^2) + (2+\tau^2)E_1 E_3 \\ &\quad - 2\chi_2[(1 + \tau + \tau^2 + \tau^3)E_1{}^2 + (2\tau + \tau^3)E_3{}^2 \\ &\quad + (2 + 4\tau^2 + \tau^4)E_1 E_3], \\ {}_{\mathrm{E}}t_{12} &= {}_{\mathrm{E}}t_{23} = 0, \end{aligned} \tag{7.11.15}$$

[9] Eringen [1962, Sect. 115] and [1963].

$$
\begin{aligned}
P_1 &= [\chi + (1+\tau^2)\chi_1 + (1+3\tau^2+\tau^4)\chi_2]E_1 \\
&\quad + [\tau\chi_1 + (2\tau+\tau^3)\chi_3]E, \\
P_2 &= 0, \\
P_3 &= [\chi_1\tau + (2\tau+\tau^3)\chi_2]E_1 + [\chi + \chi_1 + (1+\tau^2)\chi_2]E_3.
\end{aligned}
\tag{7.11.16}
$$

The boundary conditions (7.11.5) lead to

$$
\begin{aligned}
t_{(3)1} &= {}_{\mathrm{E}}t_{31} + E_1(w_e - P_3), \\
t_{(3)2} &= 0, \\
t_{(3)3} &= {}_{\mathrm{E}}t_{33} + \tfrac{1}{2}\left[\left(\overset{+}{E}_3 + w_e - P_3\right)^2 - \left(\overset{+}{E}_3\right)^2\right], \\
E_3 + P_3 &= \overset{+}{E}_3 + w_e, \qquad E_1 = \overset{+}{E}_1, \qquad E_2 = 0.
\end{aligned}
\tag{7.11.17}
$$

From (7.11.14) it is clear that the free-energy function Σ depends on E_1, E_3, and τ, i.e.,

$$
\Sigma = \Sigma(E_1, E_3, \tau). \tag{7.11.18}
$$

Given the energy function, the constitutive moduli b_{-1}, b_0, b_1, χ, χ_1, and χ_2 are determined from (7.2.14), as functions of E_1, E_3, and τ. The electric field is determined from the last three equations of (7.11.17). Consequently, (7.11.15) and (7.11.16) give the stresses and polarizations as a function of the applied field $\overset{+}{E}_1$, $\overset{+}{E}_3$, and the shear angle τ.

The mechanical surface tractions to be applied on the slab consist of the tensions $t_{(3)3}$ and the shears $t_{(3)1}$ given by (7.11.17). Thus, the simple shear cannot be maintained without a prescribed tension which will keep the width of the slab unchanged. This is the *Poynting effect* encountered in many nonlinear problems. This effect continues to exist even when the elastic parts of the traction vanish (i.e., $\tau = 0$). The lowest-order term in E_1 and E_3 is of the second degree. We also need hydrostatic pressure to keep the volume unchanged. This is clear from the existence of ${}_{\mathrm{E}}t_{22}$. This is the *Kelvin effect.*

The body couple is given by

$$
{}_{\mathrm{M}}\mathbf{c} = \mathbf{P} \times \mathbf{E}, \tag{7.11.19}
$$

which has the components

$$
{}_{\mathrm{M}}c_1 = {}_{\mathrm{M}}c_3 = 0, \qquad {}_{\mathrm{M}}c_2 = P_3E_1 - P_1E_3, \tag{7.11.20}
$$

which indicate that the slab tends to rotate about the x_2-axis. This tendency persists even if the applied electric field is perpendicular to the surfaces of the slab $\left(\overset{+}{E}_1 = 0\right)$.

For a vanishing shear angle ($\tau = 0$), the tractions are of second degree in the electric field. This is the well-known *electrostrictive effect* which does not exist in the linear theory. The shear angle and the electric field produce other cross effects, as can be seen from (7.11.15) and (7.11.16).

7.12. Cylindrically Symmetric Deformation of a Tube Subject to a Radial Field

Consider a thick tube of infinite length, with inner radius R_1 and outer radius R_2, in the natural state. For the undeformed state and the deformed state, we select the cylindrical coordinates (R, Θ, Z) and (r, θ, z), respectively. The Z- and z-axes are taken along the cylinder axis (Figure 7.12.1). The deformation is described by

$$r = r(R), \qquad \theta = C\Theta + DZ, \qquad z = E\Theta + FZ, 0 \leq \theta, \quad \Theta < 2\pi, \quad (7.12.1)$$

where r is a function of R only and C, D, E, and F are constants.

The deformation tensors in cylindrical coordinates are calculated from

$$\overset{-1}{c}{}^{kl} = G^{KL} x^k{}_{,K} x^l{}_{,L}, \qquad c_{kl} = G_{KL} X^K{}_{,k} X^L{}_{,l}, \quad (7.12.2)$$

where G_{KL} is the metric tensor and G^{KL} is its reciprocal in the material coordinates X^K. We denote the metric tensor and its reciprocal in the spatial coordinates by g_{kl} and g^{kl}, respectively. The metric tensors are given by

$$G_{KL} = \begin{pmatrix} 1 & 0 & 0 \\ 0 & R^2 & 0 \\ 0 & 0 & 1 \end{pmatrix}, \qquad G^{KL} = \begin{pmatrix} 1 & 0 & 0 \\ 0 & R^{-2} & 0 \\ 0 & 0 & 1 \end{pmatrix},$$
$$g_{kl} = \begin{pmatrix} 1 & 0 & 0 \\ 0 & r^2 & 0 \\ 0 & 0 & 1 \end{pmatrix}, \qquad g^{kl} = \begin{pmatrix} 1 & 0 & 0 \\ 0 & r^{-2} & 0 \\ 0 & 0 & 1 \end{pmatrix}. \quad (7.12.3)$$

We identify $X^1 = R$, $X^2 = \Theta$, $X^3 = Z$, $x^1 = r$, $x^2 = \theta$, and $x^3 = z$. Using (7.12.1) in (7.12.3) we obtain c_{kl} and $\overset{-1}{c}_{kl}$. The mixed components of these tensors are then obtained by raising and lowering indices with g_{kl}, and g^{kl}, i.e.,

$$\overset{-1}{c}{}^k{}_l = g_{lm} \overset{-1}{c}{}^{km}, \qquad c^k{}_l = g^{km} c_{ml}. \quad (7.12.4)$$

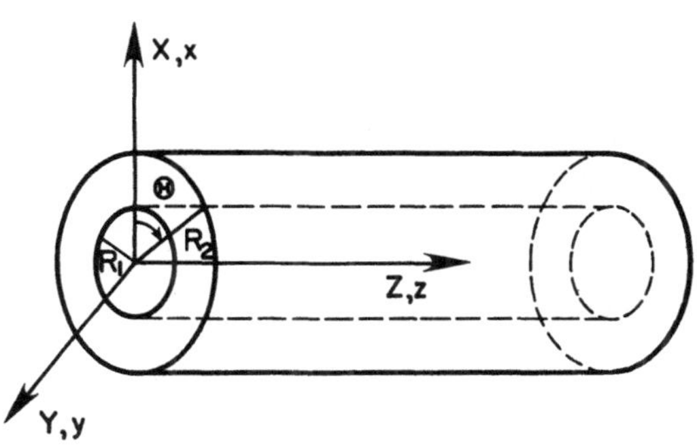

Figure 7.12.1. Cylindrical deformations of a tube.

The physical components of vectors and second-order tensors are given by

$$a^{(k)} = a^k \sqrt{g_{\underline{kk}}}, \qquad a^{(k)}{}_{(l)} = a^k{}_l \sqrt{g_{\underline{kk}}/g_{\underline{ll}}} \tag{7.12.5}$$

The physical components of $\overset{-1}{\mathbf{c}}$ and $\mathbf{c}$, we then find to be

$$\overset{-1}{c}{}^{(k)}{}_{(l)} = \begin{pmatrix} r'^2 & 0 & 0 \\ \cdot & r^2\left(\dfrac{C^2}{R^2} + D^2\right) & r\left(\dfrac{CE}{R^2} + DF\right) \\ \cdot & \cdot & \dfrac{E^2}{R^2} + F^2 \end{pmatrix},$$

$$c^{(k)}{}_{(l)} = \begin{pmatrix} 1/r'^2 & 0 & 0 \\ \cdot & \dfrac{1}{r^2}\dfrac{R^2F^2 + E^2}{(CF - DE)^2} & -\dfrac{1}{r}\dfrac{R^2DF + CE}{(CF - DE)^2} \\ \cdot & \cdot & \dfrac{R^2D^2 + C^2}{(CF - DE)^2} \end{pmatrix}. \tag{7.12.6}$$

From the first one of these we have

$$\mathrm{III} = (r'^2 r^2/R^2)(CF - DE)^2. \tag{7.12.7}$$

The deformation is isochoric if III = 1. For III = const. we have

$$r = (AR^2 + B)^{1/2}, \tag{7.12.8}$$

where A and B are constants. For

$$A = \pm(CF - DE)^{-1}, \tag{7.12.9}$$

the deformation is isochoric. Suppose that the deformation carries R_α to r_α, Θ_α (and of course $\Theta_\alpha + 2n\pi$) to θ_α, and Z_α to z_α. From (7.12.1) we have

$$\begin{aligned} r_\alpha &= (AR_\alpha{}^2 + B)^{1/2}, && \alpha = 1, 2, \\ \theta_\alpha &= C\Theta_\alpha + DZ + 2n\pi C\Theta_\alpha, && 0 \le \Theta_\alpha < 2\pi, \quad 0 \le \theta_\alpha < 2\pi, \\ z_\alpha &= E\Theta + FZ_\alpha + 2n\pi E\Theta_\alpha, && 0 \le \Theta_\alpha < 2\pi. \end{aligned} \tag{7.12.10}$$

The first of these determines A and B in terms of r_α and R_α (deformed and undeformed, and inner and outer radii of the cylinder). The second indicates that supplementary material must be added, if $C < 1$, and material must be deleted, if $C > 1$. If $C = 1$, the tube is complete, but it must be severed to perform the deformation, if $E \neq 0$, the following few special cases are included in the displacement fields described by (7.12.1):

(a) $A > 0$, $C = 1$, $D = E = 0$, inflation,
(b) $A < 0$, $C = 1$, $D = E = 0$, eversion,
(c) $A = 1$, $B = 0$, $C = 1$, $E = F = 0$, simple shear,
(d) $A = 1$, $B = 0$, $C = 1$, $D = F = 0$, screw dislocation.

Suppose that a radial field of flux is applied to the tube by placing the tube between the plates of a coaxial cylindrical condenser

$$\mathbf{E} = [E(r), 0, 0]. \tag{7.12.11}$$

The flux of field is determined by solving

$$-\varepsilon_0 \nabla^2 \phi + \nabla \cdot \mathbf{P} = 0, \qquad \mathbf{E} = -\nabla \phi, \tag{7.12.12}$$

in the regions $r < r_1$, $r_1 < r < r_2$, and $r > r_2$, subject to boundary conditions where the components of the electric field, tangential to the boundary surfaces, are continuous and where the surface charge is given, i.e.,

$$\mathbf{n} \times [\mathbf{E}] = \mathbf{0}, \qquad [\varepsilon_0 \nabla \phi - \mathbf{P}] \cdot \mathbf{n} = -w_e. \tag{7.12.13}$$

In the region $r < r_1$, we have $\mathbf{P} = \mathbf{0}$. At $r = 0$, ϕ is bounded. Hence the appropriate solution of (7.12.12) is

$$\phi = A_1 = \text{const.}, \qquad r < r_1. \tag{7.12.14}$$

In the region $r_1 < r < r_2$, $\mathbf{P}$ is given by (7.2.13), i.e.,

$$\mathbf{P} = \chi \mathbf{E} + \chi_1 \overset{-1}{\mathbf{c}} \mathbf{E} + \chi_2 \overset{-2}{\mathbf{c}} \mathbf{E}, \tag{7.12.15}$$

where χ, χ_1, and χ_2 are functions of the invariants I, II, III, I_4, I_6, and I_8 listed in (7.2.14). In view of (7.12.11), this is expressible as

$$P_r = \bar{\chi}(r'^2, r'^2 E^2) E_r, \qquad P_\theta = P_z = 0. \tag{7.12.16}$$

For the deformation field (7.12.8), r' is a function of r, so that

$$\varepsilon_0 E_r + P(r, E_r) = b_1/r, \qquad r_1 < r < r_2, \tag{7.12.17}$$

where

$$P(r, E_r) = \bar{\chi}(r'^2, r'^2 E_r^{\,2}) E_r. \tag{7.12.18}$$

In the region $r > r_2$, $\mathbf{P} = \mathbf{0}$, and the potential is given by

$$\varepsilon_0 \phi = C_1 \ln r + C_2, \qquad r > r_2, \tag{7.12.19}$$

where C_1 and C_2 are constants.

Employing the boundary conditions $(7.12.13)_2$ at $r = r_1$ and $r = r_2$, we determine B_1 and C_1 to be

$$B_1 = -r_1 w_e, \qquad C_1 = (r_1 - r_2) w_e. \tag{7.12.20}$$

The electric field is given by

$$\begin{aligned} \mathbf{E} &= \mathbf{0}, & r &< r_1, \\ \varepsilon_0 E_r + P(r, E_r) &= -(r_1/r) w_e, & r_1 &\le r \le r_2, \\ \varepsilon_0 E_r &= (r_2 - r_1) w_e r^{-1}, & r &> r_2. \end{aligned} \tag{7.12.21}$$

The second of these is a nonlinear equation to be solved for $E(r)$, since

$$P = [\chi + (\chi_1 + \chi_2 r'^2) r'^2] E_r, \tag{7.12.22}$$

where χ, χ_1, and χ_2 are calculated from their expressions, as given in (7.2.14). Since the strain and electric fields are functions of r only, their invariants also depend on r only. The **P** and $_{\mathrm{E}}$**t**-fields given by the constitutive equations will again be functions of r only. Here we treat the case of an incompressible cylinder for which (7.12.9) is valid. The physical components of the stress and the polarization fields are found by using (7.2.16) and (7.2.17):

$$
\begin{aligned}
{}_{\mathrm{E}}\bar{t}_{rr} &= {}_{\mathrm{E}}t_{rr} + p = \{b_{-1} - 2[\chi_1 + 2\chi_2(AR/r)^2]E_r^{\,2}\}(AR/r)^2 + b_1(r/AR)^2,\\
{}_{\mathrm{E}}\bar{t}_{\theta\theta} &= {}_{\mathrm{E}}t_{\theta\theta} + p = b_{-1}r^2\left(\frac{C^2}{R^2} + D^2\right) + b_1(A^2/r^2)(R^2F^2 + E^2),\\
{}_{\mathrm{E}}\bar{t}_{zz} &= {}_{\mathrm{E}}t_{zz} + p = b_{-1}\left(\frac{E^2}{R^2} + F^2\right) + b_1A^2(R^2D^2 + C^2), \qquad (7.12.23)\\
{}_{\mathrm{E}}t_{\theta z} &= b_{-1}r\left(\frac{CE}{R^2} + DF\right) - \frac{b_1}{r}A^2(R^2DF + CE),\\
{}_{\mathrm{E}}t_{r\theta} &= {}_{\mathrm{E}}t_{rz} = 0,
\end{aligned}
$$

$$P_r = P = \{\chi + [\chi_1 + \chi_2(AR/r)^2](AR/r)^2\}E_r, \qquad P_\theta = P_z = 0. \quad (7.12.24)$$

At equilibrium, eq. (7.2.2) reduce to

$$-\frac{dp}{dr} + \frac{d}{dr}({}_{\mathrm{E}}\bar{t}_{rr}) + \frac{1}{r}({}_{\mathrm{E}}\bar{t}_{rr} - {}_{\mathrm{E}}\bar{t}_{\theta\theta}) - \frac{1}{r}E_r\frac{d}{dr}(rP) = 0. \qquad (7.12.25)$$

Integration gives

$$-p + {}_{\mathrm{E}}\bar{t}_{rr} + \int^r \left[{}_{\mathrm{E}}\bar{t}_{rr} - {}_{\mathrm{E}}\bar{t}_{\theta\theta} - E_r\frac{d}{dr}(rP)\right]\frac{dr}{r} - D_1 = 0. \qquad (7.12.26)$$

The constant D_1 can be determined by clearing the inner or outer surface of the deformed cylinder from the tractions. We clear the outer surface, $r_2 = r(R_2)$, so that

$$p = {}_{\mathrm{E}}\bar{t}_{rr}(r) - \tfrac{1}{2}E_r^{\,2}(r_2) + \int_{r_2}^r \left[{}_{\mathrm{E}}\bar{t}_{rr} - {}_{\mathrm{E}}\bar{t}_{\theta\theta} - E_r\frac{d}{d\rho}(\rho P)\right]\frac{d\rho}{\rho} \qquad (7.12.27)$$

determines the pressure $p(r)$. Since, through the expression (7.12.22), E_r is determined as a function r, ${}_{\mathrm{E}}\bar{t}_{rr}$, ${}_{\mathrm{E}}\bar{t}_{\theta\theta}$, ${}_{\mathrm{E}}\bar{t}_{zz}$, and ${}_{\mathrm{E}}\bar{t}_{\theta z}$ are all known as functions of r. Consequently, (7.12.27) can be evaluated to give the unknown pressure, thus determining the stress and polarization fields. Note, however, that the inner surface $r = r(R_1)$ of the cylinder cannot be cleared of normal tractions. Thus, to maintain the deformation prescribed by (7.12.1) of an incompressible cylinder, prescribed normal tractions must be supplied to the inner surface of the cylinder. In addition, normal tractions t_{zz} and shear $t_{z\theta}$ will act at the end section of the cylinder. This situation is well known to us from nonlinear elasticity (see Eringen [1962, p. 186]). However, even when the elastic deformations are absent, we still have a radial stress t_{rr} due to the electric field. This is again the well-known *electrostrictive effect*.

The case of the uniform extension of a tube was solved by Eringen [1961]. The general solution presented here was outlined by Singh and Pipkin [1966], who also discussed other problems concerning controllable states of elastic dielectrics.

7.13. Axisymmetric Oscillations of a Tube

In the case of radial deformations of an incompressible tube of unit length, we have $A = C = F = 1$, $D = E = 0$. In cylindrical coordinates (Figure 7.13.1), the stress and electric fields given by (7.12.23) and (7.12.24) are reduced to

$$
\begin{aligned}
{}_{E}t_{rr} &= -p + b_{-1}u^{-1} + b_1 u - 2\kappa E^2, \\
{}_{E}t_{\theta\theta} &= -p + b_{-1}u + b_1 u^{-1}, \qquad\qquad (7.13.1)\\
{}_{E}t_{zz} &= -p + b_{-1} + b_1, \qquad t_{r\theta} = t_{\theta z} = t_{zr} = 0,
\end{aligned}
$$

$$P = (\chi + \kappa - \chi_2 u^{-2})E, \qquad (7.13.2)$$

where we introduced

$$u = (r/R)^2, \qquad \kappa = \chi_1 u^{-1} + 2\chi_2 u^{-2}, \qquad E_r = E. \qquad (7.13.3)$$

The incompressibility condition (7.12.9) is satisfied, and by (7.12.8) we have

$$r^2(R, t) = {r_1}^2 + R^2 - {R_1}^2 = ({r_1}^2 - {R_1}^2)u(u-1)^{-1}. \qquad (7.13.4)$$

The constitutive moduli b_{-1}, b_1, χ, and κ are given by (7.2.18) and (7.2.15), in which Σ is a function of the invariants

$$\mathrm{I} = \mathrm{II} = u^{-1} + u + 1, \qquad \mathrm{I}_4 = E^2, \qquad \mathrm{I}_6 = E^2u^{-1}, \qquad \mathrm{I}_8 = E^2u^{-2}. \quad (7.13.5)$$

For frequencies far below optical frequencies, the time rate of change of

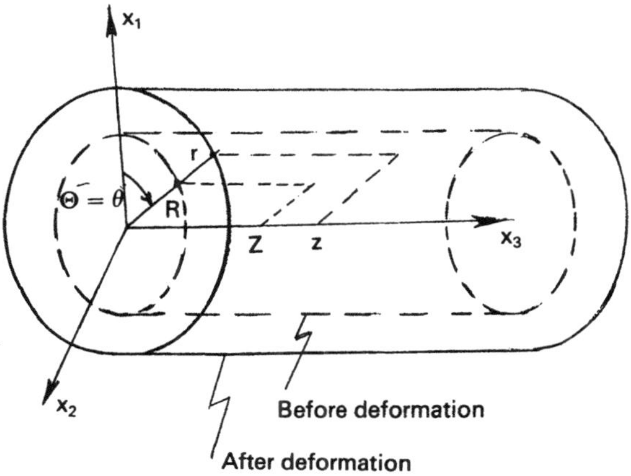

Figure 7.13.1. Axi-symmetric deformation.

magnetic induction may be discarded. For nonmagnetizable materials, the effect of the magnetic field arising from the motion is nil. Thus, for such dielectrics, we may neglect the magnetic effect entirely. In this case, then, all we need is to consider the electrostatics of the dielectrics, which have already been treated in Section 7.12 for a tube under radial polarization only (eqs. (7.12.22)). The radial oscillations of the tube are governed by the equation of motion

$$\frac{\partial_{\mathrm{E}} t_{rr}}{\partial r} + \frac{1}{r}({}_{\mathrm{E}}t_{rr} - {}_{\mathrm{E}}t_{\theta\theta}) - \frac{1}{r}\frac{\partial}{\partial r}(rP)E = \rho_0 \ddot{r}. \tag{7.13.6}$$

It is not difficult to verify that $\ddot{r}$, computed from (7.13.4), has a potential ζ, i.e.,

$$\ddot{r} = -\frac{\partial \zeta}{\partial z}, \tag{7.13.7}$$

where

$$-\zeta = (r_1\ddot{r} + \dot{r}_1^{\,2}) \ln r + \tfrac{1}{2} r^{-2} r_1^{\,2} \dot{r}_1^{\,2}. \tag{7.13.8}$$

Hence, the integral of (7.13.6), upon recalling $t_{rr} = {}_{\mathrm{E}}t_{rr} - PE$, gives

$$t_{rr} + \int^{r} \left({}_{\mathrm{E}}t_{rr} - {}_{\mathrm{E}}t_{\theta\theta} - PE + rP\frac{\partial E}{\partial r} \right)\frac{dr}{r} + \rho_0 \zeta = D_1(t). \tag{7.13.9}$$

The only allowable traction boundary conditions are the uniform tractions applied on the surfaces of the cylinder, i.e.,

$$\begin{aligned} t_{rr} &= -p_1(t), \qquad r = r_1, \\ t_{rr} &= -p_2(t), \qquad r = r_2. \end{aligned} \tag{7.13.10}$$

Using these conditions, we eliminate $D_1(t)$, to obtain

$$\rho_0(\zeta_1 - \zeta_2) + \int_{r_2}^{r_1} \left({}_{\mathrm{E}}t_{rr} - {}_{\mathrm{E}}t_{\theta\theta} - PE + rP\frac{\partial E}{\partial r} \right)\frac{dr}{r} = p_1(t) - p_2(t), \tag{7.13.11}$$

where

$$\zeta_\alpha = -(r_1\ddot{r}_1 + \dot{r}_1^{\,2}) \ln r_\alpha - \tfrac{1}{2} r_1^{\,2} \dot{r}_1^{\,2} r_\alpha^{\,-2} \tag{7.13.12}$$

If we set

$$x(t) = r_1(t)/R_1, \qquad \gamma = R_2^{\,2}/R_1^{\,2} - 1 > 0, \tag{7.13.13}$$

then (7.13.3) gives

$$\begin{aligned} r_2^{\,2}/R_2^{\,2} &= (\gamma + x^2)(1+\gamma)^{-1}, \qquad r_2^{\,2}/r_1^{\,2} = 1 + \gamma x^2, \\ u(R, t) &= 1 + (R_1^{\,2}/R^2)[x^2(t) - 1]. \end{aligned} \tag{7.13.14}$$

Introducing this into (7.13.11) and (7.13.12), we obtain

$$\begin{aligned} x\ddot{x} \ln(1 + \gamma x^{-2}) + [\ln(1 + \gamma x^{-2}) - \gamma(\gamma + x^2)^{-1}]\dot{x}^2 + f(x, \gamma) \\ = (2/\rho_0 R_1^{\,2})[p_1(t) - p_2(t)], \end{aligned} \tag{7.13.15}$$

where

$$f(x, \gamma) = \frac{1}{\rho_0 R_1^2} \int_{(\gamma+x^2)/(1+\gamma)}^{x^2} \left[(b_{-1} - b_1)(1+u)u^{-2} - \frac{2\kappa E^2}{u(1-u)} \right.$$
$$\left. - (\chi + \kappa - \chi_2 u^{-2}) \left(\frac{E^2}{u - u^2} - \frac{\partial E^2}{\partial u} \right) \right] du. \qquad (7.13.16)$$

Given the free energy function

$$\Sigma(\mathrm{I}, \mathrm{II}, \mathrm{I}_4, \mathrm{I}_6, \mathrm{I}_8) = \hat{\Sigma}(u, \mathrm{I}_4), \qquad (7.13.17)$$

χ and κ are determined as functions of u and E. In fact, by differentiation of (7.13.17), we have

$$b_{-1} - b_1 = \left(2\frac{\partial \hat{\Sigma}}{\partial u} u - \chi_1 \mathrm{I}_6 - 2\chi_2 \mathrm{I}_8 \right)(u - u^{-1})^{-1},$$
$$\chi + \kappa - \chi_2 u^{-2} = -2\frac{\partial \hat{\Sigma}}{\partial \mathrm{I}_4}, \qquad (7.13.18)$$
$$\chi_1 = -2\frac{\partial \Sigma}{\partial \mathrm{I}_6}, \qquad \chi_2 = -2\frac{\partial \Sigma}{\partial \mathrm{I}_8}.$$

Employing these in (7.13.16), we obtain

$$f(x, \gamma) = \frac{2}{\rho_0 R_1^2} \int_{(\gamma+x^2)/(1+\gamma)}^{x^2} \left\{ \frac{\partial \Sigma}{\partial u} u - \frac{\partial \Sigma}{\partial \mathrm{I}_4} \left[\mathrm{I}_4 + (u^2 - u)\frac{\partial E^2}{\partial u} \right] \right.$$
$$\left. - \tfrac{1}{2}\chi_1 \mathrm{I}_6 - \chi_2 \mathrm{I}_8 + \kappa \mathrm{I}_4 \right\} \frac{du}{u^2 - u}. \qquad (7.13.19)$$

The electric field is determined by eliminating P between (7.13.2) and $(7.12.21)_2$, i.e.,

$$\left(\varepsilon_0 - 2\frac{\partial \hat{\Sigma}}{\partial \mathrm{I}_4} \right) E = -w_e x (x^2 - 1)^{-1/2} (1 - u^{-1})^{1/2}, \qquad (7.13.20)$$

where we used (7.13.13), (7.13.14), and $(7.13.18)_2$. This functional equation must be solved for E. Once this is done, the integrand in (7.13.19) will be determined as a function of x and u, and the integral (7.13.19) can be evaluated, at least, in principle.

It may be verified that (7.13.15) can be put into the form

$$\frac{d}{dt}[\tfrac{1}{2}x^2\dot{x}^2 \ln(1 + \gamma x^{-2})] + x f(x, \gamma)\frac{dx}{dt} = 2[p_1(t) - p_2(t)](x/\rho_0 R_1^2)\frac{dx}{dt}. \qquad (7.13.21)$$

In the case where the applied loads vanish, $p_1(t) = p_2(t) = 0$, and the tube will have free radial oscillations. We define the function

$$F(x, \gamma) = \int_1^x \xi f(\xi, \gamma)\, d\xi. \qquad (7.13.22)$$

Clearly, this function satisfies the relations

$$\frac{dF}{dx} = xf(x, \gamma), \qquad F(1, \gamma) = 0. \tag{7.13.23}$$

We assume that in the undeformed state $x = 1$, the stored energy Σ vanishes. The integration of (7.13.21), for the case $p_1 = p_2 = 0$, gives

$$\tfrac{1}{2}v^2x^2 \ln(1 + \gamma x^{-2}) + F(x, \gamma) = C. \tag{7.13.24}$$

To determine C, we need the initial conditions

$$x(0) = x_0, \qquad \dot{x}(0) = v_0, \tag{7.13.25}$$

which imply that the motion is imparted to the inner wall of the tube by displacing it to the position $r_1 = x_0 R_1$, with velocity $\dot{r}_1 = v_0 R$, at time $t = 0$. Employing (7.13.24), we obtain

$$C = \tfrac{1}{2}{v_0}^2{x_0}^2 \ln(1 + \gamma {x_0}^{-2}) + F(x_0, \gamma). \tag{7.13.26}$$

From (7.13.24) we find the velocity of oscillations of the inner surface of the tube

$$v = \pm\left[\frac{2C - 2F(x, \gamma)}{x^2 \ln(1 + \gamma x^{-2})}\right]^{1/2}. \tag{7.13.27}$$

For the existence of a real velocity, we must have

$$F(x, \gamma) \leq C. \tag{7.13.28}$$

The period of oscillations is given by

$$T = 2\int_{x_1}^{x_2} v^{-1}\, dx, \tag{7.13.29}$$

where x_1 and x_2 are two simple roots of

$$F(x, \gamma) = C. \tag{7.13.30}$$

It is well known that the motion is periodic if trajectories, described by (7.13.24), are closed curves in the (v, x)-plane. Thus, such a closed curve intersects the x-axis at two points, $x = x_1 < 1$ and $x = x_2 > 1$, where $x = 1$ corresponds to the undeformed state of the tube. Similar to the vibrations of the unpolarized tube discussed by Knowles [1960], [1962] (see also Eringen and Suhubi [1974, Sect. 3.3]), x_1 and x_2 represent minimum and maximum radii of the deformed inner surface of the tube.

To carry the analysis further, we now consider a special case:

(1) *Special Material.* Extending the concept of the Mooney material to dielectrics, we consider a stored energy of the form

$$\Sigma = \tfrac{1}{2}\Sigma_1(\mathrm{I} - 3) + \tfrac{1}{2}\Sigma_2(\mathrm{II} - 3) + \tfrac{1}{2}\Sigma_3 E^2 + \tfrac{1}{2}\Sigma_4 E^2 u^{-1} + \tfrac{1}{2}\Sigma_5 E^2 u^{-2}. \tag{7.13.31}$$

Hence, we have

$$b_{-1} = \Sigma_1, \quad b_1 = -\Sigma_2, \quad \chi = -\Sigma_3, \quad \chi_1 = -\Sigma_4, \quad \chi_2 = -\Sigma_5, \quad \kappa = -\Sigma_4 u^{-1} - 2\Sigma_5 u^{-2}, \tag{7.13.32}$$

(7.13.1) and (7.13.2) give

$$\begin{aligned} t_{rr} &= -p + (\Sigma_1 + 2\Sigma_4 E^2)u^{-1} - \Sigma_2 u + 4\Sigma_5 u^{-2}E^2, \\ t_{\theta\theta} &= -p + \Sigma_1 u - \Sigma_2 u^{-1}, \\ t_{zz} &= -p + \Sigma_1 - \Sigma_2, \qquad t_{r\theta} = t_{\theta z} = t_{zz} = 0, \\ P &= -(\Sigma_3 + \Sigma_4 u^{-1} + \Sigma_5 u^{-2})E. \end{aligned} \tag{7.13.33}$$

For the electric field we have, from (7.13.20),

$$E = -w_e x[(1 - u^{-1})(x^2 - 1)^{-1}]^{1/2}(\varepsilon_0 - \Sigma_3 - \Sigma_4 u^{-1} - \Sigma_5 u^{-2})^{-1}. \tag{7.13.34}$$

With these (7.13.16) takes the form

$$\begin{aligned} f(x, \gamma) = \frac{1}{\rho_0 R_1{}^2} \int_{(\gamma+x^2)/(1+\gamma)}^{x^2} \bigg[&(\Sigma_1 + \Sigma_2)(u^{-1} + u^{-2}) - \Sigma_3 E^2 \\ &- (\Sigma_4 + 2\Sigma_5) E^2 (u^3 - u^2)^{-1} - 5\Sigma_5 E^2 (u^4 - u^3)^{-1} \\ &- (\Sigma_3 + \Sigma_4 u^{-1} + \Sigma_5 u^{-2}) \frac{\partial E^2}{\partial u} \bigg] \, du. \end{aligned} \tag{7.13.35}$$

This tedious integral will have to be evaluated. In the special case $\Sigma_4 = \Sigma_5 = 0$, using (7.13.34), the integration yields

$$\begin{aligned} f(x, \gamma) = f_0 \bigg[&\left(1 + \sigma_3 \frac{x^2}{x^2 - 1}\right) \ln \frac{(1 + \gamma)x^2}{\gamma + x^2} + \frac{\gamma(x^2 - 1)}{x^2(\gamma + x^2)} \\ &+ \sigma_3 \left(\frac{\gamma}{\gamma + x^3} - \frac{\gamma x^2}{1 + \gamma} \right) \bigg], \end{aligned} \tag{7.13.36}$$

where

$$f_0 = \frac{\Sigma_1 + \Sigma_2}{\rho_0 R_1{}^2}, \qquad \sigma_3 = \frac{\Sigma_3 w_e^2}{(\varepsilon_0 - \Sigma_3)^2(\Sigma_1 + \Sigma_2)}. \tag{7.13.37}$$

$F(x, \gamma)$ can be evaluated from (7.13.22) by integration.

The velocity of oscillations of the inner surface of the tube is now given by (7.13.27). The period is then calculated by (7.13.29).

Nonlinear vibrations of the electrostrictive tube were studied by Nowinski and Wu [1968]. There exist studies on nonlinear waves and shock waves in nonlinear elastic dielectrics (see McCarthy [1965], [1966a] and McCarthy and Green [1966] for acceleration waves and Chen and McCarthy [1974a, b] and Collet [1984] for shock waves).

7.14. Small Deformations and Fields Superimposed on Large Static Deformations and Fields

As we have seen, obtaining solutions for nonlinear problems is difficult, and often is not possible with the mathematical techniques presently available. However, in certain classes of problems, the final state of the body can be reached in two steps: initial static deformations and electromagnetic fields; followed by superposed infinitesimal deformations and electromagnetic fields from this *static biased state*. In this case, it is possible to obtain the final state of the body by solving a set of linear partial differential equations. The coefficients of these differential equations are expressible in terms of the deformations and fields describing the initial deformation and fields. Many technologically important classes of problems can be dealt with in this way. Among these are the photoelastic, electroelastic, and magneto-optical effects; propagation of discontinuities; and problems of instability.

The foregoing method of approach has been employed by several workers to discuss some of these problems. Both the method of approach and the solution of several problems, in the case of a purely mechanical system, have been discussed by Eringen and Suhubi [1974, Chap. IV]. A discussion of the nonmagnetizable dielectrics was given by Toupin [1963].[10] Other works on the subject are to be found in Dunkin and Eringen [1963] on electromagnetic waves under bias electric and magnetic fields and waves in plates; in Dixon and Eringen [1965b] on waves in polar elastic dielectrics; in Boulanger et al. [1973], Baumhauer and Tiersten [1973], Hutter [1975], Maugin and Pouget [1980], and Prechtl [1983] on electroelasticity; and in Maugin [1979a], and his coworkers, [1982], [1984], [1985], on magnetoelasticity.

In this section we derive the basic equations of elastic dielectrics for infinitesimal deformations and fields superposed on an initial static deformation and field. In the remaining parts of this chapter, we discuss the solution of these equations for various classes of problems involving different types of biases.

Essential to this formulation are three different configurations of the body (Figure 7.14.1):

(i) *The Reference Configuration B*: At time $t = 0$, the body is undeformed and free of all fields. A generic point at this state is denoted by $\mathbf{X}$ having rectangular coordinates X_K, $K = 1, 2, 3$.

---

[10] Toupin considers nonmagnetizable dielectrics. Boulanger et al. [1973] include magnetic fields, but their ponderomotive force and constitutive equations are different from ours. The same remarks apply to the work of Hutter, which considers the effect of conduction. Baumhauer and Tiersten employ a somewhat different approach in quasi-electrostatics.

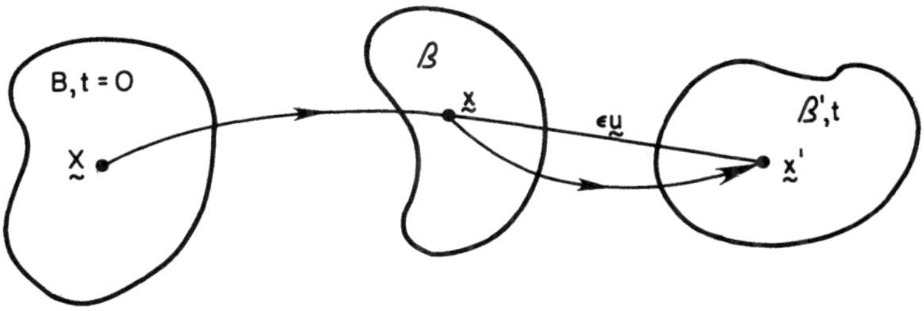

Figure 7.14.1. Reference configuration B, initial configuration $\mathscr{B}$, and the present configuration $\mathscr{B}'$.

(ii) *The Initial Configuration*, $\mathscr{B}$: In this state the body is deformed statically and carries static fields. The deformation is described by

$$x_k = x_k(\mathbf{X}), \tag{7.14.1}$$

where x_k are the rectangular coordinates, in $\mathscr{B}$, of the materials point $\mathbf{X}$ of B. The mechanical and electromagnetic fields developed in $\mathscr{B}$ are denoted by a superscript (^{0}), e.g.,

$$_{\mathrm{E}}T^0{}_{kl}(\mathbf{x}), \qquad P^0{}_k(\mathbf{x}), \qquad M^0{}_k(\mathbf{x}), \qquad E^0{}_k(\mathbf{x}), \qquad B^0{}_k(\mathbf{x}), \qquad \theta^0(\mathbf{x}). \tag{7.14.2}$$

Referring to the material frame in B, we have the following representation:

$$\begin{aligned} &_{\mathrm{E}}T^0{}_{KL}(\mathbf{X}) = J_0\, {}_{\mathrm{E}}T^0{}_{kl}(\mathbf{x}) X_{K,k} X_{L,l} = {}_{\mathrm{E}}T^0{}_{Kl} X_{K,k}, \\ &\Pi^0{}_K(\mathbf{X}) = J_0 P^0{}_k(\mathbf{x}) X_{K,k}, \qquad M^0{}_K(\mathbf{X}) = J_0 M^0{}_k(\mathbf{x}) X_{K,k}, \\ &E^0{}_K(\mathbf{X}) = E^0{}_k(\mathbf{x}) x_{k,K}, \qquad B^0{}_K(\mathbf{X}) = B^0{}_k(\mathbf{x}) x_{k,K}, \end{aligned} \tag{7.14.3}$$

where

$$J_0 = \rho/\rho_0 = \det(x_{k,K}). \tag{7.14.4}$$

The field (7.14.2) satisfies the following static field equations at a regular point in $\mathscr{B}$:

$$_{\mathrm{E}}T^0{}_{kl,k} + (B^0{}_{k,l} - B^0{}_{l,k})M^0{}_k - M^0{}_{k,k}B^0{}_l + (q_e^0 - P^0{}_{k,k})E^0{}_l + \rho f^0{}_l = 0, \tag{7.14.5}$$

$$\begin{aligned} \nabla\cdot(\mathbf{E}^0 + \mathbf{P}^0) = q_e^0, \qquad & \nabla \times \mathbf{E}^0 = \mathbf{0}, \\ \nabla\cdot\mathbf{B}^0 = 0, \qquad & \nabla \times (\mathbf{B}^0 - \mathbf{M}^0) = \mathbf{0}, \end{aligned} \tag{7.14.6}$$

and the jump conditions at a static singular surface σ^0 with unit normal $\mathbf{n}$ in $\mathscr{B}$.

$$\begin{aligned} &[T^0{}_{kl} + {}_{\mathrm{M}}T^0{}_{kl}]n_k = 0, \\ &\mathbf{n} \times [\mathbf{E}^0] = \mathbf{0}, \qquad [\mathbf{D}^0]\cdot\mathbf{n} = w_e^0, \\ &\mathbf{n} \times [\mathbf{H}^0] = \mathbf{0}, \qquad [\mathbf{B}^0]\cdot\mathbf{n} = 0. \end{aligned} \tag{7.14.7}$$

Here ρ_0 is the mass density in B, and ρ, q_e^0, $\mathbf{f}^0$, are, respectively, the mass density, the charge density, and the body force density in $\mathscr{B}$, while w_e^0 is the electric change per unit surface on σ^0. We also have the usual definitions

$$\begin{aligned} t^0{}_{kl} &\equiv T^0{}_{kl} = {}_{\mathrm{E}}T^0{}_{kl} - P^0{}_k E^0{}_l - M^0{}_k B^0{}_l, \qquad {}_{\mathrm{E}}t^0{}_{kl} \equiv {}_{\mathrm{E}}T^0{}_{kl}, \\ {}_{\mathrm{M}}t^0{}_{kl} &\equiv {}_{\mathrm{M}}T^0{}_{kl} = P^0{}_k E^0{}_l - B^0{}_k M^0{}_l + E^0{}_k E^0{}_l + B^0{}_k B^0{}_l \\ &\quad - \tfrac{1}{2}(E^{0^2} + B^{0^2} - 2\mathbf{M}^0 \cdot \mathbf{B}^0)\delta_{kl}, \\ D^0{}_k &= E^0{}_k + P^0{}_k, \qquad H^0{}_k = B^0{}_k - M^0{}_k. \end{aligned} \tag{7.14.8}$$

The constitutive equations for the fields in $\mathscr{B}$ are

$$\begin{aligned} {}_{\mathrm{E}}T^0{}_{kl} &= J_0{}^{-1}\left(\frac{\partial \Sigma}{\partial E_{KL}}\right)_0 x_{k,K} x_{l,K} = J_0{}^{-1}\left(\frac{\partial \Sigma}{\partial E_{kl}}\right)_0, \\ P^0{}_k &= -J_0{}^{-1}\left(\frac{\partial \Sigma}{\partial \mathscr{E}_K}\right)_0 x_{k,K} = -J_0{}^{-1}\left(\frac{\partial \Sigma}{\partial \mathscr{E}_k}\right)_0, \\ M^0{}_k &= -J_0{}^{-1}\left(\frac{\partial \Sigma}{\partial B_K}\right)_0 x_{k,K} = -J_0{}^{-1}\left(\frac{\partial \Sigma}{\partial B_k}\right)_0. \end{aligned} \tag{7.14.9}$$

Equations (7.14.5) may also be expressed in the material frame as

$$\begin{aligned} {}_{\mathrm{E}}T^0{}_{Kl,K} &+ (B^0{}_{k,l} - B^0{}_{l,k}) x_{k,K} M^0{}_K - M^0{}_{K,K} B^0{}_l \\ &+ (J_0 q_e^0 - \Pi^0{}_{K,K}) E^0{}_l + \rho_0 f^0{}_l = 0. \end{aligned} \tag{7.14.10}$$

(iii) *The Present Configuration* $\mathscr{B}'$: To the deformed body at the initial configuration, time- dependent small deformations and fields are applied, resulting in the present configuration $\mathscr{B}'$. The quantities referred to $\mathscr{B}'$ will carry a prime. For example, the final position $\mathbf{x}'$ of $\mathbf{x}$ is given by

$$\mathbf{x}'(\mathbf{X}, t) = \mathbf{x}(\mathbf{X}) + \varepsilon \mathbf{u}(\mathbf{X}, t), \tag{7.14.11}$$

where $\varepsilon \mathbf{u}$ is the displacement vector, $\varepsilon \ll 1$ being a perturbation parameter. Referring to the coordinates x_k and $x_{k'}$, (7.14.11) reads

$$x_{k'} = [x_k + \varepsilon u_k(\mathbf{x}, t)]\delta_{kk'}, \tag{7.14.12}$$

where $u_k(\mathbf{x}, t)$ is the displacement vector referred to the coordinates x_k. Since the *same* rectangular coordinates are being used, $\delta_{kk'}$ are Kronecker deltas.

In the present configuration $\mathscr{B}'$ the fields are denoted by

$$\begin{aligned} &{}_{\mathrm{E}}t_{k'l'}, \qquad P_{k'}, \qquad \mathscr{M}_{k'} = M_{k'} + \frac{1}{c}(\mathbf{v} \times \mathbf{P})_{k'}, \\ &\mathscr{E}_{k'} = E_{k'} + \frac{1}{c}(\mathbf{v} \times \mathbf{B})_{k'}, \qquad B_{k'}, \qquad \theta_{,k'}. \end{aligned} \tag{7.14.13}$$

These have images in $\mathscr{B}$ given by

$${}_{\mathrm{E}}T_{kl'} = {}_{\mathrm{E}}T_{kl} x_{l',l} = J\, {}_{\mathrm{E}}t_{k'l'} x_{k,k'},$$

$$P_k = JP_{k'}x_{k,k'}, \qquad \mathscr{M}_k = J\mathscr{M}_{k'}x_{k,k'},$$
$$\mathscr{E}_k = \mathscr{E}_{k'}x_{k',k}, \qquad B_k = B_{k'}x_{k',k}, \tag{7.14.14}$$

where

$$J = \rho/\rho' = \det(x_{k',k}). \tag{7.14.15}$$

The dynamical field equations at regular points of $\mathscr{B}'$ are

$$\begin{aligned} &{}_{\mathrm{E}}t_{k'l',k'} + (B_{k',l'} - B_{l',k'})\mathscr{M}_{k'} - \mathscr{M}_{k',k'}B_{l'} + (q'_e - \nabla'\cdot\mathbf{P}')\mathscr{E}_{l'} \\ &\qquad + \frac{1}{c}\left(\overset{*}{\mathbf{P}} \times \mathbf{B}\right)_{l'} + \rho'(f_{l'} - \dot{v}_{l'}) = 0, \\ &(E_{k'} + P_{k'})_{,k'} = q'_e, \qquad \nabla' \times \mathbf{E}' + \frac{1}{c}\frac{\partial \mathbf{B}'}{\partial t} = \mathbf{0}, \\ &B_{k',k'} = 0, \qquad \nabla' \times \mathbf{H}' - \frac{1}{c}\frac{\partial \mathbf{D}'}{\partial t} = \frac{1}{c}q'_e\mathbf{v}'. \end{aligned} \tag{7.14.16}$$

If we now use the identities

$$(J^{-1}x_{k',k})_{,k'} = 0, \qquad \overset{*}{P}_{k'} = J^{-1}\dot{P}_k x_{k',k}, \tag{7.14.17}$$

and (7.14.14), we can transform $(7.14.16)_1$ and $(7.14.16)_5$ to the forms

$$\begin{aligned} &{}_{\mathrm{E}}T_{kl',k} + (B_{k',l'} - B_{l',k'})x_{k',k}\mathscr{M}_k - \mathscr{M}_{k,k}B_{l'} + (Jq'_e - P_{k,k})\mathscr{E}_{l'} \\ &\qquad + \frac{1}{c}\varepsilon_{l'm'n'}x_{m',k}\dot{P}_k B_{n'} + \rho(f_{l'} - \dot{v}_{l'}) = 0, \end{aligned} \tag{7.14.18}$$

$$\nabla' \times \mathbf{B}' - \frac{1}{c}\frac{\partial \mathbf{E}'}{\partial t} = \frac{1}{c}[(q'_e - \nabla'\cdot\mathbf{P}')\mathbf{v}' + \overset{*}{\mathbf{P}}' + c\nabla' \times \mathscr{M}']. \tag{7.14.19}$$

From (7.14.12) we find that

$$\begin{aligned} &x_{k',k} = (\delta_{lk} + \varepsilon u_{l,k})\delta_{lk'}, \\ &x_{k,k'} = (\delta_{lk} - \varepsilon u_{k,l})\delta_{lk'}, \qquad J^{-1} = 1 - \varepsilon u_{k,k}. \end{aligned} \tag{7.14.20}$$

By means of these equations, we can express the dynamical fields as the sum of the bias static fields and incremental dynamical fields (linear in ε). To this end, we use (7.14.14)

$$\begin{aligned} {}_{\mathrm{E}}t_{k'l'} &= ({}_{\mathrm{E}}t^0{}_{kl} + \varepsilon\, {}_{\mathrm{E}}\tilde{t}_{kl})\delta_{kk'}\delta_{ll'}, \\ {}_{\mathrm{E}}T_{kl'} &= ({}_{\mathrm{E}}T^0{}_{kl} + \varepsilon\, {}_{\mathrm{E}}\bar{T}_{kl})\delta_{ll'}, \\ E_{k'} &= (E^0{}_k + \varepsilon e_k)\delta_{kk'}, \\ B_{k'} &= (B^0{}_k + \varepsilon b_k)\delta_{kk'}, \\ P_{k'} &= (P^0{}_k + \varepsilon p_k)\delta_{kk'}, \\ \mathscr{M}_{k'} &= \left(M^0{}_k + \varepsilon m_k + \frac{\varepsilon}{c}(\dot{\mathbf{u}} \times \mathbf{P}^0)_k\right)\delta_{kk'}, \end{aligned} \tag{7.14.21}$$

where

$$p_k = d_k - e_k, \qquad m_k = b_k - h_k. \tag{7.14.22}$$

We also set

$$\begin{aligned}
{}_{\mathrm{E}}T_{kl}(\mathbf{x}, t) &= {}_{\mathrm{E}}T^0{}_{kl}(\mathbf{x}) + \varepsilon\, {}_{\mathrm{E}}\tilde{T}_{kl}(\mathbf{x}, t), \qquad {}_{\mathrm{E}}T^0{}_{kl} = {}_{\mathrm{E}}t^0{}_{kl},\\
E_k(\mathbf{x}, t) &= E^0{}_k(\mathbf{x}) + \varepsilon \tilde{E}_k(\mathbf{x}, t),\\
B_k(\mathbf{x}, t) &= B^0{}_k(\mathbf{x}) + \varepsilon \tilde{B}_k(\mathbf{x}, t),\\
P_k(\mathbf{x}, t) &= P^0{}_k(\mathbf{x}) + \varepsilon \tilde{P}_k(\mathbf{x}, t),\\
\mathscr{M}_k(\mathbf{x}, t) &= M^0{}_k(\mathbf{x}) + \varepsilon \tilde{\mathscr{M}}_k(\mathbf{x}, t),\\
\mathscr{E}_k &= E^0{}_k + \varepsilon \tilde{e}_k,
\end{aligned} \tag{7.14.23}$$

and

$$\begin{aligned}
\rho' &= \rho - \varepsilon\tilde{\rho}, \qquad f_{k'} = (f^0{}_k + \varepsilon \tilde{f}_k)\delta_{kk'},\\
q'_{\mathrm{e}} &= q^0_{\mathrm{e}} + \varepsilon \tilde{q}_{\mathrm{e}}, \qquad w'_{\mathrm{e}} = w^0_{\mathrm{e}} + \varepsilon \tilde{w}_{\mathrm{e}}.
\end{aligned} \tag{7.14.24}$$

Using (7.14.20)–(7.14.23) and $(7.14.24)_1$ in (7.14.14) and (7.14.15), we identify the coefficients of ε

$$\begin{aligned}
{}_{\mathrm{E}}\bar{T}_{kl} &= {}_{\mathrm{E}}T^0{}_{kr}u_{l,r} + {}_{\mathrm{E}}\tilde{T}_{kl},\\
{}_{\mathrm{E}}\tilde{t}_{kl} &= {}_{\mathrm{E}}\tilde{T}_{kl} - u_{r,r}\, {}_{\mathrm{E}}T^0{}_{kl} + {}_{\mathrm{E}}T^0{}_{kr}u_{l,r} + {}_{\mathrm{E}}T^0{}_{rl}u_{k,r},\\
e_k &= \tilde{E}_k - E^0{}_r u_{r,k},\\
b_k &= \tilde{B}_k - B^0{}_r u_{r,k},\\
\tilde{e}_k &= \tilde{E}_k + \frac{1}{c}(\dot{\mathbf{u}} \times \mathbf{B}^0)_k,\\
p_k &= \tilde{P}_k + P^0{}_r u_{k,r} - P^0{}_k u_{r,r},\\
m_k &= \tilde{\mathscr{M}}_k + M^0{}_r u_{k,r} - M^0{}_k u_{r,r},\\
\tilde{\rho} &= \rho u_{k,k}.
\end{aligned} \tag{7.14.25}$$

We also have

$$\mathscr{E}_{k'} = \left\{E^0{}_k + \varepsilon\left[e_k + \frac{1}{c}(\dot{\mathbf{u}} \times \mathbf{B}^0)_k\right]\right\}\delta_{kk'}, \tag{7.14.26}$$

where we used $\mathbf{v} = \varepsilon\dot{\mathbf{u}}$.

Field Equations for Incremental Fields

To obtain the field equations satisfied by the incremental quantities u_k, e_k, b_k, p_k, and m_k, we substitute (7.14.21) and (7.14.26) into (7.14.18), (7.14.19), $(7.14.16)_2$, $(7.14.16)_3$, and $(7.14.16)_4$. Upon using (7.14.5) and (7.14.6), we obtain, to the first order in ε,

$$
\begin{aligned}
{}_{\mathrm{E}}\tilde{T}_{kl,k} &+ {}_{\mathrm{E}}T^0{}_{kr,k}u_{l,r} + {}_{\mathrm{E}}T^0{}_{kr}u_{l,rk} + (b_{k,l} - b_{l,k})M^0{}_k \\
&+ (B^0{}_{k,l} - B^0{}_{l,k})(m_k - M^0{}_r u_{k,r} + M^0{}_k u_{r,r}) + B^0{}_{k,l}u_{k,r}M^0{}_r \\
&- B^0{}_{r,k}M^0{}_r u_{k,l} - B^0{}_l(m_{k,k} - M^0{}_{r,k}u_{k,r} + M^0{}_{k,k}u_{r,r}) - M^0{}_{k,k}b_l \\
&+ (q^0_{\mathrm{e}}u_{r,r} + \tilde{q}_{\mathrm{e}} - p_{k,k} + P^0{}_{r,k}u_{k,r} - P^0{}_{k,k}u_{r,r})E^0{}_l \\
&+ (q^0_{\mathrm{e}} - P^0{}_{k,k})\left[e_l + \frac{1}{c}(\dot{\mathbf{u}} \times \mathbf{B}^0)_l\right] + \frac{1}{c}\varepsilon_{lmn}(\dot{p}_m - P^0{}_r\dot{u}_{m,r} \\
&+ P^0{}_m\dot{u}_{r,r})B^0{}_n + \rho_0 J_0^{-1}(\tilde{f}_l - \ddot{u}_l) = 0,
\end{aligned}
\tag{7.14.27}
$$

$$
d_{k,k} - D^0{}_{r,l}u_{l,r} = \tilde{q}_{\mathrm{e}},
\tag{7.14.28}
$$

$$
\varepsilon_{klm}(e_{m,l} - E^0_{m,r}u_{r,l}) + \frac{1}{c}\dot{b}_k = 0,
\tag{7.14.29}
$$

$$
b_{k,k} - B^0{}_{r,l}u_{l,r} = 0,
\tag{7.14.30}
$$

$$
\begin{aligned}
\varepsilon_{klm}(h_{m,l} &- H^0{}_{m,r}u_{r,l}) - \frac{1}{c}\dot{d}_k \\
&+ \frac{1}{c}(P^0{}_r\dot{u}_{k,r} - P^0{}_k\dot{u}_{r,r} + P^0{}_{r,r}\dot{u}_k - \dot{u}_r{}^{P0}{}_{k,r}) = \frac{1}{c}q^0_{\mathrm{e}}\dot{u}_k.
\end{aligned}
\tag{7.14.31}
$$

In these equations a superposed dot indicates the partial time derivative.

Jump Conditions

To obtain the relevant jump conditions for the incremental fields we need to calculate the unit normal $\mathbf{n}'$, to the discontinuity surface σ', in $\mathscr{B}'$. The elements of area $da_{k'}$ in $\mathscr{B}'$, and da_k in $\mathscr{B}$, are related by

$$
da_{k'} = J x_{k,k'}\, da_k.
\tag{7.14.32}
$$

The unit normal of σ' is given by

$$
n_{k'} = da_{k'}/da', \qquad da' = (da_{k'}\, da_{k'})^{1/2}.
\tag{7.14.33}
$$

Employing (7.14.32) and (7.14.20), this gives

$$
n_{k'} = (n_k + \varepsilon\tilde{n}_k)\delta_{kk'},
\tag{7.14.34}
$$

where n_k is the unit normal, to the image σ of σ', in $\mathscr{B}$ and the increment $\tilde{n}_k$ is given by

$$
\tilde{n}_k = (\tilde{E}_{mn}n_m n_n\delta_{kl} - u_{l,k})n_l;
\tag{7.14.35}
$$

and

$$
\tilde{E}_{mn} = \tfrac{1}{2}(u_{m,n} + u_{n,m})
\tag{7.14.36}
$$

is the linear strain tensor.

Replacing $\mathbf{v}$ by $\varepsilon\mathbf{v}$, compatible with the velocity $\mathbf{v}$, from the jump conditions (3.14.16), (3.14.17), and (3.14.11)–(3.14.14), we derive the jump conditions for the incremental fields

$$
\begin{aligned}
&[\rho_0(\dot{\mathbf{u}} - \mathbf{v})]\cdot\mathbf{n} = 0,\\
&\left[\tilde{t}_{kl} + {}_{\mathrm{M}}\tilde{t}_{kl} - \frac{1}{c}v_k(\mathbf{E}^0 \times \mathbf{B}^0)_l\right]n_k + [T^0{}_{kl} + {}_{\mathrm{M}}T^0{}_{kl}]\tilde{n}_k = 0,\\
&[\mathbf{d}]\cdot\mathbf{n} + [\mathbf{D}^0]\cdot\tilde{\mathbf{n}} = \tilde{w}_e,\\
&\mathbf{n} \times [\mathbf{e}] + \tilde{\mathbf{n}} \times [\mathbf{E}^0] - \frac{1}{c}\mathbf{v}\cdot\mathbf{n}[\mathbf{B}^0] = \mathbf{0},\\
&[\mathbf{B}^0]\cdot\tilde{\mathbf{n}} + [\mathbf{b}]\cdot\mathbf{n} = 0,\\
&\mathbf{n} \times [\mathbf{h}] + \tilde{\mathbf{n}} \times [\mathbf{H}^0] + \frac{1}{c}\mathbf{v}\cdot\mathbf{n}[\mathbf{D}^0] = \frac{1}{c}w_e^0\mathbf{v},
\end{aligned}
\tag{7.14.37}
$$

where $\tilde{\mathbf{n}}$ is given by (7.14.35), and the incremental stress tensors $\tilde{t}_{kl}$ and ${}_{\mathrm{M}}\tilde{t}_{kl}$ are given

$$
\tilde{t}_{kl} = {}_{\mathrm{E}}\tilde{t}_{kl} - \left(P^0{}_k\tilde{e}_l + p_kE^0{}_l - P^0{}_kE^0{}_ru_{r,l} + M^0{}_kb_l + m_kB^0{}_l + \frac{1}{c}B^0{}_l(\dot{\mathbf{u}} \times \mathbf{P}^0)_k\right),
\tag{7.14.38}
$$

$$
\begin{aligned}
{}_{\mathrm{M}}\tilde{t}_{kl} = {}& p_kE^0{}_l + P^0{}_k\left[e_l + \frac{1}{c}(\dot{\mathbf{u}} \times \mathbf{B}^0)_l\right] - b_kM^0{}_l\\
&- B^0{}_k\left[m_l + \frac{1}{c}(\dot{\mathbf{u}} \times P^0)_l\right] + E^0{}_ke_l + E^0{}_le_k + B^0{}_kb_l + B^0{}_lb_k\\
&- (E^0{}_re_r + H^0{}_rb_r - m_rB^0{}_r)\delta_{kl}.
\end{aligned}
\tag{7.14.39}
$$

Henceforth we drop the small terms containing $(\dot{\mathbf{u}} \times \mathbf{P}^0)/c$. The jump conditions (7.14.37) can be used to derive the boundary conditions at the surface $\partial\mathscr{V}$ of the body. To this end, all we need is to observe that $v = \mathbf{v}$,

$$[A] = A^+ - A^-$$

and take A^+ as the field applied to the surface of the body from outside $\partial\mathscr{V}$. For a stationary surface, $\mathbf{v} = \mathbf{0}$.

Constitutive Equations

To complete the formulation we need to obtain constitutive equations for the incremental fields. For the isothermal case, the energy Σ is given by

$$\Sigma = \Sigma(E_{kl}, \mathscr{E}_k, B_k, \theta_0), \tag{7.14.40}$$

where

$$E_{kl} = E^0{}_{kl} + \varepsilon\tilde{E}_{kl}, \qquad \mathscr{E}_k = E^0{}_k + \varepsilon\tilde{e}_k, \qquad B_k = B^0{}_k + \varepsilon\tilde{B}_k. \tag{7.14.41}$$

We express Σ into a power series of ε, retaining only the first-order terms in ε, and then substituting into the constitutive equations

$$
\begin{aligned}
{}_{\mathrm{E}}t_{k'l'} &= J'^{-1}\frac{\partial\Sigma}{\partial E_{KL}}x_{k',K}x_{l',L} = (J_0 J)^{-1}\frac{\partial\Sigma}{\partial E_{kl}}x_{k',k}x_{l',l},\\
P_{k'} &= -J'^{-1}\frac{\partial\Sigma}{\partial \mathscr{E}_K}x_{k',K} = -(J_0 J)^{-1}\frac{\partial\Sigma}{\partial \mathscr{E}_k}x_{k',k},\\
\mathscr{M}_{k'} &= -J'^{-1}\frac{\partial\Sigma}{\partial B_K}x_{k',K} = -\frac{\partial\Sigma}{\partial B_k}x_{k',k}.
\end{aligned}
\tag{7.14.42}
$$

This leads to the constitutive equations for the bias fields

$$
\begin{aligned}
{}_{\mathrm{E}}t^0{}_{kl} &= J_0{}^{-1}\left(\frac{\partial\Sigma}{\partial E_{KL}}\right)_0 x_{k,K}x_{l,L} = (J_0)^{-1}\left(\frac{\partial\Sigma}{\partial E_{kl}}\right)_0,\\
P^0{}_k &= -J_0{}^{-1}\left(\frac{\partial\Sigma}{\partial E_K}\right)_0 x_{k,K} = -J_0{}^{-1}\left(\frac{\partial\Sigma}{\partial \mathscr{E}_k}\right)_0,\\
\mathscr{M}^0{}_k &= -J_0{}^{-1}\left(\frac{\partial\Sigma}{\partial B_K}\right)_0 x_{k,K} = -J_0{}^{-1}\left(\frac{\partial\Sigma}{\partial B_k}\right)_0,
\end{aligned}
\tag{7.14.43}
$$

and the incremental fields

$$
\begin{aligned}
{}_{\mathrm{E}}\tilde{t}_{kl} &= -u_{r,r}\,{}_{\mathrm{E}}t^0{}_{kl} + {}_{\mathrm{E}}t^0{}_{km}u_{l,m} + {}_{\mathrm{E}}t^0{}_{lm}u_{k,m} + \sigma_{klmn}\tilde{E}_{mn} - e_{mkl}\tilde{e}_m\\
&\quad - h_{mkl}\tilde{B}_m,\\
P_k &= -P^0{}_k u_{r,r} + P^0{}_l u_{k,l} + \chi^{\mathrm{E}}{}_{kl}\tilde{e}_l + e_{klm}\tilde{E}_{lm} + \lambda_{kl}\tilde{B}_l,\\
m_k &= -M^0{}_k u_{r,r} + M^0{}_l u_{k,l} + \chi^{\mathrm{B}}{}_{kl}\tilde{B}_l + h_{klm}\tilde{E}_{lm} + \lambda_{lk}\tilde{e}_l.
\end{aligned}
\tag{7.14.44}
$$

Alternatively,

$$
\begin{aligned}
{}_{\mathrm{E}}\tilde{T}_{kl} &= \sigma_{klmn}\tilde{E}_{mn} - e_{mkl}\left[e_m + E^0{}_r u_{l,m} + \frac{1}{c}(\mathbf{u}\times\mathbf{B}^0)_m\right]\\
&\quad - h^0{}_{mkl}(b_m + B^0{}_r u_{r,m}),\\
p_k &= -P^0{}_k u_{r,r} + P^0{}_l u_{k,l} + \chi^{\mathrm{E}}{}_{kl}\left[e_l + E^0{}_r u_{r,l} + \frac{1}{c}(\dot{\mathbf{u}}\times\mathbf{B}^0)_l\right]\\
&\quad + e_{klm}\tilde{E}_{lm} + \lambda_{kl}(b_l + B^0{}_r u_{r,l}),\\
m_k &= -M^0{}_k u_{r,r} + M^0{}_l u_{k,l} + \chi^{\mathrm{B}}{}_{kl}(b_l + B^0{}_r u_{r,l}) + h_{klm}\tilde{E}_{lm}\\
&\quad + \lambda_{lk}\left[e_l + E^0{}_r u_{r,l} + \frac{1}{c}(\dot{\mathbf{u}}\times\mathbf{B}^0)_l\right],
\end{aligned}
\tag{7.14.45}
$$

where we have introduced the following material moduli at $\mathscr{B}$:

$$
\sigma_{klmn} = J_0{}^{-1}\left(\frac{\partial^2\Sigma}{\partial E_{kl}\,\partial E_{mn}}\right)_0, \qquad e_{mkl} = -(J_0)^{-1}\left(\frac{\partial^2\Sigma}{\partial \mathscr{E}_m\,\partial E_{kl}}\right)_0,
$$

$$h_{mkl} = -J_0^{-1}\left(\frac{\partial^2\Sigma}{\partial B_m\,\partial E_{kl}}\right)_0, \qquad \lambda_{kl} = -J_0^{-1}\left(\frac{\partial^2\Sigma}{\partial B_l\,\partial \mathscr{E}_k}\right)_0,$$

$$\chi^{\mathrm{E}}{}_{kl} = -J_0^{-1}\left(\frac{\partial^2\Sigma}{\partial \mathscr{E}_k\,\partial \mathscr{E}_l}\right)_0, \qquad \chi^{\mathrm{B}}{}_{kl} = -J_0^{-1}\left(\frac{\partial^2\Sigma}{\partial B_k\,\partial B_l}\right)_0, \quad (7.14.46)$$

where, and in (7.14.43), a (0) attached to parentheses indicates that the quantity enclosed within the parentheses is evaluated at $\varepsilon = 0$.

From (7.14.46) it is clear that the material moduli possess the following symmetry regulations:

$$\begin{gathered} \sigma_{klmn} = \sigma_{lkmn} = \sigma_{klnm} = \sigma_{mnkl}, \\ e_{mkl} = e_{mlk}, \qquad h_{mkl} = h_{mlk}, \\ \chi^{\mathrm{E}}{}_{kl} = \chi^{\mathrm{E}}{}_{lk}, \qquad \chi^{\mathrm{B}}{}_{kl} = \chi^{\mathrm{B}}{}_{lk}. \end{gathered} \quad (7.14.47)$$

Consequently, in general, there are 21 independent components of σ_{klmn}, 18 independent components of each of e_{mkl} and h_{mkl}, and 6 independent components of each of $\chi^{\mathrm{E}}{}_{kl}$ and $\chi^{\mathrm{B}}{}_{kl}$.

Isotropic Materials

For isotropic materials Σ is a function of θ_0 and 14 invariants of E_{kl}, $\mathscr{E}_k$, and B_k, i.e.,

$$\Sigma = \Sigma(\mathrm{I}_\alpha, \theta_0), \qquad \alpha = 1, 2, \ldots, 14, \quad (7.14.48)$$

where

$$\begin{gathered} \bar{\mathrm{I}}_1 = \operatorname{tr}\mathbf{E}, \quad \bar{\mathrm{I}}_2 = \operatorname{tr}\mathbf{E}^2, \quad \bar{\mathrm{I}}_3 = \operatorname{tr}\mathbf{E}^3, \quad \bar{\mathrm{I}}_4 = \boldsymbol{\mathscr{E}}\cdot\boldsymbol{\mathscr{E}}, \quad \bar{\mathrm{I}}_5 = \mathbf{B}\cdot\mathbf{B}, \\ \bar{\mathrm{I}}_6 = \boldsymbol{\mathscr{E}}\cdot\mathbf{E}\boldsymbol{\mathscr{E}}, \quad \bar{\mathrm{I}}_7 = \mathbf{B}\cdot\mathbf{E}\mathbf{B}, \quad \bar{\mathrm{I}}_8 = \boldsymbol{\mathscr{E}}\cdot\mathbf{E}^2\boldsymbol{\mathscr{E}}, \quad \bar{\mathrm{I}}_9 = \mathbf{B}\cdot\mathbf{E}^2\mathbf{B}, \\ \bar{\mathrm{I}}_{10} = (\boldsymbol{\mathscr{E}}\cdot\mathbf{B})^2, \quad \bar{\mathrm{I}}_{11} = \boldsymbol{\mathscr{E}}\cdot[\mathbf{B}\times(\mathbf{E}\boldsymbol{\mathscr{E}})], \quad \bar{\mathrm{I}}_{12} = \boldsymbol{\mathscr{E}}\cdot[\mathbf{B}\times(\mathbf{E}^2\boldsymbol{\mathscr{E}})], \\ \bar{\mathrm{I}}_{13} = \mathbf{B}\cdot[(\mathbf{E}\mathbf{B})\times(\mathbf{E}^2\mathbf{B})], \quad \bar{\mathrm{I}}_{14} = (\boldsymbol{\mathscr{E}}\cdot\mathbf{B})\boldsymbol{\mathscr{E}}\cdot[\mathbf{B}\times(\mathbf{E}\mathbf{B})]. \end{gathered} \quad (7.14.49)$$

When the deformation and the fields are determined, by solving the field equations for the bias fields, then $t^0{}_{kl}$, $P^0{}_k$, and $M^0{}_k$ will be known. The constitutive equations for the incremental fields are then determined from (7.14.43) and (7.14.44). Now the problems for the incremental fields can be tackled by employing Cauchy's equations (7.14.27) and Maxwell's equations (7.14.28)–(7.14.31), under the appropriate boundary and initial conditions. Note that, because of the initial bias fields, the field equations are now partial differential equations with *space-dependent* variable coefficients. Moreover, it is also clear from (7.14.43) and (7.14.44) that a material that is isotropic in the reference state is no longer isotropic in the initial state $\mathscr{B}$. Thus, the bias fields produce both inhomogeneity and anisotropy, so that, upon small superposed fields on the statically deformed body, a body which is not piezoelectric and piezomagnetic will, in general, acquire both piezoelectricity and piezomagnetism. This is the phenomenon of *symmetry breaking*.[11]

The equations developed in this section constitute the bases of various types of interactions of electromagnetic fields and elastic deformations. Photoelasticity, electro-optical effects, and magneto-optical effects are among some of the cross phenomena which have found extensive technological applications. These are studied in the following sections.[12] It is important to note that the entire field of stability theory can be constructed on the foundations laid here. These important problems are, as yet, to be tackled.

7.15. Photoelastic Effect

In this section we consider the inducement of optical anisotropy in an elastic solid by the application of finite strains. We consider solids that are paramagnetic (or equivalently soft ferromagnetic) dielectrics[13] which are isotropic in the natural configuration B.

We assume only an initial deformation field

$$\mathbf{x} = \mathbf{x}(\mathbf{X}), \qquad \mathbf{P}^0 = \mathbf{M}^0 = \mathbf{0}, \qquad \mathbf{E}^0 = \mathbf{B}^0 = \mathbf{0}, \qquad q_e^0 = 0, \quad (7.15.1)$$

according to which a point $\mathbf{X}$ in B moves to a point $\mathbf{x}$ in $\mathscr{B}$ with no electromagnetic fields. Maxwell's equations (7.14.6) are trivially satisfied, and the constitutive equations for the initial stress reduce to the classical equations of the theory of nonlinear elasticity in the absence of electromagnetic fields. From (7.15.1) and $(7.14.9)_{2,3}$ it follows that

$$\left(\frac{\partial \Sigma}{\partial \mathscr{E}_k}\right)_0 = 0, \qquad \left(\frac{\partial \Sigma}{\partial B_k}\right)_0 = 0. \quad (7.15.2)$$

For isotropic materials Σ is a function of the invariants listed in (7.14.49). Thus, for example, for $\chi^{E}{}_{kl}$ given by (7.14.46), we have

$$\chi^{E}{}_{kl} = -J_0{}^{-1}\left(\frac{\partial \Sigma}{\partial \bar{\mathrm{I}}_\alpha}\frac{\partial^2 \bar{\mathrm{I}}_\alpha}{\partial \mathscr{E}_k\,\partial \mathscr{E}_l} + \frac{\partial^2 \Sigma}{\partial \bar{\mathrm{I}}_\alpha\,\partial \bar{\mathrm{I}}_\beta}\frac{\partial \bar{\mathrm{I}}_\alpha}{\partial \mathscr{E}_k}\frac{\partial \bar{\mathrm{I}}_\beta}{\partial \mathscr{E}_l}\right)_0.$$

But $(\partial \bar{\mathrm{I}}_\alpha/\partial \mathscr{E}_k)_0 = 0$, because $\mathbf{P}^0 = \mathbf{0}$. Therefore, we obtain

$$\chi^{E}{}_{kl} = -2J_0{}^{-1}\left[\left(\frac{\partial \Sigma}{\partial \bar{\mathrm{I}}_4}\right)_0 \delta_{kl} + \left(\frac{\partial \Sigma}{\partial \bar{\mathrm{I}}_6}\right)_0 E^0{}_{kl} + \left(\frac{\partial \Sigma}{\partial \bar{\mathrm{I}}_8}\right)_0 E^0{}_{km}E^0{}_{ml}\right]. \quad (7.15.3)$$

---

[11] This is examined in a general manner by Maugin [1984a], and in magnetoelasticity by Turov [1983].

[12] These effects were studied by Boulanger et al [1973]. However, because of the difference in the constitutive independent variable used and the electromagnetic force law, our analysis and results are different from theirs.

[13] In soft ferromagnetic materials, we may consider that there is no spontaneous magnetization and, therefore, the nonlinear constitutive equations can be shown to be the same as those of paramagnetic materials.

Similarly, from (7.14.46), we obtain

$$\chi^{\mathrm{B}}{}_{kl} = -2J_0{}^{-1}\left[\left(\frac{\partial\Sigma}{\partial\bar{\mathrm{I}}_5}\right)_0\delta_{kl} + \left(\frac{\partial\Sigma}{\partial\bar{\mathrm{I}}_7}\right)_0 E^0{}_{kl} + \left(\frac{\partial\Sigma}{\partial\bar{\mathrm{I}}_9}\right)_0 E^0{}_{km}E^0{}_{ml}\right], \tag{7.15.4}$$

$$\lambda_{kl} = 0, \qquad e_{klm} = h_{klm} = 0.$$

Constitutive equations for the electromagnetic perturbations are given by (7.14.45), so that

$$p_k = \chi^{\mathrm{E}}{}_{kl}e_l, \qquad m_k = \chi^{\mathrm{B}}{}_{kl}b_l. \tag{7.15.5}$$

The stress field $\tilde{T}_{kl}$, given by $(7.14.45)_1$, is not affected by the electromagnetic field, so that, to the first order, electromagnetic perturbations do not give rise to any displacement. Therefore, the photoelastic effect can be investigated by means of Maxwell's equations and the constitutive equations (7.15.5). The complete set of electromagnetic field equations, as deduced from (7.14.28)–(7.14.31), are

$$\begin{aligned} &\nabla\cdot\mathbf{d} = 0, \qquad \nabla\times\mathbf{e} + \frac{1}{c}\dot{\mathbf{b}} = 0,\\ &\nabla\cdot\mathbf{b} = 0, \qquad \nabla\times\mathbf{h} - \frac{1}{c}\dot{\mathbf{d}} = 0, \end{aligned} \tag{7.15.6}$$

where $\mathbf{d}$ and $\mathbf{h}$ are given by

$$d_k = e_k + p_k = \varepsilon_{kl}e_l, \qquad h_k = b_k - m_k = \overset{-1}{\mu}_{kl}\, b_l, \tag{7.15.7}$$

with ε_{kl} and $\overset{-1}{\mu}_{kl}$ defined by

$$\varepsilon_{kl} = \delta_{kl} + \chi^{\mathrm{E}}{}_{kl}, \quad \overset{-1}{\mu}_{kl} = \delta_{kl} - \chi^{\mathrm{B}}{}_{kl}. \tag{7.15.8}$$

Birefringence

We seek plane-wave solutions for the system (7.15.6) and (7.15.7), in the form

$$\mathbf{a} = \mathrm{Re}\left\{\bar{\mathbf{a}}\exp\left[i\omega\left(t - \frac{n}{c}\mathbf{s}\cdot\mathbf{x}\right)\right]\right\}, \qquad \mathbf{s}\cdot\mathbf{s} = 1, \tag{7.15.9}$$

where $\bar{\mathbf{a}}$ is a constant complex vector and n is the refractive index. Substituting (7.15.9) for the four vectors $\mathbf{d}$, $\mathbf{e}$, $\mathbf{b}$, and $\mathbf{h}$ in (7.15.6) and (7.15.7), we obtain

$$\begin{aligned} &\bar{d}_k s_k = 0, \qquad nS_{kl}\bar{e}_l - \bar{b}_k = 0,\\ &\bar{b}_k s_k = 0, \qquad nS_{kl}\bar{h} + \bar{d}_k = 0,\\ &\bar{d}_k = \varepsilon_{kl}\bar{e}_l, \qquad \bar{h}_l = \overset{-1}{\mu}_{kl}\, b_l, \end{aligned} \tag{7.15.10}$$

where

$$S_{kl} = -\varepsilon_{klm}s_m, \qquad S_{kl}s_l = 0. \tag{7.15.11}$$

Eliminating $\bar{\mathbf{b}}$, $\bar{\mathbf{d}}$, and $\bar{\mathbf{h}}$ among (7.15.10), we obtain

$$(\varepsilon_{kl} - n^2 K_{kl})\bar{e}_l = 0, \tag{7.15.12}$$

where

$$K_{kl} = -S_{km} \overset{-1}{\mu}_{mn} S_{nl}. \tag{7.15.13}$$

When the medium is not magnetizable, then $\overset{-1}{\mu}_{kl} = \delta_{kl}$

$$K_{kl} = -S_{km}S_{ml} = P_{kl} = \delta_{kl} - s_k s_l. \tag{7.15.14}$$

The operator $\mathbf{P} = -\mathbf{S}^2$ is none other than the *projection operator*, which projects a vector onto the plane normal to the direction of propagation.

The refractive indices are obtained by using the conditions of the existence of the solution of (7.15.12) for $\bar{\mathbf{e}}$, i.e.,

$$\det(\varepsilon_{kl} - n^2 K_{kl}) = 0. \tag{7.15.15}$$

Since $\det \mathbf{K} = 0$, upon expanding (7.15.15), we obtain

$$N(\varepsilon, \mathbf{K})n^4 - N(\mathbf{K}, \varepsilon)n^2 + \det \boldsymbol{\varepsilon} = 0, \tag{7.15.16}$$

where

$$\mathbf{N}(\mathbf{u}, \mathbf{v}) = \tfrac{1}{2} \operatorname{tr} \mathbf{u}[(\operatorname{tr} \mathbf{v})^2 - \operatorname{tr} \mathbf{v}^2] + \operatorname{tr}(\mathbf{u}\mathbf{v}^2) - \operatorname{tr} \mathbf{v} \operatorname{tr}(\mathbf{u}\mathbf{v}), \tag{7.15.17}$$

for two second-order tensors u_{ij} and v_{ij}. In the case where $\overset{-1}{\mu}_{kl} = \delta_{kl}$, (7.15.17) reduces to

$$\det[\varepsilon_{kl} - n^2(\delta_{kl} - s_k s_l)] = 0. \tag{7.15.18}$$

Unless the strain tensor $E^0{}_{kl}$ is proportional to the unit tensor, as in a uniform dilatation, the tensor K_{kl} will have unequal eigenvalues and the material will be birefringent. Suppose that $n_1{}^2$ and $n_2{}^2$ are two distinct solutions of (7.15.16). Then they give rise to two real directions of $\bar{\mathbf{e}}$. We then have two polarization states, with corresponding amplitudes $\bar{\mathbf{e}}^{(1)}$ and $\bar{\mathbf{e}}^{(2)}$, such that

$$\begin{aligned} \varepsilon_{kl}\bar{e}^{(1)}{}_l &= -n_1{}^2 K_{kl}\bar{e}^{(1)}{}_l, \\ \varepsilon_{kl}\bar{e}^{(2)}{}_l &= -n_2{}^2 K_{kl}\bar{e}^{(2)}{}_l. \end{aligned} \tag{7.15.19}$$

Multiplying the first of these by $\bar{e}^{(2)}{}_k$, th second by $\bar{e}^{(1)}{}_k$, and subtracting, we obtain

$$K_{kl}\bar{e}^{(1)}{}_k\bar{e}^{(2)}{}_l = 0. \tag{7.15.20}$$

But this is equivalent to

$$A^{(1)}{}_k B_{kl} A^{(2)}{}_l = 0, \tag{7.15.21}$$

where

$$A^{(\alpha)}{}_k = S_{kl}\bar{e}^{(\alpha)}{}_l, \qquad B_{kl} = P_{km} \overset{-1}{\mu}_{mn} P_{nl}, \qquad \alpha = 1, 2, \tag{7.15.22}$$

as can be verified by substituting (7.15.22) into (7.15.21) and using (7.15.14) and (7.15.11). Expression (7.15.21) states that the vectors $A^{(1)}{}_k$ and $A^{(2)}{}_k$, in the plane orthogonal to $\mathbf{s}$, are conjugate with respect to the conic defined by B_{kl}.

From (7.15.10) it follows that $\bar{\mathbf{b}}$ is in the direction of $s_{kl}e_l$, and $\bar{\mathbf{d}}$ is given by

$$d_k = nS_{kl}\bar{h}_l = n^2 K_{kl}\bar{e}_l. \tag{7.15.23}$$

Because of (7.15.20),

$$\bar{\mathbf{d}}^{(1)} \cdot \bar{\mathbf{e}}^{(2)} = \bar{\mathbf{d}}^{(2)} \cdot \bar{\mathbf{e}}^{(1)} = 0, \tag{7.15.24}$$

so that $\bar{\mathbf{d}}^{(1)}$ is orthogonal to $\bar{\mathbf{e}}^{(2)}$, and $\bar{\mathbf{d}}^{(2)}$ is orthogonal to $\bar{\mathbf{e}}^{(1)}$. For $\bar{\mathbf{h}}$ we have

$$\bar{h}_k = \overset{-1}{\mu}_{kl}\, \bar{b}_l = n\, \overset{-1}{\mu}_{kl}\, S_{lm} \bar{e}_m. \tag{7.15.25}$$

Finally, the vector $P_{kl}\bar{h}_l$ is along the vector $S_{kl}\bar{d}_l$ and $S_{kl}\bar{h}_l$ is along $\bar{d}_k$,

$$\begin{aligned} P_{kl}\bar{h}_l &= nS_{kl}K_{lm}\bar{e}_m = \frac{1}{n} S_{kl}\, \bar{d}_l, \\ S_{kl}\bar{h}_l &= nK_{kl}\bar{e}_l = \frac{1}{n}\, \bar{d}_k. \end{aligned} \tag{7.15.26}$$

Using $(7.14.10)_2$, we also obtain

$$\bar{\mathbf{b}}^{(1)} \cdot \bar{\mathbf{b}}^{(2)} = n^2 S_{kl} S_{lm} \bar{e}^{(1)}{}_m \bar{e}^{(2)}{}_k. \tag{7.15.27}$$

When the medium is not magnetizable, $K_{kl} = P_{kl}$, and because of (7.15.20), we see that the magnetic inductions $\bar{\mathbf{b}}^{(1)}$ and $\bar{\mathbf{b}}^{(2)}$, associated with $\bar{\mathbf{e}}^{(1)}$ and $\bar{\mathbf{e}}^{(2)}$, become orthogonal. In the general case, the polarization modes would be orthogonal, if and only if the sections of the quadrics defined by $\overset{-1}{\varepsilon}_{kl}$ and μ_{kl}, with the plane normal to $\mathbf{s}$, had common principal directions.

For anisotropic solids, Nelson and Lax [1971] discuss a possible dependence of the photoelastic effect on rotations.

7.16. Electro-Optical Effect

In this section we discuss the effect of an initial uniform electric polarization in an isotropic elastic dielectric ($q_e = 0$, $\mathbf{f} = \mathbf{0}$). The dielectric is unstrained and nonmagnetized at the initial state. We have at the state $\mathscr{B}$

$$x_k = \delta_{kK} X_K, \qquad \mathbf{B}^0 = \mathbf{M}^0 = \mathbf{0}, \qquad \mathbf{E}^0 \text{ and } \mathbf{P}^0 \text{ uniform}. \tag{7.16.1}$$

The equations of motion (7.14.5), and Maxwell's equations (7.14.6), for the initial fields are satisfied trivially. Equations of motion and Maxwell's equations for the perturbations follow from (7.14.27)–(7.14.31)

$${}_{\mathrm{E}}\tilde{T}_{kl,k} + {}_{\mathrm{E}}T^0{}_{kr} u_{l,rk} - p_{k,k} E^0{}_l - \rho_0 \ddot{u}_l = 0, \tag{7.16.2}$$

$$\begin{aligned} \nabla \cdot \mathbf{d} = 0, &\qquad \nabla \times \mathbf{e} + \frac{1}{c}\dot{\mathbf{b}} = \mathbf{0}, \\ \nabla \cdot \mathbf{b} = 0, &\qquad \nabla \times \mathbf{h} - \frac{1}{c}\dot{\mathbf{d}} - \frac{1}{c}[\mathbf{P}^0(\nabla \cdot \dot{\mathbf{u}}) - (\mathbf{P}^0 \cdot \nabla)\dot{\mathbf{u}}] = \mathbf{0}. \end{aligned} \tag{7.16.3}$$

The constitutive equations are given by (7.14.43) and (7.14.45). For isotropic solids Σ is a function of the invariants $\bar{\mathrm{I}}_\alpha$ listed in (5.14.49). Consequently, we find

$$ {}_{E}T^{0}{}_{kl} = \sigma_0 \delta_{kl} - e_1 E^0{}_k E_l, \qquad P^0{}_k = \chi^{E}{}_1 E^0{}_k \qquad B^0{}_k = 0, \tag{7.16.4} $$

$$ \begin{aligned} {}_{E}\tilde{T}_{kl} &= \sigma_{klmn} u_{m,n} - e_{mkl}(E^0{}_r u_{r,m} + e_m) - h_{mkl} b_m, \\ p_k &= d_k - e_k = -P^0{}_k u_{r,r} + P^0{}_l u_{k,l} + \chi^{E}{}_{kl}(E^0{}_r u_{r,l} + e_l) + e_{klm} u_{l,m}, \\ m_h &= b_k - h_k = \chi^{B}{}_{kl} b_l + h_{klm} u_{l,m}, \end{aligned} \tag{7.16.5} $$

where the constitutive moduli are given by

$$ \begin{aligned} \sigma_{klmn} &= \sigma_1 \delta_{kl}\, d_{mn} + \sigma_2(\delta_{km}\delta_{ln} + \delta_{kn}\delta_{lm}) + \sigma_3(E^0{}_k E^0{}_l \delta_{mn} + E^0{}_m E^0{}_n \delta_{kl}) \\ &\quad + \sigma_4(E^0{}_k E^0{}_n \delta_{lm} + E^0{}_l E_n{}^0 \delta_{km} + E^0{}_k E^0{}_m \delta_{ln} + E^0{}_l E^0{}_m \delta_{kn}) \\ &\quad + \sigma_5 E^0{}_k E^0{}_l E^0{}_m, \\ e_{mkl} &= e_1(E^0{}_k \delta_{lm} + E^0{}_l \delta_{km}) + e_2 \delta_{kl} E^0{}_m + e_3 E^0{}_k E^0{}_l E^0{}_m, \\ h_{mkl} &= h_1(\varepsilon_{krm} E^0{}_l + \varepsilon_{lrm} E^0{}_k) E^0{}_r, \\ \chi^{E}{}_{kl} &= \chi^{E}{}_1 \delta_{kl} + \chi^{E}{}_2 E^0{}_k E^0{}_l, \qquad \chi^{B}{}_{kl} = \chi^{B}{}_1 \delta_{kl} + \chi^{B}{}_2 E^0{}_k E^0{}_l, \\ \lambda_{kl} &= 0, \end{aligned} \tag{7.16.6} $$

with

$$ \begin{aligned} &\sigma_0 = J_0{}^{-1}\left(\frac{\partial \Sigma}{\partial \bar{I}_1}\right)_0, \qquad \sigma_1 = J_0{}^{-1}\left(\frac{\partial^2 \Sigma}{\partial \bar{I}_1{}^2}\right)_0, \qquad \sigma_2 = J_0{}^{-1}\left(\frac{\partial \Sigma}{\partial \bar{I}_2}\right)_0, \\ &\sigma_3 = \tfrac{1}{2} J_0{}^{-1}\left(\frac{\partial^2 \Sigma}{\partial \bar{I}_1\, \partial \bar{I}_6}\right)_0, \quad \sigma_4 = \tfrac{1}{4} J_0{}^{-1}\left(\frac{\partial \Sigma}{\partial \bar{I}_8}\right)_0, \qquad \sigma_5 = J_0{}^{-1}\left(\frac{\partial^2 \Sigma}{\partial \bar{I}_6{}^2}\right)_0, \\ &e_1 = J_0{}^{-1}\left(\frac{\partial \Sigma}{\partial \bar{I}_6}\right)_0, \qquad e_2 = -2J_0{}^{-1}\left(\frac{\partial^2 \Sigma}{\partial \bar{I}_1\, \partial \bar{I}_4}\right)_0, \quad e_3 = -2J_0{}^{-1}\left(\frac{\partial^2 \Sigma}{\partial \bar{I}_4\, \partial \bar{I}_6}\right)_0, \\ &h_1 = -\tfrac{1}{2} J_0{}^{-1}\left(\frac{\partial \Sigma}{\partial \bar{I}_{11}}\right)_0, \quad \chi^{E}{}_1 = -2J_0{}^{-1}\left(\frac{\partial \Sigma}{\partial \bar{I}_4}\right)_0, \qquad \chi^{E}{}_2 = -J_0{}^{-1}\left(\frac{\partial^2 \Sigma}{\partial \bar{I}_4{}^2}\right)_0, \\ &\chi^{B}{}_1 = -2J_0{}^{-1}\left(\frac{\partial \Sigma}{\partial \bar{I}_5}\right)_0, \quad \chi^{B}{}_2 = -2J_0{}^{-1}\left(\frac{\partial \Sigma}{\partial \bar{I}_{10}}\right)_0. \end{aligned} \tag{7.16.7} $$

From (7.16.4) it is clear that the initial stress possesses transverse isotropy with respect to the direction of the initial polarization $\mathbf{P}^0 = \chi^{E}{}_1 \mathbf{E}^0$ where $\chi^{E}{}_1$ is the static susceptibility.

Substituting (7.16.6) into (7.16.5) we obtain the explicit forms of the constitutive equations of perturbations

$$ \begin{aligned} {}_{E}\tilde{T}_{kl} &= [\sigma_1 u_{r,r} + (\sigma_3 - e_2) E^0{}_m E^0{}_n u_{m,n}]\delta_{kl} + \sigma_2(u_{k,l} + u_{l,k}) \\ &\quad + \sigma_3 E^0{}_k E^0{}_l u_{r,r} + (\sigma_4 - e_1) E^0{}_k E^0{}_r u_{r,l} + (\sigma_4 - e_1) E^0{}_k E^0{}_r u_{r,k} \\ &\quad + \sigma_4 E^0{}_k E^0{}_r u_{l,r} + \sigma_4 E^0{}_l E^0{}_r u_{k,r} \\ &\quad + (\sigma_5 - e_3) E^0{}_k E^0{}_l E^0{}_m E^0{}_n u_{m,n} - e_1(E^0{}_k e_l + E^0{}_l e_k) \end{aligned} $$

$$
\begin{aligned}
&- e_2 E^0{}_r e_r \delta_{kl} - e_3 E^0{}_k E^0{}_l E^0{}_r e_r \\
&- h_1(\varepsilon_{kmn} E^0{}_m E^0{}_l + \varepsilon_{lmn} E_m E^0{}_k) b_n, \qquad (7.16.8)
\end{aligned}
$$

$$
\begin{aligned}
p_k = d_k - e_k = {} & (\chi^{\mathrm{E}}{}_1 + e_1) E^0{}_1 u_{k,l} + (\chi^{\mathrm{E}}{}_1 + e_1) E^0{}_1 u_{l,k} \\
& + (e_2 - \chi^{\mathrm{E}}{}_1) E^0{}_k u_{r,r} + \chi^{\mathrm{E}}{}_2 E^0{}_k E^0{}_l E^0{}_r u_{r,l} + e_3 E^0{}_k E^0{}_l E^0{}_r u_{l,r} \\
& + (\chi^{\mathrm{E}}{}_1 \delta_{kl} + \chi^{\mathrm{E}}{}_2 E^0{}_k E^0{}_l) e_l, \\
m_k = b_k - h_k = {} & (\chi^{\mathrm{B}}{}_1 \delta_{kl} + \chi^{\mathrm{B}}{}_2 E^0{}_k E^0{}_l) b_l \\
& + h_1(\varepsilon_{klr} E^0{}_m E^0{}_r + \varepsilon_{kmr} E^0{}_l E^0{}_r) u_{l,m}.
\end{aligned}
$$

The field equations (7.16.2) and (7.16.3) take the forms

$$
\begin{aligned}
(\sigma_1 + \sigma_2) u_{k,lk} & + (\sigma_0 + \sigma_2) u_{l,kk} + (\sigma_3 + \sigma_4 - e_1 - e_2) E^0{}_k E^0{}_r u_{k,lr} \\
& + (\sigma_3 + \sigma_4 - e_1 - e_2) E^0{}_k E^0{}_l u_{r,rk} + (\sigma_4 - e_1) E^0{}_k E^0{}_r u_{l,rk} \\
& + (\sigma_4 - 2e_1 - \chi^{\mathrm{E}}{}_1) E^0{}_l E^0{}_r u_{r,kk} \\
& + (\sigma_5 - 2e_3 - \chi^{\mathrm{E}}{}_2) E^0{}_k E_l{}^0 E^0{}_m E^0{}_n u_{m,nk} - e_1 E^0{}_k e_{l,k} \\
& - (e^0{}_1 + \chi^{\mathrm{E}}{}_1) E^0{}_l e_{k,k} - e_2 E^0{}_k e_{k,l} - (e_3 + \chi^{\mathrm{E}}{}_2) E^0{}_k E^0{}_l E^0{}_r e_{r,k} \\
& - h_1(\varepsilon_{kmn} E^0{}_l + \varepsilon_{lmn} E^0{}_k) E^0{}_m b_{n,k} - \rho \ddot{u}_l = 0, \qquad (7.16.9)
\end{aligned}
$$

$$
\begin{aligned}
(e_1 + e_2) E^0{}_l u_{k,lk} & + (\chi^{\mathrm{E}}{}_1 + e_1) E^0{}_l u_{l,kk} + (\chi^{\mathrm{E}}{}_2 + e_3) E^0{}_k E^0{}_l E^0{}_r u_{l,rk} \\
& + (1 + \chi^{\mathrm{E}}{}_1) e_{k,k} + \chi^{\mathrm{E}}{}_2 E^0{}_k E^0{}_l e_{l,k} = 0, \qquad (7.16.10)
\end{aligned}
$$

$$
\varepsilon_{klm} e_{m,l} + \frac{1}{c} \dot{b}_k = 0, \qquad (7.16.11)
$$

$$
b_{k,k} = 0, \qquad (7.16.12)
$$

$$
\begin{aligned}
& h_1(E^0{}_k E^0{}_r u_{l,rl} + E^0{}_k E^0{}_l u_{l,rr} - E^0{}_l E^0{}_r u_{k,lr} - E^0{}_l E^0{}_r u_{l,kr}) \\
& \quad - \frac{1}{c} [e_1 E^0{}_l \dot{u}_{k,l} + (\chi^{\mathrm{E}}{}_1 + e_1) E^0{}_l \dot{u}_{l,k} + e_2 E^0{}_k \dot{u}_{l,l} \\
& \quad + (\chi^{\mathrm{E}}{}_2 + e_3) E^0{}_k E^0{}_l E^0{}_r \dot{u}_{l,r}] + (1 - \chi^{\mathrm{B}}{}_1) \varepsilon_{klr} b_{r,l} - \chi^{\mathrm{B}}{}_2 \varepsilon_{klr} E^0{}_r E^0{}_p b_{p,l} \\
& \quad - \frac{1}{c} (1 + \chi^{\mathrm{E}}{}_1) \dot{e}_k - \frac{1}{c} \chi^{\mathrm{E}}{}_2 E^0{}_k E^0{}_l \dot{e}_l = 0. \qquad (7.16.13)
\end{aligned}
$$

These equations are adequate to determine u_k, b_k, and e_k under the appropriate boundary and initial conditions.

In order to place the electro-optical effect in evidence, we investigate the propagation of plane harmonic waves in a body initially polarized under a uniform electric field $\mathbf{E}^0$, ($\mathbf{P}^0 \neq \mathbf{0}$). Various vector fields are represented by

$$
\begin{aligned}
& (\mathbf{u}, \mathbf{e}, \mathbf{b}) = \operatorname{Re}(\bar{\mathbf{u}}, \bar{\mathbf{e}}, \bar{\mathbf{b}}) \exp\left[i\omega\left(t - \frac{n}{c} \mathbf{s} \cdot \mathbf{x} \right) \right], \\
& \mathbf{s} \cdot \mathbf{s} = 1.
\end{aligned} \qquad (7.16.14)
$$

Substituting these into (7.16.9)–(7.16.13), and eliminating $\bar{\mathbf{b}}$ from (7.16.9) and (7.16.13) by means of (7.16.11), we obtain

$$
\begin{aligned}
(\bar{c}_{lm} - \rho_0 v^2 \delta_{lm})\bar{u}_m - (iv/\omega)\bar{e}_{lm}\bar{e}_m &= 0,\\
\bar{p}_m \bar{u}_m + (iv/\omega)\bar{\chi}_m \bar{e}_m &= 0,\\
\bar{d}_{km}\bar{u}_m - (ic/\omega)\bar{\kappa}_{km}\bar{e}_m &= 0,
\end{aligned}
\tag{7.16.15}
$$

and

$$
\begin{aligned}
\varepsilon_{klm} s_l \bar{e}_m - (v/c)\bar{b}_k &= 0,\\
s_k \bar{b}_k &= 0,
\end{aligned}
\tag{7.16.16}
$$

where

$$
\begin{aligned}
\bar{c}_{lm} &= [\sigma_0 + \sigma_2 + (\sigma_4 - e_1)(\mathbf{E}^0\cdot\mathbf{s})^2]\delta_{lm} + (\sigma_1 + \sigma_2)s_l s_m\\
&\quad + (\sigma_3 + \sigma_4 - e_1 - e_2)(\mathbf{E}^0\cdot\mathbf{s})s_l E^0{}_m\\
&\quad + (\sigma_3 + \sigma_4 - e_1 - e_2)(\mathbf{E}^0\cdot\mathbf{s})E^0{}_l s_m\\
&\quad + (\sigma_4 - 2e_1 - \chi^{\mathrm{E}}{}_1)E^0{}_l E^0{}_m\\
&\quad + (\sigma_5 - 2e_2 - \chi^{\mathrm{E}}{}_2)(\mathbf{E}^0\cdot\mathbf{s})^2 E^0{}_l E^0{}_m,\\
\bar{e}_{lm} &= [(\bar{\mathbf{E}}\cdot\mathbf{s})e_1 - n(\mathbf{E}^0\cdot\mathbf{s})^2 h_1]\delta_{lm} + (e_1 + \chi^{\mathrm{E}}{}_1)E^0{}_l s_m + e_2 s_l E^0{}_m\\
&\quad + (e_3 + \chi^{\mathrm{E}}_2)(\mathbf{E}^0\cdot\mathbf{s})E^0{}_l E^0{}_m + nh_1(\mathbf{E}^0\cdot\mathbf{s})s_l E^0{}_m + nh_1 E^0{}_l E^0{}_m\\
&\quad + -nh_1(\mathbf{E}^0\cdot\mathbf{s})E^0{}_l s_m,
\end{aligned}
\tag{7.16.17}
$$

$$
\begin{aligned}
\bar{P}_m &= (e_1 + e_2)(\mathbf{E}^0\cdot\mathbf{s})s_m + (\chi^{\mathrm{E}}{}_1 + e_1)E^0{}_m + (\chi^{\mathrm{E}}{}_1 + e_3)(\mathbf{E}^0\cdot\mathbf{s})^2 E^0{}_m,\\
\bar{\chi}_m &= (1 + \chi^{\mathrm{E}}{}_1)s_m + \chi^{\mathrm{E}}{}_2(\mathbf{E}^0\cdot\mathbf{s})E^0{}_m,\\
\bar{d}_{km} &= h_1[(\mathbf{E}^0\cdot\mathbf{s})^2\delta_{km} - (\mathbf{E}^0\cdot\mathbf{s})E^0{}_k s_m + (\mathbf{E}^0\cdot\mathbf{s})E^0{}_m s_k - E^0{}_k E^0{}_m]\\
&\quad - (v/c)[e_1(\mathbf{E}^0\cdot\mathbf{s})\delta_{km} + (\chi^{\mathrm{E}}{}_1 + e_1)s_k E^0{}_m + e_2 E^0{}_k s_m\\
&\quad + (\chi^{\mathrm{E}}{}_2 + e_3)(\mathbf{E}^0\cdot\mathbf{s})E^0{}_k E^0{}_m],\\
\kappa_{km} &= (v/c)^2[(1 + \chi^{\mathrm{E}}{}_1)\delta_{km} + \chi^{\mathrm{E}}{}_2 E^0{}_k E^0{}_m] + (1 - \chi^{\mathrm{B}}{}_1)(s_k s_m - \delta_{km})\\
&\quad + \chi^{\mathrm{B}}{}_2(\mathbf{E}^0 \times \mathbf{s})_k(\mathbf{E}^0 \times \mathbf{s})_m,\\
v &= c/n.
\end{aligned}
$$

By means of the set of equations (7.16.15), we determine $\bar{\mathbf{u}}$, $\bar{\mathbf{e}}$, and v. Equations (7.16.16) then serve to obtain $\bar{\mathbf{b}}$. In the following, we consider two special cases.

(i) *The Transverse Kerr Effect*

Suppose that the bias electric field is in the transverse direction to the direction of propagation $\mathbf{s}$, i.e.,

$$\mathbf{E}^0\cdot\mathbf{s} = 0. \tag{7.16.18}$$

With this, the coefficients given by (7.16.17) reduce to

$$\begin{aligned}
\bar{c}_{lm} &= (\sigma_0 + \sigma_2)\delta_{lm} + (\sigma_1 + \sigma_2)s_l s_m + (\sigma_4 - 2e_1 - \chi^{E}{}_1)E^0{}_l E^0{}_m, \\
\bar{e}_{lm} &= (e_1 + \chi^{E}{}_1)E^0{}_l s_m + e_2 s_l E^0{}_m + nh_1 E^0{}_l E^0{}_m, \\
\bar{P}_m &= (\chi^{E}{}_1 + e_1)E^0{}_m, \\
\bar{\chi}_m &= (1 + \chi^{E}{}_1)s_m, \\
\bar{d}_{km} &= -h_1 E^0{}_k E^0{}_m - (v/c)[(\chi^{E}{}_1 + e_1)s_k E^0{}_m + e_2 E^0{}_k s_m], \\
\kappa_{km} &= (v/c)^2[(1 + \chi^{E}{}_1)\delta_{km} + \chi^{E}{}_2 E^0{}_k E^0{}_m] + (1 - \chi^{B}{}_1)(s_k s_m - \delta_{km}) \\
&\quad + \chi^{B}{}_2(\mathbf{E}^0 \times \mathbf{s})_k(\mathbf{E}^0 \times \mathbf{s})_m.
\end{aligned} \tag{7.16.19}$$

To obtain the solution of the system (7.16.15), we project the first and the third equations in the direction of $\mathbf{s}$, $\mathbf{E}^0$, and $\mathbf{E}^0 \times \mathbf{s}$.

The projection in the direction of $\mathbf{s}$ gives

$$\begin{aligned}
(\sigma_0 + \sigma_1 + 2\sigma_2 - \rho_0 v^2)(\mathbf{s}\cdot\bar{\mathbf{u}}) - (iv/\omega)e_2(\mathbf{E}^0\cdot\bar{\mathbf{e}}) &= 0, \\
(\chi^{E}{}_1 + e_1)(\mathbf{E}^0\cdot\bar{\mathbf{u}}) + (iv/\omega)(1 + \chi^{E}{}_1)(\mathbf{s}\cdot\bar{\mathbf{e}}) &= 0.
\end{aligned} \tag{7.16.20}$$

The projection in the direction of $\mathbf{E}^0$ gives

$$\begin{aligned}
&[\sigma_0 + \sigma_2 + (\sigma_4 - 2e_1 - \chi^{E}{}_1)(\mathbf{E}^0\cdot\mathbf{E}^0) - \rho_0 v^2](\mathbf{E}^0\cdot\bar{\mathbf{u}}) \\
&\qquad - (iv/\omega)(\mathbf{E}^0\cdot\mathbf{E}^0)(e_1 + \chi^{E}{}_1)(\mathbf{s}\cdot\bar{\mathbf{e}}) - (ic/\omega)h_1(\mathbf{E}^0\cdot\mathbf{E}^0)(\mathbf{E}^0\cdot\bar{\mathbf{e}}) = 0, \\
&h_1(\mathbf{E}^0\cdot\bar{\mathbf{u}}) + (v/c)e_2(\mathbf{s}\cdot\bar{\mathbf{u}}) \\
&\qquad + (ic/\omega)(\mathbf{E}^0\cdot\mathbf{E}^0)^{-1}[(v/c)^2(1 + \chi^{E}{}_1 + \chi^{E}{}_2\mathbf{E}^0\cdot\mathbf{E}^0) - 1 + \chi^{B}{}_1](\mathbf{E}^0\cdot\bar{\mathbf{e}}) = 0,
\end{aligned} \tag{7.16.21}$$

and the projection in the direction $\mathbf{E}^0 \times \mathbf{s}$ results in

$$\begin{aligned}
(\sigma_0 + \sigma_2 - \rho_0 v^2)(\mathbf{E}^0 \times \mathbf{s})\cdot\bar{\mathbf{u}} &= 0, \\
[(1 + \chi^{E}{}_1)(v/c)^2 - 1 + \chi^{B}{}_1 + \chi^{B}{}_2(\mathbf{E}^0\cdot\mathbf{E}^0)](\mathbf{E}^0 \times \mathbf{s})\cdot\bar{\mathbf{e}} &= 0.
\end{aligned} \tag{7.16.22}$$

From (7.16.22) it follows that

$$(\mathbf{E}^0 \times \mathbf{s})\cdot\bar{\mathbf{u}} = 0 \quad \text{or} \quad v^2 = (\sigma_0 + \sigma_2)/\rho_0, \tag{7.16.23}$$

$$(\mathbf{E}^0 \times \mathbf{s})\cdot\bar{\mathbf{e}} = 0 \quad \text{or} \quad (v/c)^2 = (1 - \chi^{B}{}_1 - \chi^{B}{}_2\mathbf{E}^0\cdot\mathbf{E}^0)/(1 + \chi^{E}{}_1). \tag{7.16.24}$$

The first of (7.16.23) indicates that $\mathbf{u}$ is in the plane of $\mathbf{E}^0$ and $\mathbf{s}$ and, of course, $\mathbf{E}^0\cdot\mathbf{s}$ and $\mathbf{E}^0 \times \mathbf{s}$ constitute an orthogonal triad. Similarly, the first of (7.16.24) expresses that $\bar{\mathbf{e}}$ is in the plane of $\mathbf{E}^0$ and $\mathbf{s}$. If this is not the case for $\mathbf{u}$, then there will be an elastic wave traveling with the speed given by $(7.16.24)_2$. This is a transverse wave whose displacement field is normal to both $\mathbf{E}^0$ and $\mathbf{s}$, with the electric vector directed along $\mathbf{s}$. If $(7.16.24)_1$ is not valid, then there will be an electromagnetic fast wave with the displacement field along the bias field and the induced electric field perpendicular to both $\mathbf{E}^0$ and $\mathbf{s}$ (Figure 7.16.1(a), (b)).

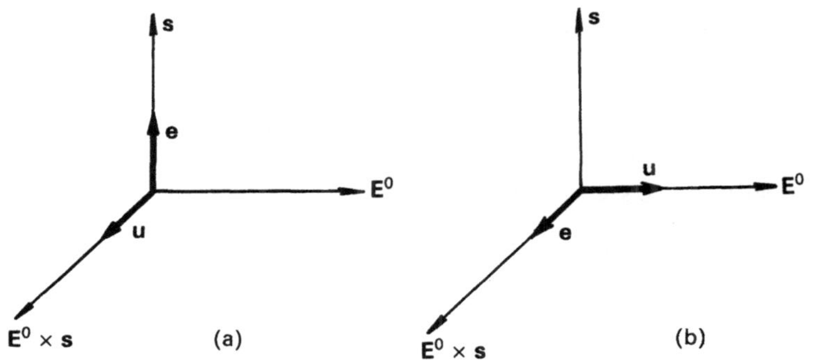

Figure 7.16.1. Transverse Kerr effect: (a) Transverse wave **u** with velocity given by $(7.6.24)_2$; (b) electromagnetic fast wave **u**.

From $(7.16.20)_1$ and $(7.16.20)_2$ we solve for

$$\begin{aligned} \mathbf{E}^0 \cdot \bar{\mathbf{e}} &= -(i\omega/ve_2)(\sigma_0 + \sigma_1 + 2\sigma_2 - \rho_0 v^2)\mathbf{s} \cdot \mathbf{u}, \\ \mathbf{s} \cdot \bar{\mathbf{e}} &= (i\omega/v)(e_1 + \chi^E{}_1)(1 + \chi^E{}_1)^{-1}\mathbf{E}^0 \cdot \bar{\mathbf{u}}. \end{aligned} \tag{7.16.25}$$

Upon substituting these into (7.16.21) we obtain two equations

$$\begin{aligned} (c/ve_2)(\alpha_1 - \alpha_2 v^2)\mathbf{s} \cdot \bar{\mathbf{u}} + (\beta_1 - \rho_0 v^2)\mathbf{E}^0 \cdot \bar{\mathbf{u}} = 0, \\ (c/ve_2)[-\gamma_1 + (\gamma_2 c^2 + \gamma_3)(v/c)^2 - \gamma_4(v^4/c^2)]\mathbf{s} \cdot \mathbf{u} + h_1 \mathbf{E}^0 \cdot \bar{\mathbf{u}} = 0, \end{aligned} \tag{7.16.26}$$

where

$$\begin{aligned} \alpha_1 &= h_1 \mathbf{E}^0 \cdot \mathbf{E}^0(\sigma_0 + \sigma_1 + 2\sigma_2), \\ \alpha_2 &= \rho_0 h_1 \mathbf{E}^0 \cdot \mathbf{E}^0, \\ \beta_1 &= \sigma_0 + \sigma_2 + [\sigma_4 - 2e_1 - \chi^E{}_1 + (e_1 + \chi^E{}_1)^2(1 + \chi^E{}_1)^{-1}]\mathbf{E}^0 \cdot \mathbf{E}^0, \\ \gamma_1 &= (1 - \chi^B{}_1)(\sigma_0 + \sigma_1 + 2\sigma_2)/\mathbf{E}^0 \cdot \mathbf{E}^0, \\ \gamma_2 &= \rho_0(1 - \chi^B{}_1)/\mathbf{E}^0 \cdot \mathbf{E}^0, \\ \gamma_3 &= e_2^2 + [(1 + \chi^E{}_1)(\mathbf{E}^0 \cdot \mathbf{E}^0)^{-1} + \chi^E{}_2](\sigma_0 + \sigma_1 + 2\sigma_2), \\ \gamma_4 &= \rho_0[(1 + \chi^E{}_1)(\mathbf{E}^0 \cdot \mathbf{E}^0)^{-1} + \chi^E{}_2]. \end{aligned} \tag{7.16.27}$$

For (7.16.26) to possess nontrivial solutions, the determinant of the coefficients of $\mathbf{s} \cdot \bar{\mathbf{u}}$ and $\mathbf{E}^0 \cdot \bar{\mathbf{u}}$ must vanish. Hence,

$$\begin{aligned} v^6 - [(\beta_1/\rho_0) + (\gamma_3/\gamma_4) + (\gamma_2/\gamma_4)c^2]v^4 + [(\beta_1\gamma_3/\rho_0\gamma_4) \\ + (\beta_1\gamma_2 + \rho_0\gamma_1 - \alpha_2 h_1)c^2(\rho_0\gamma_4)^{-1}]v^2 + (\alpha_1 h_1 - \beta_1\gamma_1)c^2(\rho_0\gamma_4)^{-1} = 0. \end{aligned} \tag{7.16.28}$$

The three roots of this cubic equation in v^2 can be determined by solving it in the usual way. However, more light can be shed on the nature of these three roots by means of a perturbation process. Accordingly, we try

$$v = v_0 \qquad \text{and} \qquad v = cv_1, \tag{7.16.29}$$

where v_0 and v_1 are moderate-order quantities, independent of c which is very large. Substituting (7.16.29) into (7.16.28), and after dividing them by c^2 and c^6, respectively, we see that the lowest-order terms give

$$v_0{}^4 - 2\lambda v_0{}^2 + \mu = 0, \qquad v_1{}^2 = \gamma_2/\gamma_4, \tag{7.16.30}$$

where

$$\lambda = (\beta_1\gamma_2 + \rho_0\gamma_1 - \alpha_2 h_1)/2\rho_0\gamma_2, \qquad \mu = (\alpha_1 h_1 - \beta_1\gamma_1)/\rho_0\gamma_2. \tag{7.16.31}$$

Hence, there exist two slow waves with phase velocities given by

$$v_0{}^2 = \lambda \pm (\lambda^2 - \mu)^{1/2}, \tag{7.16.32}$$

and one fast wave with the phase velocity determined by

$$(v/c)^2 = \rho_0(1 - \chi^{\mathrm{B}}{}_1)/(1 + \chi^{\mathrm{E}}{}_1 + \chi^{\mathrm{E}}{}_2 \mathbf{E}^0 \cdot \mathbf{E}^0). \tag{7.16.33}$$

One of the slow waves is a longitudinal elastic wave and the other is a transverse wave. The fast wave is a light wave. Consequently, the continuum model explains the Kerr effect.

(ii) *Longitudinal Pockels Effect*

We now investigate the propagation of waves along the direction of the bias field $\mathbf{E}^0$, i.e., we set

$$\mathbf{E}^0 = E_0 \mathbf{s}. \tag{7.16.34}$$

With this, (7.16.17) reduces to

$$\begin{aligned}
\bar{c}_{lm} &= \bar{c}_0\delta_{lm} + \bar{c}_1 s_l s_m, \\
\bar{e}_{lm} &= \bar{e}_0\delta_{lm} + \bar{e}_1 s_l s_m + (c/v)\bar{e}_2(s_l s_m - \delta_{lm}), \\
\bar{P}_m &= \bar{P}_0 s_m, \qquad \bar{\chi}_m = \bar{\chi}_0 s_m, \\
\bar{d}_{km} &= \bar{d}_0(\delta_{km} - s_k s_m) - (v/c)(\bar{d}_1\delta_{km} + \bar{d}_2 s_k s_m), \\
\bar{\kappa}_{km} &= \bar{\kappa}_0(s_k s_m - \delta_{km}) + (v/c)^2(\bar{\kappa}_1\delta_{km} + \bar{\kappa}_2 s_k s_m),
\end{aligned} \tag{7.16.35}$$

where

$$\begin{aligned}
\bar{c}_0 &= [\sigma_0 + \sigma_2 + (\sigma_4 - e_1)E_0{}^2] \\
\bar{c}_1 &= \sigma_1 + \sigma_2 + (2\sigma_3 + 3\sigma_4 - 4e_1 - 2e_2 - \chi^{\mathrm{E}}{}_1)E_0{}^2 \\
&\quad + (\sigma_5 - 2e_2 - \chi^{\mathrm{E}}{}_2)E_0{}^4 \\
\bar{e}_0 &= E_0 e_1, \qquad \bar{e}_1 = E_0[e_1 + e_2 + \chi^{\mathrm{E}}{}_1 + (e_3 + \chi^{\mathrm{E}}{}_2)E_0{}^2], \\
\bar{e}_2 &= h_1 E_0{}^2, \qquad \bar{\chi}_0 = 1 + \chi^{\mathrm{E}}{}_1 + \chi^{\mathrm{E}}{}_2 E_0{}^2, \\
\bar{P}_0 &= (2e_1 + e_2 + \chi^{\mathrm{E}}{}_1)E_0 + (\chi^{\mathrm{E}}{}_2 + e_3)E_0{}^3, \\
\bar{d}_0 &= h_1 E^2{}_0, \qquad \bar{d}_1 = e_1 E_0,
\end{aligned}$$

$$\bar{d}_2 = (e_1 + e_2 + \chi^{E}{}_1)E_0 + (\chi^{E} + e_3)E^3{}_0,$$
$$\bar{\kappa}_0 = 1 - \chi^{E}{}_1, \qquad \bar{\kappa}_1 = 1 + \chi^{E}{}_1, \qquad \bar{\kappa}_2 = \chi^{E}{}_2 E_0{}^2. \tag{7.16.36}$$

The scalar products of $(7.16.15)_1$ and $(7.16.15)_3$ with $\mathbf{s}$ gives

$$\begin{aligned}(\bar{c}_0 + \bar{c}_1 - \rho_0 v^2)\mathbf{s}\cdot\bar{\mathbf{u}} - (iv/\omega)(\bar{e}_0 + \bar{e}_1)(\mathbf{s}\cdot\bar{\mathbf{e}}) &= 0,\\ (\bar{d}_1 + \bar{d}_2)(\mathbf{s}\cdot\bar{\mathbf{u}}) + (iv/\omega)(\bar{\kappa}_1 + \bar{\kappa}_2)(\mathbf{s}\cdot\bar{\mathbf{e}}) &= 0.\end{aligned} \tag{7.16.37}$$

These equations may possess a nontrivial solution for $\mathbf{s}\cdot\bar{\mathbf{u}}$ and $\mathbf{s}\cdot\bar{\mathbf{e}}$, if the determinant of the coefficients vanishes, i.e.,

$$v^2 = [\bar{c}_0 + \bar{c}_1 + (\bar{e}_0 + \bar{e}_1)(\bar{d}_1 + \bar{d}_2)(\bar{\kappa}_1 + \bar{\kappa}_2)^{-1}]/\rho_0. \tag{7.16.38}$$

This is the phase velocity of an elastic wave propagating in the direction of the bias field.

If we take the vector products of $(7.16.15)_1$ and $(7.16.15)_3$ with $\mathbf{s}$, we obtain

$$\begin{aligned}(\bar{c}_0 - \rho_0 v^2)(\mathbf{s}\times\bar{\mathbf{u}}) - (iv/\omega)[\bar{e}_0 - (c/v)\bar{e}_2](\mathbf{s}\times\mathbf{e}) &= \mathbf{0},\\ [\bar{d}_0 - (v/c)\bar{d}_1](\mathbf{s}\times\bar{\mathbf{u}}) - (ic/\omega)[-\bar{\kappa}_0 + (v/c)^2\kappa_1](\mathbf{s}\times\bar{\mathbf{e}}) &= \mathbf{0}.\end{aligned} \tag{7.16.39}$$

The vanishing determinant of the coefficients $\mathbf{s}\times\bar{\mathbf{u}}$ and $\mathbf{s}\times\bar{\mathbf{e}}$ leads to

$$\begin{aligned}&(v/c)^4 - (\kappa_1\bar{c}_0 + \bar{e}_0\,\bar{d}_1 + \rho_0\bar{\kappa}_0 c^2)(\rho_0\bar{\kappa}_1 c^2)^{-1}(v/c)^2\\ &\quad + (\bar{e}_0\,\bar{d}_0 - \bar{e}_2\,\bar{d}_1)(\rho_0\bar{\kappa}_1 c^2)^{-1}(v/c) + (\bar{c}_0\bar{\kappa}_0 - \bar{e}_2\,\bar{d}_0)(\rho_0\bar{\kappa}_1 c^2)^{-1} = 0.\end{aligned} \tag{7.16.40}$$

Approximate solutions of (7.16.40) can be obtained by a particular process using c^{-2} as a perturbation parameter. A slow wave having the phase velocity

$$v^2 = \rho_0{}^{-1}[\sigma_0 + \sigma_2 + (\sigma_4 - e_1)E_0{}^2] - h_1 E_0{}^4[\rho_0(1 - \chi^{B}{}_1)]^{-1} \tag{7.16.41}$$

represents an elastic wave. By writing $v/c = \nu$, we obtain a fast electromagnetic wave with phase velocity

$$(v/c)^2 = (1 - \chi^{B}{}_1)/(1 + \chi^{E}{}_1). \tag{7.16.42}$$

7.17. Magneto-Optical Effects

Anisotropy can be induced by means of an initially applied magnetic field on an isotropic dielectric. This is known as the magneto-optical effect. This section studies briefly two such effects, namely:

(i) *The Voigt–Cotton–Mouton Effect*: The propagation of waves in a direction normal to the bias magnetic field; and
(ii) *The Faraday Effect*: Propagation in the direction of the bias field.

The method followed is similar to that pursued in the previous section. We set

$$\mathbf{x} = \mathbf{X}, \qquad \mathbf{E}^0 = \mathbf{P}^0 = \mathbf{0}, \qquad \mathbf{B}^0 \text{ and } \mathbf{M}^0 \text{ uniform}. \tag{7.17.1}$$

The equations of motion (7.14.5), and Maxwell's equations (7.14.6) for the initial fields, are satisfied identically with $q_e^0 = 0$ and $\mathbf{f} = \mathbf{0}$. The perturbation fields, (7.14.27)–(7.14.31), reduce to

$$
{}_{\mathrm{E}}\tilde{T}_{kl,k} + {}_{\mathrm{E}}T^0{}_{kr}u_{l,rk} + (b_{k,l} - b_{l,k})M^0{}_k - B^0{}_l m_{k,k} + \frac{1}{c}\varepsilon_{lmn}\dot{p}_m B^0{}_n - \rho_0 \ddot{u}_l = 0, \tag{7.17.2}
$$

$$
\begin{aligned}
\nabla\cdot\mathbf{d} &= 0,\\
\nabla\times\mathbf{e} + \frac{1}{c}\dot{\mathbf{b}} &= 0,\\
\nabla\cdot\mathbf{b} &= 0,\\
\nabla\times\mathbf{h} - \frac{1}{c}\dot{\mathbf{d}} &= 0.
\end{aligned} \tag{7.17.3}
$$

Constitutive equations are given by (7.14.43) and (7.14.45), i.e.,

$$
\begin{aligned}
&{}_{\mathrm{E}}T^0{}_{kl} = \sigma_0\delta_{kl} - h_2 B^0{}_k B^0{}_l, \qquad M^0{}_k = \chi^{\mathrm{B}}{}_1 B^0{}_k,\\
&{}_{\mathrm{E}}\tilde{T}_{kl} = \sigma_{klmn}u_{m,n} - h_{mkl}(B^0{}_r u_{r,m} + b_m),
\end{aligned} \tag{7.17.4}
$$

$$
\begin{aligned}
d_k &= e_k + p_k = e_k + \chi^{\mathrm{E}}{}_{kl}\left[e_l + \frac{1}{c}(\dot{\mathbf{u}}\times\mathbf{B}^0)_l\right],\\
m_k &= b_k - h_k = -M^0{}_k u_{r,r} + M^0{}_l u_{k,l} + \chi^{\mathrm{B}}{}_{kl}(B^0{}_r u_{r,l} + b_l) + h_{klm}u_{l,m}.
\end{aligned} \tag{7.17.5}
$$

The constitutive moduli are obtained by using (7.14.43), (7.14.45), and (7.14.46)

$$
\begin{aligned}
\sigma_{klmn} &= \sigma_1\delta_{kl}\delta_{mn} + \sigma_2(\delta_{km}\delta_{ln} + \delta_{kn}\delta_{lm}) + \sigma_3(B^0{}_k B^0{}_l\delta_{mn} + B^0{}_m B^0{}_n\delta_{kl})\\
&\quad + \sigma_4(B^0{}_k B^0{}_m\delta_{ln} + B^0{}_l B^0{}_m\delta_{kn} + B^0{}_k B^0{}_n\delta_{lm} + B^0{}_l B^0{}_n\delta_{km})\\
&\quad + \sigma_5 B^0{}_k B^0{}_l B^0{}_m B^0{}_n,\\
e_{klm} &= 0,\\
h_{klm} &= h_1 B^0{}_k\delta_{lm} + h_2(B^0{}_m\delta_{kl} + B^0{}_l\delta_{km}) + h_3 B^0{}_k B^0{}_l B^0{}_m,\\
\lambda_{kl} &= 0,\\
\chi^{\mathrm{E}}{}_{kl} &= \chi^{\mathrm{E}}{}_1\delta_{kl} + \chi^{\mathrm{E}}{}_2 B^0{}_k B^0{}_l,\\
\chi^{\mathrm{B}}{}_{kl} &= \chi^{\mathrm{B}}{}_1\delta_{kl} + \chi^{\mathrm{B}}{}_2 B^0{}_k B^0{}_l,
\end{aligned} \tag{7.17.6}
$$

where

$$
\sigma_0 = J_0{}^{-1}\left(\frac{\partial\Sigma}{\partial\bar{\mathrm{I}}_1}\right)_0, \qquad \sigma_1 = J_0{}^{-1}\left(\frac{\partial^2\Sigma}{\partial\bar{\mathrm{I}}_1{}^2}\right)_0, \qquad \sigma_2 = J_0{}^{-1}\left(\frac{\partial\Sigma}{\partial\bar{\mathrm{I}}_2}\right)_0,
$$

$$
\sigma_3 = \tfrac{1}{2}J_0{}^{-1}\left(\frac{\partial^2\Sigma}{\partial\bar{\mathrm{I}}_1\,\partial\bar{\mathrm{I}}_7}\right)_0, \qquad \sigma_4 = \tfrac{1}{2}J_0{}^{-1}\left(\frac{\partial\Sigma}{\partial\bar{\mathrm{I}}_4}\right)_0, \qquad \sigma_5 = J_0{}^{-1}\left(\frac{\partial^2\Sigma}{\partial\bar{\mathrm{I}}_7{}^2}\right)_0,
$$

$$h_1 = -2J_0{}^{-1}\left(\frac{\partial^2\Sigma}{\partial\bar{I}_1\,\partial\bar{I}_5}\right)_0, \quad h_2 = -J_0{}^{-1}\left(\frac{\partial\Sigma}{\partial\bar{I}_7}\right)_0, \quad h_3 = -2J_0{}^{-1}\left(\frac{\partial^2\Sigma}{\partial\bar{I}_5\,\partial\bar{I}_7}\right)_0,$$

$$\chi^{E}{}_1 = -2J_0{}^{-1}\left(\frac{\partial\Sigma}{\partial\bar{I}_4}\right)_0, \qquad \chi^{E}{}_2 = -2J_0{}^{-1}\left(\frac{\partial\Sigma}{\partial\bar{I}_{10}}\right)_0,$$

$$\chi^{B}{}_1 = -2J_0{}^{-1}\left(\frac{\partial\Sigma}{\partial\bar{I}_5}\right)_0, \qquad \chi^{B}{}_2 = -4J_0{}^{-1}\left(\frac{\partial^2\Sigma}{\partial\bar{I}_5{}^2}\right)_0. \tag{7.17.7}$$

Substituting (7.17.4)–(7.17.6) into (7.17.2) and (7.17.3), we obtain the field equations

$$\begin{aligned}
&(\sigma_1 + \sigma_2)u_{k,kl} + (\sigma_0 + \sigma_2)u_{l,kk} + (\sigma_3 + \sigma_4 - h_2 - h_1)B^0{}_kB^0{}_lu_{r,rk} \\
&\quad + (\sigma_3 + \sigma_4 - h_1 - h_2)B^0{}_kB^0{}_ru_{k,rl} + (\sigma_4 - \chi^{B}{}_1 - 2h_2)\mathbf{B}^0{}_lB^0{}_ru_{r,kk} \\
&\quad + (\sigma_4 - h_2)B^0{}_kB^0{}_ru_{l,rk} + (\sigma_5 - 2h_3 - \chi^{B}{}_2)B^0{}_kB^0{}_lB^0{}_mB^0{}_nu_{m,nk} \\
&\quad + (-h_1 + \chi^{B}{}_1)B^0{}_kb_{k,l} - (h_2 + \chi^{B}{}_1)B^0{}_kb_{l,k} - (h_2 + \chi^{B}{}_1)B^0{}_lb_{k,k} \\
&\quad - (h_3 + \chi^{B}{}_2)B^0{}_kB^0_lB^0{}_mb_{m,k} + \frac{1}{c}\chi^{E}{}_1\varepsilon_{lmn}B^0{}_n\left[\dot{e}_m + \frac{1}{c}(\ddot{\mathbf{u}} \times \mathbf{B}^0)_m\right] \\
&\quad - \rho_0\ddot{u}_l = 0,
\end{aligned} \tag{7.17.8}$$

$$\begin{aligned}
(1 + \chi^{E}{}_1)e_{k,k} + \chi^{E}{}_2B^0{}_kB^0{}_le_{l,k} + \frac{1}{c}\chi^{E}{}_1(\dot{\mathbf{u}} \times \mathbf{B}^0)_{k,k} \\
+ \frac{1}{c}\chi^{E}{}_2B^0{}_kB^0{}_l(\dot{\mathbf{u}} \times \mathbf{B}^0)_{l,k} = 0
\end{aligned} \tag{7.17.9}$$

$$\nabla \times \mathbf{e} + \frac{1}{\mathbf{c}}\dot{\mathbf{b}} = 0, \tag{7.17.10}$$

$$\nabla \cdot \mathbf{b} = 0, \tag{7.17.11}$$

$$\begin{aligned}
&\varepsilon_{lmk}[(1 - \chi^{B}_1)b_{k,m} - \chi^{B}{}_2B^0{}_kB^0{}_nb_{n,m} - (\chi^{B}{}_1 + h_2)B^0{}_ru_{r,km} \\
&\quad - (\chi^{B}{}_1 + h_2)B^0{}_ru_{k,rm} - h_1B^0{}_ku_{r,rm} - (\chi^{B}{}_2 + h_3)B^0{}_kB^0{}_nB^0{}_ru_{n,mr}] \\
&\quad - \frac{1}{c}\left\{(1 + \chi^{E}{}_1)\dot{e}_l + \chi^{E}{}_2B^0{}_kB^0{}_l\dot{e}_k + \frac{1}{c}\chi^{E}{}_1(\ddot{\mathbf{u}} \times \mathbf{B}^0)_l\right\} = 0.
\end{aligned} \tag{7.17.12}$$

Equations (7.17.8)–(7.17.12) are the field equations for the determination of the displacement field **u** and the electromagnetic fields **e** and **b**, given the constant bias field $\mathbf{B}^0$.

(i) *Propagation of Plane Harmonic Waves*

We seek a plane-wave solution of the system of equations (7.17.8)–(7.17.12) in the form

$$\{\mathbf{u}, \mathbf{e}, \mathbf{b}\} = \operatorname{Re}\{\bar{\mathbf{u}}, \bar{\mathbf{e}}, \bar{\mathbf{b}}\} \exp\left[i\omega\left(t - \frac{n}{c}\mathbf{s}\cdot\mathbf{x}\right)\right], \tag{7.17.13}$$

where $\mathbf{s}$ is the unit vector directed along the propagation and c/n is the propagation velocity

$$v = c/n, \qquad \mathbf{s}\cdot\mathbf{s} = 1. \tag{7.17.14}$$

Substituting (7.17.13) into (7.17.8)–(7.17.12) we will have

$$(S_{lk} - \rho_0 v^2 \delta_{lk})\bar{u}_k + (iv/\omega)B_{lk}\bar{b}_k - (iv^2/c\omega)\chi^{E}{}_1 \varepsilon_{lmn}\bar{e}_m B^0{}_n = 0, \tag{7.17.15}$$

$$(1 + \chi^{E}{}_1)\bar{\mathbf{e}}\cdot\mathbf{s} + \chi^{E}{}_2(\mathbf{B}^0\cdot\mathbf{s})(\mathbf{B}^0\cdot\bar{\mathbf{e}}) + (i\omega/c)\chi^{E}{}_1(\bar{\mathbf{u}}\times\mathbf{B}^0)\cdot\mathbf{s} = 0, \tag{7.17.16}$$

$$\mathbf{s}\times\bar{\mathbf{e}} - (v/c)\bar{\mathbf{b}} = \mathbf{0}, \tag{7.17.17}$$

$$\mathbf{s}\cdot\mathbf{b} = 0, \tag{7.17.18}$$

$$\begin{aligned}(\chi^{B}{}_1 + h_2)(\mathbf{B}^0\cdot\mathbf{s})\mathbf{s}\times\bar{\mathbf{u}} &+ [(h_1 - \chi^{B}{}_1)\mathbf{s}\cdot\bar{\mathbf{u}} + (\chi^{B}{}_2 + h_3)(\mathbf{B}^0\cdot\mathbf{s})(\mathbf{B}^0\cdot\bar{\mathbf{u}})]\mathbf{s}\times\mathbf{B}^0\\ &- (iv/\omega)[(1 - \chi^{B}{}_1)\mathbf{s}\times\bar{\mathbf{b}} - \chi^{B}{}_2(\mathbf{s}\times\mathbf{B}^0)\mathbf{B}^0\cdot\bar{\mathbf{b}}]\\ &- (iv^2/\omega c)[(1 - \chi^{E}{}_1)\bar{\mathbf{e}} + \chi^{E}{}_2(\mathbf{B}^0\cdot\bar{\mathbf{e}})\mathbf{B}^0] + (v/c)^2\chi^{E}{}_1(\bar{\mathbf{u}}\times\mathbf{B}^0) = \mathbf{0},\end{aligned} \tag{7.17.19}$$

where

$$\begin{aligned} S_{lk} &= (\sigma_1 + \sigma_2)s_l s_k + (\sigma_3 + \sigma_4 - h_1 - h_2)(\mathbf{B}^0\cdot\mathbf{s})B^0{}_l s_k\\ &\quad + (\sigma_3 + \sigma_4 - h_1 - h_2)(\mathbf{B}^0\cdot\mathbf{s})s_l B^0{}_k\\ &\quad + [\sigma_4 - 2h_2 - \chi^{B}{}_1 + (\sigma_5 - 2h_3 - \chi^{E}{}_2)(\mathbf{B}^0\cdot\mathbf{s})^2]B^0{}_l B^0{}_k\\ &\quad + [\sigma_0 + \sigma_2 + (\sigma_4 - h_2)(\mathbf{B}_0\cdot\mathbf{s})^2]\delta_{lk},\\ B_{lk} &= (\chi^{B}{}_1 - h_1)s_l B^0{}_k - (\chi^{B}{}_1 + h_2)B^0{}_1 s_k\\ &\quad - (\chi^{B}{}_2 + h_3)(\mathbf{B}^0\cdot\mathbf{s})B^0{}_k B^0{}_l - (\chi^{B}{}_1 + h_2)(\mathbf{B}^0\cdot\mathbf{s})\delta_{kl}. \end{aligned} \tag{7.17.20}$$

The set of equations (7.17.15)–(7.17.19) is adequate to determine the phase velocity v, and the wave amplitudes $\bar{\mathbf{u}}$, $\bar{\mathbf{b}}$, and $\bar{\mathbf{e}}$. Below, we study two special cases.

(ii) *The Longitudinal Faraday Effect*

Consider the case in which the direction of propagation coincides with that of the bias field, i.e.,

$$\mathbf{B}^0 = B_0\mathbf{s}, \qquad B_0 = |\mathbf{B}^0|. \tag{7.17.21}$$

In this case, equations (7.17.15)–(17.17.19) take the following forms

$$\begin{aligned} S_2(\mathbf{s}\cdot\bar{\mathbf{u}})\mathbf{s} + (S_0 - \rho_0 v^2)\bar{u} &- (iv/\omega)[B_2(\mathbf{s}\cdot\bar{\mathbf{b}})\mathbf{s} + (\chi^{B}{}_1 + h_2)B_0\bar{\mathbf{b}}]\\ &- (iv^2/c\omega)\chi^{E}{}_1 B_0(\bar{\mathbf{e}}\times\mathbf{s}) = \mathbf{0}, \end{aligned} \tag{7.17.22}$$

$$(1 + \chi^{E}{}_1 + \chi^{E}{}_2 B^2{}_0)(\mathbf{s}\cdot\bar{\mathbf{e}}) = 0, \tag{7.17.23}$$

$$\mathbf{s}\times\bar{\mathbf{e}} - (v/c)\bar{\mathbf{b}} = \mathbf{0}, \tag{7.17.24}$$

$$\mathbf{s}\cdot\bar{\mathbf{b}} = 0, \tag{7.17.25}$$

$$B_0[\chi^B{}_1 + h_2 - (v/c)^2\chi^E{}_1](\mathbf{s} \times \bar{\mathbf{u}}) - (iv/\omega)(1 - \chi^B{}_1)(\mathbf{s} \times \bar{\mathbf{b}})$$
$$- (iv^2/c\omega)[(1 + \chi^E{}_1)\bar{\mathbf{e}} + \chi^E{}_2 B_0{}^2(\mathbf{s}\cdot\bar{\mathbf{e}})\mathbf{s}] = \mathbf{0}, \quad (7.17.26)$$

where

$$S_0 = \sigma_0 + \sigma_2 + (\sigma_4 - h_2)B_0{}^2,$$
$$S_2 = \sigma_1 + \sigma_2 + (2\sigma_3 + 3\sigma_4 - 2h_1 - 4h_2 - \chi^B{}_1)B_0{}^2 + (\sigma_5 - 2h_3 - \chi^B{}_2)B_0^4,$$
$$B_2 = B_0[h_1 + h_2 + (\chi^B{}_2 + h_3)B^2{}_0]. \quad (7.17.27)$$

If we multiply (7.17.22) and (7.17.24) by s, and use (7.17.23), we find that

$$\mathbf{s}\cdot\bar{\mathbf{e}} = 0, \qquad \mathbf{s}\cdot\bar{\mathbf{b}} = 0, \quad (7.17.28)$$

and

$$v^2 = (S_0 + S_2)/\rho_0. \quad (7.17.29)$$

This is the phase velocity of an elastic wave propagating in the direction of the bias magnetic field. Next, we take the vector product of (7.17.22) and (7.17.24) with s, and upon using (7.17.23), we obtain

$$(S_0 - \rho_0 v^2)(\mathbf{s} \times \bar{\mathbf{u}}) - (iv/\omega)(\chi^B{}_1 + h_2)B_0(\mathbf{s} \times \bar{\mathbf{b}}) - (iv^2/c\omega)\chi^E{}_1 B_0\bar{\mathbf{e}} = \mathbf{0},$$
$$(v/c)\mathbf{s} \times \bar{\mathbf{b}} + \bar{\mathbf{e}} = \mathbf{0}. \quad (7.17.30)$$

From the last equations we have $\mathbf{s}\cdot\bar{\mathbf{e}} = 0$, so that (7.17.26) reduces to

$$B_0[\chi^B{}_1 + h_2 - (v/c)^2\chi^E{}_1](\mathbf{s} \times \bar{\mathbf{u}}) - (iv/\omega)(1 - \chi^B{}_1)(\mathbf{s} \times \bar{\mathbf{b}})$$
$$- (iv^2/c\omega)(1 + \chi^E{}_1)\bar{\mathbf{e}} = \mathbf{0}. \quad (7.17.31)$$

Equations (7.17.30) and (7.17.31) form a linear system for $\mathbf{s} \times \bar{\mathbf{u}}$, $\mathbf{s} \times \bar{\mathbf{b}}$, and $\bar{\mathbf{e}}$. For nontrivial solutions to exist, the determinant of the coefficient must vanish. This gives

$$(v/c)^4 - \alpha(v/c)^3 - \beta(v/c)^2 + \gamma(v/c) + \delta = 0, \quad (7.17.32)$$

where

$$\alpha = (B_0\chi^E{}_1)^2/\rho_0 c^2(1 + \chi^E{}_1),$$
$$\beta = [S_0(1 + \chi^E{}_1) + B_0{}^2\chi^E{}_1(\chi^B{}_1 + h_2) + \rho_0 c^2(1 - \chi^B{}_1)]/\rho_0 c^2(1 + \chi^E{}_1),$$
$$\gamma = B_0{}^2\chi^E{}_1(\chi^B{}_1 + h_2)/\rho_0 c^2(1 + \chi^E{}_1),$$
$$\delta = [S_0(1 - \chi^B{}_1) + (\chi^B{}_1 + h_2)^2 B^2{}_0]/\rho_0 c^2(1 + \chi^E{}_1). \quad (7.17.33)$$

If we use $1/c$ as the perturbation parameter, we can extract two wave velocities from (7.17.32), one of which corresponds to the phase velocity of an elastic wave given by

$$v^2 = \frac{S_0}{\rho_0} + B_0{}^2\frac{\chi^B{}_1 + h_2}{1 - \chi^B{}_1}, \quad (7.17.34)$$

and the other to that of a fast transverse wave of electromagnetic origin

$$\left(\frac{v}{c}\right)^2 = \frac{1 - \chi^B{}_1}{1 - \chi^E{}_1}. \tag{7.17.35}$$

This elastic model does not exhibit any Faraday rotation.

(iii) *Transverse Birefringence Effect*

In the case where the direction of propagation is perpendicular to the bias magnetic field, i.e.,

$$\mathbf{B}^0 \cdot \mathbf{s} = 0, \tag{7.17.36}$$

the field equations (7.17.15)–(7.17.19) become

$$\begin{aligned}(\sigma_1 + \sigma_2)(\mathbf{s}\cdot\bar{\mathbf{u}})\mathbf{s} - \rho_0 v^2 \bar{\mathbf{u}} + (\sigma_4 - 2h_2 - \chi^B{}_1)(\mathbf{B}^0\cdot\bar{\mathbf{u}})\mathbf{B}^0 \\ + (iv/\omega)(\chi^B{}_1 - h_1)(\mathbf{B}^0\cdot\bar{\mathbf{b}})\mathbf{s} - (iv/\omega)(\chi^B{}_1 + h_2)(\mathbf{s}\cdot\bar{\mathbf{b}})\mathbf{B}^0 \\ - (iv^2/c\omega)\chi^E{}_1(\bar{\mathbf{e}} \times \mathbf{B}^0) = \mathbf{0},\end{aligned} \tag{7.17.37}$$

$$(1 + \chi^E{}_1)(\bar{\mathbf{e}}\cdot\mathbf{s}) + (i\omega/c)\chi^E{}_1(\bar{\mathbf{u}} \times \mathbf{B}^0)\cdot\mathbf{s} = \mathbf{0}, \tag{7.17.38}$$

$$\mathbf{s} \times \bar{\mathbf{e}} - (v/c)\bar{\mathbf{b}} = \mathbf{0}, \tag{7.17.39}$$

$$\mathbf{s}\cdot\bar{\mathbf{b}} = 0, \tag{7.17.40}$$

$$\begin{aligned}(h_1 - \chi^B{}_1)(\mathbf{s}\cdot\bar{\mathbf{u}})(\mathbf{s} \times \mathbf{B}^0) + (v/c)^2\chi^E{}_1(\bar{\mathbf{u}} \times \mathbf{B}^0) \\ - (iv/\omega)(1 - \chi^B{}_1)(\mathbf{s} \times \bar{\mathbf{b}}) + (iv/\omega)\chi^B{}_2(\mathbf{s} \times \mathbf{B}^0)(\mathbf{B}^0\cdot\bar{\mathbf{b}}) \\ - (iv^2/\omega c)(1 + \chi^E{}_1)\bar{\mathbf{e}} - (iv^2/\omega c)\chi^E{}_2(\mathbf{B}^0\cdot\bar{\mathbf{e}})\mathbf{B}^0 = \mathbf{0}.\end{aligned} \tag{7.17.41}$$

We take the scalar product of (7.17.37), (7.17.39), and (7.17.41) by $\mathbf{s}$, $\mathbf{B}^0$, and $\mathbf{s} \times \mathbf{B}^0$. Upon using (7.17.39) to eliminate $\bar{\mathbf{e}}$, we obtain

$$\begin{aligned}(\sigma_1 + \sigma_2 - \rho_0 v^2)(\mathbf{s}\cdot\bar{\mathbf{u}}) + (iv/\omega)[\chi^B{}_1 - h_1 - \chi^E{}_1(v/c)^2]\mathbf{B}^0\cdot\bar{\mathbf{b}} = 0, \\ \mathbf{s}\cdot\bar{\mathbf{b}} = \mathbf{s}\cdot\bar{\mathbf{e}} = \mathbf{s}\cdot(\bar{\mathbf{u}} \times \mathbf{B}^0) = 0, \\ [(\sigma_4 - 2h_2 - \chi^B{}_1)B^{0^2} - \rho_0 v^2](\mathbf{B}^0\cdot\bar{\mathbf{u}}) = 0, \\ [1 - \chi^B{}_1 - (v/c)^2(1 + \chi^E{}_1 + \chi^E{}_2 B^{0^2}]\mathbf{B}^0\cdot\bar{\mathbf{e}} = 0, \\ [(h_1 - \chi^B{}_1) + (v/c)^2\chi^E{}_1]B^{0^2}(\mathbf{s}\cdot\bar{\mathbf{u}}) \\ - (iv/\omega)[1 - \chi^B{}_1 - \chi^B{}_2 B^{0^2} - (v/c)^2(1 + \chi^E{}_1)](\mathbf{B}^0\cdot\bar{\mathbf{b}}) = 0.\end{aligned} \tag{7.17.42}$$

From the third and fourth equations it follows that $\mathbf{B}^0\cdot\bar{\mathbf{u}}$ and $\mathbf{B}^0\cdot\bar{\mathbf{e}}$ may not vanish if

$$v^2 = (\sigma_4 - 2h_2 - \chi^B{}_1)B^{0^2}/\rho_0, \tag{7.17.43}$$

$$(v/c)^2 = (1 - \chi^B{}_1)/(1 + \chi^E{}_1 + \chi^E{}_2 B^{0^2}). \tag{7.17.44}$$

The first and the last of (7.17.42) may possess nontrivial solutions, if the determinant of the coefficient of $\mathbf{s}\cdot\bar{\mathbf{u}}$ and $\mathbf{B}^0\cdot\bar{\mathbf{b}}$ vanishes. Hence,

$$(v/c)^4 - 2\beta(v/c)^2 + \gamma = 0, \tag{7.17.45}$$

where

$$\beta = \tfrac{1}{2}[\rho_0 c^2(1 - \chi^B{}_1 - \chi^B{}_2 B^{02}) + (\sigma_1 + \sigma_2)(1 + \chi^E{}_1) + 2(h_1 - \chi^B{}_1)\chi^E{}_1 B^{02} - \chi^B{}_1\chi^E{}_1 B^{02}][\rho_0 c^2(1 + \chi^E{}_1) - (\chi^E{}_1 B^0)^2]^{-1},$$

$$\gamma = [(\sigma_1 + \sigma_2)(1 - \chi^B{}_1 - \chi^B{}_2 B^{02}) - (\chi^B{}_1 - h_1)^2 B^{02}][\rho_0 c^2(1 + \chi^E{}_1) - (\chi^E{}_1 B^0)^2]^{-1}.$$

The solution of (7.17.45) is

$$(v/c)^2 = \beta \pm (\beta^2 - \gamma)^{1/2}. \tag{7.17.47}$$

Since γ is very small, because of the appearance of c^{-2}, by a binonial expansion we obtain

$$(v/c)^2 = (1 - \chi^B{}_1 - \chi^B{}_2 B^{02})/(1 + \chi^E{}_1), \tag{7.17.48}$$

$$v^2 = \frac{\sigma_1 + \sigma_2}{\rho_0} - \frac{(\chi^B{}_1 - h_1)^2 B^{02}}{\rho_0(1 - \chi^B{}_1 - \chi^B{}_2 B^{02})}. \tag{7.17.49}$$

The phase velocity given by (7.17.49) is that of a longitudinal elastic wave, with the displacement directed along the direction of propagation. The phase velocity (7.17.43) corresponds to that of a transverse elastic wave propagating in the direction of the bias field. This wave comes to exist solely by the presence of the bias magnetic field. The medium exhibits an *acoustical transverse birefringence* which vanishes with the bias field.

The remaining phase velocities given by (7.17.44) and (7.17.48) are fast waves of electromagnetic origin. The first of these is a transverse wave with the electric vector along the bias field $\mathbf{B}^0$, while the second one is a transverse wave with the induced magnetic field vector along $\mathbf{B}^0$. These two waves account for the *Voigt–Cotton–Mouton* effect. Thus, the medium exhibits optical transverse linear birefringence.

The relationships of magnetization, magnetic susceptibility, and the Voigt–Cotton–Mouton effect were investigated extensively during the period 1930–1940. Experimental data are reported by Beams [1932] and Schultz [1963]. Most dielectrics are observed to be weakly magnetizable. The value of the classical Cotton–Mouton constant is very small (of the order 10^{-12} g^{-1} sec^2). This is in accordance with the theory which shows that the difference between the two velocities (7.17.43) and (7.17.48) is very small for weakly magnetizable dielectrics. In special dielectrics (e.g., colloidal suspensions of ferrites in a dielectric jelly), the Cotton–Mouton constant is much larger (about 10^7 times). The velocity of the longitudinal elastic waves is known to increase some 0.53%, at 300 K, in a transverse magnetic field (Truell and Elbaum [1965]). This is a very small change. Data published by Goodrich and Lange [1971] leads to similar conclusions. Acoustical birefringence, induced by a transverse magnetic field, was also observed by Watkins and Feher [1962] ad Wang and Crow [1970].

7.18. Elastic Ionic Crystals, Ferroelectrics,[14] and Other Polarizable Solids

The linear piezoelectricity theory (7.2.19), (7.2.20) is not equipped to describe the electromechanical behavior of *ionic crystals* such as the alkali halides (NaCl, NaI, etc.) in a sufficiently *fine* manner. The same holds true of the nonlinear theory (7.2.8)–(7.2.10) as regards the electromechanical behavior of *ferroelectric crystals* such as barium titanate ($BaTiO_3$) and sodium nitrite ($NaNO_2$).

Ionic Crystals

In this case, the deficiency of the theory (7.2.19), (7.2.20) is that subtler interactions, such as those occurring between the inner electronic components of each molecule and the other electronic layer (the "shell"), must be explained. Through a lattice model of solid ionic crystals (Askar et al. [1974], [1970] and Mindlin [1972]) and the so-called "long wave" limit, it can be shown that these effects can be incorporated in a continuum theory, by introducing the gradient $\nabla\mathbf{P}$ of the polarization vector, as an additional independent variable in the constitutive theory (compare (5.4.30) where $\mathscr{E}_{K,L}$ is present). This is a type of *weak nonlocality*. The corresponding continuum models were presented by Mindlin [1968] for infinitesimal strains, and by Şuhubi [1969] for finite strains. They have shown that the additional variables allow us to exhibit interesting surface and boundary-layer properties in ionic crystals (see Askar et al. [1971], Mindlin [1972], and Nowacki [1983, Chap. 2]). The nonlinear wave propagation studies, mentioned in Section 7.13, can be extended to this more involved model which will feature many interesting phenomena (Collet [1981], [1982], [1984]).

Ferroelectric Crystals

In the case of ferroelectric crystals of the *displacive* type, such as barium titanate ($BaTiO_3$), a permanent electric dipole moment exists in the crystal in its natural configuration, as a result of the relative displacement of the central atom, with respect to the other atoms of the crystalline structure. In the case of ferroelectric crystals of the *molecular group* type, such as sodium nitrite ($NaNo_2$) a dipole is attached to the central molecular group, and both the rotational inertia of this group and the electrostatic interaction with neighboring atomic components must be taken into account. To cope with this situation, a fully nonlinear theory, including both polarization inertia and gradients, is required. Such a theory was proposed by Collet and Maugin [1974] and, in a more general form (describing antiferroelectrics as well), by

[14] The electromechanical study of these crystals is a field closely related to crystal lattice dynamics and solid-state physics. See Maugin [1988, Chap. 7]. (Note added at proof)

Maugin [1977]. The Cauchy stress tensor is *not* symmetric in this theory, while the ponderomotive couple plays a prominent role. The corresponding electroacoustic equations are developed by Maugin and Pouget [1980].

The dynamics of such dielectrics, studied with the method developed in Section 7.14 above, show resonance couplings between acoustic and ferroelectric modes. With temperature approaching the ferroelectric transition temperature the ferroelectric modes disappear just as in the case of magnetoacoustic resonance coupling in ferromagnets (see Chapter 3) (Pouget and Maugin [1980]). Hence they are called *soft*. Piezoelectric Rayleigh modes and antiplane surface waves, similar to the Bleustein waves of Section 7.7 (but with a *strong dispersion*), can also be exhibited in such bodies (Pouget and Maugin [1981a, b]). All these results are confirmed by lattice models of ferroelectric solids, and experimental investigations using inelastic neutron scattering and Raman spectroscopy (Askar et al. [1984] and Pouget et al. [1985a, b]). In the nonlinear case, electromechanical interactions in such dielectrics may be shown to be disturbed by electroacoustic solitary waves which behave like *solitons* under collision, and may serve as a dynamic model for the structure in domains and walls observed in ferroelectric crystals (Pouget and Maugin [1984], [1985a]). Such crystals may be considered as good examples of the *micropolar elastic continua* of Eringen [1966b, c, 1970c].

Piezoelectric Powders

These are granular media, such as sugar, made of grains, which are piezoelectric. Thus, the linear piezoelectricity theory (7.2.19)–(7.2.22) applies to the constituents of the medium; but the latter has, overall, a much more complex electromechanical behavior since it is an example of the *micromorphic continuum* of Eringen [1964], [1966d], [1972d, e], i.e., a continuum with a deformable microstructure. Each grain in a deformable piezoelectric powder is able *to flow*, exhibiting an overall fluid behavior (Pouget and Maugin [1983a]). This complex behavior allows the production of strange echo phenomena, according to which piezoelectric powders can be used for the storage of information, like a computer memory (Pouget and Maugin [1983b]), and for the correlation of three successfully applied electric signals (Pouget [1982]). More on these phenomena are contained in the reviews by Pouget [1984], and Smolensky and Yushin [1984].

Piezoelectric Semiconductors

Heretofore, all materials considered in this chapter have exhibited electromechanical interactions such as piezoelectricity or electrostrictive effects—but they are *dielectrics* from the electric point of view. In particular, they do not conduct electricity, or do so only negligibly. Moreover, some good piezoelectrics such as cadmium sulfide CdS and germanium are, in fact, semiconductors. In that case two phenomena greatly complicate the phenomenological description, namely electric conduction and the diffusion of charges. The study of harmonic waves on a simple model then shows the possibility of spatial *amplification* of electromechanical waves (Huston and White [1962]). Continuum models, in the spirit of this book, have been tentatively developed by deLorenzi and Tiersten [1975], Ancona and Tiersten [1983], Daher [1984], and Daher and Maugin [1984]. Shock waves in such materials have been considered by McCarthy and Tiersten [1977]. This is too cumbersome to be repeated here. We refer the reader to the review by Tiersten [1984] and the papers by Maugin and Daher [1986] and Daher and Maugin [1985].

In conclusion, we note that the principle of virtual power (3.13.2), when extended to include new interactions and new effects, appears to be a safe rational method for constructing the involved phenomenological theories of the electromagnetic matter mentioned in this section.

Problems

7.1. Determine the acoustic phase velocity in an infinite piezoelectric solid.

7.2. Show that the eigenvectors $\mathbf{b}^\alpha$ of the displacement field satisfy

$$(\bar{\sigma}_{klmn}n_k n_n - \rho v_\alpha^2 \delta_{lm})b_m^\alpha = 0,$$

where $\bar{\sigma}$ is the piezoelectrically stiffened stiffness tensor. Prove that $\mathbf{b}^\alpha$ are orthonormal vectors.

7.3. Determine the reflection angle and the amplitude of an acoustic wave from the free surface of a half-space.

7.4. Obtain the Rayleigh surface wave velocity in a piezoelectric half-space.

7.5. A thick isotropic spherical shell carries large uniform surface charge on its outer surface. Determine the electric and displacement fields.

7.6. The spherical shell described in Problem 7.5 is subject to a uniform internal pressure. Determine the displacement and electric fields.

7.7. A circular cylinder is subject to an initial uniform electric field in the direction of its axis. Determine the longitudinal wave velocity.

7.8. A circular cylinder is subject to an initial uniform magnetic field in the axial direction. Determine the longitudinal wave velocity.

7.9. Research the literature and write a report on electroacoustic *solitary waves and solitons* in elastic ferroelectric crystals (e.g., $NaNO_2$).

7.10. (Short term paper.) Research the literature and write a report on the *amplification* of elastic waves in piezoelectric semiconductors.

7.11. (Short term paper.) Make a literature search and write a report on the propagation of *one-dimensional shock waves* in nonlinear elastic dielectrics.

7.12. (Short term paper.) Research the literature and write a report on the *shock compression technique* used to measure electroelastic coefficients in nonlinear elastic piezoelectrics such as $LiNbO_3$.

CHAPTER 8

Magnetoelasticity

8.1. Scope of the Chapter

This chapter is concerned with the interaction of magnetic fields with elastic solids. The theory of magnetoelasticity deals with the deformation and magnetization of nonferrous elastic materials under external effects, which consist generally of applied loads and/or deformations and thermal and magnetic fields. For these materials the effect of charge and polarization can be neglected. Hysteresis effects, micromagnetism, and spin waves are also excluded here—they will be discussed later in Chapter 9.

The theory of magnetoelasticity thus delineated is useful in several branches of the physical sciences and technology. For instance, in geophysics, the magnetoelastic wave motion in amorphous materials has been employed in various studies relevant to the earth's mantle and core. One result is that the magnetic field—which may be extremely intense near the mantle-core boundary, an essentially infinitely conducting layer—adduces a magnetoelastic explanation for the velocity gradient of compression and shear waves (Keilisborok and Munin [1959] and Knopoff [1955]). Recent studies, related to astrophysics, emphasize the interest in magnetoelasticity for the study of the crustal disturbances in the exotic stellar objects that are pulsars—"solid" stars which provide the best example of a perfect magnetoelastic medium (Bazer and Eriscon [1974] and Maugin [1978c]). Metal forming by magnetic forces (Birdsall et al. [1961], Baines et al. [1965–66], Hillier and Lal [1968], and Al-Hassani et al. [1974]) and the design of extremely powerful electromagnets, for magnetic levitation and thermonuclear fusion devices, are technologies which require the use of magnetoelasticity theory (Moon [1980], [1984]).

In Section 8.2 we collect equations of balance, jump conditions, and constitutive equations for nonlinear and linear theories. Anisotropic, isotropic, and incompressible solids are considered and field equations are obtained involving the magnetic field $\mathbf{H}$, the displacement vector $\mathbf{u}$, and the temperature change T. Boundary and initial conditions are summarized.

In Section 8.3 we present a method of solution for the static magnetoelastic case using the Stokes–Helmholtz resolution. To illustrate this method of solution, the case of a uniformly magnetized sphere in a nonpermeable medium is discussed in Section 8.4. In Section 8.5 we develop the theory of static magnetoelasticity in two dimensions. The powerful complex-function theory is presented and applied to the case of a circular cavity in a plate subject to magnetic fields. In Section 8.6 stress concentrations at a cavity surface, due to applied magnetic fields, are determined, by means of the complex-function technique.

Dynamical problems are much more difficult to deal with. Radial motions of a cylinder under an axial magnetic field, which are presented in Section 8.7, represent an exact "inverse solution" for a nonlinear problem. The propagation of plane waves, in an isotropic elastic solid carrying a bias magnetic field (Section 8.8), is another example of coupled thermo-magnetoelasticity revealing the effect of a magnetic field on the nature of waves. Experimental verification of the theoretical results are in full support of the theory. Section 8.9 discusses the problem of magnetoelastic waves produced by the application of a thermal shock to the surface of a perfectly conducting half-space which carries a constant bias magnetic field. Electromagnetic and thermoelastic fields are determined and illustrated.

In Section 8.10 we start to consider the large static deformations of elastic materials subject to magnetic fields. Homogeneous strains are discussed in this section, simple shear in Section 8.11, and the torsion of a circular cylinder subject to an axial magnetic field in Section 8.12. In all these cases, magnetostrictive effects, Poynting and Kelvin effects (relevant to finite deformations) are clearly revealed. These effects disappear in the linear theory.

Electric conduction in a cylinder undergoing large twist is the subject of Section 8.13, where we find that under large twist the current flows in spiral lines and the magnetic field tends to shorten the cylinder. Poynting and Kelvin effects contain magnetostrictive interactions.

Structural members (beams and plates) subject to strong fields are encountered in many technological applications, e.g., thermonuclear reactors, magnetically levitating trains, and insulating materials that are subject to strong transverse magnetic fields. The theory of magnetoelastic plates developed in Section 8.14 and Section 8.15 is used in Section 8.16 to discuss buckling of a plate under a strong magnetic field. This field of research is now being developed with some vigor, with the design of high field magnets. Other problems concerning the nonlinear dynamics of magnetoelasticity are mentioned in Section 8.17 by way of conclusion. Most of the topics presented in this chapter (Sections 8.4, 8.6, 8.7, 8.10–8.13) are new, and are not to be found elsewhere.

8.2. Résumé of Basic Equations

The theory of magnetoelasticity is concerned with the deformation and magnetic behavior of *nonferrous materials*. For these materials the effects of charge, displacement current, and polarization can be neglected. The basic equations of magnetoelastic solids, obtained from (5.2.1)–(5.2.18) by dropping these terms, are listed below:

Equations of Balance (in $\mathscr{V} - \sigma$)[1]

$$\rho_0 = \rho \mathrm{III}_C^{1/2} \qquad \text{or} \qquad \dot{\rho} + \rho \nabla \cdot \mathbf{v} = 0, \tag{8.2.1}$$

$$_{\mathrm{E}}t_{kl,k} + \rho(f_l - \dot{v}_l) + \frac{1}{c}(\mathbf{J} \times \mathbf{B})_l + B_{k,l}M_k - B_{l,k}M_k - M_{k,k}B_l = 0, \tag{8.2.2}$$

$$_{\mathrm{E}}t_{kl} = {}_{\mathrm{E}}t_{lk}, \tag{8.2.3}$$

$$\rho\theta\dot{\eta} - \nabla \cdot \mathbf{q} - \mathbf{J} \cdot \mathscr{E} - \rho h = 0, \tag{8.2.4}$$

$$\rho\gamma \equiv \frac{1}{\theta}\mathbf{q} \cdot \nabla\theta + \mathbf{J} \cdot \mathscr{E} \geq 0, \tag{8.2.5}$$

$$\nabla \cdot \mathbf{E} = 0, \tag{8.2.6}$$

$$\nabla \times \mathbf{E} + \frac{1}{c}\frac{\partial \mathbf{B}}{\partial t} = \mathbf{0}, \tag{8.2.7}$$

$$\nabla \cdot \mathbf{B} = 0, \tag{8.2.8}$$

$$\nabla \times \mathbf{H} = \frac{1}{c}\mathbf{J}. \tag{8.2.9}$$

Equations of Jump (on σ)

$$[\rho(\mathbf{v} - \boldsymbol{\nu})] \cdot \mathbf{n} = 0, \tag{8.2.10}$$

$$\left[\rho v_l(v_k - \nu_k) - {}_{\mathrm{E}}t_{kl} - {}_{\mathrm{M}}t_{kl} + M_k B_l - \frac{1}{c}\nu_k(\mathbf{E} \times \mathbf{B})_l\right] n_k = 0, \tag{8.2.11}$$

$$[\{\rho\varepsilon + \tfrac{1}{2}\rho v^2 + \tfrac{1}{2}(E^2 + B^2)\}(v_k - \nu_k)$$
$$- ({}_{\mathrm{E}}t_{kl} + {}_{\mathrm{M}}t_{kl} + \nu_k G_l)v_l - q_k + \mathscr{S}_k] n_k = 0, \tag{8.2.12}$$

$$\mathbf{n} \cdot [\mathbf{E}] = 0, \tag{8.2.13}$$

$$\mathbf{n} \times \left[\mathbf{E} + \frac{1}{c}\boldsymbol{\nu} \times \mathbf{B}\right] = \mathbf{0}, \tag{8.2.14}$$

$$\mathbf{n} \cdot [\mathbf{B}] = 0, \tag{8.2.15}$$

$$\mathbf{n} \times \left[\mathbf{H} - \frac{1}{c}\boldsymbol{\nu} \times \mathbf{D}\right] = \mathbf{0}. \tag{8.2.16}$$

Constitutive Equations

Anisotropic Solids

$$\rho_0 \Psi = \Sigma(C_{KL}, B_K, \theta, \mathbf{X}), \tag{8.2.17}$$

$$\eta = -\frac{1}{\rho_0}\frac{\partial \Sigma}{\partial \theta}, \tag{8.2.18}$$

[1] Equation (8.2.2) was obtained by Maugin and Eringen [1977, p. 139].

$$M_K = -\frac{\partial \Sigma}{\partial B_K}, \tag{8.2.19}$$

$$_{\mathrm{E}}t_{kl} = \frac{2\rho}{\rho_0}\frac{\partial \Sigma}{\partial C_{KL}} x_{k,K} x_{l,L} = t_{kl} + M_k B_l, \tag{8.2.20}$$

$$q_k = q_k(\mathbf{C}, \mathscr{E}, \mathbf{B}, \theta_{,K}, \theta, \mathbf{X}), \tag{8.2.21}$$

$$\mathscr{J}_k = \mathscr{J}_k(\mathbf{C}, \mathscr{E}, \mathbf{B}, \theta_{,K}, \theta, \mathbf{X}). \tag{8.2.22}$$

For the material symmetry regulations, see Section 5.10.

Isotropic Solids (see Section 5.9)

For simplicity, we drop the dependence on the high order invariant I_{13}.

$$\Sigma = \Sigma(\mathrm{I}, \mathrm{II}, \mathrm{III}, \mathrm{I}_5, \mathrm{I}_7, \mathrm{I}_9, \theta, \mathbf{X}), \tag{8.2.23}$$

$$\eta = -\frac{1}{\rho}\frac{\partial \Sigma}{\partial \theta}, \tag{8.2.24}$$

$$_{\mathrm{E}}\mathbf{t} = b_{-1}\overset{-1}{\mathbf{c}} + b_0 \mathbf{1} + b_1 \mathbf{c} - 2\chi_1^{\mathrm{B}}[\mathbf{B} \otimes (\overset{-1}{\mathbf{c}}\mathbf{B})]_{\mathrm{S}} - 4\chi_2^{\mathrm{B}}[\mathbf{B} \otimes (\overset{-2}{\mathbf{c}}\mathbf{B})]_{\mathrm{S}}, \tag{8.2.25}$$

where a subscript S means symmetrization, e.g.,

$$[T_{kl}]_{\mathrm{S}} = \tfrac{1}{2}(T_{kl} + T_{lk}).$$

$$\mathscr{M} = \chi^{\mathrm{B}}\mathbf{B} + \chi_1^{\mathrm{B}}\overset{-1}{\mathbf{c}}\mathbf{B} + \chi_2^{\mathrm{B}}\overset{-2}{\mathbf{c}}\mathbf{B}, \tag{8.2.26}$$

$$\begin{aligned}\mathbf{q} = {}& \kappa_1 \nabla\theta + \kappa_2 \mathscr{E} + \kappa_3 \overset{-1}{\mathbf{c}}\nabla\theta + \kappa_4 \overset{-1}{\mathbf{c}}\mathscr{E} + \kappa_5 \nabla\theta \times \mathbf{B} + \kappa_6 \mathscr{E} \times \mathbf{B} \\ & + \kappa_7 \overset{-2}{\mathbf{c}}\nabla\theta + \kappa_8 \overset{-2}{\mathbf{c}}\mathscr{E} + \kappa_9(\mathbf{B}\cdot\nabla\theta)\mathbf{B} + \kappa_{10}(\mathbf{B}\cdot\mathscr{E})\mathbf{B} \\ & + \kappa_{11}[\overset{-1}{\mathbf{c}}(\nabla\theta \times \mathbf{B}) - (\overset{-1}{\mathbf{c}}\nabla\theta) \times \mathbf{B}] \\ & + \kappa_{12}[\overset{-1}{\mathbf{c}}(\mathscr{E} \times \mathbf{B}) - (\overset{-1}{\mathbf{c}}\mathscr{E}) \times \mathbf{B}],\end{aligned} \tag{8.2.27}$$

$$\begin{aligned}\mathscr{J} = {}& \sigma_1 \mathscr{E} + \sigma_2 \nabla\theta + \sigma_3 \overset{-1}{\mathbf{c}}\mathscr{E} + \sigma_4 \overset{-1}{\mathbf{c}}\nabla\theta + \sigma_5 \mathscr{E} \times \mathbf{B} + \sigma_6 \nabla\theta \times \mathbf{B} \\ & + \sigma_7 \overset{-2}{\mathbf{c}}\mathscr{E} + \sigma_8 \overset{-2}{\mathbf{c}}\nabla\theta + \sigma_9(\mathbf{B}\cdot\mathscr{E})\mathbf{B} + \sigma_{10}(\mathbf{B}\cdot\nabla\theta)\mathbf{B} \\ & + \sigma_{11}[\overset{-1}{\mathbf{c}}(\mathscr{E} \times \mathbf{B}) - (\overset{-1}{\mathbf{c}}\mathscr{E}) \times \mathbf{B}] \\ & + \sigma_{12}[\overset{-1}{\mathbf{c}}(\nabla\theta \times \mathbf{B}) - (\overset{-1}{\mathbf{c}}\nabla\theta) \times \mathbf{B}],\end{aligned} \tag{8.2.28}$$

where

$$b_{-1} = 2\frac{\rho}{\rho_0}\frac{\partial \Sigma}{\partial \mathrm{I}}, \qquad b_0 = 2\frac{\rho}{\rho_0}\left(\mathrm{II}\frac{\partial \Sigma}{\partial \mathrm{II}} + \mathrm{III}\frac{\partial \Sigma}{\partial \mathrm{III}}\right), \qquad b_1 = -2\frac{\rho}{\rho_0}\mathrm{III}\frac{\partial \Sigma}{\partial \mathrm{II}},$$

$$\chi^{\mathrm{B}} = -2\frac{\rho}{\rho_0}\frac{\partial \Sigma}{\partial \mathrm{I}_5}, \qquad \chi_1^{\mathrm{B}} = -2\frac{\rho}{\rho_0}\frac{\partial \Sigma}{\partial \mathrm{I}_7}, \qquad \chi_2^{\mathrm{B}} = -2\frac{\rho}{\rho_0}\frac{\partial \Sigma}{\partial \mathrm{I}_9},$$

$$\mathrm{I} = \operatorname{tr}\overset{-1}{\mathbf{c}}, \qquad \mathrm{II} = \tfrac{1}{2}[(\operatorname{tr}\overset{-1}{\mathbf{c}})^2 - \operatorname{tr}\overset{-2}{\mathbf{c}}], \qquad \mathrm{III} = \det\overset{-1}{\mathbf{c}},$$

$$\mathrm{I}_5 = \mathbf{B}\cdot\mathbf{B}, \qquad \mathrm{I}_7 = \mathbf{B}\overset{-1}{\mathbf{c}}\mathbf{B}, \qquad \mathrm{I}_9 = \mathbf{B}\overset{-2}{\mathbf{c}}\mathbf{B},$$

$$\overset{-1}{c}_{kl} = x_{k,K} x_{l,K}, \qquad c_{kl} = X_{K,k} X_{K,l}, \qquad C_{KL} = x_{k,K} x_{k,L}, \tag{8.2.29}$$

$$t_{kl} = {}_{\mathrm{E}}t_{kl} - M_k B_l, \qquad \mathbf{D} = \mathbf{E}, \qquad \mathbf{B} = \mathbf{H} + \mathbf{M},$$

$$_{\mathrm{M}}t_{kl} = B_k H_l + E_k E_l - \tfrac{1}{2}(E^2 + B^2 - 2\mathbf{M}\cdot\mathbf{B})\delta_{kl},$$

$$\mathbf{G} = \frac{1}{c}\mathbf{E} \times \mathbf{B}, \qquad \mathscr{S} = c\boldsymbol{\mathscr{E}} \times \boldsymbol{\mathscr{H}}, \qquad \boldsymbol{\mathscr{E}} = \mathbf{E} + \frac{1}{c}\mathbf{v} \times \mathbf{B},$$

$$\boldsymbol{\mathscr{H}} = \mathbf{H} - \frac{1}{c}\mathbf{v} \times \mathbf{D}.$$

and κ_α and σ_α are functions of the joint invariants of $\overset{-1}{\mathbf{c}}$, $\boldsymbol{\mathscr{E}}$, $\mathbf{B}$, and $\nabla\theta$. These invariants can be obtained from the Table E1. The Clausius–Duhem (C–D) inequality (8.2.5) and the axiom of time reversal place restrictions on these coefficients.

Incompressible Solids. For incompressible materials $\rho = \rho_0$ and therefore $\mathrm{III} = 1$. Σ is no longer a function of III and b_0 is replaced by an unknown pressure $-p(\mathbf{x}, t)$. With these modifications (8.2.25) takes the form

$$_{\mathrm{E}}\mathbf{t} = -p\mathbf{1} + 2\frac{\partial \Sigma}{\partial \mathrm{I}}\overset{-1}{\mathbf{c}} - 2\frac{\partial \Sigma}{\partial \mathrm{II}}\mathbf{c} - 2\chi_1^{\mathrm{B}}[\mathbf{B} \otimes \overset{-1}{\mathbf{c}}\mathbf{B}]_{\mathrm{S}} - 4\chi_2^{\mathrm{B}}[\mathbf{B} \otimes \overset{-2}{\mathbf{c}}\mathbf{B}]_{\mathrm{S}}. \tag{8.2.30}$$

Linear Theory. Constitutive equations of the linear theory of magnetoelastic solids follow from (5.11.13)–(5.11.19) by dropping the effect of polarization. We assume that the initial fields are absent.

$$\Sigma = \sigma_0 - \rho_0 \eta_0 T - \frac{\rho_0 \gamma}{2T_0}T^2 - \beta_{kl} T \tilde{e}_{kl} + \tfrac{1}{2}\sigma_{klmn}\tilde{e}_{kl}\tilde{e}_{mn} - h_{klm} B_k \tilde{e}_{lm}$$
$$- \gamma_k T B_k - \tfrac{1}{2}\chi^{\mathrm{B}}_{kl} B_k B_l, \tag{8.2.31}$$

$$\eta = \eta_0 + \frac{\gamma}{T_0}T + \frac{1}{\rho_0}\beta_{kl}\tilde{e}_{kl} + \frac{\gamma_k}{\rho_0}B_k, \tag{8.2.32}$$

$$_{\mathrm{E}}t_{kl} = -\beta_{kl} T + \sigma_{klmn}\tilde{e}_{mn} - h_{mkl} B_m, \tag{8.2.33}$$

$$M_k = \gamma_k T + \chi^{\mathrm{B}}_{kl} B_l + h_{klm}\tilde{e}_{lm}, \tag{8.2.34}$$

$$q_k = \kappa_{kl} T_{,l} + \kappa^{\mathrm{E}}_{kl}\mathscr{E}_l, \tag{8.2.35}$$

$$J_k = \sigma_{kl}\mathscr{E}_l + \sigma^{\theta}_{kl} T_{,l}, \tag{8.2.36}$$

where $\tilde{e}_{kl}$ is the linear strain measure which is expressed in terms of the gradients of the displacement vector $\mathbf{u}$ by

$$\tilde{\mathbf{e}}_{kl} = \tfrac{1}{2}(u_{k,l} + u_{l,k}). \tag{8.2.37}$$

The heat and electric conduction coefficients κ_{kl}, κ^{E}_{kl}, σ_{kl}, and σ^{θ}_{kl} are subject to the restrictions arising from the entropy inequality (8.2.5) (see Section 5.11).

Linear Isotropic Solids. For isotropic materials the odd-order material tensors vanish and the even-order tensors are expressed in terms of Kronecker deltas

$$\{\beta_{kl}, \chi^{\mathrm{B}}_{kl}, \kappa_{kl}, \kappa^{\mathrm{E}}_{kl}, \sigma_{kl}\} = \{\beta_{\mathrm{e}}, \chi^{\mathrm{B}}, \kappa, \kappa^{\mathrm{E}}, \sigma\}\delta_{kl},$$
$$\sigma_{klmn} = \lambda_{\mathrm{e}}\delta_{kl}\delta_{mn} + \mu_{\mathrm{e}}(\delta_{km}\delta_{ln} + \delta_{kn}\delta_{lm}). \tag{8.2.38}$$

The constitutive equations are therefore given by

$$\Sigma = \sigma_0 - \rho_0\eta_0 T - \frac{\rho_0\gamma}{2T_0}T^2 - \beta_{\mathrm{e}}T \operatorname{tr}\tilde{\mathbf{e}} + \tfrac{1}{2}\lambda_{\mathrm{e}}(\operatorname{tr}\tilde{\mathbf{e}})^2 + \mu_{\mathrm{e}}\operatorname{tr}\tilde{\mathbf{e}}^2$$
$$- \tfrac{1}{2}\chi^{\mathrm{B}}\mathbf{B}\cdot\mathbf{B}, \tag{8.2.39}$$

$$\eta = \eta_0 + \frac{\gamma}{T_0}T + \frac{1}{\rho_0}\beta_{\mathrm{e}}\operatorname{tr}\tilde{\mathbf{e}}, \tag{8.2.40}$$

$${}_{\mathrm{E}}\mathbf{t} = (-\beta_{\mathrm{e}}T + \lambda_{\mathrm{e}}\tilde{e}_{rr})\mathbf{1} + 2\mu_{\mathrm{e}}\tilde{\mathbf{e}}, \tag{8.2.41}$$

$$\mathbf{M} = \chi^{\mathrm{B}}\mathbf{B} \qquad \text{or} \qquad \mathbf{B} = \mu\mathbf{H}, \tag{8.2.42}$$

$$\mathbf{q} = \kappa\nabla T + \kappa^{\mathrm{E}}\left(\mathbf{E} + \frac{1}{c}\dot{\mathbf{u}}\times\mathbf{B}\right), \tag{8.2.43}$$

$$\mathbf{J} = \sigma\left(\mathbf{E} + \frac{1}{c}\dot{\mathbf{u}}\times\mathbf{B}\right) + \sigma^{\theta}\nabla T, \tag{8.2.44}$$

where

$$\mu = (1 - \chi^{\mathrm{B}})^{-1}. \tag{8.2.45}$$

The nonequilibrium material moduli κ, κ^{E}, σ, and σ^{θ} are subject to the restrictions arising from the C–D inequality (8.2.5) (see (5.11.41))

$$\kappa \geq 0, \qquad \sigma \geq 0, \qquad 4\kappa\sigma T_0^{-1} - (\kappa^{\mathrm{E}}T_0^{-1} + \sigma^{\theta})^2 \geq 0. \tag{8.2.46}$$

If we impose the condition that the strain energy density $\varepsilon = \rho_0^{-1}\Sigma + T_0\eta$ be nonnegative for all $\tilde{\mathbf{e}}$ and $\mathbf{B}$ (the *material stability condition*), then we obtain the usual conditions on the elastic moduli λ_{e}, μ_{e}, and on χ^{B}:

$$3\lambda_{\mathrm{e}} + 2\mu_{\mathrm{e}} \geq 0, \qquad \mu_{\mathrm{e}} \geq 0, \qquad \chi^{\mathrm{B}} \leq 0. \tag{8.2.47}$$

Field Equations. We take the curl of (8.2.44) and use (8.2.7), (8.2.9), and $(8.2.42)_2$ to eliminate **E** and **J**. This gives

$$\nu_{\mathrm{H}}\nabla\times\nabla\times\mathbf{H} + \frac{\partial\mathbf{H}}{\partial t} - \nabla\times(\dot{\mathbf{u}}\times\mathbf{H}) = \mathbf{0},$$
$$\nu_{\mathrm{H}} \equiv c^2/\sigma\mu. \tag{8.2.48}$$

Next we carry (8.2.37) into (8.2.41) and the result into (8.2.2). Upon using $(8.2.42)_2$ and **J**, solved from (8.2.9), in (8.2.2) we obtain

$$(\lambda_{\mathrm{e}} + 2\mu_{\mathrm{e}})\nabla\nabla\cdot\mathbf{u} - \mu_{\mathrm{e}}\nabla\times\nabla\times\mathbf{u} - \beta_{\mathrm{e}}\nabla T + \mu_1(\nabla\times\mathbf{H})\times\mathbf{H} + \rho(\mathbf{f} - \ddot{\mathbf{u}}) = \mathbf{0},$$
$$\mu_1 \equiv \mu(2 - \mu). \tag{8.2.49}$$

To obtain the equation of heat conduction, we first eliminate $\mathbf{E} + \overset{-1}{c}\mathbf{v} \times \mathbf{B}$ between (8.2.43) and (8.2.44). We then substitute $\mathbf{q}$ so obtained, and (8.2.9) and (8.2.40) into (8.2.4). Upon linearization of the term $\rho\theta\dot{\eta}$, by introducing $\theta = T_0 + T$, $T_0 \gg |T| > 0$, we obtain

$$-\rho\gamma\dot{T} - \beta_e T_0 \nabla\cdot\dot{\mathbf{u}} + \kappa_e \nabla^2 T + (c^2/\sigma)(\nabla \times \mathbf{H})\cdot(\nabla \times \mathbf{H}) - (c\sigma^\theta/\sigma)(\nabla \times \mathbf{H})\cdot\nabla T + \rho h = 0, \quad (8.2.50)$$

$$\kappa_e \equiv \kappa - (\sigma^\theta \kappa^E/\sigma).$$

Equations (8.2.48)–(8.2.50) are the basic field equations of magneto-thermoelasticity for isotropic solids to determine $\mathbf{u}$, $\mathbf{H}$, and T. The fields $\mathbf{B}$, $\mathbf{E}$, and $\mathbf{J}$ follow from (8.2.42), (8.2.44), and (8.2.9).

Boundary Conditions. Neglecting electromagnetic momentum, radiation, and dragging effects, and making the discontinuity surface σ coincide with the boundary $\partial\mathscr{V}$, and from (8.2.11) and (8.2.16), we deduce the boundary conditions

$$\begin{aligned} [{}_E t_{kl} - \mu(\mu - 1)H_k H_l + {}_M t_{kl}^- - {}_M t_{kl}^+] n_k &= t_{(\mathbf{n})l} \quad \text{on } \partial\mathscr{V}_\mathbf{t}, \\ u_k = \bar{u}_k \quad \text{on } \partial\mathscr{V}_\mathbf{u} &= \partial\mathscr{V} - \partial\mathscr{V}_\mathbf{t}, \end{aligned} \quad (8.2.51)$$

$$\begin{aligned} \mathbf{n}\cdot[\mathbf{E}] &= 0, \qquad \mathbf{n} \times [\mathbf{E}] = \mathbf{0}, \\ \mathbf{n}\cdot[\mathbf{B}] &= 0, \qquad \mathbf{n} \times [\mathbf{H}] = \mathbf{0}, \quad \text{on } \partial\mathscr{V}, \end{aligned} \quad (8.2.52)$$

$$\begin{aligned} \mathbf{q}\cdot\mathbf{n} &= q_{(\mathbf{n})} \quad \text{on } \partial\mathscr{V}_\mathbf{q}, \\ T &= T_s \quad \text{on } \partial\mathscr{V}_T = \partial\mathscr{V} - \partial\mathscr{V}_\mathbf{q}, \end{aligned} \quad (8.2.53)$$

where $t_{(\mathbf{n})l}$, $\bar{u}_k$, $q_{(\mathbf{n})}$, and T_s are prescribed, and the superscripts $(-)$ and $(+)$ denote, respectively, the values on the interior and exterior sides of $\partial\mathscr{V}$.

According to the expression of ${}_M t_{kl}$ given by (8.2.29) and (8.2.42) we have

$${}_M t_{kl} = \mu H_k H_l + E_k E_l - \tfrac{1}{2}[E^2 + \mu\mathbf{H}\cdot(\mathbf{H} - \mathbf{M})]\delta_{kl}, \quad (8.2.54)$$

so that

$$\begin{aligned} {}_M t_{kl}^- &= \mu H_k H_l + E_k E_l - \tfrac{1}{2}[E^2 + \mu(2 - \mu)H^2]\delta_{kl}, \\ {}_M t_{kl}^+ &= H_k H_l + E_k E_l - \tfrac{1}{2}(E^2 + H^2)\delta_{kl}. \end{aligned} \quad (8.2.55)$$

Initial Conditions. These generally consist of the Cauchy data

$$\mathbf{u}(\mathbf{x}, 0) = \mathbf{u}^0(\mathbf{x}), \qquad \dot{\mathbf{u}}(\mathbf{x}, 0) = \mathbf{v}^0(\mathbf{x}), \quad (8.2.56)$$

$$\mathbf{E}(\mathbf{x}, 0) = \mathbf{E}^0(\mathbf{x}), \qquad \mathbf{H}(\mathbf{x}, 0) = \mathbf{H}^0(\mathbf{x}), \quad (8.2.57)$$

$$T(\mathbf{x}, 0) = T^0(\mathbf{x}). \quad (8.2.58)$$

Formulation In Terms of the Stress Tensor. In some cases it is useful to express basic equations in terms of the stress tensor. This is achieved by eliminating $\mathbf{u}$ from equations (8.2.48)–(8.2.50) by means of the equations of compatibility (similar to the Beltrami–Mitchell equations). Here we only give the simple static case for which (8.2.48) and (8.2.50) remain valid with $\dot{\mathbf{u}} = \mathbf{0}$. The equations of equilibrium (8.2.2) take the form

$$({}_{\mathrm{E}}t_{kl} + \hat{t}_{kl})_{,k} + \rho f_l = 0, \tag{8.2.59}$$

where

$$\begin{aligned} \hat{t}_{kl} &\equiv \mu_1(H_k H_l - \tfrac{1}{2}H^2\delta_{kl}), \\ \mu_1 &= \mu(2-\mu). \end{aligned} \tag{8.2.60}$$

Equation (8.2.59) is supplemented by the compatibility conditions

$$\tilde{e}_{kl,mn} + \tilde{e}_{mn,kl} - \tilde{e}_{km,ln} - \tilde{e}_{ln,km} = 0, \tag{8.2.61}$$

and the strain–stress relations

$$\begin{aligned} \tilde{e}_{kl} &= \frac{1}{2\mu}\left({}_{\mathrm{E}}t_{kl} - \frac{\nu}{1+\nu}\,{}_{\mathrm{E}}t_{rr}\delta_{kl}\right), \qquad \text{(3-dimensional)}, \\ \tilde{e}_{kl} &= \frac{1}{2\mu}({}_{\mathrm{E}}t_{kl} - \nu\,{}_{\mathrm{E}}t_{rr}\delta_{kl}), \qquad \text{(2-dimensional)}, \end{aligned} \tag{8.2.62}$$

where $\nu = \lambda_e/2(\lambda_e + \mu_e)$, is Poisson's ratio.

In this case the boundary conditions on the tractions read

$$[{}_{\mathrm{E}}t_{kl} + \hat{t}_{kl}]n_k = 0 \quad \text{on } \partial\mathscr{V}_t. \tag{8.2.63}$$

Weakly Magnetizable Solids. These are solids for which $|\chi^{\mathrm{B}}| \ll 1$, so that μ is not very different from unity. In this case we have

$$\mu_1 \simeq \mu. \tag{8.2.64}$$

8.3. Static Magnetoelastic Fields

When the dynamical effects and heat conduction are negligible the field equations of magnetoelasticity are simplified a great deal. For the homogeneous and isotropic materials, with $\mathbf{f} = \mathbf{0}$, (8.2.8), (8.2.48), and (8.2.49) reduce to

$$\nabla \cdot \mathbf{H} = 0, \tag{8.3.1}$$

$$\nabla \times \nabla \times \mathbf{H} = \mathbf{0}, \tag{8.3.2}$$

$$(\lambda_e + 2\mu_e)\nabla\nabla\cdot\mathbf{u} - \mu_e\nabla \times \nabla \times \mathbf{u} + \mu_1(\nabla \times \mathbf{H}) \times \mathbf{H} = \mathbf{0}. \tag{8.3.3}$$

The magnetic field is thus uncoupled from $\mathbf{u}$ and can be determined by solving (8.3.1) and (8.3.2). Once this is done the last term in (8.3.3) will be known, so that the displacement field can be solved from (8.3.3) by treating

$$\mathbf{f}^{\mathrm{H}} = \mu_1(\nabla \times \mathbf{H}) \times \mathbf{H} \tag{8.3.4}$$

as a body-force distribution. To this end, the Lamé potentials (based on the Stokes–Helmholtz resolution) may be used conveniently. Accordingly, every sufficiently smooth vector field $\mathbf{f}(\mathbf{x})$ may be decomposed as

$$\mathbf{f} = \nabla f + \nabla \times \mathbf{F}, \qquad \nabla \cdot \mathbf{F} = 0 \tag{8.3.5}$$

If we introduce a vector field $\mathbf{U}$ by

$$f = \nabla \cdot \mathbf{U}, \qquad \mathbf{F} = -\nabla \times \mathbf{U}, \tag{8.3.6}$$

eq. (8.3.5) becomes

$$\nabla^2 \mathbf{U} = \mathbf{f}, \tag{8.3.7}$$

where we used the identity for the Laplacian

$$\nabla^2 \mathbf{U} = \nabla\nabla \cdot \mathbf{U} - \nabla \times \nabla \times \mathbf{U}. \tag{8.3.8}$$

Equation (8.3.7) has the well-known solution

$$\mathbf{U}(\mathbf{x}) = -\frac{1}{4\pi}\int_R r^{-1}\mathbf{f}(\mathbf{x}')\,dv(\mathbf{x}'), \qquad r = |\mathbf{x} - \mathbf{x}'|. \tag{8.3.9}$$

By carrying this into (8.3.6), we obtain

$$\begin{aligned} f(\mathbf{x}) &= -\frac{1}{4\pi}\int_R r^{-1}\nabla' \cdot \mathbf{f}(\mathbf{x}')\,dv(\mathbf{x}'), \\ \mathbf{F}(\mathbf{x}) &= \frac{1}{4\pi}\int_R r^{-1}\nabla' \times \mathbf{f}(\mathbf{x}')\,dv(\mathbf{x}'), \end{aligned} \tag{8.3.10}$$

where R is a region of space outside which $\mathbf{f}$ vanishes, and ∇' denotes the gradient operator acting at the point $\mathbf{x}'$. Given $\mathbf{f} = \mathbf{f}^{\mathrm{H}}$ by (8.3.4), the fields $f(\mathbf{x})$ and $F(\mathbf{x})$ are determined by (8.3.10).

We now apply the Stokes–Helmholtz resolution to the displacement field and write

$$\mathbf{u} = \nabla\phi + \nabla \times \boldsymbol{\psi}, \qquad \nabla \cdot \boldsymbol{\psi} = 0, \tag{8.3.11}$$

where ϕ and $\boldsymbol{\psi}$ are, respectively, scalar- and vector-valued functions of $\mathbf{x}$. Introducing (8.3.11) into (8.3.3), we will have (8.3.3) satisfied identically if

$$(\lambda_e + 2\mu_e)\nabla^2\phi + f = 0, \qquad \mu_e\nabla^2\boldsymbol{\psi} + \mathbf{F} = \mathbf{0}. \tag{8.3.12}$$

Since f and $\mathbf{F}$ are now known through (8.3.10), Poisson's equations (8.3.12) will have a solution similar to (8.3.9). Therefore, the *particular integral* of (8.3.3) can be expressed as

$$\begin{aligned} \mathbf{u}(\mathbf{x}) = \frac{\lambda_e + \mu_e}{8\pi\mu_e(\lambda_e + 2\mu_e)}\int_{\mathscr{V}} \mu_1 \Bigg[&\frac{\lambda_e + 3\mu_e}{\lambda_e + \mu_e}\frac{\mathbf{f}^{\mathrm{H}}(\mathbf{x}')}{r} \\ &+ \frac{\mathbf{f}'^{\mathrm{H}}(\mathbf{x}')\cdot(\mathbf{x} - \mathbf{x}')}{r^3}(\mathbf{x} - \mathbf{x}')\Bigg]\,dv(\mathbf{x}'). \end{aligned} \tag{8.3.13}$$

To satisfy the boundary conditions to this particular solution we must add the general solution of Navier's equation.

According to (8.3.8), (8.3.1), and (8.3.2) each rectangular component of the magnetic field **H** satisfies Laplace's equation

$$\nabla^2 \mathbf{H} = \mathbf{0}, \tag{8.3.14}$$

whose solutions have been studied extensively under various boundary conditions in potential theory (see Chapter 6).

Hard Ferromagnet

A common situation occurs when the magnetization is independent of applied fields at moderate field strengths. Such is the case for hard ferromagnets for which the magnetization $\mathbf{M}(\mathbf{x})$ is fixed. In the case of $\mathbf{J} = 0$, we may introduce a scalar potential Φ by

$$\mathbf{H} = -\nabla\Phi. \tag{8.3.15}$$

Then (8.2.9) is satisfied identically and (8.2.8) gives

$$\nabla^2\Phi = -\nabla\cdot\mathbf{M}. \tag{8.3.16}$$

The solution of this equation is given by

$$\Phi(\mathbf{x}) = -\frac{1}{4\pi}\int_{\mathscr{V}} \frac{\nabla'\cdot\mathbf{M}(\mathbf{x}')}{|\mathbf{x}-\mathbf{x}'|}\,dv(\mathbf{x}') + \frac{1}{4\pi}\int_{\partial\mathscr{V}} \frac{\mathbf{n}'\cdot\mathbf{M}(\mathbf{x}')}{|\mathbf{x}-\mathbf{x}'|}\,da(\mathbf{x}'). \tag{8.3.17}$$

Alternatively, it is possible to employ a vector potential

$$\mathbf{B} = \nabla\times\mathbf{A}, \qquad \nabla\cdot\mathbf{A} = 0, \tag{8.3.18}$$

which satisfies (8.2.8), and (8.2.9) gives

$$\nabla^2\mathbf{A} = -\nabla\times\mathbf{M}, \tag{8.3.19}$$

whose integral is

$$\mathbf{A}(\mathbf{x}) = \frac{1}{4\pi}\int_{\mathscr{V}} \frac{\nabla'\times\mathbf{M}(\mathbf{x}')}{\mathbf{x}-\mathbf{x}'|}\,dv(\mathbf{x}') + \frac{1}{4\pi}\int_{\partial\mathscr{V}} \frac{\mathbf{M}(\mathbf{x}')\times\mathbf{n}'}{|\mathbf{x}-\mathbf{x}'|}\,da(\mathbf{x}'). \tag{8.3.20}$$

Either of the results (8.3.17) or (8.3.20) may be used to determine the magnetic field in hard magnets.

8.4. Uniformly Magnetized Sphere in a Nonpermeable Medium

Here we illustrate the method of solution, discussed in the previous section, for the case of a spherical hard magnet, with a uniform permanent magnetization $\mathbf{M}_0$ parallel to the z-axis, imbedded in a nonpermeable material of infinite extent. We consider the surface magnetic charge density given by

$$\sigma = \mathbf{n}\cdot\mathbf{M} = M_0\cos\theta, \tag{8.4.1}$$

where θ is the polar angle in the spherical polar coordinates (r, θ, ϕ) (Figure 8.4.1). According to (8.3.17), only the surface integral survives and

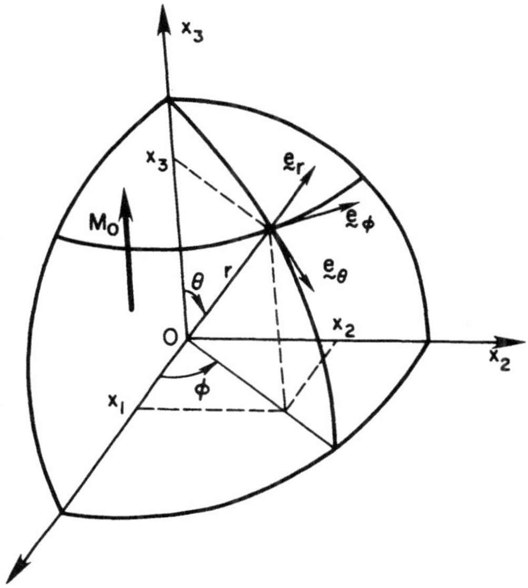

Figure 8.4.1. Uniformly magnetized sphere (after Jackson [1975, p. 196]).

we have

$$\Phi(r, \theta) = \frac{1}{4\pi} M_0 a^2 \int_0^{\pi} d\theta' \int_0^{2\pi} \frac{\cos\theta' \sin\theta'}{|\mathbf{x}' - \mathbf{x}|} d\phi', \tag{8.4.2}$$

where

$$|\mathbf{x}' - \mathbf{x}| = r'^2 + r^2 - 2rr'[\cos(\phi' - \phi) \sin\theta' \sin\theta + \cos\theta' \cos\theta]. \tag{8.4.3}$$

By expanding in the inverse of distance, the integrals of (8.4.2) can be evaluated to obtain

$$\begin{aligned} \Phi(r, \theta) &= \tfrac{1}{3} M_0 r \cos\theta, \qquad & r < a, \\ \Phi(r, \theta) &= \tfrac{1}{3} M_0 (a^3/r^2) \cos\theta, \qquad & r > a, \end{aligned} \tag{8.4.4}$$

where $r = a$ is the radius of the sphere. Equation $(8.4.4)_2$ is the potential of a dipole with dipole moment

$$\mathbf{m} = \tfrac{1}{3} a^3 \mathbf{M}. \tag{8.4.5}$$

The magnetic field $\mathbf{H}$ is given by (eq. (8.3.15))

$$\mathbf{H} = -\nabla\Phi = -\left(\frac{\partial\Phi}{\partial r}\mathbf{e}_r + \frac{1}{r}\frac{\partial\Phi}{\partial\theta}\mathbf{e}_\theta\right),$$

which has symmetry about the polar axis

$$\begin{aligned} \mathbf{H} &= \tfrac{1}{3} M_0 (\cos\theta \mathbf{e}_r - \sin\theta \mathbf{e}_\theta), \qquad & r < a, \\ \mathbf{H} &= \tfrac{1}{3} M_0 (a/r)^3 (2\cos\theta \mathbf{e}_r + \sin\theta \mathbf{e}_\theta), \qquad & r > a. \end{aligned} \tag{8.4.6}$$

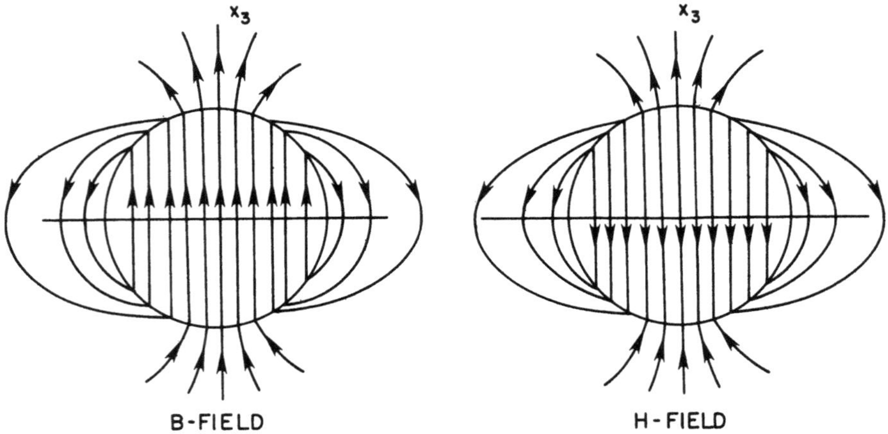

Figure 8.4.2. Uniformly magnetized sphere. Reprinted with permission from J.D. Jackson, *Classical Electrodynamics* (Second Edition), © 1975, John Wiley and Sons.

The magnetic induction **B** follows from

$$\mathbf{B} = \mathbf{H} + \mathbf{M}, \tag{8.4.7}$$

$$\begin{aligned} \mathbf{B} &= \mathbf{H} + M_0\mathbf{i}_3 = \tfrac{2}{3}M_0(\cos\theta\mathbf{e}_r - \sin\theta\mathbf{e}_\theta), \qquad r < a, \\ \mathbf{B} &= \mathbf{H} = \tfrac{1}{3}M_0(a/r)^3(2\cos\theta\mathbf{e}_r + \sin\theta\mathbf{r}_\theta), \qquad r > a. \end{aligned} \tag{8.4.8}$$

The lines of constants **B** and **H** are shown in Figure 8.4.2. We notice that the directions of the **B**-lines are reversed in the core magnet. The lines of **B** are closed curves, but the lines of **H** originate on the surface of the sphere.

To determine the displacement field in the surrounding medium we employ Navier's equation (8.3.3), in which the magnetic force (the last term) vanishes. Thus, the deformation is purely due to the surface tractions. We now calculate the surface tractions due to the magnetic field.

The magnetic stress tensor is given by

$$_{\mathrm{M}}\mathbf{t} = -\tfrac{1}{2}\mathbf{H}\cdot\mathbf{H}\mathbf{1} + \mathbf{H}\otimes\mathbf{H}. \tag{8.4.9}$$

Hence the surface tractions are

$$\begin{aligned} {}_{\mathrm{M}}t_{rr} &= \tfrac{1}{18}M_0^2\cos 2\theta, \\ {}_{\mathrm{M}}t_{r\theta} &= -\tfrac{1}{18}M_0^2\sin 2\theta, \qquad {}_{\mathrm{M}}t_{rz} = 0, \end{aligned} \qquad r < a, \tag{8.4.10}$$

$$\begin{aligned} {}_{\mathrm{M}}t_{rr} &= \tfrac{1}{36}M_0^2(a/r)^6(3 + 5\cos 2\theta), \\ {}_{\mathrm{M}}t_{r\theta} &= \tfrac{1}{9}M_0^2(a/r)^6\sin 2\theta, \qquad {}_{\mathrm{M}}t_{rz} = 0, \end{aligned} \qquad r > a, \tag{8.4.11}$$

The jump conditions $(8.2.47)_1$, or surface tractions, are given by

$$[t_{kl} + {}_{\mathrm{M}}t_{kl}]n^k = 0$$

which can be used to obtain the boundary conditions.

If we suppose that the core magnet is loose so that the cavity surface is free to deform, then the cavity stress at the surface of the cavity is given by

$$\begin{aligned} t_{rr} &= {}_{\mathrm{M}}t_{rr}^{+} - {}_{\mathrm{M}}t_{rr}^{-} = -\frac{M_0^2}{12}(1 + \cos 2\theta), \\ t_{r\theta} &= {}_{\mathrm{M}}t_{r\theta}^{+} - {}_{\mathrm{M}}t_{r\theta}^{-} = -\frac{M_0^2}{6}\sin 2\theta. \end{aligned} \tag{8.4.12}$$

The problem is now reduced to solving Navier's equations for $\mathbf{u}$

$$(\lambda_e + 2\mu_e)\nabla\nabla\cdot\mathbf{u} - \mu_e\nabla \times \nabla \times \mathbf{u} = \mathbf{0}, \tag{8.4.13}$$

subject to (8.4.10) at the cavity surface and vanishing $\mathbf{u}$ at $r = \infty$. The cavity stress tensor is then calculated by

$$\mathbf{t} = {}_{\mathrm{E}}\mathbf{t} - \mathbf{M} \otimes \mathbf{B}, \tag{8.4.14}$$

where ${}_{\mathrm{E}}\mathbf{t}$ is given by (8.2.41). Alternatively, in terms of ${}_{\mathrm{E}}\mathbf{t}$, the boundary conditions read

$$\left.\begin{aligned} {}_{\mathrm{E}}t_{rr} &= \tfrac{1}{4}M_0^2(1 + \cos 2\theta), \\ {}_{\mathrm{E}}t_{r\theta} &= 0, \qquad {}_{\mathrm{E}}t_{r\phi} = 0 \end{aligned}\right\} \quad \text{at } r = a. \tag{8.4.15}$$

Thus, the problem is one of classical elasticity for an elastic solid, containing a spherical cavity whose surface is subjected to the prescribed tractions (8.4.15), and at $r = \infty$, the displacement field vanishes.

8.5. Two-Dimensional Magnetoelasticity

For two-dimensional problems the vectors $\mathbf{u}$ and $\mathbf{H}$ have the components $(u, v, 0)$ and $(H_x, H_y, 0)$. For boundary-value problems in which surface tractions are prescribed, it is more convenient to use the stress-function technique. From (8.2.59), and (8.2.60), with $f_l = 0$, we have

$$\begin{aligned} &\frac{\partial}{\partial x_k}[{}_{\mathrm{E}}t_{kl} + \mu_1(H_kH_l - \tfrac{1}{2}H^2\delta_{kl})] = 0, \\ &\mu_1 = \mu(2 - \mu). \end{aligned} \tag{8.5.1}$$

We introduce the stress function Φ by

$$\begin{aligned} {}_{\mathrm{E}}t_{xx} &= \Phi_{,yy} - \tfrac{1}{2}\mu_1(H_x^2 - H_y^2), \\ {}_{\mathrm{E}}t_{yy} &= \Phi_{,xx} + \tfrac{1}{2}\mu_1(H_x^2 - H_y^2), \\ {}_{\mathrm{E}}t_{xy} &= -\Phi_{,xy} - \mu_1 H_x H_y. \end{aligned} \tag{8.5.2}$$

The equations of equilibrium (8.5.1) are satisfied identically. The compatibility conditions (8.2.61) for the plane strain components may be expressed as

$$\tilde{e}_{xx,yy} + \tilde{e}_{yy,xx} - 2\tilde{e}_{xy,xy} = 0. \tag{8.5.3}$$

Substituting $\tilde{e}_{kl}$ from $(8.2.62)_2$ into (8.5.3) we obtain

$$[(1-\nu)\,{}_{\mathrm{E}}t_{xx} - \nu\,{}_{\mathrm{E}}t_{yy}]_{,yy} + [(1-\nu)\,{}_{\mathrm{E}}t_{yy} - \nu\,{}_{\mathrm{E}}t_{xx}]_{,xx} - 2\,{}_{\mathrm{E}}t_{xy,xy} = 0. \quad (8.5.4)$$

Upon carrying (8.5.2) into (8.5.4) we obtain

$$\nabla^4\Phi + \frac{\mu_1}{2(1-\nu)}\left[\left(\frac{\partial^2}{\partial x^2} - \frac{\partial^2}{\partial y^2}\right)(H_x^2 - H_y^2) + 4\frac{\partial^2}{\partial x\,\partial y}(H_x H_y)\right] = 0. \quad (8.5.5)$$

The magnetic field is determined by (8.3.2), which in two dimensions reads

$$\nabla^2 H_x = 0, \qquad \nabla^2 H_y = 0. \quad (8.5.6)$$

The use of the complex variables $z = x + iy$ and $\bar{z} = x - iy$ proves to be useful.

Transformation of (8.5.6) gives

$$\frac{\partial^2 H_x}{\partial z\,\partial \bar{z}} = 0, \qquad \frac{\partial^2 H_y}{\partial z\,\partial \bar{z}} = 0, \quad (8.5.7)$$

whose solution, subject to $\nabla \cdot \mathbf{H} = 0$, is

$$2H_x = f_1(z) + \overline{f_1(z)}, \qquad 2H_y = i[f_1(z) - \overline{f_1(z)}], \quad (8.5.8)$$

where $f_1(z)$ is an analytic function and $\overline{f_1(z)}$ is its complex conjugate. Using (8.5.8) and transforming (8.5.5) we obtain[2]

$$\frac{\partial^4 \Phi}{\partial z^2\,\partial \bar{z}^2} = 0. \quad (8.5.9)$$

The general solution of (8.5.9) is

$$\Phi = \tfrac{1}{2}[f_2(z) + \overline{f_2(z)} + \bar{z} f_3(z) + z\overline{f_3(z)}], \quad (8.5.10)$$

where $f_2(z)$ and $f_3(z)$ are analytic functions which, together with $f_1(z)$, are to be determined by use of the boundary conditions.

Employing (8.5.1) and (8.5.10) in (8.5.2) we obtain

$$\begin{aligned} {}_{\mathrm{E}}t_{xx} + {}_{\mathrm{E}}t_{yy} &= 2\varphi(z) + 2\overline{\varphi(z)}, \\ {}_{\mathrm{E}}t_{yy} - {}_{\mathrm{E}}t_{xx} + 2i\,{}_{\mathrm{E}}t_{xy} &= 2\bar{z}\varphi'(z) + 2\psi(z) + \mu_1 f_1^2(z), \end{aligned} \quad (8.5.11)$$

where we set

$$f_3'(z) = \varphi(z), \qquad f_2''(z) = \psi(z). \quad (8.5.12)$$

In polar coordinates (r, θ) the stress components are given by

$$\begin{aligned} {}_{\mathrm{E}}t_{rr} + {}_{\mathrm{E}}t_{\theta\theta} &= {}_{\mathrm{E}}t_{xx} + {}_{\mathrm{E}}t_{yy}, \\ {}_{\mathrm{E}}t_{\theta\theta} - {}_{\mathrm{E}}t_{rr} + 2i\,{}_{\mathrm{E}}t_{r\theta} &= ({}_{\mathrm{E}}t_{yy} - {}_{\mathrm{E}}t_{xx} + 2i\,{}_{\mathrm{E}}t_{xy})e^{2i\theta}, \end{aligned} \quad (8.5.13)$$

---

[2] In Paria's work [1967], his equation corresponding to (8.5.9) erroneously contains a term due to the magnetic field. The present section and the following two sections are new and different from Paria's account.

from which we calculate

$$_{\rm E}t_{rr} - i\,_{\rm E}t_{r\theta} = \varphi + \bar{\varphi} - (\bar{z}\varphi' + \psi + \tfrac{1}{2}\mu_1 f_1^2)e^{2i\theta}. \tag{8.5.14}$$

The boundary conditions for the surface tractions require the satisfaction of the jump conditions

$$[_{\rm E}t_{kl} + \mu_1(H_k H_l - \tfrac{1}{2}H^2\delta_{kl})]n_k = 0 \quad \text{on } \partial\mathscr{V}. \tag{8.5.15}$$

Note that outside the body $\mu_1 = 1$. These are expressible in terms of the potentials by use of (8.5.8) and (8.5.11). For straight boundaries (8.5.11) is useful, and for circular boundaries (8.5.14) is useful.

It is possible to use the method of conformal mapping to treat boundary-value problems concerning other types of curvilinear boundaries and mixed boundary-value problems. This extends the complex-function technique of Muskhelishvili and Kolossov. However, the development of this method is too lengthy to be presented here. Interested readers may consult the books by Muskhelishvili [1953], Green and Zerna [1954], and others, for an exposure to this powerful method. In the next section we present solutions of two two-dimensional problems to illustrate the use of the present development.

8.6. Stress Concentration at a Circular Hole

An infinite plate with a circular hole is under the influence of a primary magnetic field of constant magnitude at infinity, and the circular boundary of the hole is subjected to a prescribed magnetic field (Figure 8.6.1). It is required that we determine the stress field. This problem was treated by Paria [1967][3] for weakly magnetized bodies. The magnetic field strength at the surface $r = a$ is given by

$$H_x = H_0 + h_1 \cos\theta, \qquad H_y = h_1 \sin\theta, \tag{8.6.1}$$

[3] We do not agree with Paria's treatment.

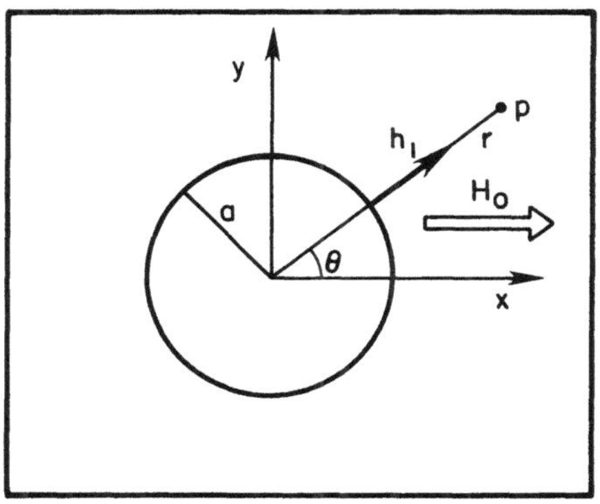

Figure 8.6.1. Circular cavity under bias magnetic field.

where H_0 and h_1 are constant. Boundary conditions on **H** are satisfied by the complex potential

$$f_1(z) = H_0\left(1 + c\frac{a}{z}\right), \qquad c \equiv h_1/H_0. \tag{8.6.2}$$

Substituting this into (8.5.14) we obtain

$${}_{\mathrm{E}}t_{rr} - i\,{}_{\mathrm{E}}t_{r\theta} = \varphi + \bar{\varphi} - z\varphi' - \frac{z}{\bar{z}}\psi - \tfrac{1}{2}\mu_1 H_0^2\left(\frac{z}{\bar{z}} + 2c\frac{a}{\bar{z}} + \frac{c^2a^2}{z\bar{z}}\right), \tag{8.6.3}$$

$${}_{\mathrm{E}}t_{rr} + {}_{\mathrm{E}}t_{\theta\theta} = 2\varphi + 2\bar{\varphi}. \tag{8.6.4}$$

Boundary conditions on the circular hole require that

$$[{}_{\mathrm{E}}t_{kl} + \mu_1(H_kH_l - \tfrac{1}{2}H^2\delta_{kl})]n_k = 0, \qquad r = a, \tag{8.6.5}$$

where $\mu_1 = 2\mu - \mu^2$ inside the body and $\mu_1 = 1$ outside it. It follows that

$$[{}_{\mathrm{E}}t_{rr} - i\,{}_{\mathrm{E}}t_{r\theta} + \tfrac{1}{2}\mu_1(H_r^2 - H_\theta^2 - 2iH_rH_\theta)] = 0, \qquad r = a. \tag{8.6.6}$$

The continuity conditions for the magnetic field at $r = a$ require that

$$\mu H_r^- = H_r^+, \qquad H_\theta^- = H_\theta^+, \qquad r = a. \tag{8.6.7}$$

Using these and the fact that ${}_{\mathrm{E}}t_{rr}^+ = {}_{\mathrm{E}}t_{r\theta}^+ = 0$ at $r = a$, from (8.6.6), we have

$${}_{\mathrm{E}}t_{rr} - i\,{}_{\mathrm{E}}t_{r\theta} + \tfrac{1}{2}\mu_1(H_r^2 - H_\theta^2 - 2iH_rH_\theta) = \tfrac{1}{2}(\mu^2H_r^2 - H_\theta^2 - 2i\mu H_rH_\theta), \tag{8.6.8}$$

where we dropped the superscripts $(-)$ from the quantities, e.g., we wrote $H_r^- = H_r$, ${}_{\mathrm{E}}t_{rr}^- = {}_{\mathrm{E}}t_{rr}$.

The radial and tangential components of the magnetic field on the boundary are given by

$$\begin{aligned} H_r &= H_0\cos\theta + h_1 = \frac{H_0}{2}(\zeta + \zeta^{-1} + 2c), \\ H_\theta &= -H_0\sin\theta = i\frac{H_0}{2}(\zeta - \zeta^{-1}), \end{aligned} \tag{8.6.9}$$

where

$$c = h_1/H_0, \qquad \zeta = e^{i\theta} = z/a. \tag{8.6.10}$$

Substituting (8.6.9) into (8.6.8) and using (8.6.3) we obtain

$$\begin{aligned} \varphi(\zeta) + \bar{\varphi}(\zeta) - \zeta\varphi'(\zeta) - \zeta^2\psi(\zeta) = \frac{H_0^2}{8}[&(\mu+1)^2\zeta^2 + (\mu-1)^2\zeta^{-2} \\ &+ 4\mu(\mu+1)c\zeta + 4\mu(\mu-1)c\zeta^{-1} \\ &+ 2\mu(1+2c^2) - 2]. \end{aligned} \tag{8.6.11}$$

To determine $\varphi(\zeta)$ and $\psi(\zeta)$, we substitute polynomials in ζ^{-1} for $\varphi(\zeta)$ and $\psi(\zeta)$ and determine their coefficients

$$\varphi(z) = \frac{H_0^2}{4}\left[\mu(\mu-1)c\frac{a}{z} + \tfrac{1}{2}(\mu+1)^2\frac{a^2}{z^2}\right],$$

$$\psi(z) = \frac{H_0^2}{4}\left[-(\mu^2 + 3\mu)c\frac{a}{z} - (\mu^2 + 2c^2\mu^2 - 1)\frac{a^2}{z^2} - (\mu^2 + 4\mu + 1)\frac{a^4}{z^4}\right]. \tag{8.6.12}$$

The stress field $_{\mathrm{E}}\mathbf{t}$ is determined by substituting these into (8.6.3) and (8.6.4)

$$\begin{aligned}
{}_{\mathrm{E}}t_{rr} &= \frac{H_0^2}{4}\bigg[8\mu(\mu - 1)c\frac{a}{r}\cos\theta + 2(\mu + 1)^2\frac{a^2}{r^2}\cos 2\theta \\
&\quad - (\mu^2 + 4\mu + 1)\frac{a^4}{r^4}\cos 2\theta + 2\mu(\mu - 2)\cos 2\theta \\
&\quad + (\mu - 1)(\mu + 1 + 4\mu c^2)\frac{a^2}{r^2}\bigg], \\
{}_{\mathrm{E}}t_{\theta\theta} &= \frac{H_0^2}{4}\bigg[6\mu(\mu - 1)c\frac{a}{r}\cos\theta + (\mu^2 + 4\mu + 1)\frac{a}{r^4}\cos 2\theta \\
&\quad + 2\mu(2 - \mu)\cos 2\theta + (1 - \mu)(\mu + 1 + 4\mu c^2)\frac{a^2}{r^2}\bigg], \\
{}_{\mathrm{E}}t_{r\theta} &= \frac{H_0^2}{4}\bigg[4\mu(1 - \mu)c\frac{a}{r}\sin\theta + (\mu + 1)^2\frac{a^2}{r^2}\sin 2\theta \\
&\quad - (\mu^2 + 4\mu + 1)\frac{a^4}{r^4}\sin 2\theta + 2\mu(2 - \mu)\sin 2\theta\bigg].
\end{aligned} \tag{8.6.13}$$

The total magnetoelastic stress field is given by

$$\begin{aligned}
T_{kl} &= {}_{\mathrm{E}}t_{kl} + \hat{t}_{kl} \\
\hat{t}_{rr} &= -\hat{t}_{\theta\theta} = \frac{\mu_1}{2}(H_r^2 - H_\theta^2), \qquad \hat{t}_{r\theta} = \mu_1 H_r H_\theta.
\end{aligned} \tag{8.6.14}$$

Consequently,

$$\begin{aligned}
T_{rr}(r, \theta) &= \frac{H_0^2}{4}\bigg[4\mu^2 c\frac{a}{r}\cos\theta + 2(\mu + 1)^2\frac{a^2}{r^2}\cos 2\theta \\
&\quad - (\mu^2 + 4\mu + 1)\frac{a^4}{r^4}\cos 2\theta + (\mu^2 - 1 + 2\mu^2 c^2)\frac{a^2}{r^2}\bigg], \\
T_{\theta\theta}(r, \theta) &= \frac{H_0^2}{4}\bigg[(10\mu^2 - 14\mu)c\frac{a}{r}\cos\theta + (\mu^2 + 4\mu + 1)\frac{a^4}{r^4}\cos 2\theta \\
&\quad + (1 - \mu^2 - 2\mu^2 c^2)\frac{a^2}{r^2}\bigg], \\
T_{r\theta}(r, \theta) &= \frac{H_0^2}{4}\bigg[-4\mu c\frac{a}{r}\sin\theta + (\mu + 1)^2\frac{a^2}{r^2}\sin 2\theta - (\mu^2 + 4\mu + 1)\frac{a^4}{r^4}\sin 2\theta\bigg].
\end{aligned} \tag{8.6.15}$$

The total hoop stress at the periphery of the circle is maximum for $\theta = 0$,

$\pi/2$, and π. We have

$$T_{\theta\theta}(a, 0) = \frac{H_0^2}{2}[(5\mu^2 - 7\mu)c - \mu^2c^2 + 2\mu + 1],$$

$$T_{\theta\theta}(a, \pi/2) = -\frac{\mu}{2}H_0^2(\mu + 2 + \mu c^2), \tag{8.6.16}$$

$$T_{\theta\theta}(a, \pi) = \frac{H_0^2}{2}[(7\mu - 5\mu^2)c - \mu^2c^2 + 2\mu + 1],$$

$T_{\theta\theta}(a, \theta)$ is stationary at $\theta = 0$ and

$$\theta = \theta_c = \frac{\mu(7 - 5\mu)c}{2(\mu^2 + 4\mu + 1)}. \tag{8.6.17}$$

For weakly magnetized bodies and $|c| \ll 1$, this last expression gives $\theta_c \simeq 0$. Thus, the maximum stress is at $\theta = 0$. Note that the hoop stress at $r = a$, $\theta = \pi/2$, is compressive. Consequently, by means of a magnetic field so established, it is possible to reduce the elastic stress concentration at $r = a$, $\theta = \pi/2$. As is well known, under a constant tension T_0 at $y = 0$, $x = \pm\infty$, the purely elastic stress (without magnetic field) is given by (see Timoshenko and Goddier [1955, Sect. 32]).

$$_{\mathrm{E}}t_{rr} = \tfrac{1}{2}T_0\left(1 - \frac{a^2}{r^2}\right) + \tfrac{1}{2}T_0\left(1 - \frac{4a^2}{r^2} + \frac{3a^4}{r^4}\right)\cos 2\theta,$$

$$_{\mathrm{E}}t_{\theta\theta} = \tfrac{1}{2}T_0\left(1 + \frac{a^2}{r^2}\right) - \tfrac{1}{2}T_0\left(1 + \frac{3a^4}{r^4}\right)\cos 2\theta, \tag{8.6.18}$$

$$_{\mathrm{E}}t_{r\theta} = -\tfrac{1}{2}\left(1 + \frac{2a^2}{r^2} - \frac{3a^4}{r^4}\right)\sin 2\theta.$$

The hoop stress at $r = a$, $\theta = \pi/2$, is maximum

$$_{\mathrm{E}}t_{\theta\theta} = 3T_0. \tag{8.6.19}$$

Thus we see that the total hoop stress, when the magnetic field is present, is

$$T_{\theta\theta}(a, \pi/2) + 3T_0 = 3T_0 - \frac{\mu H_0^2}{2}(2 + \mu + \mu c^2). \tag{8.6.20}$$

Consequently, by means of the applied magnetic field, we can reduce the stress concentration. Note, however, that the stress field at $r = a$ is increased by the application of the magnetic field. This is permissible since the purely elastic stress field here is small $(-T_0)$.

Magnetic Field h_1 Tangential to the Cavity Surface

If the magnetic field $\mathbf{h}_1$ is taken tangential to the cavity surface, but $\mathbf{H}_0$ is in the x-direction (Figure 8.6.2), we will have

$$\begin{aligned} H_x &= H_0(1 - c\sin\theta), & H_y &= H_0 c\cos\theta \\ H_r &= H_0\cos\theta, & H_\theta &= H_0(-\sin\theta + c). \end{aligned} \tag{8.6.21}$$

Figure 8.6.2. Magnetic field tangential to the circular cavity.

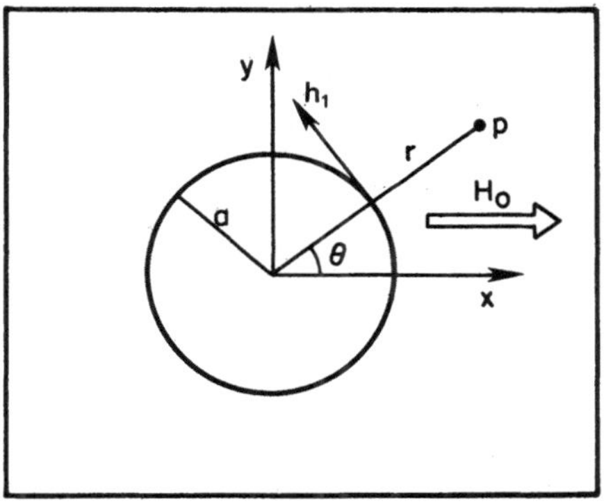

The magnetic potential has the form

$$f_1(z) = H_0\left(1 - ic\frac{a}{z}\right). \tag{8.6.22}$$

The magnetic field $\mathbf{h}_1$ corresponds to a field created by a steady current winding around the circular cavity. Such a situation occurs in practical applications.

Comparing (8.6.22) with (8.6.2) we see that we can obtain the complex potentials φ and ψ from (8.6.12) by merely replacing c by $-ic$.

$$\varphi = \frac{H_0^2}{4}\left[-i\mu(\mu - 1)c\frac{a}{z} + \tfrac{1}{2}(\mu + 1)^2\frac{a^2}{z^2}\right],$$

$$\psi = \frac{H_0^2}{4}\left\{i(\mu^2 + 3\mu)c\frac{a}{z} - [\mu^2(1 + 2c^2) - 1]\frac{a^2}{z^2}\right. \tag{8.6.23}$$

$$\left. + (\mu^2 + 4\mu + 1)\frac{a^4}{z^4}\right\}.$$

Consequently,

$${}_{\mathrm{E}}t_{rr} = \frac{H_0^2}{4}\left[2\mu(\mu - 2)c\frac{a}{r}\sin\theta + 2(\mu + 1)^2\frac{a^2}{r^2}\cos 2\theta\right.$$

$$\left. - (\mu^2 + 4\mu + 1)\frac{a^4}{r^4}\cos 2\theta + 2\mu(\mu - 2)\cos 2\theta + (\mu^2 - 1 + 4\mu c^2)\frac{a^2}{r^2}\right],$$

$${}_{\mathrm{E}}t_{\theta\theta} = \frac{H_0^2}{4}\left[2\mu(4 - 3\mu)c\frac{a}{r}\sin\theta + (\mu^2 + 4\mu + 1)\frac{a^4}{r^4}\cos 2\theta\right.$$

$$\left. + 2\mu(2 - \mu)\cos 2\theta - (\mu^2 - 1 + 4\mu c^2)\frac{a^2}{r^2}\right] \tag{8.6.24}$$

$${}_{\mathrm{E}}t_{r\theta} = \frac{H_0^2}{4}\left[6\mu(\mu - 1)c\frac{a}{r}\cos\theta + (\mu + 1)^2\frac{a^2}{r^2}\sin 2\theta\right.$$

$$\left. - (\mu^2 + 4\mu + 1)\frac{a^4}{r^4}\sin 2\theta + 2\mu(2 - \mu)\sin 2\theta\right].$$

The total stress field is given by

$$
\begin{aligned}
T_{rr} &= \frac{H_0^2}{4}\left[2\mu(2-\mu)\frac{ca}{r}\sin\theta + 2(\mu+1)^2\frac{a^2}{r^2}\cos 2\theta \right.\\
&\quad \left. - (\mu^2+4\mu+1)\frac{a^4}{r^4}\cos 2\theta + (\mu^2 - 1 + 2\mu^2c^2)\frac{a^2}{r^2}\right],\\
T_{\theta\theta} &= \frac{H_0^2}{4}\left[-2\mu^2\frac{ca}{r}\sin\theta + (\mu^2+4\mu+1)\frac{a^4}{r^4}\cos 2\theta \right.\\
&\quad \left. + (1-\mu^2-2\mu^2c^2)\frac{a^2}{r^2}\right],\\
T_{r\theta} &= \frac{H_0^2}{4}\left[2\mu(\mu+1)\frac{ca}{r}\cos\theta + (\mu+1)^2\frac{a^2}{r^2}\sin 2\theta \right.\\
&\quad \left. - (\mu^2+4\mu+1)\frac{a^4}{r^4}\sin 2\theta\right].
\end{aligned}
\tag{8.6.25}
$$

The total hoop stresses at $r = a$, $\theta = 0$, $\pi/2$, and π are obtained to be

$$
\begin{aligned}
T_{\theta\theta}(a, 0) = T_{\theta\theta}(a, \pi) &= \frac{H_0^2}{2}(1 + 2\mu - \mu^2c^2),\\
T_{\theta\theta}(a, \pi/2) &= -\frac{\mu H_0^2}{2}(2 + \mu + \mu c + \mu c^2).
\end{aligned}
\tag{8.6.26}
$$

The stationary value of $T_{\theta\theta}$ at $r = a$ occurs at $\theta = \pi/2, 3\pi/2$, and $\theta = \theta_c$ given by

$$
\sin\theta_c = -\frac{\mu^2 c}{2(\mu^2+4\mu+1)}. \tag{8.6.27}
$$

8.7. Radial Motions of a Cylinder Under an Axial Magnetic Field

A long circular cylindrical core is subjected to an axial magnetic field by means of a current carried in its winding. We assume that the system is a perfect conductor and investigate the radial motions of the core. The current density $\mathbf{J}$ and the displacement field are each assumed to have single components that are functions of the radial coordinate r and time t. In cylindrical coordinates we write

$$
\mathbf{J} = J(r, t)\mathbf{e}_\theta, \qquad \mathbf{u} = u(r, t)\mathbf{e}_r. \tag{8.7.1}
$$

For a perfect conductor $\sigma = \infty$ and $\mathscr{E} = \mathbf{0}$. Assuming also that the heat conduction is negligible and that $\mathbf{f} = \mathbf{0}$, the field equations (8.2.48) and (8.2.49) reduce to

$$
\frac{\partial H}{\partial t} + \frac{1}{r}\left(r\frac{\partial u}{\partial t}H\right) = 0, \tag{8.7.2}
$$

$$c_1^2 \frac{\partial}{\partial r}\left[\frac{1}{r}\frac{\partial}{\partial r}(ru)\right] - \frac{\mu_1}{\rho} H \frac{\partial H}{\partial r} - \ddot{u} = 0, \tag{8.7.3}$$

where

$$c_1^2 = (\lambda_e + 2\mu_e)/\rho, \qquad \mathbf{J} = -c\frac{\partial H}{\partial r}\mathbf{e}_\theta. \tag{8.7.4}$$

The second equation is the result of (8.2.9) with $\mathbf{H} = H(r, t)\mathbf{e}_z$. The nonlinear equations (8.7.2) and (8.7.3) will have to be solved to determine $H(r, t)$ and $u(r, t)$. By a transformation of variables, as used by Ladikov [1961] and Paria [1967] in connection with a problem concerning unsteady motions of a sphere, we can obtain an exact solution of this system. Thus, we introduce

$$\xi = r\tau, \qquad \tau = \tau(t). \tag{8.7.5}$$

With this (8.7.2) is transformed into

$$\left(\frac{\xi\tau'}{\tau} + \tau\frac{\partial u}{\partial t}\right)\frac{\partial H}{\partial \xi} + \frac{\partial H}{\partial \tau}\tau' + \frac{H}{r}\frac{\partial}{\partial \xi}\left(\xi\frac{\partial u}{\partial t}\right) = 0. \tag{8.7.6}$$

If we choose

$$\frac{\partial u}{\partial t} = -\xi\tau'/\tau^2 \tag{8.7.7}$$

this equation has the general solution

$$H = H_0(\xi)\tau^2 \tag{8.7.8}$$

provided $\tau' \neq 0$. Excluding the static case $\tau' = 0$ (8.7.8), with $H_0(\xi)$ as an arbitrary function, determines the functional form of H. From (8.7.7), by integration,

$$u(r, t) = -r \ln \tau + u_0(r), \tag{8.7.9}$$

where $u_0(r)$ is an arbitrary function of r. If the displacement is zero, initially we must have $\tau(0) = 1$ and $u_0(r) = 0$. Utilizing (8.7.8) and (8.7.9) in (8.7.3) we will have

$$\frac{\mu_1}{2\rho}\frac{1}{\xi}\frac{dH_0^2}{d\xi} = \tau^{-6}\frac{d^2 \ln \tau}{dt^2} = 3k^2, \tag{8.7.10}$$

where k^2 is a constant. Integration gives

$$\begin{aligned} H_0^2 &= 3\frac{\rho}{\mu_1}k^2\xi^2 + H_1^2, \\ \tau'^2 &= k^2\tau^8 + k_1\tau^2, \end{aligned} \tag{8.7.11}$$

where H_1^2 and k_1 are also constants of integration.

The stress tensor ${}_{\mathrm{E}}\mathbf{t}$, in cylindrical coordinates, is given by

$${}_{\mathrm{E}}t_{rr} = \frac{\lambda_e}{r}\frac{\partial}{\partial r}(ru) + 2\mu_e\frac{\partial u_r}{\partial r} = -2(\lambda_e + \mu_e)\ln \tau,$$

$$
\begin{aligned}
{}_{\mathrm{E}}t_{\theta\theta} &= \frac{\lambda_e}{r}\frac{\partial}{\partial r}(ru) + 2\mu_e\frac{u}{r} = -2(\lambda_e + \mu_e)\ln\tau, \\
{}_{\mathrm{E}}t_{zz} &= \frac{\lambda_e}{r}\frac{\partial}{\partial r}(ru) = -2\lambda_e \ln\tau, \\
{}_{\mathrm{E}}t_{rz} &= {}_{\mathrm{E}}t_{\theta z} = {}_{\mathrm{E}}t_{r\theta} = 0.
\end{aligned}
\tag{8.7.12}
$$

The electric field is obtained from $\mathscr{E} = 0$, hence

$$
\mathbf{E} = -\frac{1}{c}\dot{\mathbf{u}} \times \mu\mathbf{H} = -\frac{\mu}{c}\xi\tau' H_0(\xi)\mathbf{e}_\theta. \tag{8.7.13}
$$

If the initial electric field is not present we must have $\tau' = 0$ for $t = 0$. This, through $(8.7.11)_2$, implies that $k_1 = -k^2$. Integrating $(8.7.11)_2$, under the conditions that $\tau = 1$ and $\tau' = 0$ for $t = 0$, we obtain

$$
\tau^6 = 1 \pm \tan(3kt). \tag{8.7.14}
$$

Retaining the $(+)$ sign we see that τ is a continuous, increasing function of t in the range $0 \le 3kt < \pi/2$, and the solution is valid, at least in this range.

The electromagnetic fields $\mathbf{E}^0$ and $\mathbf{H}^0$ outside the cylinder satisfy Maxwell's equations in vacuum

$$
\nabla \times \mathbf{H}^0 = \mathbf{0}, \qquad \nabla \times \mathbf{E}^0 = -\frac{1}{c}\frac{\partial \mathbf{H}^0}{\partial t}, \qquad \nabla \cdot \mathbf{H}^0 = 0. \tag{8.7.15}
$$

Under the conditions of axial symmetry, the first and third of these equations give

$$
\mathbf{H}^0 = C_1(t) r^{-1}\mathbf{e}_\theta + C_2(t)\mathbf{e}_z. \tag{8.7.16}
$$

With this $(8.7.15)_2$ gives

$$
E^0_\theta = -C_2'\frac{r}{2c} + \frac{1}{r}C_3(t), \qquad E^0_z = -\frac{1}{c}C_1' \ln r + C_4(t), \tag{8.7.17}
$$

where $C_1(t)$ and $C_4(t)$ are functions of integrations. The boundary conditions (8.2.13) and (8.2.16), at $r = a$, lead to

$$
\begin{aligned}
C_1 = C_4 = 0, \qquad C_3 &= C_2' a^2/2c, \\
C_2 = H_0(a\tau)\tau^2[1 &- (\mu/c^2)a^2k^2(\tau^6 - 1)].
\end{aligned}
\tag{8.7.18}
$$

Thus, the nonvanishing components of the external electromagnetic fields are given by

$$
\begin{aligned}
H^0_z &= \tau^2 H_0(a\tau)[1 - (\mu/c^2)a^2k^2(\tau^6 - 1)], \\
E^0_\theta &= \frac{a}{2c}\left(\frac{a}{r} - \frac{r}{a}\right)\frac{dH^0_z}{dt}.
\end{aligned}
\tag{8.7.19}
$$

The electromagnetic stress tensor ${}_M\mathbf{t}$ is calculated by

$$
{}_{\mathrm{M}}t^k{}_l = \mu H^k H_l + E^k E_l - \tfrac{1}{2}[E^2 + \mu(2 - \mu)H^2]\delta^k{}_l, \tag{8.7.20}
$$

where we also used $\mathbf{B} = \mu\mathbf{H}$ and $\mathbf{M} = (\mu - 1)\mathbf{H}$. For the external field, we have $\mu = 1$. By use of (8.7.20), we find that the only nonvanishing electromagnetic traction on any cylindrical surface $r = \text{const.}$ is

$$\begin{aligned} {}_{\mathrm{M}}t_{rr} &= -\tfrac{1}{2}\tau^4 H_0^2(r\tau)[(\mu/c)^2k^2r^2(\tau^6 - 1) + \mu(2-\mu)], \qquad 0 \le r \le a \\ {}_{\mathrm{M}}t_{zz} &= -\tfrac{1}{2}\left[(a/2c)^2\left(\frac{a}{r} - \frac{r}{a}\right)^2\left(\frac{dH_z^0}{dt}\right)^2 + (H_z^0)^2\right]. \end{aligned} \tag{8.7.21}$$

The boundary condition on tractions at $r = a$, according to (8.2.11), is

$$\left[{}_{\mathrm{E}}t_{rr} + {}_{\mathrm{M}}t_{rr} + \frac{\mu}{c}\dot{u}E_\theta H_z\right] = 0. \tag{8.7.22}$$

Employing (8.7.8), (8.7.12), (8.7.13), (8.7.19), (8.7.21) in (8.7.22) we arrive at an applied traction given by

$$t_{rr}^0 = -2(\lambda_e + \mu_e)\ln\tau + \tfrac{1}{2}\tau^4 H_0^2(a\tau)\left[\frac{\mu^2a^2k^2}{c^2}(\tau^6 - 1) - \mu(2-\mu)\right]. \tag{8.7.23}$$

From this result it is clear that the radial stress of the magnitude given by (8.7.23) must be applied to the surface of the cylinder to provide the particular motion described by (8.7.9) and the electromagnetic fields obtained. The constants H_1^2 and k can be used to adjust the initially applied magnetic field or the stress. For example, by taking $H_1^2 = -3\rho k^2a^2/\mu$ we can make the initial surface tractions vanish. However, the surface tractions subsequent to the application of the magnetic field are specified as functions of time and cannot be made to vanish.[4] In this sense then the solution is an "inverse solution." Nevertheless, it provides a method of exact solution for a nonlinear problem which, aside from its pedagogical value, may be valuable in dealing with large electromagnetic fields.

A metal-forming technology exists that employs high-intensity currents to form thin sheets. Under high-intensity discharges, electric wires are known to explode. In such situations it is necessary to obtain solutions of the nonlinear field equations. Lightening damage also falls into this category of problems. It must be mentioned, however, that thermal effects also become very significant requiring us to treat the full problem of magneto-thermoelasticity.

8.8. Propagation of Plane Waves

In a large class of dynamical problems we are concerned with the propagation of disturbances in a medium that carries a bias field. If the magnitude of the bias field is large, compared to the field created by the disturbance superposed on the medium, we can linearize the field equations. Let $\mathbf{H}^0$ denote a constant

[4] Paria [1967], in his analysis of a similar problem for a sphere, imposes vanishing surface tractions for all time. However, it appears that he has ignored the condition of continuity for the tangential electric field which, if used, would determine his arbitrary function $\psi_1(t)$. This makes it impossible to have the boundary traction vanish for all times.

bias field and write

$$\mathbf{H} = \mathbf{H}^0 + \mathbf{h}(\mathbf{x}, t), \qquad |\mathbf{h}| \ll |\mathbf{H}^0|. \tag{8.8.1}$$

Substituting (8.8.1) into (8.2.8), (8.2.48)–(8.2.50) and dropping products of **u**, **h**, and ∇T we will have

$$\nabla \cdot \mathbf{h} = 0, \tag{8.8.2}$$

$$\nu_H \nabla \times \nabla \times \mathbf{h} + \frac{\partial \mathbf{h}}{\partial t} - \nabla \times (\dot{\mathbf{u}} \times \mathbf{H}^0) = \mathbf{0}, \tag{8.8.3}$$

$$(\lambda_e + 2\mu_e)\nabla\nabla \cdot \mathbf{u} - \mu_e \nabla \times \nabla \times \mathbf{u} - \beta_e \nabla T + \mu_1(\nabla \times \mathbf{h}) \times \mathbf{H}^0 + \rho(\mathbf{f} - \ddot{\mathbf{u}}) = \mathbf{0}, \tag{8.8.4}$$

$$-\rho\gamma\dot{T} - \beta_e T_0 \nabla \cdot \dot{\mathbf{u}} + \kappa_e \nabla^2 T + \rho h = 0, \tag{8.8.5}$$

where

$$\nu_H = c^2/\sigma\mu, \qquad \kappa_e = \kappa - (\sigma^\theta \kappa^e/\sigma). \tag{8.8.6}$$

From (8.8.5) it is clear that in the linear theory the equation of heat conduction is not affected by the magnetic field.

In the case of infinite conductivity $\sigma = \infty$ and (8.8.3) simplifies to

$$\frac{\partial \mathbf{h}}{\partial t} - \nabla \times (\dot{\mathbf{u}} \times \mathbf{H}^0) = \mathbf{0}.$$

Integration of this under the initial conditions that $\mathbf{h} = \mathbf{u} = \mathbf{0}$ for $t = 0$ gives

$$\mathbf{h} = \nabla \times (\mathbf{u} \times \mathbf{H}^0). \tag{8.8.7}$$

With this (8.8.3) takes the form

$$(\lambda_e + 2\mu_e)\nabla\nabla \cdot \mathbf{u} - \mu_e \nabla \times \nabla \times \mathbf{u} - \beta_e \nabla T + \mu_1[\nabla \times \nabla \times (\mathbf{u} \times H^0)] \times \mathbf{H}^0 + \rho(\mathbf{f} - \ddot{\mathbf{u}}) = \mathbf{0}, \tag{8.8.8}$$

so that **h** no longer appears. For a perfect conductor therefore (8.8.4) and (8.8.7) must be solved to determine **u** and T. Afterwards (8.8.7) gives the magnetic field.

For plane harmonic waves we write

$$\{\mathbf{u}, \mathbf{h}, T\} = \{\bar{\mathbf{u}}, \bar{\mathbf{h}}, \bar{T}\} \exp[i(\boldsymbol{\xi} \cdot \mathbf{x} - \omega t)], \tag{8.8.9}$$

where $\bar{\mathbf{u}}$ and $\bar{\mathbf{h}}$ are constant vectors and $\bar{T}$ is a constant scalar. Substituting this into (8.8.2)–(8.8.5), with $h = 0$, $\mathbf{f} = \mathbf{0}$, we obtain

$$\boldsymbol{\xi} \cdot \bar{\mathbf{h}} = 0, \tag{8.8.10}$$

$$\omega(\boldsymbol{\xi} \cdot \bar{\mathbf{u}})\mathbf{H}^0 - \omega(\boldsymbol{\xi} \cdot \mathbf{H}^0)\bar{\mathbf{u}} + (\nu_H \xi^2 - i\omega)\bar{\mathbf{h}} = \mathbf{0}, \tag{8.8.11}$$

$$(c_2^2 - c_1^2)(\boldsymbol{\xi} \cdot \bar{\mathbf{u}})\boldsymbol{\xi} + (\omega^2 - c_2^2 \xi^2)\bar{\mathbf{u}} - i\beta_e \boldsymbol{\xi} \bar{T} + i(\mu_1/\rho)[(\boldsymbol{\xi} \cdot \mathbf{H}^0)\bar{\mathbf{h}} - (\mathbf{H}^0 \cdot \bar{\mathbf{h}})\boldsymbol{\xi}] = \mathbf{0}, \tag{8.8.12}$$

$$(i\rho\gamma\omega - \kappa_e \xi^2)\bar{T} - \beta_e T_0 \omega \boldsymbol{\xi} \cdot \bar{\mathbf{u}} = 0. \tag{8.8.13}$$

The scalar product of (8.8.12) and (8.8.13) with $\mathbf{H}^0 \times \xi$ gives two equations which cannot be satisfied for arbitrary values of $\boldsymbol{\xi}\cdot\mathbf{H}^0$ unless

$$\bar{\mathbf{h}}\cdot(\mathbf{H}^0\times\boldsymbol{\xi})=0,\qquad \bar{\mathbf{u}}\cdot(\mathbf{H}^0\times\boldsymbol{\xi})=0, \tag{8.8.14}$$

provided $\nu_H\xi^2 - i\omega \neq 0$ and $\omega^2 - c_2^2\xi^2 \neq 0$. Of these $\nu_H\xi^2 - i\omega = 0$ corresponds to a purely electromagnetic wave and the second to a purely elastic equivoluminal wave. Equation (8.8.14) implies that $\mathbf{h}$ and $\mathbf{u}$ are in the planes of $\mathbf{H}^0$ and $\boldsymbol{\xi}$. We can select this plane as the (x_1, x_2)-plane so that all vectors will have only two components. In fact, we can also select the x_1-axis in the direction of $\boldsymbol{\xi}$ so that

$$\boldsymbol{\xi}=\xi\mathbf{i}_1. \tag{8.8.15}$$

With this (8.8.10) implies that $h_1 = 0$ and therefore

$$\mathbf{h}=h_2\mathbf{i}_2=\omega\xi(H_1^0\bar{u}_2-H_2^0\bar{u}_1)(\nu_H\xi^2-i\omega)^{-1}\mathbf{i}_2. \tag{8.8.16}$$

Substituting this into (8.8.12), with the $\bar{T}$ replaced by (8.8.13), leads to two equations for the two unknowns $\bar{u}_1$ and $\bar{u}_2$

$$\begin{aligned}&\left[\omega^2-c_1^2\xi^2-i\beta_e\alpha\xi^2+i\frac{\mu_1}{\rho}H_2^{0^2}\omega\xi^2(\nu_H\xi^2-i\omega)^{-1}\right]\bar{u}_1\\&\qquad\qquad-i\frac{\mu_1}{\rho}H_1^0H_2^0\omega\xi^2(\nu_H\xi^2-i\omega)^{-1}\bar{u}_2=0,\\&-i\frac{\mu_1}{\rho}H_1^0H_2^0\omega\xi^2(\nu_H\xi^2-i\omega)^{-1}\bar{u}_1\\&\qquad\qquad+[\omega^2-c_2^2\xi^2+i\frac{\mu_1}{\rho}H_1^{0^2}\omega\xi^2(\nu_H\xi^2-i\omega)^{-1}]\bar{u}_2=0,\end{aligned} \tag{8.8.17}$$

where

$$\alpha=\beta_e T_0\omega(i\rho\gamma\omega-\kappa_e\xi^2)^{-1}. \tag{8.8.18}$$

Equation (8.8.17) constitutes two equations for the determination of two unknowns $\bar{u}_1$ and $\bar{u}_2$.

For perfect conductors $\nu_H = 0$, and the system (8.8.17) reduces to

$$\begin{aligned}&\left[\omega^2-\xi^2\left(c_1^2+i\beta_e\alpha-\frac{\mu_1}{\rho}H_2^{0^2}\right)\right]\bar{u}_1+\frac{\mu_1}{\rho}H_1^0H_2^0\bar{u}_2=0,\\&\frac{\mu_1}{\rho}H_1^0H_2^0\xi^2\bar{u}_1+\left[\omega^2-\xi^2\left(c_2^2-\frac{\mu_1}{\rho}H_1^{0^2}\right)\right]\bar{u}_2=0.\end{aligned} \tag{8.8.19}$$

In the case of negligible heat conduction (or isothermal state) $\alpha = 0$.

The system (8.8.17) may possess nontrivial solutions if the determinant of the coefficients of $\bar{u}_1$ and $\bar{u}_2$ vanishes, i.e.,

$$\begin{vmatrix}[\omega^2-\xi^2(c_1^2+i\beta_e\alpha)](\nu_H\xi^2-i\omega)+i\frac{\mu_1}{\rho}H_2^{0^2}\omega\xi^2 & -i\frac{\mu_1}{\rho}H_1^0H_2^0\omega\xi^2\\ -i\frac{\mu_1}{\rho}H_1^0H_2^0\omega\xi^2 & (\omega^2-\xi^2c_2^2)(\nu_H\xi^2-i\omega)+i\frac{\mu_1}{\rho}H_1^{0^2}\omega\xi^2\end{vmatrix}=0. \tag{8.8.20}$$

Several special cases are of interest:

(i) *Bias Field Is Absent*: In this case $\mathbf{H}^0 = \mathbf{0}$ and (8.8.20) possesses three roots

$$\begin{aligned} \omega + i\nu_H\xi^2 &= 0, \\ \omega^2 - c_2^2\xi^2 &= 0, \\ \omega^2 - \xi^2(c_1^2 + i\beta_e\alpha) &= 0. \end{aligned} \tag{8.8.21}$$

Of these, the first one corresponds to diffusion of the magnetic field with $\mathbf{u} = \mathbf{0}$. The second one gives the phase velocity of the transverse elastic waves not coupled with either thermal or electromagnetic fields. The last one gives the phase velocity of longitudinal elastic waves coupled with the thermal fields.

(ii) *Bias Field Is in the x_2-direction*: Hence $H_1^0 = 0$, and we can have electromagnetic waves $(8.8.21)_1$, or transverse waves $(8.8.21)_2$, or longitudinal waves with the phase velocity determined by

$$(\omega^2 - c_1^2\xi^2 - i\beta_e\alpha\xi^2)(\nu_H\xi^2 - i\omega) + i\frac{\mu_1}{\rho}H_2^{0^2}\omega\xi^2 = 0. \tag{8.8.22}$$

(iii) *Bias Field Is in the Direction of Propagation* ($H_2^0 = 0$): In this case we have electromagnetic waves described by $(8.8.21)_1$, longitudinal waves described by $(8.8.21)_3$, and transverse waves with phase velocity determined by

$$(\omega^2 - c_2^2\xi^2)(\nu_H\xi^2 - i\omega) + i\frac{\mu_1}{\rho}H_1^{0^2}\omega\xi^2 = 0. \tag{8.8.23}$$

Returning to the general case, we introduce the following nondimensional quantities:

$$\begin{aligned} &\chi = \omega/\overset{*}{\omega}, \qquad \eta = \xi c_1/\overset{*}{\omega}, \qquad \varepsilon_T = \beta_e^2 T_0/\rho\gamma c_1^2, \\ &\varepsilon_H = \nu_H\overset{*}{\omega}/c_1^2, \qquad R_H = \mu_1 H_2^{0^2}/\rho c_1^2, \qquad R_L = \mu_1 H_1^{0^2}/\rho c_1^2, \\ &s^2 = c_2^2/c_1^2, \end{aligned} \tag{8.8.24}$$

where $\overset{*}{\omega}$ is the characteristic frequency defined by

$$\overset{*}{\omega} = \rho\gamma c_1^2/\kappa_e. \tag{8.8.25}$$

The dispersion relation (8.8.20) now becomes

$$\begin{aligned} (s^2\eta^2 - \chi^2)\{[(\eta^2 - \chi^2)(\chi + i\varepsilon_H\eta^2) + R_H\eta^2\chi](\chi + i\eta^2) + \varepsilon_T\eta^2\chi(\chi + i\varepsilon_H\eta^2)\} \\ + R_L\eta^2\chi[(\eta^2 - \chi^2)(\chi + i\eta^2) + \varepsilon_T\eta^2\chi] = 0. \end{aligned} \tag{8.8.26}$$

This is a quartic equation in η^2 with complex coefficients. In the case $R_L = 0$, (8.8.26) factorizes into two parts: $s^2\eta^2 - \chi^2 = 0$ which corresponds to a transverse elastic wave, and another factor given by

$$[(\eta^2 - \chi^2)(\chi + i\varepsilon_H\eta^2) + R_H\eta^2\chi](\chi + i\eta^2) + \varepsilon_T\eta^2\chi(\chi + i\varepsilon_H\eta^2) = 0. \tag{8.8.27}$$

This is the dispersion relation corresponding to the longitudinal waves. Therefore, in the general case (8.8.26) both the longitudinal and transverse waves are coupled. For the special case of $R_L = R_H = \varepsilon_H = 0$, we have the dispersion relation of purely thermoelastic waves

$$(\eta^2 - \chi^2)(\chi + i\eta^2) + \varepsilon_T \eta^2 \chi^2 = 0. \tag{8.8.28}$$

The coupling between thermoelasticity and the magnetic field is, therefore, described by the factors R_L, R_H, and ε_H.

In the *low-frequency* region following Wilson [1963], we can obtain the roots of (8.8.26). The characteristic frequency $\overset{*}{\omega}$ for various metals at 20 °C was provided by Chadwick [1960].

	Aluminum	Copper	Iron	Lead
$\overset{*}{\omega}$ in $\sec^{-1}$	4.66×10^{11}	1.73×10^{11}	1.75×10^{11}	1.9×10^{11}

To determine the phase velocity of waves we set

$$\chi = \nu\eta, \qquad \nu = \omega/\xi c_1. \tag{8.8.29}$$

The phase velocity is then given by $v = c_1 \operatorname{Re} \nu$. Substituting (8.8.29) into (8.8.26) we will have

$$\begin{aligned} \nu^6 &+ i\eta(1 + \varepsilon_H)\nu^5 - (1 + R_H + \varepsilon_T + \varepsilon_H \eta^2 + s^2 + R_L)\nu^4 \\ &- i\eta[1 + \varepsilon_H + R_H + \varepsilon_H \varepsilon_T + (1 + \varepsilon_H)s^2 + R_L]\nu^3 \\ &+ [\varepsilon_H \eta^2 + s^2(1 + R_H + \varepsilon_T + \varepsilon_H \eta^2) + (1 + \varepsilon_T)R_L]\nu^2 \\ &+ i\eta[s^2(1 + \varepsilon_H + R_H + \varepsilon_H \varepsilon_T + R_L]\nu - \varepsilon_H s^2 \eta^2 = 0. \end{aligned} \tag{8.8.30}$$

This is a six-order equation with complex coefficients. Since the coefficients contain η, it is clear that the phase velocity will depend on the wavelength and therefore these waves are dispersive.

If $\omega \ll \overset{*}{\omega}$, i.e., $\chi \ll 1$, we see that

$$\operatorname{Re} \nu = v/c_1 = O(\chi^{1/2}) \ll 1 \tag{8.8.31}$$

and we have *slow waves*. In this case, if we retain the lowest-order terms in ν, we obtain

$$\nu = -i\eta[1 + \varepsilon_T + (1 + R_H + R_L s^{-2})\varepsilon_H^{-1}]^{-1}. \tag{8.8.32}$$

For the purely thermoelastic case $\varepsilon_H \to \infty$, and we have

$$\nu = -i\eta(1 + \varepsilon_T)^{-1}. \tag{8.8.33}$$

It is clear that for low frequencies the magnetic field increases the thermoelastic coupling by $(1 + R_H + R_L s^{-2})\varepsilon_H^{-1}$. Otherwise, the nature of purely thermoelastic waves is not altered. Thus we may wish to examine higher-order terms. For example, for the quadratic approximation we would retain the last three terms. This corresponds to the approximation discussed by Wilson

[1963]. In fact, for this approximation, we set

$$\eta^2 = i\phi\chi + (\text{term of order } \chi^2),$$
$$v = \chi/i\phi, \tag{8.8.34}$$

and we obtain Wilson's equation

$$\phi^2 - \phi[(1 + R_H + \varepsilon_H + \varepsilon_H\varepsilon_T)\varepsilon_H^{-1} + (R_L/\varepsilon_H s^2)] + [(1 + \varepsilon_T + R_H)\varepsilon_H^{-1} + R_L(1 + \varepsilon_T)\varepsilon_H^{-1}s^{-2}] = 0. \tag{8.8.35}$$

For zero conductivity $\varepsilon_H \to \infty$ and this gives $\phi = 1 + \varepsilon_T$ which leads to the same result as (8.8.33). For finite conductivity, but vanishing bias field, $R_H = R_L = 0$ and we obtain

$$\phi_1 = 1/\varepsilon_H, \qquad \phi_2 = 1 + \varepsilon_T. \tag{8.8.36}$$

Of these, the first one corresponds to a purely electromagnetic wave, while the second corresponds to a purely thermoelastic wave. The complex polarization $\bar{u}_1/\bar{u}_2$ can be solved from (8.8.17)

$$\bar{u}_1/\bar{u}_2 = (H_1^0/H_2^0)[1 + (\chi + i\varepsilon_H\eta^2)(s^2\eta^2 - \chi^2)(R_L\eta^2\chi)^{-1}]. \tag{8.8.37}$$

If the approximation (8.8.32) is used this gives

$$\bar{u}_1/\bar{u}_2 = (H_1^0/H_2^0)[1 + (1 - \varepsilon_H\phi)s^2 R_L^{-1}]. \tag{8.8.38}$$

The value of the complex polarization can be obtained from (8.8.37) and (8.8.38), under the approximation (8.8.32), by substituting the value of η^2 (or ϕ) calculated from (8.8.34) or (8.8.35).

(iv) *Heat Conduction Is Not Present*: In this case $\varepsilon_T = 0$. After the cancellation of a factor $\chi + i\eta^2$, (8.8.26) reduces to

$$(s^2\eta^2 - \chi^2)(\eta^2 - \chi^2)(\chi + i\varepsilon_H\eta^2) + R_H\eta^2\chi(s^2\eta^2 - \chi^2) + R_L\eta^2\chi(\eta^2 - \chi^2) = 0. \tag{8.8.39}$$

This expression shows that the longitudinal and transverse waves are coupled. Upon using (8.8.29) in (8.8.39), we obtain the following dispersion relation:

$$(s^2 - v^2)(1 - v^2)(v + i\varepsilon_H\eta) + R_H v(s^2 - v^2) + R_L v(1 - v^2) = 0. \tag{8.8.40}$$

This is a fifth-degree polynomial in v with complex coefficients. It is clear that longitudinal and transverse waves are coupled. The uncoupling occurs only in the case when the bias field is directed along either the x_1- or x_2-axes, since in these cases either $R_L = 0$ or $R_H = 0$, and (8.8.40) reduces to a third-degree polynomial. An estimate of the phase velocities indicates that, in moderate to reasonably high fields (10^4 Oersted or so), the effect of the magnetic field on the phase velocity is small (about one part in a million). Thus, we may use a perturbation process to obtain the longitudinal and transverse wave velocities. To this end, we write

$$v = 1 + w, \qquad |w| \ll 1, \tag{8.8.41}$$

in (8.8.40) and retain the linear terms in w to obtain

$$w = \frac{R_H}{2}\left(1 - R_H\frac{3 - s^2}{2(1 - s^2)} - \frac{R_L}{1 - s^2} + i\varepsilon_H\eta\right)^{-1}. \tag{8.8.42}$$

Since η can be a complex number, writing $\eta = \eta_1 + i\eta_2$ we obtain the real and imaginary parts of (8.8.42)

$$\begin{aligned} w_1 &= \operatorname{Re} w = \frac{R_H}{2}[1 - r_L + \varepsilon_H^2\eta_1^2(1 - r_L)^{-1}]^{-1}, \\ w_2 &= \operatorname{Im} w = -\frac{R_H}{2}\varepsilon_H\eta_1(1 - r_L)^{-1}[1 - r_L + \varepsilon_H^2\eta_1^2(1 - r_L)^{-1}]^{-1}, \end{aligned} \tag{8.8.43}$$

where

$$r_L = \frac{R_H}{2}(3 - s^2)(1 - s^2)^{-1} + R_L(1 - s^2)^{-1} + \varepsilon_H\eta_2. \tag{8.8.44}$$

The phase velocity of the longitudinal waves now follows from (8.8.41), by writing $\xi = \xi_1 + i\xi_2$ and

$$v_L/c_1 = \omega/\xi_1 c_1 = (\xi_1 + i\xi_2)(1 + w_1 + iw_2)T_1.$$

The real and imaginary parts of this expression give

$$\begin{aligned} v_L/c_1 &= 1 + w_1 - (\xi_2/\xi_1)w_2 = 1 + w_1 + w_2^2(1 + w_1)^{-1}, \\ \xi_2/\xi_1 &= -w_2(1 + w_1)^{-1} = \eta_2/\eta_1. \end{aligned} \tag{8.8.45}$$

An examination of R_H and R_L shows that these quantities are very small compared to unity for moderate to fairly large fields (e.g., 10^4 Oersted) Therefore, we may neglect R_H^2 compared to R_H, and R_L compared to unity. This latter approximation becomes clear after taking $(8.8.45)_2$ into (8.8.44). With this, eqs. (8.8.45) reduce to

$$\begin{aligned} \frac{v_L}{c_1} &= 1 + \frac{\mu_1 H^2 \sin^2\theta}{2\rho c_1^2(1 + v_H^2\xi_1^2 c_1^{-2})}, \\ \frac{\xi_2}{\xi_1} &= \frac{\eta_2}{\eta_1} = \frac{\mu_1 H^2 \sin^2\theta}{2\rho c_1^2}\cdot\frac{v_H\xi_1 c_1^{-1}}{1 + v_H^2\xi_1^2 c_1^{-2}}, \end{aligned} \tag{8.8.46}$$

where we used (8.8.24).

To obtain the transverse wave velocity and attenuation factor, we substitute

$$v = sv_T = s(1 + w_T) \tag{8.8.47}$$

into (8.8.40). This gives

$$\begin{aligned} w_{T1} &= \operatorname{Re} w_T = \frac{R_L}{2s^2}\frac{1 - r_T}{(1 - r_T)^2 + \varepsilon_H^2\eta_1^2 s^{-2}}, \\ w_{T2} &= \operatorname{Im} w_T = -\frac{R_L}{2s^2}\frac{\varepsilon_H\eta_1 s^{-1}}{(1 - r_T)^2 + \varepsilon_H^2\eta_1^2 s^{-2}}, \end{aligned} \tag{8.8.48}$$

where

$$r_T = \varepsilon_H \eta_2 s^{-1} - [R_H s^2 + \tfrac{1}{2}(3s^2 - 1)R_L](1 - s^2)^{-1} s^{-2}. \tag{8.8.49}$$

The transverse phase velocity and the attenuation factor are given by

$$\begin{aligned} v_T/c_1 &= 1 + w_{T1} + w_{T2}^2(1 + w_{T1})^{-1}, \\ \xi_2/\xi_1 &= -w_{T2}(1 + w_{T1})^{-1} = \eta_2/\eta_1. \end{aligned} \tag{8.8.50}$$

The approximation leading to (8.8.46), in this case, gives

$$\begin{aligned} \frac{v_T}{c_2} &= 1 + \frac{\mu_1 H^2 \cos^2\theta}{2\rho c_2^2(1 + v_H^2 \xi_1^2 c_2^{-2})}, \\ \frac{\xi_2}{\xi_1} &= \frac{\mu_1 H^2 \cos^2\theta}{2\rho c_2^2} \cdot \frac{v_H \xi_1 c_2^{-1}}{1 + v_H^2 \xi_1^2 c_2^{-2}}. \end{aligned} \tag{8.8.51}$$

The phase velocities (8.8.46) and (8.8.51) also follow from (8.8.40) by taking R_L and R_H equal to zero, respectively, i.e., when the bias field is directed along the x_2-axis for v_L and along the x_1-axis for v_T. We therefore conclude that, even for fairly large fields, the couplings between longitudinal and transverse waves are weak and (8.8.46) and (8.8.51) represent very good approximations.

We notice that $v_H \xi_1$ can be expressed as

$$v_H \xi_1 = 2\pi^2 \delta^2/\lambda^2, \qquad \delta = c(\lambda/\sigma\mu_1)^{1/2}, \tag{8.8.52}$$

where $\lambda = 2\pi/\text{Re}\,\xi$ is the wavelength and δ is the skin depth that an electromagnetic wave, with the same frequency as the elastic wave, would have in the material. With this, (8.8.46) and (8.8.51) become identical to the corresponding expressions announced by Alers and Fleury [1963], if we remember to convert our electromagnetic units to theirs by replacing c and μ by $c/4\pi$ and $\mu/4\pi$. Experiments performed by these authors on gold have shown that (8.8.46) and (8.8.51) give excellent agreement with observations in the range from zero to 21×10^4 Oersted bias magnetic fields. They found that the phase velocity is linear in H^2 and variations with the angle follow $\sin^2\theta$ and $\cos^2\theta$

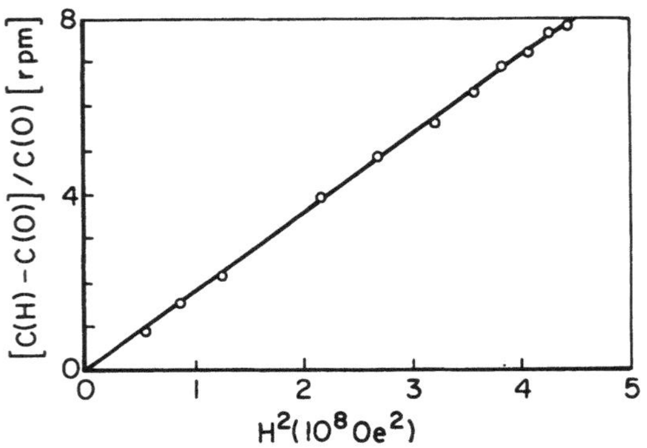

Figure 8.8.1. Phase velocity versus H^2 (after Alers and Fleury [1963]). Reprinted with permission of the authors and the American Physical Society.

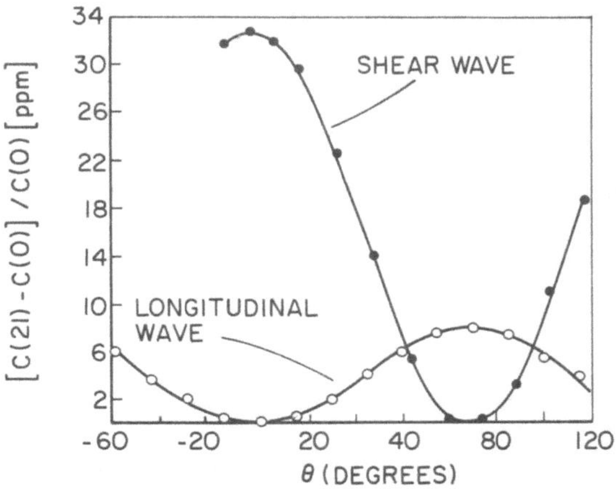

Figure 8.8.2. Variation of the change in sound velocity accompanying the application of a 21-kOe magnetic field with the angle between the field direction and the [110] propagation direction. The solid lines are the angular dependence predicted by the theory (after Alers and Fleury [1963]). Reprinted with permission of the authors and the American Physical Society.

(see Figures 8.8.1 and 8.8.2). The longitudinal wave was propagated along the [110] crystal axis. It was also observed that small temperature changes were very important in the phase velocity. The attenuation factor also agrees very well with observation (see Figure 8.8.3). These results are, of course, valid in the macroscopic range where the wavelength is large compared to the penetration depth. Even for a wavelength comparable to the penetration depth, the theory predicts a correction factor which is slightly smaller than needed for perfect agreement with experiments.

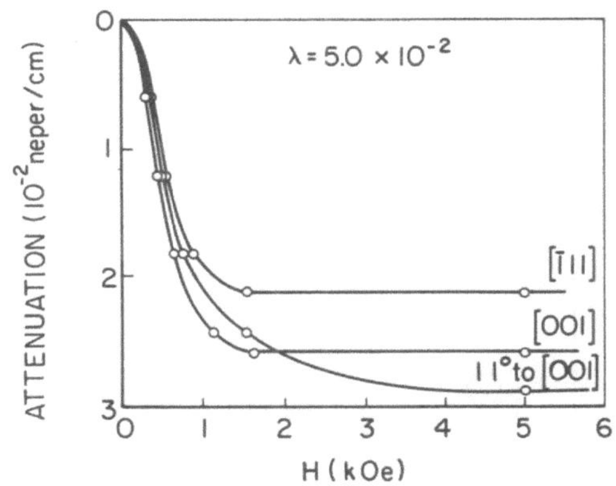

Figure 8.8.3. Attenuation factor versus H (after Alers and Fleury [1963]). Reprinted with permission of the authors and the American Physical Society.

Problems concerning harmonic magnetoelastic waves can also be found in Nowacki [1975] and Paria [1962]. When wavelengths are very close to the skin depth, *anomalous* dispersion occurs. For this case, see Chapter 13. New problems to be tackled include the diffraction of magnetoelastic waves on defects (Chattopadhyay and Maugin [1985]), inhomogeneities (Selezov [1984]), the propagation of magnetoelastic waves in random media (Dhar [1979]), and also the coupling between the magnetoelasticity of structures and an ambient electrically conducting gas which may arise in aero (space) elasticity (Librescu [1977]).

8.9. Waves Induced by a Thermal Shock

Here we consider the propagation of one-dimensional elastic waves produced by a thermal shock applied to the boundary $x_3 = 0$ of a half-space $x_3 \geq 0$. For simplicity, we assume that the medium is a perfect conductor.[5] The medium is biased with an initial constant magnetic field $\mathbf{H}_0$ parallel to the surface of the half-space. We choose the x_1-axis of the rectangular coordinates in the direction of $\mathbf{H}_0$ (Figure 8.9.1). We consider that all fields are functions of x_3 and t so that the problem is one-dimensional. The magnetic field is expressed as

$$\mathbf{H} = H_0 \mathbf{i}_1 + \mathbf{h}(x_3, t), \qquad |\mathbf{h}| \ll H_0. \tag{8.9.1}$$

The medium is assumed to be a perfect conductor, i.e., $\sigma = \infty$, and therefore $\nu_H = 0$. Dropping the higher-order terms involving products of h_k and $\dot{u}_l$ in the linearization, the field equations (8.2.48)–(8.2.50) reduce to

$$\frac{\partial h_1}{\partial t} + H_0 \frac{\partial^2 u_3}{\partial x_3 \, \partial t} = 0, \qquad h_2 = h_3 = 0, \tag{8.9.2}$$

[5] This problem and the problem for a medium with finite conductivity were treated by Kaliski and Nowacki [1962a, b].

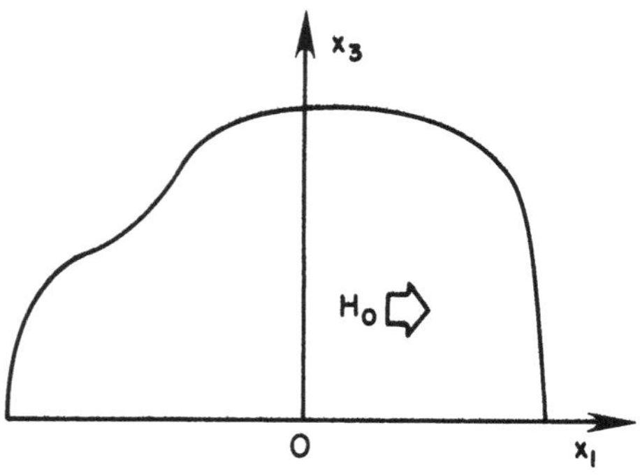

Figure 8.9.1. Half-space subject to thermal shock.

$$c_1^2 \frac{\partial^2 u_3}{\partial x_3^2} - \frac{\beta_e}{\rho} \frac{\partial T}{\partial x_3} - \frac{\mu_1}{\rho} \frac{\partial h_1}{\partial x_3} - \frac{\partial^2 u_3}{\partial t^2} = 0, \tag{8.9.3}$$

$$c_2^2 \frac{\partial^2 u_\alpha}{\partial x_3^2} - \frac{\partial^2 u_\alpha}{\partial t^2} = 0, \qquad (\alpha = 1, 2),$$

$$-\rho_0 \gamma \frac{\partial T}{\partial t} - \beta_e T_0 \frac{\partial^2 u_3}{\partial x_3 \, \partial t} + \kappa_e \frac{\partial^2 T}{\partial x_3^2} = 0. \tag{8.9.4}$$

For infinite conductivity, the electric field follows from (8.2.44) by dividing by $\sigma = \infty$:

$$\mathbf{E} = -\frac{1}{c} \dot{\mathbf{u}} \times \mathbf{B} = \frac{\mu_1}{c} H_0 (\dot{u}_2 \mathbf{i}_3 - \dot{u}_3 \mathbf{i}_2). \tag{8.9.5}$$

From $\nabla \cdot \mathbf{E} = 0$, it follows that $\dot{u}_{2,3} = 0$. Since $u_2 = \dot{u}_2 = 0$ initially, it follows that u_2 will vanish for all times. In fact, because u_1 and u_2 are uncoupled from u_3 in (8.9.3), we may also set $u_1 = 0$ henceforth. Thus

$$E_1 = E_3 = 0, \qquad E_2 = -\frac{\mu_1}{c} H_0 \dot{u}_3. \tag{8.9.6}$$

In vacuum electromagnetic fields $\mathbf{E}^0$ and $\mathbf{h}^0$ satisfy Maxwell's equations which reduce to

$$\begin{aligned} -\frac{\partial E_2^0}{\partial x_3} + \frac{1}{c} \frac{\partial h^0}{\partial t} &= 0, \\ \frac{\partial h_1^0}{\partial x_3} - \frac{1}{c} \frac{\partial E_2^0}{\partial t} &= 0, \qquad E_1^0 = \text{const.} \end{aligned} \tag{8.9.7}$$

Boundary conditions on traction, electromagnetic, and thermal fields at $x_3 = 0$ follow from (8.2.51)–(8.2.53)

$$\begin{aligned} t_{(3)k} &= {}_{\mathrm{E}}t_{3k} + {}_{\mathrm{M}}t_{3k}^{-} - {}_{\mathrm{M}}t_{3k}^{+} - M_3 B_k, \\ E_k = E_k^0, \qquad h_1 &= h_1^0, \qquad H_2^0 = H_3^0 = 0, \\ T &= T_1 H(t) \quad \text{at } x_3 = 0. \end{aligned} \tag{8.9.8}$$

In addition, all fields vanish at $x_3 = \pm\infty$. In (8.9.8), $H(t)$ is the Heaviside unit function and ${}_{\mathrm{M}}t_{3k}^{-}$ and ${}_{\mathrm{M}}t_{3k}^{+}$ are, respectively, electromagnetic tractions at $x_3 = 0-$ and $x_3 = 0+$. We calculate ${}_{\mathrm{E}}t_{lk}$ and ${}_{\mathrm{M}}t_{lk}$ by using (8.2.41) and (8.2.55). In fact, the only surviving components of these are ${}_{\mathrm{E}}t_{33}$ and ${}_{\mathrm{M}}t_{33}$ which are given by

$$\begin{aligned} {}_{\mathrm{E}}t_{33} &= -\beta_e T + (\lambda_e + 2\mu_e) u_{3,3}, \\ {}_{\mathrm{M}}t_{33} &= -\frac{\mu_1}{2}(H_0^2 + 2H_0 h_1), \qquad \mu_1 \equiv \mu(2 - \mu). \end{aligned} \tag{8.9.9}$$

We must remember to set $\mu_1 = 1$ for ${}_{\mathrm{M}}t_{33}^{+}$.

Under the initial bias field H_0, the body is assumed to be unstrained and

$T = 0$, but there is a Maxwell stress of magnitude

$$ {}_{M}t^{0}_{33} = \frac{1-\mu_1}{2} H_0^2. \tag{8.9.10}$$

We subtract this from the surface tractions $t_{(3)k}$ given by (8.9.8), in order to express the boundary conditions on the vanishing mechanical tractions at $x_3 = 0$ as

$$c_1^2 u_{3,3} - \frac{\beta_e}{\rho} T + \frac{1}{\rho}(\mu - 1)^2 H_0 h_1 = 0, \qquad x_3 = 0. \tag{8.9.11}$$

We keep the thermal shock condition on T but, using $(8.9.7)_2$ and $(8.9.6)_2$, recast the boundary condition on E_2. Hence

$$\begin{aligned} &h^0_{1,3} + (\mu_1 H_0/c^2)\ddot{u}_3 = 0, \\ &T = T_1 H(t), \qquad x_3 = 0. \end{aligned} \tag{8.9.12}$$

Initial conditions are conditions of rest at $t = 0$, i.e.,

$$u_3(x_3, 0) = \dot{u}_3(x_3, 0) = 0, \qquad h_1(x_3, 0) = 0. \tag{8.9.13}$$

The problem is now well-posed, namely: determine the solutions of $(8.9.2)_1$, $(8.9.3)_1$, and (8.9.4) for $u_3(x_3, t)$, $h_1(x_3, t)$, and $T(x_3, t)$ under the boundary and initial conditions (8.9.11)–(8.9.13).

It must be remembered, however, that to the total stress field obtained in this way we must add the static magnetic stress field due to the bias field H_0. Consequently, at the surface $x_3 = 0$ of the body, there will be a residual magnetic stress ${}_{M}t^{0}_{33}$ given by (8.9.10).

From equation (8.9.2) and the initial condition $(8.9.13)_3$ it follows that

$$h_1 = -H_0 \frac{\partial u_3}{\partial x_3}. \tag{8.9.14}$$

Substituting this into $(8.9.3)_1$, we eliminate h_1 leading to

$$a^2 \frac{\partial^2 u_3}{\partial x_3^2} - \frac{\partial^2 u_3}{\partial t^2} - \frac{\beta_e}{\rho}\frac{\partial T}{\partial x_3} = 0, \qquad a^2 \equiv c_1^2 + \frac{\mu_1}{\rho} H_0^2. \tag{8.9.15}$$

We simplify (8.9.4) by neglecting the thermomechanical coupling $\beta_e T_0 \dot{u}_{3,3}$. It is well known that the contribution of this term is generally very small (see Boley and Weiner [1960]).

$$\frac{\partial^2 T}{\partial x_3^2} - \frac{1}{\eta}\frac{\partial T}{\partial t} = 0, \qquad \eta \equiv \kappa_0/\rho\gamma. \tag{8.9.16}$$

We introduce the notation

$$\begin{aligned} &u_3 = u, \qquad h_1^0 = h^0, \qquad h_1 = h, \qquad E_2^0 = E^0, \\ &z = a x_3/\eta, \qquad \tau = a^2 t/\eta, \qquad m = \beta_e \eta/\rho a^3, \\ &m_0 = m a^2/c_1^2, \qquad n = \mu_1 H_0 a^3/c^2 \eta, \qquad \beta_0 = \eta(\mu - 1)^2 H_0/c_1^2 a\rho. \end{aligned} \tag{8.9.17}$$

Equations (8.9.14). (8.9.15), and (8.9.7) are then transformed to

$$\begin{aligned} u_{,zz} - u_{,\tau\tau} - mT_{,z} &= 0, \\ T_{,zz} - T_{,\tau} &= 0, \\ h^0_{,zz} - (a/c)^2 h^0_{,\tau\tau} &= 0, \qquad z > 0, \\ E^0_{,zz} - (a/c)^2 E^0_{,\tau\tau} &= 0, \end{aligned} \tag{8.9.18}$$

where we replaced (8.9.7) by its equivalent wave equations. The boundary conditions (8.9.11) and (8.9.12) are expressed as

$$\begin{aligned} u_{,z} - m_0 T + \beta_0 h^0 &= 0, \\ h^0_{,z} + nu_{,\tau\tau} &= 0, \\ T = T_1 H(\tau), &\qquad z = 0. \end{aligned} \tag{8.9.19}$$

In addition, at $z = \infty$ the fields u and T must vanish, and at $z = -\infty$ the fields h^0 and E^0 must vanish. Initially, the medium is assumed to be at rest.

We proceed to obtain the solution of the set (8.9.18) by means of the Laplace transform, defined by

$$\bar{f}(z, p) = \int_0^\infty f(z, \tau) e^{-p\tau}\, d\tau. \tag{8.9.20}$$

Applying the Laplace transform to (8.9.18) and (8.9.19) we will have

$$\begin{aligned} \bar{u}_{,zz} - p^2\bar{u} - m\bar{T}_{,z} = 0, &\qquad \bar{T}_{,zz} - p\bar{T} = 0, \\ \bar{h}^0_{,zz} - (ap/c)^2\bar{h}^0 = 0, &\qquad \bar{E}^0_{,zz} - (ap/c)^2\bar{E}^0 = 0, \end{aligned} \tag{8.9.21}$$

and

$$\bar{u}_{,z} - m_0\bar{T} + \beta_0\bar{h}^0 = 0, \qquad \bar{h}^0_{,z} + np^2\bar{u} = 0, \qquad \bar{T} = T_1/p. \tag{8.9.22}$$

Integrating (8.9.21) using boundary conditions at $z = \pm\infty$, and integrating (8.9.22) using boundary conditions at $z = 0$, we obtain

$$\begin{aligned} \bar{u} = \frac{mT_1}{p^2(1-p)(1+\beta)} &\{[1 - \theta(1-p) + \beta\sqrt{p}]e^{-pz} \\ &- (1+\beta)\sqrt{p}e^{-\sqrt{p}z}\}, \end{aligned} \tag{8.9.23}$$

$$\bar{T} = (T_1/p)e^{-\sqrt{p}z}, \qquad z \geq 0, \tag{8.9.24}$$

$$\bar{h}^0 = -(npc/a)\bar{u}(0, p)\exp(paz/c), \qquad z \leq 0, \tag{8.9.25}$$

where

$$\theta = m_0/m = a^2/c_1^2 > 1, \qquad \beta = \beta_0 cn/a. \tag{8.9.26}$$

The Laplace transform of the mechanical stress ${}_{\mathrm{E}}t_{33}$ is given by

$$\begin{aligned} {}_{\mathrm{E}}\bar{t}_{33} &= \rho c_1^2 \bar{u}_{3,3} - \beta_e \bar{T} = s(\bar{u}_{,z} - m_0\bar{T}), \\ s &= \rho c_1^2 a/\eta. \end{aligned} \tag{8.9.27}$$

Substituting $\bar{u}$ and $\bar{T}$ from (8.9.23) and (8.9.24) into (8.9.27), we obtain

$$
\begin{aligned}
{}_{\mathrm{E}}\bar{t}_{33} = &-\frac{M}{(1+\beta)\theta(1-p)}\left[\left(\frac{1-\theta}{p}+\theta+\frac{\beta}{\sqrt{p}}\right)e^{-pz}\right.\\
&\left.-(1+\beta)\left(\frac{1-\theta}{p}+\theta\right)e^{-\sqrt{p}z}\right],
\end{aligned} \tag{8.9.28}
$$

$$M = \beta_e T_1.$$

The transform $\bar{h}^0$, given by (8.9.25), has the explicit form

$$
\begin{aligned}
\bar{h}^0 &= -\frac{N}{1-p}\left(\frac{1-\theta}{p}+\theta-\frac{1}{\sqrt{p}}\right)\exp(paz/c), \qquad z<0,\\
N &= H_0\mu_1 a^3 m T_1/\eta c(1+\beta),
\end{aligned} \tag{8.9.29}
$$

and the transform of $\bar{h}$ follows from (8.9.14)

$$
\begin{aligned}
\bar{h} &= \frac{P}{1-p}\left[\left(\frac{1-\theta}{p}+\theta+\frac{\beta}{\sqrt{p}}\right)e^{-pz}-\frac{1+\beta}{p}e^{-\sqrt{p}z}\right],\\
P &= mH_0T_1a/(1+\beta)\eta.
\end{aligned} \tag{8.9.30}
$$

The inverse transforms of (8.9.28) and (8.9.29) are found to be

$$
\begin{aligned}
{}_{\mathrm{E}}t_{33}(z,\tau) = &-\frac{M}{(1+\beta)\theta}\{(1-\theta)f_1(z,\tau)-f_2(z,\tau)-\beta f_3(z,\tau)\\
&-(1+\beta)[(1-\theta)g_1(z,\tau)-g_2(z,\tau)]\},
\end{aligned} \tag{8.9.31}
$$

$$h^0(z,\tau) = N[\theta-1+f_2^0(z,\tau)-f_3^0(z,\tau)], \qquad z<0, \tag{8.9.32}$$

$$
\begin{aligned}
h(z,\tau) = P\{&(1-\theta)f_1(z,\tau)-f_2(z,\tau)-\beta f_3(z,\tau)\\
&-(1+\beta)[g_1(z,\tau)-g_2(z,\tau)]\},
\end{aligned} \tag{8.9.33}
$$

where

$$
\begin{aligned}
f_1(z,\tau) &= H(\tau-z), \qquad f_2(z,\tau) = H(\tau-z)\exp(\tau-z),\\
f_2^0(z,\tau) &= H[\tau+(az/c)]\exp[\tau+(a/c)z],\\
f_3(z,\tau) &= H(\tau-z)\exp(\tau-z)\,\mathrm{Erf}(\tau-z)^{1/2},\\
f_3^0(z,\tau) &= H[\tau+(az/c)]\exp[\tau+(a/c)z]\,\mathrm{Erf}[\tau+(a/c)z],\\
g_1(z,\tau) &= \mathrm{Erfc}(z/\sqrt{4\tau}),\\
g_2(z,\tau) &= \tfrac{1}{2}\exp(\tau+z)\,\mathrm{Erfc}[(z/\sqrt{4\tau})+\sqrt{\tau}]\\
&\quad+\tfrac{1}{2}\exp(\tau-z)\,\mathrm{Erfc}[(z/\sqrt{4\tau})-\sqrt{\tau}].
\end{aligned} \tag{(8.9.34)}
$$

When the magnetic field is absent ($H_0 = 0$), we have the special case of a thermoelastic problem for which t_{33} reduces to

$$t_{33}(z,\tau) = -M[g_2(z,\tau)-f_2(z,\tau)], \qquad h^0 = h = 0, \tag{8.9.35}$$

a result which was obtained by Danilovskaya [1950]. At $z = 0$, the mechanical stress ${}_{\mathrm{E}}t_{33}$ is given by

$$ {}_{\mathrm{E}}t_{33}(0, \tau) = -\frac{M\beta}{(1+\beta)\theta}[e^{\tau}(1 - \operatorname{Erf}\sqrt{\tau}) - 1 + \theta]. \qquad (8.9.36)$$

When this is added to the electromagnetic stresses ${}_{\mathrm{M}}t_{33} - {}_{\mathrm{M}}t_{33}^{+}$ we obtain the total stress which vanishes, as it should.

The solutions (8.9.31)–(8.9.33) are expressed in terms of the functions f_1, f_2, and f_3 which represent propagating waves, and g_1 and g_2 which are diffusional in character. At plane $z = \text{const.}$, a term of stress represented by g_1 and g_2 appears immediately. This stress is modified with the arrival of a wave represented by f_1, f_2, and f_3 at the time $\tau = z$, i.e., $t = x_3/a$. The wave propagates with the phase velocity a which is larger than the phase velocity of the medium when the bias field is absent. When the wave arrives the stress undergoes a jump by the amount $M/(1 + \beta)$.

When the magnetic field is absent ($\beta = 0$, $\theta = 1$) the wave propagates with the velocity c_1 and the jump in stress, upon arrival of the wave, is M. For $z > \tau$, the stress decreases rapidly approaching zero as $z \to \infty$.

The situation for the induced magnetic field h is similar to that of the normal stress t_{33} and can be seen from (8.9.33). The modified electromagnetic wave propagates with the Alfvèn velocity

$$a = [c_1^2 + (\mu H_0^2/\rho)]^{1/2}. \qquad (8.9.37)$$

Upon arrival of the wave $\tau = z$, the magnetic field $h(z, \tau)$ undergoes a jump by the amount $P\theta$. The externally induced magnetic field has the form

$$\begin{aligned} h^0(z, \tau) = N\{\theta - 1 + \exp[\tau + (az/c)] \operatorname{Erfc}[\tau + (az/c)] \\ \cdot H[\tau + (az/c)]\}, \qquad z < 0. \end{aligned} \qquad (8.9.38)$$

This wave propagates in the vacuum with the velocity of light. At the wave front $\tau = -x_3/c$, the vacuum field h^0 acquires a jump $N\theta$. At the boundary $z = 0$, we have

$$h^0(0, \tau) = N(\theta - 1 + e^{\tau} \operatorname{Erfc}\sqrt{\tau}) > \theta N. \qquad (8.9.39)$$

8.10. Homogeneous Strain in a Magnet

In order to obtain some idea of the nonlinear effects in magnetoelasticity, in this section and in Sections 8.11–8.13, we explore several problems concerned with finite deformations and fields. To understand the coupling between the deformation and the magnetic field, we consider here the homogeneous deformation of a cuboid subject to a magnetic field.

Referring to the same rectangular coordinates, a point in the underformed body is located by X_K. After deformation this point takes the spatial place x_k

$$x_1 = \lambda_1 X_1, \qquad x_2 = \lambda_2 X_2, \qquad x_3 = \lambda_3 X_3,$$

where λ_1, λ_2, λ_3 are, respectively, the stretches in the x_1-, x_2-, and x_3-directions. The deformation tensors $\overset{-1}{\mathbf{c}}$ and $\mathbf{c}$ are given by

$$\| \overset{-1}{c}_{kl} \| = \begin{bmatrix} \lambda_1^2 & 0 & 0 \\ 0 & \lambda_2^2 & 0 \\ 0 & 0 & \lambda_3^2 \end{bmatrix}, \qquad \| c_{kl} \| = \begin{bmatrix} \lambda_1^{-2} & 0 & 0 \\ 0 & \lambda_2^{-2} & 0 \\ 0 & 0 & \lambda_3^{-2} \end{bmatrix}. \tag{8.10.1}$$

The joint invariants of $\overset{-1}{\mathbf{c}}$ and $\mathbf{B}$ are (see (8.2.29))

$$\begin{aligned} &\mathrm{I} = \lambda_1^2 + \lambda_2^2 + \lambda_3^2, \quad &&\mathrm{II} = \lambda_1^2\lambda_2^2 + \lambda_2^2\lambda_3^2 + \lambda_3^2\lambda_1^2, \quad &&\mathrm{III} = \lambda_1^2\lambda_2^2\lambda_3^2, \\ &\mathrm{I}_5 = B^2, &&\mathrm{I}_7 = \sum_k B_k\lambda_k^2 B_k, &&\mathrm{I}_9 = \sum_k B_k\lambda_k^4 B_k. \end{aligned} \tag{8.10.2}$$

In the absence of the electric field, $\mathbf{E} = \mathbf{0}$, $\mathbf{J} = \mathbf{0}$, and the constant fields $\mathbf{B}$ and $\mathbf{H}$ satisfy Maxwell's equations. The equations of equilibrium (8.2.2) are also trivially satisfied. We calculate the stress field ${}_{\mathrm{E}}\mathbf{t}$, the magnetization $\mathbf{M}$, and the magnetic field $\mathbf{H}$ by use of (8.2.25) and (8.2.26)

$${}_{\mathrm{E}}t_{kl} = \tau_{\underline{k}}\delta_{\underline{k}l} - (\gamma_{\underline{k}} + \gamma_{\underline{l}})B_{\underline{k}}B_{\underline{l}}, \tag{8.10.3}$$

$$M_k = \chi_{\underline{k}}B_{\underline{k}}, \tag{8.10.4}$$

$$H_k = (1 - \chi_{\underline{k}})B_{\underline{k}}, \tag{8.10.5}$$

where underbarred indices are not summed, and

$$\begin{aligned} \tau_k &= b_{-1}\lambda_k^2 + b_0 + b_1\lambda_k^{-2}, \\ \chi_k &= \chi^{\mathrm{B}} + \chi_1^{\mathrm{B}}\lambda_k^2 + \chi_2^{\mathrm{B}}\lambda_k^4, \\ \gamma_k &= \chi_1^{\mathrm{B}}\lambda_k^2 + 2\chi_2^{\mathrm{B}}\lambda_k^4. \end{aligned} \tag{8.10.6}$$

Constitutive moduli $b_{-1}, b_0, b_1, \chi^{\mathrm{B}}, \chi_1^{\mathrm{B}}$, and χ_2^{B} are functions of the invariants listed in (8.10.2). Since λ_k and B_k are constants these moduli and, consequently, τ_k, χ_k, and γ_k are also constants.

Cauchy's stress tensor is given by

$$t_{kl} = {}_{\mathrm{E}}t_{kl} - M_k B_l = \tau_{\underline{k}}\delta_{\underline{k}l} - (\gamma_{\underline{k}} + \gamma_{\underline{l}} + \chi_{\underline{k}})B_{\underline{k}}B_{\underline{l}}. \tag{8.10.7}$$

The surface traction $\mathbf{t}_{(\mathbf{n})}$ on a surface, with exterior unit normal $\mathbf{n}$, is given by

$$t_{(\mathbf{n})l} = (t_{kl}^{-} + {}_{\mathrm{M}}t_{kl}^{-} - {}_{\mathrm{M}}t_{kl}^{+})n_k, \tag{8.10.8}$$

where a superscript $(-)$ indicates the value taken from the negative side of the normal (interior of the body), and $(+)$ indicates the value taken from the positive side of the normal (exterior of the body). The Maxwell stresses are given by

$$\begin{aligned} {}_{\mathrm{M}}t_{kl}^{-} &= B_k H_l - \tfrac{1}{2}\mathbf{B}\cdot(\mathbf{H} - \mathbf{M})\delta_{kl} \\ &= (1 - \chi_{\underline{l}})B_{\underline{k}}B_{\underline{l}} - \tfrac{1}{2}\delta_{kl}\sum_r (1 - 2\chi_r)B_r^2, \\ {}_{\mathrm{M}}t_{kl}^{+} &= H_k^0 H_l^0 - \tfrac{1}{2}\mathbf{H}^0\cdot\mathbf{H}^0\delta_{kl}, \end{aligned} \tag{8.10.9}$$

where $\mathbf{H}^0$ is the magnetic field outside of the body, adjacent to the surface.

Substituting (8.10.7) and (8.10.9) into (8.10.8) we obtain

$$t_{(\mathbf{n})l} = \sum_k [\tau_k \delta_{kl} - (\gamma_k + \gamma_{\underline{l}} + \chi_k + \chi_{\underline{l}} - 1) B_k B_{\underline{l}}] n_k - \mathbf{H}^0 \cdot \mathbf{n} H_l^0$$
$$- \tfrac{1}{2}\left(\mathbf{B}\cdot\mathbf{B} - \mathbf{H}^0\cdot\mathbf{H}^0 - 2\sum_k \chi_k B_k^2\right) n_l. \tag{8.10.10}$$

For example, on the surface $x_2 = b$ we have the traction

$$t_{(2)l} = \left(\tau_2 - \tfrac{1}{2}\mathbf{B}\cdot\mathbf{B} + \tfrac{1}{2}\mathbf{H}^0\cdot\mathbf{H}^0 + \sum_k \chi_k B_k^2\right)\delta_{2l}$$
$$- (\chi_2 + \chi_{\underline{l}} + \gamma_2 + \gamma_{\underline{l}} - 1) B_2 B_{\underline{l}} - H_2^0 H_l^0. \tag{8.10.11}$$

It is clear that, unlike the purely elastic case, the shear components of the tractions do not vanish unless B_1, B_3 and H_1^0, H_3^0 vanish.

Boundary conditions on the surfaces $x_2 = \pm b$ and $x_3 = \pm c$ require that

$$\begin{aligned} B_2 &= H_2^0, \quad & H_1 &= H_1^0, \quad & H_3 &= H_3^0, \quad & x_2 &= \pm b,\\ B_3 &= H_3^0, \quad & H_1 &= H_1^0, \quad & H_2 &= H_2^0, \quad & x_3 &= \pm c. \end{aligned} \tag{8.10.12}$$

For the special case $H_2^0 = H_3^0 = 0$, (8.10.10) reduces to

$$t_{(\mathbf{n})l} = \sum_k (\tau_k n_k \delta_{kl}) - (\gamma_1 + \gamma_{\underline{l}} + \chi_{\underline{l}} - \chi_1 + \chi_1^2) B_1^2 n_1 \delta_{\underline{l}1}$$
$$- \tfrac{1}{2}[(1 - 2\chi_1) B_1^2 - H_1^{0^2}] n_l. \tag{8.10.13}$$

Therefore, on the surface $x_2 = b$, we have the traction

$$t_{(2)l} = [\tau_2 + \tfrac{1}{2}(\chi_1 B_1)^2]\delta_{2l}. \tag{8.10.14}$$

This indicates that the shear components of traction vanish. In fact, this is also the case on the surfaces $x_3 = c$ and $x_1 = a$, so that, in this case, only normal tractions act on the surfaces of the cuboid.

We consider several special cases.

(i) *Simple Extension.* For the particular case of simple extension, we have $\lambda_2 = \lambda_3$, where $t_{(2)2} = t_{(3)3} = 0$ and $t_{(1)1} = t_1$ is the applied traction. Thus

$$\begin{aligned} t_1 &= \tau_1 - (2\gamma_1 + \tfrac{1}{2}\chi_1^2) B_1^2,\\ t_2 &= t_3 = \tau_2 + \tfrac{1}{2}(\chi_1 B_1)^2 = 0, \end{aligned} \tag{8.10.15}$$

where, upon using the expression (8.2.29) of b_{-1}, b_0, b_1, $\rho/\rho_0 = 1/\lambda_1\lambda_2\lambda_3$, and (8.10.6)

$$\begin{aligned} \tau_1 &= 2\lambda_1\left(\frac{\partial\Sigma}{\partial \mathrm{I}}\lambda_2^{-2} + 2\frac{\partial\Sigma}{\partial \mathrm{II}} + \frac{\partial\Sigma}{\partial \mathrm{III}}\lambda_2^2\right),\\ \tau_2 &= \tau_3 = \frac{2}{\lambda_1}\left\{\frac{\partial\Sigma}{\partial \mathrm{I}} + (\lambda_1^2 + \lambda_2^2)\frac{\partial\Sigma}{\partial \mathrm{II}} + \lambda_1^2\lambda_2^2\frac{\partial\Sigma}{\partial \mathrm{III}}\right\},\\ \chi_1 &= \chi^{\mathrm{B}} + \chi_1^{\mathrm{B}}\lambda_1^2 + \chi_2^{\mathrm{B}}\lambda_1^4,\\ \gamma_1 &= \chi_1^{\mathrm{B}}\lambda_1^2 + 2\chi_2^{\mathrm{B}}\lambda_1^4. \end{aligned} \tag{8.10.16}$$

Equations (8.10.15) constitute two transcendental equations to determine the stretches λ_1 and λ_2 in terms of the applied magnetic field and axial traction. In the case $B_1 = 0$, the second equation determines λ_2 as a function of λ_1 which, in the linear theory, leads to the contraction ratio known as the Poisson ratio (see Eringen [1962, p. 177.)] In the case $t_1 = 0$ but $B_1 \neq 0$, (8.10.2) may be used to calculate the stretches as functions of the applied magnetic field. To obtain an explicit form we assume that the extensions are small compared to unity (as in the case of metals), and retain only the lowest-order terms in extensions and in the applied field. To this end, we substitute

$$\lambda_k = 1 + e_k, \qquad |e_k| \ll 1,$$

and the linear stress–strain relations expressed in terms of e_k read

$$\begin{aligned} \tau_1 &= (\lambda_e + 2\mu_e)e_1 + 2\lambda_e e_2, \\ \tau_2 &= \lambda_e e_1 + 2(\lambda_e + \mu_e)e_2. \end{aligned} \tag{8.10.17}$$

Then we obtain

$$\begin{aligned} (\lambda_e + 2\mu_e)e_1 + 2\lambda_e e_2 &= t_1 - K_1 B_1^2, \\ \lambda_e e_1 + 2(\lambda_e + \mu_e)e_2 &= -K_2 B_1^2, \end{aligned} \tag{8.10.18}$$

where

$$\begin{aligned} K_1 &= 2\chi_1^B + 4\chi_2^B, \\ K_2 &= \tfrac{1}{2}(\chi^B + \chi_1^B + \chi_2^B).^2 \end{aligned} \tag{8.10.19}$$

Hence

$$\begin{aligned} e_1 &= \frac{1}{3\lambda_e + 2\mu_e}\left\{\left(\frac{\lambda_e}{\mu_e} + 1\right)t_1 - \left[\left(\frac{\lambda_e}{\mu_e} + 1\right)K_1 + \frac{\lambda_e}{\mu_e}K_2\right]B_1^2\right\}, \\ e_2 &= -\frac{1}{3\lambda_e + 2\mu_e}\left\{\frac{\lambda_e}{2\mu_e}t_1 + \left[\frac{\lambda_e}{2\mu_e}K_1 - \left(\frac{\lambda_e}{2\mu_e} + 1\right)K_2\right]B_1^2\right\}. \end{aligned} \tag{8.10.20}$$

Extensions e_1 and e_2 are therefore proportional to B_1^2 for $t_1 = 0$. Consequently, a bar would undergo deformation in proportion to the square of the magnetic field, even when there is no applied load. This effect is known as *magnetostriction* and can be used to deform elastic solids.

(ii) *Incompressible Solids*: For the incompressible media we have $\lambda_1\lambda_2\lambda_3 = 1$. Cauchy's stress tensor (8.10.7) retains its general form, however, τ_k takes the form

$$\tau_k = -p + 2\frac{\partial\Sigma}{\partial \mathrm{I}}\lambda_k^2 - 2\frac{\partial\Sigma}{\partial \mathrm{II}}\lambda_2^{-2}. \tag{8.10.21}$$

With the replacement $\lambda_3 = 1/\lambda_1\lambda_2$ in this equation, and in the expression (8.10.6) of χ_k and γ_k, from (8.10.7) we obtain

$$t_{\underline{kk}} = -p + 2\frac{\partial\Sigma}{\partial \mathrm{I}}\lambda_k^2 - 2\frac{\partial\Sigma}{\partial \mathrm{II}}\lambda_k^{-2} - (\chi^B + 3\chi_1^B\lambda_{\underline{k}}^2 + 5\chi_2^B\lambda_{\underline{k}}^4)B_{\underline{k}}B_{\underline{k}}. \tag{8.10.22}$$

In the case of simple extension we have $\lambda_2 = \lambda_3$, $t_{22} = t_{33}$. Compatible with this we also take $B_2 = B_3 = 0$. It then follows that the pressure is given by

$$p = \frac{2}{\lambda_1}\frac{\partial\Sigma}{\partial \mathrm{I}} - 2\lambda_1\frac{\partial\Sigma}{\partial \mathrm{II}}, \tag{8.10.23}$$

where we used $\lambda_1\lambda_2^2 = 1$. Substituting this into (8.10.22) for $k = 1$ we will have

$$t_{11} = 2\left(\lambda_1^2 - \frac{1}{\lambda_1}\right)\left(\frac{\partial\Sigma}{\partial \mathrm{I}} + \frac{1}{\lambda_1}\frac{\partial\Sigma}{\partial \mathrm{II}}\right) - (\chi^{\mathrm{B}} + 3\chi_1^{\mathrm{B}}\lambda_1^2 + 5\chi_2^{\mathrm{B}}\lambda_1^4)B_1^2. \tag{8.10.24}$$

Again we see that the magnetostrictive effect causes a stress in proportion to the square of the applied field.

(iii) *Hydrostatic Case*: If the stretches are the same, i.e., $\lambda_1 = \lambda_2 = \lambda_2$, Cauchy's stress tensor (8.10.7) does not reduce to hydrostatic pressure unless $B_1 = B_2 = B_3 = B$. In this case, then we have

$$t_{kl} = -\bar{p}\delta_{kl}, \tag{8.10.25}$$

where

$$\bar{p} = -2\left(\frac{1}{K}\frac{\partial\Sigma}{\partial \mathrm{I}} + 2K\frac{\partial\Sigma}{\partial \mathrm{II}} + K^3\frac{\partial\Sigma}{\partial \mathrm{III}}\right) - (\chi^{\mathrm{B}} + 3\chi_1^{\mathrm{B}}K^2 + 5\chi_2^{\mathrm{B}}K^4)B^2. \tag{8.10.26}$$

Since $K = (\rho_0/\rho)^{1/3}$, it follows that we have an *equation of state* of the form $\bar{p} = f(\rho, B)$. In fact, this is used in the case of magnetic stars.

8.11. Simple Shear

In a simple shear, an aggregate of parallel planes moves, remaining parallel to their original direction by an amount proportional to their distance from one of these planes. Thus, a rectangular cross section $Z =$ const. (OACB) of a cuboid after deformation becomes a parallelogram (OAcb) (Figure 8.11.1).

Referring to the same rectangular coordinates, the deformation is prescribed by

$$x = X + SY, \qquad y = Y, \qquad z = Z. \tag{8.11.1}$$

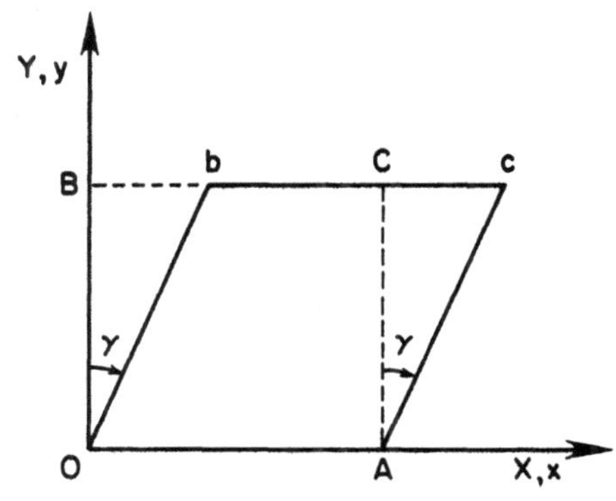

Figure 8.11.1. Simple shear.

The deformation tensors $\overset{-1}{\mathbf{c}}$ and $\mathbf{c}$ are given by

$$\|\overset{-1}{c}_{kl}\| = \begin{bmatrix} 1+S^2 & S & 0 \\ S & 1 & 0 \\ 0 & 0 & 1 \end{bmatrix}, \qquad \|c_{kl}\| = \begin{bmatrix} 1 & -S & 0 \\ -S & 1+S^2 & 0 \\ 0 & 0 & 1 \end{bmatrix}. \tag{8.11.2}$$

According to (8.2.29), the invariants of $\overset{-1}{\mathbf{c}}$ and $\mathbf{B}$ are

$$\begin{aligned} &\mathrm{I} = \mathrm{II} = 3 + S^2, \qquad \mathrm{III} = 1, \qquad \mathrm{I}_5 = B^2, \\ &\mathrm{I}_7 = (1 + S^2)B_1^2 + 2SB_1B_2 + B_2^2 + B_3^2, \\ &\mathrm{I}_9 = (1 + 3S^2 + S^4)B_1^2 + 2S(2 + S^2)B_1B_2 + (1 + S^2)^2B_2^2 + B_3^2. \end{aligned} \tag{8.11.3}$$

The deformation is therefore isochoric. In the absence of the electric field, $\mathbf{E} = \mathbf{0}$, $\mathbf{J} = \mathbf{0}$, and the constant fields $\mathbf{B}$ and $\mathbf{H}$ satisfy Maxwell's equations. The equations of equilibrium (8.2.2) are also satisfied trivially.

We consider the case of a magnetic field directed along the x-axis, i.e., $H_2 = H_3 = 0$. The stress field ${}_{\mathrm{E}}\mathbf{t}$ for incompressible solids, magnetization $\mathbf{M}$, and the magnetic field $\mathbf{H}$ are given by

$$\begin{aligned} {}_{\mathrm{E}}t_{kl} &= -p\delta_{kl} + 2\frac{\partial\Sigma}{\partial \mathrm{I}}\overset{-1}{c}_{kl} - 2\frac{\partial\Sigma}{\partial \mathrm{II}}c_{kl} - (\gamma_{kl} + \gamma_{lk})B_1^2, \\ M_k &= \chi_k B_1, \\ H_k &= (\delta_{k1} - \chi_k)B_1, \end{aligned} \tag{8.11.4}$$

where p is an unknown pressure and

$$\begin{aligned} \gamma_{kl} &= \chi_1^{\mathrm{B}}\delta_{k1}\overset{-1}{c}_{l1} + 2\chi_2^{\mathrm{B}}\delta_{k1}\overset{-1}{c}_{lm}\overset{-1}{c}_{m1}, \\ \chi_k &= \chi^{\mathrm{B}}\delta_{k1} + \chi_1^{\mathrm{B}}\overset{-1}{c}_{k1} + \chi_2^{\mathrm{B}}\overset{-1}{c}_{km}\overset{-1}{c}_{m1}. \end{aligned} \tag{8.11.5}$$

Here Σ, χ^{B}, χ_1^{B}, and χ_2^{B} are functions of invariants (8.11.3), i.e.,

$$\begin{aligned} &\mathrm{I} = \mathrm{II} = 3S^2, \qquad \mathrm{III} = 1, \qquad \mathrm{I}_5 = B_1^2, \\ &\mathrm{I}_7 = (1 + S^2)B_1^2, \qquad \mathrm{I}_9 = [(1 + S^2)^2 + S^2]B_1^2. \end{aligned} \tag{8.11.6}$$

Hence they are functions of S^2 and B_1^2, which are constants. Cauchy's stress tensor is given by

$$\begin{aligned} t_{kl} = {}_{\mathrm{E}}t_{kl} - M_kB_l = -p\delta_{kl} + 2\frac{\partial\Sigma}{\partial \mathrm{I}}\overset{-1}{c}_{kl} - 2\frac{\partial\Sigma}{\partial \mathrm{II}}c_{kl} \\ - (\gamma_{kl} + \gamma_{lk} + \chi_k\delta_{l1})B_1^2. \end{aligned} \tag{8.11.7}$$

The surface traction $\mathbf{t}_{(\mathbf{n})}$ on a surface with exterior unit normal $\mathbf{n}$ is given by

$$t_{(\mathbf{n})l} = (t_{kl}^- + {}_{\mathrm{M}}t_{kl}^- - {}_{\mathrm{M}}t_{kl}^+)n_k. \tag{8.11.8}$$

Maxwell's stress tensor ${}_{\mathrm{M}}\mathbf{t}$ is calculated by

$$\begin{aligned} {}_{\mathrm{M}}t_{kl}^- &= (\delta_{k1}\delta_{l1} - \tfrac{1}{2}\delta_{kl} - \chi_l\delta_{k1} + \chi_1\delta_{kl})B_1^2, \\ {}_{\mathrm{M}}t_{kl}^+ &= H_k^0H_l^0 - \tfrac{1}{2}\mathbf{H}^0\cdot\mathbf{H}^0\delta_{kl}, \end{aligned} \tag{8.11.9}$$

where $\mathbf{H}^0$ is the magnetic field on the surface from outside the body. On the surface $x_2 = b$, we have

$$H_2^0 = 0,$$
$$H_1^0 = (1 - \chi_1)B_1, \qquad x_2 = b. \tag{8.11.10}$$

Combining (8.11.7)–(8.11.10) we obtain tractions. On the surface $x_2 = b$, we have $n_k = \delta_{k2}$ and thus

$$t_{(2)l} = -p\delta_{2l} + 2\frac{\partial \Sigma}{\partial \mathrm{I}} \overset{-1}{c}_{2l} - 2\frac{\partial \Sigma}{\partial \mathrm{II}} c_{2l}$$
$$- [\gamma_{l2} + \gamma_{2l} + \chi_2\delta_{l1} - \tfrac{1}{2}\chi_1^2\delta_{2l}]B_1^2, \tag{8.11.11}$$

or in explicit components

$$t_{(2)1} = 2S\left(\frac{\partial \Sigma}{\partial \mathrm{I}} + \frac{\partial \Sigma}{\partial \mathrm{II}}\right) - 2S'[\chi_1^{\mathrm{B}} + 2\chi_2^{\mathrm{B}}(2 + S^2)]B_1^2,$$
$$t_{(2)2} = -p + 2\frac{\partial \Sigma}{\partial \mathrm{I}} - 2(1 + S^2)\frac{\partial \Sigma}{\partial \mathrm{II}} + \tfrac{1}{2}\chi_1^2 B_1^2, \tag{8.11.12}$$
$$t_{(2)3} = 0,$$

where

$$\chi_1 = \chi^{\mathrm{B}} + (1 + S^2)\chi_1^{\mathrm{B}} + [(1 + S^2)^2 + S^2]\chi_2^{\mathrm{B}}. \tag{8.11.13}$$

From (8.11.12) it is clear that the magnetic field exerts a shear and normal traction on the surface bc, proportional to B_1^2. To maintain the simple shear we must therefore apply a normal traction $t_{(2)2}$, in addition to a shear, to the surfaces OA and bc of the body. For the linear strain theory the elastic part of $t_{(2)2}$ vanishes but the magnetic part remains unless B_1^2 is small. This is the *Poynting effect* arising from the magnetostriction. In order to prohibit the volume change it is also necessary to apply a normal traction $t_{(3)3}$. This is the so-called *Kelvin effect*. These nonlinear effects are not present in the linear theory where both the strains and the magnetic field are small. The universal relation

$$t_{11} - t_{22} = St_{12},$$

that is valid in the case of $B = 0$, is no longer valid when the magnetic field is present.

8.12. Torsion of a Cylindrical Magnet

A uniform twist of a circular cylinder may be represented by

$$r = R, \qquad \theta = \Theta + SZ, \qquad z = Z, \tag{8.12.1}$$

where (R, Θ, Z) and (r, θ, z) are, respectively, the cylindrical coordinates of a point in the undeformed and deformed bodies (Figure 8.12.1). The constant S

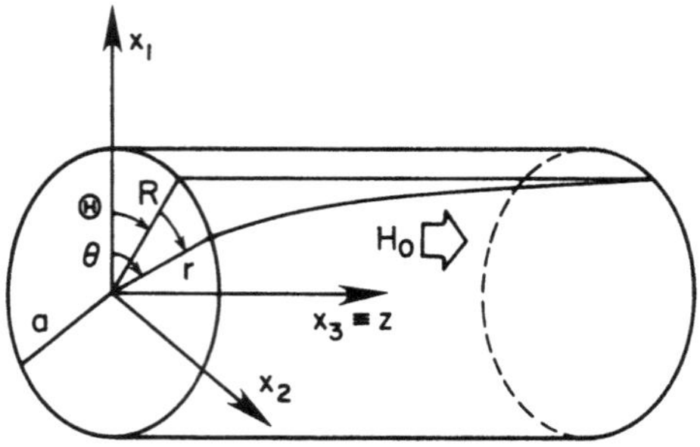

Figure 8.12.1. Torsion of a cylindrical magnet.

is the angle of twist per unit length. The metric tensors are given by

$$\|G_{KL}\| = \begin{bmatrix} 1 & 0 & 0 \\ 0 & R^2 & 0 \\ 0 & 0 & 1 \end{bmatrix}, \qquad \|g_{kl}\| = \begin{bmatrix} 1 & 0 & 0 \\ 0 & r^2 & 0 \\ 0 & 0 & 1 \end{bmatrix}. \tag{8.12.2}$$

The deformation tensors $\overset{-1}{\mathbf{c}}$ and $\mathbf{c}$ in cylindrical coordinates are calculated by

$$\overset{-1}{c}{}^{kl} = G^{KL} x^k_{,K} x^l_{,L}, \qquad c_{kl} = G_{KL} X^K_{,k} X^L_{,l}, \tag{8.12.3}$$

where $X^1 = R$, $X^2 = \Theta$, $X^3 = Z$ and $x^1 = r$, $x^2 = \theta$, $x^3 = z$. By means of the metric tensors (8.12.2) and their reciprocals we raise and lower indices so that

$$\|\overset{-1}{c}{}^{kl}\| = \begin{bmatrix} 1 & 0 & 0 \\ 0 & r^2 + S^2 & 0 \\ 0 & S & 1 \end{bmatrix}, \qquad \|c^{kl}\| = \begin{bmatrix} 1 & 0 & 0 \\ 0 & r^{-2} & -S \\ 0 & -S & 1 + S^2 r^2 \end{bmatrix}, \tag{8.12.4}$$

$$\|\overset{-1}{c}{}^k{}_l\| = \begin{bmatrix} 1 & 0 & 0 \\ 0 & 1 + S^2 r^2 & S \\ 0 & Sr^2 & 1 \end{bmatrix}, \qquad \|c^k{}_l\| = \begin{bmatrix} 1 & 0 & 0 \\ 0 & 1 & -S \\ 0 & -Sr^2 & 1 + S^2 r^2 \end{bmatrix}, \tag{8.12.5}$$

$$\|\overset{-2}{c}{}^k{}_m\| = \begin{bmatrix} 1 & 0 & 0 \\ 0 & 1 + 3S^2 r^2 + S^4 r^4 & S(2 + S^2 r^2) \\ 0 & Sr^2(2 + S^2 r^2) & 1 + S^2 r^2 \end{bmatrix}.$$

We assume that the cylinder carries a uniform magnetic field in the axial direction, i.e., $H_r = H_\theta = 0$ and $H_z = H$ and the material is incompressible. It then follows that

$${}_{\mathrm{E}}t^k{}_l = -p\delta^k{}_l + 2\frac{\partial \Sigma}{\partial \mathrm{I}} \overset{-1}{c}{}^k{}_l - 2\frac{\partial \Sigma}{\partial \mathrm{II}} c^k{}_l - \beta^k{}_{lmn} B^m B^n, \tag{8.12.6}$$

and

$$\begin{aligned} M^k &= \kappa^k{}_l B^l, & B^1 &= 0, \\ H^k &= (\delta^k{}_l - \kappa^k{}_l) B^l, & H^1 &= H^2 = 0, \end{aligned} \tag{8.12.7}$$

where p is an unknown pressure

$$\begin{aligned}\beta^k{}_{lmn} &= \chi_1^{\mathrm{B}}(\delta^k{}_n \overset{-1}{c}_{lm} + g_{ln} \overset{-1}{c}{}^k{}_m) + 2\chi_2^{\mathrm{B}}(\delta^k{}_n \overset{-2}{c}_{lm} + g_{ln} \overset{-2}{c}{}^k{}_m),\\ \kappa^k{}_l &= \chi^{\mathrm{B}}\delta^k{}_l + \chi_1^{\mathrm{B}} \overset{-1}{c}{}^k{}_l + \chi_2^{\mathrm{B}} \overset{-2}{c}{}^k{}_l,\end{aligned} \tag{8.12.8}$$

and Σ, χ^{B}, χ_1^{B}, and χ_2^{B} are functions of the invariants of $\overset{-1}{\mathbf{c}}$ and $\mathbf{B}$, i.e.,

$$\begin{aligned} &\mathrm{I} = \mathrm{II} = 3 + S^2r^2, \qquad \mathrm{III} = 1,\\ &\mathrm{I}_5 = \mathbf{B}\cdot\mathbf{B}, \qquad \mathrm{I}_7 = \mathbf{B}\cdot \overset{-1}{\mathbf{c}}\mathbf{B}, \qquad \mathrm{I}_9 = \mathbf{B}\cdot \overset{-2}{\mathbf{c}}\mathbf{B}.\end{aligned} \tag{8.12.9}$$

With (8.12.7), Maxwell's equations and the boundary conditions on $\mathbf{H}$ are satisfied trivially. Cauchy's stress tensor is given by

$$t^k{}_l = {}_{\mathrm{E}}t^k{}_l - \kappa^k{}_m B^m B_l. \tag{8.12.10}$$

Maxwell's stress tensor is given by

$${}_{\mathrm{M}}t^k{}_l = {}_{\mathrm{M}}\beta^k{}_{lmn} B^m B^n, \qquad r < a, \tag{8.12.11}$$

$${}_{\mathrm{M}}t^k{}_l = (\delta^k{}_3 g_{l3} - \tfrac{1}{2}\delta^k{}_l)(H)^2, \qquad r > a, \tag{8.12.12}$$

where

$${}_{\mathrm{M}}\beta^k{}_{lmn} = (g_{lm} - \kappa_{lm})\delta^k{}_n - \tfrac{1}{2}(g_{mn} - 2\kappa_{mn})\delta^k{}_l. \tag{8.12.13}$$

We observe that $(8.12.7)_1$ consists of only two equations, since for $k = 1$ it is trivially satisfied. These two equations are nonlinear in B^2, B^3, and r^2 (superscripts on B are not powers). We assume that solutions exist for B^2 and B^3 (not necessarily unique)

$$B^2 = B^2(r^2, H), \qquad B^3 = B^3(r^2, H). \tag{8.12.14}$$

With this then $\beta^k{}_{lmn}$, $\kappa^k{}_l$, and ${}_{\mathrm{M}}\beta^k{}_{lmn}$ are fully determined as functions of r^2 and H.

Equations of equilibrium (8.2.2), with $\mathbf{f} = \dot{\mathbf{v}} = \mathbf{0}$ in cylindrical coordinates, reduce to

$$\begin{aligned}\frac{\partial}{\partial r}\left(-p + 2\frac{\partial \Sigma}{\partial \mathrm{I}} - 2\frac{\partial \Sigma}{\partial \mathrm{II}}\right) - 2S^2 r\frac{\partial \Sigma}{\partial \mathrm{I}} + \frac{\beta_2}{r}B_\theta^2 + \beta_3 S B_\theta B_z = 0,\\ \frac{\partial p}{\partial \theta} = 0, \qquad \frac{\partial p}{\partial z} = 0,\end{aligned} \tag{8.12.15}$$

where

$$\begin{aligned}\beta_2 &= 2\chi_1^{\mathrm{B}}(1 + S^2r^2) + 4\chi_2^{\mathrm{B}}(1 + 3S^2r^2 + S^4r^4),\\ \beta_3 &= 2\chi_1^{\mathrm{B}} + 4\chi_2^{\mathrm{B}}(2 + S^2r^2).\end{aligned} \tag{8.12.16}$$

In deriving (8.12.16) we replaced vectors and tensors by their physical components, e.g., we wrote

$$\begin{aligned} &B^k = B^{(k)}/\sqrt{g_{\underline{kk}}}, \qquad t^k{}_l = t^{(k)}{}_{(l)}\sqrt{g_{\underline{ll}}}/\sqrt{g_{\underline{kk}}},\\ &B^{(1)} = B_r, \qquad B^{(2)} = B_\theta, \qquad B^{(3)} = B_z,\\ &t^{(1)}{}_{(1)} = t_{rr}, \qquad t^{(2)}{}_{(2)} = t_{\theta\theta}, \qquad t^{(2)}{}_{(3)} = t_{\theta z}, \ldots .\end{aligned} \tag{8.12.17}$$

We also find that

$$ {}_{\mathrm{E}}t_{r\theta} = {}_{\mathrm{E}}t_{rz} = 0. \tag{8.12.18}$$

Hence p is a function of r and it is obtained by integrating (8.12.15)

$$p = 2\frac{\partial \Sigma}{\partial \mathrm{I}} - 2\frac{\partial \Sigma}{\partial \mathrm{II}} + \int^r \left(\frac{\beta_2}{r} B_\theta^2 + \beta_3 S B_\theta B_z - 2S^2 r \frac{\partial \Sigma}{\partial \mathrm{I}}\right) dr + C_1. \tag{8.12.19}$$

The constant C_1 is determined by setting to zero the radial component of the surface traction

$$t_{(\mathbf{n})}{}^l = (t^-{}_k{}^l + {}_{\mathrm{M}}t^-{}_k{}^l - {}_{\mathrm{M}}t^+{}_k{}^l) n^k \tag{8.12.20}$$

at $r = a$. From (8.12.10)–(8.12.13) and (8.12.15) we find that

$$t_{(r)r} = {}_{\mathrm{E}}t_{rr} + p_{\mathrm{B}}(r), \tag{8.12.21}$$

where

$$p_{\mathrm{B}} = \lambda^m{}_n B_m B^n = \lambda^2{}_2 B_\theta^2 + (r\lambda^2{}_3 + r^{-1}\lambda^3{}_2) B_\theta B_z + \lambda^3{}_3 B_z^2, \tag{8.12.22}$$

with

$$\begin{aligned} \lambda^m{}_n &= \kappa^m{}_n + \tfrac{1}{2}(\kappa^{3m}\kappa^3{}_n + \delta^m{}_n), \\ \kappa^{mn} &= g^{nr}\kappa^m{}_r. \end{aligned} \tag{8.12.23}$$

Setting $t_{(r)z} = 0$ at $r = a$, through (8.12.19) and (8.12.21) we determine C_1 so that

$$p = 2\frac{\partial \Sigma}{\partial \mathrm{I}} - 2\frac{\partial \Sigma}{\partial \mathrm{II}} + \mathscr{I}(B_\theta, B_z) + p_{\mathrm{B}}(a), \tag{8.12.24}$$

where

$$\mathscr{I}(B_\theta, B_z) = \int_a^r \left(\frac{\beta_2}{r} B_\theta^2 + \beta_3 B_\theta B_z - 2S^2 r \frac{\partial \Sigma}{\partial \mathrm{I}}\right) dr. \tag{8.12.25}$$

Here $p_{\mathrm{B}}(a)$ is the pressure on $r = a$, solely due to the magnetic field. We note that on $r = a$, $t_{(r)\theta} = t_{(r)z} = 0$, so that the surface $r = a$ is now cleared from all tractions. Note, however, that there is no other arbitrary constant left to clear the end surfaces $z = 0, L$, of the cylinder from normal tractions.

Cauchy's stress tensor, given by (8.12.10), reads

$$\begin{aligned} t_{rr} &= -p_{\mathrm{B}} - \mathscr{I}(B_\theta, B_z), \\ t_{\theta\theta} &= -p_{\mathrm{B}} + 2S^2 r^2 \frac{\partial \Sigma}{\partial \mathrm{I}} - \mathscr{I}(B_\theta, B_z) - b_{22} B_\theta^2 - b_{23} S r B_\theta B_z, \\ t_{zz} &= -p_{\mathrm{B}} - 2S^2 r^2 \frac{\partial \Sigma}{\partial \mathrm{II}} - \mathscr{I}(B_\theta, B_z) - b_{32} S r B_\theta B_z - b_{33} B_z^2, \\ t_{z\theta} &= 2Sr\left(\frac{\partial \Sigma}{\partial \mathrm{I}} + \frac{\partial \Sigma}{\partial \mathrm{II}}\right) - b_{42} B_\theta^2 - b_{43} B_\theta B_z - \frac{Sr}{2}\beta_3 B_z^2, \\ t_{z\theta} &= t_{\theta z} = (\kappa_{32}/r)(B_z^2 - B_\theta^2 + S r B_\theta B_z), \\ t_{zr} &= t_{rz} = t_{r\theta} = t_{\theta r} = 0, \end{aligned} \tag{8.12.26}$$

where

$$\begin{aligned}
b_{22} &= \chi^{B} + 3\chi_1^{B}(1 + S^2r^2) + 5\chi_2^{B}(1 + 3S^2r^2 + S^4r^4),\\
b_{23} &= b_{32} = 3\chi_1^{B} + 5\chi_2^{B}(2 + S^2r^2),\\
b_{33} &= \chi^{B} + 3\chi_1^{B} + 5\chi_2^{B}(1 + S^2r^2)\\
b_{42} &= Sr[2\chi_1^{B} + 3\chi_2^{B}(2 + S^2r^2)]\\
b_{43} &= \chi^{B} + \chi_1^{B}(3 + S^2r^2) + \chi_2^{B}(5 + 9S^2r^2 + S^4r^4).
\end{aligned} \tag{8.12.27}$$

The jump conditions at the surface $z = L$ require that $B_z = H_z^0$, $H_\theta = H_\theta^0$, $H_r = H_r^0$, $z = L$. Since Maxwell's equations give $H_z^0 = \text{const.}$ and $H_\theta^0 = C_1/r$ we see that $H_\theta^0 = H_r^0 = 0$, but

$$B_z(r, H) = H^0 = \text{const.} \tag{8.12.28}$$

This cannot be satisfied since B_z is a function of r, as can be seen from $(8.12.14)_2$. However, the integral of B_z over the cross section will give a constant which determines H^0, the situation here is similar to the end problem encountered in elasticity. In fact, the tractions $t_{(z)z}$ and $t_{(z)\theta}$, acting at the end cross sections, cannot be balanced *pointwise* with zero axial load and linearly distributed shear in radial coordinates.

For the tractions at $z = 0, L$, we find

$$\begin{aligned}
t_{(z)r} &= 0,\\
t_{(z)\theta} &= t_{z\theta} + B_\theta B_z - \kappa^2{}_2 B_\theta B_z - r\kappa^2{}_3 B_z^2,\\
t_{(z)z} &= t_{zz} + \tfrac{1}{2}(2\kappa^2{}_2 - \kappa^2{}_3 - 1)B_\theta^2 - 2r\kappa^2{}_3(2 - \kappa^3{}_3)B_\theta B_z\\
&\quad + [2\kappa^3{}_3 - (\kappa^3{}_3)^2 - 1]B_z^2,
\end{aligned} \tag{8.12.29}$$

where $t_{z\theta}$ and t_{zz} are given by (8.12.26) and

$$\begin{aligned}
\kappa^2{}_2 &= \chi^{B} + \chi_1^{B}(1 + S^2r^2) + \chi_2^{B}(1 + 3S^2r^2 + S^4r^4),\\
\kappa^2{}_3 &= S\chi_1^{B} + \chi_2^{B}S(2 + S^2r^2),\\
\chi^3{}_3 &= \chi + \chi_1^{B} + \chi_2^{B}(1 + S^2r^2).
\end{aligned} \tag{8.12.30}$$

Therefore, the *Poynting effect appears as a normal traction $t_{(z)z}$ which must be supplied to the end sections in order to maintain the surface $r = a$ of the cylinder free of tractions.* Even when the shear is small, so that S^2 can be set equal to zero compared to S, we will have magnetic tractions. The surface tractions distributed at the end sections are equipollent to a torque T and an axial load N which can be calculated by

$$T = 2\pi \int_0^a r^2 t_{(z)\theta}\, dr, \qquad N = 2\pi \int_0^a r t_{(z)z}\, dr. \tag{8.12.31}$$

If this axial load is not applied, the cylinder will undergo an axial deformation. In the absence of the magnetic field, the cylinder is known to elongate (see Eringen [1962, p. 182]).

Special Case (Axial Magnetic Field Only: $B_\theta = 0$)

In this case great simplifications are achieved. From (8.12.22) and (8.12.25) we see that

$$p_{\mathrm{B}} = \lambda^3{}_3 B_z^2 = [\kappa^3{}_3 + \tfrac{1}{2}(\kappa^3{}_3)^2 + \tfrac{1}{2}]B_z^2,$$
$$\mathscr{I} = -\int_a^r 2S^2 r \frac{\partial \Sigma}{\partial \mathrm{I}}\, dr \equiv \mathscr{I}_0. \tag{8.12.32}$$

The stress field (8.12.26) reads

$$t_{rr} = -p_{\mathrm{B}} + \mathscr{I}_0,$$
$$t_{\theta\theta} = -p_{\mathrm{B}} + 2S^2 r^2 \frac{\partial \Sigma}{\partial \mathrm{I}} - \mathscr{I}_0,$$
$$t_{zz} = -p_{\mathrm{B}} - 2S^2 r^2 \frac{\partial \Sigma}{\partial \mathrm{II}} - \mathscr{I}_0 - b_{33} B_z^2,$$
$$t_{z\theta} = 2Sr\left(\frac{\partial \Sigma}{\partial \mathrm{I}} + \frac{\partial \Sigma}{\partial \mathrm{II}}\right) - \frac{Sr}{2}\beta_3 B_z^2,$$
$$t_{z\theta} = t_{\theta z} = \kappa^2{}_3 r B_z^2,$$
$$t_{zr} = t_{rz} = t_{r\theta} = t_{\theta r} = 0. \tag{8.12.33}$$

Consequently, the direct effect of the magnetic field (excluding the dependence of Σ on B) appears in t_{zz}, $t_{z\theta}$, and $t_{\theta z}$. The traction an $z = 0, L$, now reduces to

$$t_{(z)r} = 0,$$
$$t_{(z)\theta} = t_{z\theta} - r\kappa^2{}_3 B_z^2, \tag{8.12.34}$$
$$t_{(z)z} = t_{zz} + [2\kappa^3{}_3 - (\kappa^3{}_3)^2 - 1]B_z^2.$$

The Poynting effect due to the magnetic field is again present.

There is no difficulty in examining the case of $B_z = 0$, but $B_\theta \neq 0$. In this case, $t_{\theta\theta}$, $t_{z\theta}$, and $t_{\theta z}$ are affected by the magnetic pressure B_θ^2 and the tractions at $z = 0, L$, are reduced to

$$t_{(z)\theta} = t_{z\theta},$$
$$t_{(z)z} = t_{zz} + \tfrac{1}{2}(2\kappa^2{}_2 - \kappa^2{}_3 - 1)B_\theta^2. \tag{8.12.35}$$

8.13. Electric Conduction in a Twisted Cylinder

In this section we discuss the steady electric conduction in a twisted circular cylinder. All fields are assumed to depend only on the radial coordinate r in the cylindrical coordinates (r, θ, z), with z taken along the axis of the cylinder. The cylindrical coordinates of the undeformed body are denoted by (R, Θ, Z)

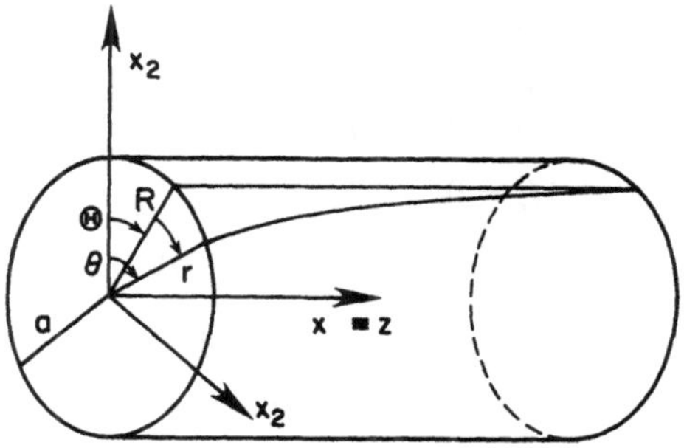

Figure 8.13.1. Twist of a magneto-elastic cylinder.

(Figure 8.13.1). Maxwell's equations read

$$\begin{aligned} \nabla \times \mathbf{E} &= \mathbf{0}, \qquad \nabla \cdot \mathbf{E} = 0, \\ \nabla \times \mathbf{H} &= \frac{1}{c}\mathbf{J}, \qquad \nabla \cdot \mathbf{B} = 0. \end{aligned} \tag{8.13.1}$$

In cylindrical coordinates we have

$$\nabla \times \mathbf{E} = -\frac{\partial E_z}{\partial r}\mathbf{e}_\theta + \frac{1}{r}\frac{\partial}{\partial r}(rE_\theta)\mathbf{e}_z. \tag{8.13.2}$$

From the first two of (8.13.1), it then follows that

$$E_r = 0, \qquad E_z = E, \qquad E_\theta = 0, \tag{8.13.3}$$

where E is constant and we used the regularity condition at $r = 0$.

From the second set of (8.13.1), we obtain

$$\begin{gathered} B_r = 0, \\ \frac{dH_z}{dr} = -\frac{1}{c}J_\theta, \qquad \frac{1}{r}\frac{d}{dr}(rH_\theta) = \frac{1}{c}J_z. \end{gathered} \tag{8.13.4}$$

The constitutive equations (8.2.28) for the current give

$$\begin{aligned} J_r &= -(\sigma_5 B_\theta + \sigma_{11} SrB_z)E_z, \\ J_\theta &= [\sigma_3 Sr + \sigma_7 Sr(2 + r^2S^2) + \sigma_9 B_\theta B_z]E_z, \\ J_z &= [\sigma_1 + \sigma_3 + \sigma_7(1 + r^2S^2) + \sigma_9 B_z^2]E_z, \end{aligned} \tag{8.13.5}$$

where σ_α are functions of the joint invariants of $\overset{-1}{\mathbf{c}}$, $\mathbf{E}$, and $\mathbf{B}$. These invariants depend on r only.

From (8.13.5) it is clear that the conduction is three dimensional and current will flow in all directions. However, we observe that the term with coefficients σ_5 in the expression of $\mathbf{J}$ represents the Hall current which does not contribute to the entropy production. The coefficient of σ_{11} in $\mathbf{J}$ changes the sign of the entropy production $\mathbf{J} \cdot \mathbf{E}$, so that according to the axiom of time reversal σ_{11} must be an odd function of $\mathbf{B}$. Thus, the term containing σ_{11} in (8.2.26) is at least of fourth degree in the strain and electromagnetic variables. Both this term and the Hall current are generally negligible. In this case then $J_r = 0$ and the current flows in helices.

Constitutive equations (8.2.26) for the magnetization gives

$$
\begin{aligned}
M_r &= 0,\\
M_\theta &= [\chi^{\mathrm{B}} + (1 + S^2r^2)\chi_1^{\mathrm{B}} + (1 + S^2r^2)^2\chi_2^{\mathrm{B}} + S^2r^2\chi_2^{\mathrm{B}}]B_\theta \\
&\quad + Sr[\chi_1^{\mathrm{B}} + (2 + S^2r^2)\chi_2^{\mathrm{B}})B_z,\\
M_z &= Sr[\chi_1^{\mathrm{B}} + \chi_2^{\mathrm{B}}(2 + S^2r^2)]B_\theta + [\chi^{\mathrm{B}} + \chi_1^{\mathrm{B}} + (1 + S^2r^2)\chi_2^{\mathrm{B}}]B_z.
\end{aligned}
\tag{8.13.6}
$$

For the incompressible cylinder we obtain constitutive equations of ${}_{\mathrm{E}}\mathbf{t}$

$$
\begin{aligned}
{}_{\mathrm{E}}t^{11} &= {}_{\mathrm{E}}t_{rr} = -p + 2\frac{\partial\Sigma}{\partial \mathrm{I}} - 2\frac{\partial\Sigma}{\partial \mathrm{II}},\\
r^2\,{}_{\mathrm{E}}t^{22} &= {}_{\mathrm{E}}t_{\theta\theta} = -p + 2(1 + S^2r^2)\frac{\partial\Sigma}{\partial \mathrm{I}} - 2\frac{\partial\Sigma}{\partial \mathrm{II}} - B_\theta^2[\tfrac{1}{2}(1 + S^2r^2)\chi_1^{\mathrm{B}}\\
&\quad + 2(1 + 3S^2r^2 + S^4r^4)\chi_2^{\mathrm{B}} - B_\theta B_z Sr[\tfrac{1}{2}\chi_1^{\mathrm{B}} + 2(2 + S^2r^2)\chi_2^{\mathrm{B}}],\\
{}_{\mathrm{E}}t^{33} &= {}_{\mathrm{E}}t_{zz} = -p + 2\frac{\partial\Sigma}{\partial \mathrm{I}} - 2(1 + S^2r^2)\frac{\partial\Sigma}{\partial \mathrm{II}}\\
&\quad - B_\theta B_z Sr[\tfrac{1}{2}\chi_1^{\mathrm{B}} + 2\chi_2^{\mathrm{B}}(2 + S^2r^2)]\\
&\quad - B_z^2[\tfrac{1}{2}\chi_1^{\mathrm{B}} + 2(1 + S^2r^2)\chi_2^{\mathrm{B}}],\\
r\,{}_{\mathrm{E}}t^{23} &= {}_{\mathrm{E}}t_{\theta z} = 2Sr\left(\frac{\partial\Sigma}{\partial \mathrm{I}} + \frac{\partial\Sigma}{\partial \mathrm{II}}\right) - B_\theta^2 Sr[\tfrac{1}{4}\chi_1^{\mathrm{B}} + (2 + S^2r^2)\chi_2^{\mathrm{B}}]\\
&\quad - B_\theta B_z[\tfrac{1}{4}\chi_1^{\mathrm{B}}(2 + S^2r^2) + (2 + 4S^2r^2 + S^4r^4)\chi_2^{\mathrm{B}}]\\
&\quad - B_z^2 Sr[\tfrac{1}{4}\chi_1^{\mathrm{B}} + (2 + S^2r^2)\chi_2^{\mathrm{B}}],\\
{}_{\mathrm{E}}t^{31} &= {}_{\mathrm{E}}t^{12} = 0.
\end{aligned}
\tag{8.13.7}
$$

Equations of equilibrium take the form

$$
\begin{aligned}
&\frac{\partial\,{}_{\mathrm{E}}t_{rr}}{\partial r} + \frac{1}{r}({}_{\mathrm{E}}t_{rr} - {}_{\mathrm{E}}t_{\theta\theta}) + \frac{1}{c}(J_\theta B_z - J_z B_\theta) + M_\theta\frac{\partial B_\theta}{\partial r} + M_z\frac{\partial B_z}{\partial r} = 0,\\
&\qquad -\frac{\partial p}{\partial\theta} - \frac{1}{c}J_r B_z = 0, \qquad -\frac{\partial p}{\partial z} + \frac{1}{c}J_r B_\theta = 0.
\end{aligned}
\tag{8.13.8}
$$

Now we see that unless $J_r = 0$ the pressure will depend on θ and z. Setting $\sigma_5 = \sigma_{11} = 0$ leads to $J_r = 0$ and the pressure p becomes a function of r alone.

Writing $\mathbf{H} = \mathbf{B} - \mathbf{M}$ in (8.2.9) we obtain

$$
\begin{aligned}
\frac{d}{dr}(B_z - M_z) &= -\frac{1}{c}J_\theta,\\
\frac{1}{r}\frac{d}{dr}(rB_\theta - rM_\theta) &= \frac{1}{c}J_z.
\end{aligned}
\tag{8.13.9}
$$

Recalling that, according to (8.13.5) and (18.13.6), M_θ, M_z, J_θ, and J_z are functions of B_z and B_θ, eqs. (8.13.9) constitute a set of two nonlinear equations for two unknown functions for B_z and B_θ. Upon integration and using the boundary conditions

$$H_\theta(a) = H_\theta^0, \qquad H_z(a) = H_z^0,$$

we determine B_z and B_θ as functions of r, where H_θ^0 and H_z^0 are the constant magnetic fields outside the body near $r = a$. Once $B_\theta(r)$ and $B_z(r)$ are determined, $(8.13.8)_1$ may be integrated, with the boundary conditions on tractions on $r = a$, to determine the pressure p. The stress field will then follow from (8.13.7). Unfortunately, the solution of (8.13.9) is not possible until explicit forms of the constitutive coefficients appearing in (8.13.6) and (8.13.7) are known. Even then integration of (8.13.9) may be difficult. Therefore we consider certain approximations.

For a conducting cylinder, magnetization is generally small so that we may retain only the linear terms in (8.2.26). This means that in (8.13.6) all terms other than the factor of χ^B are negligible, i.e.,

$$M_r = 0, \qquad M_\theta = \chi^B B_\theta, \qquad M_z = \chi^B B_z, \tag{8.13.10}$$

where χ^B is a constant. Next we assume that the effect of the magnetic induction **B** on the current is small. This is also expected to be the case for conductors. With this (8.13.5) reduces to

$$\begin{aligned} J_r &= 0, \qquad J_\theta = \sigma_3 S r E_z = k_\theta(r^2) r, \\ J_z &= [\sigma_1 + \sigma_3 + \sigma_7(1 + S^2 r^2)]E_z = k_z(r^2), \end{aligned} \tag{8.13.11}$$

where k_θ and k_z are functions of r^2, since σ_α depend on r^2 and E_z^2. Equations (8.13.9) now take the form

$$\begin{aligned} \frac{dB_z}{dr} &= k_\theta(r^2) r / c(1 - \chi^B), \\ \frac{1}{r}\frac{d}{dr}(rB_\theta) &= k_z(r^2)/c(1 - \chi^B). \end{aligned} \tag{8.13.12}$$

The right-hand sides of these equations are functions of r^2, so that integration gives

$$B_z = -\beta_z(r^2) + C_1, \qquad rB_\theta = \beta_\theta(r^2) + C_2, \tag{8.13.13}$$

where C_1 and C_2 are constants of integration and

$$\begin{aligned} \beta_z &= \int^r [k_\theta(r^2) r / c(1 - \chi^B)]\, dr, \\ \beta_\theta &= \int^r [k_z(r^2) r / c(1 - \chi^B)]\, dr. \end{aligned} \tag{8.13.14}$$

Maxwell's equations for the outside of the cylinder give

$$H_\theta^0 = C_3/r, \qquad H_z^0 = C_4. \tag{8.13.15}$$

We set $C_4 = 0$ and use the continuity of H_θ and H_z, at $r = a$ and the regularity of fields at $r = 0$. Assuming that $B_\theta(r^2)/r$ vanishes for $r = 0$, this gives

$$B_\theta = \frac{1}{r}\beta_\theta(r^2), \qquad B_z = \beta_z(a^2) - \beta_z(r^2),$$
$$H_\theta^0 = (1 - \chi^B)\beta_\theta(a^2)/r. \tag{8.13.16}$$

Since χ_1^B and χ_2^B are taken as zero, the magnetostriction terms in ${}_E\mathbf{t}$ disappear and the strain energy Σ depends on I, II, and B^2, i.e.,

$$\Sigma = \Sigma[3 + S^2r^2, B^2(r^2)] = \hat{\Sigma}(r^2). \tag{8.13.17}$$

The equation of equilibrium (8.13.8) becomes

$$\frac{d}{dr}\left[-p + 2\frac{\partial\Sigma}{\partial \mathrm{I}} - 2\frac{\partial\Sigma}{\partial \mathrm{II}} + \tfrac{1}{2}\chi^B(B_\theta^2 + B_z^2)\right] - 2S^2r\frac{\partial\Sigma}{\partial \mathrm{I}}$$
$$+ \frac{1}{c}\left\{k_\theta(r^2)r[\beta_z(a^2) - \beta_z(r^2)] - k_z(r^2)\frac{1}{r}\beta_\theta(r^2)\right\} = 0. \tag{8.13.18}$$

Integration gives p as

$$p = 2\frac{\partial\Sigma}{\partial \mathrm{I}} - 2\frac{\partial\Sigma}{\partial \mathrm{II}} + \tfrac{1}{2}\chi^B(B_\theta^2 + B_z^2) - 2S^2\int^r r\frac{\partial\Sigma}{\partial \mathrm{I}}\,dr$$
$$+ \frac{1}{c}\int^r \left\{k_\theta(r^2)r[\beta_z(a^2) - \beta_z(r^2)] - k_z(r^2)\frac{1}{r}\beta_\theta(r^2)\right\} dr + C. \tag{8.13.19}$$

On the surface ${}_Et_{rz} = {}_Et_{r\theta} = 0$. To clear the surface $r = a$ from the normal traction we set

$${}_Et_{rr}(a) + {}_Mt_{rr}(a) - {}_Mt_{rr}^+(a) = 0, \tag{8.13.20}$$

Since $M_r = B_r = 0$. The Maxwell stresses ${}_Mt_{rr}$ and ${}_Mt_{rr}^+$ are given by

$${}_Mt_{rr} = -\tfrac{1}{2}[E_z^2 + (1 - 2\chi^B)(B_\theta^2 + B_z^2)], \qquad 0 \le r \le a,$$
$${}_Mt_{rr}^+ = -\tfrac{1}{2}(H_\theta^{0^2} + E_z^2), \qquad r \ge a. \tag{8.13.21}$$

Employing these and (8.13.19) in (8.13.20), we obtain the unknown constant C

$$C = -2S^2\int^a r\frac{\partial\Sigma}{\partial \mathrm{I}}\,dr - \frac{1}{c}\int^a \left\{k_\theta(r^2)r[\beta_z(a^2) - \beta_z(r^2)] - \frac{1}{r}k_z(r^2)\beta_\theta(r^2)\right\} dr$$
$$+ \tfrac{1}{2}\chi^B(\chi^B - 1)[\beta_\theta(a^2)/a]^2. \tag{8.13.22}$$

The pressure is now given by

$$p = {}_Mp(r) + 2\frac{\partial\Sigma}{\partial \mathrm{I}} - 2\frac{\partial\Sigma}{\partial \mathrm{II}} - 2S^2\int_r^a r\frac{\partial\Sigma}{\partial \mathrm{I}}\,dr, \tag{8.13.23}$$

where ${}_Mp(r)$ is the pressure due to electromagnetic effects, defined by

$${}_Mp(r) = \frac{1}{c}\int_r^a \left\{rk_\theta(r^2)[\beta_z(a^2) - \beta_z(r^2)] - \frac{1}{r}k_z(r^2)\cdot\beta_\theta(r^2)\right\} dr$$
$$+ \tfrac{1}{2}\chi^B(\chi^B - 1)[\beta_\theta(a^2)/a]^2. \tag{8.13.24}$$

The stress field follows from (8.13.7) by substituting (8.13.23) for p and setting $\chi_1^B = \chi_2^B = 0$.

$$
\begin{aligned}
{}_{\mathrm{E}}t_{rr} &= -{}_{\mathrm{M}}p(r) - 2S^2 \int_r^a r\frac{\partial \Sigma}{\partial \mathrm{I}}\,dr, \\
{}_{\mathrm{E}}t_{\theta\theta} &= -{}_{\mathrm{M}}p(r) + 2S^2\left(r^2\frac{\partial \Sigma}{\partial \mathrm{I}} - \int_r^a r\frac{\partial \Sigma}{\partial \mathrm{I}}\,dr\right), \\
{}_{\mathrm{E}}t_{zz} &= -{}_{\mathrm{M}}p(r) - 2S^2\left(r^2\frac{\partial \Sigma}{\partial \mathrm{II}} + \int_r^a r\frac{\partial \Sigma}{\partial \mathrm{I}}\,dr\right), \\
{}_{\mathrm{E}}t_{\theta z} &= 2Sr\left(\frac{\partial \Sigma}{\partial \mathrm{I}} + \frac{\partial \Sigma}{\partial \theta}\right), \qquad {}_{\mathrm{E}}t_{zr} = {}_{\mathrm{E}}t_{r\theta} = 0.
\end{aligned}
\tag{8.13.25}
$$

The electrical conduction therefore causes pressure in the cylinder given by (8.13.24). At the end sections $z = 0, L$, of the cylinder there will be an extra axial load due to the electric conduction. Poynting and Kelvin effects are present as usual. Electrical conduction in deformed elastic materials was studied by Pipkin and Rivlin [1960a], [1961a, b], and [1962].

8.14. Theory of Magnetoelastic Plates

The classical plate theory must be modified in order to take into account the effect of the applied magnetic field.

We consider a thin plate P, of thickness $2h$, with a median surface $x_3 = 0$. The upper and lower surfaces of the plate are denoted, respectively, by $S^+(x_3 = h)$ and $S^-(x_3 = -h)$, and its contour by $\mathscr{C}$ (Figure 8.14.1). The classical plate theory of Love and Kirchhoff assumes that the displacement field is

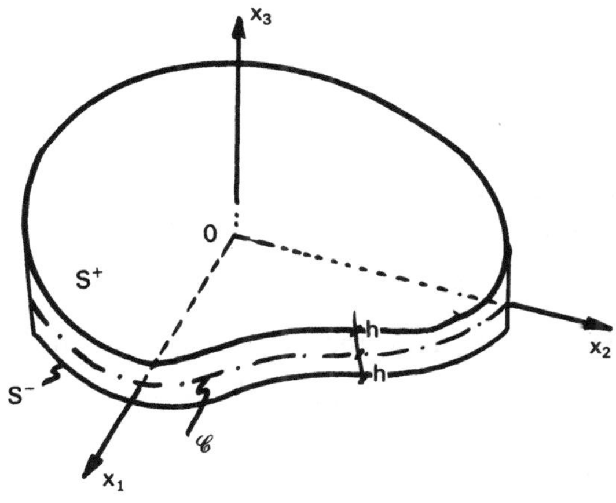

Figure 8.14.1. Elastic plate.

of the form

$$u_\alpha = v_\alpha(x_1, x_2, t) - x_3 w_{,\alpha}(x_1, x_2, t),$$
$$u_3 = w(x_1, x_2, t), \tag{8.14.1}$$

where, and henceforth, Greek indices are assumed to have the range (1, 2), while Latin indices have the range (1, 2, 3). For the electric and magnetic fields, we write

$$\mathbf{H} = \mathbf{H}^0 + \mathbf{h}, \qquad \mathbf{E} = \mathbf{e}. \tag{8.14.2}$$

Here $\mathbf{H}^0$ is a static bias magnetic field, while $\mathbf{h}$ and $\mathbf{e}$ are, respectively, the induced magnetic and electric fields which are considered to be small, e.g.,

$$|\mathbf{h}|/|\mathbf{H}^0| = O(\varepsilon), \tag{8.14.3}$$

where ε is a perturbation parameter. Likewise, the displacement fields u_α and u_3 are considered to be small. The deformation is measured from the initial configuration which carries the $\mathbf{H}^0$-field, but no initial stress, i.e., ${}_{\mathrm{E}}T^0_{kl} = 0$. The bias field satisfies Maxwell's equations

$$\nabla \times \mathbf{H}^0 = \mathbf{0}, \qquad \nabla \cdot \mathbf{B}^0 = 0. \tag{8.14.4}$$

Differential equations for the incremental fields follow from (7.14.27)–(7.14.31)

$${}_{\mathrm{E}}\tilde{T}_{kl,k} + {}_{\mathrm{M}}\tilde{F}_l - \rho_0 \ddot{u}_l = \mathbf{0}, \tag{8.14.5}$$

$$\nabla \cdot \mathbf{d} = 0, \tag{8.14.6}$$

$$\nabla \times \mathbf{e} + \frac{1}{c}\dot{\mathbf{b}} = \mathbf{0}, \tag{8.14.7}$$

$$\nabla \cdot \mathbf{b} - B^0_{r,l} u_{l,r} = 0, \tag{8.14.8}$$

$$\varepsilon_{klm}(h_{m,l} - H^0_{m,r} u_{r,l}) - \frac{1}{c}\dot{d}_k = 0, \tag{8.14.9}$$

where a superposed dot denotes a partial time derivative and

$${}_{\mathrm{M}}\tilde{F}_l = (b_{k,l} - b_{l,k})M^0_k + (B^0_{k,l} u_{k,r} - B^0_{r,k} u_{k,l})M^0_r$$
$$- B^0_l(m_{k,k} - M^0_{r,k} u_{k,r}) + \frac{1}{c}\varepsilon_{lmn}\dot{p}_m B^0_n \tag{8.14.10}$$

is the magnetic body force.

The constitutive equation for the bias field is given by

$$\mathbf{B}^0 = \mathbf{H}^0 + \mathbf{M}^0 = \mu_0 \mathbf{H}^0 \tag{8.14.11}$$

while the incremental fields, in isotropic media, follow from (7.14.45)

$${}_{\mathrm{E}}\tilde{T}_{kl} = \lambda_e \tilde{E}_{rr}\delta_{kl} + 2\mu_e \tilde{E}_{kl}; \qquad \tilde{E}_{kl} = \tfrac{1}{2}(u_{k,l} + u_{l,k}),$$
$$\mathbf{p} = \mathbf{d} - \mathbf{e} = (\varepsilon - 1)\left[\mathbf{e} + \frac{1}{c}(\dot{\mathbf{u}} \times \mathbf{B}^0)\right], \tag{8.14.12}$$
$$m_k = (1 - \mu_0^{-1})(b_k + 2B^0_r \tilde{E}_{kr}) - M^0_k u_{r,r}.$$

Jump conditions for the bias field are

$$\mathbf{n} \times [\mathbf{H}^0] = \mathbf{0}, \qquad \mathbf{n} \cdot [\mathbf{B}^0] = 0. \tag{8.14.13}$$

For incremental fields they are listed in (7.14.37)

$$\begin{aligned}[{}_{\mathrm{E}}\tilde{T}_{kl} - M_k^0 b_l - m_k B_l^0 + {}_{\mathrm{M}}\tilde{t}_{kl}]n_k + [{}_{\mathrm{M}}T_{kl}^0 - M_k^0 B_l^0]\tilde{n}_k &= 0,\\ [\mathbf{d}]\cdot\mathbf{n} &= 0,\\ \mathbf{n}\times[\mathbf{e}] &= \mathbf{0},\\ [\mathbf{b}]\cdot\mathbf{n} + [\mathbf{B}^0]\cdot\tilde{\mathbf{n}} &= \mathbf{0},\\ \mathbf{n}\times[\mathbf{h}] + \tilde{\mathbf{n}}\times[\mathbf{H}^0] &= \mathbf{0},\end{aligned} \tag{8.14.14}$$

where

$$\begin{aligned}{}_{\mathrm{M}}\tilde{t}_{kl} &= B_k^0 h_l + H_l^0 b_k - (\mathbf{H}^0\cdot\mathbf{b} - \mathbf{m}\cdot B^0)\delta_{kl},\\ {}_{\mathrm{M}}T_{kl}^0 &= B_k^0 H_l^0 - \tfrac{1}{2}(\mathbf{B}^0\cdot\mathbf{B}^0 - 2\mathbf{M}_0\cdot\mathbf{B}^0)\delta_{kl}.\end{aligned} \tag{8.14.15}$$

As usual, boldface brackets [ ] indicate the difference in values of its enclosure, obtained by approaching the surface from the outside and inside, i.e.,

$$[A] = A^+ - A^-.$$

For the outside fields, we must remember to replace $\mathbf{B}^{0+}$, by $\mathbf{H}^{0+}$; b^+ by h^+, i.e., $\mathbf{M}^{0+} = \mathbf{m}^+ = \mathbf{0}$.

Equations (8.14.4)–(8.14.14) are valid for the three-dimensional isotropic magnetoelastic solids. In order to obtain equations for thin plates, we employ (8.14.1) and perform the following two sets of operations on (8.14.5) and (8.14.10), Eringen [1988]:

(a) We integrate these equations across the thickness of the plate.
(b) We multiply the first two components ($l = 1, 2$) of these equations by x_3 and then integrate them across the thickness of the plate.

In the spirit of the plane stress we first set ${}_{\mathrm{E}}\tilde{T}_{33} = 0$, which gives $\tilde{E}_{33} = -\lambda\tilde{E}_{\gamma\gamma}/(\lambda_e + 2\mu_e)$. Using this, the stress–strain relations $(8.14.12)_1$ become

$$\begin{aligned}{}_{\mathrm{E}}\tilde{T}_{\alpha\beta} &= \frac{E}{1-\nu^2}[\nu\tilde{E}_{\gamma\gamma}\delta_{\alpha\beta} + (1-\nu)\tilde{E}_{\alpha\beta}],\\ {}_{\mathrm{E}}\tilde{T}_{\alpha 3} &= {}_{\mathrm{E}}\tilde{T}_{33} = 0, \qquad (\alpha, \beta, \gamma) = 1, 2,\end{aligned} \tag{8.14.16}$$

where

$$\tilde{E}_{\alpha\beta} = \tfrac{1}{2}(v_{\alpha,\beta} + v_{\beta,\alpha}) - x_3 w_{,\alpha\beta}. \tag{8.14.17}$$

Operations (a) and (b) introduce the planar forces $N_{\alpha\beta}$, vertical shears Q_α, and couples $M_{\alpha\beta}$, as defined by (Figure 8.14.2)

$$\begin{aligned}N_{\alpha\beta} &= \int_{-h}^{h} {}_{\mathrm{E}}\tilde{T}_{\alpha\beta}\,dx_3, \qquad Q_\alpha = \int_{-h}^{h} {}_{\mathrm{E}}\tilde{T}_{\alpha 3}\,dx_3,\\ M_{\alpha\beta} &= \int_{-h}^{h} {}_{\mathrm{E}}\tilde{T}_{\alpha\beta}x_3\,dx_3.\end{aligned} \tag{8.14.18}$$

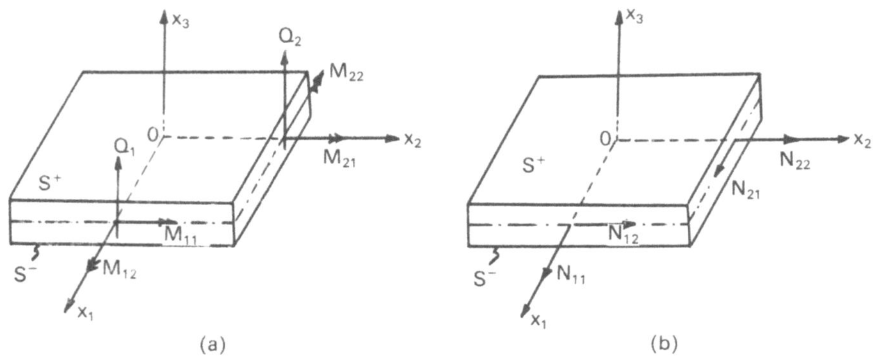

Figure 8.14.2. (a) Bending moments and transverse shears. (b) Planar tractions.

Using (8.14.16) and (8.14.17), we obtain

$$\begin{aligned} N_{\alpha\beta} &= \frac{Eh}{1-\nu^2}[2\nu v_{\gamma,\gamma}\delta_{\alpha\beta} + (1-\nu)(v_{\alpha,\beta} + v_{\beta,\alpha})], \\ M_{\alpha\beta} &= -D[\nu w_{,\gamma\gamma}\delta_{\alpha\beta} + (1-\nu)w_{,\alpha\beta}], \end{aligned} \tag{8.14.19}$$

where D is the flexural rigidity defined by

$$D = \frac{2Eh^3}{3(1-\nu^2)}. \tag{8.14.20}$$

We note that Q_α vanishes. However, we retain the above definition for later use, since the magnetic field contributes a nonvanishing transverse shear along the contour of the plate. This contradiction in the classical plate theory is resolved by considering the effect of the shear deformations which bring additional small terms to the expressions (8.14.1) (see Reissner [1944] and Green and Zerna [1954, p. 224]). These effects become important for thick plates.

The operations (a) and (b) on (8.14.5) give

$$\frac{Eh}{1-\nu}\left(v_{\alpha,\alpha\beta} + \frac{1-\nu}{1+\nu}v_{\beta,\alpha\alpha}\right) + q_\beta + \tilde{f}_\beta - 2\rho_0 h\ddot{v}_\beta = 0, \tag{8.14.21}$$

$$Q_{\alpha,\alpha} + [{}_{\mathrm{E}}\tilde{T}_{33}]^{h}_{-h} + \tilde{f}_3 - 2\rho_0 h\ddot{w} = 0, \tag{8.14.22}$$

$$M_{\alpha\beta,\alpha} - Q_\beta + [x_3\, {}_{\mathrm{E}}\tilde{T}_{3\beta}]^{h}_{-h} + \tilde{l}_\beta + \tfrac{2}{3}\rho_0 h^3\ddot{w}_{,\beta} = 0, \tag{8.14.23}$$

where

$$\begin{aligned} q_\beta &= [{}_{\mathrm{E}}\tilde{T}_{3\beta}]^{h}_{-h}, \qquad \tilde{f}_k = \int_{-h}^{h} {}_{\mathrm{M}}\tilde{F}_k\, dx_3, \\ l_\beta &= \int_{-h}^{h} {}_{\mathrm{M}}\tilde{F}_\beta x_3\, dx_3. \end{aligned} \tag{8.14.24}$$

From equations (8.14.21)–(8.14.23), $[{}_{\mathrm{M}}\tilde{T}_{3\beta}]_{-h}^{h}$ and $[{}_{\mathrm{E}}\tilde{T}_{33}]_{-h}^{h}$ were not dropped. This is because on the surface of the body there may be surface tractions.

Eliminating Q_α between (8.14.22) and (8.14.23), and using $(8.14.19)_2$, we obtain

$$D\nabla^4 w - p - f + 2\rho_0 h\ddot{w} - \tfrac{2}{3}\rho_0 h^3 \nabla^2 \ddot{w} = 0, \tag{8.14.25}$$

where ∇^4 is the two-dimensional biharmonic operator and

$$\begin{aligned} p &\equiv [x_3\, {}_{\mathrm{E}}\tilde{T}_{3\beta,\beta} + {}_{\mathrm{E}}\tilde{T}_{33}]_{-h}^{h}, \\ f &= \int_{-h}^{h} ({}_{\mathrm{M}}\tilde{F}_3 + x_3\, {}_{\mathrm{M}}\tilde{F}_{\beta,\beta})\, dx_3. \end{aligned} \tag{8.14.26}$$

Usually, the rotatory inertia (the last term in (8.14.25)) is also neglected, so that

$$D\nabla^4 w - p - f + 2\rho_0 h\ddot{w} = 0. \tag{8.14.27}$$

Equation (8.14.21) is the differential equation for the planar motions of the plate, and (8.14.27) is that for the flexural motions. We note that these equations are coupled with the magnetic fields through $\tilde{f}_\beta$, q_β, p, f, and the boundary conditions.

Boundary conditions on the tractions and fields follow from (8.14.13) and (8.14.14). In these equations, $\mathbf{n}$ represents the unit normal of the undeformed surface under consideration, and $\tilde{\mathbf{n}}$ represents its increment due to deformation. For $\tilde{\mathbf{n}}$ we have, from (7.14.35),

$$\tilde{n}_k = (u_{i,j} n_i n_j \delta_{kl} - u_{l,k}) n_l. \tag{8.14.28}$$

For example, on the upper and lower surfaces $x_3 = \pm h$ of the plate, we have

$$\mathbf{n} = \pm \mathbf{i}_3, \qquad \tilde{n}_\alpha = \mp w_{,\alpha}, \qquad \tilde{n}_3 = 0. \tag{8.14.29}$$

On the contour surface $\mathscr{C}$ of the plate, we have

$$\begin{aligned} &\mathbf{n} = (n_\alpha, 0), \\ &\tilde{n}_\alpha = (\tilde{n}_{\alpha\beta} - x_3 \tilde{m}_{\alpha\beta}) n_\beta, \qquad \tilde{n}_3 = w_{,\beta} n_\beta, \end{aligned} \tag{8.14.30}$$

where

$$\begin{aligned} \tilde{n}_{\alpha\beta} &= v_{\gamma,\delta} n_\gamma n_\delta \delta_{\alpha\beta} - v_{\beta,\alpha}, \\ \tilde{m}_{\alpha\beta} &= w_{,\gamma\delta} n_\gamma n_\delta \delta_{\alpha\beta} - w_{,\beta\alpha}. \end{aligned} \tag{8.14.31}$$

Initial conditions usually consist of Cauchy data, namely, the prescription of v, w, $\dot{v}$, $\dot{w}$, and of electromagnetic fields throughout the body, at time $t = 0$.

8.15. Plate Under Transverse Magnetic Field

Here we consider a special case of the theory developed in Section 8.14, namely, the plate under a static transverse magnetic field. The bias field $\mathbf{H}^0$ is assumed to be constant throughout the plate P, and it is directed perpendicu-

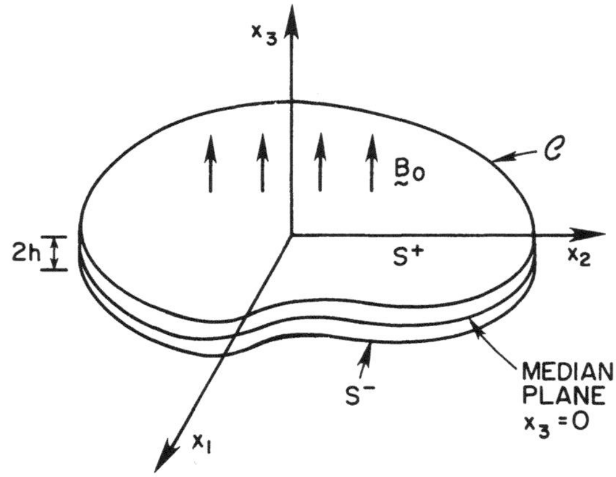

Figure 8.15.1. Elastic plate under constant transverse magnetic field.

lar to the median plane $x_3 = 0$ of the plate, Figure 18.15.1. For soft ferromagnetic materials, the electric field can be ignored, so that we have

$$\begin{aligned} \mathbf{B} &= B^0 \mathbf{i}_3, \qquad |x_3| < h, \\ \mathbf{H} &= B^0 \mathbf{i}_3, \qquad |x_3| > h. \end{aligned} \tag{8.15.1}$$

This approximation is valid almost everywhere in P and outside P, except near the boundaries ∂P of the plate, where in a small boundary layer, $\mathbf{B}$ differs from $B^0 \mathbf{i}_3$. However, this difference is generally negligible.[6]

Differential equations (8.14.8) and (8.14.9) reduce to

$$\nabla \times \mathbf{h} = \mathbf{0}, \qquad \nabla \cdot \mathbf{b} = 0. \tag{8.15.2}$$

Since the effect of the electric field can be ignored, (8.14.6) and (8.14.7) are satisfied automatically.

The constitutive equation (8.14.12) for $\mathbf{m}$ reduces to

$$\mathbf{b} = \mu_0 \mathbf{h} - (\mu_0 - 1)\mathbf{B}^0 \nabla \cdot \mathbf{u}. \tag{8.15.3}$$

As a result of this and (8.15.1), the magnetic force (8.14.10) vanishes

$${}_{\mathrm{M}}\tilde{F}_l = 0. \tag{8.15.4}$$

Relevant equations of jump (8.14.14), at $x_3 = \pm h$, reduce to

$${}_{\mathrm{E}}\tilde{T}^+_{3l} - {}_{\mathrm{E}}\tilde{T}^-_{3l} + (1 - \mu_0^{-1})B^0 b_l^- + B^0(h_l^+ - h_l^-) + B^{0^2}(1 - \mu_0^{-1})w_{,l} = 0 \tag{8.15.5}$$

$$h_3^+ - \mu_0 h_3^- \mp (\mu_0 - 1)B^0 h \nabla^2 w = 0, \tag{8.15.6}$$

$$h_\alpha^+ - h_\alpha^- + (1 - \mu_0^{-1})B^0 w_{,\alpha} = 0. \tag{8.15.7}$$

---

[6] For an account of the boundary layer effect, see Van de Ven [1975], [1984]. See also Goudjo and Maugin [1983].

For flexural motions, we have

$$u_{\alpha,\beta} = -x_3 w_{,\alpha\beta}, \qquad u_{3,\alpha} = -u_{\alpha,3} = w_{,\alpha}, \qquad u_{3,3} = 0.$$

Using (8.15.3) and (8.15.5) in (8.14.26), and setting ${}_{\mathrm{E}}\tilde{T}^{+}_{3l} = 0$, we obtain

$$p = p_{\mathrm{H}} + 2hB^{0^2}(\mu_0 - 1)\nabla^2 w, \tag{8.15.8}$$

where

$$p_{\mathrm{H}} = B^0[(\mu_0 - 2)(x_3 h^{-}_{\alpha,\alpha} + h^{-}_3) + x_3 h^{+}_{\alpha,\alpha} + h^{+}_3]^{h}_{-h}. \tag{8.15.9}$$

Differential equations of flexural motion of the plate read

$$D\nabla^4 w - 2hB^{0^2}(\mu_0 - 1)\nabla^2 w - p_{\mathrm{H}} + 2\rho_0 h\ddot{w} = 0. \tag{8.15.10}$$

Accompanying this equation are boundary conditions on the contour of the plate. The plate may be free, clamped, simply supported, elastically supported, or elastically clamped along $\mathscr{C}$. These conditions are similar to the classical ones. We list below some of these conditions.

(i) *Clamped Plate*

$$w = 0, \qquad \frac{\partial w}{\partial n} = 0 \quad \text{on } \mathscr{C}, \tag{8.15.11}$$

where $\partial/\partial n$ denotes the normal derivative along the unit normal $\mathbf{n}$, i.e.,

$$\frac{\partial w}{\partial n} = w_{,\alpha} n_\alpha.$$

(ii) *Simply Supported Plate*

$$w = 0, \qquad M_{nn} = 0 \quad \text{on } \mathscr{C}, \tag{8.15.12}$$

where the subscript n denotes the direction of $\mathbf{n}$.

(iii) *Free Plate*

For the plate with free contour $\mathscr{C}$, the magnetic moment vanishes. The classical Kirchhoff boundary condition is modified by the vertical magnetic shear force. The magnetic couple is given by

$${}_{\mathrm{M}}\mathbf{C} = \mathbf{M}^0 \times \mathbf{b} + \mathbf{m} \times \mathbf{B}^0.$$

As discussed in Section 3.10, the jump conditions on the magnetic couple are satisfied identically. But there is a contribution to the vertical shears from the magnetic field. This is calculated by

$${}_{\mathrm{M}}Q = \int_{-h}^{h} {}_{\mathrm{E}}\tilde{T}^{-}_{\alpha 3} n_\alpha \, dx_3 \quad \text{along } \mathscr{C}, \tag{8.15.13}$$

where ${}_{\mathrm{E}}\tilde{T}^{-}_{\alpha 3} = {}_{\mathrm{E}}\tilde{T}^{-}_{3\alpha}$ follows from (8.15.5) by setting ${}_{\mathrm{E}}\tilde{T}^{+}_{\alpha 3} = 0$. Hence

$${}_{\mathrm{E}}\tilde{T}^{-}_{\alpha 3} n_\alpha = [B^0 h^{+}_\alpha + B^0(\mu_0 - 2)h^{-}_\alpha + (1 - \mu_0^{-1})B^{0^2} w_{,\alpha}] n_\alpha. \tag{8.15.14}$$

Consequently,

$$ {}_{\mathrm{M}}Q = 2h(1-\mu_0^{-1})B^{0^2}w_{,\alpha}n_\alpha + B^0\int_{-h}^{h}[h_\alpha^+ + (\mu_0 - 2)h_\alpha^-]n_\alpha\,dx_3. \quad (8.15.15) $$

The mechanical vertical shear Q is given by (8.14.23)

$$ Q = Q_\beta n_\beta = (M_{\alpha\beta,\alpha} + [x_3\,{}_{\mathrm{E}}\tilde{T}_{3\beta}]_{-h}^{h})n_\beta, \quad (8.15.16) $$

where the rotatory inertia term was dropped. The second term in (8.15.16) is calculated by using (8.15.5) and setting ${}_{\mathrm{E}}\tilde{T}_{3\beta}^{+} = 0$. Hence,

$$ \begin{aligned} Q = \{&M_{\alpha\beta,\alpha} + 2h(1-\mu_0^{-1})B^{0^2}w_{,\beta} \\ &+ B^0[x_3h_\beta^+ + x_3(\mu_0-2)h_\beta^-]_{-h}^{h}\}n_\beta. \end{aligned} \quad (8.15.17) $$

The boundary conditions for the free contour are then expressed as

$$ M_{nn} = 0, \qquad Q + {}_{\mathrm{M}}Q - \frac{\partial M_{nT}}{\partial T} = 0, \quad (8.15.18) $$

where M_{nT} is the twisting moment at $\mathscr{C}$ and $\partial/\partial T$ represents the tangential derivative along $\mathscr{C}$ (Figure 8.15.2).

It should be noted that, in a boundary layer along $\mathscr{C}$, the magnetic field is nonuniform and magnetic saturation may occur. This effect is neglected here.

Equation (8.15.2) suggests that we introduce magnetic potentials

$$ \begin{aligned} h^+ &= -\nabla\psi, \qquad |x_3| > h, \\ h^- &= -\nabla\phi, \qquad |x_3| < h, \end{aligned} \quad (8.15.19) $$

so that ψ and ϕ must satisfy

$$ \begin{aligned} \nabla^2\psi &= 0, \\ \nabla^2\phi - (1-\mu_0^{-1})B^0\nabla^2 w &= 0, \end{aligned} \quad (8.15.20) $$

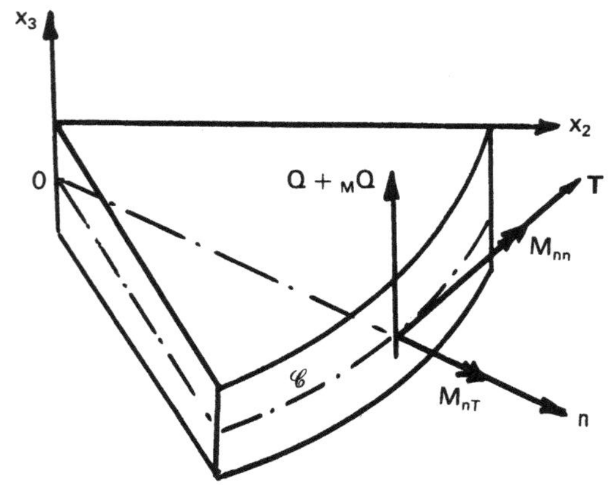

Figure 8.15.2. Bending moment M_{nn}, twisting moment M_{nT}, and vertical shear $Q + {}_{\mathrm{M}}Q$ at the edge surface of the plate.

subject to the jump conditions (8.15.6), namely,

$$\begin{aligned} \psi_{,3} - \mu_0\phi_{,3} \pm (\mu_0 - 1)B^0 h\nabla^2 w &= 0, \\ \psi_{,\alpha} - \phi_{,\alpha} - (1 - \mu_0^{-1})B^0 w_{,\alpha} &= 0, \end{aligned} \qquad x_3 = \pm h, \tag{8.15.21}$$

$$\psi \to 0 \qquad \text{as} \quad |x_3| \to \infty.$$

For the flexural motion we have (8.15.9), where

$$p_{\mathrm{H}} = B^0[x_3\psi_{33} - \psi_{,3} + (\mu_0 - 2)(x_3\phi_{,33} - \phi_{,3})]_{-h}^{h}. \tag{8.15.22}$$

Thus, the problem is reduced to solving the coupled differential equations (8.15.10) and (8.15.20) under the boundary conditions (8.15.21) and the plate edge support conditions. The initial conditions on w usually consist of Cauchy data which prescribe w and $\dot{w}$ at time $t = 0$.

8.16. Magnetoelastic Buckling of a Circular Plate

To illustrate the solutions of the equations developed in Section 8.15, we consider here the buckling of a circular plate of radius R and thickness $2h$. We consider only the lowest rotationally symmetric buckling under the transverse magnetic induction B^0.

The differential equations to be solved are (8.15.20) and (8.15.10) with vanishing inertia, i.e.,

$$\nabla^2\phi + \phi_{,zz} - (1 - \mu_0^{-1})B^0\nabla^2 w = 0, \qquad |z| < h, \tag{8.16.1}$$

$$\nabla^2\psi + \psi_{,zz} = 0, \qquad |z| > h, \tag{8.16.2}$$

$$D\nabla^4 w - h(1 - \mu_0^{-1})B^{0^2}\nabla^2 w - p_{\mathrm{H}} = 0, \qquad r < R, \quad z = 0, \tag{8.16.3}$$

where

$$p_{\mathrm{H}} = B^0[x_3\psi_{,33} - \psi_{,3} + (\mu_0 - 2)(x_3\phi_{,33} - \phi_{,3})]_{-h}^{h} \tag{8.16.4}$$

Boundary conditions for (8.16.1) and (8.16.2), at $|z| = h$, are

$$\psi_{,z} - \mu_0\phi_{,z} = \mp(\mu_0 - 1)B^0 h\nabla^2 w, \qquad z = \pm h, \tag{8.16.5}$$

$$\psi_{,r} - \phi_{,r} = (1 - \mu_0^{-1})B^0 w_{,r}, \tag{8.16.6}$$

$$\psi \to 0 \qquad |z| \to \infty. \tag{8.16.7}$$

For the plate equation (8.16.3), we consider the following boundary conditions:

(i) *A Clamped Plate*:

$$w = w_{,r} = 0, \qquad r = R. \tag{8.16.8}$$

(ii) *A Simply Supported Plate*:

$$w = M_{rr} = 0, \qquad r = R. \tag{8.16.9}$$

We try a solution of the form

$$
\begin{aligned}
w_{,r} &= AJ_1(\alpha r), \\
\phi(r, z) &= B^0 C(z) J_0(\alpha r) - (1 - \mu_0^{-1})\alpha^{-1} B^0 A J_0(\alpha r), \\
\psi(r, z) &= B^0 D(z) J_0(\alpha r),
\end{aligned}
\tag{8.16.10}
$$

where $J_n(\alpha r)$ are the Bessel functions of nth order, A and α are constants. From the first of these and $(8.16.8)_1$, it follows that

$$
w = A\alpha^{-1}[J_0(\alpha R) - J_0(\alpha r)]. \tag{8.16.11}
$$

For the clamped plate the second expression of (8.16.8) gives

$$
J_1(\alpha R) = 0, \tag{8.16.12}
$$

whose smallest positive root is

$$
\alpha R \simeq 3.832. \tag{8.16.13}
$$

For the simply supported plate, from $(8.16.9)_2$, it follows that

$$
w_{,rr}(R) + \nu R^{-1} w_{,r}(R) = 0.
$$

Consequently,

$$
J_0(\alpha R) - \frac{1 - \nu}{\alpha R} J_1(\alpha R) = 0 \tag{8.16.14}
$$

whose lowest root, for $\nu = 0.3$, is

$$
\alpha R \simeq 2.049. \tag{8.16.15}
$$

In order to determine $C(z)$ and $D(z)$, we substitute $(8.16.10)_{2,3}$, into (8.16.1) and (8.16.2), and use $\nabla^2 J_0 = -\alpha^2 J_0$. Hence,

$$
C_{,zz} - \alpha^2 C = 0, \qquad D_{,zz} - \alpha^2 D = 0. \tag{8.16.16}
$$

From the symmetry of ϕ with respect to z, and the asymmetric condition (8.16.7), we conclude that

$$
\begin{aligned}
C(z) &= C_0 \cosh(\alpha z), \qquad |z| < h, \\
D(z) &= D_0 \exp(-\alpha |z|), \qquad |z| > h,
\end{aligned}
\tag{8.16.17}
$$

The boundary conditions (8.16.5) and (8.16.6) give

$$
\begin{aligned}
C_0 &= (\mu_0 - 1)(\alpha\Lambda)^{-1} A(h\alpha + 2\mu_0^{-1}), \\
D_0 &= (\mu_0 - 1)(\alpha\Lambda)^{-1} A[h\alpha \cosh(\alpha h) - 2\sinh(\alpha h)],
\end{aligned}
\tag{8.16.18}
$$

where

$$
\Lambda \equiv \cosh(\alpha h) + \mu_0 \sinh(\alpha h). \tag{8.16.19}
$$

Consequently, ϕ, ψ, and p_H are obtained to be

$$
\phi = B^0(\mu_0 - 1)(\alpha\Lambda)^{-1} A J_0(\alpha r)[(h\alpha + 2\mu_0^{-1})\cosh(\alpha z) - \mu_0 \Lambda],
$$

$$\psi = B^0(\mu_0 - 1)(\alpha\Lambda)^{-1} A J_0(\alpha r)[h\alpha \cosh(\alpha h) - 2\sinh(\alpha h)] \cdot \exp[\alpha(h - |z|)], \tag{8.16.20}$$

$$p_H = B^{0^2}(\mu_0 - 1) 2h\alpha\Lambda^{-1} A J_0(\alpha r)\left\{[h\alpha \cosh(\alpha h) - 2\sinh(\alpha h)]\left(1 + \frac{1}{h\alpha}\right) + (\mu_0 - 2)(h\alpha + 2\mu_0^{-1})\cdot[\cosh(\alpha h) - (\alpha h)^{-1}\sinh(\alpha h)]\right\}. \tag{8.16.21}$$

Substituting this and (8.16.11) into (8.16.3), we obtain the critical value of the magnetic induction that causes buckling

$$B_{\mathrm{cr}}^{0^2} = \frac{E(h\alpha)^2}{3(1-\nu^2)}\left\{-2(1-\mu_0^{-1}) - (\mu_0 - 1)\Lambda^{-1}[h\alpha\cosh(\alpha h) - 2\sinh(\alpha h)] \cdot \left(1 + \frac{1}{\alpha h}\right) - (\mu_0 - 1)(\mu_0 - 2)\Lambda^{-1} \cdot (h\alpha + 2\mu_0^{-1})[\cosh(\alpha h) - (\alpha h)^{-1}\sinh(\alpha h)]\right\}^{-1}. \tag{8.16.22}$$

We note that αh is small compared to unity, since $h/R \ll 1$. If we write $\cosh(\alpha h) \simeq 1$ and $\sinh(\alpha h) \simeq \alpha h$, (8.16.22) is simplified to

$$B_{\mathrm{cr}}^{0^2} = \frac{E(h\alpha)^2}{3(1-\nu^2)} \cdot \frac{\mu_0(\mu_0\alpha h + 1)}{(\mu_0 - 1)(\mu_0 - 2\mu_0\alpha h - 2)}. \tag{8.16.23}$$

We note that for buckling to occur, we must have

$$\mu_0 > 1, \qquad \mu_0 > \mu_{\mathrm{cr}}; \qquad \mu_{\mathrm{cr}} \equiv 2(1 - 2\alpha h)^{-1}. \tag{8.16.24}$$

The first of these conditions ($\mu_0 > 1$) is fulfilled, since $\mu_0 = 1$ means nonexistence of the plate. Since $\alpha h \ll 1$, we must have $\mu_0 > 2$. With this condition we find that

$$B_{\mathrm{cr}}^{0^2} = \frac{E(\alpha h)^3}{3(1-\nu^2)}\left(1 + \frac{1}{\mu_0\alpha h}\right) \quad \text{if } \mu_0\alpha h \gg 1, \tag{8.16.25}$$

$$B_{\mathrm{cr}}^{0^2} = \frac{E(\alpha h)^2}{3(1-\nu^2)\mu_0} \quad \text{if } \mu_0\alpha h \ll 1, \tag{8.16.26}$$

$$B_{\mathrm{cr}}^{0^2} = \frac{2E(\alpha h)^3}{3(1-\nu^2)} \quad \text{if } \mu_0\alpha h = 1. \tag{8.16.27}$$

In these equations we must substitute α from (8.16.13) for the clamped plates, and from (8.16.15) for the simply supported plates. Thus, for example, for the case (8.16.25) we will have

$$\begin{aligned} B_{\mathrm{cr}}^0 &= \left(\frac{E}{1-\nu^2}\right)^{1/2} 4.33\left(\frac{h}{R}\right)^{3/2}, \quad \text{(clamped)},\\ B_{\mathrm{cr}}^0 &= \left(\frac{E}{1-\nu^2}\right)^{1/2} 1.78\left(\frac{h}{R}\right)^{3/2}, \quad \text{(simply supported, } \nu = 0.3). \end{aligned} \tag{8.16.28}$$

These results are identical to those obtained by Van de Ven [1978] in a different way. In various previous literature B_{cr}^0 is obtained to be proportional to $(h/R)^{3/2}$ (Van de Ven [1978] and Goudjo and Maugin [1983]). Also, similar results for the beam plate were given by Moon [1984], Wallerstein and Peach [1972], Poplar [1972], Yeh [1971] and Miya et al. [1978], where experimental results were quoted and various additional effects were considered in order to bring the theoretical results into proximity with experimental observation. In all of these works the effects of piezomagnetism were ignored and various simplified force laws were used. The present result (8.16.22) employes the exact force law and includes piezomagnetic effects. Therefore, it is believed to be the correct theoretical prediction. While this result depends on the permeability μ_0, in the thin plate approximation and for $\mu_0 \alpha h \gg 1$, this dependence disappears.

According to Moon [1984, p. 166] experimental data (for the cantilever beam plate) fall below the theoretical value of B_{cr}^0, given by him and others, up to a factor of 2. The theoretical and experimental curves displayed in his Figure 5.27, using his data, show that $\mu_0 \alpha h \simeq 0.7$. If this is used in (8.16.25), the present result would be much closer to his experimental curve. Also, it is without other considerations such as imperfection sensitivity (Popular [1972]), pole effects (Yeh [1971]), and magnetic hysteresis (Miya et al. [1978]).

It is also important to note in the case $\mu_0 \alpha h \ll 1$ (not noticed before) that B_{cr}^0 is linear in αh and depends on μ_0, as displayed by (8.16.26).

8.17. Other Works: Dynamic Buckling and Nonlinear Waves

The case of buckling examined in Section 8.16 is but one simple example of the study of static magnetoelastic buckling. The works of Alblas [1978], Ambartsumian et al. [1977], Dalrymple et al. [1974], Moon [1978], [1984], Maugin and Goudjo [1982], Goudjo and Maugin [1983], Parkus [1979], Van de Ven [1975], [1978], and Wallerstein and Peach [1972] contain other examples of static magnetoelastic buckling for various structures (beams, beam plates, plates of various forms, shells) for both insulating and conducting (or superconducting) magnetoelastic materials. Many relevant references are to be found in Ambartsumian [1982], Green and Naghdi [1983], and Moon [1980], [1984], the latter developing several engineering aspects of the equation. For the development of the general theory of plates, subject to electromagnetic fields and buckling of plates under an applied electric field, see Eringen [1989].

Elements of dynamic magnetoelastic stability and parametric excitations may be found in Moon and Pao [1969], Das et al. [1981], and Goudjo and Maugin [1983]. Plane waves in an infinite elastic solid, in the presence of a large magnetic field, were studied by Dunkin and Eringen [1963]. They gave dispersion relations for a bias magnetic field in an arbitrary direction and a current vector, involving the Hall effect. In the same work, vibrations of an infinite plate in a large, static, magnetic field were analyzed. Dispersion relations were obtained for the first and second modes, and the damping caused by the magnetic field was obtained as a function of the wave number. A discussion of the nonlinear, rigid, annular, dielectric cylinder, subject to an axial magnetic field, may be found in Jordan and Eringen [1964b].

Finally, corresponding to nonlinear wave propagation in pure nonlinear elasticity (Eringen and Şuhubi [1974]) and magnetohydrodynamics (see Chapter 10 below), we have *nonlinear wave propagation in magnetoelasticity*. For this advanced subject, we refer the reader to original works such as: Bazer [1971] for the magnetoelastic analogue of geometrical optics; Bazer and Karal [1971] for the study of simple waves; Bazer and Ericson [1974] and Maugin [1981a], [1988] for shock wave problems; McCarthy [1966a, b], [1967], [1968] and Chen and McCarthy [1975] for the propagation of weak discontinuities and their growth; Maugin [1978c] for exact results in relativistic frameworks; and Maugin [1981b] for the formation of shocks.

Problems

8.1. Obtain constitutive equations that are second degree in the magnetic field.

8.2. Express field equations in terms of the stress tensor and the magnetic field for isothermal solids.

8.3. A spherical shell of high permeability μ is placed in a uniform field B_0 in the x_1-direction. Show that the magnetic potential Φ can be expressed in the forms

$$\Phi = -B_0 r \cos\theta + \sum_{k=0}^{\infty} a_k r^{-k-1} P_k(\cos\theta), \qquad r > b,$$

$$\Phi = \sum_{k=0}^{a} (b_k r^k + c_k r^{-k}) P_k(\cos\theta), \qquad a < r < b,$$

$$\Phi = \sum_{k=0}^{a} d_k r^k P_k(\cos\theta) \qquad r < a,$$

where a and b are, respectively, the inner and outer radius of the shell and P_k is the Legendre polynomial. Determine constants a_k, b_k, c_k, and d_k. Show that the inside of the spherical shell is shielded.

8.4. A circular cylinder of radius a carries a uniform surface charge distribution w_e. The cylinder is rotating with constant angular velocity Ω about its axis. Find the magnetic flux density.

8.5. Determine the stress field in a plate with an elliptic hole under a bias magnetic field directed along the minor axis of the ellipse.

8.6. A sphere of radius a carries a uniform surface charge. The sphere is undergoing small amplitude radial vibrations isothermally. Determine the phase velocity.

8.7. A half-plane is under a constant bias field $\mathbf{H}_0$. Determine the surface wave velocity for small amplitude waves.

8.8. A thick, incompressible spherical shell carrying a uniform charge on its outer surface is inflated. Determine the stress and the magnetic fields.

8.9. A circular cylinder carries steady electric current. If the cylinder is stretched without volume change, what will be the stress and the magnetic fields in the cylinder?

8.10. Determine the frequency of the axisymmetric vibrations of a circular plate clamped along its outer edge. The plate is under a transverse uniform magnetic field.

8.11. By appropriate replacement of the flexural rigidity D one can obtain the equations of motion of beams from (8.14.25). For a simply supported beam obtain the buckling load under the constant transverse field B_0.

8.12. Derive the equations of motion of a beam when it is subjected to a bias axial magnetic field B_0.

8.13. Obtain the frequency of a cantilever beam:
(a) under a transverse magnetic field;
(b) under an axial magnetic field.

APPENDIX A

Crystallographic Point Groups

In this appendix we examine briefly the symmetry properties of nonmagnetic crystals. According to the axiom of material invariance, the macroscopic symmetry of all nonmagnetic crystals may be described by an isotropy group $\{\mathbf{S}\}$. Accordingly, under the transformations of the material frame of reference

$$\bar{\mathbf{X}} = \mathbf{S}\mathbf{X}, \qquad \mathbf{S}\mathbf{S}^{\mathrm{T}} = \mathbf{S}^{\mathrm{T}}\mathbf{S} = \mathbf{1}, \qquad \det \mathbf{S} = \pm 1, \tag{A.1}$$

the constitutive functionals must remain form-invariant for all members of the symmetry group $\{\mathbf{S}\}$. Local properties of crystals are restricted by the *point group*.

The symmetry operators, that act at a fixed point O and leave invariant all distances and angles in a three-dimensional space, are called the *point group*. The symmetry operators that have these properties are rotations about axes through O, and products (combinations) of rotations and inversions. Of course, such products include reflections in planes through O.

If the group contains only rotations, it is called a *proper rotation group*. This is isomorphic with the group $O^+(3)$ of all 3×3 orthogonal matrices. Operators, whose matrices have determinant (-1), are called *improper rotations*. They are products of proper rotations and inversion. We note that the inversion commutes with all rotations.

Every subgroup of $O^+(3)$ is a *proper point group*. Proper point groups of *finite* order are classified as: Cyclic ($C_n = n$); Dihedral ($D_n = n22$, n even, $D_n = n2$, n odd); Tetrahedral ($T = 23$); and Octahedral ($O = 432$).

A crystallographic point group is restricted by a requirement that an operator must be compatible with the translational symmetry of a crystalline solid. Hence, the appropriate symmetry operations are

$$\text{identity} = E, \qquad \text{inversion} = C,$$
$$\text{reflection in certain planes} = \sigma, \qquad \text{rotations} = C_{nr}.$$

The rotation C_{nr} is an anticlockwise rotation through $2\pi/n$ radians about the axis indicated by r.

The eleven proper point groups are listed in Table A.1, together with their

Table A.1. Crystallographic pure rotation groups.

Cyclic groups	Symmetry elements
$C_1 = 1$	E
$C_2 = 2$	E, C_{2z}
$C_3 = 3$	E, C_{3z}, C_{3z}^-
$C_4 = 4$	$E, C_{4z}, C_{4z}^-, C_{2z}$
$C_6 = 6$	$E, C_{6z}, C_{6z}^-, C_{3z}, C_{3z}^-, C_{2z}$
Dihedral groups	
$D_2 = 222$	$E, C_{2x}, C_{2y}, C_{2z}$
$D_3 = 32$	$E, C_{3z}, C_{3z}^-, C'_{21}, C'_{22}, C'_{23}$
$D_4 = 422$	$E, C_{4z}, C_{4z}^-, C_{2x}, C_{2y}, C_{2z}, C_{2a}, C_{2b}$
$D_6 = 622$	$E, C_{6z}, C_{6z}^-, C_{3z}, C_{3z}^-, C_{2z}, C'_{2r}, C''_{2r}$
Tetrahedral group	
$T = 23$	$E, C_{2m}, C_{3j}, C_{3j}^-$
Octahedral group	
$O = 432$	$E, C_{2m}, C_{3j}, C_{3j}^-, C_{2p}, C_{4m}, C_{4m}^-$

symmetry elements. In this table the first column ($C_1, C_2, \ldots, 0$) denotes the Schönflies notation, and the second column ($1, 2, \ldots, 432$) denotes the international notation.

In addition to purely rotational symmetry, the space lattice possesses symmetries of reflections in various planes ($\det \mathbf{S} = -1$). In order to include such symmetry operations, we multiply the proper point group $\{P\}$ by $\{E, C\}$. This produces a new set of eleven point groups that are subgroups of $O(3)$.

If the point group $\{P\}$ has an invariant subgroup $\{H\}$ of index[1] 2, then

$$\{P\} = \{H\} + C\{P - H\} \tag{A.2}$$

is also a point group. This process gives ten more point groups. The possible crystallographic point groups are 32 in number, as listed in Table A.2.

By examination of the metrical properties, crystal classes are divided into seven *crystal systems*. Each system possesses one and the same metrical property. If $\mathbf{h}_i$ denotes the lattice bases then the length of lattice bases $|\mathbf{h}_1| = a$, $|\mathbf{h}_2| = b$, $|\mathbf{h}_3| = c$, and angles $\alpha = \text{angle}(\mathbf{h}_2, \mathbf{h}_3)$, $\beta = \text{angle}(\mathbf{h}_3, \mathbf{h}_1)$, and $\gamma = \text{angle}(\mathbf{h}_1, \mathbf{h}_3)$, for each crystal system, are the same. This is called a *holohedry* of the space lattices.

In Table A.1, $j = 1, 2, 3, 4$; $m = x, y, z$; $p = a, b, c, d, e, f$; and $r = 1, 2, 3$; and the labels of the symmetry operations can be identified from Figures A.1–A.3. In Figures A.1 and A.2 the labels of the symmetry operations are placed on the figure in the position to which the letter E is taken by that operation.

[1] The index of a subgroup is the integer obtained by dividing the order of the group by that of the subgroup.

Figure A.1. Symmetry elements: triclinic, monoclinic, rhombic, and tetragonal systems.

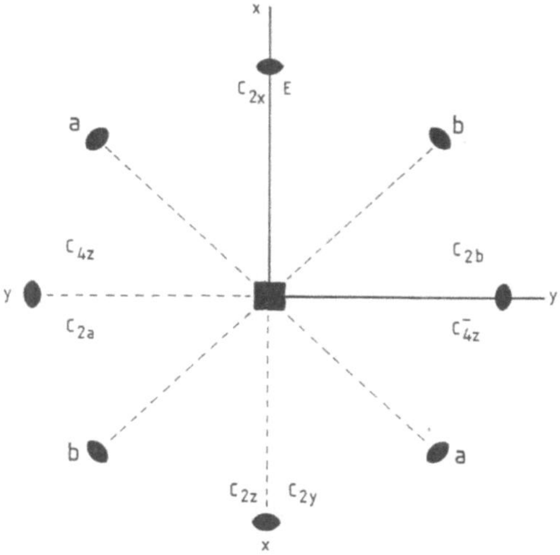

Figure A.2. Symmetry elements: trigonal and hexagonal systems.

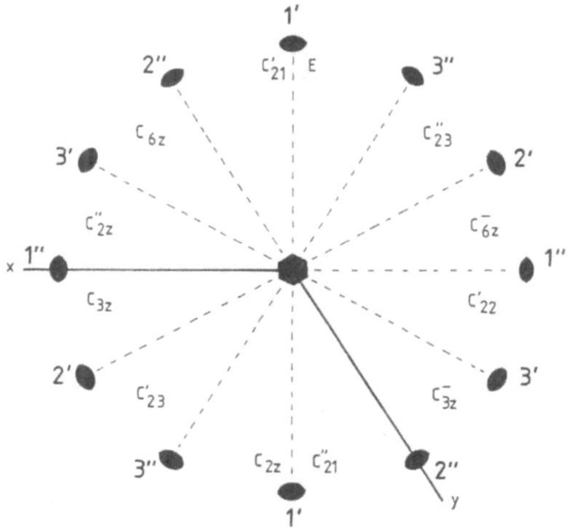

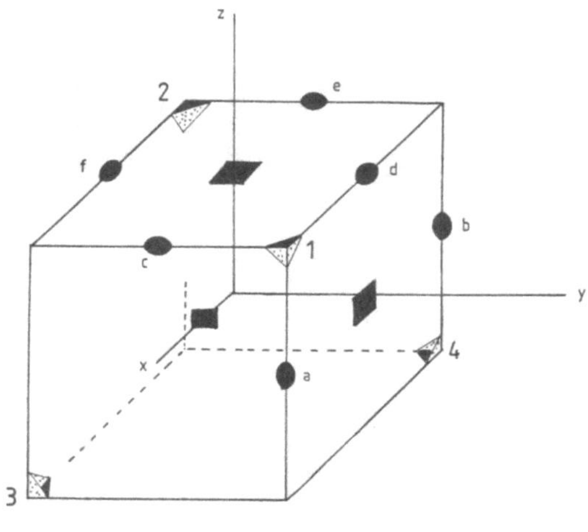

Figure A.3. Symmetry elements: cubic system.

Table A.2. The 32 conventional crystal classes.

System	Class number	Class name		Symmetry transformations	Order
Triclinic	1	Pedial	C_1 1	I	1
	2	Pinacoidal	C_i $\bar{1}$	I, C	2
Monoclinic	3	Sphenoidal	C_2 2	I, D_3	2
	4	Domatic	C_s m	I, R_3	2
	5	Prismatic	C_{2h} $2/m$	I, C, R_3, D_3	4
Orthorhombic	6	Rhombic–disphenonidal	D_2 222	I, D_1, D_2, D_3	4
	7	Rhombic–pyromidal	C_{2v} $2mm$	I, R_1, R_2, D_3	4
	8	Rhombic–dipyramidal	D_{2h} mmm	$I, C, D_1, D_2, D_3, R_1, R_2, R_3$	8
Tetragonal	9	Tetragonal–pyramidal	C_4 4	I, D_3, R_1T_3, R_2T_3	4
	10	Tetragonal–disphenoidal	$\bar{C}_2$ $\bar{4}$	I, D_3, D_1T_3, D_2T_3	4
	11	Tetragonal–dipyramidal	C_{4h} $4/m$	$I, D_3, D_1T_3, D_2T_3, R_1T_3, R_2T_3, C, R_3$	8
	12	Tetragonal–trapezahedral	D_4 422	$I, D_1, D_2, D_3, CT_3, R_1T_3, R_2T_3, R_3T_3$	8
	13	Ditetragonal–pyramidal	C_{4v} $4mm$	$I, R_1, R_2, D_3, T_3, R_1T_3, R_2T_3, D_3T_3$	8
	14	Tetragonal–scalenohedral	D_{2v} $\bar{4}2m$	$I, D_1, D_2, D_3, T_3, D_1T_3, D_2T_3, D_3T_3$	8
	15	Ditetragonal–dipyramidal	D_{4h} $4/mmm$	$I, D_1, D_2, D_3, CT_3, R_1T_3, R_2T_3, R_3T_3, C, R_1, R_2, R_3, T_3, D_1T_3,$	16

Trigonal	16	Trigonal–pyramidal	C_3 3	I, S_1, S_2	3
	17	Rhombohedral	$\bar{C}_3$ $\bar{3}$	$I, S_1, S_2, C, CS_1, CS_2$	6
	18	Trigonal–trapezohedral	D_3 32	$I, S_1, S_2, D_1, D_1S_1, D_1S_2$	6
	19	Ditrigonal–pyramidal	C_{3v} 3_m	$I, S_1, S_2, R_1, R_1S_1, R_1S_2$	6
	20	Hexagonal–scalenohedral	D_{3v} $\bar{3}_m$	$I, S_1, S_2, C, CS_1, CS_2, R_1, R_1S_1, R_1S_2, D_1, D_1S_1, D_1S_2$	12
Hexagonal	21	Hexagonal–pyramidal	C_6 6	$I, S_1, S_2, D_3, D_3S_1, D_3S_2$	6
	22	Trigonal–dipyramidal	C_{3h} $\bar{6}$	$I, S_1, S_2, R_3, R_3S_1, R_3S_2$	6
	23	Hexagonal–dipyramidal	C_{6h} $6/m$	$I, S_1, S_2, R_3, R_3S_1, R_3S_2, C, CS_1, CS_2, D_3, D_3S_1, D_3S_2$	12
	24	Hexagonal–trapezohedral	D_6 622	$I, S_1, S_2, D_3, D_3S_1, D_3S_2, D_1, D_1S_1, D_1S_2, D_2S_1, D_2S_2, D_2$	12
	25	Dihexagonal–pyramidal	C_{6v} $6mm$	$I, S_1, S_2, D_3, D_3S_1, D_3S_2, R_1, R_1S_1, R_1S_2, R_2, R_2S_1, R_2S_2$	12
	26	Ditrigonal–dipyramidal	D_{3h} $\bar{6}2m$	$I, S_1, S_2, R_3, R_3S_1, R_3S_2, R_1, R_1S_1, R_1S_2, D_2, D_2S_1, D_2S_2$	12
	27	Dihexagonal–dipyramidal	D_{6h} $6/mm$	$I, S_1, S_2, C, CS_1, CS_2, D_1, D_1S_1, D_1S_2, D_2, D_2S_1, D_2S_2, R_1, R_1S_1, R_1S_2, R_2, R_2S_1, R_2S_2, R_3, R_3S_1, R_3S_2, D_3, D_3S_1, D_3S_2$	24
Cubic	28	Tetartoidal	T 23	$I, D_1, D_2, D_3, C_{3j}, C_{3j}^-$	12
	29	Diploidal	T_h $m3$	$I, D_1, D_2, D_3, C, R_1, R_2, R_3, C_{3j}, C_{3j}^-, S_{6j}, S_{6j}^-$	24
	30	Gyroidal	O 432	$I, D_1, D_2, D_3, C_{2p}, C_{3j}, C_{3j}^-, C_{4m}, C_{4m}^-$	24
	31	Hextetrahedral	Td $43m$	$I, D_1, D_2, D_3, \sigma_{dp}, C_{3j}, C_{3j}^-, S_{4m}, S_{4m}^-$	24
	32	Hexoctohedral	O_h $m3m$	$I, D_1, D_2, D_3, C_{2p}, C_{3j}, C_{3j}^-, C_{4m}, C_{4m}^-, C, R_1, R_2, R_3, \sigma_{dp}, S_{6j}, S_{6j}^-, S_{4m}, S_{4m}^-$	48

The transformation matrices are given by

$$\mathbf{I}=\begin{pmatrix}1&0&0\\0&1&0\\0&0&1\end{pmatrix},\qquad \mathbf{C}=\begin{pmatrix}-1&0&0\\0&-1&0\\0&0&-1\end{pmatrix},$$

$$\mathbf{R}_1=\begin{pmatrix}-1&0&0\\0&1&0\\0&0&1\end{pmatrix},\ \mathbf{R}_2=\begin{pmatrix}1&0&0\\0&-1&0\\0&0&1\end{pmatrix},\quad \mathbf{R}_3=\begin{pmatrix}1&0&0\\0&1&0\\0&0&-1\end{pmatrix},$$

$$\mathbf{D}_1=\begin{pmatrix}1&0&0\\0&-1&0\\0&0&-1\end{pmatrix},\quad \mathbf{D}_2=\begin{pmatrix}-1&0&0\\0&1&0\\0&0&-1\end{pmatrix},\ \mathbf{D}_3=\begin{pmatrix}-1&0&0\\0&-1&0\\0&0&1\end{pmatrix},$$

$$\mathbf{T}_1=\begin{pmatrix}1&0&0\\0&0&1\\0&1&0\end{pmatrix},\quad \mathbf{T}_2=\begin{pmatrix}0&0&1\\0&1&0\\1&0&0\end{pmatrix},\quad \mathbf{T}_3=\begin{pmatrix}0&1&0\\1&0&0\\0&0&1\end{pmatrix},$$

$$\mathbf{M}_1=\begin{pmatrix}0&1&0\\0&0&1\\1&0&0\end{pmatrix},\quad \mathbf{M}_2=\begin{pmatrix}0&0&1\\1&0&0\\0&1&0\end{pmatrix}, \tag{A.3}$$

$$\mathbf{S}_1=\begin{pmatrix}-1/2&\sqrt{3}/2&0\\-\sqrt{3}/2&-1/2&0\\0&0&1\end{pmatrix},\qquad \mathbf{S}_2=\begin{pmatrix}-1/2&-\sqrt{3}/2&0\\\sqrt{3}/2&-1/2&0\\0&0&1\end{pmatrix},$$

where **I** is the identity and **C** is the central inversion. $\mathbf{R}_1, \mathbf{R}_2, \mathbf{R}_3$ are reflections in the planes whose normals are along the $x_1 = x$-, $x_2 = y$-, and $x_3 = z$-directions, respectively. $\mathbf{D}_1, \mathbf{D}_2, \mathbf{D}_3$ are rotations through π radians about the x_1-, x_2-, and x_3-axes, respectively. $\mathbf{T}_1$ is a reflection through a plane which bisects the x_2- and x_3-axes and contains the x_1-axis. $\mathbf{T}_2$ and $\mathbf{T}_3$ are analogously defined. $\mathbf{M}_1$ and $\mathbf{M}_2$ are rotations through $2\pi/3$ clockwise and anticlockwise, about an axis making equal acute angles with the axes x_1, x_2, and x_3. $\mathbf{S}_1$ and $\mathbf{S}_2$ are rotations through $2\pi/3$ clockwise and anticlockwise, respectively, about the $x_3 = z$-axis.

APPENDIX B

Crystallographic Magnetic Groups

As noted in Section 5.4, the symmetry properties of magnetic materials must include a time-inversion operator which reverses the spin of each atom. The situation is visualized simply by considering a chain of equally spaced atoms on a line (Figure B.1). Disregarding their spin, we see that the X_2-axis is a twofold symmetry axis, and in addition, the X_2X_3-plane is a reflection plane (Figure B.1(a)). Now if the spins are as shown in Figure B.1(b), then the situation is the same. However, if the spins are oppositely directed (Figure B.1(c)), then X_2 is no longer a twofold rotation axis. Moreover, the X_2X_3-plane is *not* a reflection plane. Thus, the full characterization of the magnetic properties of crystals requires the incorporation of the symmetry property of the individual atoms constituting the lattice points to the symmetry of the lattice. This means the consideration of spin or, interpreted as an orbital angular momentum, time reversal. Atoms of certain materials do not possess magnetic moments and in some other materials the spin is randomly distributed. The first of these two classes of materials is called *diamagnetic* and the second *paramagnetic*. These materials may therefore be referred to as *nonmagnetic*, and the point group of 32 classes discussed in Appendix A constitutes their symmetry group.

However, there exist large classes of other materials which exhibit magnetic properties. These are the *ferromagnetic*, *antiferromagnetic*, and *ferrimagnetic* materials. In ferromagnetic materials (e.g., Fe, Zn, Co) the adjacent lattice sites possess parallel spins so that, in the absence of an external field, the material posseses net magnetization (Figure B.2(a)). In antiferromagnetic materials (e.g., CoF_2, MnF_2, Cr_2O_3) the spin distribution is in a periodic arrangement, alternating parallel and antiparallel motifs, that results in zero magnetization in the absence of an external field (Figure B.2(b)). The ferrimagnetic materials (e.g., $MnFe_2O_4$, $NiFe_2O_4$) also contain antiparallel spin arrangements, however, the cancellation is incomplete and the body possesses magnetic dipole density. All three types of materials have highly nonlinear **B**–**H** relationships. Ferromagnetic, antiferromagnetic, and ferrimagnetic materials are called *magnetic materials.*

The arrangement of atomic magnetic moments can be affected in all mag-

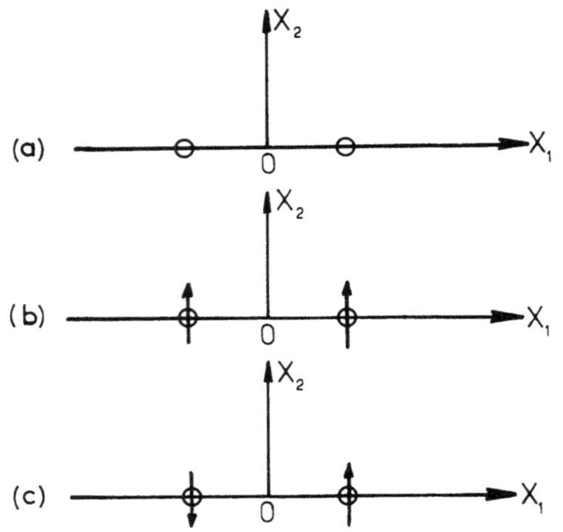

Figure B.1. Magnetic symmetry.

netic classes to produce antiferromagnetism. This includes even those that exhibit ferromagnetism. For example, NiF_2 in its crystallized magnetic symmetry $m\underline{m}\underline{m}$ (a ferromagnetic class), exhibits antiferromagnetism. Conversely, by applying a small rotation to the spins of antiferromagnetic materials we can obtain weak ferromagnetism. This phenomenon has been observed for several substances, among which are α-Fe_2O_3 above 250 K, NiF_2, $MnCO_3$, and $CoCO_3$.

For *magnetic materials*, as discussed before, the spin symmetry can be incorporated into the crystal symmetry group by means of the time-reversal operator R. Alternatively, we can use a four-dimensional formalism involving 4×4 matrices, in Minkowski space, as the members of the symmetry group. Here, for the sake of simplicity, we briefly discuss the use of the time-reversal operator R. It is conventional to denote the time reversal by an underscore, e.g., if $(E, S^1, S^2, \ldots)$ denote the elements of the nonmagnetic group G. The reversal of the atomic magnetic moment for an element S^α of G is denoted by $\underline{S}^\alpha$ and is called the *complement* of S^α. If the product rule of matrices being applied to the elements of S^α is $S^1S^2 = S^3$, then we can easily see that the product rule for the complement group is $\underline{S}^1\underline{S}^2 = S^3$, $\underline{S}^1S^2 = S^1\underline{S}^2 = \underline{S}^3$. In this way, from the symmetry elements of $G = \{\mathbf{S}\}$, we obtain complementary elements by replacing some of these symmetry operations by their complements, such that the resulting set of operations form a group under the product rule defined above. By exhausting all possibilities for the 32 elements of the

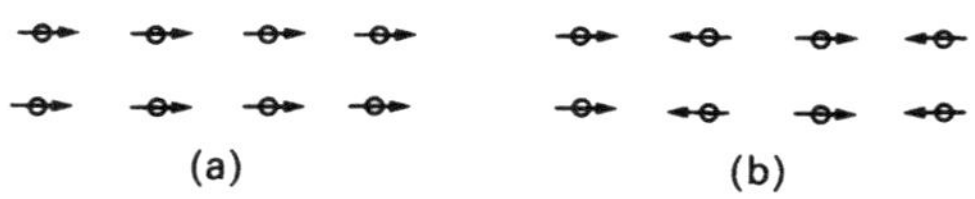

Figure B.2. Magnetic materials: (a) ferromagnetic; (b) antiferromagnetic.

nonmagnetic crystal group, we find that there are only 58 distinct groups which are of magnetic origin. A systematic way of determining the magnetic group is given by Tavger and Zaitsev [1956]. The 32 nonmagnetic point groups, of course, do not contain the time reversal R. The remaining 58 groups, called *additional magnetic groups*, contain R in combination with the spatial symmetry operators. Thus, if H is a subgroup of index 2 of the nonmagnetic group $G \equiv \{\mathbf{S}\}$, then the elements of the additional magnetic group are of two types:

(a) $\mathbf{S}^\alpha \in H \subset G$;
(b) $R\mathbf{S}^\beta$ such that $\mathbf{S}^\beta \in (G - H)$.

Birss [1964] proves that $\mathbf{S}^\alpha$ and $\mathbf{S}^\beta$ are disjoint, and therefore it is possible to represent a magnetic point group $\{\mathbf{M}\}$ in the form

$$\{\mathbf{M}\} = \{H + R(G - H)\}, \tag{B.1}$$

or

$$\{\mathbf{M}\} = \{H + R\mathbf{S}^\beta H\}, \tag{B.2}$$

where $\mathbf{S}^\beta$ is a particular element of the set $(G - H)$. From (B.2) it is clear that magnetic point groups can be generated as follows:

(i) For any particular class, one group of magnetic symmetry is identical to the nonmagnetic class G.
(ii) From G select all subgroups H of index 2.
(iii) Replace all elements $\mathbf{S}^\beta$ of $(G - H)$ (which do not belong to H) by $\underline{\mathbf{S}}^\beta = R\mathbf{S}^\beta$.
(iv) Reject all groups $\{\mathbf{M}\} = \{H + R(-H)\}$ for which any element $\mathbf{S}^\beta$ is of odd order. This is because a magnetic group with an element $R\mathbf{S}^\beta$ is to be rejected if $\mathbf{S}^\beta$ is of odd order, since $(R\mathbf{S}^\beta)^n = R$ (n = odd) is not a magnetic symmetry group.

EXAMPLE. To illustrate, consider the prismatic class $C_{2h} = 2/m = 2{:}m$ whose symmetry elements are $\mathbf{I}$, $\mathbf{D}_1$, $\mathbf{C}$, and $\mathbf{R}_1$. It has three subgroups with index 2, namely,

$$m = \{\mathbf{I}, \mathbf{R}_1\} = C_5,$$

$$2 = \{\mathbf{I}, \mathbf{D}_1\} = C_2,$$

$$\bar{1} = \{\mathbf{I}, \mathbf{C}\} = C_i.$$

We thus have

$$\left\{\frac{2}{m} - m\right\} = \{\mathbf{D}_1, \mathbf{C}\},$$

$$\left\{\frac{2}{m} - 2\right\} = \{\mathbf{C}, \mathbf{R}_1\},$$

$$\left\{\frac{2}{m} - \bar{1}\right\} = \{\mathbf{D}_1, \mathbf{R}_1\}.$$

Hence, the three magnetic groups originating from $2/m$ are

$$\underline{2}/m\colon\quad m + R\left\{\frac{2}{m} - m\right\} = \mathbf{I}, \mathbf{R}_1, R\mathbf{D}_1, R\mathbf{C},$$

$$2/\underline{m}\colon\quad 2 + R\left\{\frac{2}{m} - 2\right\} = \mathbf{I}, \mathbf{D}_1, R\mathbf{C}, R\mathbf{R}_1,$$

$$\underline{2}/\underline{m}\colon\quad \bar{1} + R\left\{\frac{2}{m} - \bar{1}\right\} = \mathbf{I}, \mathbf{C}, R\mathbf{D}_1, R\mathbf{R}_1.$$

Note that none of the elements of these three classes are of odd order. They constitute 8 to 11 classes out of the 90 magnetic groups in Table B.1.

Table B.1. Magnetic point groups.

No.	Magnetic point group $\{M\}$	Classical subgroup $\{H\}$ International	Classical subgroup $\{H\}$ Schönflies	G-H
1	$\underline{\bar{1}}$	1	C_1	C
2	$\underline{2}$	1	C_1	D_3
3	$\underline{m}$	1	C_1	R_3
4	$2/\underline{m}$	2	C_2	C, R_3
5	$\underline{2}/m$	m	$C_{1h} = C_s$	C, D_3
6	$\underline{2}/\underline{m}$	$\bar{1}$	C_i	D_3, R_3
7	$\underline{2}\underline{2}2$	2	C_2	D_1, D_2
8	$2\underline{m}\underline{m}$	2	C_2	R_1, R_2
9	$\underline{2}m\underline{m}$	m	C_s	D_3, R_1
10	$\underline{m}\underline{m}\underline{m}$	222	D_2	C, R_1, R_2, R_3
11	$\underline{m}mm$	$2mm$	C_{2v}	C, D_1, D_2, R_3
12	$m\underline{m}\underline{m}$	$2/m$	C_{2h}	D_1, D_2, R_1, R_2
13	$\underline{4}$	2	C_2	$R_2 T_3, R_1 T_3$
14	$\underline{\bar{4}}$	2	C_2	$D_2 T_3, D_1 T_3$
15	$4\underline{2}\underline{2}$	4	C_4	$D_1, D_2, CT_3, R_3 T_3$
16	$\underline{4}2\underline{2}$	222	D_2	$R_2 T_3, R_1 T_3, CT_3, R_3 T_3$
17	$4/\underline{m}$	4	C_4	$C, R_3, D_2 T_3, D_1 T_3$
18	$\underline{4}/\underline{m}$	$\bar{4}$	S_4	$C, R_3, R_2 T_3, R_1 T_3$
19	$\underline{4}/m$	$2/m$	C_{2h}	$R_2 T_3, R_1 T_3, D_2 T_3, D_1 T_3$
20	$4\underline{m}\underline{m}$	4	C_4	$R_1, R_2, T_3, D_3 T_3$
21	$\underline{4}m\underline{m}$	$2mm$	C_{2v}	$R_2 T_3, R_1 T_3, T_3, D_3 T_3$
22	$\bar{4}\underline{2}\underline{m}$	$\bar{4}$	S_4	$D_1, D_2, T_3, D_3 T_3$
23	$\underline{\bar{4}}2\underline{m}$	222	D_2	$D_2 T_3, D_1 T_3, T_3, D_3 T_3$
24	$\underline{\bar{4}}\underline{2}m$	$2mm$	C_{2v}	$D_1, D_2, D_2 T_3, D_1 T_3$
25	$4/\underline{m}\underline{m}\underline{m}$	422	D_4	$C, R_1, R_2, R_3, D_2 T_3, D_1 T_3, T_3, D_3 T_3$
26	$4/\underline{m}mm$	$4mm$	C_{4v}	$C, R_3, D_2 T_3, D_1 T_3, D_1, D_2, CT_3, R_3 T_3$
27	$\underline{4}/mm\underline{m}$	mmm	D_{2h}	$R_2 T_3, R_1 T_3, CT_3, R_3 T_3, D_2 T_3, D_1 T_3, T_3, D_3 T_3$
28	$\underline{4}/\underline{m}m\underline{m}$	$\bar{4}2m$	D_{2d}	$C, R_1, R_2, R_3, R_2 T_3, R_1 T_3, CT_3, R_3 T_4$
29	$\underline{4}/m\underline{m}\underline{m}$	$4/m$	C_{4h}	$D_1, D_2, R_1, R_2, CT_3, R_3 T_3, T_3, D_3 T_3$
30	$3\underline{2}$	3	C_3	$D_1, D_1 S_1, D_1 S_2$
31	$3\underline{m}$	3	C_3	$R_1, R_1 S_1, R_1 S_2$

(continued)

Table B.1 (*continued*)

No.	Magnetic point group $\{M\}$	Classical subgroup $\{H\}$ International	Classical subgroup $\{H\}$ Schönflies	*G-H*
32	$\underline{\bar{6}}$	3	C_3	R_3, R_3S_1, R_3S_2
33	$\bar{6}\underline{m}\underline{2}$	$\bar{6}$	C_{3h}	$D_2, D_2S_1, D_2S_2, R_1, R_1S_1, R_2S_2$
34	$\underline{\bar{6}}m\underline{2}$	$3m$	C_{3v}	$D_2, D_2S_1, D_2S_2, R_3, R_3S_1, R_3S_2$
35	$\underline{\bar{6}}\underline{m}2$	32	D_3	$R_3, R_3S_2, R_3S_1, R_1, R_1S_1, R_1S_2$
36	$\underline{6}$	3	C_3	D_3, D_3S_2, D_3S_1
37	$\underline{\bar{3}}$	3	C_3	C, CS_1, CS_2
38	$\bar{3}\underline{m}$	$\bar{3}$	C_{3i}	$D_1, D_1S_1, D_1S_2, R_1, R_1S_1, R_1S_2$
39	$\underline{\bar{3}}m$	$3m$	C_{3v}	$D_1, D_1S_1, D_1S_2, C, CS_1, CS_2$
40	$\underline{\bar{3}}\underline{m}$	32	D_3	$C, CS_1, CS_2, R_1, R_1S_1, R, S_2$
41	$6\underline{2}\underline{2}$	6	C_6	$D_1, D_1S_1, D_1S_2, D_2, D_2S_1, D_2S_2$
42	$\underline{6}2\underline{2}$	32	D_3	$D_3, D_3S_2, D_3S_1, D_2, D_2S_1, D_2S_2$
43	$6/\underline{m}$	6	C_6	$C, CS_1, CS_2, R_3, R_3S_2, R_3S_1$
44	$\underline{6}/\underline{m}$	$\bar{3}$	C_{3i}	$D_3, D_3S_2, D_3S_1, R_3, R_3S_2, R_3S_1$
45	$6/m$	$\bar{6}$	C_{3h}	$C, CS_1, CS_2, D_3, D_3S_2, D_3S_1$
46	$6\underline{m}\underline{m}$	6	C_6	$R_1, R_1S_1, R_1S_2, R_2, R_2S_1, R_2S_2$
47	$\underline{6}m\underline{m}$	$3m$	C_{3v}	$D_3, D_3S_2, D_3S_1, R_2, R_2S_1, R_2S_2$
48	$\underline{6}/mm\underline{m}$	$\bar{6}2m$	D_{3h}	$C, CS_1, CS_2, D_3, D_3S_2, D_3S_1, D_2, D_2S_1,$ $D_2S_2, R_1, R_1S_1, R_1S_2$
49	$\underline{6}/\underline{m}m\underline{m}$	$\bar{3}m$	D_{3d}	$D_3, D_3S_2, D_3S_1, D_2, D_2S_1, D_2S_2, R_3, R_3S_1,$ $R_3S_2, R_2, R_2S_1, R_2S_2$
50	$6/\underline{m}\underline{m}\underline{m}$	622	D_6	$C, CS_1, CS_2, R_3, R_3S_1, R_3S_2, R_1, R_1S_1,$ $R_1S_2, R_2, R_2S_1, R_2S_2$
51	$6/\underline{m}mm$	$6mm$	C_{6v}	$D_1, D_1S_1, D_1S_2, D_2, D_2S_1, D_2S_2, C, CS_1,$ $CS_2, R_3, R_3S_1, R_3S_2$
52	$6/m\underline{m}\underline{m}$	$6/m$	C_{6h}	$D_1, D_1S_1, D_1S_2, D_2, D_2S_2, R_1, R_1S_1, R_1S_2,$ $R_2, R_2S_1, R_2S_2, D_2S_1$
53	$\underline{m}3$	23	T	$C, S_{6j}, S_{6j}^-, R_1, R_2, R_3$
54	$\underline{\bar{4}}3\underline{m}$	23	T	$\sigma_{dp}, S_{4m}, S_{4m}^-$
55	$\underline{4}3\underline{2}$	23	T	C_{2p}, C_{4m}, C_{4m}^-
56	$\underline{m}3\underline{m}$	432	O	$C, S_{6j}, S_{6j}^-, R_1, R_2, R_3, \sigma_{dp}, S_{4m}, S_{4m}^-$
57	$\underline{m}3m$	$\bar{4}3m$	T_d	$C, S_{6j}, S_{6j}^-, R_1, R_2, R_3, C_{2p}, C_{4m}, C_{4m}^-$
58	$m3\underline{m}$	$m3$	T_h	$C_{2p}, C_{4m}, C_{4m}^-, \sigma_{dp}, S_{4m}, S_{4m}^-$

APPENDIX C

Integrity Bases of Crystallographic Groups

Tables C1.1–C1.16 give the linear combinations of the components of an absolute (polar) vector p_i, an axial vector a_i, and a symmetric second-order tensor S_{ij}, which form the carrier spaces for the irreducible representations Γ_1, $\Gamma_2, \ldots$ associated with various crystal classes. The notation Γ_3: ϕ, ψ indicates that ϕ and ψ are the basic quantities associated with the representation Γ_3 of degree one. The notation Γ_5: $(a_1, a_2), (b_1, b_2)$ indicates that (a_1, a_2) and (b_1, b_2) are basic quantities associated with the representation Γ_5 of degree two, and so on. Typical elements of the integrity basis for a crystal class are listed following the tables, given by Kiral [1972] and Kiral and Smith [1974]. The complete set of integrity basis would be obtained from these by using the format (5.5.18).

Table C1. Basic quantities. For a symmetric second-order tensor S_{ij}, a polar vector p_i, and an axial vector a_i that form the carrier space for the irreducible representations $\Gamma_1, \Gamma_2, \ldots$ associated with various conventional crystal classes.

Table C1.1

$C_1 \cdot \Gamma_1$: $a_1, a_2, a_3, S_{11}, S_{22}, S_{33}, S_{13}, S_{23}, S_{12}$;	Γ_2: p_1, p_2, p_3;
$C_2 \cdot \Gamma_1$: $p_2, p_3, a_1, S_{11}, S_{22}, S_{33}, S_{23}$;	Γ_2: $p_1, a_2, a_3, S_{12}, S_{13}$;
$C_2 \cdot \Gamma_1$: $p_1, a_1, S_{11}, S_{22}, S_{33}, S_{23}$;	Γ_2: $p_2, p_3, a_2, a_3, S_{12}, S_{13}$;

Table C1.2

$C_{2h} \cdot \Gamma_1$: $a_1, S_{11}, S_{22}, S_{33}, S_{23}$;	Γ_2: p_1;	Γ_3: p_2, p_3;	Γ_4: a_2, a_3, S_{12}, S_{13};
$C_{2v} \cdot \Gamma_1$: $p_1, S_{11}, S_{22}, S_{33}$;	Γ_2: a_1, S_{23};	Γ_3: p_2, a_3, S_{12};	Γ_4: p_3, a_2, S_{13};
$D_2 \cdot \Gamma_1$: S_{11}, S_{22}, S_{33};	Γ_2: p_1, a_1, S_{23};	Γ_3: p_2, a_2, S_{13};	Γ_4: p_3, a_3, S_{12};

Table C1.3

$D_{2h} \cdot \Gamma_1$: S_{11}, S_{22}, S_{33};	Γ_2: a_1, S_{23};	Γ_3: a_2, S_{13};	Γ_4: a_3, S_{12};	Γ_2': p_1;	Γ_3': p_2;	Γ_4': p_3;

Table C1.4

$\bar{C}_2 \cdot \Gamma_1$: $a_3, S_{33}, S_{11} + S_{22}$; Γ_2: $p_3, S_{12}, S_{11} - S_{22}$;
Γ_3: $p_1 - ip_2, a_1 + ia_2, S_{13} + iS_{23}$; Γ_4: $p_1 + ip_2, a_1 - ia_2, S_{13} - iS_{23}$;
$C_4 \cdot \Gamma_1$: $a_3, S_{33}, S_{11} + S_{22}, p_3$; Γ_2: $S_{12}, S_{11} - S_{22}$;
Γ_3: $p_1 + ip_2, a_1 + ia_2, S_{13} + iS_{23}$; Γ_4: $p_1 - ip_2, a_1 - ia_2, S_{13} - iS_{23}$;

Table C1.5

$C_{4h} \cdot \Gamma_1$: $a_3, S_{33}, S_{11} + S_{22}$; Γ_2: $S_{12}, S_{11} - S_{22}$; Γ_3: $S_{13} + iS_{23}, a_1 + ia_2$;
Γ_4: $a_1 - ia_2, S_{13} - iS_{23}$; Γ_1': p_3; Γ_3': $p_1 + ip_2$; Γ_4': $p_1 - ip_2$;

Table C1.6

$C_{4v} \cdot \Gamma_1$: $p_1, S_{33}, S_{11} + S_{22}$; Γ_2: a_3; Γ_3: S_{12}; Γ_4: $S_{11} - S_{22}$;
Γ_5: $(p_1, p_2), (a_2, -a_1), (S_{13}, S_{23})$;
$D_4 \cdot \Gamma_1$: $S_{33}, S_{11} + S_{22}$; Γ_2: p_3, a_3; Γ_3: S_{12}; Γ_4: $S_{11} - S_{22}$
Γ_5: $(p_1, p_2), (a_1, a_2), (S_{23}, -S_{13})$;
$D_{2v} \cdot \Gamma_1$: $S_{33}, S_{11} + S_{22}$; Γ_2: a_3; Γ_3: p_3, S_{12}; Γ_4: $S_{11} - S_{22}$;
Γ_5: $(p_1, p_2), (a_1, -a_2), (S_{23}, S_{13})$;

Table C1.7

$D_{4h} \cdot \Gamma_1$: $S_{33}, S_{11} + S_{22}$; Γ_2: a_3; Γ_3: S_{12}; Γ_4: $S_{11} - S_{22}$;
Γ_5: $(a_1, a_2), (S_{23}, -S_{13})$; Γ_2': p_3; Γ_5': (p_1, p_2);

Table C1.8

$C_3 \cdot \Gamma_1$: $p_3, a_3, S_{33}, S_{11} + S_{22}$; Γ_2: $p_1 - ip_2, a_1 - ia_2, S_{13} - iS_{23}$,
$S_{11} - S_{22} + 2iS_{12}$; Γ_3: $p_1 + ip_2, a_1 + ia_2, S_{13} + iS_{23}, S_{11} - S_{22} - 2iS_{12}$;

Table C1.9

$C_{3v} \cdot \Gamma_1$: $p_3, S_{33}. S_{11} + S_{22}$; Γ_2: a_3;
Γ_3: $(p_1, p_2), (a_2, -a_1), (S_{13}, S_{23}), (2S_{12}, S_{11} - S_{22})$;
$D_3 \cdot \Gamma_1$: $S_{33}, S_{11} + S_{22}$; Γ_2: p_3, a_3;
Γ_3: $(p_2, -p_1), (a_2, -a_1), (S_{13}, S_{23}), (2S_{12}, S_{11} - S_{22})$;

Table C1.10

$\bar{C}_3 \cdot \Gamma_1$: $a_1, S_{33}, S_{11} + S_{22}$; Γ_2: $a_1 - ia_2, S_{13} - iS_{23}, S_{11} - S_{22} + 2iS_{12}$;
Γ_3: $a_1 + ia_2, S_{11} + iS_{21}, S_{11} - S_{22} - 2iS_{12}$; Γ_4: p_1; Γ_5: $p_1 - ip_2$; Γ_6: $p_1 + ip_2$;
$C_{3h} \cdot \Gamma_1$: $a_3, S_{33}, S_{11} + S_{22}$; Γ_2: $p_1 - ip_2, S_{11} - S_{22} + 2iS_{12}$; Γ_3: $p_1 + ip_2$,
$S_{11} - S_{22} - 2iS_{12}$; Γ_4: p_3; Γ_5: $a_1 - ia_2, S_{13} - iS_{23}$; Γ_6: $a_1 + ia_2, S_{13} + iS_{23}$;
$C_6 \cdot \Gamma_1$: $p_3, a_3, S_{33}, S_{11} + S_{22}$; Γ_2: $S_{11} - S_{22} + 2iS_{12}$; Γ_3: $S_{11} - S_{22} - 2iS_{12}$;
Γ_5: $p_1 - ip_2, a_1 - ia_2, S_{13} - iS_{23}$; Γ_6: $p_1 + ip_2, a_1 + ia_2, S_{13} + iS_{23}$;

Table C1.11

$D_{3h}\cdot\Gamma_1$: $S_{33}, S_{11}+S_{22}$; Γ_2: a_3; Γ_3: p_3; Γ_5: $(a_1, a_2), (S_{23}, -S_{13})$;
Γ_6: $(p_1, p_2), (2S_{12}, S_{11}-S_{22})$;
$D_{3v}\cdot\Gamma_1$: $S_{33}, S_{11}+S_{22}$; Γ_2: a_3; Γ_4: p_3; Γ_5: (p_1, p_2);
Γ_6: $(a_2, -a_1), (S_{13}, S_{23}), (2S_{12}, S_{11}-S_{22})$;
$D_6\cdot\Gamma_1$: $S_{33}, S_{11}+S_{22}$; Γ_2: a_3, p_3; Γ_5: $(p_1, p_2), (a_1, a_2)$,
$(S_{23}, -S_{13})$; Γ_6: $(2S_{12}, S_{11}-S_{22})$;
$C_{6v}\cdot\Gamma_1$: $p_3, S_{33}, S_{11}+S_{22}$; Γ_2: a_3; Γ_5: $(p_1, p_2), (a_2, -a_1)$,
(S_{13}, S_{23}); Γ_6: $(2S_{12}, S_{11}-S_{22})$;

Table C1.12

$C_{6h}\cdot\Gamma_1$: $a_3, S_{33}, S_{11}+S_{22}$; Γ_2: $S_{11}-S_{22}+2iS_{12}$; Γ_3: $S_{11}-S_{22}-2iS_{12}$;
Γ_5: $a_1-ia_2, S_{13}-iS_{23}$; Γ_6: $a_1+ia_2, S_{13}+iS_{23}$; Γ_1': p_3;
Γ_5': p_1-ip_2; Γ_6': p_1+ip_2;

Table C1.13

$T\cdot\Gamma_1$: $S_{11}+S_{22}+S_{33}$;
Γ_2: $S_{11}+\omega^2 S_{22}+\omega S_{33}$; Γ_3: $S_{11}+\omega S_{22}+\omega^2 S_{33}$;
Γ_4: $(p_1, p_2, p_3), (a_1, a_2, a_3), (S_{23}, S_{13}, S_{12})$;

Table C1.14

$T_h\cdot\Gamma_1$: $S_{11}+S_{22}+S_{33}$;
Γ_2: $S_{11}+\omega^2 S_{22}+\omega S_{33}$; Γ_3: $S_{11}+\omega S_{22}+\omega^2 S_{33}$;
Γ_4: $(a_1, a_2, a_3), (S_{23}, S_{13}, S_{12})$;
$\bar{\Gamma}_4$: (p_1, p_2, p_3);

Table C1.15

$T_d\cdot\Gamma_1$: $S_{11}+S_{22}+S_{33}$; Γ_3: $(S_{11}-S_{33}, \sqrt{3}/3(2S_{22}-S_{11}-S_{33}))$;
Γ_4: $(S_{23}, S_{13}, S_{12}), (p_1, p_2, p_3)$; Γ_5: (a_1, a_2, a_3);
$O\cdot\Gamma_1$: $S_{11}+S_{22}+S_{33}$; Γ_3: $(S_{11}-S_{33}, \sqrt{3}/3(2S_{22}-S_{11}-S_{33}))$;
Γ_4: (S_{23}, S_{13}, S_{12}); Γ_5: $(p_1, p_2, p_3), (a_1, a_2, a_3)$;

Table C1.16

$O_h\cdot\Gamma_1$: $S_{11}+S_{22}+S_{33}$; Γ_3: $(S_{11}-S_{33}, \sqrt{3}/3(2S_{22}-S_{11}-S_{33}))$;
Γ_4: (S_{23}, S_{13}, S_{12}); Γ_5: (a_1, a_2, a_3); $\bar{\Gamma}_5$: (p_1, p_2, p_3);

where

$$\omega = -\frac{1}{2} + i\frac{\sqrt{3}}{2}, \qquad \omega^2 = -\frac{1}{2} - i\frac{\sqrt{3}}{2}.$$

For each crystal class there is listed a table of the form

$R\backslash\alpha$	$\mathbf{S}^1$	$\mathbf{S}^2$	$\dots$	$\mathbf{S}^N$	Basic quantities
Γ_1	$\mathbf{T}_1^1$	$\mathbf{T}_1^2$	$\dots$	$\mathbf{T}_1^N$	$\psi, \psi', \dots$
Γ_2	$\mathbf{T}_2^1$	$\mathbf{T}_2^2$		$\mathbf{T}_2^N$	$a,\ b, \dots$
$\vdots$	$\vdots$	$\vdots$		$\vdots$	$\vdots\ \vdots$
Γ_r	$\mathbf{T}_r^1$	$\mathbf{T}_r^2$		$\mathbf{T}_r^N$	$\mathbf{A}, \mathbf{B}, \dots$

representing the unequivalent irreducible representations $\Gamma_1, \Gamma_2, \dots, \Gamma_r$ of the crystallographic group $\{\mathbf{S}\}$. Tables C2.1–C2.14 display these irreducible representations for various crystal classes. These classes are identified by name and also by listing their Hermann–Mauguin, Schönflies, and Shubnikov symbols. The basic quantities that form the carrier spaces for irreducible representations $\Gamma_1, \Gamma_2, \dots, \Gamma_r$ are denoted by

$$\psi, \psi', \psi''\psi''', \dots,$$

$$a, b, c, d, \dots,$$

$$\mathbf{A} = \begin{bmatrix} A_1 \\ A_2 \end{bmatrix}, \quad \mathbf{B} = \begin{bmatrix} B_1 \\ B_2 \end{bmatrix}, \quad \mathbf{C} = \begin{bmatrix} C_1 \\ C_2 \end{bmatrix}, \quad \mathbf{D} = \begin{bmatrix} D_1 \\ D_2 \end{bmatrix}.$$

The irreducible representations Γ_r are either of degree one or two. Those of degree one are either real or complex numbers, and those of degree two are expressed in terms of the matrices $\mathbf{E}, \mathbf{A}, \dots, \mathbf{L}$, listed below

$$\mathbf{E} = \begin{bmatrix} 1 & 0 \\ 0 & 1 \end{bmatrix}, \quad \mathbf{A} = \begin{bmatrix} -\frac{1}{2} & \sqrt{3}/2 \\ -\sqrt{3}/2 & -\frac{1}{2} \end{bmatrix}, \quad \mathbf{B} = \begin{bmatrix} -\frac{1}{2} & -\sqrt{3}/2 \\ \sqrt{3}/2 & -\frac{1}{2} \end{bmatrix},$$

$$\mathbf{F} = \begin{bmatrix} 1 & 0 \\ 0 & -1 \end{bmatrix},$$

$$\mathbf{G} = \begin{bmatrix} -\frac{1}{2} & \sqrt{3}/2 \\ \sqrt{3}/2 & \frac{1}{2} \end{bmatrix}, \quad \mathbf{H} = \begin{bmatrix} -\frac{1}{2} & -\sqrt{3}/2 \\ -\sqrt{3}/2 & -\frac{1}{2} \end{bmatrix}, \quad \mathbf{K} = \begin{bmatrix} 0 & 1 \\ 1 & 0 \end{bmatrix},$$

$$\mathbf{L} = \begin{bmatrix} \mathbf{0} & \mathbf{1} \\ \mathbf{1} & \mathbf{0} \end{bmatrix}.$$

A superposed bar indicates complex conjugate. The generic elements of the integrity basis are listed following Tables C2.1–C2.14 (from Kiral and Smith [1974] and Kiral [1972]).

Pedial class. No symmetry. Hence all independent components of vectors and tensors constitute basic quantities.

Pinacoidal class, C_1, $\bar{1}$, $\bar{2}$. *Domatic class,* C_v, *m*, *m*. *Sphenoidal class,* C_2, 2, 2

Table C2.1

C_1	**I**	**C**	
C_v	**I**	$\mathbf{R}_1$	Basic
C_2	**I**	$\mathbf{D}_1$	quantities
Γ_1	1	1	$a, a', \ldots$
Γ_2	1	-1	$b, b', \ldots$

Application of Theorem D.6 (Appendix D) immediately yields the result that the typical multilinear elements of the integrity bases for C_1, C_s, and C_2 are given by

$$\begin{aligned} &1.\ a; \\ &2.\ bb'. \end{aligned} \tag{C2.1}$$

Prismatic class, C_{2h}, $2/m$, $2{:}m$. *Rhombic–pyramidal class,* C_{2v}, $mm2$, $2\cdot m$. *Rhombic–disphenoidal class,* D_2, 222, 2:2

Table C2.2

C_{2h}	**I**	$\mathbf{D}_1$	$\mathbf{R}_1$	**C**	
C_{2v}	**I**	$\mathbf{D}_1$	$\mathbf{R}_3$	$\mathbf{R}_2$	Basic
D_2	**I**	$\mathbf{D}_1$	$\mathbf{D}_2$	$\mathbf{D}_3$	quantities
Γ_1	1	1	1	1	$a, a', \ldots$
Γ_2	1	1	-1	-1	$b, b', \ldots$
Γ_3	1	-1	1	-1	$c, c', \ldots$
Γ_4	1	-1	-1	1	$d, d', \ldots$

The typical multilinear elements of the integrity bases for C_{2h}, C_{2v}, and D_2 are given by

$$\begin{aligned} &1.\ a; \\ &2.\ bb',\ cc',\ dd'; \\ &3.\ bcd. \end{aligned} \tag{C2.2}$$

Rhombic–dipyramidal class, D_{2h}, *mmm*, $m\cdot 2{:}m$.
Repeated application of Theorem D.6 yields the result that the typical multilinear elements of the integrity basis for D_{2h} are given by

$$\begin{aligned} &1.\ a; \\ &2.\ bb',\ cc',\ dd',\ AA',\ BB',\ CC',\ DD'; \\ &3.\ bcd,\ bAB,\ bCD,\ cAC,\ cBD,\ dAD,\ dBC; \\ &4.\ bcBC,\ bcAD,\ bdBD,\ bdAC,\ cdCD,\ cdAB,\ ABCD. \end{aligned} \tag{C2.3}$$

Table C2.3

D_{2h}	$\mathbf{I}$	$\mathbf{D}_1$	$\mathbf{D}_2$	$\mathbf{D}_3$	$\mathbf{C}$	$\mathbf{R}_1$	$\mathbf{R}_2$	$\mathbf{R}_3$	Basic quantities
Γ_1	1	1	1	1	1	1	1	1	$a, a', \ldots$
Γ_2	1	1	−1	−1	1	1	−1	−1	$b, b', \ldots$
Γ_3	1	−1	1	−1	1	−1	1	−1	$c, c', \ldots$
Γ_4	1	−1	−1	1	1	−1	−1	1	$d, d', \ldots$
Γ'_1	1	1	1	1	−1	−1	−1	−1	$A, A', \ldots$
Γ'_2	1	1	−1	−1	−1	−1	1	1	$B, B', \ldots$
Γ'_3	1	−1	1	−1	−1	1	−1	1	$C, C', \ldots$
Γ'_4	1	−1	−1	1	−1	1	1	−1	$D, D', \ldots$

Tetragonal–disphenoidal class, $\bar{C}_2$, $\bar{4}$, $\bar{4}$. Tetragonal–pyramidal class, C_4, 4, 4.

Table C2.4

$\bar{C}_2$ C_4	$\mathbf{I}$ $\mathbf{I}$	$\mathbf{D}_3$ $\mathbf{D}_3$	$\mathbf{D}_1\mathbf{T}_3$ $\mathbf{R}_1\mathbf{T}_3$	$\mathbf{D}_2\mathbf{T}_3$ $\mathbf{R}_2\mathbf{T}_3$	Basic quantities
Γ_1	1	1	1	1	$\varphi, \varphi', \ldots$
Γ_2	1	1	−1	−1	$\psi, \psi', \ldots$
Γ_3	1	−1	i	$-i$	$a, b, \ldots$
Γ_4	1	−1	$-i$	i	$a, \bar{b}, \ldots$

In Table C2.4, the quantities $\bar{a}, \bar{b}, \ldots$ denote the complex conjugates of the quantities $a, b, \ldots$, respectively. The typical multilinear elements of the integrity basis for $\bar{C}_2$ and C_4 are given by

1. φ;
2. $a\bar{b}$, $\Psi\Psi'$;
3. ψab;
4. $abcd$.

(C2.4)

Note that the presence of the complex invariants ab, Ψab, $abcd$ in (C2.4) indicates that both the real and imaginary parts $a\bar{b} \pm \bar{a}b$, $\Psi ab \pm \Psi\bar{a}b$, $abcd \pm \bar{a}\bar{b}\bar{c}\bar{d}$ of $a\bar{b}$, Ψab, $abcd$ are typical multilinear elements of the integrity basis.

Tetragonal–dipyramidal class, C_{4h}, $4/m$, $4{:}m$.

Table C2.5

C_{4h}	$\mathbf{I}$	$\mathbf{D}_3$	$\mathbf{R}_1\mathbf{T}_3$	$\mathbf{R}_2\mathbf{T}_3$	$\mathbf{C}$	$\mathbf{R}_3$	$\mathbf{D}_1\mathbf{T}_3$	$\mathbf{D}_2\mathbf{T}_3$	Basic quantities
Γ_1	1	1	1	1	1	1	1	1	$\varphi, \varphi', \ldots$
Γ_2	1	1	−1	−1	1	1	−1	−1	$\Psi, \Psi', \ldots$
Γ_3	1	−1	i	$-i$	1	−1	i	$-i$	$a, b, \ldots$
Γ_4	1	−1	$-i$	i	1	−1	$-i$	i	$\bar{a}, \bar{b}, \ldots$
Γ'_1	1	1	1	1	−1	−1	−1	−1	$\xi, \xi', \ldots$
Γ'_2	1	1	−1	−1	−1	−1	1	1	$\eta, \eta', \ldots$
Γ'_3	1	−1	i	$-i$	−1	1	$-i$	i	$A, B, \ldots$
Γ'_4	1	−1	$-i$	i	−1	1	i	$-i$	$\bar{A}, \bar{B}, \ldots$

In Table C2.5, the quantities $\bar{a}$, $\bar{b}$, ..., $\bar{A}$, $\bar{B}$, ... denote the complex conjugates of a, b, ..., A, B, ..., respectively. We find upon repeated application of Theorem D6 that the typical multilinear elements of the integrity basis for C_{4h} are given by

$$\begin{array}{l} 1.\ \varphi; \\ 2.\ a\bar{b},\ A\bar{B},\ \Psi\Psi',\ \xi\xi',\ \eta\eta'; \\ 3.\ \Psi ab,\ \Psi AB,\ \xi a\bar{A},\ \eta aA,\ \Psi\xi\eta; \\ 4.\ abcd,\ abAB,\ ab\bar{A}\bar{B},\ ABCD,\ \Psi\xi aA,\ \Psi\eta aA,\ \xi\eta ab,\ \xi\eta AB; \\ 5.\ \xi aABC,\ \xi Aabc,\ \eta\bar{a}ABC,\ \eta\bar{A}abc. \end{array} \tag{C2.5}$$

The presence of the complex invariants $a\bar{b}$, $A\bar{B}$, ..., $\eta\bar{A}abc$ in (C2.5) indicates that both the real and imaginary parts of these invariants are typical multilinear elements of the integrity basis.

Ditetragonal–pyramidal class, C_{4v}, 4mm, $4\cdot m$. Tetragonal–trapezohedral class, D_4, 422, 4:2. Tetragonal–scalenohedral class, D_{2v}, $\bar{4}2m$, $\bar{4}\cdot m$

Table C2.6

C_{4v} D_4 D_{2v}	I I I	$\mathbf{R}_2$ $\mathbf{D}_1$ $\mathbf{D}_1$	$\mathbf{R}_1$ $\mathbf{D}_2$ $\mathbf{D}_2$	$\mathbf{D}_3$ $\mathbf{D}_3$ $\mathbf{D}_3$	$\mathbf{T}_3$ $\mathbf{R}_2\mathbf{T}_3$ $\mathbf{T}_3$	$\mathbf{R}_2\mathbf{T}_3$ $\mathbf{R}_2\mathbf{T}_3$ $\mathbf{D}_1\mathbf{T}_3$	$\mathbf{R}_1\mathbf{T}_3$ $\mathbf{R}_1\mathbf{T}_3$ $\mathbf{D}_2\mathbf{T}_3$	$\mathbf{D}_1\mathbf{T}_3$ $\mathbf{CT}_3$ $\mathbf{D}_1\mathbf{T}_3$	Basic quantities
Γ_1	1	1	1	1	1	1	1	1	$\varphi, \varphi', \ldots$
Γ_2	1	−1	−1	1	−1	1	1	−1	$\psi, \psi', \ldots$
Γ_3	1	−1	−1	1	1	−1	−1	1	$\nu, \nu', \ldots$
Γ_4	1	1	1	1	−1	−1	−1	−1	$\tau, \tau', \ldots$
Γ_5	E	F	−F	−E	K	L	−L	−K	$\mathbf{a}, \mathbf{b}, \ldots$

Repeated application of Theorem D.6 yields the result that the typical multilinear elements of the integrity basis for C_{4v}, D_4, and D_{2v}, are given by

$$\begin{array}{l} 1.\ \varphi; \\ 2.\ a_1b_1 + a_2b_2,\ \Psi\Psi',\ \nu\nu',\ \tau\tau'; \\ 3.\ \Psi(a_1b_2 - a_2b_1),\ \nu(a_1b_2 + a_2b_1),\ \tau(a_1b_1 - a_2b_2),\ \Psi\nu\tau; \\ 4.\ a_1b_1c_1d_1 + a_2b_2c_2d_2,\ \Psi\nu(a_1b_1 - a_2b_2),\ \Psi\tau(a_1b_2 + a_2b_1), \\ \quad \nu\tau(a_1b_2 + a_2b_1); \\ 5.\ \Psi(a_1b_1c_1d_2 + a_1b_1d_1c_2 + a_1c_1d_1b_2 + b_1c_1d_1a_2 - a_2b_2c_2d_1 - a_2b_2d_2c_1 \\ \quad - a_2c_2d_2b_1 - b_2c_2d_2a_1). \end{array} \tag{C2.6}$$

Ditetragonal–dipyramidal class, D_{4h}, 4/mmm, m·4:m

Table C2.7

D_{4h}	I	D_1	D_2	D_3	CT_3	R_1T_3	R_2T_3	R_3T_3	Basic quantities
Γ_1	1	1	1	1	1	1	1	1	$\varphi, \varphi', \ldots$
Γ_2	1	−1	−1	1	−1	1	1	−1	$\Psi, \Psi', \ldots$
Γ_3	1	−1	−1	1	1	−1	−1	1	$\nu, \nu', \ldots$
Γ_4	1	1	1	1	−1	−1	−1	−1	$\tau, \tau', \ldots$
Γ_5	**E**	**F**	−**F**	−**E**	−**K**	−**L**	**L**	**K**	**a**, **b**, ...
Γ_1'	1	1	1	1	1	1	1	1	$\xi, \xi', \ldots$
Γ_2'	1	−1	−1	1	−1	1	1	−1	$\eta, \eta', \ldots$
Γ_3'	1	−1	−1	1	1	−1	−1	1	$\theta, \theta', \ldots$
Γ_4'	1	1	1	1	−1	−1	−1	−1	$\gamma, \gamma', \ldots$
Γ_5'	**E**	**F**	−**F**	−**E**	−**K**	−**L**	**L**	**K**	**A**, **B**, ...

D_{4h}	C	R_1	R_2	R_3	T_3	D_1T_3	D_2T_3	D_3T_3	Basic quantities
Γ_1	1	1	1	1	1	1	1	1	$\varphi, \varphi', \ldots$
Γ_2	1	−1	−1	1	−1	1	1	−1	$\Psi, \Psi', \ldots$
Γ_3	1	−1	−1	1	1	−1	−1	1	$\nu, \nu', \ldots$
Γ_4	1	1	1	1	−1	−1	−1	−1	$\tau, \tau', \ldots$
Γ_5	**E**	**F**	−**F**	−**E**	−**K**	−**L**	**L**	**K**	**a**, **b**, ...
Γ_1'	−1	−1	−1	−1	−1	−1	−1	−1	$\xi, \xi', \ldots$
Γ_2'	−1	1	1	−1	1	−1	−1	1	$\eta, \eta', \ldots$
Γ_3'	−1	1	1	−1	−1	1	1	−1	$\theta, \theta', \ldots$
Γ_4'	−1	−1	−1	−1	1	1	1	1	$\gamma, \gamma', \ldots$
Γ_5'	−**E**	−**F**	**F**	**E**	**K**	**L**	−**L**	−**K**	**A**, **B**, ...

Repeated application of Theorem D.6 yields the result that the typical multilinear elements of the integrity basis for D_{4h} are given by

1. φ;
2. $a_1b_1 + a_2b_2$, $A_1B_1 + A_2B_2$, $\Psi\Psi'$, $\nu\nu'$, $\tau\tau'$, $\xi\xi'$, $\eta\eta'$, $\theta\theta'$, $\gamma\gamma'$;
3. $\Psi(a_1b_2 - a_2b_1)$, $\Psi(A_1B_2 - A_2B_1)$, $\nu(a_1b_2 + a_2b_1)$, $\nu(A_1B_2 + A_2B_1)$, $\tau(a_1b_1 - a_2b_2)$, $\tau(A_1B_1 - A_2B_2)$, $\xi(a_1A_1 + a_2A_2)$, $\eta(a_1A_2 - a_2A_1)$, $\theta(a_1A_2 + a_2A_1)$, $\gamma(a_1A_1 - a_2A_2)$, $\Psi\nu\tau$, $\Psi\theta\gamma$, $\Psi\xi\eta$, $\nu\eta\gamma$, $\tau\xi\gamma$, $\tau\eta\theta$, $\nu\xi\theta$;
4. $a_1b_1c_1d_1 + a_2b_2c_2d_2$, $A_1B_1C_1D_1 + A_2B_2C_2D_2$, $(a_1b_2 + a_2b_1)(A_1B_2 + A_2B_1)$, $(a_1b_2 - a_2b_1)(A_1B_2 + A_2B_1)$, $(a_1b_1 - a_2b_2)(A_1B_1 - A_2B_2)$, $(\Psi\nu, \xi\gamma, \eta\theta)(a_1b_1 - a_2b_2)$, $(\Psi\tau, \xi\theta, \eta\gamma)(a_1b_2 + a_2b_1)$, $(\nu\tau, \xi\eta, \theta\gamma)(a_1b_2 - a_2b_1)$, $(\Psi\nu, \xi\gamma, \eta\theta)(A_1B_1 - A_2B_2)$, $(\Psi\tau, \xi\theta, \eta\gamma)(A_1B_2 + A_2B_1)$, $(\nu\tau, \xi\eta, \theta\gamma)(A_1B_2 - A_2B_1)$, $(\Psi\eta, \nu\theta, \tau\gamma)(a_1A_1 + a_2A_2)$, $(\Psi\theta, \nu\eta, \tau\xi)(a_1A_1 - a_2A_2)$, $(\Psi\gamma, \nu\xi, \tau\eta)(a_1A_2 + a_2A_1)$, $(\Psi\xi, \nu\gamma, \tau\theta)(a_1A_2 - a_2A_1)$, $\Psi\nu\xi\gamma$, $\Psi\nu\eta\theta$, $\Psi\tau\xi\theta$, $\Psi\tau\eta\gamma$, $\nu\tau\xi\eta$, $\nu\tau\theta\gamma$, $\xi\eta\theta\gamma$;
5. $\Psi(a_1b_1c_1d_2 + a_1b_1d_1c_2 + a_1c_1d_1b_2 + b_1c_1d_1a_2 - a_2b_2c_2d_1 - a_2b_2d_2c_1 - a_2c_2d_2b_1 - b_2c_2d_2a_1)$,
   $\Psi(A_1B_1C_1D_2 + A_1B_1D_1C_2 + A_1C_1D_1B_2 + B_1C_1D_1A_2 - A_2B_2C_2D_1 - A_2B_2D_2C_1 - A_2C_2D_2B_1 - B_2C_2D_2A_1)$,

$$\begin{aligned}
&\Psi(a_1b_2 + a_2b_1)(A_1B_1 - A_2B_2),\ \Psi(a_1b_1 - a_2b_2)(A_1B_2 + A_2B_1),\\
&\nu(a_1b_2 - a_2b_1)(A_1B_1 - A_2B_2),\ \nu(a_1b_1 + a_2b_2)(A_1B_2 - A_2B_1),\\
&\tau(a_1b_2 - a_2b_1)(A_1B_2 + A_2B_1),\ \tau(a_1b_2 + a_2b_1)(A_1B_2 - A_2B_1),\\
&\xi(a_1b_1c_1A_1 + a_2b_2c_2A_2),\ \xi(A_1B_1C_1a_1 + A_2B_2C_2a_2),\\
&\eta(a_1b_1c_1A_2 - a_2b_2c_2A_2),\ \eta(A_1B_1C_1a_2 - A_2B_2C_2a_1),\\
&\theta(a_1b_1c_1A_2 + a_2b_2c_2A_1),\ \theta(A_1B_1C_1a_2 + A_2B_2C_2a_1),\\
&\gamma(a_1b_1c_1A_1 - a_2b_2c_2A_2),\ \gamma(A_1B_1C_1a_1 - A_2B_2C_2a_2),\\
&(\Psi\xi\theta, \Psi\eta\gamma, \nu\xi\eta, \nu\theta\gamma)(a_1b_1 - a_2b_2),\ (a_1b_2 + a_2b_1)(\Psi\xi\gamma, \Psi\eta\theta, \tau\xi\eta, \tau\theta\gamma),\\
&(\nu\xi\gamma, \nu\eta\theta, \tau\xi\theta, \tau\eta\gamma)(a_1b_2 - a_2b_1),\ (A_1B_1 - A_2B_2)(\Psi\xi\theta, \Psi\eta\gamma, \nu\xi\eta, \nu\theta\gamma),\\
&(\Psi\xi\gamma, \Psi\eta\theta, \tau\xi\eta, \tau\theta\gamma)(A_1B_2 + A_2B_1),\ (A_1B_2 - A_2B_1)(\nu\xi\gamma, \nu\eta\theta, \tau\xi\theta, \tau\eta\gamma),\\
&(\Psi\nu\gamma, \Psi\tau\theta, \nu\tau\eta, \eta\theta\gamma)(a_1A_1 + a_2A_2),\ (a_1A_1 - a_2A_2)(\Psi\eta\tau, \Psi\nu\xi, \nu\tau\theta, \xi\eta\theta),\\
&(\Psi\nu\eta, \Psi\tau\xi, \nu\tau\gamma, \xi\eta\gamma)(a_1A_2 + a_2A_1),\ (a_1A_2 - a_2A_1)(\Psi\nu\theta, \Psi\tau\gamma, \nu\tau\xi, \xi\theta\gamma);
\end{aligned} \tag{C2.7}$$

6. $\xi\eta(a_1b_1 - a_2b_2)(A_1B_2 + A_2B_1),\ \theta\gamma(a_1b_1 - a_2b_2)(A_1B_2 + A_2B_1),$

$$\begin{aligned}
&\Psi\xi(a_1b_1c_1A_2 - a_2b_2c_2A_1),\ \Psi\xi(A_1B_1C_1a_2 - A_2B_2C_2a_1),\\
&\Psi\eta(a_1b_1c_1A_1 + a_2b_2c_2A_2),\ \Psi\eta(A_1B_1C_1a_1 + A_2B_2C_2a_2),\\
&\Psi\theta(a_1b_1c_1A_1 - a_2b_2c_2A_2),\ \Psi\theta(A_1B_1C_1a_1 - A_2B_2C_2a_2),\\
&\Psi\gamma(a_1b_1c_1A_2 + a_2b_2c_2A_1),\ \Psi\gamma(A_1B_1C_1a_2 + A_2B_2C_2a_1),\\
&(\xi\eta, \theta\gamma)(a_1b_1c_1d_2 + a_1b_1d_1c_2 + a_1c_1d_1b_2 + b_1c_1d_1a_2 - a_2b_2c_2d_1\\
&\qquad - a_2b_2d_2c_1 - a_2c_2d_2b_1 - b_2c_2d_2a_1),\\
&(\xi\eta, \theta\gamma)(A_1B_1C_1D_2 + A_1B_1D_1C_2 + A_1C_1D_1B_2 + B_1C_1D_1A_2\\
&\qquad - A_2B_2C_2D_1 - A_2B_2D_2C_1 - A_2C_2D_2B_1 - B_2C_2D_2A_1),\\
&(a_1b_2 - a_2b_1)(A_1B_1C_1D_2 + A_1B_1D_1C_2 + A_1C_1D_1B_2 + B_1C_1D_1A_2\\
&\qquad - A_2B_2C_2D_1 - A_2B_2D_2C_1 - A_2C_2D_2B_1 - B_2C_2D_2A_1),\\
&(A_1B_2 - A_2B_1)(a_1b_1c_1d_2 + a_1b_1d_1c_2 + a_1c_1d_1b_2 + b_1c_1d_1a_2\\
&\qquad - a_2b_2c_2d_1 - a_2b_2d_2c_1 - a_2c_2d_2b_1 - b_2c_2d_2a_1).
\end{aligned}$$

Trigonal–pyramidal class, $\mathbf{C}_3$, 3, 3.

Table C2.8

C_3	$\mathbf{I}$	$\mathbf{S}_1$	$\mathbf{S}_2$	Basic quantities
Γ_1	1	1	1	$\varphi, \varphi', \ldots$
Γ_2	1	ω	ω^2	$a, b, \ldots$
Γ_3	1	ω^2	ω	$\bar{a}, \bar{b}, \ldots$

The quantities ω and ω^2 in Table C2.8 are defined by

$$\omega = -1/2 + i\sqrt{3}/2, \qquad \omega^2 = -1/2 - i\sqrt{3}/2. \tag{C2.8}$$

We note that $\omega^3 = 1$ and that $\bar{a}, \bar{b}, \ldots$ denote the complex conjugates of the quantities $a, b, \ldots$, respectively. The typical multilinear elements of the integrity basis for C_3 are given by

$$\begin{aligned}
&1.\ \varphi;\\
&2.\ a\bar{b};\\
&3.\ abc.
\end{aligned} \tag{C2.9}$$

The presence of the complex invariants $a\bar{b}$ and abc in (C2.9) indicates that both the real and imaginary parts $a\bar{b} \pm \bar{a}b$ and $abc \pm \bar{a}\bar{b}\bar{c}$ of these invariants are typical multilinear elements of the integrity basis.

Ditrigonal–pyramidal class, C_{3v}, 3m, 3 · m. Trigonal–trapezohedral class, D_3, 32, 3:2.

Table C2.9

C_{3v}	I	S_1	S_2	R_1	R_1S_1	R_1S_2	Basic
D_3	I	S_1	S_2	D_1	D_1S_1	D_1S_2	quantities
Γ_1	1	1	1	1	1	1	$\varphi, \varphi', \ldots$
Γ_2	1	1	1	-1	-1	-1	Ψ, Ψ'
Γ_3	E	A	B	$-$F	$-$G	$-$H	**a**, **b**, ...

The typical multilinear elements of the integrity basis for C_{3v} and D_3 are given by

$$\begin{aligned}
&1.\ \varphi;\\
&2.\ a_1b_1 + a_2b_2,\ \Psi\Psi';\\
&3.\ a_2b_2c_2 - a_1b_1c_2 - b_1c_1a_2 - c_1a_1b_2,\ \Psi(a_1b_2 - a_2b_1);\\
&4.\ \Psi(a_1b_1c_1 - a_2b_2c_1 - b_2c_2a_1 - c_2a_2b_1).
\end{aligned} \tag{C2.10}$$

Rhombohedral class, $\bar{C}_3$, $\bar{3}$, $\bar{6}$. Trigonal–dipyramidal class, C_{3h}, $\bar{6}$, 3:*m*. *Hexagonal–pyramidal class, C_6,* 6, 6.

Table C2.10

$\bar{C}_3$	I	S_1	S_2	C	CS_1	CS_2	
C_{3h}	I	S_1	S_2	R_3	R_3S_1	R_3S_2	Basic
C_6	I	S_1	S_2	D_3	D_3S_1	D_3S_2	quantities
Γ_1	1	1	1	1	1	1	$\varphi, \varphi', \ldots$
Γ_2	1	ω	ω^2	1	ω	ω^2	$a, b, \ldots$
Γ_3	1	ω^2	ω	1	ω^2	ω	$\bar{a}, \bar{b}, \ldots$
Γ_4	1	1	1	-1	-1	-1	$\xi, \xi', \ldots$
Γ_5	1	ω	ω^2	-1	$-\omega$	$-\omega^2$	$A, B, \ldots$
Γ_6	1	ω^2	ω	$-\omega^1$	$-\omega^2$	$-\omega$	$\bar{A}, \bar{B}, \ldots$

The quantities ω and ω^2 appearing in Table C2.10 are defined by (C2.8). The quantities $\bar{a}, \bar{b}, \ldots, \bar{A}, \bar{B}, \ldots$ denote the complex conjugates of $a, b, \ldots, A, B, \ldots$, respectively. Let P be a polynomial function of the quantities $\varphi, \ldots, a, \bar{a}, b, \bar{b}, \ldots, \xi, \ldots, A, \bar{A}, B, \bar{B}, \ldots$ which is invariant under the first three transformations of Table C2.10. Then it is seen from the results for the group C_3 that P is expressible as a polynomial in the quantities obtained from the typical multilinear quantities

$$\varphi,\ a\bar{b},\ abc,\ A\bar{B},\ aAB \tag{C2.11}$$

and

$$\xi,\ a\bar{A},\ abA,\ ABC. \tag{C2.12}$$

The quantities (C2.11) remain invariant under the final three transformations of Table C2.10, and the quantities (C2.12) all change sign under any of the last three transformations of Table C2.10. With Theorem D.6 we then see that the typical multilinear elements of the integrity basis for $\bar{C}_3$, C_{3h}, and C_6 are given by

$$\begin{aligned} &1.\ \varphi; \\ &2.\ a\bar{b},\ A\bar{B},\ \xi\xi'; \\ &3.\ abc,\ aAB,\ \xi\alpha\bar{A}; \\ &4.\ ab\bar{A}\bar{B},\ \xi abA,\ \xi ABC; \\ &5.\ \bar{a}ABCD; \\ &6.\ ABCDEF. \end{aligned} \tag{C2.13}$$

The presence of the complex invariants $a\bar{b}$, $A\bar{B}$, $\ldots$, $ABCDEF$ in (C2.13) indicates that both the real and imaginary parts $a\bar{b} \pm \bar{a}b$, $A\bar{B} \pm \bar{A}B$, $\ldots$, $ABCDEF \pm \bar{A}\bar{B}\bar{C}\bar{D}\bar{E}\bar{F}$ of these invariants are typical multilinear elements of the integrity basis.

Ditrigonal–dipyramidal class, D_{3h}, $\bar{6}m2$, $m \cdot 3{:}m$. Hexagonal–scalenohedral class, D_{3v}, $\bar{3}m$, $\bar{6} \cdot m$. Hexagonal–trapezohedral class, D_6, 622, 6:2. Dihexagonal–pyramidal class, C_{6v}, 6mm, $6 \cdot m$.

Table C2.11

D_{3h} D_{3v} D_6 C_{6v}	I I I I	S_1 S_1 S_1 S_1	S_2 S_2 S_2 S_2	R_3 C D_3 D_3	R_3S_1 CS_1 D_3S_1 D_3S_1	R_3S_2 CS_2 D_3S_2 D_3S_2	Basic quantities
Γ_1	1	1	1	1	1	1	$\varphi, \varphi', \ldots$
Γ_2	1	1	1	1	1	1	$\Psi, \Psi', \ldots$
Γ_3	1	1	1	−1	−1	−1	$\xi, \xi', \ldots$
Γ_4	1	1	1	−1	−1	−1	$\eta, \eta', \ldots$
Γ_5	**E**	**A**	**B**	−**E**	−**A**	−**B**	**A**, **B**, ...
Γ_6	**E**	**A**	**B**	**E**	**A**	**B**	**a**, **b**, ...
D_{3h} D_{3v} D_6 C_{6v}	R_1 D_1 D_1 R_2	R_1S_1 D_1S_1 D_1S_1 R_2S_1	R_1S_2 D_1S_2 D_1S_2 R_2S_2	D_2 R_1 D_2 R_1	D_2S_1 R_1S_1 D_2S_1 R_1S_1	D_2S_2 R_1S_2 D_2S_2 R_1S_2	Basic quantities
Γ_1	1	1	1	1	1	1	$\varphi, \varphi', \ldots$
Γ_2	−1	−1	−1	−1	−1	−1	$\Psi, \Psi', \ldots$
Γ_3	1	1	1	−1	−1	−1	$\xi, \xi', \ldots$
Γ_4	−1	−1	−1	1	1	1	$\eta, \eta', \ldots$
Γ_5	**F**	**G**	**H**	−**F**	−**G**	−**H**	**A**, **B**, ...
Γ_6	−**F**	−**G**	−**H**	−**F**	−**G**	−**H**	**a**, **b**, ...

We note that the basic quantities φ, Ψ, ξ, η, **A**, **a** associated with Table C2.11 transform under transformations 1, 2, 3, 10, 11, 12 of Table C2.11 in the same manner as do the quantities φ, Ψ, Ψ', φ', **a**, **b** under the transformations of Table C2.9 associated with the crystal classes C_{3v} and D_3. Let us employ the notation

$$\begin{aligned} &A = A_1 + iA_2, \quad \bar{A} = A_1 - iA_2, \quad B = B_1 + iB_2, \quad \bar{B} = B_1 - iB_2, \ldots, \\ &a = a_1 + ia_2, \quad \bar{a} = a_1 - ia_2, \quad b = b_1 + ib_2, \quad \bar{b} = b_1 - ib_2, \ldots . \end{aligned} \tag{C2.14}$$

With the notation (C2.14), we see from the results for the groups C_{3v} and D_3 that the typical multilinear elements of the integrity basis for polynomial functions of the basic quantities $\varphi, \varphi', \ldots, \Psi, \Psi', \ldots, \xi, \xi', \ldots, \eta, \eta', \ldots$, **A**, **B**, ..., **a**, **b**, ... which are invariant under the group of transformations 1, 2, 3, 10, 11, 12 of Table C2.11 are given by

$$\begin{gathered} \varphi,\ \Psi\Psi',\ \xi\xi',\ A\bar{B} + \bar{A}B,\ a\bar{b} + \bar{a}b,\ \Psi(A\bar{B} - \bar{A}B), \\ \Psi(a\bar{b} - \bar{a}b),\ \xi(a\bar{A} - \bar{a}A),\ abc - \bar{a}\bar{b}\bar{c},\ aAB - \bar{a}\bar{A}\bar{B}, \\ \Psi(abc + \bar{a}\bar{b}\bar{c}),\ \Psi(aAB + \bar{a}\bar{A}\bar{B}), \\ \xi(abA + \bar{a}\bar{b}\bar{A}),\ \xi(ABC + \bar{A}\bar{B}\bar{C}) \end{gathered} \tag{C2.15}$$

and

$$\begin{gathered} \eta,\ \Psi\xi,\ a\bar{A} + \bar{a}A,\ \xi(a\bar{b} - \bar{a}b),\ \Psi(a\bar{A} - \bar{a}A),\ \xi(A\bar{B} - \bar{A}B), \\ abA - \bar{a}\bar{b}\bar{A},\ ABC - \bar{A}\bar{B}\bar{C},\ \xi(abc + \bar{a}\bar{b}\bar{c}),\ \Psi(abA + \bar{a}\bar{b}\bar{A}), \\ \xi(aAB + \bar{a}\bar{A}\bar{B}),\ \Psi(ABC + \bar{A}\bar{B}\bar{C}). \end{gathered} \tag{C2.16}$$

The quantities (C2.15) remain invariant under the remaining transformations of Table C2.11 whereas the quantities (C2.16) change sign under all of the remaining transformations of Table C2.11. Application of Theorem D.6 then yields the result upon elimination of the redundant terms that the typical multilinear elements of the integrity basis for D_{3v}, D_{3h}, D_6, and C_{6v} are given by

$$\begin{aligned} &1.\ \varphi; \\ &2.\ a\bar{b} + \bar{a}b,\ A\bar{B} + \bar{A}B,\ \Psi\Psi',\ \xi\xi',\ \eta\eta'; \\ &3.\ abc - \bar{a}\bar{b}\bar{c},\ aAB - \bar{a}\bar{A}\bar{B}, \\ &\quad \Psi(a\bar{b} - \bar{a}b),\ \Psi(A\bar{B} - \bar{A}B),\ \xi(a\bar{A} - \bar{a}A),\ \eta(a\bar{A} + \bar{a}A),\ \Psi\xi\eta; \\ &4.\ ab\bar{A}\bar{B} + \bar{a}\bar{b}AB,\ (a\bar{b} - \bar{a}b)(A\bar{B} - \bar{A}B),\ \Psi(abc + \bar{a}\bar{b}\bar{c}), \\ &\quad \Psi(aAB + \bar{a}\bar{A}\bar{B}),\ \xi(abA + \bar{a}\bar{b}\bar{A}),\ \xi(ABC + \bar{A}\bar{B}\bar{C}), \\ &\quad \eta(abA - \bar{a}\bar{b}\bar{A}),\ \eta(ABC - \bar{A}\bar{B}\bar{C}),\ \Psi\xi(a\bar{A} + \bar{a}A),\ \Psi\eta(a\bar{A} - \bar{a}A), \\ &\quad \xi\eta(a\bar{b} - \bar{a}b),\ \xi\eta(A\bar{B} - \bar{A}B); \\ &5.\ (abc + \bar{a}\bar{b}\bar{c})(A\bar{B} - \bar{A}B),\ a\bar{A}\bar{B}\bar{C}\bar{D} - \bar{a}ABCD, \\ &\quad \Psi(ab\bar{A}\bar{B} - \bar{a}\bar{b}AB),\ \Psi\xi(abA - \bar{a}\bar{b}\bar{A}),\ \Psi\xi(ABC - \bar{A}\bar{B}\bar{C}), \\ &\quad \Psi\eta(ABC + \bar{A}\bar{B}\bar{C}),\ \Psi\eta(abA + \bar{a}\bar{b}\bar{A}),\ \xi\eta(abc + \bar{a}\bar{b}\bar{c}),\ \xi\eta(aAB + \bar{a}\bar{A}\bar{B}); \end{aligned} \tag{C2.17}$$

6. $ABCDEF + \bar{A}\bar{B}\bar{C}\bar{D}\bar{E}\bar{F}, \Psi(a\bar{A}\bar{B}\bar{C}\bar{D} + \bar{a}ABCD)$;
7. $\Psi(ABCDEF - \bar{A}\bar{B}\bar{C}\bar{D}\bar{E}\bar{F})$.

We recall that the quantities $A, \bar{A}, \ldots, F, \bar{F}, a, \bar{a}, \ldots, c, \bar{c}$ appearing in (C2.17) are defined as in (C2.14)

Hexagonal–dipyramidal class, C_{6h}, $6/m$, $6{:}m$.

Table C2.12

C_{6h}	$\mathbf{I}$	$\mathbf{S}_1$	$\mathbf{S}_2$	$\mathbf{D}_3$	$\mathbf{D}_3\mathbf{S}_1$	$\mathbf{D}_3\mathbf{S}_2$	Basic quantities
Γ_1	1	1	1	1	1	1	$\varphi, \varphi', \ldots$
Γ_2	1	ω	ω^2	1	ω	ω^2	$a, b, \ldots$
Γ_3	1	ω^2	ω	1	ω^2	ω	$a, b, \ldots$
Γ_4	1	1	1	-1	-1	-1	$\xi, \xi', \ldots$
Γ_5	1	ω	ω^2	-1	$-\omega$	$-\omega^2$	$A, B, \ldots$
Γ_6	1	ω^2	ω	-1	$-\omega^2$	$-\omega$	$\bar{A}, \bar{B}, \ldots$
Γ_1'	1	1	1	1	1	1	$\pi, \pi', \ldots$
Γ_2'	1	ω	ω^2	1	ω	ω^2	$X, Y, \ldots$
Γ_3'	1	ω^2	ω	1	ω^2	ω	$\bar{X}, \bar{Y}, \ldots$
Γ_4'	1	1	1	-1	-1	-1	$\delta, \delta', \ldots$
Γ_5'	1	ω	ω^2	-1	$-\omega$	$-\omega^2$	$x, y, \ldots$
Γ_6'	1	ω^2	ω	-1	$-\omega^2$	$-\omega$	$\bar{x}, \bar{y}, \ldots$
C_{6h}	$\mathbf{C}$	$\mathbf{CS}_1$	$\mathbf{CS}_2$	$\mathbf{R}_3$	$\mathbf{R}_3\mathbf{S}_1$	$\mathbf{R}_3\mathbf{S}_2$	Basic quantities
Γ_1	1	1	1	1	1	1	$\varphi, \varphi', \ldots$
Γ_2	1	ω	ω^2	1	ω	ω^2	$a, b, \ldots$
Γ_3	1	ω^2	ω	1	ω^2	ω	$a, b, \ldots$
Γ_4	1	1	1	-1	-1	-1	$\xi, \xi', \ldots$
Γ_5	1	ω	ω^2	-1	$-\omega$	$-\omega^2$	$A, B, \ldots$
Γ_6	1	ω^2	ω	-1	$-\omega^2$	$-\omega$	$\bar{A}, \bar{B}, \ldots$
Γ_1'	-1	-1	-1	-1	-1	-1	$\pi, \pi', \ldots$
Γ_2'	-1	$-\omega$	$-\omega^2$	-1	$-\omega$	$-\omega^2$	$X, Y, \ldots$
Γ_3'	-1	$-\omega^2$	$-\omega$	-1	$-\omega^2$	$-\omega$	$\bar{X}, \bar{Y}, \ldots$
Γ_4'	-1	-1	-1	1	1	1	$\delta, \delta', \ldots$
Γ_5'	-1	$-\omega$	$-\omega^2$	1	ω	ω^2	$x, y, \ldots$
Γ_6'	-1	$-\omega^2$	$-\omega$	1	ω^2	ω	$\bar{x}, \bar{y}, \ldots$

The quantities ω and ω^2 appearing in Table C2.12 are defined by (C2.8). We note that the quantities φ, ξ, π, δ and a, A, X, x associated with Table C2.12 transform under the first three transformations of Table C2.8 in the same manner as do the quantities φ and a associated with Table C2.8 (crystal class C_3) under the transformations of Table C2.8. We see from the results for the group C_3 that the typical multilinear elements of the integrity basis for polynomial functions of the basic quantities φ, ξ, π, δ, a, A, X, x, which are invariant under the first three transformations of Table C2.12

are given by

$$\begin{gathered}\varphi,\ \xi,\ \pi,\ \delta,\ a\bar{b},\ a\bar{A},\ a\bar{X},\ a\bar{x},\ A\bar{B},\ A\bar{X},\ A\bar{x},\\ X\bar{Y},\ X\bar{x},\ x\bar{y},\ abc,\ abA,\ abX,\ abx,\ ABC,\\ ABa,\ ABX,\ ABx,\ XYa,\ XYA,\ XYZ,\ XYx,\ xya,\\ xyA,\ xyX,\ xyz,\ aAX,\ aAx,\ aXx,\ AXx.\end{gathered} \tag{C2.18}$$

Under any of the remaining nine transformations of Table C2.12, certain of the quantities (C2.18) remain invariant and the others change sign. Then, repeated application of Theorem D6 will yield the result that the typical multilinear elements of the integrity basis for the crystal class C_{6h} are given by

$$\begin{array}{ll}
1. & \varphi;\\
2. & a\bar{b},\ A\bar{B},\ X\bar{Y},\ x\bar{y},\ \pi\pi',\ \xi\xi',\ \delta\delta';\\
3. & abc,\ ABa,\ XYa,\ xya,\ AXx,\ \xi a\bar{A},\ \xi x\bar{X},\\
 & \delta a\bar{x},\ \delta A\bar{X},\ \pi A\bar{x},\ \pi\xi\delta;\\
4. & ab\bar{A}\bar{B},\ ab\bar{X}\bar{Y},\ ab\bar{x}\bar{y},\ AB\bar{X}\bar{Y},\ AB\bar{x}\bar{y},\ XY\bar{x}\bar{y},\\
 & \quad a\bar{A}X\bar{x},\ a\bar{A}\bar{X}x,\ aA\bar{X}\bar{x},\\
 & \quad \pi abX,\ \pi ABX,\ \pi XYZ,\ \pi xyX,\ \pi aAx,\\
 & \quad \xi abA,\ \xi ABC,\ \xi XYA,\ \xi xyA,\ \xi aXx,\\
 & \quad \delta abx,\ \delta ABx,\ \delta XYx,\ \delta xyz,\ \delta aAX,\\
 & \quad \pi\xi a\bar{x},\ \pi\xi A\bar{X},\ \pi\delta a\bar{A},\ \xi\delta x\bar{X},\ \xi\delta A\bar{x};\\
5. & \bar{a}ABCD,\ \bar{a}ABXY,\ \bar{a}ABxy,\ \bar{a}XYZU,\ \bar{a}XYxy,\ \bar{a}xyzu,\\
 & ABC\bar{X}x,\ ABCX\bar{x},\ XYZ\bar{A}x,\ XYZA\bar{x},\ abxA\bar{X},\\
 & abx\bar{A}X,\ ab\bar{x}AX,\ xyzA\bar{X},\ \mathrm{xyz}\bar{A}X,\\
 & \pi ab\bar{A}\bar{x},\ \pi AB\bar{a}\bar{X},\ \pi xy\bar{a}\bar{X},\ \pi XY\bar{A}\bar{x},\\
 & \xi ab\bar{X}\bar{x},\ \xi AB\bar{X}\bar{x},\ \xi XY\bar{A}\bar{a},\ \xi xy\bar{A}\bar{a},\\
 & \delta ab\bar{A}\bar{X},\ \delta AB\bar{a}\bar{x},\ \delta XY\bar{a}\bar{x},\ \delta xy\bar{A}\bar{X},\\
 & \pi\xi abx,\ \pi\xi ABx,\ \pi\xi XYx,\ \pi\xi xyz,\ \pi\xi aAX,\\
 & \pi\delta abA,\ \pi\delta ABC,\ \pi\delta XYA,\ \pi\delta xyA,\ \pi\delta aXx,\\
 & \xi\delta abX,\ \xi\delta ABX,\ \xi\delta XYZ,\ \xi\delta xyX,\ \xi\delta aAx;\\
6. & ABCDEF,\ ABCDXY,\ ABXYZU,\ XYZUVW,\ ABCDxy,\\
 & ABxyzu,\ xyzuvw,\ XYZUxy,\ XYxyzu,\ aAx\bar{X}\bar{Y}\bar{Z},\\
 & aAX\bar{x}\bar{y}\bar{z},\ aXx\bar{A}\bar{B}\bar{C},\\
 & \pi\bar{a}ABCx,\ \pi\bar{a}xyzA,\ \pi ABCD\bar{X},\ \pi ABxy\bar{X},\ \pi xyzu\bar{X},\\
 & \xi\bar{a}XYZx,\ \xi\bar{a}xyzX,\ \xi XYZU\bar{A},\ \xi XYxy\bar{A},\ \xi xyzu\bar{A},\\
 & \delta\bar{a}ABCX,\ \delta\bar{a}XYZA,\ \delta XYZU\bar{x},\ \delta XYAB\bar{x},\ \delta ABCD\bar{x};\\
7. & ABCDE\bar{X}\bar{x},\ XYZUV\bar{A}\bar{x},\ xyzuv\bar{A}\bar{X},\\
 & \pi ABCDEx,\ \pi ABCxyz,\ \pi Axyzuv,\\
 & \xi XYZUVx,\ \xi XYZxyz,\ \xi Xxyzuv,\\
 & \delta ABCDEX,\ \delta ABCXYZ,\ \delta AXYZUV.
\end{array} \tag{C2.19}$$

The presence of the complex invariants $a\bar{b}$, $A\bar{B}$, ..., $\delta AXYZUV$ in (C2.19) indicates that both the real and imaginary parts of these invariants are typical multilinear elements of the integrity basis.

Table C2.13

T	$\mathbf{I}$	$\mathbf{D}_1$	$\mathbf{D}_2$	$\mathbf{D}_3$	$\mathbf{D}_1\mathbf{M}_1$	$\mathbf{M}_1$	Basic quantities
Γ_1	1	1	1	1	1	1	$\psi, \psi', \ldots$
Γ_2	1	1	1	1	ω	ω	$\psi, \psi', \ldots$
Γ_3	1	1	1	1	ω^2	ω^2	$\tau, \tau', \ldots$
Γ_4	$\mathbf{I}$	$\mathbf{D}_1$	$\mathbf{D}_2$	$\mathbf{D}_3$	$\mathbf{D}_1\mathbf{M}_1$	$\mathbf{M}_1$	$\mathbf{x}, \mathbf{y}, \ldots$

T	$\mathbf{D}_2\mathbf{M}_1$	$\mathbf{D}_3\mathbf{M}_1$	$\mathbf{D}_2\mathbf{M}_2$	$\mathbf{D}_3\mathbf{M}_2$	$\mathbf{D}_1\mathbf{M}_2$	$\mathbf{M}_2$	Basic quantities
Γ_1	1	1	1	1	1	1	$\psi, \psi', \ldots$
Γ_2	ω	ω	ω^2	ω^2	ω^2	ω^2	$\psi, \psi', \ldots$
Γ_3	ω^2	ω^2	ω	ω	ω	ω	$\tau, \tau', \ldots$
Γ_4	$\mathbf{D}_2\mathbf{M}_1$	$\mathbf{D}_3\mathbf{M}_1$	$\mathbf{D}_2\mathbf{M}_2$	$\mathbf{D}_3\mathbf{M}_2$	$\mathbf{D}_1\mathbf{M}_2$	$\mathbf{M}_2$	$\mathbf{x}, \mathbf{y}, \ldots$

Diploidal class, T_h, m3, $\overline{6}/2$
(Table follows from that of T, since $T_h = T \times S_2$)

Hextetrahedral class, T_d, $\overline{4}3m$, $3/\overline{4}$
Gyroidal class, O, 432, 3/4

Table C2.14

T_d O	$\mathbf{E}$ $\mathbf{E}$	$\mathbf{D}_1$ $\mathbf{D}_1$	$\mathbf{D}_2$ $\mathbf{D}_2$	$\mathbf{D}_3$ $\mathbf{D}_3$	$\mathbf{D}_1\mathbf{T}_2$ $\mathbf{R}_1\mathbf{T}_2$	$\mathbf{D}_1\mathbf{T}_3$ $\mathbf{R}_1\mathbf{T}_3$	$\mathbf{D}_2\mathbf{T}_1$ $\mathbf{R}_2\mathbf{T}_1$	$\mathbf{D}_2\mathbf{T}_3$ $\mathbf{R}_2\mathbf{T}_3$	$\mathbf{D}_3\mathbf{T}_1$ $\mathbf{R}_3\mathbf{T}_1$	$\mathbf{D}_3\mathbf{T}_2$ $\mathbf{R}_3\mathbf{T}_2$	$\mathbf{T}_1$ $\mathbf{CT}_1$	$\mathbf{T}_2$ $\mathbf{CT}_2$	$\mathbf{T}_3$ $\mathbf{CT}_3$	Basic quantities
Γ_1	1	1	1	1	1	1	1	1	1	1	1	1	1	$\psi, \psi', \ldots$
Γ_2	1	1	1	1	−1	−1	−1	−1	−1	−1	−1	−1	−1	$\gamma, \gamma', \ldots$
Γ_3	$\mathbf{I}$	$\mathbf{I}$	$\mathbf{I}$	$\mathbf{I}$	$\mathbf{F}$	$\mathbf{H}$	$\mathbf{G}$	$\mathbf{H}$	$\mathbf{G}$	$\mathbf{F}$	$\mathbf{G}$	$\mathbf{F}$	$\mathbf{H}$	$\begin{pmatrix}\tau_1\\\tau_2\end{pmatrix}, \begin{pmatrix}\tau_1'\\\tau_2'\end{pmatrix}, \ldots$
Γ_4	$\mathbf{I}$	$\mathbf{D}_1$	$\mathbf{D}_2$	$\mathbf{D}_3$	$\mathbf{D}_1\mathbf{T}_2$	$\mathbf{D}_1\mathbf{T}_3$	$\mathbf{D}_2\mathbf{T}_1$	$\mathbf{D}_2\mathbf{T}_3$	$\mathbf{D}_3\mathbf{T}_1$	$\mathbf{D}_3\mathbf{T}_2$	$\mathbf{T}_1$	$\mathbf{T}_2$	$\mathbf{T}_3$	$\mathbf{x}, \mathbf{y}, \ldots$
Γ_5	$\mathbf{I}$	$\mathbf{D}_1$	$\mathbf{D}_2$	$\mathbf{D}_3$	$\mathbf{R}_1\mathbf{T}_2$	$\mathbf{R}_1\mathbf{T}_3$	$\mathbf{R}_2\mathbf{T}_1$	$\mathbf{R}_2\mathbf{T}_3$	$\mathbf{R}_3\mathbf{T}_1$	$\mathbf{R}_3\mathbf{T}_2$	$\mathbf{CT}_1$	$\mathbf{CT}_2$	$\mathbf{CT}_3$	, , ...

T_d O	$\mathbf{D}_1\mathbf{T}_1$ $\mathbf{R}_1\mathbf{T}_1$	$\mathbf{D}_2\mathbf{T}_2$ $\mathbf{R}_2\mathbf{T}_3$	$\mathbf{D}_3\mathbf{T}_3$ $\mathbf{R}_3\mathbf{T}_3$	$\mathbf{M}_1$ $\mathbf{M}_1$	$\mathbf{M}_2$ $\mathbf{M}_2$	$\mathbf{D}_1\mathbf{M}_1$ $\mathbf{D}_1\mathbf{M}_1$	$\mathbf{D}_2\mathbf{M}_2$ $\mathbf{D}_1\mathbf{M}_2$	$\mathbf{D}_3\mathbf{M}_1$ $\mathbf{D}_2\mathbf{M}_1$	$\mathbf{D}_2\mathbf{M}_2$ $\mathbf{D}_2\mathbf{M}_2$	$\mathbf{D}_3\mathbf{M}_1$ $\mathbf{D}_3\mathbf{M}_1$	$\mathbf{D}_3\mathbf{M}_2$ $\mathbf{D}_3\mathbf{M}_2$	Basic quantities
Γ_1	1	1	1	1	1	1	1	1	1	1	1	$\psi, \psi', \ldots$
Γ_2	−1	−1	−1	1	1	1	1	1	1	1	1	$\gamma, \gamma', \ldots$
Γ_3	$\mathbf{G}$	$\mathbf{F}$	$\mathbf{H}$	$\mathbf{B}$	$\mathbf{A}$	$\mathbf{B}$	$\mathbf{A}$	$\mathbf{B}$	$\mathbf{A}$	$\mathbf{B}$	$\mathbf{A}$	$\begin{pmatrix}\tau_1\\\tau_2\end{pmatrix}, \begin{pmatrix}\tau_1'\\\tau_2'\end{pmatrix}, \ldots$
Γ_4	$\mathbf{D}_1\mathbf{T}_1$	$\mathbf{D}_2\mathbf{T}_2$	$\mathbf{D}_3\mathbf{T}_3$	$\mathbf{M}_1$	$\mathbf{M}_2$	$\mathbf{D}_1\mathbf{M}_1$	$\mathbf{D}_1\mathbf{M}_2$	$\mathbf{D}_2\mathbf{M}_1$	$\mathbf{D}_2\mathbf{M}_2$	$\mathbf{D}_3\mathbf{M}_1$	$\mathbf{D}_3\mathbf{M}_2$	$\mathbf{x}, \mathbf{y}, \ldots$
Γ_5	$\mathbf{R}_1\mathbf{T}_1$	$\mathbf{R}_2\mathbf{T}_3$	$\mathbf{R}_3\mathbf{T}_3$	$\mathbf{M}_1$	$\mathbf{M}_2$	$\mathbf{D}_1\mathbf{M}_1$	$\mathbf{D}_1\mathbf{M}_2$	$\mathbf{D}_2\mathbf{M}_1$	$\mathbf{D}_2\mathbf{M}_2$	$\mathbf{D}_3\mathbf{M}_1$	$\mathbf{D}_3\mathbf{M}_2$	, , ...

Hexoctahedral class, O_h, m3m, $\overline{6}/4$
(Table follows from that of O, since $O_h = O \times S_2$)

APPENDIX D

Some Theorems on Symmetric Polynomial Functions

Here we give some basic theorems without proof (see Weyl [1946, pp. 36, 53, 276]), that provide a systematic method for constructing the integrity basis of polynomials from the typical multilinear elements. The abbreviation $\sum x_i y_j \dots z_k$ is understood to denote the sum of quantities obtained by permuting the subscripts in the summant cyclically, e.g.,

$$\sum x_1 = \sum x_2 = \sum x_3 \equiv x_1 + x_2 + x_3,$$
$$\sum x_1 y_2 = \sum x_2 y_3 = \sum x_3 y_1 \equiv x_1 y_2 + x_2 y_3 + x_3 y_1.$$

Theorem 1. *A set of typical multilinear elements of the integrity basis for polynomials* $P(x_1^{(1)}, x_2^{(1)}, \dots, x_1^{(n)}, x_2^{(n)})$, *which are invariant under interchange of subscripts* 1 *and* 2 *on* $x^{(1)}, x^{(2)}, \dots, x^{(n)}$, *is formed by the quantities*

$$x_1 + x_2, \qquad x_1 y_2 + x_2 y_1. \tag{D1}$$

To obtain the multilinear elements we form n sets of quantities by substituting $x^{(1)}, \dots, x^{(n)}$ for x in $(\text{D1})_1$ and $n(n-1)/2$ quantities by substituting $x^{(i)}$ for x and $x^{(j)}$ for y $(i, j = 1, 2, \dots, n; i < j)$ in $(\text{D1})_2$.

Theorem 2. *A set of typical multilinear elements of the integrity basis for polynomials* $P(x_1^{(1)}, x_2^{(1)}, x_3^{(1)}, \dots, x_1^{(n)}, x_2^{(n)}, x_3^{(n)})$, *which are invariant under all permutations of the subscripts* 1, 2, *and* 3, *is formed by the quantities*

$$x_1, \qquad \sum x_1 y_1, \qquad \sum x_1 y_1 z_1. \tag{D2}$$

Thus the multilinear elements consist of n sets of quantities obtained by substituting $x^{(1)}, x^{(2)}, \dots, x^{(n)}$ in $(\text{D2})_1$; the $n(n-1)/2$ sets are obtained by substituting $x^{(i)}$ for x and $x^{(j)}$ for y $(i, j = 1, 2, \dots, n; i < j)$ in $(\text{D2})_2$, and the $n(n-1)(n-2)/2$ quantities are obtained by substituting $x^{(i)}$ for x, $x^{(j)}$ for y and $x^{(k)}$ for z $(i, j, k = 1, 2, \dots, n; i < j < k)$ in $(\text{D2})_3$. For example, for $n = 3$, we have

$$\sum x_1 \colon \quad x_1^{(i)} + x_2^{(i)} + x_3^{(i)}, \qquad i = 1, 2, 3,$$

$$\begin{aligned}\sum x_1 y_1\colon\quad & x_1^{(1)}x_1^{(2)} + x_2^{(1)}x_2^{(2)} + x_3^{(1)}x_3^{(2)},\\ & x_1^{(2)}x_1^{(3)} + x_2^{(2)}x_2^{(3)} + x_3^{(2)}x_3^{(3)},\\ & x_1^{(3)}x_1^{(1)} + x_2^{(3)}x_2^{(1)} + x_3^{(3)}x_3^{(1)},\\ \sum x_1 y_1 z_1\colon\quad & x_1^{(1)}x_1^{(2)}x_1^{(3)} + x_2^{(1)}x_2^{(2)}x_2^{(3)} + x_3^{(1)}x_3^{(2)}x_3^{(3)}.\end{aligned}$$

Theorem 3. *A set of typical multilinear elements of the integrity basis for polynomials* $P(x_1^{(1)}, x_2^{(1)}, x_3^{(1)}, \ldots, x_1^{(n)}, x_2^{(n)}, x_3^{(n)})$, *which are invariant under cyclic permutations of the subscripts* 1, 2, *and* 3, *is formed by the quantities*

$$\begin{aligned}&\sum x_1 & & \\ &\sum x_1 y_1, & &\sum x_1(y_2 - y_3), \\ &\sum x_1 y_1 z_1, & &\sum x_1 y_1 (z_2 - z_3).\end{aligned} \tag{D3}$$

Theorem 4. *A set of multilinear elements of the integrity basis for polynomials* $P(x_1^{(1)}, x_2^{(1)}, x_3^{(1)}, \ldots, x_1^{(n)}, x_2^{(n)}, x_3^{(n)}, L_1, \ldots, L_m)$, *which are invariant under all odd permutations of the subscripts* 1, 2, *and* 3 (*i.e.*, (12), (13), (23)) *on the* $x_i^{(1)}, \ldots, x_i^{(n)}$ ($i = 1, 2, 3$), *with simultaneous changes of sign of the quantities* $L_1, L_2, \ldots, L_m$, *i.e.*,

$$\begin{aligned}&P(x_i^{(1)}, x_j^{(1)}, x_k^{(1)}, \ldots, x_i^{(n)}, x_j^{(n)}, x_k^{(n)}, L_1, \ldots, L_m)\\ &\quad = P(x_i^{(1)}, x_k^{(1)}, x_j^{(1)}, \ldots, x_i^{(n)}, x_k^{(n)}, x_j^{(n)}, -L_1, \ldots, -L_m)\\ &\quad = P(x_k^{(1)}, x_i^{(1)}, x_j^{(1)}, \ldots, x_k^{(n)}, x_i^{(n)}, x_j^{(n)}, L_1, \ldots, L_m)\end{aligned}$$

where i, j, k *is any permutation of* 1, 2, 3, *is formed by the quantities*:

(i) $L_i L_j$ ($i, j = 1, \ldots, m$; $i < j$);
(ii) *the typical multilinear integrity basis in* $x_1^{(i)}, x_i^{(n)}$ ($i = 1, 2, 3$) *which are invariant under all permutations of the subscripts* 1, 2, *and* 3 (*Theorem* 2);
(iii) $L_i M_j$ ($i = 1, \ldots, m$; $j = 1, 2, \ldots$) *where* M_j *are given by*

$$x_1(y_2 - y_3) \qquad \text{and} \qquad x_1 y_1 (z_2 - z_3).$$

Theorem 5. *An integrity basis for polynomials in the variables* $x_1, \ldots, x_p$, $I_1, \ldots, I_q$, *which are invariant under a group of transformations for which* $I_1, \ldots, I_q$ *are invariants, is formed by adjoining to the quantities* $I_1, \ldots, I_q$ *an integrity basis for polynomials in the variables* $x_1, \ldots, x_p$ *which are invariant under the same group of transformations.*

Theorem 6. *If* P *is a polynomial function of the complex quantities* $\alpha_1, \ldots, \alpha_n$, $\beta_1, \ldots, \beta_m$, *which satisfy the relation* $P(\alpha_1, \ldots, \alpha_n; \beta_1, \ldots, \beta_m) = P(\alpha_1, \ldots, \alpha_n; -\beta_1, \ldots, -\beta_m)$, *then* P *is expressible as a polynomial in the quantities*

$$\begin{aligned}&\alpha_i \ (i = 1, 2, \ldots, n,\\ &\beta_j \beta_k, \qquad \beta_j \bar{\beta}_k \qquad (j, k = 1, \ldots, m),\end{aligned} \tag{D4}$$

where $\bar{\beta}_k$ *is the complex conjugate of* β_k.

Theorem 7. *A polynomial integrity basis in the n vectors* $x^{(r)} = (x_1^{(r)}, \ldots, x_n^{(r)})$ $(r = 1, 2, \ldots, n)$ *in n-dimensional space, which is invariant under all proper orthogonal transformations, is formed by the scalar products*

$$\alpha_i^{(r)}\alpha_i^{(s)} \tag{D5}$$

and the determinant

$$|\alpha_s^{(r)}| \tag{D6}$$

where $i, r, s = 1, 2, \ldots, n$.

APPENDIX E

Representations of Isotropic, Scalar, Vector, and Tensor Functions

The representations for isotropic, scalar, vector, and tensor-valued functions were studied by Wang [1–4] and Smith [5, 6] using different procedures. The results given by them were not identical. After the modifications discussed by Boehler [7] both representations are made identical. For ease of reference the results are reproduced here, where Wang's notation is used.

References

[1] Wang, C.C., *Arch. Rational Mech. Anal.*, **33**, 249 (1969).
[2] Wang, C.C., *Arch. Rational Mech. Anal.*, **33**, 268 (1969).
[3] Wang, C.C., *Arch. Rational Mech. Anal.*, **36**, 166 (1970).
[4] Wang, C.C., *Arch. Rational Mech. Anal.*, **43**, 392 (1971).
[5] Smith, G.F., *Arch. Rational Mech. Anal.*, **36**, 161 (1970).
[6] Smith, G.F., *Internat. J. Engng. Sci.* **19**, 899 (1971).
[7] Boehler, J.P., *Z. Angew. Math. Mech.* **57**, 323 (1977).

Table E.1. Complete and irreducible sets of invariants of symmetric tensors **A**, vectors **v**, and skew-symmetric tensors **W**.

I. Invariants depending on one variable	
Variable	Invariants
$\mathbf{A}$	$\mathrm{tr}\,\mathbf{A}$, $\mathrm{tr}\,\mathbf{A}^2$, $\mathrm{tr}\,\mathbf{A}^3$
$\mathbf{v}$	$\mathbf{v}, \mathbf{v}$
$\mathbf{W}$	$\mathrm{tr}\,\mathbf{W}^2$
II. Invariants depending on two variables when I is assumed	
Variables	Invariants
$\mathbf{A}_1, \mathbf{A}_2$	$\mathrm{tr}\,\mathbf{A}_1\mathbf{A}_2$, $\mathrm{tr}\,A_1^2A_2$, $\mathrm{tr}\,\mathbf{A}_1\mathbf{A}_2^2$, $\mathrm{tr}\,A_1^2A_2^2$
$\mathbf{A}, \mathbf{v}$	$\mathbf{v}\cdot\mathbf{A}\mathbf{v}$, $\mathbf{v}\cdot\mathbf{A}^2\mathbf{v}$
$\mathbf{A}, \mathbf{W}$	$\mathrm{tr}\,\mathbf{A}\mathbf{W}^2$, $\mathrm{tr}\,\mathbf{A}^2\mathbf{W}^2$, $\mathrm{tr}\,\mathbf{A}^2\mathbf{W}^2A\mathbf{W}$
$\mathbf{v}_1, \mathbf{v}_2$	$\mathbf{v}_1\cdot\mathbf{v}_2$
$\mathbf{v}, \mathbf{W}$	$\mathbf{v}\cdot\mathbf{W}^2\mathbf{v}$
$\mathbf{W}_1, \mathbf{W}_2$	$\mathrm{tr}\,\mathbf{W}_1\mathbf{W}_2$

(continued)

Table E.1 (*continued*)

III. Invariants depending on three variables when II is assumed

Variables	Invariants
$\mathbf{A}_1, \mathbf{A}_2, \mathbf{A}_3$	$\operatorname{tr} \mathbf{A}_1\mathbf{A}_2\mathbf{A}_3$
$\mathbf{A}_1\mathbf{A}_2, \mathbf{v}$	$\mathbf{v}\cdot\mathbf{A}_1\mathbf{A}_2\mathbf{v}$
$\mathbf{A}, \mathbf{v}_1, \mathbf{v}_2$	$\mathbf{v}_1\cdot\mathbf{A}\mathbf{v}_2, \mathbf{v}_1\cdot\mathbf{A}^2\mathbf{v}_2$
$\mathbf{A}, \mathbf{W}_1, \mathbf{W}_2$	$\operatorname{tr} \mathbf{A}\mathbf{W}_1\mathbf{W}_2, \operatorname{tr} \mathbf{A}\mathbf{W}_1\mathbf{W}_2^2, \operatorname{tr} \mathbf{A}\mathbf{W}_1^2\mathbf{W}_2$
$\mathbf{A}_1, \mathbf{A}_2, \mathbf{W}$	$\operatorname{tr} \mathbf{A}_1\mathbf{A}_2\mathbf{W}, \operatorname{tr} \mathbf{A}_1^2\mathbf{A}_2\mathbf{W}, \operatorname{tr} \mathbf{A}_1\mathbf{W}^2\mathbf{A}_2\mathbf{W}, \operatorname{tr} \mathbf{A}_1\mathbf{A}_2^2\mathbf{W},$
$\mathbf{W}_1, \mathbf{W}_2, \mathbf{W}_3$	$\operatorname{tr} \mathbf{W}_1\mathbf{W}_2\mathbf{W}_3$
$\mathbf{v}_1, \mathbf{v}_2, \mathbf{W}$	$\mathbf{v}_1\cdot\mathbf{W}\mathbf{v}_2, \mathbf{v}_1\cdot\mathbf{W}^2\mathbf{v}_2$
$\mathbf{v}, \mathbf{W}_1, \mathbf{W}_2$	$\mathbf{v}\cdot\mathbf{W}_1\mathbf{W}_2\mathbf{v}, \mathbf{v}\cdot\mathbf{W}_1^2\mathbf{W}_2\mathbf{v}, \mathbf{v}\cdot\mathbf{W}_1\mathbf{W}_2^2\mathbf{v}$
$\mathbf{A}, \mathbf{v}, \mathbf{W},$	$\mathbf{v}\cdot\mathbf{A}\mathbf{W}\mathbf{v}, \mathbf{v}\cdot\mathbf{A}^2\mathbf{W}\mathbf{v}, \mathbf{v}\cdot\mathbf{W}\mathbf{A}\mathbf{W}^2\mathbf{v}$

IV. Invariants depending on four variables when III is assumed

Variables	Invariants
$\mathbf{A}_1, \mathbf{A}_2, \mathbf{v}_1, \mathbf{v}_2$	$\mathbf{v}_1\cdot(\mathbf{A}_1\mathbf{A}_2 - \mathbf{A}_2\mathbf{A}_1)\mathbf{v}_2$
$\mathbf{A}, \mathbf{v}_1, \mathbf{v}_2, \mathbf{W}$	$\mathbf{v}_1\cdot(\mathbf{A}\mathbf{W} - \mathbf{W}\mathbf{A})\mathbf{v}_2$
$\mathbf{v}_1, \mathbf{v}_2, \mathbf{W}_1, \mathbf{W}_2$	$\mathbf{v}_1\cdot(\mathbf{W}_1\mathbf{W}_2 - \mathbf{W}_2\mathbf{W}_1)\mathbf{v}_2$

Table E.2. Generators for vector-valued isotropic functions

I. Generators depending on one variable

Variable	Generator
$\mathbf{v}$	$\mathbf{v}$

II. Generators depending on two variables

Variables	Generators
$\mathbf{A}, \mathbf{v}$	$\mathbf{A}\mathbf{v}, \mathbf{A}^2\mathbf{v}$
$\mathbf{W}, \mathbf{v}$	$\mathbf{W}\mathbf{v}, \mathbf{W}^2\mathbf{v}$

III. Generators depending on three variables

Variables	Generators
$\mathbf{A}_1, \mathbf{A}_2, \mathbf{v}$	$(\mathbf{A}_1\mathbf{A}_2 - \mathbf{A}_2\mathbf{A}_1)\mathbf{v}$
$\mathbf{W}_1, \mathbf{W}_2, \mathbf{v}$	$(\mathbf{W}_1\mathbf{W}_2 - \mathbf{W}_2\mathbf{W}_1)\mathbf{v}$
$\mathbf{A}, \mathbf{v}, \mathbf{W},$	$(\mathbf{A}\mathbf{W} - \mathbf{W}\mathbf{A})\mathbf{v}$

Table E.3. Generators for symmetric tensor-valued isotropic functions.

I. Generator depending on no variable $\mathbf{I}$

II. Generators depending on one variable

Variable	Generators
$\mathbf{A}$	$\mathbf{A}, \mathbf{A}^2$
$\mathbf{v}$	$\mathbf{v} \otimes \mathbf{v}$
$\mathbf{W}$	$\mathbf{W}^2$

III. Generators depending on two variables

Variables	Generators
$\mathbf{A}_1, \mathbf{A}_2$	$\mathbf{A}_1\mathbf{A}_2 + \mathbf{A}_2\mathbf{A}_1, \mathbf{A}_1^2\mathbf{A}_2 + \mathbf{A}_2\mathbf{A}_1^2, \mathbf{A}_1\mathbf{A}_2^2 + \mathbf{A}_2^2\mathbf{A}_1$
$\mathbf{A}, \mathbf{v}$	$\mathbf{v} \otimes \mathbf{A}\mathbf{v} + \mathbf{A}\mathbf{v} \otimes \mathbf{v}, \mathbf{v} \otimes \mathbf{A}^2\mathbf{v} + \mathbf{A}^2\mathbf{v} \otimes \mathbf{v}$
$\mathbf{A}, \mathbf{W}$	$\mathbf{AW} - \mathbf{WA}, \mathbf{WAW}, \mathbf{A}^2\mathbf{W} - \mathbf{WA}^2, \mathbf{WAW}^2 - \mathbf{W}^2\mathbf{AW}$
$\mathbf{v}_1, \mathbf{v}_2$	$\mathbf{v}_1 \otimes \mathbf{v}_2 + \mathbf{v}_2 \otimes \mathbf{v}_1$
$\mathbf{v}, \mathbf{W}$	$\mathbf{Wv} \otimes \mathbf{Wv}, \mathbf{v} \otimes \mathbf{Wv} + \mathbf{Wv} \otimes \mathbf{v}, \mathbf{Wv} \otimes \mathbf{W}^2\mathbf{v} + \mathbf{W}^2\mathbf{v} \otimes \mathbf{Wv}$
$\mathbf{W}_1, \mathbf{W}_2$	$\mathbf{W}_1\mathbf{W}_2 + \mathbf{W}_2\mathbf{W}_1, \mathbf{W}_1\mathbf{W}_2^2 - \mathbf{W}_2^2\mathbf{W}_1, \mathbf{W}_1^2\mathbf{W}_2 - \mathbf{W}_2\mathbf{W}_1^2$

IV. Generators depending on three variables

Variables	Generators
$\mathbf{A}, \mathbf{v}_1, \mathbf{v}_2$	$\mathbf{A}(\mathbf{v}_1 \otimes \mathbf{v}_2 - \mathbf{v}_2 \otimes \mathbf{v}_1) - (\mathbf{v}_1 \otimes \mathbf{v}_2 - \mathbf{v}_2 \otimes \mathbf{v}_1)\mathbf{A}$
$\mathbf{W}, \mathbf{v}_1, \mathbf{v}_2$	$\mathbf{W}(\mathbf{v}_1 \otimes \mathbf{v}_2 - \mathbf{v}_2 \otimes \mathbf{v}_1) + (\mathbf{v}_1 \otimes \mathbf{v}_2 - \mathbf{v}_2 \otimes \mathbf{v}_1)\mathbf{W}$

Table E.4. Generators for skew-symmetric tensor-valued isotropic functions.

I. Generator depending on one variable

Variable	Generator
$\mathbf{W}$	$\mathbf{W}$

II. Generators depending one two variables

Variables	Generators
$\mathbf{A}_1, \mathbf{A}_2$	$\mathbf{A}_1\mathbf{A}_2 - \mathbf{A}_2\mathbf{A}_1, \mathbf{A}_1^2\mathbf{A}_2 - \mathbf{A}_2\mathbf{A}_1^2, \mathbf{A}_1\mathbf{A}_2^2 - \mathbf{A}_2^2\mathbf{A}_1,$ $\mathbf{A}_1\mathbf{A}_2\mathbf{A}_1^2 - \mathbf{A}_1^2\mathbf{A}_2\mathbf{A}_1, \mathbf{A}_2\mathbf{A}_1\mathbf{A}_2^2 - \mathbf{A}_2^2\mathbf{A}_1\mathbf{A}_2$
$\mathbf{A}, \mathbf{v}$	$\mathbf{v} \otimes \mathbf{Av} - \mathbf{Av} \otimes \mathbf{v}, \mathbf{v} \otimes \mathbf{A}^2\mathbf{v} - \mathbf{A}^2\mathbf{v} \otimes \mathbf{v}, \mathbf{Av} \otimes \mathbf{A}^2\mathbf{v} - \mathbf{A}^2\mathbf{v} \otimes \mathbf{Av}$
$\mathbf{A}, \mathbf{W}$	$\mathbf{AW} + \mathbf{WA}, \mathbf{AW}^2 - \mathbf{W}^2\mathbf{A}$
$\mathbf{W}, \mathbf{v}$	$\mathbf{v} \otimes \mathbf{Wv} - \mathbf{Wv} \otimes \mathbf{v}, \mathbf{v} \otimes \mathbf{W}^2\mathbf{v} - \mathbf{W}^2\mathbf{v} \otimes \mathbf{v}$
$\mathbf{v}_1, \mathbf{v}_2$	$\mathbf{v}_1 \otimes \mathbf{v}_2 - \mathbf{v}_2 \otimes \mathbf{v}_1$
$\mathbf{W}_1, \mathbf{W}_2$	$\mathbf{W}_1\mathbf{W}_2 - \mathbf{W}_2\mathbf{W}_1$

III. Generators depending on three variables

Variables	Generators
$\mathbf{A}_1, \mathbf{A}_2, \mathbf{A}_3$	$\mathbf{A}_1\mathbf{A}_2\mathbf{A}_3 + \mathbf{A}_2\mathbf{A}_3\mathbf{A}_1 + \mathbf{A}_3\mathbf{A}_1\mathbf{A}_2 - \mathbf{A}_2\mathbf{A}_1\mathbf{A}_3 - \mathbf{A}_1\mathbf{A}_3\mathbf{A}_2 - \mathbf{A}_3\mathbf{A}_2\mathbf{A}_1$
$\mathbf{A}_1, \mathbf{A}_2, \mathbf{v}$	$\mathbf{A}_1\mathbf{v} \otimes \mathbf{A}_2\mathbf{v} - \mathbf{A}_2\mathbf{v} \otimes \mathbf{A}_1\mathbf{v} + \mathbf{v} \otimes (\mathbf{A}_1\mathbf{A}_2 - \mathbf{A}_2\mathbf{A}_1)\mathbf{v} - (\mathbf{A}_1\mathbf{A}_2$ $- \mathbf{A}_2\mathbf{A}_1)\mathbf{v} \otimes \mathbf{v}$
$\mathbf{A}, \mathbf{v}_1, \mathbf{v}_2$	$\mathbf{A}(\mathbf{v}_1 \otimes \mathbf{v}_2 - \mathbf{v}_2 \otimes \mathbf{v}_1) + (\mathbf{v}_1 \otimes \mathbf{v}_2 - \mathbf{v}_2 \otimes \mathbf{v}_1)\mathbf{A}$
$\mathbf{W}, \mathbf{v}_1, \mathbf{v}_2$	$\mathbf{W}(\mathbf{v}_1 \otimes \mathbf{v}_2 - \mathbf{v}_2 \otimes \mathbf{v}_1) - (\mathbf{v}_1 \otimes \mathbf{v}_2 - \mathbf{v}_2 \otimes \mathbf{v}_1)\mathbf{W}$

APPENDIX F

Maxwell's Equations in Various Systems of Units

Three formulations of Maxwell's equations in matter (all in Lorentz–Heaviside units). After Maugin [1978a, p. 17].

Equation	(A) Four-vector covariant formulation*	(B) Three-dimensional Galilean formulation in a co-moving frame $R_c(x, t)$†	(C) Three-dimensional formulation in a fixed laboratory frame R_G
Gauss	$\overset{\perp}{\boldsymbol{\nabla}} \cdot \mathscr{D} + \frac{2}{c}\boldsymbol{\omega} \cdot \mathscr{H} = q$	$\boldsymbol{\nabla} \cdot \mathbf{D} = q$	$\boldsymbol{\nabla} \cdot \mathbf{D} = q_e$
Ampère	$(\overset{\perp}{\boldsymbol{\nabla}} + c^{-2}\mathscr{C}) * \mathscr{H} - \frac{1}{c}\overset{*}{\mathscr{D}} = \frac{1}{c}\mathscr{J}$	$\boldsymbol{\nabla} \times \mathscr{H} - \frac{1}{c}\overset{*}{\mathscr{D}} = \frac{1}{c}\mathscr{J}$	$\boldsymbol{\nabla} \times \mathbf{H} - \frac{1}{c}\frac{\partial \mathbf{D}}{\partial t} = \frac{1}{c}\mathbf{J}$
Faraday	$(\overset{\perp}{\boldsymbol{\nabla}} + c^{-2}\mathscr{C}) * \mathscr{E} + \frac{1}{c}\overset{*}{\mathscr{B}} = \mathbf{0}$	$\boldsymbol{\nabla} \times \mathscr{E} + \frac{1}{c}\overset{*}{\mathbf{B}} = \mathbf{0}$	$\boldsymbol{\nabla} \times \mathbf{E} + \frac{1}{c}\frac{\partial \mathbf{B}}{\partial t} = \mathbf{0}$
Conservation of magnetic flux	$\overset{\perp}{\boldsymbol{\nabla}} \cdot \mathscr{B} - \frac{2}{c}\boldsymbol{\omega} \cdot \mathscr{E} = 0$	$\boldsymbol{\nabla} \cdot \mathbf{B} = 0$	$\boldsymbol{\nabla} \cdot \mathbf{B} = 0$
Conservation of charge	$\overset{*}{q} + \nabla_\alpha \mathscr{J}^\alpha = 0$	$\overset{*}{q} + \boldsymbol{\nabla} \cdot \mathscr{J} = 0$	$\frac{\partial q}{\partial t} + \boldsymbol{\nabla} \cdot \mathbf{J} = 0$
Potentials	$\mathscr{B} = \overset{\perp}{\boldsymbol{\nabla}} * \mathscr{A} + \frac{2\phi}{c}\boldsymbol{\omega}$	$\mathbf{B} = \boldsymbol{\nabla} \times \mathbf{A}$	$\mathbf{B} = \boldsymbol{\nabla} \times \mathbf{A}$
	$\mathscr{E} = -(\overset{\perp}{\boldsymbol{\nabla}} + c^{-2}\mathscr{C})\phi - \frac{1}{c}\left(\underset{u}{\pounds} \mathscr{A}\right)_\perp$	$\mathscr{E} = -\boldsymbol{\nabla}\phi - \frac{1}{c}\left[\frac{d\mathbf{A}}{dt} + (\boldsymbol{\nabla}\mathbf{A}) \cdot \mathbf{v}\right]$	$\mathbf{E} = -\boldsymbol{\nabla}\phi - \frac{1}{c}\frac{\partial \mathbf{A}}{\partial t}$

* All four-vectors (boldface type) in formulation (A) are spatial. See Chapter 15, Vol. II.

† In formulation (B): $\mathscr{E} = \mathbf{E} + c^{-1}\mathbf{v} \times \mathbf{B}$, $\mathscr{H} = \mathbf{H} - c^{-1}\mathbf{v} \times \mathbf{D}$, $\mathscr{J} = \mathbf{J} - q\mathbf{v}$, $d/dt \equiv \partial/\partial t + \mathbf{v} \cdot \boldsymbol{\nabla}$, $\overset{*}{\mathbf{D}} \equiv d\mathbf{D}/dt - (\mathbf{D} \cdot \boldsymbol{\nabla})\mathbf{v} + \mathbf{D}(\boldsymbol{\nabla} \cdot \mathbf{v})$ where $\mathbf{v}$ is the three-velocity.

Macroscopic Maxwell's equations and Lorentz force in various systems of units. After Jackson [1962]. The Heaviside–Lorentz system is used throughout the book in theoretical considerations.

System	ε_0	μ_0	D, H	Macroscopic Maxwell's equations				Lorentz force per unit charge
Electrostatic (esu)	1	c^{-2} (t^2l^{-2})	$\mathbf{D} = \mathbf{E} + 4\pi\mathbf{P}$ $\mathbf{H} = c^2\mathbf{B} - 4\pi\mathbf{M}$	$\nabla\cdot\mathbf{D} = 4\pi q_e$	$\nabla\times\mathbf{H} = 4\pi\mathbf{J} + \dfrac{\partial\mathbf{D}}{\partial t}$	$\nabla\times\mathbf{E} + \dfrac{\partial\mathbf{B}}{\partial t} = 0$	$\nabla\cdot\mathbf{B} = 0$	$\mathbf{E} + \mathbf{v}\times\mathbf{B}$
Electromagnetic (emu)	c^{-2} (t^2l^{-2})	1	$\mathbf{D} = \dfrac{1}{c^2}\mathbf{E} + 4\pi\mathbf{P}$ $\mathbf{H} = \mathbf{B} - 4\pi\mathbf{M}$	$\nabla\cdot\mathbf{D} = 4\pi q_e$	$\nabla\times\mathbf{H} = 4\pi\mathbf{J} + \dfrac{\partial\mathbf{D}}{\partial t}$	$\nabla\times\mathbf{E} + \dfrac{\partial\mathbf{B}}{\partial t} = 0$	$\nabla\cdot\mathbf{B} = 0$	$\mathbf{E} + \mathbf{v}\times\mathbf{B}$
Gaussian	1	1	$\mathbf{D} = \mathbf{E} + 4\pi\mathbf{P}$ $\mathbf{H} = \mathbf{B} - 4\pi\mathbf{M}$	$\nabla\cdot\mathbf{D} = 4\pi q_e$	$\nabla\times\mathbf{H} = \dfrac{4\pi}{c}\mathbf{J} + \dfrac{1}{c}\dfrac{\partial\mathbf{D}}{\partial t}$	$\nabla\times\mathbf{E} + \dfrac{1}{c}\dfrac{\partial\mathbf{B}}{\partial t} = 0$	$\nabla\cdot\mathbf{B} = 0$	$\mathbf{E} + \dfrac{\mathbf{v}}{c}\times\mathbf{B}$
Heaviside–Lorentz	1	1	$\mathbf{D} = \mathbf{E} + \mathbf{P}$ $\mathbf{H} = \mathbf{B} - \mathbf{M}$	$\nabla\cdot\mathbf{D} = q_e$	$\nabla\times\mathbf{H} = \dfrac{1}{c}\left(\mathbf{J} + \dfrac{\partial\mathbf{D}}{\partial t}\right)$	$\nabla\times\mathbf{E} + \dfrac{1}{c}\dfrac{\partial\mathbf{B}}{\partial t} = 0$	$\nabla\cdot\mathbf{B} = 0$	$\mathbf{E} + \dfrac{\mathbf{v}}{c}\times\mathbf{B}$
Rationalized MKS	$\dfrac{10^7}{4\pi c^2}$ $(q^2t^2m^{-1}l^{-2})$	$4\pi\times 10^{-2}$ (mlq^{-2})	$\mathbf{D} = \varepsilon_0\mathbf{E} + \mathbf{P}$ $\mathbf{H} = \dfrac{1}{\mu_0}\mathbf{B} - \mathbf{M}$	$\nabla\cdot\mathbf{D} = q_e$	$\nabla\times\mathbf{H} = \mathbf{J} + \dfrac{\partial\mathbf{D}}{\partial t}$	$\nabla\times\mathbf{E} + \dfrac{\partial\mathbf{B}}{\partial t} = 0$	$\nabla\cdot\mathbf{B} = 0$	$\mathbf{E} + \mathbf{v}\times\mathbf{B}$

Where necessary the dimensions of quantities are given in parentheses. The symbol c stands for the velocity of light in vacuum with dimension (l/t).

References

ABLOWICZ, M.J. and SEGUR, H. [1981]: *Solitons and the Inverse Scattering Transform*, SIAM, Philadelphia.

ABRAHAM, M. [1909]: Zur Elektrodynamik bewegter Körper, *Rend. Circ. Mat. Palermo*, **28**, 1–28.

ABRAHAM, M. [1910]: Sull'elettrodinamica di Minkowski, *Rend. Circ. Mat. Palermo*, **30**, 33, 46. See also *Theorie der Electrizitat*, Vol. II, Teubner, Leipzig, 1923, p. 300.

AGRANOVICH, V.M. and Ginzburg, V.L. [1984]: *Crystal Optics with Spatial Dispersion and Excitons*, Springer-Verlag, New York.

AKHIEZER, I.A. and BOLOTIN, Yu, L. [1967]: Theory of scattering of electromagnetic waves in ferromagnetic substances, *Soviet Phys.* JETP, **25**, 925–933.

AKHIEZER, A.I., BAR'YAKHTAR, V.G., and PELETMINSKII, S.V. [1958]: Coupled magnetoelastic waves in ferromagnetic media and ferroacoustic resonance, *Zhur. Eksper. Teoret. Fiz.* (in Russian), **35**, 228–239.

AKHIEZER, A.I., BAR'YAKHTAR, V.G., and PELETMINSKII, S.V. [1968]: *Spin Waves* (translation from Russian), North-Holland, Amsterdam.

AKULOV, N. [1936]: Zur Quantentheorie der Temperaturabhangigkeit der Magnetisierungskurve, *Zeit. Phys.*, **100**, 197–202.

ALBLAS, J.B. [1968]: Continuum mechanics of media with internal structure, *Symposia Mathematica*, I, pp. 229–251, Inst. Naz. di Alta Mat., Academic Press, New York.

ALBLAS, J.B. [1974]: Electro-magneto-elasticity, in *Topics in Applied Mechanics*, pp. 71–114, eds. J.L. Zeman and F. Ziegler, Springer-Verlag, Wien.

ALBLAS, J.B. [1978]: Magneto-elastic stability of some composite structures, in *Continuum Models of Discrete Systems*, pp. 283–312, ed. J.W. Provan, University of Waterloo Press, Waterloo, Ontario, Canada.

ALERS, P. and FLEURY, P.A. [1963]: Modification of the velocity of sound in metals by magnetic fields, *Phys. Rev.*, **129**, 2428–2429.

ALFVÉN, H. and FÄLTHAMMAR, C. [1963]: *Cosmical Electrodynamics*, 2nd edition, Oxford University Press, New York, London.

AL-HASSANI, S.T.S., DUNCAN, J.L., and JOHNSON W. [1974]; On the parameters of the magnetic forming process, *J. Mech. Engng. Sci.*, **16**, 1–9.

AMBARTSUMIAN, S.A. [1982]; Magneto-elasticity of thin plates and shells, *Appl. Mech. Reviews*, **35**, 1–5.

AMBARTSUMIAN, S.A., BAGDASARIAN G.E., and BELUBEKIAN M.V. [1977]: *Magneto-Elasticity of Thin Shells and Plates* (in Russian), Nauka, Moscow.

AMENT, W. and RADO, G. [1955]: Electromagnetic effect of spin–wave resonance in ferromagnetic metals, *Phys. Rev.*, **97**, 1558–1566.

American Institute of Physics (The) [1957]: *American Institute of Physics Handbook*, McGraw-Hill, New York.

ANCONA, M.G. and TIERSTEN, H.F. [1983]: Fully macroscopic description of electrical conduction in metal–insulator–semiconductor structures, *Phys. Rev.*, **B27**, 7018–7045.

ANDERSON, J.C. [1968]: *Magnetism and Magnetic Materials*, Chapman and Hall, London.

ANDERSON, J.E. [1963]: *Magnetohydrodynamics Shock Waves*, M.I.T. Press, Cambridge, MA.

ARENZ, R.J., FERGUSON, C.W., and WILLIAMS M.L. [1967]: The mechanical and optical characterization of Solithane 113 composition, *Exp. Mech*, **7**, 183–188.

ARI, N. and ERINGEN, A.C. [1983]: Nonlocal stress field at Griffith crack, *Cryst. Lattice Def. Amorph. Mat.*, **10**, 937–945.

ARIMAN, T., TURK, M.A., and SYLVESTER, N.D. [1973]: Microcontinuum fluid mechanics—A review, *Int. J. Engng. Sci.*, **11**, 905–930.

ARIMAN, T., TURK, M.A., and SYLVESTER, N.D. [1974]: Application of microcontinuum fluid mechanics, *Int. J. Engng. Sci.*, **12**, 273–293.

ARP, P.A., FOISTER, R.T., and MASON, S.G. [1980]: Some electrohydrodynamic effects in fluid dispersion. *Adv. Colloid Interface Sci.*, **12**, 295–356.

ASKAR, A. and LEE, P.C.Y. [1974]: Lattice dynamics approach to the theory of diatomic elastic dielectrics, *Phys. Rev.*, **B9**, 5291–5299.

ASKAR, A., LEE, P.C.Y., and ÇAKMAK, A.S. [1970]: Lattice dynamics approach to the theory of elastic dielectrics with polarization gradients, *Phys. Rev.*, **B1**, 3525–3537.

ASKAR, A., LEE, P.C.Y., and ÇAKMAK, A.S. [1971]: The effects of surface curvature and a discontinuity on the surface energy and energy density and other induced fields in elastic dielectrics with polarization gradients, *Int. J. Solids and Structures*, **7**, 523–537.

ASKAR, A., POUGET, J., and MAUGIN, G.A. [1984]: Lattice models for elastic ferroelectrics and related continuum theories, in *The Mechanical Behavior of Electromagnetic Solid Continua*, pp. 151–156, ed. G.A. Maugin, North-Holland, Amsterdam.

ASTROV, D.H. [1960]: On the magneto-electric effect in antiferromagnetics, *Soviet Phys. JETP*, **11**, 708–709.

ASTROV, D.H. [1961]: Magnetoelectric effect in chromium oxide, *Soviet Phys. JETP*, **13**, 729–733.

ATTEN, P. [1974]: Electrohydrodynamic stability of dielectric liquids during transient regime of space-charge-limited injection, *Phys. Fluids*, **17**, 1822–1827.

ATTEN, P. [1975]: Stabilité électrohydrodynamique des liquides de faible conductivité, *J. Mécanique*, **14**, 461–495.

ATTEN, P. and MOREAU, R. [1972]: Stabilité électrohydrodynamique des liquides isolants soumis à une injection unipolaire, *J. Mécanique*, **11**, 471–520.

AVSEC, D. and LUNTZ, M. [1936]: Tourbillons électroconvectifs, *C. R. Acad. Sci. Paris*, **203**, 1104–1144.

BAILEY, R.L. [1983]: Lesser known applications of ferrofluids, *J. Magnetism and Magnetic Materials*, **39**, 173–177.

BAINES, K., DUNCAN, J.L., and JOHNSON, W. [1965–1966]: Electromagnetic metal forming, *Proc. Inst. Mech. Engrs*, **180**, Paper 37, Part 31, 348–362.

BAKIRTAS, I. and MAUGIN, G.A. [1982]: Ondes. de surface SH pures en élasticité inhomogène, *J. Méc. Théor. Appl.* **1**, 995–1013.

BARNETT, S.J. [1931]: Electron–inertia effect and the determination of m/c for the free electrons in copper, *Phil. Mag.* (7th Series), **12**, 349–360.

BARRABES, C. [1975]: Elastic and thermoelastic media in general relativity, *Nuovo Cimento*, **28B**, 377–394.

BARRABES, C. [1985]: Transient thermodynamics of electromagnetic media in general relativity, *J. Math. Phys.*, **26**, 798–803.

BAR'YAKHTAR, V.G. and MAKHMUDOV, Z.Z. [1964]: Coupled magnetoelastic waves in antiferromagnets with a magnetic structure of the $MnCO_3$ type, *Zhur. Èksper. Teoret. Fiz.* (in Russian), **47**, 1989–1994.

BAR'YAKHTAR, V.G., SAVCHENKO, M.A., and TARASENKO, V.V. [1965]: Coupled magnetoelastic waves in antiferromagnets in strong magnetic fields, *Zhur. Èksper. Teoret. Fiz.* (in Russian), **49**, 944–952.

BASHTOVOI, V.G., BERKOVSKY, B. and VISLOVICH, A. [1988]: *Introduction to thermodynamics of Magnetic Fluids*, Hemisphere, Washington.

BASHTOVOI, V., REX, A., and FOIGUEL, R. [1983]: Some nonlinear wave processes in magnetic fluids, *J. Magnetism and Magnetic Materials*, **39**, 115–118.

BATEMAN, G. [1978]: *MHD Instabilities*, M.I.T. Press, Cambridge, MA.

BAUMHAUER, J.C. and TIERSTEN, H.F. [1973]: Nonlinear electrostatics equations for small fields superimposed on a bias, *J. Acoust. Soc. Amer.*, **54**, 1017–1034.

BAZER, J. [1971]: Geometrical magnetoelasticity, *Geophys. J. Roy. Astron.*, **25**, 203–237.

BAZER, J. and ERICSON, W.B. [1959]: Hydromagnetic shocks, *Astrophys. J.*, **129**, 758.

BAZER, J. and ERICSON, W.B. [1974]: Nonlinear motion in magnetoelasticity, *Arch. Rat. Mech. Anal.*, **55**, 124–192.

BAZER, J. and KARAL, F. [1971]: Simple wave motion in magnetoelasticity, *Geophys. J. Roy. Astron.*, **25**, 127–156.

BEAMS, J.W. [1932]: Electric and magnetic double refraction, *Rev. Mod. Phys.*, **4**, 133.

BEDNORZ, J.G. and MÜLLER, K.A. [1986]: Possible high T_c super conductivity in the Ba–La–Cu–O system, *Z. Phys.* **B**, **64**, 189–193.

BENACH, R. [1974]: Toward a rational dynamics of plasmas, Ph. D. Thesis, Technical University of Eindhoven, The Netherlands.

BENACH, R. and MULLER, I. [1974]: Thermodynamics and the description of magnetizable dielectric mixtures of fluids, *Arch. Rat. Mech. Anal.*, **53**, 312–346.

BERGMANN, S. [1971]: *Integral Operators in the Theory of Linear Partial Differential Equations*, Springer-Verlag, Berlin, Heidelberg, New York.

BERGMANN, S. and SCHIFFER, M. [1953]: *Kernel Functions and Elliptic Differential Equations in Mathematical Physics*, Academic Press, New York.

BERKOVSKY, B. (editor) [1978]: *Thermomechanics of Magnetic Fluids*, Hemisphere, Washington.

BERLINCOURT, D.A. [1981]: Piezoelectric ceramics: Characteristics and applications, *J. Acoust. Soc. Amer.*, **70**, 1506–1595.

BERLINCOURT, D.A., CURRAN, D.R., and JAFFE, H. [1964]: Piezoelectric and piezomagnetic materials and their functions as transducers, in *Physical Acoustics*, Vol. IA, pp. 170–267, ed. W.P. Mason, Academic Press, New York.

BERNSTEIN, I.B., FRIEMAN, E.A., KRUSKAL, M.A., and KULSRUD, R.M. [1958]: An energy principle for hydromagnetic stability problems, *Proc. Roy. Soc. London*, **A244**, 17–40.

BHAGAVANTAM, S. [1966]: *Crystal Symmetry and Physical Properties*, Academic Press, New York.

BIRDSALL, D.H., FORD, F.C., FURTH, H.D., and RILEY, R.E. [1961]: Magnetic forming, *Amer. Mach*, **105**, 117–121.

BIRSS, R.R. [1964]: *Symmetry and Magnetism*, North-Holland, Amsterdam.

BLEUSTEIN, J.L. [1968]: A new surface wave in piezoelectric materials, *Appl. Phys. Lett.*, **13**, 412–413.

BLOCH, F. [1930]: Zur theorie des ferromagnetismus, *Zeit. Physik*, **61**, 206–219.

BLOEMBERGEN, N. [1965]: *Nonlinear Optics*, Benjamin, New York.

BOARDMAN, A.D. and COOPER, G.S. [1984]: Nonlinear wave propagation in optical waveguide sciences, Kweilin, People's Republic of China, June 20–23, 1983. Guest Editors: Huang Hung-Chia and Allan W. Snyder. *Applied Scientific Research* **41**, 384.

BOEHLER, J.P. [1977]: On irreducible representations for isotropic scalar functions, *Zeit. angew. Math. Mech.*, **57**, 323–327.

BOGARDUS, H., KRUEGER, D.A., and THOMPSON, D. [1978]: Dynamic magnetization in ferrofluids, in *Thermomechanics of Magnetic Fluids*, pp. 75–85, ed. B. Berkovsky, Hemisphere, Washington.

BOLEY, B.A. and WEINER, J.M. [1960]: *Theory of Thermal Stresses*, Wiley, New York.

BORN, M. [1972]: *Optik*, 3rd edition, Springer-Verlag, Berlin.

BORN, M. and HUANG, K. [1954]: *Dynamical Theory of Crystal Lattices* Oxford University Press, New York, Sect. 8.

BOROVICK-ROMANOV, A.S. [1959]: Piezomagnetism in the antiferromagnetic fluorides of cobalt and manganese, *Soviet Phys. JETP*, **36**, 1954–1955.

BÖTTCHER, C.J.F. [1952]: *Theory of Electric Polarization*, Elsevier, New York.

BOULANGER, Ph. and MAYNE, G. [1971]: Tenseur impulsion-énergie d'un milieu soumis à des effets thermiques et électro-magnétiques, *Bull. Acad. Belg. Roy., Cl. Sci.*, **57**, 872–890.

BOULANGER, Ph., MAYNE G., and VAN GEEN, R. [1973]: Magneto-optical, electro-optical and photoelastic effects in an elastic polarizable and magnetizable isotropic continuum, *Int. J. Solids and Structures*, **9**, 1439–1464.

BOULANGER, Ph., MAYNE, G., HERMANNE, A., KESTENS, J., and VAN GEEN, R. [1971]: L'effet photoélastique dans le cadre de la mécanique rationnelle des milieux continus, *Revue de l'Industrie Minérale–Mines*, June issue, 1–35.

BRADLEY, R. [1978]: Overstable electroconvective instabilities, *J. Mech. Appl. Math.*, **31**, 381–390.

BRANCHER, J.P. [1980a]: Existence et stabilité d'une aimantation constante dans un ferrofluide en mouvement, *C. R. Acad. Sci. Paris*, 290B, 457–459.

BRANCHER, J.P. [1980b]: Sur l'hydrodynamique des ferrofluides, Doctoral Thesis, Université de Nancy, France.

BRANCHER, J.P. and DENIS, J.P. [1981]: Phénomène de relaxation dans les ferrofluides, *C. R. Acad. Sci. Paris*, **292-II**, 1247–1250.

BRANOVER, H. (editor) [1976]: *MHD Flows and Turbulence*, Wiley, New York and Israel University Press, Jerusalem.

BRENNER, H. [1970]: Rheology of a dilute suspension of dipolar spherical particles in an external field, *J. Colloid and Interface Sci.*, **32**, 141–158.

BRESSAN, A. [1963]: Cinematica dei sistemi continui in relatività generale, *Ann. Mat. Pura Appl.*, **62**, 99–148.

BRESSAN, A. [1978]: *Relativistic Theories of Materials*, Springer Tracts in Natural Philosophy, Vol. 29, Springer-Verlag, Berlin, Heidelberg, New York.

BROWN, C.S., KELL, R.C., TAYLOR, R., and THOMAS, L. A. [1962]: Piezoelectric materials, *Proc. Inst. Elect. Engrs.* (London), **109**, P.E.B. No. 43, 99–114.

BROWN, W.F. Jr. [1963]: *Micromagnetics*, Wiley–Interscience, New York.

BROWN, W.F. Jr. [1965]: Theory of magnetoelastic effects in ferromagnetism, *J. Appl. Phys.*, **36**, 994–1000.

BROWN, W.F. Jr. [1966]: *Magnetoelastic Interactions*, Springer-Verlag, New York.

BURFOOT, J.C. [1967]: *Ferroelectrics*, Van Nostrand, Princeton, NJ.

CABANNES, H. [1970]: *Theoretical Magnetofluiddynamics* (translated from the French), Academic Press, New York.

CADY, W.G. [1946]: *Piezoelectricity*, McGraw-Hill, New York.

CALOGERO, F. and DEGASPERIS, A. [1982]: *Spectral Transform and Solitons*, Vol. I, North-Holland, Amsterdam.

CARTER, B. [1980]: Rheometric structure theory, convective differentiation and continuum electrodynamics, *Proc. Roy. Soc. London*, **A372**, 169–200.

CARTER, B. and QUINTANA, H. [1972]: Foundations of general relativistic high-pressure elasticity theory, *Proc. Roy. Soc. London*, **A331**, 58–71.

CATTANEO, C. [1962]: *Formulation relativiste des lois physiques en relativité générale*, Multigraphed Lecture Notes, Collège de France, Paris.

CHADWICK, P. [1960]: Thermoelasticity, the dynamical theory, in *Progress in Solid Mechanics*, North-Holland, Amsterdam.

CHANDRASEKHAR, S. [1961]: *Hydrodynamic and Hydromagnetic Stability*, Oxford University Press, London.

CHATTOPADHYAY, A. and MAUGIN, G.A. [1985]: Diffraction of magnetoelastic waves by a rigid strip, *J. Acoust. Soc. Amer.*, **78**, 217–222.

CHELKOWSKI, A. [1980]: *Dielectric Physics*, Elsevier, Amsterdam (translation from Polish).

CHEN, P.J. and MCCARTHY, M.F. [1974a]: One-dimensional shock waves in elastic dielectrics, *Istit. Lombardo Accad. Sci. Lett. Rend.*, **107**, 715–727.

CHEN, P.J. and MCCARTHY, M.F. [1974b]: Thermodynamic influences on the behavior of one-dimensional shock waves in elastic dielectrics, *Int. J. Solids and Structures*, **10**, 1229–1242.

CHEN, P.J. and MCCARTHY, M.F. [1975]: The behavior of plane shock waves in deformable magnetic materials, *Acta Mechanica*, **23**, 91–102.

CHIKAZUMI, S. [1966]: *Physics of Magnetism*, Wiley, New York.

CHRISTENSEN, R.M. [1971]: *Theory of Viscoelasticity*, Academic Press, New York, London.

CLARK, A.E. and STRAKNA, R.E. [1960]: Elastic constants of single crystals YIG, *J. Appl. Phys.*, **32** 1172–1173.

COLEMAN, B.D. [1964]: Thermodynamics of materials with memory, *Arch. Rat. Mech. Anal.*, **17**, 1–46.

COLEMAN, B.D. and DILL, E.H. [1971a]: On the thermodynamics of electromagnetic fields in materials with memory, *Arch. Rat. Mech. Anal.*, **41** 132–162.

COLEMAN, B.D. and DILL, E.H. [1971b]: Thermodynamical restrictions on the constitutive equations of electromagnetic theory, *Zeit. angew. Math. Phys.*, **22**, 691–702.

COLEMAN, B.D. and DILL, E.H. [1975]: Photoviscoelasticity: Theory and practice, in *The Photoelastic Effects and its Applications*, pp. 455–505, ed. J. Kestens, Springer-Verlag, Berlin.

COLEMAN, B.D. and NOLL, W. [1961]: Foundations of linear viscoelasticity, *Rev. Mod. Phys.*, **33**, 239–249.

COLEMAN, B.D., DILL, E.H., and TOUPIN, R.A. [1970]: A phenomenological theory of streaming birefringence, *Arch. Rat. Mech. Anal.*, **39**, 358–399.

COLLET, B. [1978]: Higher-order surface couplings in elastic ferromagnets, *Int. J. Engng. Sci.*, **16**, 349–364.

COLLET, B. [1981]: One-dimensional acceleration waves in deformable dielectrics with polarization gradients, *Int. J. Engng. Sci.*, **19**, 389–407.

COLLET, B. [1982]: Shock waves in deformable dielectrics with polarization gradients, *Int. J. Engng. Sci.*, **20**, 1145–1160.

COLLET, B. [1984]: Shock waves in deformable ferroelectric materials, in *The Mechanical Behavior of Electromagnetic Solid Continua*, pp. 157–163, ed. G.A. Maugin, North-Holland, Amsterdam.

COLLET, B. and MAUGIN, G.A. [1974]: Sur l'électrodynamique des milieux continus avec interactions, *C. R. Acad. Sci. Paris*, **279B**, 379–382.

COLLET, B. and MAUGIN, G.A. [1975]: Couplage magnétoélastique de surface dans les matériaux ferromagnétiques, *C. R. Acad. Sci. Paris*, **280A**, 1641–1644.

COMSTOCK, R.L. [1964]: Parametric coupling of the magnetization and strain field in a ferromagnet—I, II, *J. Appl. Phys.*, **34**, 1461–1464, 1465–1468.

COMSTOCK, R.L. [1965]: Parallel pumping of magnetoelastic waves in ferromagnets, *J. Appl. Phys.*, **35**, 2427–2431.

COOK, W.R. [1962]: Ferroelectric and piezoelectric materials, in *Digest of the Literature on Dielectrics*, National Academy of Sciences, Washington.

COQUIN, G.A. and TIERSTEN, H.F. [1967]: Rayleigh waves in linear elastic dielectrics, *J. Acoust. Soc. Amer.*, **41**, 921–939.

COURANT, R. [1965]: *Methods of Mathematical Physics*, Vol. II, Interscience, New York.

COWLEY, M.D. and ROSENSWEIG, R.E. [1967]: The interfacial stability of a ferromagnetic fluid, *J. Fluid Mech.*, **30**, 671–688.

CROSIGNANI, B. and DI PORTO, P. [1981]: Soliton propagation in multimode optical fibers, *Optic Letters*, **6**, 329–330.

CROSIGNANI, B., PAPAS, C.H., and DI PORTO, P. [1981]: Coupled-mode theory approach to nonlinear pulse propagation in optical fibers, *Optics Letters*, **6**, 61–63.

CULICK, F.E.C. [1964]: Compressible magnetogasdynamics flow, *Zeit. angew. Math. Phys.*, **15**, 129–143.

CURIE, P. [1908]: *Oeuvres de Pierre Curie*, Société Française de Physique, Paris.

CURTIS, H.D. and LIANIS, G. [1971]: Relativistic thermodynamics of deformable electromagnetic materials with memory, *Int. J. Engng. Sci.*, **9**, 451–468.

CURTIS, R.A. [1971]: Flow and wave propagation in ferrofluids, *Phys. Fluids*, **14**, 2096–2102.

DAHER, N. [1984]: Waves in elastic semiconductors in a bias electric field, in *The Mechanical Behavior of Electromagnetic Solid Continua*, pp. 115–120, ed. G.A.

Maugin, North-Holland, Amsterdam.
DAHER, N. and MAUGIN, G.A. [1984]: Modèle phénoménologique de semi-conducteurs piézoélectriques, *C. R. Acad. Sci. Paris*, **299-II**, 999–1002.
DAHER, N. and MAUGIN, G.A. [1986]: Waves in elastic semiconductors in a bias electric field, *Int. J. Engng. Sci.*, **24**, 733–754.
DALRYMPLE, J.M., PEACH, M.O., and VLIGELAHN, G.L. [1974]: Magnetoelastic buckling of thin magnetically soft plates in a cylindrical Mode, *J. Appl. Mech. Trans. ASME*, **41**, 145–150.
DANIEL, I.M. [1964]: Static and dynamic stress analysis in viscoelastic materials, Ph. D. Thesis, Illinois Institute of Technology, Urbana, IL.
DANIEL, I.M. [1965]: Experimental methods for dynamic stress analysis in viscoelastic materials, *J. Appl. Mech.*, **32**, 598–606.
DANIEL, I.M. [1966]: Dynamical properties of a photoviscoelastic material, *Exp. Mech.*, **5**, 83–89.
DANILOVSKAYA, V.I. [1950]: Thermal stresses in an elastic half-space under an instantaneous heating of the surface (in Russian), *Prik. Mat. Mech.*, **14**, 316–318.
DAS, N.C., BATTACHARYA, S.K., and DAS, S.N. [1981]: Quasi-static magnetoelastic vibration of an infinite ferromagnetic plate in a transverse magnetic field, *Mech. Res. Commun.*, **8**, 153–160.
DE GENNES, P.G. [1966]: *Superconductivity of Metals and Alloys*, Benjamin, New York.
DE GENNES, P.G. [1974]: *The Physics of Liquid Crystals*, Oxford University Press, Oxford.
DE LORENZI, H. and TIERSTEN, H.F. [1975]: On the interaction of the electromagnetic field with heat conducting deformable semiconductors, *J. Math. Phys.*, **16**, 938–957.
DEMIRAY, H. and ERINGEN, A.C., [1973a]: Constitutive equations of a plasma with bound charges, *Plasma Physics*, **15**, 889–901.
DEMIRAY, H. and ERINGEN, A.C., [1973b]: Continuum theory of a slightly ionized plasma, diagmagnetic effects, *Plasma Physics*, **15**, 903–920.
DEMIRAY, H. and EFINGEN, A.C. [1973c]: On the constitutive equations of polar elastic dielectrics, *Letts. Appl. Engng. Sci.*, **1**, No. 6, 517–527.
DEMIRAY, H. and ERINGEN, A.C. [1974]: Motion of electron gas in conducting solids, *Plasma Physics*, **16**, 589–602.
DHAR, P.K. [1979]: Coupled electromagnetic and elastic waves in random media, *Int. J. Engng. Sci.*, **17**, 145–150.
DIEULESAINT, E. and ROYER, D. [1980]: *Elastic Waves in Solids: Applications to Signal Processing* (translation from the French), Wiley, New York.
DILL, E.H. [1975]: Simple materials with fading memory, in *Continuum Physics*, Vol. II. Chap. 4, ed. A.C. Eringen, Academic Press, New York.
DILL, E.H. and FOWLKES, C.W. [1966]: Photoviscoelasticity, NASA CR-44, National Aeronautics and Space Administration, U.S.A.
DIRAC, P.A.M. [1929]: Quantum mechanics of many-electron systems, *Proc. Roy. Soc. London*, **A123**, 714–733.
DIXON, R.C. and ERINGEN, A.C. [1965a]: A dynamical theory of polar elastic dielectrics—I, *Int. J. Engng. Sci.*, **3**, 359–377.
DIXON, R.C. and ERINGEN, A.C. [1965b]: A dynamical theory of polar elastic dielectrics—II, *Int. J. Engng. Sci.*, **3**, No. 3, 379–398.
DORING, W. [1966]: Ferromagnetisms, in *Handbuch der Physik*, Bd. XVIII/2, ed. S. Flügge, Springer-Verlag, Berlin, Heidelberg, New York.

DRAGOS, L. [1975]: *Magnetofluid Dynamics* (translation from Romanian edition), Abacus Press, Tunbridge Wells, U.K.

DROUOT, R. and MAUGIN, G.A. [1985]: Continuum modeling of polyelectrolytes in solution, *Rheologica Acta*, **24**, 474–487.

DUNKIN, J.W. and ERINGEN, A.C. [1963]: On the propagation of waves in an electromagnetic elastic solid, *Int. J. Engng. Sci.*, **1**, No. 4, 461–495.

DUVAUT, G. and LIONS, J.L. [1972]: *Les inéquations en mécanique et en physique*, Dunod, Paris.

EASTMAN, D.E. [1966]: Second-order magnetoelastic properties of yttrium–iron–garnet, *J. Appl. Phys.*, **37**, 996–997; see also *Phys. Rev.*, **148**, 2, 530–542.

ECKART, C. [1940]: The thermodynamics of irreversible processes: III, Relativitic theory of the simple fluid, *Phys. Rev.*, **58**, 919–924.

EINSTEIN, A. [1956]: *The Meaning of Relativity*, Princeton University Press, Princeton, NJ.

EINSTEIN, A. and DE HAAS, W.J. [1915]: Experimenteller Nachweis der Ampèreschen Molekullarstrome, *Verh. d. Deutschen Phys. Gesellschaft*, **17**, 152–170.

EINSTEIN, A. and LAUB, J. [1908]: Uber die elektromagnetischen Felde auf ruhende Körper ausgeübten ponderomotorischen Kräfte, *Ann. Phys.* (Leipzig), **26**, 541–550.

ELLIOT, R.S. [1966]: *Electromagnetics*, McGraw-Hill, New York.

EMTAGE, P.R. [1976]: Nonreciprocal attenuation of magnetoelastic surface waves, *Phys. Rev.*, **B13**, 3063–3070.

ERINGEN, A.C. [1954]: The finite Sturm–Liouville transform, *Quart. J. Math.*, **5**, 120–129.

ERINGEN, A.C. [1955]: The solution of a class of mixed–mixed boundary value problems in plane elasticity, *Proceedings of the 2nd National Congress of Applied Mechanics*, pp. 142–144, ASME, New York.

ERINGEN, A.C. [1962]: *Nonlinear Theory of Continua*, McGraw-Hill, New York.

ERINGEN, A.C. [1963]: On the foundations of electrostatics, *Int. J. Engng. Sci.*, **1**, 127–153.

ERINGEN, A.C. [1964]: Simple micro-fluids, *Int. J. Engng. Sci.*, **2**, No. 2, 205–217.

ERINGEN, A.C. [1966a]: A unified theory of thermomechanical materials, *Int. J. Engng. Sci.*, **4**, 179–202.

ERINGEN, A.C. [1966b]: Linear theory of micropolar elasticity, *J. Math Mech.*, **15**, No. 6, 909–923.

ERINGEN, A.C. [1966c]: Theory of micropolar fluids, *J. Math Mech.*, **16**, No. 1, 1–18.

ERINGEN, A.C. [1966d]: Mechanics of micromorphic materials, *Proceedings of the 11th International Congress of Applied Mechanics* (held 1964, Munich, Germany), pp. 131–138. ed. H. Görtler, Springer-Verlag, Berlin.

ERINGEN, A.C. [1967]: *Mechanics of Continua*, Wiley, New York.

ERINGEN, A.C. [1967a]: Linear theory of micropolar viscoelasticity, *Int. J. Engng. Sci.*, **5**, No. 2, 191–204.

ERINGEN, A.C. [1967b]: Theory of micropolar continua, in *Developments in Mechanics*, Vol. 3, Part I—Solid Mechanics and Materials, *Proceedings of the 9th Midwestern Conference, University of Wisconsin*, August, 1965, pp. 23–40, eds. T.C. Huang and M.W. Johnson, Jr., Wiley, New York.

ERINGEN, A.C. [1968]: Theory of micropolar elasticity, in *Fracture*, Vol. II, Chap. 7, pp. 621–729, ed. H. Liebowitz, Academic Press, New York.

ERINGEN, A.C. [1969a]: Micropolar fluids with stretch, *Int. J. Engng. Sci.*, **7**, No. 1, 115–127.

ERINGEN, A.C. [1969b]: Mechanics of micropolar continua, in *Contributions to Mechanics*, pp. 23–40, ed. D. Abir, Pergamon Press, London.

ERINGEN, A.C. [1970a]: On a theory of general relativistic thermodynamics and viscous fluids, in *A Critical Review of Thermodynamics* (*Proceedings of Symposium on A Critical Review of the Foundations of Relativistic and Classical Thermodynamics, University of Pittsburgh, April*, 1969), pp. 483–503, eds. E.B. Stuart, B. Gal-Or, and A.J. Brainard, Mono Book Corp., Baltimore.

ERINGEN, A.C. [1970b]: Balance laws of micromorphic mechanics, *Int. J. Engng. Sci.*, **8**, No. 10, 819–828.

ERINGEN, A.C. [1970c]: *Foundations of Micropolar Thermoelasticity*, Springer-Verlag, Wien.

ERINGEN, A.C. [1971a]: Tensor analysis, in *Continuum Physics*, Vol. I, pp. 1–154, ed. A.C. Eringen, Academic Press, New York.

ERINGEN, A.C. (editor) [1971b]: *Continuum Physics*, Vol. 1, Academic Press, New York.

ERINGEN, A.C. [1971c]: Micromagnetism and superconductivity, *J. Math. Phys.*, **12**, No. 7, 1353–1358.

ERINGEN, A.C. [1972a]: Linear theory of nonlocal elasticity and dispersion of plane waves, *Int. J. Engng. Sci.*, **10**, 561–575.

ERINGEN, A.C. [1972b]: Nonlocal polar elastic continua, *Int. J. Engng. Sci.*, **10**, No. 1, 1–16.

ERINGEN, A.C. [1972c]: On nonlocal fluid mechanics, *Int. J. Engng. Sci.*, **10**, No. 6, 561–575.

ERINGEN, A.C. [1972d]: Theory of thermomicrofluids, *J. Math. Anal. Appl.*, **38**, No. 2, 480–496.

ERINGEN, A.C. [1972e]: Theory of micromorphic materials with memory, *Int. J. Engng. Sci.*, **10**, No. 7, 623–641.

ERINGEN, A.C. [1973a]: On nonlocal microfluid mechanics, *Int. J. Engng. Sci.*, **11**, No. 2, 291–306.

ERINGEN, A.C. [1973b]: Theory of nonlocal electromagnetic elastic solids, *J. Math. Phys.*, **14**, No. 6, 733–740.

ERINGEN, A.C. [1974a]: On nonlocal continuum thermodynamics, in *Modern Developments in Thermodynamics*, pp. 121–142. ed. B. Gal-Or, Wiley, New York.

ERINGEN, A.C. [1974b]: Memory-dependent nonlocal elastic solids, *Letts. Appl. Engng. Sci.*, **2**, No. 3. 145–159.

ERINGEN, A.C. [1974c]: Nonlocal elasticity and waves, in *Continuum Mechanics Aspects of Geodynamics and Rock Fracture Mechanics*, pp. 81–105, ed. P. Thoft-Christensen, Reidel, Dordrecht, Holland (*Proceedings of the NATO Advanced Study Institute Held in Iceland*, August, 1974).

ERINGEN, A.C. [1975a]: *Continuum Physics*, Vol. II, Secs. 1.1–1.4, ed. A.C. Eringen, Academic Press, New York.

ERINGEN, A.C. [1975b]: Polar and nonlocal theories of continua and applications (Twenty Lectures Given at Bogazici University, Turkey), Bogazici University Publications, 75-35/01, Spring 1975.

ERINGEN, A.C. [1976a]: Polar field theories, in *Continuum Physics*, Vol. 4, pp. 1–73, ed. A.C. Eringen, Academic Press, New York.

ERINGEN, A.C. [1976b]: Nonlocal polar field theories, in *Continuum Physics*, Vol. 4,

Part III, pp. 205–267, ed. A.C. Eringen, Academic Press, New York.
ERINGEN, A.C. [1977a]: Fundamentals of continuum field theories, in *Topics in Mathematical Physics*, Colorado University Press (papers presented at the International Symposium at Bogazici University 1975).
ERINGEN, A.C. [1977b]: Continuum mechanics at the atomic scale, in *Crystal Lattice Defects*, Vol. 7, pp. 109–130,
ERINGEN, A.C. [1978a]: Nonlocal continuum mechanics and some applications, in *Nonlinear Equations in Physics and Mathematics*, pp. 271–318, ed. A.O. Barut, Reidel, Dordrecht, Holland.
ERINGEN, A.C. [1978b]: Micropolar theory of liquid crystals, in *Liquid Crystals and Ordered Fluids*, Vol. 3, pp. 443–474, eds. J.F. Johnson and R.S. Porter, Plenum, New York.
ERINGEN, A.C. [1978c]: Line crack subject to shear, *Int. J. Fracture*, **14**, No. 4, 367–379.
ERINGEN, A.C. [1979a]: Electrodynamics of cholesteric liquid crystals, *Mol. Cryst. Liq. Cryst.*, **54**, 21–44.
ERINGEN, A.C. [1979b], Continuum theory of nematic liquid crystals subject to electromagnetic fields, *J. Math. Phys.*, **20**, 2671–2681.
ERINGEN, A.C. [1980]: *Mechanics of Continua* (2nd enlarged edition), Krieger, New York.
ERINGEN, A.C. [1983]: On differential equations of nonlocal elasticity and solutions of screw dislocation and surface waves, *J. Appl. Phys.*, **54** (9), 4703–4710.
ERINGEN, A.C. [1984a]: Nonlocal stress fields of dislocations and crack, in *Modelling Problems in Crack Tip Mechanics*, ed. J.T. Pindera, from *Proceedings of the 10th Canadian Fracture Conference*, pp. 113–130, Martinus Nijhoff, University of Waterloo, Canada.
ERINGEN, A.C. [1984b]: On continuous distributions of dislocations in nonlocal elasticity, *J. Appl. Phys.*, **56** (10), 2675–2680.
ERINGEN, A.C. [1984c]: Theory of nonlocal piezoelectricity, *J. Math. Phys.*, **25**, 717–727.
ERINGEN, A.C. [1984d]: Electrodynamics of memory-dependent nonlocal elastic continua, *J. Math. Phys.*, **25** (11), 3235–3249.
ERINGEN, A.C. [1984e]: A continuum theory of rigid suspensions, *Int. J. Engng. Sci.*, **22**, 1373–1388.
ERINGEN, A.C. [1985a]: Nonlocal continuum theory for dislocations and fracture, in *The Mechanics of Dislocations Proceedings of an International Symposium* pp. 101–110, American Society for Metals, Michigan, 1983.
ERINGEN, A.C. [1985b]: Rigid suspensions in viscous fluid, *Int. J. Engng. Sci.*, **23**, 491–495.
ERINGEN, A.C. [1987]: Theory of nonlocal elasticity and some applications, *Res. Mechanica*, **21**, 313–342.
ERINGEN, A.C. [1988]: Theory of electromagnetic elastic plates, *Int. J. Engng. Sci.* (1989) **27**, 363–375. (Reference added at proof.)
ERINGEN, A.C. and EDELEN, D.G.B. [1972]: On nonlocal elasticity, *Int. J. Engng. Sci.*, **10**, 233–248.
ERINGEN, A.C. and INGRAM, J.D. [1966]: A continuum theory of chemically reacting media—I, *Int. J. Engng. Sci.*, **3**, 197–212.
ERINGEN, A.C. and KAFADAR, C.B. [1970]: Relativistic theory of microelectromagnetism, *J. Math. Phys.*, **11**, 1984–1991.
ERINGEN, A.C. and KAFADAR, C.B. [1976]: Nonlocal polar field theories, in *Continuum*

Physics, Vol. 4, Part III, pp. 205–267, ed. A.C. Eringen Academic Press, New York.

ERINGEN, A.C. and KIM, B.S. [1974]: On the problem of crack tip in nonlocal elasticity, in *Continuum Mechanics Aspects of Geodynamics and Rock Fracture Mechanics*, pp. 107–113, ed. P. Thoft-Christensen, Reidel, Dordrecht, Holland.

ERINGEN, A.C. and KIM, B.S. [1977]: Relations between nonlocal elasticity and lattice dynamics, in *Crystal Lattice Defects*, Vol. 1, pp. 51–57,

ERINGEN, A.C., SPEZIALE, C.G., and KIM, B.S. [1977]: Crack-tip problem in nonlocal elasticity, *J. Mech. Phys. Solids*, **25**, 339–355.

ERINGEN, A.C. and ŞUHUBI, E.S. [1964]: Nonlinear theory of simple micro-elastic solids—I, *J. Engng. Sci.*, **2**, No. 2, 189–203.

ERINGEN, A.C. and ŞUHUBI, E.S. [1974]: *Elastodynamics*, Vol. I, Academic Press, New York.

ERINGEN, A.C. and ŞUHUBI, A.S. [1975]: *Elastodynamics*, Vol. II, Academic Press, New York.

ERSOY, Y. [1979]: Plane waves in electrically conducting and magnetizable viscoelastic isotropic solids subjected to a uniform magnetic field, *Int. J. Engng. Sci.*, **17**, 193–214.

ESHBACK, J.R. [1963]: Spin–Wave propagation and the magnetoelastic interactions in yttrium–iron–garnet, *J. Appl. Phys.*, Suppl. **34**, 1298–1304.

EZEKIEL, F.D. [1974]: The broad new applications of ferrolubricants, A.S.M.E., 74-DE-21 paper.

FANO, R.M., CHU, J.J., and ADLER, R.B. [1960]: *Electromagnetic Fields, Forces and Energy*, Wiley, New York.

FARNELL, G.W. [1978]: Types and properties of surface waves, in *Acoustic Surface Waves*, ed. A.A. Oliner, Vol. 24 of *Topics in Applied Physics*, pp. 13–60, Springer-Verlag, Berlin.

FATTUZO, E. and MERZ, W.J. [1967]: Ferroelectricity, in *Selected Topics in Solid State Physics*, Vol. 7, ed. E.P. Wohlfarth, Wiley, New York.

FEDOROV, F.I. [1968]: *Theory of Elastic Waves in Crystals* (translated from the Russian), Plenum, New York.

FELICI, N.J. [1969]: Phénomènes hydro et aérodynamiques dans la conduction des diélectriques fluides, *Revue Générale d'Electricité* (Paris), **78**, 717–734.

FELICI, N.J. [1972]: DC conduction in liquid dielectrics—II. Electrohydrodynamic phenomena, *Direct Current*, **2**, 147–165.

FILLIPINI, J.C., LACROIX, J.C., and TOBAZEON R. [1970]: Quelques remarques sur les phénomènes électrohydrodynamiques transitoires et stationnaires en régime d'injection unipolaire de porteurs de charges dans les diélectriques liquides, *C. R. Acad. Sci. Paris*, **271B**, 73–76.

FIZEAU, H. [1859]: Sur les hypothèses relatives à l'éther lumineux et sur une expérience qui paraît démontrer que le mouvement des corps change la vitesse à laquelle la lumière se propage dans leur intérieur, *Ann. Chimie Phys.*, **57** (3), 385–404.

FOLEN, V.G., RADO, G.T., and STOLDER, F.W. [1961]: Anisotropy of the magneto-electric effect in Cr_2O_3, *Phys. Rev. Lett.*, **6**, 607–608.

FOMÉTHE, A. and MAUGIN G.A. [1982]: Influence of dislocations on magnon–phonon couplings—A phenomenological approach, *Int. J. Engng. Sci.*, **20**, 1125–1144.

FORSBERGH, P.W. [1956]: Piezoelectricity, electrostriction, and ferroelectricity, in

Handbuch der Physik, Bd. XVII, p. 264, ed. S. Flügge, Springer-Verlag, Berlin, Heidelberg, New York.

FOSTER, N.F. [1981]: Piezoelectricity in thin film materials, *J. Acoust. Soc. Amer.*, **70**, 1609–1614.

FOWLKES, C.W. [1969]: Photoviscoelastic model testing, NASA CR-1289, National Aeronautics and Space Administration, U.S.A.

FRIEDMAN, N. and KATZ, M. [1966]: A representation theorem for additive functionals, *Arch. Rat. Mech. Anal.*, **21**, 49–57.

FRIEDRICHS, K.O. [1974]: On the laws of relativistic electro-magnetofluid dynamics, *Commun. Pure Appl. Math.*, **27**, 749–808.

FRIEDRICHS, K.O. and KRANZER, H. [1958]: Notes on magnetohydrodynamics, VIII. Nonlinear wave propagation, N.Y.U. Institute of Mathematical Science Report, NYO-6486, New York University, New York.

FRÖHLICH, H. [1958]: *Theory of Dielectrics*, Oxford University Press, London.

GALEEV, A.A. and SUDAN, R.N. [1983]: *Plasma Physics*, I, North-Holland, Amsterdam.

GALEEV, A.A., and SUDAN, R.N. [1984]: *Plasma Physics*, II, North-Holland, Amsterdam.

GANGULY, A.K., DAVIS, K.L., and WEBB, D.C. [1978]: Magnetoelastic surface waves on the (110) plane of highly magnetostrictive cubic crystals, *J. Appl. Phys.*, **49**, 759–767.

GERMAIN, P. [1959]: Contribution à l'étude des ondes de choc en magnétodynamique des fluides, Publ. ONERA, no. 97, Office National d'Etude et de Recherches Aéronautiques, Paris.

GERMAIN, P. [1960]: Shock waves and shock wave structure in magneto fluid dynamics, *Rev. Mod. Phys.*, **32**, 951–958.

GERMAIN, P. [1972]: Shock waves, jump relations and structure, in *Advances in Applied Mechanics*, Vol. 12, pp. 131–194, ed. C.S. Yih, Academic Press, New York.

GERMAIN, P. [1973]: La méthode des puissances virtuelles en mécanique des milieux continus—I, *J. Mécanique*, **12**, 235–274.

GERSDORFF, R. [1960]: Uniform and non-uniform effect in magnetostriction, *Physica*, **26**, 553–574.

GILBERT, T.L. [1955]: in *Proceedings of the Pittsburgh Conference on Magnetism and Magnetic Materials*, AIEE Publ. no. T78, p. 253, AIEE, New York.

GILBERT, T.L. [1956]: A phenomenological theory of ferromagnetism, Ph. D. Thesis, Illinois Institute of Technology, Chicago.

GOLDSTEIN, H. [1950]: *Classical Mechanics*, Addison-Wesley, Reading, Mass.

GOODRICH, G.W. and LANGE, J.N., [1971] Longitudinal and shear magnetoelastic behavior of metals, *J. Acoust. Soc. Amer.*, **50**, 869–874.

GOUDJO, C. and MAUGIN, G.A. [1983]: On the static and dynamic stability of soft-ferromagnetic elastic plates, *J. Méc. Théor. Appl.*, **2**, 947–975.

GREEN, A.E. and NAGHDI, P.M. [1983]: On electromagnetic effects in the theory of shells and plates, *Phil. Trans. Roy. Soc. London*, **A309**, 559–610.

GREEN, A.E. and ZERNA, W. [1954]: *Theoretical Elasticity*, Oxford University Press, London.

DE GROOT, S.R. and MAZUR, P. [1962]: *Non-Equilibrium Thermodynamics*, North-Holland, Amsterdam.

DE GROOT, S.R. and SUTTORP, L.G. [1972]: *Foundations of Electrodynamics*, North-Holland, Amsterdam.

GROSSMAN, W., HAMEIRI, E., and WEITZNER, H. [1983]: Magnetohydrodynamic and double adiabatic stability of compact toroid plasmas, *Phys. Fluids*, **26**, 508–519.

GROT, R.A. [1976]: Relativistic continuum physics: Electromagnetic interactions in *Continuum Physics*, Vol. 3, pp. 129–219, ed. A.C. Eringen, Academic Press, New York.

GROT, R.A. and ERINGEN, A.C. [1966a]: Relativistic continuum mechanics—I. Mechanics and thermodynamics, *Int. J. Engng. Sci.*, **4**, 611–638.

GROT, R.A. and ERINGEN, A.C. [1966b]: Relativistic continuum mechanics—II. Electromagnetic interactions with matter, *Int. J. Engng. Sci.*, **4**, 639–670.

GROT, R.A. and ERINGEN, A.C. [1966c]: Continuum theory of nonlinear viscoelasticity, in *Mechanics and Chemistry of Solid Propellants*, pp. 157–201, eds. A.C. Eringen, H. Liebowitg, S. Koh, and J. Crowley, Pergamon Press, London.

GUREVICH, A.G. [1973]: *Magnetic Resonance in Ferrites and Antiferromagnets* (in Russian), Nauka, Moscow.

HACKETT, R.M. and KROKOSKY, E.M. [1968]: A photoviscoelastic analysis of time-dependent stresses in polyphase system, *Exp. Mech.*, **8**, 537–547.

HAJDO, L. and ERINGEN, A.C. [1979a]: Theory of light reflection by cholesteric liquid crystals possessing a pitch gradient, *J. Opt. Soc. Amer.*, **69**, No. 7, 1017–1023.

HAJDO, L. and ERINGEN, A.C. [1979b]: Theory of light reflection by cholesteric liquid crystals possessing a tilted structure, *J. Opt. Soc. Amer.*, **69**, No. 11, 1509–1513.

HAJDO, L. and ERINGEN, A.C. [1979c]: Application of nonlocal theory to electromagnetic dispersion, *Lett. Appl. Engng. Sci.*, **17**, 785–791.

HALL, W.F. and BUSENBERG, S.N. [1969]: Viscosity of magnetic suspensions, *J. Chem. Phys.*, **51**, 137–144.

HAMEIRI, E. [1983]: The equilibrium and stability of rotating plasmas, *Phys. Fluids*, **26**, 230–237.

HARTMANN, J. [1937]: Hydrodynamics—I. Theory of the laminar flow of an electrically conductive liquid in a homogeneous magnetic field, *Kgl. Danske Viden. Selskab. Math. Fys. Med.*, **15**, no. 6.

HASEGAWA, A. and BRINKMAN, W.F. [1980]: Tunable coherent IR and FIR sources utilizing modulational instability, *IEEE J. Quantum Electronics*, **QE-16**, 694–697.

HASEGAWA, A. and KODAMA, Y. [1981]: Signal transmission by optical solitons in monomode fiber, *Proc. IEEE*, 69, 1145–1150.

HASEGAWA, A. and KODAMA, Y. [1982]: Amplification and reshaping of optical solitons in a glass fiber I, *Optics Letters*, **7**, 285–287.

HASEGAWA, A. and TAPPERT, F. [1973]: Transmission of stationary optical pulses in dispersive dielectric fibers, Parts 1 and 2, *Appl. Phys. Lett.*, **23**, 142–144, 146–149.

HEISENBERG, W. [1928]: Zur theorie des ferromagnetismus, *Zeit. Physik*, **49**, 619–636.

HELLIWELL, R.A. [1965]: *Whistlers and Related Ionospheric Phenomena*, Stanford University Press, Stanford, CA.

HILLIER, M.J. and LAL, G.K. [1968]: The electrodynamics of electromagnetic forming, *Int. J. Mech. Sci.*, **10**, 491–500.

HUGHES, W.F. and YOUNG, F.J. [1966]: *The Electromagnetodynamics of Fluids*, Wiley, New York.

HUSTON, A.R. and WHITE, D.L. [1962]: Elastic wave propagation in piezoelectric semiconductors, *J. Appl. Phys.*, **33**, 40–47.

HUTTER, K. [1975]: Wave propagation and attenuation in paramagnetic and soft-ferromagnetic materials, *Int. J. Engng. Sci.*, **14**, 883–894.

HUTTER, K. and VAN DE VEN, A.A.F. [1978]: *Field Matter Interactions in Thermoelastic Solids*, Lecture Notes in Physics, Springer-Verlag, Berlin, Heiderberg, New York.

Institute of Radio Engineers (The) [1949]: Standards on piezoelectric crystals, *Proc. Inst. Radio Engineers*, **37**, 1378–1395.

Institute of Radio Engineers (The) [1958]; Standards on piezoelectric materials, *Proc. Inst. Radio Engineers* **46**, 764–778.

IRVING, J.H. and KIRKWOOD, J.G. [1950]: The statistical mechanical theory of transport processes IV, *J. Chem. Phys.*, **18**, 817–829.

ISRAEL, W. and STEWART, J.M. [1980]: Progress in relativistic thermodynamics and electrodynamics of continuous media, in *General Relativity and Gravitation*, Vol. 2, pp. 491–525, ed. A. Held, Plenum, New York.

JACKSON, J.D. [1975]: *Classical Electrodynamics*, 2nd edition, Wiley, New York.

JEFFREY, A. [1966]: *Magnetohydrodynamics*, Oliver and Boyd, Edinburgh.

JEFFREY, A. and TANIUTI, T. [1964]: *Nonlinear Wave Propagation*, Academic Press, New York.

JEFFREYS, H. and JEFFREYS, B.S. [1950]: *Methods of Mathematical Physics*, 2nd edition, Cambridge University Press, London.

JENKINS, J.T. [1971]: Some simple flows of a paramagnetic fluid, *J. Physique*, **32**, 931–938.

JENKINS, J.T. [1972]: A theory of magnetic fluids, *Arch. Rat. Mech. Anal.*, **46**, 42–60.

JENKINS, J.T. [1975]: Steady jets of a magnetic fluid, in *Recent Advances in Engineering Science*, Vol. 6, pp. 373–379, Scientific Publishers, Boston.

JESSOP, H.T. [1958]: Photoelasticity, in *Handbuch der Physik*, Bd. VI, ed. S. Flügge, Springer-Verlag, Berlin, Heidelberg, New York.

JOFFRE, G., PRUNET-FOCH, B., BERTHOMME, S., and CLOUPEAU, M. [1980]: Deformation of liquid menisci under the action of an electric field, *J. Electrostatics*, **13**, 151–165.

JOHNSON, C.W. and GOLDSMITH, W. [1969]: Optical and mechanical properties of birefringent polymers, *Exp. Mech.*, **9**, 263–268.

JOHNSON, T.A., FOWLKES, C.W., and DILL, E.H. [1968]: An experiment on creep at varying temperature, in *Proc. Fifth Intern. Cong. Rheology*, Vol. 3, pp. 349–355. University Park Press, Maryland.

JONA, F. and SHIRANE, G. [1962]: *Ferroelectric Crystals*, Pergamon Press, New York.

JORDAN, N.F. and ERINGEN, A.C. [1964a]: On the static nonlinear theory of electromagnetic thermoelastic solids—I, *Int. J. Engng. Sci.*, **2**, No. 1, 59–95.

JORDAN, N.F. and ERINGEN, A.C. [1964b]: On the static nonlinear theory of electromagnetic thermoelastic solids—II, *Int. J. Engng. Sci.*, **2**, No. 1, 97–114.

Journal of Magnetism and Magnetic Materials [1983]: Magnetic Fluid Bibliography (Literature and Patents), *J. Magnetism and Magnetic Materials*, **39**, 119–220.

KAFADAR, C.B. [1971]: The theory of multipoles in classical electromagnetism, *Int. J. Engng. Sci.*, **9**, 831–853.

KAFADAR, C.B. and ERINGEN, A.C. [1971a]: Micropolar media—I: The classical theory, *Int. J. Engng. Sci.*, **9**, No. 3, 271–305.

KAFADAR, C.B. and ERINGEN, A.C. [1971b]: Micropolar theory—II: The relativistic theory, *Int. J. Engng. Sci.*, **9**, 271–305.

KALISKI, S. [1969a]: Equations of a combined electromagnetic, elastic and spin field and coupled drift-type amplification effects—I: General equations, *Proc. Vibr. Problems*, **10**, 123–131.

KALISKI, S. [1969b]: Equations of a combined electromagnetic, elastic and spin field and coupled drift-type amplification effects—II: Drift-type Amplifiers, *Proc. Vibr. Problems*, **10**, 133–146.

KALISKY, S. and KAPALEWSKI, J. [1968]: Surface waves of the spin–elastic type in a discrete body of cubic structure, *Proc. Vibr. Problems*, **9**, 269–278.

KALISKI, S. and NOWACKI, W. [1962a, b]: Excitation of mechanical-electromagnetic waves induced by a thermal shock, I, II, *Bull. Acad. Pol. Sci. Ser. Sci. Techn.*, **10**, 25–34.

KAMBERSKY, V. and PATTON, C.E. [1975]: Spin–wave relaxation and phenomelogical damping in ferromagnetic resonance, *Phys. Rev.*, **B11**, 2668–2672.

KARPMAN, V. I. [1975]: *Nonlinear Waves in Dispersive Media* Pergamon Press, London.

KARPMAN, V. I. and KRUSHKAL, E.M. [1969]: Modulaled waves in nonlinear dispersive media *Soviet Phys. JETP*, **28**, 277–281.

KATAYEV, I.G. [1966]: *Electromagnetic Shock Waves* (translation from the Russian), Iliffe Books, London.

KAYE, G.W.C. and LABY, T.H. [1973]: *Tables of Physical and Chemical Constants*, 14th edition, Longman, London.

KAZAKIA, J.Y. and VENKATARAMAN, R. [1975]: Propagation of electromagnetic waves in a nonlinear dielectric slab, *Zeit. angew. Math. Phys.*, **26**, 61–76.

KEILIS-BOROK, V.I. and MUNIN, A.S. [1959]: Magnetoelastic waves and the boundary of the earth's core (in Russian), *Isvd. Geophys. Ser.*, 1529–1541.

KELLOG, O.D. [1929]: *Foundations of Potential Theory*, Springer-Verlag, Berlin.

KEOWN, R. [1975]: *An Introduction to Group Representation Theory*, Academic Press, New York.

KIKUCHI, H. and HIROTA, M. [1985]: Nonlinear electromagnetics in terms of quasi-particles and solitons and its application to nonlinear dispersive and dissipative media, in *Nonlinear and Environmental Electromagnetics*, ed. H. Kikuchi, Elsevier Science, Amsterdam.

KIRAL, E. [1972]: Symmetry restriction on the constitutive relations for anisotropic materials—Polynomial integrity bases for cubic crystals system, Habilitation Thesis, M.E.T.U., Ankara, Turkey.

KIRAL, E. and ERINGEN, A.C. [1976]: Nonlinear constitutive equations of magnetic crystals, Princeton University Report, Department of Civil and Geological Engineering, Princeton, NJ. Scheduled for publication by Springer-Verlag.

KIRAL, E. and SMITH, G.F. [1974]: On the constitutive relations for anisotropic materials—Triclinic, monoclinic, rhombic, tetragonal, and hexagonal crystal systems, *Int. J. Engng. Sci.*, **12**, 471–490.

KIRIUSHIN, V.V. [1983]: Mathematical model of structure phenomena in magnetic fluids, *J. Magnetism and Magnetic Materials*, **39**, 14–16.

KIRIUSHIN, V.V. NALETOVA, V.A., and CHEKANOV, V.V. [1978]: The motion of magnetizable fluid in a rotating homogeneous magnetic field, *P. M. M. J. Appl. Math. Mech.* (English translation), **42**, 710–715.

KITTEL, C. [1958a]: Interactions of spin waves and ultrasonic waves in ferromagnetic crystals, *Phys. Rev.*, **110**, 836–841.

KITTEL, C. [1958b]: Excitation of spin waves in a ferromagnetic by a uniform rf field, *Phys. Rev.*, **110**, 1295–1297.

KITTEL, C. [1971]: *Introduction to Solid State Physics*, 2nd edition, Wiley, New York.

KLEIN, M.V. [1970]: *Optics*, Wiley, New York.

KNOPOFF, L. [1955]: The interaction between elastic wave motion and a magnetic field in electrical conductors, *J. Geophys. Res.*, **73** 6527–6533.

KNOWLES, J.K. [1960]: Large amplitude oscillations of a tube of incompressible elastic material, *Quart. Appl. Math.*, **18**, 71–77.

KNOWLES, J.K. [1962]: On a class of oscillations in the finite deformation theory of elasticity, *J. Appl. Mech.*, **29**, 283–286.

KODAMA, Y. and HASEGAWA, A. [1982]: Amplification and reshaping of optical solitons in glass fiber—II, *Optics Letters*, **7**, 339–341.

KOSILOVA, V.G. KUNIN, I.A., and SOSNINA, E.G. [1968]: Interaction of point defects with allowance for spatial dispersion, *Fiz. Tverd. Tela.*, **10**, 367–374.

KÖSZEGI, L. and KRONMULLER, H. [1984]: Magnetic hysteresis loops for several amorphous alloys after various heat treatments below the curie point, *Appl. Phys.*, **A34**, 95–103.

KOTOWSKI, R. [1979]: On the Brillouin delta function of cubic and hexagonal lattices, *Z. Phys.* **B**, **33**, 321–330.

KOZHENKOV, V.I. and FUKS, N.A. [1976]: Electrohydrodynamic atomization of liquids, *Russian Chem. Rev.*, **45**, 1179.

KRANYŠ, M. [1980]: Relativistic electrodynamics of dissipative elastic media, *Can. J. Phys.*, **58**, 666–682.

KRÖNER, E. [1967]: Elasticity theory of materials with long-range cohesive forces, *Int. J. Structures*, **3**, 731–742.

KRÖNER, E. (editor) [1967]: Mechanics of generalized continua, *Proc. IUTAM Symposium*, Springer-Verlag, Berlin, Heidelberg, New York.

KUBO, R. and NAGAMIYA, T. [1969]: *Solid State Physics*, McGraw-Hill, New York.

KUNIN, I.A. [1967]: Inhomogeneous elastic medium with nonlocal interaction, *J. Appl. Mech. Tech. Phys.*, **8**, 41–44.

KUNIN, I.A. [1982, 1983]: *Elastic Media with Microstructure*, Vols. I and II, Springer-Verlag, Berlin, Heidelberg, New York.

KUSKE, A. and ROBERTSON, G. [1974]: *Photoelastic Stress Analysis*, Wiley, New York.

LACROIX, J.C. ATTEN, P., and HOPFINGER, E.J. [1975]: Electro-convection in a dielectric liquid layer subjected to unipolar induction, *J. Fluid Mech.*, **69**, 539–563.

LADIKOV, Ya. P. [1961]: Some exact solutions of the equations of non-steady motion in magneto-hydrodynamics, *Soviet Phys. Dokl.*, **6**, 198–201.

LAMB, J., G.L. [1980]: *Elements of Soliton Theory*, Wiley, New York.

LANDAU, L.D. and LIFSHITZ, E.M. [1935]: On the theory of the dispersion of magnetic permeability in ferromagnetic bodies, *Phys. Z. Sowjet*, **8**, 153.

LANDAU, L.D. and LIFSHITZ, E.M. [1960]: *Electrodynamics of Continuous Media* (translation from the Russian), Pergamon Press, Oxford.

LANDOLT-BÖRNSTEIN, [1959]: *Numerical Values and Functions*, Vol. II, 6th edition, pp. 414–448, Springer-Verlag, Berlin.

LAWSON, M.O. and DECAIRE, J.A. [1967]: Investigation on power generation using electrofluid-dynamic processes, in *Proc. Intersociety Energy Conversion Engineering Conference*, Miami Beach, Florida (August 13–17, 1967).

LAX, M. and NELSON, D.F. [1971]: Linear and nonlinear electrodynamics in elastic anisotropic dielectrics, *Phys. Rev.*, **B4**, 3694–3731.

LEE, E.W. [1955]: Magnetostriction and magnetomechanical effects, *Rep. Progr. in Physics*, **18**, 184–229.

LIANIS, G. [1973a]: The general form of constitutive equations in continuum relativistic physics, *Nuovo Cimento*, **14B**, 57–105.

LIANIS, G. [1973b]: Formulation and application of relativistic constitutive equations for deformable electromagnetic materials, *Nuovo Cimento*, **16B**, 1–43.

LIANIS, G. [1974]: Relativistic thermodynamics of viscoelastic dielectrics, *Arch. Rat. Mech. Anal.*, **55**, 300–331.

LIANIS, G. and RIVLIN, R.S. [1972]: Relativistic equations of balance in continuum mechanics, *Arch. Rat. Mech. Anal.*, **48**, 64–82.

LIANIS, G. and WHICKER, D. [1975]: Electromagnetic phenomena in rotating media, *Arch. Rat. Mech. Anal*, **57**, 325–362.

LIBRESCU, L. [1977]: Recent contributions concerning the flutter problem of elastic thin bodies in an electrically conducting gas flow, a magnetic field being present, *SM Archives*, Vol. 2, pp. 1–108, Noordhoff, Leyden.

LICHNEROWICZ, A. [1967]: *Relativistic Hydrodynamics and Magnetohydrodynamics*, Benjamin, New York.

LICHNEROWICZ, A. [1971]: Ondes de choc, ondes infinitésimales et rayons en hydrodynamique et magnétohydrodynamique relativistes, in *Relativistic Fluid Dynamics*, pp. 87–203, ed. C. Cattaneo, Cremonese, Rome.

LICHNEROWICZ, A. [1976]: Shock waves in relativistic magnetohydrodynamics under general assumptions, *J. Mat. Phys.*, **17**, 2135–2142.

LIELAUSIS, O. [1975]: Liquid metal magnetohydrodynamics, *Atomic Energy Review*, **13**, no. 3.

LINES, M.E. [1979]: Elastic properties of magnetic materials, *Phys. Rep.*, **55**, 133–181.

LIPSON, S.G. and LIPSON, H. [1970]: *Optical Physics*, Cambridge University Press, London.

LIU, I.S. and MULLER, I. [1972]: On the thermodynamics and thermostatics of fluids in electromagnetic fields, *Arch. Rat. Mech. Anal.*, **46**, 149–179.

LIVENS, G.H. [1962]: *The Theory of Electricity*, 2nd edition, Cambridge University Press, London.

LOMONT, J.S. [1959]: *Applications of Finite Groups*, Academic Press, New York.

LORENTZ, H.A. [1952]: *The Theory of Electrons*, 2nd edition, Dover, New York.

LORENTZ, H.A., EINSTEIN, A., WEYL, H. and MINKOWSKI, H. [1923]: *The Principle of Relativity* (Collection of Reprints), Dover, New York.

LUIKOV, A.V. and BERKOVSKY, B. [1974]: *Convective and Thermal Waves* (in Russian), Energya, Moscow.

MCCARTHY, M.F. [1965]: Propagation of plane acceleration discontinuities in hyperelastic dielectrics, *Int. J. Engng. Sci.*, **4**, 361–381.

MCCARTHY, M.F. [1966a]: The propagation and growth of plane acceleration waves in a perfectly electrically conducting elastic material in a magnetic field, *Int. J. Engng. Sci.*, **4**, 361–381.

MCCARTHY, M.F. [1966b]: The growth of magnetoelastic waves in a Cauchy elastic

material of finite electrical conductivity, *Arch. Rat. Mech. Anal.*, **23**, 191–217.
McCarthy, M.F. [1967]: wave propagation in nonlinear magneto-thermoelasticity. Propagation of acceleration waves, *Proc. Vibr. Problems*, **8**, 337–348.
McCarthy, M.F. [1968]: Wave propagation in nonlinear Magneto-thermoelasticity. On the variation of the amplitude of acceleration waves. *Proc. Vibr. Problems*, **9**, 367–381.
McCarthy, M.F. [1971]: Thermodynamics of electromagnetic materials with memory, *Arch. Rat. Mech. Anal.*, **41**, 333–353.
McCarthy, M.F. [1974]: Thermodynamics of deformable magnetic materials with memory, *Int. J. Engng. Sci.*, **12**, 45–60.
McCarthy, M.F. and Green, W.A. [1966]: The growth of plane acceleration discontinuities propagating into a homogeneously deformed hyperelastic dielectric material in the presence of a magnetic field, *Int. J. Engng. Sci.*, **4**, 403–422.
McCarthy, M.F. and Tiersten, H.F. [1977]: Shock waves and acoustoelectric domains in piezoelectric semiconductors, *J. Appl. Phys.*, **48**, 159.
Magnus, W. and Oberhettinger, F. [1949]: *Formulas and Theorems for the Functions of Mathematical Physics*, Chelsea, New York.
Malik, S.K. and Singh, M. [1983]: Nonlinear instability in superposed magnetic fluids, *J. Magnetism and Magnetic Materials*, **39**, 123–126.
Martinet, A. [1974]: Biréfringence et dichroïsme linéaire des ferrofluides sous champ magnétique, *Rheol. Acta*, **13**, 260–264.
Martinet, A. [1978]: Experimental evidences of static and dynamic anisotropies of magnetic colloids, in *Thermomechanics of Magnetic Fluids*, pp. 97–114, ed. B. Berkovsky, Hemisphere, Washington.
Mason, W.P. [1950]: *Piezoelectric Crystals and Their Application to Ultrasonics*, Van Nostrand, New York.
Mason, W.P. [1966]: *Crystal Physics and Interaction Processes*, Academic Press, New York.
Mason, W.P. [1981]: Piezoelectricity, its history and applications, *J. Acoust. Soc. Amer.*, **70**, 1561–1566.
Masson, M. and Weaver, W. [1929]: *The Electromagnetic Field*, Dover, New York.
Matthews, H. and LeCraw, R.G. [1962]: Acoustic Faraday rotation by magnon–phonon interaction, *Phys. Rev. Lett.*, **8**, 397–399.
Maugin, G.A. [1971a]: Magnetized deformable media in general relativity, *Ann. Inst. Henri Poincaré*, **A15**, 275–302.
Maugin, G.A. [1971b]: Micromagnetism and polar media, Ph.D. Thesis, Princeton University, Dept. of AMS, Princeton, NJ.
Maugin, G.A. [1972a]: Remarks on dissipative processes in the continuum theory of micromagnetics, *J. Phys.*, **A5**, 1550–1562.
Maugin, G.A. [1972b]: An action principle in general relativistic magnetohydrodynamics, *Ann. Inst. Henri Poincaré*, **A16**, 133–169.
Maugin, G.A. [1972c]: Relativistic theory of magnetoelastic interactions, I, *J. Phys.*, **A5**, 786–802.
Maugin, G.A. [1973a]: Relativistic theory of magnetoelastic interactions, II: Constitutive theory, *J. Phys.*, **A6**, 306–321.
Maugin, G.A. [1973b]: Relativistic theory of magnetoelastic interactions, III: Isotropic media, *J. Phys.*, **A6**, 1647–1666.
Maugin, G.A. [1973c]: Harmonic oscillations of elastic continua and detection of gravitational waves, *General Relativity Gravitat. J.*, **4**, 241–272.

MAUGIN, G.A. [1974a]: Sur la dynamique des milieux magnétisés avec spin magnétique, *J. Mécanique*, **13**, 75–96.
MAUGIN, G.A. [1974b]: Quasi-electrostatics of electrically polarized continua, *Lett. Appl. Engng. Sci.*, **2**, 293–306.
MAUGIN, G.A. [1974c]: Sur les fluides relativistes à spin, *Ann. Inst. Henri Poincaré*, **A20**, 41–68.
MAUGIN, G.A. [1974d]: Relativistic theory of magnetoelastic interactions, IV: Hereditary processes, *J. Phys.*, **A7**, 818–837.
MAUGIN, G.A. [1975]: On the spin relaxation in deformable ferromagnets, *Physica*, **81A**, 454–468.
MAUGIN, G.A. [1976a]: Micromagnetism, in *Continuum Physics*, Vol. III, pp. 213–312, ed. A.C. Eringen, Academic Press, New York.
MAUGIN, G.A. [1976b]: A continuum theory of deformable ferrimagnetic Bodies—I: Field equations, *J. Math. Phys.*, **17** 1727–1738.
MAUGIN, G.A. [1976c]: A continuum theory of deformable ferrimagnetic bodies—II: Thermodynamics, constitutive theory, *J. Math. Phys.*, **17**, 1739–1751.
MAUGIN, G.A. [1976d]: On the foundations of the electrodynamics of deformable media with interactions, *Lett. Appl. Engng. Sci.*, **4**, 3–17.
MAUGIN, G.A. [1977]: Deformable dielectrics—II, III, *Arch. Mech. Stosow.*, **29**, 143–159, 251–258.
MAUGIN, G.A. [1978a]: On the covariant formulation of Maxwell's equations in matter, *J. Franklin Inst.*, **305**, 11–26.
MAUGIN, G.A. [1978b]: Exact relativistic theory of wave propagation in prestressed elastic solids, *Ann. Inst. Henri Poincaré*, **A28**, 155–185.
MAUGIN, G.A. [1978c]: Relation between wave speeds in the crust of dense magnetic stars, *Proc. Roy. Soc. London* **A364**, 537–552.
MAUGIN, G.A. [1978d]: Sur les invariants des chocs dans les milieux continus relativistes magnétiques, *C. R. Acad. Sci. Paris*, **287A**, 171–174.
MAUGIN, G.A. [1978e]: On the covariant equations of the relativistic electrodynamics of continua—I: General equations, *J. Math. Phys.*, **19**, 1198–1205.
MAUGIN, G.A. [1978f]: On the covariant equations of the relativistic electrodynamics of continua—II: Fluids, *J. Math. Phys.*, **19**, 1206–1211.
MAUGIN, G.A. [1978g]: On the covariant equations of the relativistic electrodynamics of continua—III: Elastic solids, *J. Math. Phys.*, **19**, 1212–1219.
MAUGIN, G.A. [1978h]: On the covariant equations of the relativistic electrodynamics of continua—IV: Media with spin, *J. Math. Phys.*, **19**, 1220–1226.
MAUGIN, G.A. [1978i]: A phenomenological theory of ferroliquids, *Int. J. Engng. Sci.*, **16**, 1029–1044.
MAUGIN, G.A. [1979a]: A continuum approach to magnon–phonon couplings—I: General equations, background solution, *Int. J. Engng. Sci.*, **17**, 1073–1091.
MAUGIN, G.A. [1979b]: A continuum approach to magnon–phonon couplings—II: Wave propagation for hexagonal symmetry, *Int. J. Engng. Sci.*, **17**, 1093–1108.
MAUGIN, G.A. [1979c]: Classical magnetoelasticity in ferromagnets with defects, in *Electromagnetic Interactions in Elastic Solids*, pp. 243–324, ed. H. Parkus, Springer-Verlag, Wien.
MAUGIN, G.A. [1980a]: The method of virtual power in continuum mechanics; Application to coupled fields, *Acta Mechanica*, **35**, 1–70.
MAUGIN, G.A. [1980b]: Elastic-electromagnetic resonance couplings in electromagne-

tically ordered media, in *Theoretical and Applied Mechanics*, pp. 345–355, eds. F.P.J. Rimrott and B. Tabarrok, North-Holland, Amsterdam.
MAUGIN, G.A. [1980c]: Further comments on the equivalence of Abraham's, Minkowski's, and others' electrodynamics, *Can. J. Phys.*, **58**, 1163–1170.
MAUGIN, G.A. [1981a]: Wave motion in magnetizable deformable solids, *Int. J. Engng. Sci.*, **19**, 321–388.
MAUGIN, G.A. [1981b]: Ray theory and shock formation in relativistic elastic solids, *Phil. Trans. Roy. Soc. London*, **302**, 189–215.
MAUGIN, G.A. [1981c]: Dynamic magnetoelectric couplings in ferroelectric ferromagnets, *Phys. Rev.*, **B23**, 4608–4614.
MAUGIN, G.A. [1982]: Quadratic dissipative effects in ferromagnets, *Int. J. Engng. Sci.*, **20**, 295–302.
MAUGIN, G.A. [1983]: Surface elastic waves with transverse horizontal polarization, in *Advances in Applied Mechanics*, Vol. 23, pp. 373–434, ed. J.W. Hutchinson, Academic Press, New York.
MAUGIN, G.A. [1984a]: Symmetry breaking and dynamical electromagnetic–elastic couplings, in *The Mechanical Behavior of Electromagnetic Solid Continua*, pp. 35–46, ed. G.A. Maugin, North-Holland, Amsterdam.
MAUGIN, G.A. [1984b]: Attenuation of coupled waves in antiferromagnetic elastic conductors in weak magnetic fields, *Int. J. Engng. Sci.*, **22**, 1269–1290.
MAUGIN, G.A. [1985]: *Nonlinear Electromechanical Effects and Applications*, World Scientific, Singapore, New York.
MAUGIN, G.A. [1988]: *Continuum Mechanics of Electromagnetic Solids*, North-Holland, Amsterdam (Reference added at Proof).
MAUGIN, G.A. and COLLET, B. [1974]: Thermodynamique des milieux continus électromagnétiques avec interactions, *C. R. Acad. Sci. Paris*, **279**B, 379–382
MAUGIN, G.A. and DAHER, N. [1986]: Phenomenological theory of elastic semiconductors, *Int. J. Engng. Sci.*, **24**, 703–732.
MAUGIN, G.A. and DROUOT, R. [1983]: Thermomagnetic behavior of magnetically nonsaturated fluids, *J. Magnetism and Magnetic Materials*, **39**, 7–10.
MAUGIN, G.A. and ERINGEN, A.C. [1972a]: Deformable magnetically saturated media—I: Field equations, *J. Math. Phys.*, **13**, 143–155.
MAUGIN, G.A. and ERINGEN, A.C. [1972b]: Deformable magnetically saturated media—II: Constitutive theory, *J. Math. Phys.*, **13**, 1334–1347.
MAUGIN, G.A. and ERINGEN, A.C. [1972c]: Polarized elastic materials with electronic spin—A relativistic approach, *J. Math. Phys.*, **13**, 1777–1788.
MAUGIN, G.A. and ERINGEN, A.C. [1974]: Variational formulation of the relativistic theory of microelectromagnetism, *J. Math. Phys.*, **15**, 1494–1499.
MAUGIN, G.A. and ERINGEN, A.C. [1977]: On the equations of the electrodynamics of deformable bodies of finite extent, *J. Mécanique*, **16**, 101–147.
MAUGIN, G.A. and FOMÉTHE, A. [1982]: On the viscoplasticity of ferromagnetic crystals, *Int. J. Engng. Sci.*, **20**, 885–908.
MAUGIN, G.A. and GOUDJO, C. [1982]: The equations of soft-ferromagnetic elastic plates, *Int. J. Solids and Structures*, **18**, 889–912.
MAUGIN, G.A. and HAKMI, A. [1984]; Magnetoacoustic wave propagation in paramagnetic insulators exhibiting induced linear magnetoelastic couplings, *J. Acoust. Soc. Amer.*, **76**, 826–840.
MAUGIN, G.A. and HAKMI, A. [1985]: Magnetoelastic surface waves in elastic ferromagnets—I: Orthogonal setting of the bias field, *J. Acoust. Soc. Amer.* **77**, 1010–1026.

MAUGIN, G.A. and POUGET, J. [1980]: Electroacoustic equations in elastic ferroelectrics, *J. Acoust. Soc. Amer.*, **68**, 575–587.

MAUGIN, G.A. and POUGET, J. [1981]: A continuum approach to magnon–phonon couplings—III: Numerical results, *Int. J. Engng. Sci.*, **19**, 479–493.

MAUGIN, G.A. and SIOKÉ-RAINALDY, J. [1983]: Magnetoacoustic resonance in antiferromagnetic insulators in weak magnetic fields, *J. Appl. Phys.*, **54**, 1507–1518.

MAUGIN, G.A. and SIOKÉ-RAINALDY, J. [1985]: Magnetoacoustic resonance in antiferromagnetic insulators in "moderate" and strong magnetic fields, *J. Appl. Phys.*, **57**, 2131–2141.

MEDVEDEV, V.F. and KRAKOV, M.S. [1983]: Flow separation by means of magnetic fluid, *J. Magnetism and Magnetic Materials*, **39**, 119–122.

MEGAW, H.D. [1957]: *Ferroelectricity in Crystals*, Methuen, London.

MELCHER, J.R. [1963]: *Field Coupled Surface Waves*, M.I.T. Press, Cambridge, MA.

MELCHER, J.R. [1981]: *Continuum Electromechanics*, M.I.T. Press, Cambridge, MA.

MELCHER, J.R. and TAYLOR, G.I. [1969]: Electrohydrodynamics: A review of interfacial shear stresses, in *Annual Review of Fluid Mechanics*, pp. 111–146, eds. W.R. Sears and M. Van Dyke, Annual Reviews, Palo Alto, CA.

MERT, M. [1975]: Symmetry restrictions on linear and nonlinear constitutive equations for anisotropic materials—Classical and magnetic crystals classes, Ph. D. Thesis, M.E.T.U., Ankara, Turkey.

MICHELSON, A.A. and MORLEY, E.W. [1886]: Influence of motion of the medium on the velocity of light, *Amer. J. Sci.*, **31** (3), 377.

MIELNICKI, J. [1968]: Interaction of spin waves with longitudinal and transverse lattice vibrations, *Electron Technology*, **1**, 45–60.

MIELNICKI, J. [1969]: The investigation of elastic anisotropy in YIG by means of magnetoelastic interactions, *I.E.E.E. Trans.*, **SU-16**, 3, 144–146.

MIELNICKI, J. [1977]: Spin and magnetoelastic wave generation in anisotropic crystals (in Polish), *Prace Inst. Fiz. P.A.N.*, no. 63, 114 pp., Warsaw, Poland.

MINDLIN, R.D. [1968]: Polarization gradients in elastic dielectrics, *Int. J. Solids and Structures*, **4**, 637–642.

MINDLIN, R.D. [1972]: Elasticity, piezoelectricity and crystal lattice dynamics, *J. Elasticity*, **2**, 217–282.

MINKOWSKI, H. [1908]: Die Grundgleichungen für die elektromagnetischen Vorgänge in bevegten Körpern, *Gottinger Nachrichten*, 53–111.

MISNER, C.W., THORNE, K.S., and WHEELER, J.A. [1973]: *Gravitation*, Freeman, San Francisco.

MIYA, K., HARA, K., and SOMEYA, K. [1978]: Experimental and theoretical study on magnetoelastic buckling of a cantilever, *J. Appl. Mech.*, **45**, 355–360.

MOFFATT, H.K. [1976]: Generation of magnetic fields by fluid motions, in *Advances in Applied Mechanics*, Vol. 16, pp. 119–181, Academic Press, New York.

MOFFATT, H.K. [1978]: *Magnetic Field Generation in Electrically Conducting Fluids*, Cambridge University Press, Cambridge.

MOLLENAUER, L.F., STOLEN, R.H., and GORDON J.P. [1980]: Experimental observations of picosecond pulse narrowing and solitons in optical fibres, *Phys. Rev. Lett.*, **45**, 1095–1098.

MØLLER, C. [1952]: *The Theory of Relativity*, Oxford University Press, London.

MOND, M. and WEITZNER, H. [1982]: Stability of helically symmetric straight equilibria, *Phys. Fluids*, **25**, 2056–2061.

MOON, F.C. [1978]: Problems in magneto-solid mechanics, in *Mechanics Today*, Vol. 4, pp. 307–390, ed. S. Nemat-Nasser, Pergamon Press, New York.

MOON, F.C. (editor) [1980]: *Mechanics of Superconducting Structures*, A.S.M.E., A.M.D. no. 41, A.S.M.E., New York.

MOON, F.C. [1984]: *Magnetosolid Mechanics*, Wiley, New York.

MOON, F.C. and PAO, Y.H. [1969]: Vibration and dynamic instability of a beam-plate in a transverse magnetic field, *J. Appl. Mech. Trans. ASME*, **36**, 92–100.

MORGENTHALER, F.R. [1966]: Pulsed frequency and mode conversion of magneto-elastic waves, in *Ultrasonics Symposium, Cleveland*, Ohio, Paper K-6.

MORGENTHALER, F.R. [1968a]: Pulsed magnetic field conversion of thermal spin fluctuations to elastic microwave noise power, in *Ultrasonics Symposium, New York*, Paper M-7.

MORGENTHALER, F.R. [1968b]: Magnetoelastic wave propagation in time-varying magnetic fields, in *Recent Advances in Engineering Science*, pp. 117–132, ed A.C. Eringen, Gordon and Breach, New York.

MORGENTHALER, F.R. [1972]: Dynamic magnetoelastic coupling in ferromagnets and antiferromagnets, *I.E.E.E. Trans. Mag.*, **8** (1), 130–151.

MORRIS, F.E. and NARIBOLI, G.A. [1972]: Photoelastic waves, *Int. J. Engng. Sci.*, **10**, 765–774.

MORRO, A. [1973]: Su un'assiomatica per l'elettrotermodinamica relativistica di un sistema continui, *Rend. Acad. Fis. Mat. Soc. N. Sci. Nap.*, **40**, 235–243.

MORRO, A., DROUOT, R., and MAUGIN, G.A. [1985]: Thermodynamics of polyelectrolyte solutions in an electric field, *J. Non-Equilibrium Thermodynamics* **10**, 131–144.

MORSE, P.M. and FESCHBACH, H. [1953]: *Methods of Theoretical Physics*, McGraw-Hill, New York.

MOSKOWITZ, R. [1974]: Dynamic sealing with magnetic fluids, in *29th A.S.LE. Annual Meeting, Cleveland, Ohio*, Paper 74-AM-6D-2, A.S.L.E., Park Ridge, IL.

MOSKOWITZ, R. and ROSENSWEIG, R.E. [1967]: Non-mechanical torque-driven flow of a ferromagnetic fluid by an electromagnetic field, *Lett. Appl. Phys.*, **11**, 301–303.

MOTOGI, S. [1979]: Interaction between spin waves and elastic waves in one-domain ferromagnetic insulators, *Int. J. Engng. Sci.*, **17**, 889–905.

MOTOGI, S. [1982]: A phenomenological theory of hysteresis damping of vibrations in ferromagnetic insulators, *Int. J. Engng. Sci.*, **20**, 823–834.

MOTOGI, S. and MAUGIN, G.A. [1984a]: Effects of magnetostriction on vibrations of Bloch and Néel walls, *Physica Statu Solidi*, **81a**, 519–532.

MOTOGI, S. and MAUGIN, G.A. [1984b]: Magnetoelastic oscillations of a Bloch wall in ferromagnets with dissipation, *Japan J. Appl. Phys.*, **23**, 1026–1031.

MÜLLER, I. [1968]: A thermodynamic theory of mixtures of fluids, *Arch. Rat. Mech. Anal.*, **28**, 1–39.

MUSGRAVE, M.J.P. [1970]: *Crystal Acoustics*, Holden-Day, San Francisco.

MUSKHELISHVILI, N.J. [1963]: *Some Basic Problems of the Mathematical Theory of Elasticity* (translation from the Russian), Noordhoff, Groningen, Holland.

NARASIMHAMURTY, T.S. [1981]: *Photoelastic and Electro-Optic Properties of Crystals*, Plenum, New York.

NÉEL, L. [1942]: Théorie des lois d'aimantation de Lord Rayleigh, *Cahiers de Physique*, no. **12**, p. 1.

NÉEL, L. [1948]: Propriétés magnétique des ferrites, *Annal. Phys.* (Paris), **3**, 137–198.

NELSON, D.F. [1979]: *Electric, Optic and Acoustic Interactions in Dielectrics*, Wiley, New York.

NELSON, D.F. and LAX, M. [1971]: Theory of photoelastic interactions, *Phys. Rev.*, **B3**, 2778–2794.

NEURINGER, J.L. [1966]: Some viscous flows of a saturated ferrofluid under the combined influence of thermal and magnetic field gradients, *Int. J. Nonlinear Mech.*, **1**, 123–137.

NEURINGER, J.L. and ROSENSWEIG, R.E. [1964]: Ferrohydrodynamics, *Phys. Fluids*, **7**, 1927–1937.

NOWACKI, W. [1975]: *Dynamic Problems in Thermoelasticity* (translation from the Polish), Noordhoff, Leyden, and P.W.N., Warsaw.

NOWACKI, W. [1983]: *Efekty Elektromagnetyczne W Stalich Cialach Odksztalcalnych* (Polish), P.A.N., Warsaw.

NOWINSKI, J.L. and WU, T.T. [1968]: A nonlinear dynamic problem for a thick walled cylinder of electrostrictive materials, *Int. J. Engng. Sci.*, **6**, 17–26.

O'DELL, T.H. [1970]: *Electrodynamics of Magneto-electric Media*, North-Holland, Amsterdam.

OLDROYD, J.G. [1970]: Equations of state of continuous matter in general relativity, *Proc. Roy. Soc. London*, **A316**, 1–28.

OOSAWA, F. [1971]: *Polyelectrolytes*, Marcel Dekker, New York.

OSTROUMOV, G.A. [1966]: Electric convection, *J. Engng. Phys.* (translation from the Russian), **10**, 406–414.

PAI, S.I. [1962]: *Magnetogasdynamics and Plasma Dynamics*, Springer-Verlag, Wien.

PAO, Y.H. [1978]: Electromagnetic forces in deformable continua, in *Mechanics Today*, Vol. 4, pp. 209–305, ed. S. Nemat-Nasser, Pergamon Press, New York.

PAO, Y.H. and HUTTER, K. [1975]: Electrodynamics for moving elastic solids and viscous fluids, *Proc. IEEE*, **63**, 1011–1021.

PAREKH, J.P. [1972]: Magnetoelastic surface waves propagating in an arbitrary direction on a tangentially magnetized YIG substrata, in *Proc. IEEE Ultrasonic Symposium* (IEEE, Boston, 1972), p. 333, IEEE, New York.

PARIA, G. [1962]: On magneto-thermo-elastic plane waves, *Proc. Cambridge Philos. Soc.*, **58**, 527–531.

PARIA, G. [1967]: Magnetoelasticity and magneto-thermo-elasticity, in *Advances in Applied Mechanics*, Vol. 10, pp. 73–112, ed. C.S. Yih, Academic Press, New York.

PARIS, R.B. [1984]: Resistive instabilities in MHD, *Ann. Phys. Fr.*, **9**, 347–432.

PARKUS, H. [1972a]: Magneto- und elektroelastizität, *Zeit. angew. Math. Mech.*, **53**, 718–724.

PARKUS, H. [1972b]: Thermoelastic equations for ferromagnetic bodies, *Arch. Mech. Stosow.*, **24**, 819–825.

PARKUS, H. [1979]: Application of electromagnetic interaction theory, in *Electromagnetic Interactions in Elastic Solids*, pp. 363–415, ed. H. Parkus, Springer-Verlag, Wien.

PELETMINSKII, S.V. [1959]: Coupled magnetoelastic oscillations in antiferromagnets, *Zhur. Èksper. Teoret. Fiz.* (in Russian), **37**, 452–457.

PENFIELD, P. and HAUS, H.A. [1967]: *Electrodynamics of Moving Media*, M.I.T. Press, Cambridge, MA.

PERRY, M.P. [1978]: A survey of ferromagnetic liquid applications, in *Thermo-*

mechanics of Magnetic Fluids, pp. 219–230, ed. B. Berkovsky, Hemisphere, Washington.

PETTINI, G. [1970]: Sul teorema di unicita nell'elettromagnetismo nonlineare ereditaro, *Boll. Unione Mat. Ital.*, 4th Series, **3**, 55–64.

PINES, D. [1963]: *Elementary Excitations in Solids*, Benjamin, New York, Chap. 4.

PIPKIN, A.C. and RIVLIN, R.S. [1960a]: Electrical conduction in deformed isotropic materials, *J. Math. Phys.*, **1**, 127–130.

PIPKIN, A.C. and RIVLIN, R.S. [1961a]: Electrical conduction in a stretched and twisted tube, *J. Math. Phys.*, **2**, 636–638.

PIPKIN, A.C. and RIVLIN, R.S. [1961b]: Electrical conduction in a noncircular rod, *J. Math. Phys.*, **2**, 865–868.

PIPKIN, A.C. and RIVLIN, R.S. [1962]: Non-rectilinear current flow in a straight conductor, *J. Math. Phys.*, **3**, 368–371.

PIPKIN, A.C. and RIVLIN, R.S. [1966]: Electrical, thermal and magnetic constitutive equations for deformed isotropic materials, *Rend. Acad. Lincei*, **8**, 3–29.

PIPPARD, A.B. [1965]: *The Dynamics of Conduction Electrons*, Gordon and Breach, New York.

POMERANTZ, M. [1961]: Excitation of spin–wave resonance by microwave phonons, *Phys. Rev. Lett.*, **7**, 312–313.

POPLAR, C.H. [1972]: Postbuckling analysis of a magnetoelastic beam, *J. Appl. Mech.*, **39**, 207–211.

POUGET, J. [1982]: Opération de convolution au moyen d'échos électro-acoustiques, *C. R. Acad. Sci. Paris*, **II-295**, 845–848.

POUGET, J. [1984]: Electro-acoustic echoes in piezoelectric powders, in *The Mechanical Behavior of Electromagnetic Solid Continua*, pp. 177–184, ed. G.A. Maugin, North-Holland, Amsterdam.

POUGET, J., ASKAR, A., and MAUGIN, G.A. [1986a]: Lattice model for elastic ferroelectric crystals: Microscopic approach, *Phys. Rev.*, **B**, **33**, 6304–6319.

POUGET, J., ASKAR, A., and MAUGIN, G.A. [1986b]: Lattice model for elastic ferroelectric crystals: Continuum approximation, *Phys. Rev.*, **B**, **33**, 6320–6325.

POUGET, J. and MAUGIN, G.A. [1980]: Coupled acoustic-optic models in elastic ferroelectrics, *J. Acoust. Soc. Amer.*, **68**, 588–601.

POUGET, J. and MAUGIN, G.A. [1981a]: Bleustein–Gulyaev surface modes in elastic ferroelectrics, *J. Acoust. Soc. Amer.*, **69**, 1304–1318.

POUGET, J. and MAUGIN, G.A. [1981b]: Piezoelectric Rayleigh waves in elastic ferroelectrics, *J. Acoust. Soc. Amer.*, **69**, 1319–1325.

POUGET, J. and MAUGIN, G.A. [1983a]: Nonlinear electroacoustic equations for piezoelectric powders, *J. Acoust. Soc. Amer.*, **74**, 925–940.

POUGET, J. and MAUGIN, G.A. [1983b]: Electroacoustic echoes in piezoelectric powders, *J. Acoust. Soc. Amer.*, **74**, 941–954.

POUGET, J. and MAUGIN, G.A. [1984]: Solitons and electroacoustic interactions in ferroelectric crystals—I: Single soliton and domain walls, *Phys. Rev.*, **B30**, 5306–5325.

POUGET, J. and MAUGIN, G.A. [1985a]: Solitons and electroacoustic interactions in ferroelectric crystals—II: Interactions of solitons and radiations, *Phys. Rev.*, **B31**, 4633–4651.

PRECHTL, A. [1979]: Electromagnetic interactions in elastic solids: Some relativistic aspects, in *Electromagnetic Interactions in Elastic Solids*, pp. 325–362, ed. H. Parkus, Springer-Verlag, Wien.

PRECHTL, A. [1983]: Electro-elasticity with small deformations, *Zeit angew. Math. Mech.*, **63**, 419–424.

PRENDERGAST, K.H. [1956]: The equilibrium of a self-gravifating incompressible fluid sphere with a magnetic field, Part I *Astrophys. J.*, **123**, 498–508.

PRENDERGAST, K.H. [1958]: The equilibrium of a self-gravifating incompressible fluid sphere with a magnetic field, Part II *Astrophys. J.*, **128**, 361–374.

RADO, G.T. and FOLEN, V.J. [1962]: Magnetoelectric effect in antiferromagnetics, *J. Appl. Phys.*, **33S**, 1126–1132.

RAMIREZ, G.A. and LIANIS, G. [1968]: Relativistic kinematics of deformable Solids—I, *Acta Mechanica*, **6**, 326–343.

REISSNER, E. [1944]: On the theory of bending elastic plates, *J. Math. Phys.*, **23**, 184–191.

RESLER, E.L. and NEURINGER, R.E. [1964]: Magnetocaloric power, *AIAA J.*, **2**, 1418–1422.

RESLER, E.L. and SEARS, W.R. [1958]: Magneto-gasdynamics channel flow, *Zeit. angew. Math. Phys.*, **9b**, 509–518.

REUTER, G.E.H. and SONDHEIMER, E.H. [1948]: Theory of the anomalous skin effect in metals, *Proc. Roy. Soc. London* Ser A, **195**, 336–364.

REZENDE, S.M. and MORGENTHALER, F.R. [1969]: Magnetoelastic wave propagation in time-varying fields I–II, *J. Appl. Phys.*, **40**, 524–545.

RISTIC, V.M. [1983]: *Principles of Acoustic Devices*, Wiley. Interscience, New York.

RIVLIN, R.S. and SMITH, G.F. [1971]: Birefringence in viscoelastic materials, *Zeit. angew. Math. Phys.*, **22**, 325–339.

ROBDELL, D.S. [1964]: Ferromagnetic resonance absorption line width in nickel metal—Evidence for Landau–Lifshitz damping, *Phys. Rev. Lett.*, **13**, 471–474.

ROBERTS, P.H. [1967]: *An Introduction to Magnetohydrodynamics*, Longmans, London.

ROGERS, C., CERKIGE, H.M. and ASKAR, A. [1977]: Electromagnetic wave propagation in nonlinear dielectric media, *Acta Mechanica*, **26**, 59–73.

ROGULA, D. (editor) [1982]: *Nonlocal Theory of Material Media*, CISM Courses and Lectures, No. 268, Springer-Verlag, Wien, New York.

ROSENSWEIG, R.E. [1966]: Magnetic fluids, *Int. Sci. Techn.*, **55**, 48–56.

ROSENSWEIG, R.E. [1970]: Ferrohydrodynamics, in *Encyclopeadic Dictionary of Physics*, Pergamon Press, Oxford.

ROSENSWEIG, R.E. [1985]: *Ferrohydrodynamics*, Cambridge University Press, Cambridg, U.K.

ROSENSWEIG, R.A., MISKOLOGY, G., and EZEKIEL, F.E. [1968]: Magnetic fluid seals, *Machine Design*, **40**, 145–151.

ROSENSWEIG, R.E., ZAHN, M., and SHUMOVICH, D. [1983]: Labyrinthine instability in magnetic and dielectric fluids, *J. Magnetism and Magnetic Materials*, **39**, 127–134.

ROSENSWEIG, R.A., ZAHN, M., and VOGLER, T. [1978]: Stabilization of fluid penetration through a porous medium using magnetizable fluids, in *Thermomechanics of Fluids*, pp. 195–211, ed. B. Berkovsky, Hemisphere, Washington.

SANCHEZ-PALENCIA, E. [1968]: Existence de solutions de certains problèmes aux limites en magnétohydrodynamique, *J. Mécanique*, **7**, 405–426.

SANCHEZ-PALENCIA, E. [1969]: Quelques résultats d'existence et d'unicité pour des écoulements magnétohydrodynamiques non stationnaires, *J. Mécanique*, **8**, 509–541.

SCHLÖMANN, E. [1960]: Generation of phonons in high-power ferromagnetic resonance experiments, *J. Appl. Phys.*, **31**, 1647–1656.

SCHLÖMANN, E. [1961]: in *Advances in Quantum Electronics*, pp. 444–452, ed. J.R. Singer, Columbia University Press, New York.

SCHLÖMANN, E. [1964]: Generation of spin waves in nonuniform magnetic fields—I: Conversion of electromagnetic power into spin–wave power and vice-versa, *J. Appl. Phys.*, **35**, 159–166.

SCHLÖMANN, E. and JOSEPH, R.I. [1964]: Generation of spin waves in nonuniform magnetic fields—II: Calculation of the coupling length, *J. Appl. Phys.*, **35**, 167–170.

SCHNEIDER, J.M. and WATSON, P.K. [1970]: Electrohydrodynamic stability of space-charge-limited currents in dielectric liquids—I: Theoretical study, *Phys. Fluids*, **19**, 1948–1954.

SCHUBERT, M. and WILHELMI, B. [1986]: *Nonlinear Optics and Quantum Electronics*, Wiley, New York.

SCHUTZ, W. [1936]: Magnetooptik, in *Handbuch der Experimentalphysik*, Akad. Verlag, MBH, **16**, Part I, Leipzig Akad der Verlag.

SCOTT, R.Q. and MILLS, D.L. [1977]: Propagation of surface magnetoelastic waves on ferromagnetic crystal substrate, *Phys. Rev.*, **B15**, 3545–3557.

SEANOR, D.A. (editor) [1982]: *Electrical Properties of Polymers*, Academic Press, New York.

SEDOVA, G.L. [1978]: Nonlinear waves and strong discontinuities in ferromagnetics, *Izv. Akad. Nauk, SSSR, Mzh.G.*, no. 2.

SEDOVA, G.L. [1981]: Propagation of electromagnetic waves for arbitrary dependence of magnetic permeability on magnetic induction, *Prikl. Matem. Mekhan.* (English translation), **44**, 329–331.

SEDOVA, G.L. [1982]: Strong discontinuities of electromagnetic fields in magnetics, *Prikl. Matem. Mekhan.* (English translation), **45**, 718–721.

SELEZOV, I.T. [1984]: Diffraction of magnetoelastic waves by inhomogeneities, in *The Mechanical Behavior of Electromagnetic Solid Continua*, pp. 351–355, ed. G.A. Maugin, North-Holland, Amsterdam.

SESSLER, G.M. [1981]: Piezoelectricity of polyvinylidene fluoride, *J. Acoust. Soc. Amer.*, **70**, 1596–1608.

SHEN, Y.R. [1984]: *The Principle of Nonlinear Optics*, Wiley, New York.

SHTRIKMANN, S. and TREVES, D.T. [1963]: Micromagnetics, in *Magnetism*, Vol. 3, eds. G.T. Rado and H. Suhl, Academic Press, New York.

SHUBNIKOV, A.V. and BELOV, N.V. [1964]: *Colored Symmetry* (translation from the Russian), Pergamon Press, New York.

SINGH, H. and PIPKIN, A.C. [1966]: Controllable states of elastic dielectrics, *Arch. Rat. Mech. Anal.*, **21**, 169–210.

SIOKÉ-RAINALDY, J. and MAUGIN, G.A. [1983]: Magnetoelastic equations for antiferromagnetic insulators of the easy axis type, *J. Appl. Phys.*, **54**, 1490–1506.

SIROTIN, Yu. I. [1960]: Group tensor space, *Soviet Phys. Crystallography*, **5**, 157–165.

SIROTIN, Yu. I. [1961]: Plotting tensors of a given symmetry, *Soviet Phys. Crystallography*, **6**, 263–271.

SMITH, G.F. [1968]: On the generation of integrity bases, *Atti. Acad. Naz. Lincei*, series VIII, **9**, 51.

SMITH, G.F. [1970]: On a fundamental error in two papers of C.C. Wang "On Representations for Isotropic Functions, Parts I and II", *Arch. Rat. Mech. Anal.*, **36**, 166–223.

SMITH, G.F. [1971]: On isotropic functions of symmetric tensors, skew symmetric tensors and vectors, *Int. J. Engng. Sci.*, **19**, 899–916.

SMOLENSKII, G.A. [1974]: *Physics of Magnetic Dielectrics* (in Russian), Nauka, Leningrad.

SMOLENSKY, G.A. and YUSHIN, N.K. [1984]: Electroacoustic echoes in piezoelectric powders, in *The Mechanical Behavior of Electromagnetic Solid Continua*, pp. 167–176 ed. G.A. Maugin, North-Holland, Amsterdam.

SNEDDON, I.N. [1966]; *Mixed Boundary Values Problems in Potential Theory*, North-Holland, Amsterdam.

SODERHOLM, L. [1970]: A principle of objectivity in relativistic continuum mechanics, *Arch. Rat. Mech. Anal.*, **39**, 89–107.

SOOHOO, R.F. [1963]: General exchange boundary condition and surface anisotropy energy of a ferromagnet, *Phys. Rev.*, **131**, 594–601.

SPENCER, A.J.M. [1971]: Theory of invariants, in *Continuum Physics*, Vol. 1, ed. A.C. Eringen, Academic Press, New York.

STOKES, V.K. [1984]: *Theories of Fluids with Microstructure*, Springer-Verlag, Berlin, Heidelberg, New York, Tokyo.

STRATTON, J.A. [1941]: *Electromagnetic Theory*, McGraw-Hill, New York.

STRAUSS, W. [1965]: Elastic and magnetoelastic waves in yttrium–iron–garnet, *Proc. IEEE*, **53**, 1485.

STRAUSS, W. [1968]: Magnetoelastic properties of yttrium–iron–garnet, in *Physical Acoustics*, Vol. IV, Part B, pp. 2–52, ed. W.P. Mason, Academic Press, New York.

STUETZER, O.M. [1962]: Magnetohydrodynamics and electrohydrodynamics, *Phys. Fluids*, **5**, 534–544.

ŞUHUBI, E.S. [1965]: Small torsional oscillations of a circular cylinder with finite electric conductivity in a constant axial magnetic field, *Int. J. Engng. Sci.*, **2**, 441–459.

ŞUHUBI, E.S. [1969]: Elastic dielectrics with polarization gradients, *Int. J. Engng. Sci.*, **7**, 993–997.

ŞUHUBI, E.S. and ERINGEN, A.C. [1964]: Nonlinear theory of simple micro-elastic solids—II, *Int. J. Engng. Sci.*, **2**, No. 4, 389–404.

SWATIK, D.S. and HENDRICKS, C.D. [1968]: Production of ions by electrohydrodynamics spraying techniques, *AIAA J.*, **6**, 1596.

SZUSTAKOWSKI, M. [1976]: Echo of magnetoelastic waves in YIG monocrystals, *J. Techn. Physics* (Warsaw), **17**, 403–408.

TAKETOMI, S. [1985]: Equivalence between constitutive equations for magnetic fluids with an instrinsic angular momentum and those for liquid crystals, *J. Phys. Soc. Japan*, **54**, 102–107.

TAREEV, B.M. (editor) [1980]: *Electrical and Radio Engineering Materials*, MIR, Moscow (in English).

TAUB, A.H. [1948]: Relativistic Rankine–Hugoniot equations, *Phys. Rev.*, **74**, 328–334.

TAVGER, B.A., and ZAITSEV, V.M., [1956]: Magnetic symmetry of crystals, *Sov. Phys. JETP* (English transl.) **3**, 430.

TAYLER, R.J. [1958]: in *Proceedings of the 2nd Geneva Conference on the Peaceful Uses Atomic Energy*, **31**, p. 160.

TAYLOR, E.F. and WHEELER, J.A. [1966]: *Spacetime Physics*, Freeman, San Francisco.

TAYLOR, G.I. [1964]: Disintegration of water drops in an electric field, *Proc. Roy. Soc. London*, **A280**, 383.

TER HAAR, D. and WERGELAND, H. [1971]: Thermodynamics and statistical mechanics in the special theory of relativity, *Phys. Rep.*, **1**, 31–54.

TESARDI, L.R., LEVINSTEIN, H.J., and GYORGY, E.M. [1969]: Electromagnetic sound conversion by linear magnetostriction in $TIFeF_3$, *Solid State Comm.*, **7**, 1, 241–243.

TESARDI, L.R., LEVINSTEIN, N.J., GYORGY, E.M., and GUGGENHEIM, H.J. [1969]: Electromagnetic sound conversion by linear magnetostriction in $TiFeF_3$, *Solid State Comm.*, **7**, 241–243.

THEOCARIS, P.S., [1965]: A review of the rheo-optical properties of linear high polymers, *Exp. Mech.*, **5**, 105–114.

TIERSTEN, H.F. [1963]: Thickness vibrations of piezoelectric plates, *J. Acoust. Soc. Amer.*, **35**, 53–58.

TIERSTEN, H.F. [1964]: Coupled magnetomechanical equations for magnetically saturated insulators, *J. Math. Phys.*, **5**, 1298–1318.

TIERSTEN, H.F. [1965a]: Variational principle for saturated magnetoelastic insulators, *J. Math. Phys.*, **6**, 779–787.

TIERSTEN, H.F. [1965b]: Thickness vibrations of saturated magnetoelastic plates, *J. Appl. Phys.*, **36**, 2250–2259.

TIERSTEN, H.F. [1969]: *Linear Piezoelectric Plate Vibrations*, Plenum, New York.

TIERSTEN, H.F. [1981]: Electroelastic interactions and the piezoelectric equations, *J. Acoust. Soc. Amer.*, **70**, 1567–1576.

TIERSTEN, H.F. [1984]: Electric fields, deformable semiconductors and piezoelectric devices, in *The Mechanical Behavior of Electromagnetic Solid Continua*, pp. 99–114, ed. G.A. Maugin, North-Holland, Amsterdam.

TIERSTEN, H.F. and TSAI, C.F. [1972]: On the interactions of the electromagnetic field with heat conducting deformable insulators, *J. Math. Phys.*, **13**, 361–382.

TIMOSHENKO, S. and GOODIER, J.N. [1955]: *Theory of Elasticity*, McGraw-Hill, New York.

TONNELAT, M.A. [1971]: *Histoire du Principe de Relativité*, Flammarion, Paris.

TOUPIN, R.A. [1956]: The elastic dielectric, *J. Rat. Mech. Anal.*, **5**, 849–915.

TOUPIN, R.A. [1963]: A dynamical theory of dielectrics, *Int. J. Engng. Sci.*, **1**, 101–126.

TRICOMI, F.G. [1957]: *Integral Equations*, Interscience, New York.

TRUELL, R. and ELBAUM, C. [1965]: High-frequency ultrasonic stress waves, in *Handbuch der Physik*, Vol. XI, ed. S. Flügge, Springer-Verlag, Berlin.

TRUESDELL, C. and TOUPIN, R.A. [1960]: The classical field theories, in *Handbuch der Physik*, Bd. III/1, ed. S. Flügge, Springer-Verlag, Berlin, Heidelberg, New York.

TRUESDELL, C. and NOLL, W. [1965]: The nonlinear field theories of mechanics, in *Handbuch der Physik*, Bd. III/3, ed. S. Flügge, Springer-Verlag, Berlin, Heidelberg, New York.

TURNBULL, R.J. [1968]: Electroconvective instability with a stabilizing temperature gradient, I—Theory, II—Experimental results, *Phys. Fluids*, **11**, 2588–2603.

TUROV, E.A. [1983]: Symmetry breaking and magnetoacoustic effect in ferro- and antiferromagnets (in Russian), *Progress in Physical Sciences* (Uspekhi Fiz. Nauk), **140**, 429–462.

TUROV, E.A. [1984]: Magnetoacoustics of ferro- and antiferromagnetics, in *The Mechanical Behavior of Electromagnetic Solid Continua*, pp. 255–267, ed. G.A. Maugin, North-Holland, Amsterdam.

VAN DE VEN, A.A.F. [1975]: Interaction of electromagnetic and elastic fields in solids, Ph.D. Thesis, Technical University of Eindhoven, The Netherlands.

VAN DE VEN, A.A.F. [1978]: Magnetoelastic buckling of thin plates in a uniform transverse magnetic field, *J. Elasticity*, **8**, 297–312.

VAN DE VEN, A.A.F. [1984]: The influence of finite specimen dimensions on the magneto-elastic buckling of a cantilever, in *The Mechanical Behavior of Electromagnetic Solid Continua*, pp. 421–426, ed. G.A. Maugin, North-Holland, Amsterdam.

VAN VLECK, J.H. [1932]: *The Theory of Electric and Magnetic Susceptibilities*, Oxford University Press, London.

VDOVIN, V.E. and KUNIN, I.A. [1968]: Interaction of dislocations with allowance for spatial dispersion, *Fiz. Tverd. Tela.*, **10**, 375–384.

VIKTOROV, I.A. [1979]: Types of acoustic surface waves, *Soviet Phys. Acoustics*, **25**, 1–9.

VITTORIA, C., CRAIG, J.N., and BAILEY, G.C. [1974]; General dispersion law in a ferromagnetic cubic magnetoelastic conductor, *Phys. Rev.*, **B10**, 3945–3956.

VOIGT, W. [1899]: Zur Theorie der Magneto-optischen Erscheinungen, *Ann. der Phys.*, **67**, 345–365.

VOIGT, W. [1928]: *Lehrbuch der Kristallphysik*, Teubner Verlag, Leipzig.

VOLKENSHTEIN, M.V. [1983]: *Biophysics*, MIR, Moscow (in English).

VOLTERRA, V. [1959]: *Theory of Functionals and of Integral and Integro-Differential Equations*. Dover, New York, p. 21.

VON HIPPEL, H.R. [1954]: *Dielectrics and Waves*, Wiley, New York.

VONSOVSKII, S.V. [1975]: *Magnetism* (translation from the Russian), Israel University Press, Jerusalem.

WALLERSTEIN, D.V. and PEACH, M.O. [1972]: Magnetoelastic buckling of beams and thin plates of magnetically soft materials, *J. Appl. Mech. Trans. ASME*, **39**, 451–455.

WANG, C.C. [1969a]: On representations for isotropic functions—I, *Arch. Rat. Mech. Anal.*, **33**, 249–267.

WANG, C.C. [1969b]: On representations for isotropic functions—II, *Arch. Rat. Mech. Anal.*, **33**, 268–287.

WANG, C.C. [1970]: A new representation theorem for isotropic functions, Parts I and II, *Arch. Rat. Mech. Anal.*, **36**, 166–223.

WANG, S. and CROW, J. [1970]: Acoustic Faraday rotation, in *Dig. Int. Magn. Conf.*, IEEE, New York.

WATKINS, G.D. and FEHER, E. [1962]: Effect of uniaxial stress on the EPR of transition element ions in MgO, *Amer. Phys. Soc. Bull.*, **7**, 29.

WATSON, P.K., SCHNEIDER, J.M., and TILL, H.R. [1970]: Electrohydrodynamic stability of space-charge-limited currents in dielectric liquids—II: Experimental Study, *Phys. Fluids*, **13**, 1955–1961.

WEINBERG, S. [1972]: *Gravitation and Cosmology*, Wiley, New York.

WEISS, P. [1907]: L'hypothèse du champ moléculaire et la propriété ferromagnétique, *J. Physique*, **6**, 661–690.
WEYL, H. [1946]: *Classical Groups*, Princeton University Press, Princeton, NJ.
WHITTAKER, E.T. [1951]: *History of the Theories of Aether and Electricity*, 2 volumes, Nelson, London.
WHITTAKER, E.T. and WATSON, G.H. [1946]: *Modern Analysis*, Macmillan, New York.
WILLIAMS, M.L. and ARENZ, R.J. [1964]: The Engineering analysis of linear photoviscoelastic materials, *Exp. Mech.*, **4**, 249–262.
WILSON, A.H. [1953]: *The Theory of Metals*, Cambridge University Press, Cambridge.
WILSON, A.J. [1963]: The propagation of magneto-thermo-elastic waves, *Proc. Cambridge Philos. Soc.*, **59**, 483–488.
WILSON, H.A. [1905]: On the effect of rotating a dielectric in a magnetic field, *Phil. Trans. Roy. Soc.* A, **204**, 121–137.
WILSON, M. and WILSON, H.A. [1914]: On the Electric effect of rotating a magnetic insulator in a magnetic field, *Proc. Roy. Soc. London*, **A89**, 99–108.
WITHERS, R.S., MELCHER, J.R. and RICHMANN, J.W. [1978]: Charging, migration and electrohydrodynamic transport of aerosols, *J. Electrostatics*, **5**, 225–239.

YEH, C-S. [1971]: Linear theory of magnetoelasticity for soft ferromagnetic materials and magnetoelastic buckling, Ph.D. dissertation, Cornell University, Ithaca, New York.

ZAHN, M. and MELCHER, J.R. [1972]: Space charge dynamics of liquids, *Phys. Fluids*, **15**, 1197–1205; erratum *ibid*, p. 2082.
ZAKHAROV, V.E. and SHABAT, A.B. [1972]: Exact theory of two-dimensional self-focusing and one-dimensional self-modulation of waves in nonlinear media, *Soviet Phys. JETP*, **34**, 62–69.
ZEEMAN, P. [1914]: Fresnel's coefficient for light of different colours, *Proc. Acad. Sci. Amsterdam*, **17**, 445.
ZELAZO, R.E. and MELCHER, J.R. [1969]: Dynamics and stability of ferrofluids: Surface interactions, *J. Fluid Mech.*, **39**, 1–24.
ZERNIKE, F. and MIDWINTER, J.E. [1973]: *Applied Nonlinear Optics*, Wiley–Interscience, New York.
ZHELUDEV, I.S. [1971]: *Physics of Crystalline Dielectrics*, 2 volumes, (translation from the Russian), Plenum, New York.
ZIMMELS, Y. [1983]: Application of ferrofluids to separation of particulates, *J. Magnetism and Magnetic Materials*, **39**, 173–177.
ZOCHER, H. and TÖRÖK, C. [1953]: About space–time asymmetry in the realm of classical, general and crystal physics, *Proc. Nat. Acad. Sci.*, **39**, 681–686.

Index

MIX
Papier aus verantwortungsvollen Quellen
Paper from responsible sources
FSC® C105338

If you have any concerns about our products,
you can contact us on
ProductSafety@springernature.com

In case Publisher is established outside the EU,
the EU authorized representative is:
Springer Nature Customer Service Center GmbH
Europaplatz 3, 69115 Heidelberg, Germany

Printed by Libri Plureos GmbH
in Hamburg, Germany